Ecology
of World Vegetation

O.W. Archibold
Department of Geography
University of Saskatchewan
Saskatoon, Canada

CHAPMAN & HALL
London · Glasgow · Weinheim · New York · Tokyo · Melbourne · Madras

Chapman & Hall, 2-6 Boundary Row, London SE1 8HN, UK

Chapman & Hall, 2-6 Boundary Row, London SE1 8HN, UK

Blackie Academic & Professional, Wester Cleddens Road, Bishopbriggs, Glasgow G64 2NZ, UK

Chapman & Hall GmbH, Pappelallee 3, 69469 Weinheim, Germany

Chapman & Hall USA, One Penn Plaza, 41st Floor, New York NY 10119, USA

Chapman & Hall Japan, ITP-Japan, Kyowa Building, 3F, 2-2-1 Hirakawacho, Chiyoda-ku, Tokyo 102, Japan

Chapman & Hall Australia, Thomas Nelson Australia, 102 Dodds Street, South Melbourne, Victoria 3205, Australia

Chapman & Hall India, R. Seshadri, 32 Second Main Road, CIT East, Madras 600 035, India

First edition 1995

© 1995 Chapman & Hall

Typeset in Palatino 9.5 on 10 pt by Puretech Corporation, Pondicherry, India
Printed in Great Britain by St Edmundsbury Press Ltd, Bury St Edmunds, Suffolk

ISBN 0 412 44290 6 (HB)   0 412 44300 7 (PB)

A catalogue record for this book is available from the British Library

Library of Congress Catalog Card Number: 94-72008

∞ Printed on permanent acid-free text paper, manufactured in accordance with ANSI/NISO Z39.48-1992 and ANSI/NISO Z39.48-1984 (Permancence of Paper).

For David, Katherine and Gillian

Contents

Preface

The ecology of world vegetation is described in numerous books and journals, but these are usually very specialized in their scope and treatment. This book provides a synthesis of this literature. A brief introductory chapter outlines general ecological concepts and subsequent chapters examine the form and function of the major biomes of the world. A similar organization has been used for each biome type. These chapters begin with a description of environmental conditions and a brief account of floristic diversity in a regional context. The remaining pages describe characteristic adaptations and ecosystem processes.

Although there is a rapidly growing literature on ecological topics, most books are very specialized in their approach. No single volume can provide a complete overview of the diverse subject matter that is represented by 'ecology'. This book too has necessarily been selective, but it is hoped that the approach and synthesis will provide an adequate introduction to the ecology of world ecosystems. I have tried to minimize the length of this book by presenting information in a compact style. Numerous illustrations have been included to supplement this rather factual approach, and references to the original sources provide more detailed information on topics that are of interest to those for whom this book provides a first encounter with ecology and to readers with more background in this subject.

Keith Bigelow and Derek Thompson of the Department of Geography, University of Saskatchewan, completed all of the drafting and photographic work. They have spent many hours on this project and their care and skill is reflected in the consistently high quality of the illustrations throughout the book. Many friends and colleagues have provided photographs. It has not been possible to include all of them, but the 'global' perspective of the book has been greatly enhanced in this way. I wish to thank them all for the time and trouble they have taken to supply this material. I must also thank Mary Dykes and the staff of the interlibrary loans department of the Library, University of Saskatchewan, for their unfailing ability to get even the most obscure references.

Dr Clem Earle of Chapman & Hall has been involved with this book since its inception and I am grateful for his advice and encouragement at all stages of production. The nature of the book is reflected in his interest and efforts. Dr Bob Carling, Dr Jo Koster and the production staff at Chapman and Hall must also be acknowledged for their excellent work in connection with this project.

Finally, I am sincerely grateful for the assistance and constructive criticism provided by my wife Anne. Although my mind has at times been preoccupied, my thoughts have always been with her and with my children to whom I dedicate this book.

Bill Archibold
Westbury, Wiltshire, UK

CHAPTER 1

Vegetation and environment – introductory concepts

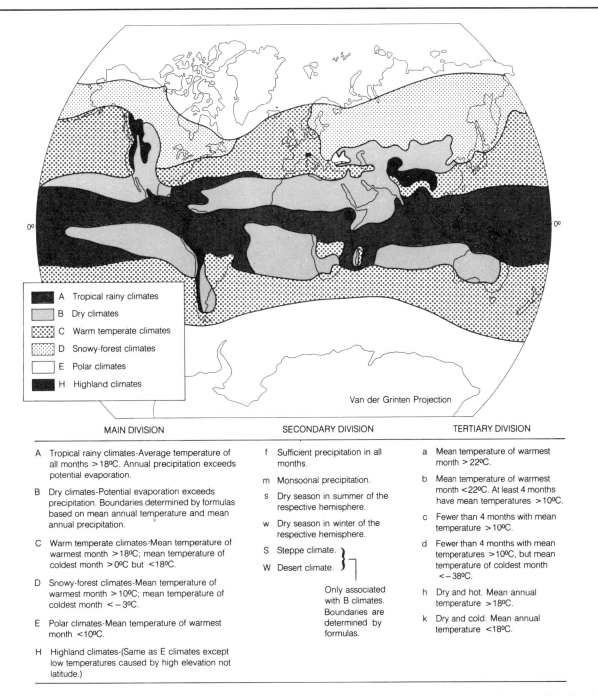

MAIN DIVISION	SECONDARY DIVISION	TERTIARY DIVISION
A Tropical rainy climates-Average temperature of all months >18ºC. Annual precipitation exceeds potential evaporation.	f Sufficient precipitation in all months.	a Mean temperature of warmest month > 22ºC.
B Dry climates-Potential evaporation exceeds precipitation. Boundaries determined by formulas based on mean annual temperature and mean annual precipitation.	m Monsoonal precipitation. s Dry season in summer of the respective hemisphere. w Dry season in winter of the respective hemisphere.	b Mean temperature of warmest month <22ºC. At least 4 months have mean temperatures >10ºC. c Fewer than 4 months with mean temperature >10ºC.
C Warm temperate climates-Mean temperature of warmest month >18ºC; mean temperature of coldest month >0ºC but <18ºC.	S Steppe climate. W Desert climate. } Only associated with B climates. Boundaries are determined by formulas.	d Fewer than 4 months with mean temperatures >10ºC, but mean temperature of coldest month <−38ºC.
D Snowy-forest climates-Mean temperature of warmest month >10ºC; mean temperature of coldest month <−3ºC.		h Dry and hot. Mean annual temperature >18ºC.
E Polar climates-Mean temperature of warmest month <10ºC.		k Dry and cold. Mean annual temperature <18ºC.
H Highland climates-(Same as E climates except low temperatures caused by high elevation not latitude.)		

Figure 1.1 The major climatic regions of the world according to the Köppen classification and associated subdivisions. (After Strahler and Strahler, 1992.) (Reproduced with permission from A. H. and A. N. Strahler, *Modern Physical Geography*, 4th edn, published by John Wiley and Sons Inc., copyright © 1992.)

Plants are a general feature of the natural landscape and grow in all but the most extreme environments. However, no species occurs everywhere in the world, each being distributed according to its own unique tolerance of the multitude of factors that comprise its environment. Species with similar ecological tolerances develop into recognizable plant formations with distinctive floristic and structural characteristics. At the broadest scale these represent the major biomes of the world. The regional extent of each biome is primarily determined by climatic conditions and this is the basis of most schemes of vegetation classification. Of historical importantance is the work of de Candolle who, in 1874, proposed that the major plant formations were distributed according to the heat requirements and drought tolerance of the plants. Although this early explanation has now been superseded, the general concepts were subsequently incorporated into the widely used Köppen system of climate classification (Table 1.1).

1.1 GLOBAL CLIMATE AND PLANT DISTRIBUTION

Climatic conditions are controlled by the amount of solar energy that is intercepted by the Earth and its atmosphere. Some energy is used in photosynthesis and becomes temporarily stored in the biosphere, but mostly it is absorbed and converted to heat. In tropical regions there is a net surplus of energy, but higher latitudes experience a net negative radiation balance because more energy is lost through re-radiation than is received. The energy surplus in lower latitudes is redistributed through the circulation of the atmosphere and oceans and this ultimately determines global temperature and precipitation patterns. In general terms the Earth's climates are warm and moist in the tropics and become colder and drier towards the poles. To this must be added the effects of mountain barriers and continentality, so that many different thermal regimes and seasonal precipitation patterns interact to form the diverse climates of the world. In this way some 25 climate types are defined in the Köppen classification scheme, each being designated by a 2- or 3-letter code (Figure 1.1). The Köppen system is an empirical classification in which climates are grouped according to predefined limits and the correspondence between the climate classes and large-scale vegetation patterns is essentially coincidental. The predominant influence of temperature and precipitation on world vegetation also forms the basis of the Holdridge method of classifying plant formations (Figure 1.2). In this scheme temperature and precipitation interact through potential evapotranspiration to define humidity provinces and so establish a functional relationship between vegetation and climate.

The general nature of each plant formation is determined by the structural character of the vegetation, and so reduces the great variation in floristic diversity to relatively few classes. The growth form of a plant is an adaptive response to its environment and provides an ecological classification that may be indicative of habitat conditions. Physiognomic characteristics such as plant height, woody or herbaceous growth and leaf type are often used in this way. The most widely used system is that proposed by Raunkiaer (1934) and is based on the arrangement of the perennating tissues of the plants growing under different climatic conditions: five principal life forms are distinguished (Figure 1.3). **Phanerophytes**, represented by trees and tall shrubs, carry their buds on the tips of their branches where they are exposed to climatic extremes: it is the predominant life form in mild, moist environments where the plants are not subject to frost and drought. **Chamaephytes** include small shrubs and herbs that grow close to the ground: they occur most frequently in regions where snow cover affords some protection during cold winter months. **Hemicryptophytes** are characteristic of moist temperate regions. These plants die back at the end of the growing season, and the buds are protected by the withered leaves and soil. **Cryptophytes** are well adapted to extreme conditions of cold or drought and persist because they regenerate from buds, bulbs and rhizomes that are completely buried in the soil. **Therophytes** are annual plants that regenerate from seed each year; they occur abundantly in desert areas. Raunkiaer's system is actually a floristic approach in that the 'biological spectrum' of a region is calculated from the number of species in each category. Different plant formations are distinguished by the proportional contribution of each category (Figure 1.4).

Table 1.1 De Candolle's plant groups as the basis of Köppen's climate classes (after Colinvaux, 1973)

De Candolle plant group	Postulated plant requirements	Formation	Köppen climate class
Megatherms	Continuous high temperature and abundant moisture	Tropical rain forest	A (tropical rainy climates)
Xerophiles	Drought and heat tolerant	Tropical desert	B (dry climates)
Mesotherms	Moderate temperature and moisture	Temperate deciduous forest	C (warm temperate climates)
Microtherms	Less heat and moisture, tolerate long cold winters	Boreal forest	D (snowy-forest climates)
Hekistotherms	Tolerate conditions beyond the tree-line	Tundra	E (polar climates)

(Reproduced with permission from P. A. Colinvaux, *Introduction to Ecology*, copyright © 1973 John Wiley & Sons Inc.)

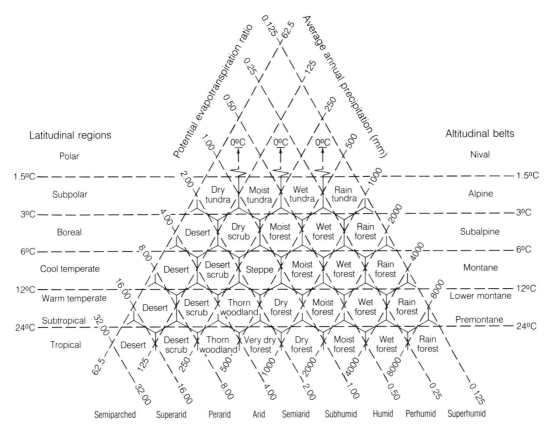

Figure 1.2 Classification of world plant formations according to mean annual temperature, precipitation and potential evapotranspiration. (After Holdridge *et al.*, 1971.) (Reproduced from Holdridge *et al.*, *Forest Environments in Tropical Life Zones*; published by Pergamon Press, 1971.)

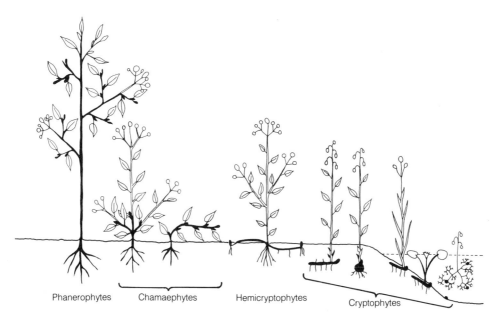

Figure 1.3 Physiognomic classification of plants according to the relative position of perennating parts. The untoned parts of the plants die back during unfavourable periods of the year but the other parts of the plant persist and give rise to new growth the following season. Therophytes, which persist only as seeds, are omitted. (After Raunkiaer, 1934.) (From C. Raunkiaer, *The Life Forms of Plants and Statistical Plant Geography*, 1934; by permission of Oxford University Press.)

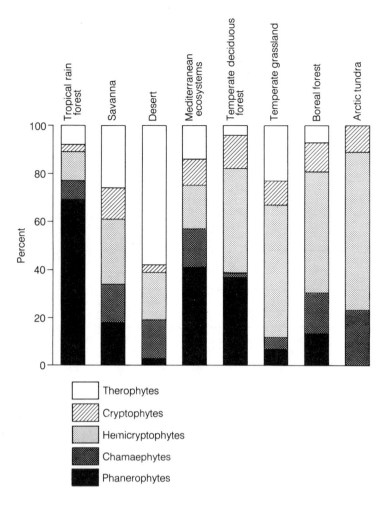

Figure 1.4 Proportional distribution of different life forms classified according to Raunkiaer (1934) in the major ecosystems of the world.

1.2 CLIMATE AND COMMUNITY STRUCTURE

Plant formations, such as tropical forest or tundra, are the largest and most complex units of vegetation and represent the level at which most world maps are compiled. Traditionally the global distribution is referred to as the '**formation-type**', with '**formation**' reserved for the regional subdivisions – for example, the American, European and East Asian formations of the temperate deciduous forest formation-type. The distribution of these complex units is generally determined by climate, although some, such as heaths, may be influenced by soil conditions. A formation is structurally similar throughout its range, even though pronounced floristic differences may occur between regions. Consequently, the formation is subdivided into **associations** on the basis of the dominant species that grow under uniform habitat conditions. For example, the clay vales of Europe support stands of pedunculate oak (*Quercus robur*) whereas beech (*Fagus sylvatica*) is the dominant species on the chalk uplands. **Faciations** are recognized where some associational dominants are locally absent, and **lociations** may

be defined according to important subdominant species. **Societies** are established within these higher classes through the inclusion of subordinate species (Polunin, 1960). In this hierarchical system it is assumed that different plant species are habitually associated with each other over wide geographic areas. This has led to the assumption that the classes are rigidly defined and thus represent discrete units in the natural landscape which have developed under the controlling influence of climate.

Such ideas are encompassed in the '**organismic**' approach developed by Clements (1916) in which communities were thought to consist of species that were so coadapted that they were most successful when they occurred together. Thus, distinct assemblages developed because all species replaced each other at discrete points along an environmental gradient (Figure 1.5). The alternative viewpoint proposed by Gleason (1926) and others emphasized the species as the essential determinant of vegetation structure and composition. This '**individualistic**' approach considers that each species responds in its own way to the physical and biotic environment. Analytical studies of community structure suggest that species are arranged so that each has an optimum envir-

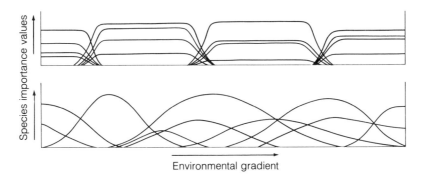

Figure 1.5 Distribution of species populations along an environmental gradient according to the organismic hypothesis (upper) proposed by Clements (1916) which suggests that formations are distinct landscape units, and the individualistic hypothesis (lower) presented by Gleason (1926) with dominant and subordinate species arranged in a non-coincidental pattern. (After Whittaker, 1970.) (Reprinted with the permission of Macmillan Publishing Company from *Communities and Ecosystems* by Robert H. Whittaker. Copyright © 1970 by Robert H. Whittaker.)

onment that does not coincide with that of a potential competitor (Whittaker, 1965). The development of well-defined communities is therefore precluded because species are distributed rather randomly and discontinuities will occur only where environmental conditions change abruptly. However, it has also been argued that there are points along environmental gradients where pronounced changes in species composition do occur, and that the appearance or disappearance of dominant species at their critical limits of tolerance can be accompanied by changes in subordinate species. For example, in the western United States the loss of shrubby sagebrush (*Artemisia tridentata*) distinguishes the *Artemisia–Festuca* association from the *Agropyron–Festuca* association dominated by bunch-grasses (Daubenmire, 1966). The loss of sagebrush not only alters the composition of the plant cover, but shrub-dependent birds also disappear, and thus the overall composition of the ecosystem is changed. This functional relationship between members of a community is an important extension of the early ideas and demonstrates the complexity of environmental interactions. Thus, whereas global vegetation patterns are related to climatic conditions, other factors such as soils, herbivory and competition account for much of the variation at a local scale.

1.3 SOIL AS AN ENVIRONMENTAL FACTOR

Soil is an integral part of the biosphere. It consists of organic and inorganic matter in the form of solids, liquids and gases, but in addition to these inert materials the soil provides a habitat for many organisms which are essential for the maintenance of the overlying plant communities. Soil is the principal source of nutrients and water for most terrestrial plants, and they in turn provide the bulk of the organic litter necessary for the main-

tenance of the living soil community. Differences in soil properties produced by the interaction of parent materials, climate, topography and vegetation over time have a profound effect on the biological systems that they support. The inorganic mineral fraction of a soil influences physical characteristics such as texture and structure, and thus affects soil moisture, aeration and other properties. Similarly, soil chemistry is partly determined by the breakdown of primary minerals through weathering processes, although the supply of plant-available nutrients is principally controlled by the amount and type of organic matter that is present. Additions of organic matter and other materials to the soil and losses from leaching, seepage and erosion, coupled with transfers and transformations within the soil itself, result in the gradual development of different soil types (Simonson, 1959). The unique combination of factors operating in different parts of the world has resulted in many distinctive soils and this has necessitated some method of classification.

One of the most widely used soil classifications is the Soil Taxonomy developed by the US Department of Agriculture (Soil Survey Staff, 1975). It uses diagnostic horizons to classify soils into 10 Orders which represent the highest class in the system. Six of these diagnostic horizons develop at the soil surface and include, for example, the dark, humus-rich mollic epipedons associated with temperate grasslands. Diagnostic subsurface horizons develop at depth within the soil: 18 are recognized including, for example, the reddish-brown, illuviated spodic horizon associated with coniferous forest and the highly weathered oxic horizon that is characteristic of many tropical soils. The presence or absence of such diagnostic horizons reflects differences in the degree and type of dominant soil-forming processes that have occurred. The soil Orders that are differentiated in this way are classified into suborders and lower classes mainly on the basis of soil moisture and temperature regimes or

Table 1.2 General characteristics of the soil orders in the Soil Taxonomy and approximate equivalence with the Great Soil Groups and higher classes of the Soil Map of the World

Soil Order	General characteristics of the Orders in the Soil Taxonomy	Approximate equivalence with the Great Soil Groups	Approximate equivalence with the highest categories in the Soil Map of the World
Entisols	Recent soils with little profile development and no diagnostic horizons.	Azonal soils, and some Low Humic Gley soils.	Fluvisols, Regosols, Rankers, Lithosols, Arenosols.
Vertisols	Soils of tropical and subtropical regions with high clay content and subject to cracking; typically associated with savanna grasslands.	Grumusols.	Vertisols.
Inceptisols	Soils having weakly developed horizons and containing weatherable minerals: often associated with arctic tundra and mountain areas.	Acid brown soils, some Brown Forest soils, Low Humic Gley and Gley soils.	Gleysols, Andosols, Cambisols.
Aridisols	Shallow stony soils of dry climates with high alkalinity and low organic matter content.	Sierozems, Solonchaks, some reddish brown soils and Solonetz.	Solochaks, Solonetz, Xerosols, Yermosols.
Mollisols	Soils of the subhumid temperate grasslands with thick dark humus-rich A horizons and high base status.	Prairie soils, Chernozems, Chestnut soils, Rendzinas, and some Brown Forest and Humic Gley soils.	Chernozems, Phaeozems, Kastanozems, Greyzems, Rendzinas.
Spodosols	Acidic eluviated soils of cool, humid areas typically associated with boreal forest; low cation exchange capacity and lacking carbonate minerals.	Podsols, Brown Podsolic soils and groundwater Podsols.	Podsols.
Alfisols	Soils of humid and subhumid regions with illuvial clay horizons; typically of high base status; associated with temperate forest, mediterranean areas and tropical grasslands.	Grey Wooded soils, Grey-Brown Podsolic, Degraded Chernozems and associated Planosols.	Luvisols, Podsoluvisols, Planosols, Nitosols.
Ultisols	Chemically weathered clayey soils of warm moist regions; relatively low base status; typically associated with subtropical broad-leaf forest and monsoon tropical forest.	Red-Yellow Podsolics, Reddish-brown Lateritic soils and associated Planosols.	Acrisols.
Oxisols	Deep, highly weathered and leached soils of the humid tropics; red in colour due to accumulated iron oxides; the soils of moist tropical forests.	Lateritic soils, Latosols.	Ferralsols.
Histosols	Dark organic soils of poorly drained areas; typically very acidic and of low fertility.	Bog soils.	Histosols.

distinctive physical and chemical properties associated with differences in parent materials and topography.

Although the Soil Taxonomy is a comprehensive empirical method of soil classification, it has not been adopted universally. A comparison of the 10 Orders in this scheme and the more familiar Great Soil Groups used in the earlier classification of Baldwin *et al.* (1938), and the equivalent categories used on the Soil Map of the World (FAO–Unesco, 1974–78) is therefore presented in Table 1.2.

1.4 THE INFLUENCE OF LIVING ORGANISMS

Plants seldom grow in isolation except in very extreme environments such as polar deserts or sea cliffs where the number of suitable sites are limited. Elsewhere plants grow together as members of a community and must compete for resources. The most successful competitors become common members of the community, others are less prominent, and some may be eliminated. Competition is most intense between organisms with similar ecological requirements. **Intraspecific competition** may occur between members of the same species when the supply of water, nutrients or other resources is limited. Because of this some individuals become stunted and die, and the population is ultimately thinned to its optimum density. Similar interaction between different species is termed **interspecific competition**, and the outcome depends on how effectively each competitor utilizes the scarce resources. Species with identical resource requirements cannot coexist in a stable community. The floristic diversity of a community is thus a reflection of the degree of resource specialization that has occurred between species in that particular habitat. The specialized functional interaction between an organism and its environment defines that species niche. Competition occurs where species niches overlap, and because of this many species are necessarily restricted to particular microhabitats for which they are best adapted.

Competition can reduce a plant population through death of established individuals or by altering the reproductive success of a species. Intraspecific competition typically operates as a density-dependent process in which mortality is regulated by the size of the population. In

Table 1.3 Growth and antiherbivore characteristics of inherently fast-growing and slow-growing plant species (after Coley et al., 1985)

	Fast-growing species	Slow-growing species
Growth characteristics:		
Resource availability in preferred habitat	High	Low
Maximum photosynthetic rates	High	Low
Dark respiration rates	High	Low
Leaf protein content	High	Low
Response to resource pulses	Flexible	Inflexible
Leaf life spans	Short	Long
Successional status	Typically early	Typically late
Antiherbivore characteristics:		
Rates of herbivory	High	Low
Amount of defence metabolites	Low	High
Type of defence	Qualitative	Quantitative
Turnover rate of defence	High	Low
Flexibility of defence mechanism	More flexible	Less flexible

(Reproduced with permission from P. D. Coley, J. P. Bryant and F. S. Chapin, Resource availability and plant antiherbivore defense, *Science*, **230**, 897, copyright 1985 by the AAAS.)

this way a larger proportion of the population is killed as population density increases. The greatest losses normally occur at the seedling stage, but stands are naturally thinned as the plants mature through the suppression of smaller individuals. Conversely, density-dependent fecundity may regulate population size by the number of seeds that are produced by plants at different population densities (Silvertown, 1982). Interspecific competition operates through the imposition of stress, and successful competitors are characteristically adapted to avoid this. **Competitive species** are usually perennial plants with well-developed capacities for resource capture, and exhibit rapid growth of roots and shoots to ensure their advantage (Grime, 1979). Other plants tolerate the stresses imposed by competitors or other unfavourable environmental conditions such as drought or cold. **Stress-tolerant species** are also perennials, but their conservative growth strategy enables them to survive for long periods with little growth or reproduction. Such is the case for many tree species that must remain in the understory of a forest until a suitable gap occurs in the canopy. A third ecological strategy is associated with areas subject to disturbance from factors such as fire and grazing. Under these conditions **ruderals** are at an advantage: mostly they are annual plants with rapid growth rates and short life spans. Reproductive allocation is typically high in ruderal species, and the production of large numbers of readily dispersed seeds with prolonged dormancy is the norm. Many ruderals are fugitive weedy herbs, but shade-intolerant trees, such as jack pine (*Pinus banksiana*), also depend on disturbance for continued establishment.

Plants comprise the resource that sustains higher trophic levels in an ecosystem, and in addition to various competitive interactions they must tolerate the impact of herbivores. Each herbivorous species has a preferred diet and selects plants which are the most palatable. In natural ecosystems the palatable species are damaged and depleted through loss of photosynthetic tissue and reduced assimilation, and this may render them less competitive or limit their reproductive capacity. However, they are never eliminated by overgrazing, because complex mechanisms have evolved to ensure the optimum balance between plants and herbivores. Plants can be classified according to the probability that they will be found by grazing animals (Feeny, 1976). **Apparent species** tend to occur abundantly in specific habitats, and because they are conspicuous to herbivores they depend on non-selective, dose-dependent mechanisms to reduce grazing intensities. **Unapparent species** exploit a number of habitats, and so their distribution is patchy and unpredictable. Further protection is afforded by potent secondary metabolites which deter specific predators. Alternatively, the different methods used by plants to reduce herbivory has been attributed to resource availability and growth rates (Table 1.3). The functional life span of leaves and other organs is longer in slower growing species of plants: this favours increased investment in defensive mechanisms and makes them less palatable to herbivores. With increasing community diversity herbivore preferences become more specific so that some insects, for example, may be restricted to a single plant species. Host-specific herbivore populations are generally limited because the plants are widely dispersed: this in turn offers some protection to the plants themselves.

1.5 ENERGY AND NUTRIENT FLOW THROUGH ECOSYSTEMS

Herbivores do not consume all of the plant material that is available to them. Similarly, most organisms at higher levels in the food web escape predation. The energy and nutrients stored in their tissues is eventually released through death and decomposition. Decomposer organisms are therefore essential for the maintenance of ecosystem structure and function and complement the assimilative activities of other plants and animals. The assimilation of energy by plants in photosynthesis transforms carbon dioxide into carbohydrates such as glucose and starch: further synthesis produces other organic compounds such as fats, oils and proteins. The rate of photosynthesis varies with light intensity, temperature and moisture availability: growth may be further limited by soil nutrient levels. Consequently, there are pronounced regional differences in plant productivity which are primarily related to latitudinal changes in climate (Figure 1.6). Productivity is highest in the humid tropics where the combination of strong radiation, warm temperatures and abundant rainfall provides favourable growing conditions all year round. At higher latitudes plant growth is limited by cooler temperatures and shorter growing seasons, and in desert areas productivity is reduced by the lack of water. Productivity in oceans and lakes is primarily limited by nutrient availability: it is generally highest in the temperate regions where seasonal overturn brings nutrient-rich water to the surface, where light intensity is sufficient for phytoplankton photosynthesis.

The amount of energy that passes to higher trophic levels in an ecosystem is determined by net primary production and the efficiency with which this plant material is transformed into animal biomass. Even under the most favourable conditions, it is estimated that as little as 1–5% of the incident radiant energy is used in photosynthesis. Some of the energy assimilated by plants is used in respiration and cellular maintenance and is therefore unavailable to consumers. Net production efficiency, defined as the ratio of net to gross production, is as high as 75–80% for rapidly growing plants in temperate regions, but drops to 40–60% in the tropics (Ricklefs, 1990). Once the food is eaten the energy it contains is dissipated in various ways. Most of the ingested energy is used in metabolic activity and is lost as heat through respiration; some is excreted in waste products or is egested because it could not be digested, and the

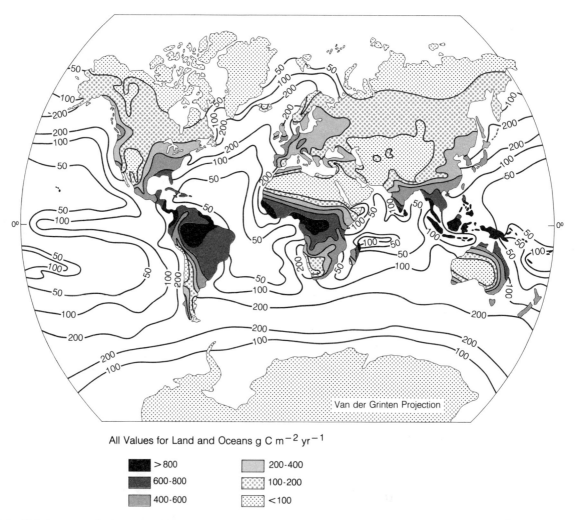

All Values for Land and Oceans g C m^{-2} yr^{-1}

- ■ >800
- ▨ 600-800
- ▩ 400-600
- ▥ 200-400
- ▒ 100-200
- ░ <100

Figure 1.6 Global pattern of primary productivity. (After Lieth, 1975.) (Redrawn from H. Lieth, 'Historical survey of primary productivity', in *Ecological Studies 14 – Primary Productivity of the Biosphere*, eds H. Lieth and R. H. Whittaker; published by Springer-Verlag N.Y. Inc., 1975.)

Table 1.4 Average residence time of energy (biomass/net primary production) in living plant biomass for representative biomes (after Ricklefs, 1990; Whittaker and Likens, 1975)

	Net primary production (g m^{-2} yr^{-1})	Biomass (g m^{-2})	Residence time (yr)
Tropical rain forest	2 200	45 000	20.5
Tropical seasonal forest	1 600	35 000	21.9
Tropical grassland	900	4 000	4.4
Temperate deciduous forest	1 200	30 000	25.0
Boreal forest	800	20 000	25.0
Temperate grassland	600	1 600	2.7
Desert	90	700	7.8
Tundra	140	600	4.3
Swamp and marsh	3 000	15 000	5.0
Lake and stream	400	20	0.05
Ocean	125	3	0.02

(Sources: R. H. Whittaker and G. E. Likens, The biosphere and Man, in *Ecological Studies 14 – Primary Productivity of the Biosphere*, ed. H. Leith and R. H. Whittaker, published by Springer-Verlag N.Y. Inc., 1975; R. E. Ricklefs, *Ecology*, 3rd edn, published by W. H. Freeman and Company, 1990.)

remainder is used for growth and reproduction. The nutritional value of vegetation depends on the amount of lignin and other indigestible materials that the foodstuffs contain. Herbivores assimilate as much as 80% of the energy in seeds and 60–70% of the energy in young vegetation: this declines to 30–40% for animals that browse on shrubby material and is only 15% for millipedes and other organisms that feed on wood. The rate at which energy is transferred from plants to herbivores is calculated as the ratio of the energy stored in biomass to that added each year in net production. In this way the average residence times of energy in living plant biomass varies from only 1 day in marine continental shelf ecosystems to 25 years in temperate and boreal forests (Table 1.4).

Solar energy assimilated by plants passes through the food web, and is dissipated as heat at each trophic level. Organic debris is similarly broken down by the activity of detritus feeders and decomposer organisms, and the energy transformed into biomass through this pathway may again pass temporarily into the food web by way of predation. All life ultimately depends on the energy that continually enters the biosphere through photosynthesis. However, nutrients are circulated within the biosphere, and the amounts available to plants depends on various gains and losses which occur within the ecosystem. Although the interchange of nutrients between the biotic and abiotic components of an ecosystem is unique for each element, two classes of biogeochemical cycles are generally recognized. In gaseous cycles elements such as nitrogen and carbon can exist as gases under normal atmospheric conditions. Alternatively, in sedimentary cycles elements such as phosphorus and potassium remain as solids or pass into solution. Gases enter the biosphere through biological fixation: in sedimentary cycles natural inputs originate from weathering of rock materials. Circulation occurs through abiotic processes such as precipitation, particulate fallout from the atmosphere and streamflow, but in many ecosystems the greatest supply of nutrients is held in organic matter and is made available to plants through decomposition. Many organisms consume detritus, but ultimately the fragmented material passes to the micro-organisms and is rendered to its inorganic components by the activities of fungi and bacteria. The residence time of dead organic matter in the litter varies from a few months in tropical areas where conditions are optimal for decomposer organisms to more than 100 years in the cool boreal forests, but here, as in most biomes, the process of mineralization is accelerated by fire. Fire not only reduces the accumulated litter, but also releases nutrients from any live plants that are consumed.

1.6 SUCCESSIONAL DEVELOPMENT IN PLANT COMMUNITIES

Disturbance from fire and other natural factors, such as disease and hurricanes, periodically alters the nature of the plant cover in an area. Habitat conditions are also changed thereby favouring the establishment of different species of plants. In **classical successions** the species in these new communities are in time replaced by others until the climax associations are once more established. Disturbance of a previously existing plant cover initiates the process of secondary succession. Alternatively, plant colonization may begin on newly formed habitats, such as river sand bars or moraine exposed by glacial retreat: this is termed primary succession. **Primary succession** begins when new sites become available. Pioneer species arrive and become established at the site, and with time alter the environment so that other species can survive. The floristic and structural complexity of these communities increases with time until environmental conditions become stabilized and a self-perpetuating climax association is formed (Clements, 1916). **Secondary succession** proceeds in a similar manner, except that the process is generally much faster because some of the previous plant cover as well as buried seeds and rootstocks will normally persist at the site, and residual soil conditions will also be more suitable. The nature of the secondary plant cover therefore reflects the degree and intensity of the disturbance and the amount time that has elapsed since it occurred (Kellman, 1970).

The appearance of a species during succession depends on the time when its propagules arrive at the site and its tolerance of existing environmental conditions. Subsequent development of the plant cover is controlled by the competitive interaction of the species and their effect on the environment. Three alternative models of successional change are suggested on this basis (Figure 1.7). The **facilitation model** incorporates the classical view of succession proposed by Clements in which the established plant cover modifies the environment so that it becomes more favourable for late successional species. Alternatively, in the **inhibition model** the initial plant cover modifies the environment so that it is less favourable for subsequent recruitment by other species, and

succession proceeds only when the inhibitory species die and are replaced. In the **tolerance model** successional development is determined by the competitive abilities and life spans of the established plant species, so that the longer-lived species characteristically associated with later stages will persist in the plant cover. Consequently, the predictable and sequential changes in vegetation and environment envisioned in classical succession theory may not occur. The development from pioneer communities through distinctive seral stages to the climax vegetation as a result of autogenic environmental modification has been termed **relay floristics** (Egler, 1954). Alternatively, viable propagules of most species may be present initially, and apparent sequential changes in the plant cover occurs because of different germination requirements and growth rates. Interspecific differences in seed dispersal and longevity are important to these concepts. **Pioneer species** are characteristically short-lived

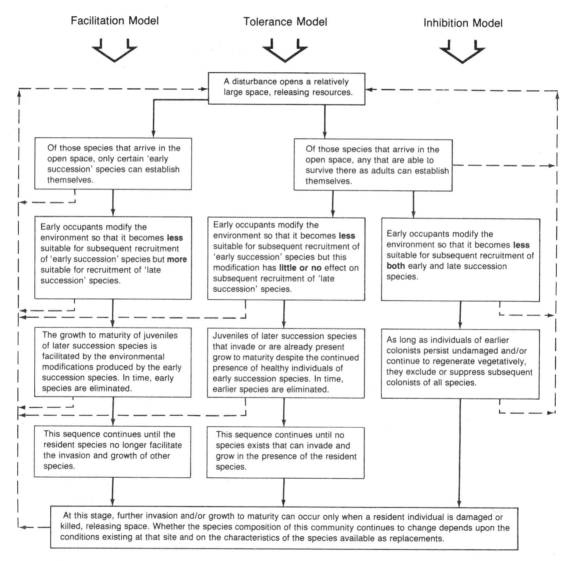

Figure 1.7 Mechanisms of succession in natural plant communities. The dashed lines indicate that secondary succession may be initiated at any stage of development. (After Connell and Slatyer, 1977.) (Reproduced with permission from J. H. Connell and R. O. Slatyer, Mechanisms of succession in natural communities and their role in community stability and organisation, *American Naturalist*, **111**, 1119–44, published by the University of Chicago Press, 1977.)

Table 1.5 General characteristics of plants associated with early and late stages of ecological succession (after Ricklefs, 1990)

	Early succession	Late succession
Adult longevity	short-lived	long-lived
Growth rate	rapid	slow
Size at maturity	small	large
Root:shoot ratio	low	high
Shade tolerance	low	high
Number of seeds	many	few
Seed size	small	large
Seed viability	long	short
Seed dispersal	long distance	short distance

(Reproduced with permission from R. E. Ricklefs, *Ecology*, 3rd edn, published by W. H. Freeman and Company, 1990.)

plants which produce abundant, readily dispersed seeds with comparatively long viability (Table 1.5) and consequently are more likely to be represented in the early plant cover in primary successions. Long-term storage of propagules only occurs beneath an established plant cover and so is more appropriately associated with secondary succession. Thus, the combination of seed availability and suitability of conditions for establishment and growth makes successional change more of a probabilistic process than originally envisioned.

The process of succession is accompanied by changes in many community attributes (Table 1.6). Early successional communities occur in newly created habitats or areas of recent disturbance in which the physical environment is characterized by extreme conditions and variable resource availability. For example, in the absence of a mature plant cover soil moisture levels may fluctuate because of direct exposure to sun, wind and rain. The nutrient capital available to pioneer plants in primary successions is typically very low and in inorganic form, although availability is generally higher in disturbed secondary sites. Characteristically the pioneer communities are composed of few species, but population densities are normally high, because of their efficient dispersal mechanisms, wide ecological tolerance and relatively unspecialized patterns of resource utilization. The rate of primary production per unit of biomass is high because the community is mainly comprised of relatively small, fast-growing species. Overall performance is greatly influenced by environmental perturbations, and overall stability is low. Later stages in succession are characterized by an increase in species diversity and biomass, more complex community structure and greater utilization of resources. Efficiency in resource use is achieved through increased specialization. However, poorly adapted species may eventually

Table 1.6 Trends in community attributes during ecological succession (after Odum, 1969)

Ecosystem attributes	Developmental stages	Mature stages
Community energetics		
1. Gross production/community respiration (P/R ratio)	greater or less than 1	approaches 1
2. Gross production/standing crop biomass (P/B ratio)	high	low
3. Biomass supported/unit energy flow (B/E ratio)	low	high
4. Net community production	high	low
5. Food chains	linear, predominantly grazing	web-like, predominantly detritus
Community structure		
6. Total organic matter	small	large
7. Inorganic nutrients	extrabiotic	intrabiotic
8. Species diversity – variety component	low	high
9. Species diversity – equitability component	low	high
10. Biochemical diversity	low	high
11. Stratification and spatial heterogeneity	poorly organized	well organized
Life history		
12. Niche specialization	broad	narrow
13. Size of organisms	small	large
14. Life cycles	short, simple	long, complex
Nutrient cycling		
15. Mineral cycles	open	closed
16. Nutrient exchange rate between organisms and environment	rapid	slow
17. Role of detritus in nutrient regeneration	unimportant	important
Selection pressure		
18. Growth form	for rapid growth (r-selection)	for feedback control (k-selection)
19. Production	quantity	quality
Overall homeostasis		
20. Internal symbiosis	undeveloped	developed
21. Nutrient conservation	poor	good
22. Stability (resistance to external perturbations)	poor	good

(Reproduced with permission from E. P. Odum, The strategy of ecosystem development, *Science*, **164**, 265, copyright 1969 by the AAAS.)

be lost from the community and consequently species diversity can decline in climax communities. Because of their intricate organization, homeostatic regulation is characteristically high in mature communities. Structure and function become increasingly self-regulated with maturity, and interaction with the external environment declines. However, the mature plant communities still exhibit seasonal rhythms as growing conditions fluctuate during the year. They must also evolve in response to long-term environmental change.

1.7 HISTORICAL PERSPECTIVES IN PLANT GEOGRAPHY

The present distribution of plant species is the product of past events, and the changing patterns which have occurred through time are preserved in the geological record. Plant microfossils extracted from lake deposits and peat bogs are most commonly used to interpret events in the Quaternary (Figure 1.8), especially the dramatic changes in plant distributions caused by the Pleistocene glaciations. Plant microfossils are mainly comprised of pollen grains and spores. They are exceptionally resistant to decay in anaerobic sediments and their unique morphologies facilitate identification. Plant macrofossils such as wood fragments, leaves and fruits may also be present, but their usefulness in palaeobotany is sometimes limited by problems of identification. It is also difficult to establish if the plants were actually growing at the fossil site (West, 1977b). The changing pattern of world vegetation is thereby projected back through time. Vascular terrestrial plants evolved in the Silurian period some 400 million years ago, and rapidly

Era	Period	Epoch	Duration (million years)	Age (million years)
CENOZOIC	Quaternary	Holocene	0.01	0.01
		Pleistocene	2	2
	Tertiary	Pliocene	3	5
		Miocene	19	24
		Oligocene	13	37
		Eocene	21	58
		Palaeocene	8	66
MESOZOIC	Cretaceous		78	144
	Jurassic		64	208
	Triassic		37	245
PALAEOZOIC	Permian		41	286
	Carboniferous		74	360
	Devonian		48	408
	Silurian		30	438
	Ordovician		67	505
	Cambrian		65	570
PRECAMBRIAN			3200	3800

Figure 1.8 The geologic time scale. (After Strahler and Strahler, 1992.) (Reproduced with permission from A. H. and A. N. Strahler, *Modern Physical Geography*, 4th edn, published by John Wiley and Sons Inc., copyright © 1992.)

increased in diversity during the Devonian period with the appearance of the first seed plants, then declined during the period of mass extinction at the end of the Permian. These early floras are distinguished by the adaptive radiation of the pteridophytes. The gymnosperms increased in diversity during the Carboniferous and remained the dominant plant group until the appearance of angiosperms in the mid-Cretaceous (Signor, 1990). Today the number of ferns and their allies totals about 12 000 species, most of which are native to the moist tropics. The gymnosperms include some 500 species of conifer, 100 species of cycad and a few other ancient plants such as the maidenhair tree (*Ginkgo biloba*), which is native to China, and *Welwitschia mirabilis*, which is restricted to the coastal fog-belt of the Namib Desert. The greatest diversity is exhibited by the angiosperms with some 250 000 species.

Biogeographic interpretation of the early fossil record has been greatly advanced by the general acceptance of the theory of continental drift. Thus, the fern-like *Glossopteris* is widely distributed in Permian rocks in India, Australia, southern Africa, South America and Antarctica, and presumably spread to these locations at a time when the southern continents formed part of Pangaea. The '*Glossopteris* flora' consists of plants with deciduous leaves and pronounced growth rings in their wood, which

suggests that they were adapted to a temperate climate (Schopf, 1970). The break up of Pangaea into the northern and southern landmasses of Laurasia and Gondwana occurred about 180 million years ago in the late-Triassic, and this early separation is still reflected in the present flora. Thus the proportion of plants with entire leaves is higher in the southern hemisphere floras than in the northern hemisphere. Similarly, the forests of high latitudes are dominated by conifers in the north but are mostly composed of angiosperms in the south, with an admixture of ancient and distinctive gymnosperms (Pielou, 1979). Rotational movements resulted in the separation of Africa, India and Madagascar from the other southern continents in the mid-Cretaceous about 100 million years ago (Figure 1.9). Australia, South America and Antarctica became separated during the Eocene, about 49 million years ago. These events are reflected in the distribution of the southern beeches (*Nothofagus* spp.) which now grow in temperate regions in South America, Australia, New Zealand and New Guinea, with fossil sites known from Antarctica. The absence of *Nothofagus* from Africa is generally attributed to the early isolation of this continent. Alternatively, the genus may have become extinct in Africa as the continent drifted northwards into tropical latitudes, or perhaps it was unable to compete with plants coming from Eurasia.

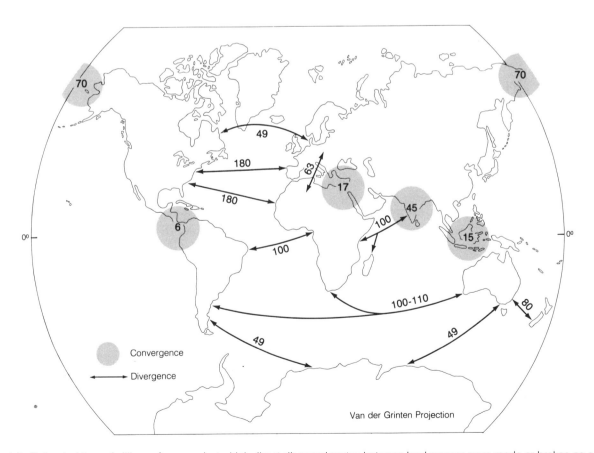

Figure 1.9 Estimated times (millions of years BP) at which direct dispersal routes between land masses were made or broken as a result of continental drift. (After Pielou, 1979.)

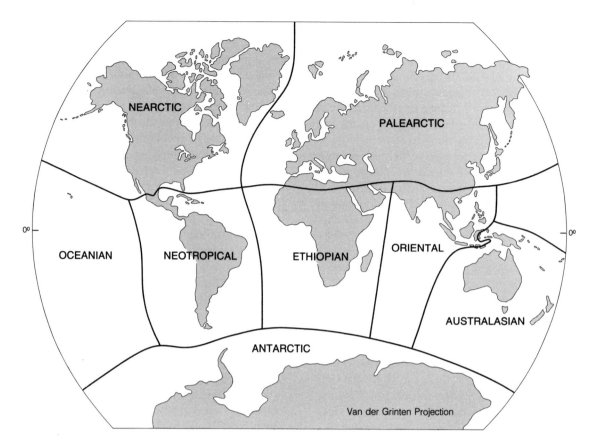

Figure 1.10 Biogeographic regions of the world combining traditional zoogeographical and phytogeographical realms. (After Pielou, 1979.)

The distribution of a plant species can thus be attributed to historical events as well as its migrational ability and adaptability to present environmental conditions. Many different types of distribution are recognized (Polunin, 1960). For example, **cosmopolitan** species are found in most parts of the world: they are mostly weedy species which have been distributed unwittingly by humans through cultivation, or are represented by mosses and other cryptogams. More restricted are the **circumpolar** plants which are found only at higher latitudes, and **pantropical** species which occur throughout the tropics. Disjunct distributions occur when locations are widely separated as is the case with some **arctic-alpine** species. In contrast **endemic** species are very restricted in their range, and may occur only in a peculiar environment, such as those associated with serpentine soils, or perhaps are limited to a single island. Some, like the red-woods (*Sequoia* spp.) may be relic species that are remnants of an earlier flora that was previously much more extensive. The geographic ranges of most species are comparatively small. Consequently, it is possible to subdivide the world into regional floras on the basis of their distinctive taxa. In this way the world is divided into 37 floristic regions (Good, 1964). These can be combined into eight biogeographic realms which account for the regional differences in phytogeography and zoogeography (Figure 1.10). Even though the biogeographic realms are taxonomically distinct because of the restricted range of most plant and animal species, adaptation to similar ecological niches has given rise to many equivalent forms. In this way the major biomes of the world are distinguished on the basis of their structural similarity and functional relationship with the environment.

CHAPTER 2

The tropical forests

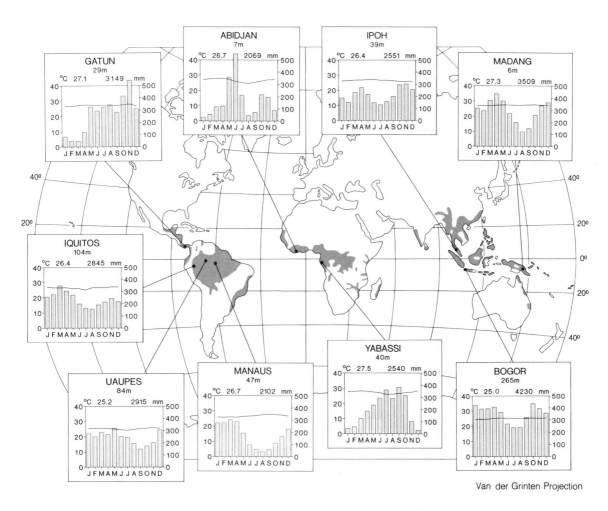

Van der Grinten Projection

Figure 2.1 Distribution of tropical forest and representative climatic conditions. Mean monthly temperatures are indicated by the line and mean precipitation for each month is shown by the bars. Station elevation, mean annual temperature and mean annual precipitation appear at the top of each climograph.

2.1 INTRODUCTION

The tropical forest ecosystem is represented by three floristically diverse formations. The American formation is the most extensive, but large tracts of tropical forest are also found in equatorial Africa and South East Asia (Figure 2.1). Throughout these regions there is a recurrent structural and physiognomic adaptation to similar environmental conditions. However, these convergent traits have arisen in very different floras and many species have extremely localized distributions. The most complex communities, described as '**rain forests**', are associated with the warm, moist tropical lowlands (Figure 2.2). Their multilayered, evergreen canopies

15

Figure 2.2 (a) Continuous cover of tropical lowland forest in Rondonia, Brazil; (b) the multi-layered structure of the canopy; (c) diversity of life forms are the characteristic features of tropical lowland forests. ((a) G. T. Prance; (b) J. T. Price; (c) G. A. J. Scott.)

provide a mosaic of niches which support a great diversity of smaller plants and animals. This is but one of many tropical forest types, the characters of which are largely determined by differences in temperature and precipitation regimes. At higher elevations where conditions are cooler the trees become smaller (Figure 2.3) and as the canopy begins to open the understory plants become increasingly abundant (Table 2.1). **Semi-deciduous forests** occur in regions where precipitation is more seasonal (Figure 2.4) and many species shed their leaves during the dry period. **Swamp forests** occur in areas that are seasonally flooded. Some, like the **várzea** that grow in areas flooded by the muddy-water tributaries of the Amazon, are physiognomically similar to lowland terra firma forests. **Igapó** forests occur in areas that are periodically inundated by clear, nutrient-poor rivers and the white sandy soils characteristically support fewer species and less biomass (Figure 2.5).

Figure 2.3 Tropical montane forests on the Serra Parima, northern Brazil. (G. T. Prance.)

Table 2.1 Classification of tropical forests (after Longman and Jeník, 1987)

A. Tropical evergreen forests: composed mostly of evergreen species with little or no bud protection against cold or drought.
   1. Lowland forest: multilayered structure with many trees exceeding 30 m in height; undergrowth sparse beneath the closed canopy; palms rare, climbers nearly absent, epiphytes uncommon.
   2. Montane forest: few trees exceed 30 m in height; tree-ferns and small palms common in undergrowth; ground layer rich in herbs and mosses, epiphytes present.
   3. Cloud forest: closed forest structure but with numerous gaps; trees often gnarled, rarely over 20 m in height; numerous woody lianas and herbaceous climbers; rich epiphytic flora, and extensive gound cover of mosses, ferns and broad-leaved herbs.
   4. Alluvial forest: multilayered closed forest with numerous gaps, growing on river banks and in areas liable to flooding; buttresses and stilt-roots characteristic of many species; palms, herbaceous undergrowth and epiphytes more common than in other lowland associations.
   5. Swamp forest: low-lying moist sites with poor soil aeration; tree flora less diverse; buttresses, stilt-roots and pneumorrhizae common; extensive growth of ferns and herbs.
   6. Peat forest: on peat covered, nutrient-poor soils; comparatively few tree species rarely growing more than 20 m in height; palms may be dominant in some sites; poorly developed ground cover mostly of ferns.
B. Tropical and subtropical seasonal forests: mainly evergreen and leaf-exchanging trees with some bud protection; trees of upper canopy moderately drought-resistant, but leaf shedding will occur in dry season; tree-ferns absent in montane forms, but climbers and evergreen shrubs common; in highest sites trees rarely exceed 10 m, and little undergrowth or ground cover develops beneath their dense canopy.
C. Tropical and subtropical semi-deciduous forests: trees of upper canopy shed leaves during dry season but smaller trees and shrubs usually evergreen; tree seedlings, saplings and shrubs common in ground cover along with grasses; herbaceous climbers present; epiphytes rare, but more abundant at higher elevations.
D. Subtropical evergreen forests: multilayered forests but less vigorous than tropical forms because of pronounced seasonal temperature regime; shrubs dominate understory.
E. Mangrove forests: single layered forests growing to 30 m in intertidal zones throughout tropics and subtropics; evergreen, halophilous trees equipped with stilt-roots, and variously adapted pneumorrhizae; little ground vegetation develops, but bryopyhtes and lichens often associated with lower stems and roots; some epiphytes may be present.

(Source: K. A. Longman and J. Jeník, *Tropical Forest and its Environment*, 2nd edn; published by Longman, 1987.)

Figure 2.4 Semideciduous forest in southern Brazil. (G. T. Prance.)

2.2 REGIONAL DISTRIBUTION

The American formation contains about 50% of the world's tropical forests. Most of the forest is found in the Amazon drainage basin of Brazil and along the Andean foothills. The high Andes effectively isolates the forests of Amazonia from those of northern Colombia. These extend northwards through Panama and Costa Rica into eastern Central America. The drier climate of central Brazil separates the Atlantic coastal forests from the Amazonian forest region. Also, small patches of forest can still be found on many islands throughout the Caribbean. Leguminous tree species are comparatively abund-

ant in neotropical forests (Table 2.2), but the distributions of many genera are often quite limited so that several floristically distinct regions are recognized (Prance, 1989).

The Indo-Malesian formation accounts for about 30% of the tropical forests. It is the most fragmented of the formations, being dispersed throughout the islands of Malesia, Melanesia and the Philippines. In mainland Asia monsoon forests and drier deciduous woodlands separate the evergreen forests of the Malay peninsula, Myanmar and Thailand from those in the coastal regions of Cambodia and Vietnam. The forests of Sri Lanka and western India are similarly isolated from the main region. The formation reaches its most southerly limit in eastern Queensland, where it intermixes with subtropical forms. The rain forests of the Far East are dominated by dipterocarps which in many stands comprise 80% of the canopy and 40% of the understorey cover (Mabberley, 1992), and in monsoon areas a single species, such as the Sal tree (*Shorea robusta*) of India, is often the sole dominant.

Most of the forested area in the African formation occurs in the Congo Basin of northern Zaire. Its eastern limit is essentially the African Rift Zone. In the west the forest reaches the Guinea coast, where it becomes restricted to the southern part of Nigeria, Benin and Togo. Drier climatic conditions in the region known as the Togo–Dahomey gap separate the forests of Ghana, the Ivory Coast, Liberia and Sierra Leone from the main part of the African formation. The relic forests of eastern Madagascar are also recognized as part of

Figure 2.5 Seasonal aspects of igapó forest which is flooded annually by the Rio Negro in Brazil. (G. T. Prance.)

this formation. Leguminous trees such as *Brachystegia laurentii*, *Gilbertiodendron dewevrei* and *Cynometra alexandri* are common canopy dominants in lowland forests which are distinctively poorer in species than the American and Indo-Malesian formations (Hamilton, 1989).

2.3 CLIMATE

In the Köppen system of climate classification, the tropical rainforest climate (Af) occurs in areas where the mean temperatures of all months exceed 18 °C, and minimum monthly precipitation is above 60 mm. Within the tropical lowland forest zone mean annual temperatures typically exceed 25 °C with an annual range of less than 5 °C (Figure 2.1): in these regions diurnal fluctuations are more pronounced than seasonal patterns. Continual daytime heating causes convectional uplift and results in a predictable daily cycle of cloud build-up and rain. Cumulus clouds begin to form as the air temperature rises; by mid-afternoon towering cumulonimbus clouds have developed and thundershower activity persists until even-

ing. Rainfall in these regions normally exceeds 2000 mm and can be even greater in highland areas. This orographic effect is particularly noticeable in places such as Central America where the moist trade winds are forced to rise over mountains. Although mountain barriers influence regional precipitation patterns, rainfall generally decreases and becomes more intermittent with distance from the equator. As a result the tropical moist evergreen forests are replaced in turn by less luxuriant seasonal forests, semi-deciduous forests and savanna woodlands.

Subdivisions of the rain forest are associated with areas of pronounced seasonal precipitation regimes. Regions such as south-eastern Brazil, West Africa, western India, Assam and Myanmar experience a monsoon climate. Abundant rain falls in the summer season as the trade winds bring moist, maritime tropical air onshore. Changes in global pressure patterns bring drier air to these regions in winter. Consequently, annual precipitation in monsoon climates usually exceeds 1500 mm, although most of this falls in the summer. For example, average annual precipitation at Rangoon is 2619 mm, but only 145 mm falls between November and April. The forests in the monsoon areas usually support fewer species, many of which, like teak (*Tectona grandis*), are deciduous or semi-deciduous. Elsewhere in the tropics a decline in total annual precipitation is normally accompanied by more prolonged dry spells and the vegetation becomes increasingly drought resistant.

In mountainous terrain there is a general cooling trend with elevation. Daily, monthly and annual temperature ranges become more pronounced and the likelihood of frost and snow increases. This results in an altitudinal zonation of tropical vegetation types. The complex lowland forests are progressively replaced by simpler communities at higher elevations. Smaller trees and denser undergrowth occur in the lower montane forests, with mossy elfin woodland above. Temperate species begin to appear in the flora above about 2500 m. In the drier zones above cloud level, xeromorphic dwarf heath may develop with alpine grasslands present at elevations of 4000–4500 m. Above 4500–5000 m these communities give way to barren terrain and permanent snow cover. However, the altitudinal limits of each zone can vary markedly. Richards (1952) suggested an approximate elevation of 1000 m for the upper boundary of the lowland forest, although this may range from 600 m to 2900 m, with a corresponding displacement of the higher zones. The lower values are normally associated with small, coastal mountains, and the zones become progressively higher inland or on larger mountain ranges. This response of vegetation to physiographic conditions is termed the '**Massenerhebung effect**'. It is most pronounced in the geographically diverse terrain of South East Asia. The physiological mechanisms which underlie these altitudinal distribution patterns is still speculative. As well as increased risk of frost damage (van Steenis,

Table 2.2 Characteristic families and genera containing dominant, abundant, conspicuous or subendemic woody plants in the rain forests of the world, with associated epiphytes and secondary forest trees (after Longman and Jeník, 1987; Mabberley, 1992)

Neotropics

Leguminosae	*Andira, Apuleia, Dalbergia, Dinizia, Hymenolobium, Mora*
Sapotaceae	*Manilkara, Pradosia*
Meliaceae	*Cedrela, Swietenia*
Euphorbiaceae	*Hevea*
Myristicaceae	*Virola*
Moraceae	*Cecropia, Ficus*
Lecythidaceae	*Bertholletia*
Epiphytes	ferns, Orchidaceae, Bromeliaceae, Cactaceae
Secondary	*Cecropia, Miconia, Vismia*

Africa

Leguminosae	*Albizia, Brachystegia, Cynometra, Gilbertiodendron*
Sapotaceae	*Afrosersalisia, Chrysophyllum*
Meliaceae	*Entandrophragma, Khaya*
Euphorbiaceae	*Macaranga, Uapaca*
Moraceae	*Chlorophora, Ficus, Musanga*
Sterculiaceae	*Cola, Triplochiton*
Ulmaceae	*Celtis*
Epiphytes	ferns, Orchidaceae
Secondary	*Harungana, Macaranga, Musanga*

Indo-Malesia

Dipterocarpaceae	*Dryobalanops, Hopea, Shorea*
Leguminosae	*Koompassia*
Meliaceae	*Aglaia, Dysoxylum*
Moraceae	*Artocarpus, Ficus*
Anacardiaceae	*Mangifera*
Dilleniaceae	*Dillenia*
Thymelaeaceae	*Gonystylus*
Epiphytes	ferns, Orchidaceae, Asclepiadaceae, Rubiaceae
Secondary	*Glochidion, Macaranga, Mallotus, Melastoma*

(Source: K. A. Longman and J. Jeník, *Tropical Forest and its Environment*, 2nd edn; published by Longman, 1987.)

1968), some species may be adversely affected by reduced assimilation in foggy, high elevation sites (Grubb and Whitmore, 1966) or by low soil fertility (Grubb, 1977).

2.4 SOILS

The continually warm, moist conditions in lowland tropical forests promotes strong chemical weathering and rapid leaching of soluble materials. The characteristic soils are **Oxisols**. These deeply weathered soils have no distinct horizons and eventually intergrade to saprolite at depths of 3 m or more. Chemical alteration of primary rock minerals such as feldspars, micas and pyroxenes through the removal of potassium, magnesium and other soluble constituents results in the accumulation of iron and aluminium sesquioxides which characteristically imparts a strong red colour to the soil. The lack of fresh, weatherable rock fragments results in a substrate that is essentially devoid of available plant nutrients. In these climates silica is also soluble, and recrystallizes with aluminium to form kaolinite, a rather inert clay mineral with a low cation exchange capacity; the low organic status of the soil further exacerbates the problem of nutrient deficiency. Most of the plant nutrients are added by the rapidly decomposed organic matter that is continually falling on to the surface, and are quickly absorbed by the growing plants. The soil itself remains acid (pH 4.5–5.5): this can result in aluminum toxicity and subsequent interference with phosphorus uptake. Despite their high clay content, most Oxisols exhibit a fine granular structure that is easily penetrated by roots, is readily drained and is resistant to erosion.

Ultisols may develop in areas with more seasonal precipitation regimes. They are generally associated with forested regions exhibiting seasonal soil-water deficits and are particularly characteristic of monsoon regions. Their distinctive property is a well-developed argillic (clay-rich) B_t horizon which is yellow-brown or reddish in colour as a result of a high concentration of iron sesquioxides. Chemically they are similar to Oxisols, being low in organic matter content and generally deficient in plant nutrients. However, Ultisols are more prone to impeded drainage and accelerated erosion, and to the formation of compact layers of iron-rich clay or laterite which with repeated wetting and drying can harden into plinthite. Laterite causes severe management problems.

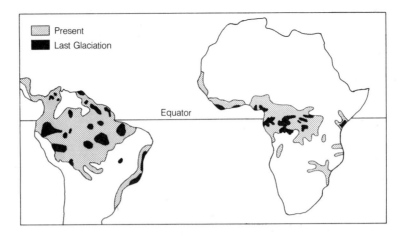

Figure 2.6 Extent of lowland tropical rain forest during the Wisconsin glacial maximum (approximately 18 000 years BP) in South America and Africa. (After Prance, 1982; Kingdon, 1970.) (Reproduced from J. Kingdon, *Island Africa. The evolution of Africa's rare animals and plants*, published by Collins, 1990 and from *Biological Diversification in the Tropics*, G. T. Prance (ed.), 1982, © Columbia University Press, New York; reprinted with permission of the publisher.)

However, it is estimated that only 7% of tropical soils are adversely affected by these conditions (Sanchez and Boul, 1975), and laterite is rare in undisturbed forests.

Volcanic activity occurs in many parts of Central America and South East Asia. Recent ash deposits quickly weather to **Andosols** with a high content of allophane. This amorphous alumino-silicate clay maintains the soils in a loose, porous condition and consequently they are well drained. However, their natural fertility is reduced because phosphorus is readily bound up. With time the allophane changes to kaolinite, and the volcanic soils are ultimately indistinguishable from Oxisols or Ultisols. In high elevation sites, incomplete decomposition of organic material produces soils which are distinctively peaty. The peat layer is best developed in the cool, ever-humid cloud forest zone. Here organic horizons may accumulate to a depth of 1 m and the saturated, acid soils are of low fertility; gleying and iron pans are common in the shallow mineral layers beneath. Intensively leached white sands are also present in some parts of the tropical forest biome. In these highly acid soils, pH can be as low as 2.8, but even though they are virtually sterile they support communities of scleromorphic plants which root in a thick layer of slowly decomposing litter (Anderson, 1981).

2.5 FLORISTIC DIVERSITY

Tropical forests are renowned for the great diversity of species they support. It is estimated that these regions contain more than 100 000 species – approximately 40% of the world's flora. Richards (1952) attributed the floristic diversity of the tropical forests to a long evolutionary history, but this is only partially supported by the fossil record (Flenley, 1979). Further evidence indicates that significant climatic changes occurred during the Pleistocene with cool, dry conditions prevailing during the

glacial maxima. Haffer (1969) postulated that these fluctuating conditions resulted in the periodic fragmentation of the forest habitat in South America (Figure 2.6). Regions of high endemism noted in present-day plant and animal distributions are considered to be a vestige of the evolutionary differentiation which occurred in these disjunct forest refugia during the drier phases (Prance, 1982; Turner, 1982). In Africa the forests were probably less fragmented with perhaps 9 or 10 refuges occuring in the Zaire Basin (Kingdon, 1990), and this may account for the relative poverty of the flora and the wide distribution of many of its species (Richards, 1973). However, Axelrod and Raven (1978) attribute Africa's unusually poor tropical flora to widespread extinction as the climate became more arid prior to the Pleistocene.

The effect of the Pleistocene in South East Asia was rather different. It is estimated that sea level was lowered by about 200 m during glacial maxima. This resulted in many of the Malesian islands becoming grouped into the land units of Sunda and Sahul which were effectively separated by an intervening region of deep water (Figure 2.7). In mountainous regions, vegetation zones were displaced by as much as 1500 m as the climate changed (Hope, 1976), which periodically restricted some elements of the flora to the higher peaks. Alternating periods of isolation and interchange has been a major factor promoting speciation in the distinctive flora of New Guinea. Geobotanical evidence indicates a similar sequence of events in East Africa (Hedberg, 1969) and in the Andes (Figure 2.8).

Geographic isolation can be maintained in animal populations which evolved in past refugia if they do not travel great distances while foraging and if they are not forced to migrate because of seasonally unfavourable conditions. Further genetic differentiation could arise from reduced contact between members of these sedentary populations. The dependency on animals for fruit

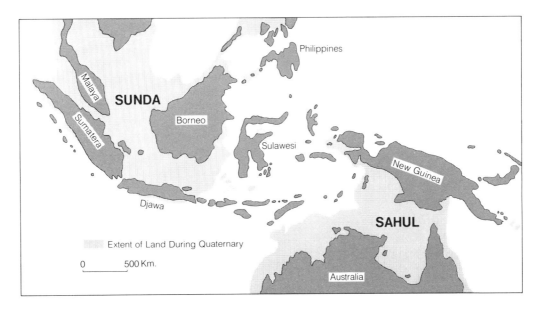

Figure 2.7 The extent of the Sunda and Sahul shelves exposed during the Quaternary as determined by the 200 m bathymetric contour at the continental limits of South East Asia and Australia. (After Walker, 1982.) (Adapted from D. Walker, Speculations on the origin and evolution of Sunda–Sahul rain forests, in *Biological Diversification in the Tropics*, G. T. Prance (ed.), 1982, © Columbia University Press, New York; reprinted with permission of the publisher.)

and seed dispersal could then have led to a parallel enrichment of the tropical forest floras. The rate of speciation may be further accelerated by high productivity in these continually favourable environments. Alternatively, greater diversity could result from lower rates of extinction because of reduced competition, or because of greater environmental stability, although Connell (1978) argued that high species diversity can only be maintained by frequent environmental disturbance. The limitation of some animal populations by a shortage of food (Smythe *et al.*, 1982), and the possibility that the scleromorphic leaf form is an adaptation to recurrent dry spells (Walter, 1985), attest to the variable conditions which can occur in these regions.

Diverse communities will develop only if species can coexist. The general uniformity of conditions which favour forest establishment conceals the variety of microhabitats beneath the canopy. Variations in light, temperature, moisture and food resources present a complex of niches each of which may offer distinct advantages to very specialized life forms. Increased rates of speciation coupled with the opportunity to succeed beneath the complex structure of the canopy has resulted in an unparalleled richness of all life forms. The Malesian flora, for example, is comprised of 25 000–30 000 species, of which 30% are trees over 10 cm in diameter (Jacobs, 1974). This great diversity of trees is one of the distinctive features of tropical lowland forests. As many as 176 species ha^{-1} can occur in parts of Malaysia (Wyatt-Smith, 1966), and Mori *et al.* (1983) have reported 178 species over 10 cm in diameter in a 0.67 ha plot in South America. In contrast only 60 species ha^{-1} may be present in Nigeria (Richards, 1952). The vascular flora in forest sites has rarely been inventoried completely. A

total of 333 species ha^{-1}, including 78 tree species, has been reported for montane forest in Java (Meijer, 1959) and in Ghana Lawson *et al.* (1970) recorded over 275 species ha^{-1}, about half of which were trees. Such floristic and structural complexities are conveniently summarized in the 'synusiae' developed by Richards (1952) which provide an ecological classification of the plants according to their form and functional role in the community (Table 2.3). Of the different classes used, the mechanically dependent plants – climbers, stranglers and epiphytes – are a very characteristic and conspicuous element in tropical forests. However, in terms of Raunkiaer's life form classification they typically represent only 2% or 3% of species present. In these mild moist climates, the life form spectrum is dominated by woody phanerophytes.

2.6 STRUCTURAL CHARACTERISTICS OF TROPICAL FORESTS

2.6.1 Canopy structure

The tree species in lowland forests can be grouped into three general height classes or strata. The tallest trees of the **A stratum** form a discontinuous layer in which there is no lateral contact between adjacent crowns (Figure 2.9). These are the emergent trees which rise above the denser canopy of the intermediate **B stratum** (Figure 2.10); members of the shortest **C stratum** are dispersed below. The height of the trees in the A stratum is typically 30–42 m, decreasing to 18–27 m in the B stratum and 8–14 m in the C stratum (Richards, 1952). However, many trees can reach heights of 50 m with emergents

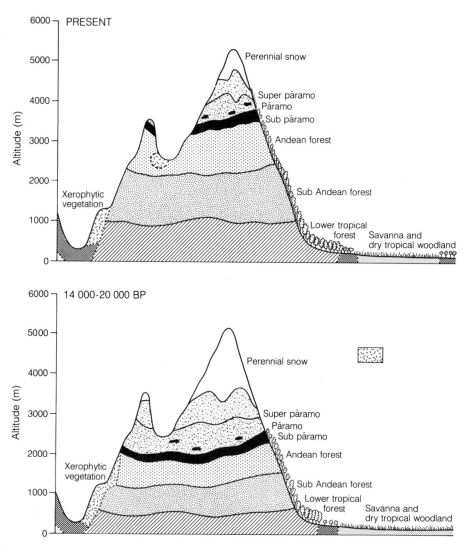

Figure 2.8 The present vegetation belts in the eastern Andes of Colombia and a tentative reconstruction of the vegetation zones during the last glacial maximum. (After van der Hammen, 1974.) (Reproduced with permission from T. van der Hammen, The Pleistocene changes of vegetation and climate in tropical South America, *Journal of Biogeography*, 1974, **1**, 5, 17.)

Table 2.3 Tropical forest synusiae (after Richards, 1952)

A. Autotrophic plants (with chlorophyll)
    1. Mechanically independent plants
        Trees and shrubs
        Herbs
    2. Mechanically dependent plants
        Climbers
        Stranglers
        Epiphytes
B. Heterotrophic plants (without chlorophyll)
    1. Saprophytes
    2. Parasites

(Source: P. W. Richards, *The Tropical Rain Forest*; published by Cambridge University Press, 1952.)

growing to 70–80 m (Bourgeron, 1983). An 84 m specimen of *Koompassia excelsa* in Sarawak is reportedly the tallest tree in the tropics (Whitmore, 1989a). The distinctiveness

of the strata is often obscured by younger individuals as they grow to their mature heights, and the strata are also less apparent in sites with high species diversity. Consequently, Brunig (1983) has argued that stratification is an artifact of sampling rather than an inherent characteristic of the forests. Beneath the trees the **D stratum** is comprised of small palms, tall herbs, large ferns and shrubs of various forms (Figure 2.11). The **E stratum** is made up of herbaceous plants, although these may be outnumbered by tree seedlings. Individuals within the D and E strata are widely dispersed, with densities usually increasing in openings or disturbed sites. A thin layer of litter 1–3 cm thick may cover the forest floor, but patches of bare soil are common.

To this structurally complex plant cover must be added the climbers and epiphytes which commonly festoon the forest. The climbing vines and lianas use different methods of attachment. Some are equipped with

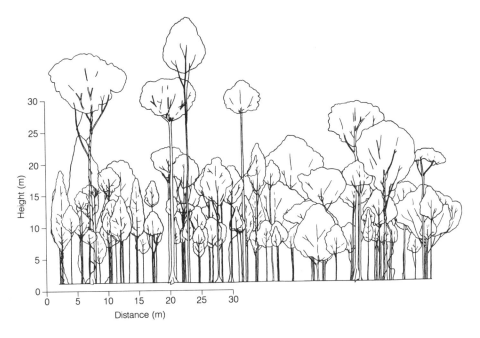

Figure 2.9 Profile diagram showing the characteristic stratification of trees in tropical rain forest. (After Richards, 1936.) (Reproduced with permission from P. W. Richards, Ecological observations on the rain forest of Mount Dulit, Sarawak. Part I, *Journal of Ecology*, 1936, **24**, 10.)

hooks or thorns, others may possess tendrils or adhesive organs, or they may simply twine around the supporting plants (Figure 2.12). Herbaceous climbers are common in the lower strata, and woody forms can reach to the highest tree crowns. The largest lianas can exceed 15 cm in diameter, their woody, rope-like stems often spreading to adjacent trees (Figure 2.13). The much-used rattan which comes from *Calamus*, a genus of climbing palm, can reach a length of 165 m. Although they represent only a small fraction of the forest biomass, some light-demanding climbers form dense tangles in larger clearings or along river banks (Figure 2.14). Most of these fast-growing plants have a high leaf-to-stem biomass

Figure 2.11 (a) *Cibotium* tree ferns in montane *Metrosideros polymorpha* forest in Hawaii; (b) *Licuala* palms in the understory tropical lowland forest in Brunei. ((a) R. A. Wright; (b) G. T. Prance.)

Figure 2.10 Emergent trees of the A stratum penetrate the continuous canopy of shorter trees. (K. A. Hudson.)

Figure 2.12 (a) A large-leaved vine encircles the trunk of a palm in Eungella National Park, Queensland; (b) most twining species climb in an anticlockwise spiral. (G. W. and R. Wilson.)

Figure 2.13 The cable-like form of woody lianas are especially abundant in tropical moist lowland forests. (G. T. Prance.)

Figure 2.14 Liana thickets often develop in clearings and along river banks, but density usually declines as the forest stand matures. (G. T. Prance.)

Epiphytes represent a second group of plants that rely on other species for mechanical support. Most are species of orchids and ferns which grow attached to trunks and branches at various heights within the canopy (Figure 2.15), although mosses become increasingly dominant in the forests at higher elevations. The epiphytic flora is best represented in the moister, montane forests where they may equal 50% of the tree leaf biomass (Edwards and Grubb, 1977). Unlike the climbers, the majority of epiphytes never come into contact with the soil, but obtain moisture and nutrients from rainwater as it percolates through the canopy. Alternatively, some species are hemiepiphytic and either begin life as vines but become epiphytic later, or develop aerial roots that extend to the ground. Notorious amongst this group are the strangling figs (*Ficus* sp.), the ever-thickening root systems of which can eventually stop the flow of fluids in the conducting tissues of the supporting trees (Figure 2.16). Subtle differences in shade and moisture within the canopy affect the distribution of the epiphytes (Johansson, 1989). Many epiphytic ferns depend on moist humus which has accumulated in crutches where branches arise from a trunk. In the lightly shaded zones further along the branch epiphytic orchids that are better

and their leaves are invariably protected by potent phytotoxins; the rattans may be further protected by biting ants which they harbour in their leaf sheaths. Despite good water conductance, the climbers are vulnerable to drought and mineral nutrient deficiencies.

(a)

(b)

Figure 2.15 (a) The spongy mantle of epiphytic *Philodendron* absorbs mineral-charged water that moves through the canopy at the onset of a rainstorm; epiphytes commonly develop free-hanging aerial roots; (b) many species of *Anthurium* are secondary hemiepiphytes in that the young plants are climbing vines that root in the soil and subsequently become epiphytic. (G. T. Prance.)

Figure 2.16 The roots of this epiphytic fig (*Ficus* sp.) originally grew entwined around a supporting tree which subsequently died and decomposed, leaving the fig supported by its own stilt roots. Plants which germinate in the tree crowns but eventually anchor in the soil are termed primary hemiepiphytes. (G. A. J. Scott.)

adapted to drier conditions are more common, while the end section of the branch supports only a few smaller epiphytes. Most epiphytes absorb nutrients through their roots, although some are equipped with hair-like trichomes which permit nutrient uptake through leaf surfaces. More unusual are the myrmecophilous epiphytes that obtain nutrients from organic litter supplied by ants which nest in specialized chambers within the plants (Huxley, 1980), or the mutualistic relationship in which arboreal ants place seeds of the epiphytic vine *Codonanthe crassifolia* in the walls of their nests (Kleinfeldt, 1978).

2.6.2 Tree architecture

The trees in each stratum have characteristic shapes to their crowns which are determined by the height at which branching occurs, the length of the branches and their angle of divergence. The taller trees possess rather long, slender trunks in comparison with their crowns which are typically flattened and umbrella-shaped. The crowns become more spherical in mature trees of the B stratum, and are distinctly elongate in individuals of the C stra-

tum (Richards, 1952). Depending on their method of growth, the trunks of the trees are relatively straight, or develop a somewhat sinuous form. In larger trees branching is generally confined to the uppermost part of the trunk, although lower branches will often develop in open sites, such as clearings, or along river banks.

All tree forms can be described by 23 architectural models according to the branching activity which occurs during growth (Figure 2.17). Straight stems, such as those of coffee (*Coffea arabica*), arise through monopodial growth with branching occurring below the terminal indeterminate meristem (Figure 2.18). The irregular trunk of cocoa (*Theobroma cacao*) is formed sympodially: in this process branches develop from the terminal meristems, and axial growth is continued from a subordinate lateral meristem. A comparison of mango (*Mangifera indica*) and rubber (*Hevea brasiliensis*) shows that similar processes control the development of the branches themselves; branching is described as orthotropic or plagiotropic depending on whether growth is mainly vertical or lateral. Branching rarely develops beyond the third order in tropical trees and there is a notable absence of fine twigs. Most trees can continue to grow despite damage to limbs, this reiterative process being controlled by the number of reserve meristems. However, the unbranched palms and cycads rely on a single meristem: to minimize damage, this indispensable tissue is usually well protected by thorns or lignified leaf sheaths.

Annual radial growth for trees in the A, B and C strata in lowland forests in the Philippines reportedly average 0.6, 0.4 and 0.3 cm, respectively (Richards, 1952). Comparable data for wet forests in Costa Rica indicate growth rates of 0.1–1.5 cm yr^{-1} for canopy species, 0.1–0.8 cm yr^{-1} for sub-canopy species and 0.1–0.3 cm yr^{-1} for understory species (Lieberman *et al.*, 1985a). In semi-deciduous forest radial growth is rarely more than 1 cm yr^{-1} (Daubenmire, 1972b). However, a considerable amount of wood is laid down during the life of these slow-growing trees. For example, Nicholson (1965) reported girths of 1.8–4.8 m for emergent species, 1.2–3 m for species in the main canopy and 0.9–1.2 m for species in the lower canopy of forests in Borneo.

2.6.3 Buttresses and root systems

Many trees in the lowland forests develop buttresses towards the base of their trunks. **Plank buttresses**, which develop as large, triangular outgrowths from lateral roots near the soil surface, are common on the taller trees and frequently extend to give a fluted appearance to the trunk. They are less common on species of the B stratum and are rare amongst the smaller species. Conversely, **stilt roots** arising from adventitious buds along the lower trunk are found almost exclusively on species of the C stratum. Buttresses and stilt roots are considered

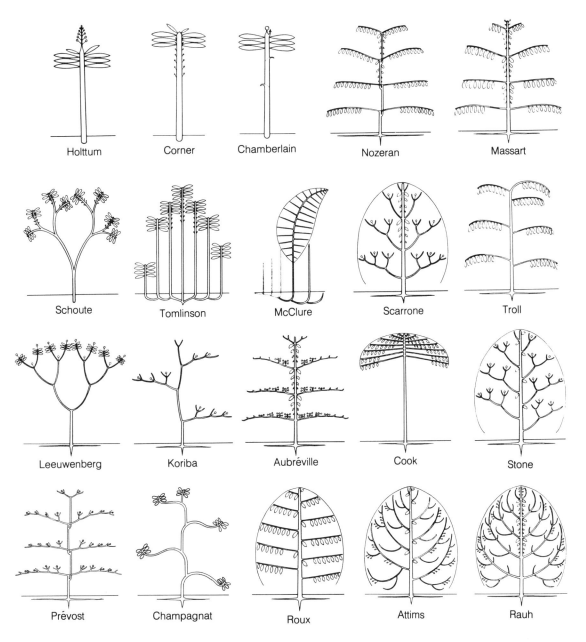

Figure 2.17 Examples of architectural models of tropical trees. (After Hallé *et al.*, 1978.) (Reproduced from F. Halle, R. A. A. Oldeman and P. B. Tomlinson, *Tropical Trees and Forests: An Architectural Analysis*, published by Springer-Verlag GmbH, 1978.)

to be adaptations for support of the trees (Figure 2.19), and they are often best developed in species which lack taproots, especially if they are growing in soils which offer little anchorage (Navez, 1930). Whitford (1906) suggested that buttressing strengthened the trunk of large-crowned trees but this relationship was not confirmed by Lowe (1963). Significant buttressing can also develop on young trees (Richards, 1952). Henwood (1973) concluded that stability would be increased in trees with irregular crowns if buttresses were developed as tension members to oppose the eccentric load, much like guy ropes on a tent: windthrow could also be reduced if they developed on the windward side of the trunk. These relationships have been demonstrated in *Ceiba petandra*

growing in West Africa (Baker, 1973b), and for the larger buttresses in *Quararibea asterolepis* from Panama (Richter, 1984). However, Lewis (1988) could find no relationship between prevailing winds and the size and orientation of buttresses in *Pterocarpus officinialis* growing in Puerto Rico.

Petch (1930) proposed that increased growth resulted from water being conducted up the stem directly above the lateral roots and subsequently Chalk and Akpalu (1963) demonstrated that the wood anatomy of species with well-developed buttresses could confine water movement to these parts of the trunk. Smith (1972), noting that buttressing is less prevalent in montane forest associations, attributed this decline to lower net energy

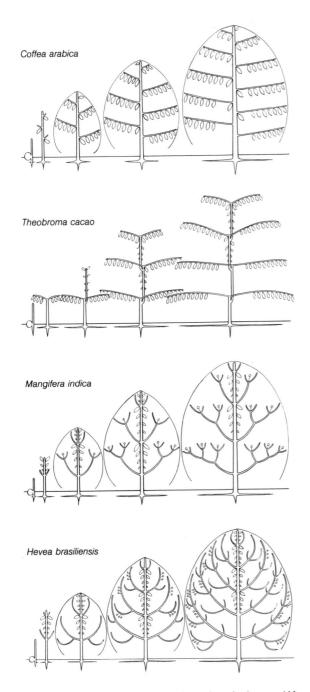

Coffea arabica

Theobroma cacao

Mangifera indica

Hevea brasiliensis

Figure 2.18 Patterns of growth in selected tropical trees. (After Hallé *et al.*, 1978.) (Reproduced from F. Halle, R. A. A. Oldeman and P. B. Tomlinson, *Tropical Trees and Forests: An Architectural Analysis*, published by Springer-Verlag GmbH, 1978.)

inputs. Outside of tropical forests, stilt roots are common on mangroves, while buttress-like appendages can develop on western red cedar (*Thuja plicata*) in Canadian west coast forests, and on elm (*Ulmus americana*) growing in poorly aerated, swampy sites: a similar form is seen in the root-knees of bald cypress (*Taxodium distichum*) growing in areas like the Florida everglades (Fowells, 1965). The higher frequency of modified trunk

forms in moist sites corresponds with the increased frequency of buttresses observed by Richards (1952) in the wetter soils of Guyana.

Lowland forest trees are generally shallow-rooted. Huttel (1975) reported that over 50% of the root biomass in West African forests was concentrated in the upper 30 cm of the soil, and that a high proportion of the finer, absorptive roots was restricted to the top 10 cm. Similarly, between 57% and 93% of the root mass occurs in the top 20 cm of Amazon forest soils (Klinge, 1973a, b). This type of root system is characteristic of many emergent species and is particularly common in soils of low fertility (Gower, 1987). However, some of the taller tropical trees also develop taproots and 'sinkers' on their lateral roots. Smaller trees usually produce fewer surface roots. Jeník (1978) has recognized 25 different rooting systems in tropical trees on the basis of morphology, development and ecological criteria. Included in this scheme are the specialized roots of **mangroves** which develop under extreme waterlogging or flooding and **aerial roots** which commonly form on trees growing in wetter soils (Figure 2.20). Some aerial roots may remain slender and free-hanging, but if they extend to the ground and become anchored they can grow into thick columnar roots which provide additional support for the branches (Figure 2.21). The banyan (*Ficus benghalensis*) is well known for this and often a grove-like clustering of trees results. The production of adventitious roots in some species can also enable crushed individuals to 'walk' away from treefall sites (Bodley and Benson, 1980).

2.6.4 Bark and wood

The bark of most tropical trees is thin, smooth and pale in colour (Richards, 1952). This is commonly the case in South America and Africa, although in South East Asia many species can be identified by the distinctive colour and texture of their bark (Whitmore, 1962). Trees with bark less than 10 mm in thickness often develop a scaly or flaky covering to the trunk, but thicker bark can be deeply fissured; spines may also be present (Figure 2.22). Wee (1978) and Grubb *et al.* (1963) have shown that bark texture can affect the degree to which epiphytes infest a tree. Species with very smooth or deeply fissured bark are less likely to harbour epiphytes than those with scaly, flaky or rougher bark. Similarly, the bark of some species of *Quercus* in Mexican cloud forests can inhibit growth of epiphytic orchids through a chemical effect (Frei and Dodson, 1972), although this may not prevent them from establishing on humus collected in branch angles.

Growth in many tropical forests is not affected by pronounced changes in day length and temperature. Consequently, annual growth rings in tropical woods are less distinct than in temperate species and rarely can be used to determine tree age. Cambial activity is closely

Figure 2.19 (a) Plank buttresses are a characteristic feature of emergent forest trees; (b) stilt roots are mostly associated with the smaller tree species (G. T. Prance.)

related to leaf development, and the most distinct growth rings occur in deciduous species in which woody growth ceases during the leafless period. However, Alvim (1964) reported that annual rings did not develop in 19 of 177 species growing in areas of Brazil with a seasonally dry climate. Conversely, distinct annual rings were observed in 21 of 60 species from the Amazon rain forest. Discontinuous development in evergreen species prior to the expansion of new leaves can produce growth rings which are incomplete or unrelated to a regular time interval. This further complicates age determinations. Maximum ages of 200–250 years were suggested by Richards (1952), but extrapolations of annual growth increments in wet forests in Costa Rica indicate a life span of 52–221 years for understory species, 78–338 years for sub-canopy species, and 78–442 years for canopy species (Lieberman *et al.*, 1985a). Trees considerably older than this are reported in South East Asia (Whitmore, 1984).

2.7 PHENOLOGICAL CHARACTERISTICS OF TROPICAL FORESTS

2.7.1 Leaf development

Although continuous growth is common for seedlings, leaf replacement in older trees may cause growth to cease periodically. This is most pronounced in decidu-

ous species which may remain leafless for several months. Other species will develop new leaves at the same time that the older foliage is shed, or retain the older leaves for some time after the new leaves have expanded. Most leaves are replaced after 15 months, although life spans of 3 months to 3 years have been reported. In palms, tree ferns and tropical conifers, continual shoot extension under favourable conditions is accompanied by new leaf formation. In regions lacking a pronounced dry season leaf development is characteristically asynchronous, even between individuals of the same species, or between different parts of the same tree (Longman and Jeník, 1987). Even in wetter regions most leaves are produced following somewhat drier periods (Fogden, 1972; Medway, 1972) and this becomes increasingly apparent in seasonal forests where leaf fall reaches a peak during the dry period (Koelmeyer, 1959; Taylor, 1960). The deciduous habit is also more characteristic of species in the upper canopy suggesting that leaf abscission may be related to water stress: some wide-ranging species become increasingly deciduous in drier parts of their range. Hence, Medina (1983) distinguished between facultative and obligate deciduous species, noting that some may remain in a leafless condition for 6 to 13 months depending on the intensity of the dry period. In contrast, some trees in Costa Rica are leafless during the wet season Janzen (1975a).

In many species the colour of the newly flushed leaves contrasts dramatically with the sombre greens

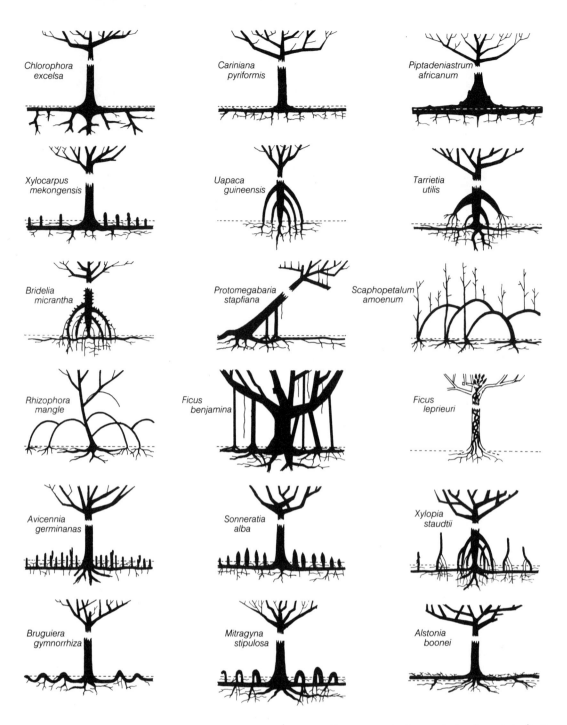

Figure 2.20 Characteristic root systems of tropical trees. (After Jeník, 1978.) (Reproduced with permission from J. Jeník, Roots and root systems in tropical trees: morphologic and ecologic aspects, in *Tropical Trees as Living Systems*, eds P. B. Tomlinson and M. H. Zimmermann; published by Cambridge University Press, 1978.)

of the mature foliage; those with high concentrations of anthocyanins are red or pink, others can be blue or white. Absorption of UV wavelengths by anthocyanins and phenols can screen young leaves from destructive shortwave radiation (Lee and Lowry, 1980); in older leaves protection is afforded by high reflectance from the cuticle, and by the development of intercellular spaces (Allen *et al.*, 1973). Leaf growth is quite rapid, and the newly expanded leaves can remain limp for a few days until mechanical tissues develop. Mature leaves in the lowland forests are generally 7–13 cm in length, somewhat elliptical in shape and of an entire, sclerophyllous type. They become progressively smaller at higher elevations (Dolph and Dilcher, 1980). The leaves of the understory trees often develop an extended **drip-tip** or acumen which allows water to drain quickly from

Figure 2.21 (a) Massive aerial root development and (b) columnar supporting roots in species of *Ficus*. ((a) O. W. Archibold; (b) G. T. Prance.)

Figure 2.22 Coarsely spinescent bark of a young kapok (*Ceiba pentandra*) tree. (G. T. Prance.)

the surface. This may increase photosynthetic efficiency of the leaf by increasing light absorption and transpiration rates, or by discouraging the growth of epiphyllous algae and lichens (Richards, 1952). The leaves are smaller, more leathery in texture, and have less pronounced drip-tips in the taller trees (Figure 2.23), although these xeromorphic characteristics typically develop only after individuals have grown beyond the juvenile stage. Leaf bud scales are generally poorly developed, being replaced by stipules, hairs or mucilaginous secretions which may reduce desiccation or deter herbivorous predation.

The varying leaf types in tropical forests may result from increased water stress in the upper canopy caused by higher insolation, stronger winds and periodic drought (Roth, 1984). However, the leaves are not particularly drought-resistant (Reich and Borchert, 1988), and in areas which are subject to periodic dry spells the tendency is for them to be shed (Frankie *et al.*, 1974). Thus, Loveless (1961) and Medina *et al.* (1990) concluded that sclerophyll anatomy is more likely an adaptation to nutrient deficiencies in the soil rather than to drought. Alternatively, a thickened leaf cuticle might deter herbivorous insects or offer protection against fungal attack (Buckley *et al.*, 1980).

Figure 2.23 (a) The entire sclerophyllous leaves of *Tepuianthus aracensis* lack drip tips; (b) lichens cover some of the leaves of understory plants despite their waxy surfaces and short drip tips. (G. T. Prance.)

Figure 2.24 Leguminous species with dissected leaves such as those of the flamboyant tree (*Delonix regia*) are common in moist tropical forests. (O. W. Archibold.)

Compound leaves are typically associated with species of drier regions (Figure 2.24), yet they are also common in the canopy of moist tropical forests (Givnish, 1978). In lowland forest in Ecuador, for example, 36% of the species have compound leaves compared with 9% of the species in adjacent montane forests (Grubb *et al.*, 1963). In lowland sites the compound leaf may give plants a competitive advantage since they can be rapidly deployed when gaps open in the canopy (Givnish, 1978), although they are more likely an adaptation for heat dissipation (Gates *et al.*, 1968). Similarly, the majority of the entire leaves in the canopy are steeply inclined to prevent overheating during high sun periods.

2.7.2 Flowering

In the wet tropics flowering is not normally restricted to a specific time of the year and some species may bloom continuously as long as conditions remain favourable. Ever-flowering species will bear flowers and fruit several times during a year, but the interval can be much longer for those that reproduce intermittently (Corner,

1956). As with leaf-flushing, individuals within a population often flower at different times, and flowering within a single canopy can also be asynchronous. However, in Malaysian dipterocarps flowering is gregarious and all individuals over a wide area blossom simultaneously. This usually occurs at intervals of 8–13 years (Wycherley, 1973). Poore (1968) considered that gregarious flowering is triggered by dry weather, but Ashton *et al.* (1988) noted that flowering is preceded by at least 3 nights when temperatures drop 2°C below normal during El Niño years. During one such episode Chan and Appanah (1980) recorded a specific flowering sequence in related *Shorea* species over a 10-week period (Figure 2.25): this staggered flowering activity reduces competition for the thrips that pollinate the blossoms. The sequential flowering pattern which occurs each year in sympatric species of *Arrabidaea* in Central America is also regarded as a strategy that ensures pollination and maintains species diversity (Gentry, 1974). The synchronous dispersal of seed which occurs following gregarious flowering in *Shorea* ensures that some seed survives predation. The importance of predation as a selective force has been demonstrated by Janzen (1975b) who noted that gregarious flowering in *Hymenaea coubaril* occurred only in areas where its principal seed predator was present. In species of *Miconia* sequential flowering results in staggered ripening of the fruit, which are eaten by birds; this may have evolved as a mechanism to reduce competition for dispersal agents (Snow, 1965).

In regions with no pronounced dry spell tropical trees can be classed into three groups according to their flowering activity (Alvim and Alvim, 1978). Some species flower at about the same time each year regardless of the rainfall regime, perhaps responding to changes in day length or temperature (Longman, 1985). Others flower at any time of the year, apparently unrelated to any external stimuli. For example, in Brazil between 10% and 30% of the trees are in bloom each month (Alvim and Alvim, 1978). The third group includes species such as coffee

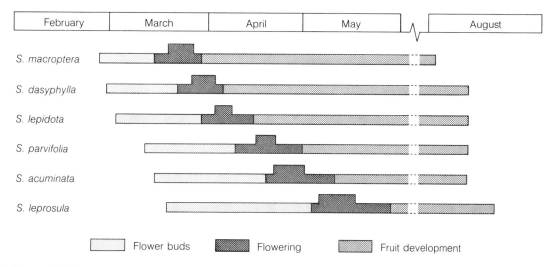

Figure 2.25 Sequential blooming and simultaneous fruiting in related species of *Shorea* during a year with gregarious flowering. Peak flowering time is indicated by the wider bar. (After Longman, 1985.) (Adapted from Fig. 2, p. 31, K. A. Longman, Tropical forest trees, in *CRC Handbook of Flowering*, Vol. 1, ed. A. H. Halevy; published by CRC Press, 1985.)

which exhibit a hydroperiodic response with the dormant buds opening only after rain interrupts a dry spell; a period of 3 days elapses before *Coffea rupestris* flowers profusely (Rees, 1964) and 7 days for *C. arabica* (Alvim, 1960). In coffee, dormancy is broken when the water potential in the subtending leaves falls below −12 bars and water flow to the flower bud ceases (Magalhaes and Angelocci, 1976).

Where the climate is more seasonal, flowering occurs mainly in the dry season. In Nigeria, for example, the number of species in flower increases from 30% in the wet season to 70% in the drier months (Njoku, 1963). This response is most apparent in deciduous species, whereas peak flowering in the evergreen species is at the start of the wet season (Figure 2.26). Similarly, in drier parts of Costa Rica 33% of the species come into flower during the wet season compared with 55% in the dry season (Frankie *et al.*, 1974): the flowering period for the remaining species lasts for an average of 25 weeks, thereby obscuring any seasonal trends. However, the smaller treelets and shrubs flower primarily during the early part of the wet season (Opler *et al.*, 1980a). At wetter sites 15–30% of the tree species are in flower each month, and many of the treelets and shrubs come into flower several times during the year, or bloom continually. A different flowering regime was noted in young secondary communities nearby. Not only did the species in the pioneer stages come into flower at a different time than the mature trees, but they also continued to flower for much longer periods and produced larger seed crops (Opler *et al.*, 1980b). With many species coming into flower during the dry period, the blossoms in seasonal forests are quite conspicuous against the leafless branches. Without this deciduous phase the subdued green of the foliage in other forest types can obscure the great profusion of flower forms and colours.

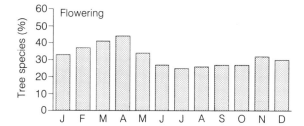

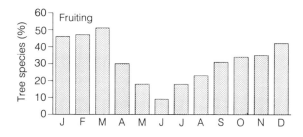

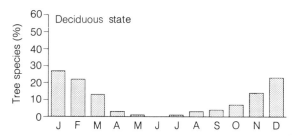

Figure 2.26 Annual cycle of flowering, fruiting and leaf fall in the forests of Ghana; the rainy season is April to September. (After Taylor, 1960.) (Source: C. J. Taylor, *Synecology and Silviculture in Ghana*; published by Thomas Nelson & Sons Ltd, 1960.)

2.8 FLORAL BIOLOGY

2.8.1 Flower development

Flowering has been observed as early as 3 months in teak seedlings (Hedegart, 1976) and several pioneer species flower after 2 to 3 years, but for many trees a long period of vegetative growth precedes flowering (Ng, 1966). In Malaysia 60 years can elapse before some of the canopy species flower (Ashton, 1969). Normally the trees will produce flowers periodically for many years. However, Foster (1977) describes *Tachigalia versicolor* as a 'suicidal neotropical tree', because after reaching heights of 30–40 m it flowers once and then dies: seedlings within the gap created by the fallen parent may benefit from higher light levels. A similar process has been reported in palms (Moore and Uhl, 1973), and the bamboo species *Phyllostachys bambusoides* flowers gregariously after 120 years and then re-establishes from seed. Such extended monocarpic flowering traits could have evolved as a mechanism to reduce seed predation (Janzen, 1976).

Although flowering may continue over several weeks, individual blossoms on most trees survive for about a day; *Passiflora foetida* is remarkable in that its flowers begin to wilt only 5–6 hours after they open (Janzen, 1968). Short-lived flowers are usually produced in large numbers which results in mass blooming: in *Casearea praecox* all plants in a region flower synchronously for a single day (Opler *et al.*, 1976). Other species extend their flowering period by producing fewer, longer-lasting blossoms, their average life span being related to flower size and the number of flowers open at any time (Stratton, 1989). Flower longevity also tends to increase at higher elevations. Species which depend on bats and other nocturnal pollinators, such as monkeys and opossums, open their flowers at night, although they often remain receptive to daytime pollinators (Janson *et al.*, 1981).

Many tropical trees are **dioecious** with male and female flowers borne on separate plants. More than 20% of the large tropical tree species in Costa Rica are dioecious compared with 12% for shrubs and small trees (Croat, 1979). A value of 26% is reported for the trees of Sarawak (Ashton, 1969) and 40% in Nigeria (Jones, 1955). This is 2–3 times greater than in most temperate forests. In semi-deciduous tropical forest the majority of dioecious species produce small flowers which blossom at the start of the wet season (Bawa and Opler, 1975). This coincides with the emergence of the small insects which are the principal pollinators and increases the possibility of cross-pollination, thereby ensuring greater seed set in an environment where seed predation is high. In dioecious species the male blossoms are invariably produced earlier and in larger numbers than the females, thereby acting as a cue for pollinators (Bawa, 1983). By producing both pollen and nectar the male flowers offer greater floral rewards, which encourages the animals to visit

them first. The female flowers of *Jacaratia dolichaula* produce no nectar; they attract pollinators by mimicking male blossoms (Bawa, 1980).

In **monoecious** species, in which male and female flowers are borne on the same plant, cross-pollination is ensured by physical, temporal or physiological mechanisms. For example, in *Ceiba acuminata* the receptive stigma is held far away from the anthers by a long style, and is brushed by hummingbirds or bats as they approach the nectaries (Baker *et al.*, 1983). Alternatively, pollen may be shed before (protandry) or after (protogyny) the stigma is receptive. Several species of bat-pollinated *Parkia* are protandrous with nectar production commencing at dusk and the stigma being exserted about midnight (Hopkins, 1984). Protogyny is less common but is found in some palms (Uhl and Moore, 1977). The small tree *Cupania guatemalensis* maximizes outcrossing in the population by producing pistillate flowers and two types of staminate flowers over a 5–7 week period (Bawa, 1977). The larger male flowers are the first to develop. Shortly after they finish blooming the pistillate flowers open, followed without interruption by the smaller male flowers. In palms the male and female flowering periods can be staggered over several weeks (Essig, 1973) and consequently flowering within a population is highly asynchronous.

Many of the smaller understory trees, such as cacao (*Theobroma* sp.), are **cauliflorous** with flowers arising directly from woody tissue (Figure 2.27). Cauliflory is rare in taller species: the Asian durian (*Durio* sp.), which bears edible fruit, is probably the best known of the canopy species which exhibit this characteristic. The Judas tree (*Cercis siliquastrum*) is one of very few cauliflorous species outside of the tropical forest environment. Richards (1952) describes two methods whereby cauliflory develops. Typically, flowering begins from buds on young wood, and these remain active despite continuing vegetative growth. Alternatively, the buds may remain dormant until the wood has increased in diameter and the flowers develop only after the buds have grown through the bark. In simple cauliflory flowers can arise indefinitely anywhere on the trunk, branches or twigs. If flowering is restricted to specific parts of the tree other terms are used – ramiflory when flowers are borne only on the branches and twigs; trunciflory when they are confined to the trunk.

2.8.2 Pollination

Wind speeds within the forest understory can be as low as 0.1 m s^{-1} (Allen *et al.*, 1972) and wind pollination is restricted to comparatively few species (Bawa and Crisp, 1980). Within the canopy, wind speeds of 1–2 m s^{-1} have been recorded and the air is generally turbulent (Baynton *et al.*, 1965), but most plants are still pollinated by animals. Janzen (1975a) suggested that wind pollination

Figure 2.27 (a) Cauliflorous blossom arising from the lichen-covered trunk of *Heterostemon* in Brazil; (b) cauliflorous fruits on the branch of a fig (*Ficus*) in Queensland. ((a) G. T. Prance; (b) G. W. and R. Wilson.)

is not efficient in species-rich forests where conspecific individuals are widely dispersed throughout the community. Even in the relatively open cover of disturbed sites, wind-pollinated species are soon at a disadvantage because of the rate at which the site becomes overgrown. The percentage of wind-pollinated species in a clearing in Costa Rica dropped from 38% to 8% in only 2 years as bees, wasps and hummingbirds became more active (Table 2.4). In all mature tropical forests pollination is mostly carried out by insects, birds and bats as they search for food (Figure 2.28).

On a regional scale, the flowering peak that is characteristic of drier periods can increase pollen transfer as the animals forage intensively for moisture. At the species level, plants might produce a large, conspicuous but short-lived flower crop that attracts relatively unspecialized pollinators, or deploy a small number of flowers which are visited continually over a longer period. Thus, solitary euglossine bees exhibit a 'trapline' strategy in which flowers are visited repeatedly over an extensive area (Janzen, 1971b). The bees can travel more than 23 km during these daily excursions, visiting the large, lightly scented flowers of trees and vines that usually bloom for a day (Frankie *et al.*, 1983). Most plants bloom in the dry season: this is the time of greatest activity as the bees search for the rich, sugary nectar. The

Table 2.4 Pollination systems for plants in different seral stages in moist lowland forest in Costa Rica. Values given as percentages except in mature forest where L = lower, S = similar, H = higher than in seral communities (after Opler *et al.*, 1980a)

Pollinator system	6 mo	1 yr	2 yr	3 yr	Mature forest
Wind	38	22	8	6	L
Small bees	34	40	40	38	S
Medium bees/wasps	7	18	23	19	L
Beetles	0	4	0	3	H
Small butterflies/moths	21	11	10	9	S
Large butterflies/moths	0	4	5	3	S
Large bees	0	0	3	6	H
Hummingbirds	0	4	10	10	L
Bats	0	0	0	3	H

(Reproduced with permission from P. A. Opler, H. G. Baker and G. W. Frankie, Plant reproductive characteristics during secondary succession in neotropical lowland forest ecosystems, *Biotropica*, **12** (supplement), 1980.)

Figure 2.28 Flower morphologies associated with pollination strategies. (a) Butterflies pollinate *Turnera ulmifolia*; (b) the brush-like flower of *Combretum* attracts birds; (c) trumpet-shaped flowers of *Galeandra* are visited by sphingid moths; (d) *Heliconia* is pollinated by hummingbirds; (e) blossoms of *Parkia* are accessible to bats; (f) the club-shaped spadix of Araceae typically attracts beetles. (G. T. Prance.)

bees are mostly opportunistic feeders, gathering nectar from as many as 20 different species. In the wet season fewer species are active, but at peak flowering times as many as 70 species of bees have been observed visiting a single tree (Frankie et al., 1976). Such a high level of activity can lead to a rivalry between visiting bees which ends in the ejection of some individuals from a flower, thereby increasing the possibility of outcrossing (Frankie and Baker, 1974). This competitive interaction is minimized when the bees forage at different heights in the canopy or on specific floral types. The larger bees generally do not visit the small flowers of shrubs and herbs which secrete little nectar: these are mostly pollinated by the smaller species of bees.

Highly specific strategies have co-evolved between many tropical plants and insects; deception and mimicry, for example, is particularly common in orchids (van der Pijl and Dodson, 1966). Likewise, solitary wasps (Blastophaga psenes) pollinate the fig Ficus carica as they deposit their eggs in the sterile ovaries of specialized gall flowers (Wiebes, 1979). Although other species do not rely on specific animals for pollination, they are nonetheless adapted to attract certain types of pollinators. Flowers which are pollinated by beetles tend to be greenish or off-white in colour but emit strong odours. Pollen does not readily adhere to the smooth bodies of beetles and their clumsy action usually restricts them to open, bowl-shaped blossoms which offer little reward of nectar or pollen (Faegri and van der Pijl, 1979). However, the spectacular Amazon lily (Victoria amazonica), like some other tropical species, traps the beetles that pollinate it and feeds them from specialized food bodies. Uhl and Moore (1977) noted similar feeding on the sepals of palms. Flies are not regular flower visitors, but the blossoms which attract them are usually pale in colour and bell-shaped; their anthers and stigmas are clearly exposed and they emit little scent. However, sapromyophilous flowers are specialized to attract flies which normally feed on rotting meat. They are typically drab brown or purple in colour, and emit a foul odour. The enormous flower of Rafflesia arnoldii deceives its pollinators in this manner. Ants are not efficient pollinators because of their small size and hard bodies, and many plants have evolved ant-repellant chemicals to prevent them from robbing the nectaries (van der Pijl, 1955) or have morphological adaptations to minimize pilferage (Guerrant and Fiedler, 1981). Some species, however, have extra-floral nectaries that are used to attract 'ant-guards' which deter other herbivores by their aggressive behaviour (Bentley, 1977).

The longer probosces of butterflies and moths provide access to the nectaries hidden within deeper, funnel-shaped flowers. Butterflies are active during the day and alight on the rim of vividly coloured flowers. Sphingid moths, on the other hand, are largely nocturnal. The flowers pollinated by these insects are usually pale in colour, but heavily scented, and may have recurved petals to prevent other insects from landing (Faegri and van der Pijl, 1979). The nectar in flowers pollinated by lepidopterans has a lower sugar concentration and is less viscous than in bee-pollinated flowers (Baker, 1978); this facilitates rapid feeding during the fleeting visits of the hovering moths.

Flowers visited by hummingbirds (Trochilidae) also produce nectars low in sugar (Pyke and Waser, 1981). Many species of hummingbird inhabit the neotropics, but they are not native to the forests of Africa and Asia: here passerine sun-birds (Nectariniidae) and honey-eaters (Meliphagidae) are active pollinators. Hummingbirds are attracted to scentless, brightly coloured tubular flowers. The corollas of some flowers can be more than 33 mm in length, and are often curved to fit the bills of non-territorial hermit hummingbirds during their 'trap-line' excursions (Stiles, 1975). In seasonal forests these species of hummingbird are active around flowers growing in open sites, but will also visit understory flowers during the wet season. In contrast, the straightbilled (non-hermit) hummingbirds will utilize canopy flowers, even those which are not adapted for hummingbird pollination. Observations of hummingbird activity around the shrub Hamelia patens and the tree Inga brenesii showed that over 90 flowers may be probed each visit during peak flowering periods (Feinsinger, 1978). The two plants differ in the amount of nectar produced by their flowers. Nectar secretions from Hamelia patens average 12.5 µl per flower with up to 300 flowers per plant, compared with an average secretion of 0.9 µl for as many as 5000 flowers per plant in Inga brenesii. In some species of hummingbird territorial behaviour occurs around shrubs bearing more than 70 flowers and around trees with over 600 flowers: this could reduce outcrossing in the plants. Conversely, in Heliconia sp. sequential blooming and clumped distribution patterns result in more active foraging and an increased probability of cross-pollination (Stiles, 1975). Hummingbirds are most active in the lower forest strata, whereas flowers in the upper canopy are more frequently visited by perching birds, especially when fruits are in short supply (Toledo, 1977).

Flowers pollinated by bats open at night to disclose flowers which are often whitish, drab green or purple in colour and emit a strong, stale odour; the blossoms are very accessible. Many species flower during the leafless phase, others are cauliflorous and produce flowers on leafless wood (Durio sp.), or are penduliflorous, suspending the flowers below the foliage on long stalks, as in species of Parkia. In Africa and Asia bats often climb around in the tree visiting numerous flowers over a 20-minute period, but in tropical America the visits usually last only a few seconds before the tree is abandoned (Baker, 1973a; Hopkins, 1984). Pollen is generally transferred to the bat from numerous brush-like anthers as they forage for nectar which, although produced in copious quantities, is generally low in sugar (Baker, 1978). Many species of bat feed solely on nectar or pollen, and must visit several plant species to sustain their high

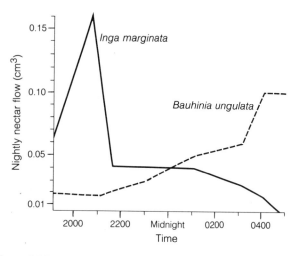

Figure 2.29 Nightly nectar flow in two bat-pollinated species of trees in Costa Rica. (Howell, 1977.) (Reproduced from D. J. Howell, Time sharing and body partitioning in bat–plant pollination systems, with permission from *Nature*, **270**, 510, copyright 1977 Macmillan Magazines Limited.)

metabolic activity. Examination of seven bat species in Costa Rica revealed that 42% of the individuals carried mixed loads of pollen, and visited, on average, 13 of the 17 available plant species (Heithaus *et al.*, 1975). Most bats are opportunistic feeders and move quickly between flowers. *Glossophaga soricina*, for example, is particularly active around *Inga marginata* trees early in the evening but later it concentrates on *Bauhinia ungulata* (Howell, 1977). This behaviour is clearly related to the availability of nectar from these two sources (Figure 2.29). Bat activity is also controlled by the slow rate at which nectar is delivered by the plant. In *Oxoxylum indicum* the flowers take up to 60 minutes to open fully (Gould, 1978). Until then the corolla is too narrow to permit access to the nectar pool and feeding is restricted to nectar adhering to a collar of hairs at the base of the anthers. Some of the flowers' nectar and pollen is removed during this early stage. Later the flowers bend downwards when the bats land. This gains them access to the nectar pool, but even then the maximum volume extracted on a single visit is only 0.05 µl. After this second peak of activity the flower becomes less attractive to the bats, and visits cease about 2 o'clock in the morning,

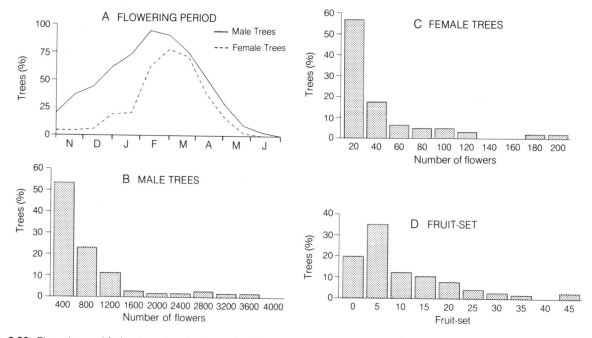

Figure 2.30 Flowering and fruit-set per tree in *Jacaratia dolichaula*, a dioecious species of the Costa Rican rain forest. (After Bullock and Bawa, 1981.) (Reprinted with permission from S. H. Bullock and K. S. Bawa, Sexual dimorphism and the annual flowering pattern in *Jacaratia dolichaula* (D. Smith) Woodson (Caricaceae) in a Costa Rican rain forest, *Ecology*, 1981, **62**, 1497, 1499.)

when they presumably search for alternative food sources. In some species nectar production may start before nightfall and continue well into the following day, thereby encouraging visits from bats, birds and insects (Crome and Irvine, 1986; Sazima and Sazima, 1978).

2.8.3 Fruit and seed production

Despite large flower crops, the number of fruits set by tropical trees is generally very low. In *Cassia grandis*, for example, fewer than 1% of the flowers produce seed pods (Janzen, 1978). Low fruit production can arise from high rates of flower abscission, or because the first pollinated flowers inhibit fruit-set in others. These types of strategies can increase the chance of pollination or ensure proper allocation of resources for successful fruit development. In addition, an unusually large number of dioecious species in tropical forests have showy male flowers that serve only to attract pollinators. Although there appears to be no consistency between numbers of trees of each sex, the males produce many more flowers than the females (Opler and Bawa, 1978). For example, in *Jacaratia dolichaula* flowering in male trees increases steadily from October to early February compared with the more abrupt rise to peak flowering later in February by the females (Figure 2.30). At this time there is approximately 1.5 times as many male trees in flower than females. Some of the male trees produce over 3600 flowers, compared with fewer than 100 flowers for most

of the females. Subsequent fruit-set is low with fewer than 10 fruits being produced on about 60% of the female trees.

Most species come into flower only after the previous fruit crop has matured, so that flowering periodicity is largely determined by the length of time required for the fruit to ripen. In *Posoqueria grandiflora*, for example, 32 months elapse before the fruit is mature. Consequently, flowering occurs only once every 3 years (Bawa, 1983). Ripening periods in Malaysia range from 3 weeks to 9 months (Ng and Loh, 1974), but although fruits mature throughout the year noticeable peaks occur in some months. A single fruiting peak occurs in Malaysia (Medway, 1972) and in Ghana (Taylor, 1960). In Panama there are two ripening peaks with seedfall for many canopy species occuring in either September or April (Figure 2.31). Species which come into flower at the start of the rainy season generally bear ripe fruit a few months later in September, whereas those species contributing to the April peak can require almost 12 months to ripen. Differences in the rates of ripening could reflect the interactive effects of energy demands on the plant and herbivory (Janzen, 1978). The predation period is reduced when fruits develop rapidly but this demands energy, often at a time when the trees are leafless. When maturation is delayed some species produce many small fruits which may be sacrificed to predators, then as the threat diminishes ripening continues in a few of the remaining fruits. Alternatively, unripe, full-sized fruit may be produced rapidly but these do not ripen until

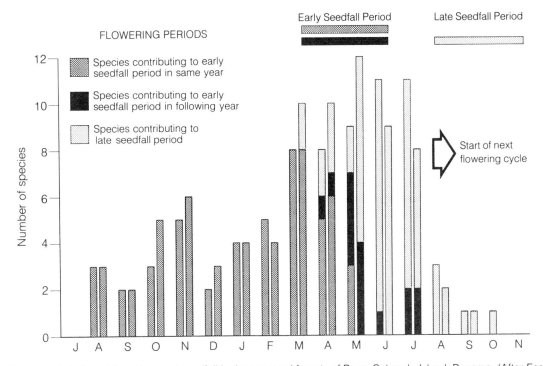

Figure 2.31 The annual rhythm of flowering and seedfall in the seasonal forests of Barro Colorado Island, Panama. (After Foster, 1982.) (Adapted from R. B. Foster, The seasonal rhythm of fruitfall on Barro Colorado Island, in *The Ecology of a Tropical Forest: Seasonal Rhythms and Long-term Changes*, eds E. G. Leigh, A. S. Rand and D. M. Windsor; published by Smithsonian Institution Press, 1982.)

Table 2.5 Seed dispersal in tropical forests (after Howe and Smallwood, 1982)

	Pptn (mm)	Plant form (and number of sp.)	Vertebrates	Wind	Water	Self	Unknown
			(Percentage of species)				
Moist forests							
Colombia	5530	trees (133)	89	3	–	–	8
Costa Rica	4000	canopy trees (79)	85	13	–	–	3
		sub-canopy trees (82)	98	2	–	–	–
Ecuador	2650	canopy trees (145)	93	4	1	2	–
		sub-canopy trees (96)	91	5	2	2	–
		lianas (83)	63	33	1	4	–
Panama	2650	canopy trees (291)	78	16	1	4	–
		sub-canopy trees (131)	87	5	3	5	–
Dry forests							
Costa Rica	1800	canopy trees (138)	64	29	1	6	–
		sub-canopy trees (60)	77	8	< 1	15	–
		lianas (49)	22	71	< 1	6	–

(H. F. Howe and J. Smallwood, Ecology of seed dispersal; reproduced, with permission, from the *Annual Review of Ecology and Systematics*, **13**, © 1982 by Annual Reviews Inc.)

several months later, following accumulation of adequate energy reserves.

2.9 SEEDING PATTERNS

2.9.1 Seed dispersal

Dispersal of mature fruits and seeds is necessary to avoid high predatory losses or to find conditions suitable for germination and growth. In wet tropical forests less than 10% of the species rely on wind for dispersal, although this may increase to 30% in more open seasonal forests (Baker *et al.*, 1983; Jackson, 1981) and can be considerably higher in specific life forms such as lianas (Table 2.5). Most of the ripe fruit is taken by bats and birds, although monkeys and rodents are important and to a lesser degree ruminants, including elephants (Figure 2.32). Fruits consumed by bats are often dull brown, green or yellow in colour, and like the bat flowers emit strong musty odours. Studies of Panamanian fruit bats have indicated that they have distinct feeding niches (Figure 2.33). Some species utilize fruits of canopy trees while others restrict their activity to understory plants with varying degrees of dietary specialization. Similarly, in seasonal forests, changes in food availability can result in distinct patterns of resource utilization (Figure 2.34). Most seeds consumed by bats tend to be deposited close to the fruiting plant, or deposited beneath their roosts (Fleming and Heithaus, 1981).

Frugivorous birds are usually attracted to brightly coloured, fleshy fruits, the seeds being quickly regurgitated or voided in wastes. Fruits taken by specialized feeders, such as toucans (Ramphastinae) and hornbills (Bucerotidae), are rich in fats and protein; the larger fruits, up to 40 × 70 mm in size, usually contain a few large, soft seeds (Snow, 1981). To produce fruits of such high quality can be costly for the plant but may ensure regular feeding visits (McKey, 1975). The smaller, many-seeded fruits characteristic of bushes and understory trees tend to be utilized by birds which usually supplement their diet with insects, and dispersal in these species is more opportunisitic. A comparison of *Tetragastris panamensis* and *Virola surinamensis* illustrates the effectiveness of different dispersal strategies (Howe, 1982). Each species is visited by a similar complement of birds. The capsules and arils of each species are of similar size, but the seeds of *T. panamensis* weigh only 0.2 g compared with 2.0 g for *V. surinamensis*; dispersal is highest for the large seeds consumed by the specialized bird population (Table 2.6). Howe and Estabrook (1977) predicted that species in which fruits are produced in small numbers over a long period are more likely to benefit from a few specialized frugivores, whereas mass fruiting, which results in a showy display for only a short time, would more likely attract opportunisitic feeders. Although this appears to be the case for most plants, Smythe (1970) and Jackson (1981) have reported that ripening of larger seeds tends to be much more seasonal than for smaller seeds.

Most birds quickly eliminate seeds and consequently their dispersal is largely determined by feeding habits. Hornbills, for example, fly off immediately after feeding (Becker and Wong, 1985), while fruit-pigeons typically rest for 20 minutes (Pratt and Stiles, 1983). In Costa Rica, parrots (Psittacidae) visit *Casearia corymbosa* trees for up to 4 hours; during this time the birds may regurgitate more than 450 seeds beneath the canopy (Howe, 1977). Masked tityras (*Tityra semifasciata*) provide much better seed dispersal for this species. After consuming the fruit they fly to neighbouring trees, regurgitate the seeds, then return to feed. However, in heavily laden trees even specialized feeders may regurgitate seeds rather than fly off with them, and often they defend the fruit aggressively. Toucans will often displace tityras in such situations (Howe, 1981).

Figure 2.32 (a) Fleshy aril of *Clusia*; (b) the large soft seeds of *Lithocarpus* are encased in a hard nut; (c) prominently displayed fruits of *Eschweilera* are consumed by bats; (d) fleshy, globose fruits of the climbing shrub *Norantea guianensis*; (e) pendulous fruit cluster of a palm. ((a)–(d) G. T. Prance; (e) K. A. Hudson.)

Feeding activity can also be conditioned by predators; in canopy trees birds tend to gather fruits quickly and fly off with them, but remain longer to feed in the denser foliage of the understory (Gautier-Hion *et al.*, 1985). Such behaviour, also common in monkeys and other frugivorous vertebrates, is the basis of the predator-avoidance

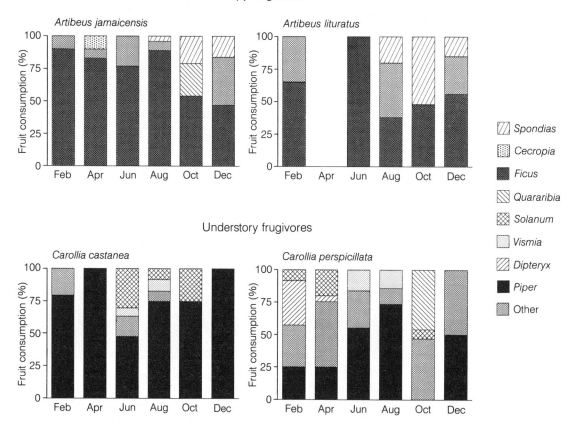

Figure 2.33 Diets of common canopy and understory frugivorous bats in the seasonal forests of Barro Colorado Island, Panama. (After Bonaccorso and Humphrey, 1984.) (Reproduced with permission from F. J. Bonaccorso and S. R. Humphrey, Fruit bat niche dynamics: their role in maintaining tropical forest diversity, in *Tropical Rain-forest: The Leeds Symposium*, eds A. C. Chadwick and S. L. Sutton; published by Leeds Philosophical and Literary Society, 1984.)

hypothesis which assumes that the high risk of predation forces frugivores to move continually while feeding (Howe, 1979). After observing the capture of a bat by a tree snake during a fleeting visit to a flower, Hopkins and Hopkins (1982) have suggested that this might encourage cross-pollination as well as enhancing seed dispersal.

The activities of monkeys, squirrels and rodents can also be important in seed dispersal (Table 2.7). Monkeys tend to be opportunisitic feeders, attracted to larger, aromatic, sugar-rich fruits often encased in brown, green or yellow husks (Janson, 1983; Leighton and Leighton, 1983). Typically, they fill their cheeks then retire some distance from the tree and spit out the seeds. Monkeys utilize many species. In Panama, for example, Oppenheimer (1982) observed capuchins (*Cebus capucinus*) feeding on 110 species, although material other than fruit was included in their diets. Squirrels and rodents often consume more seeds than they disperse unless they are protected by hard fruits, such as those of the palms *Scheelea zonensis* and *Dipteryx panamensis* taken by red-tailed squirrels (*Sciurus granatensis*) in Panama (Glantz *et al.*, 1982). Studies in West Africa suggest that

squirrels actively disperse seeds in less than 20% of the species they utilize compared with 88% for birds and 96% for monkeys (Table 2.8). Irrespective of the dispersal agent, most seeds fall very close to the source. Few of the wind-dispersed seeds of *Shorea* sp. travel further than 25 m (Whitmore, 1984). Shorter distances are common in species dispersed by monkeys, although bats and birds can carry seeds somewhat further (Hubbell, 1979).

2.9.2 Seed predation

Howe and Vande Kerckhove (1981) have reported that seeds of *Virola surinamensis* will rot if the encasing aril is not removed, while seedlings from 300-year-old *Calvaria major* trees on Mauritius have reputedly reappeared only recently now that seeds previously eaten by the extinct dodo (*Raphus cucullatus*) are artificially scarified (Temple, 1977; Owadally, 1979). However, such dependencies are uncommon in tropical species – more often the seeds are destroyed by predators. The host–predator relationships are usually very specific. In seasonal forest

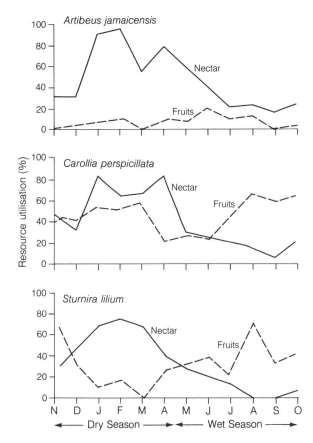

Figure 2.34 Monthly trends in utilization of nectar and fruits by bats in the seasonal tropical forests of Costa Rica. (After Heithaus *et al.*, 1975.) (Reprinted with permission from E. R. Heithaus, T. H. Fleming and P. A. Opler, Foraging patterns and resource utilization in seven species of bats in a seasonal tropical forest, *Ecology*, 1975, **56**, 846.)

in Costa Rica, for example, 110 insect species reportedly damage seeds of 100 plant species, of which 59 are attacked exclusively by a single predator species (Janzen, 1980). Janzen (1970) and Connell (1971) have argued that seed predation and seedling mortality are disproportionately high close to the parent tree, and proposed that species recruitment will be most successful some distance away (Figure 2.35). Several studies support this 'escape hypothesis'. Howe *et al.* (1985) noted a decline in weevil damage in germinating seeds and seedlings of *Virola surinamensis* at distances of 20 m from mature trees; a similar pattern was observed for bruchid beetle damage in fruits of the palm *Scheelea zonensis* (Wright, 1983). van Valen (1975) has calculated that intense predation by beetles reduces the seed production of the palm *Euterpe globosa* by more than 70% within one month of ripening, and that only 1% of the original seed population ultimately produces seedlings. Similarly, the pattern of seedling loss in *Dipteryx panamensis* illustrates the low probability that mature trees would be replaced by conspecifics (Table 2.9). Such spatially related patterns of loss are viewed as the mechanism by which high species diversity is maintained in tropical forests. However, surveys in Asian forests (Ashton, 1969; Poore, 1968) have shown that species can be distributed in clumps. The probability that some propagules will survive may be enhanced if the predators are satiated by mass fruiting, or if fruit production is delayed until the predator population has been reduced by lack of resources (Janzen, 1971a). The synchronous fruit fall in *Shorea* species following differing periods of ripening and supra-annual masting may reduce predatory losses. A subsequent modification of the escape-hypothesis model by Hubbell (1980) accommodates the possibility that some seedlings may establish close to the parent tree (Figure 2.35).

Table 2.6 Relative contributions to dispersal and waste of arillate seeds consumed by frugivores visiting *Tetragastris* and *Virola* trees in Panama (after Howe, 1982)

	Tetragastris trees		*Virola* trees	
	dispersed (%)	wasted (%)	dispersed (%)	wasted (%)
Howler monkey	6	59	–	–
White-faced monkey	1	19	–	–
Spider monkey	–	–	3	9
Coatimundi	< 1	10	–	–
Black-crested guan	–	–	9	0
Slaty-tailed trogon	< 1	0	10	< 1
Black-throated trogon	< 1	0	–	–
Rufous motmot	–	–	14	2
Collared aracari	< 1	< 1	1	0
Chestnut-mandibled toucan	< 1	0	35	2
Keel-billed toucan	< 1	< 1	8	2
Masked tityra	< 1	0	< 1	5
Fruit crow	1	0	–	–
Red-eyed vireo	0	0	–	–

(Reproduced with permission from H. F. Howe, Fruit production and animal activity in two tropical trees, in *The Ecology of a Tropical Forest: Seasonal Rhythms and Long-term Changes*, eds E. G. Leigh, A. S. Rand and D. M. Windsor; published by Smithsonian Institution Press, 1982.)

Table 2.7 Relative importance of various agents of fruit dispersal for woody species in tropical forests (after Leigh and Windsor, 1982)

| | Panama | | Costa Rica | | Ivory Coast |
	overstory (%)	understory (%)	overstory (%)	understory (%)	overstory (%)
Birds	36	66	45	68	10
Bats	13	11	6	5	1
Primates	9	2	9	2	6
Terrestrial mammals	5	7	10	5	2
Elephants	–	–	–	–	27
Wind	31	0	13	1	20
Mechanical	1	7	2	3	11
Unknown	4	7	15	15	23

(Reproduced with permission from E. G. Leigh and D. M. Windsor, Forest production and regulation of primary consumers on Barro Colorado Island, in *The Ecology of a Tropical Forest: Seasonal Rhythms and Long-term Changes*, eds E. G. Leigh, A. S. Rand and D. M. Windsor; published by Smithsonian Institution Press, 1982.)

Table 2.8 The number of fruiting species with seeds dispersed, left intact beneath the canopy (neutral), or consumed by vertebrates in a tropical rain forest in Gabon, West Africa (after Gautier-Hion *et al.*, 1985)

| | Number of species | | |
	Dispersed	Neutral	Consumed
Birds	32	0	4
Small rodents	14	0	51
Squirrels	7	0	34
Large rodents	12	4	28
Ruminants	13	8	37
Monkeys	59	10	3

(Reproduced with permission from A. Gautier-Hion *et al.*, Fruit characters as a basis of fruit choice and seed dispersal in a tropical forest vertebrate community, *Oecologia*, 1985, **65**, 328.)

2.9.3 Seedling establishment

The seeds of most tropical trees lose their viability quickly, hence rapid germination is the norm. Thus, in a sample of 335 Malaysian tree species more than 50% germinated within 6 weeks of seedfall (Ng, 1980). The seeds of a few species remain dormant for 8–12 months and occassionally germination is delayed for over 3 years (Ng, 1978). Epigeal germination, which exposes the cotyledons as the hypocotyl extends, is most common and the robust seedlings can persist for long periods in the subdued light of the forest floor. However, canopy species, at least, must ultimately gain access to light and successful regeneration is regulated by the creation of treefall gaps (Figure 2.36). The life expectancy of mature trees once they have reached the canopy may be no more than 100 years. In American tropical forests gaps recur at intervals of 114 to 159 years (Hartshorn, 1978; Putz and Milton, 1982). In tropical cloud forest an interval of 158 years is predicted (Arriaga, 1988). The incidence of treefalls increases during the wet season (Figure 2.37), possibly because of the greater weight of water in the canopy. Low cohesion in wet soils, lightning, landslips and disease might also create openings in the canopy (Whitmore, 1984), and large-

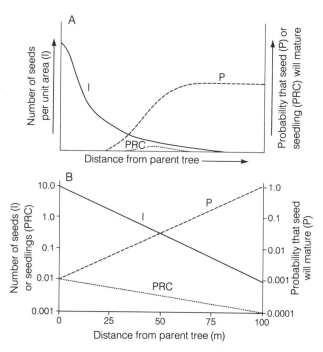

Figure 2.35 Models of seed predation and the coexistence of tree species in tropical forests. (A) The probability of seed or seedling survival is a function of the size of the seed crop, distance from the parent tree and the activity of seed or seedling predators and recruitment occurs only beyond some minimal distance from the parent tree; (B) a rescaling of Janzen's (1970) graphical model assuming that a small fraction of seeds and seedlings escape predation; in this way population recruitment is greater close to the parent tree and results in a clumped or random distribution in many species. ((a)after Janzen, 1970; (b) after Hubbell, 1980.) (Reproduced with permission from D. H. Janzen, Herbivores and the number of tree species in tropical forests, *The American Naturalist*, **104**, 501–28, published by the University of Chicago Press, 1970.)

scale devastation can result from cyclones (Dittus, 1985; Whitmore, 1974; Wood, 1970). The average size of treefall gaps is often less than 100 m², but the marked change in microclimate that occurs in these random openings is essential to the regeneration of the complex forest community.

Table 2.9 Changes in the distribution of *Dipteryx panamensis* seedlings in relation to conspecific adults over a 24-month period (after Clark and Clark, 1984)

Distance to nearest adult (m)	Seeding survival over time			
	7 mo (%)	11 mo (%)	18 mo (%)	24 mo (%)
0–8	9	9	3	0
9–13	28	23	19	13
14–21	36	38	33	40
22–46	26	30	44	47
Number of seedlings	138	81	36	15

(Reproduced with permission from D. A. Clark and D. B. Clark, Spacing dynamics of a tropical rain forest tree: evaluation of the Janzen–Connell model, *The American Naturalist*, **124**, 769–88, published by the University of Chicago Press, 1984.)

2.10 MICROCLIMATE

The difference between local microclimates and regional conditions is dependent on the structure of the forest; this ultimately determines the amount of incident radiation that penetrates the canopy (Figure 2.38). Richards (1983) used the contiguous B stratum of lowland forests to separate the upper, well illuminated euphotic zone from the very different conditions in the shaded oligophotic zone below. In mature forest less than 1% of the incident energy reaches the forest floor, and during the day air temperatures at ground level can be 3–4 °C cooler than in the overstory (Figure 2.39). Although heat is lost from the canopy during the night, the cool air does not sink through the dense foliage and so the lower strata stay warmer and temperatures are relatively uniform throughout the year. Relative humidity remains close to 100% throughout the day in the still air beneath the canopy. Reduced transpiration as a result of such humid conditions could restrict nutrient uptake to the detri-

Figure 2.36 Treefall gap in tropical lowland forest in Brunei. (G. T. Prance.)

ment of the seedlings (Longman and Jeník, 1987). Carbon dioxide, perhaps originating from the soil and litter, is also present in higher concentrations near the ground.

Structural differences within the forest create variable light conditions throughout the canopy. Hallé *et al.* (1978) have suggested that points within the canopy where light intensity is greater than expected can be combined to form ecological inversion surfaces. These

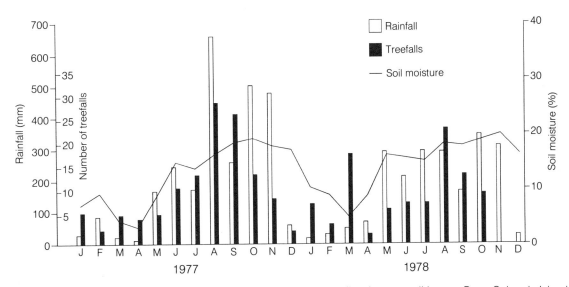

Figure 2.37 The relationship between monthly treefalls and precipitation and soil moisture conditions on Barro Colorado Island, Panama. (After Brokaw, 1982.) (Adapted from N. V. L. Brokaw, Treefalls: frequency, timing, and consequences, in *The Ecology of a Tropical Forest: Seasonal Rhythms and Long-term Changes*, eds E. G. Leigh, A. S. Rand and D. M. Windsor; published by Smithsonian Institution Press, 1982.)

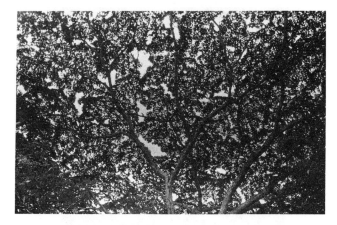

Figure 2.38 Lowland forest canopy in southern Amazonia. As in most lowland forest there is little overlap between adjacent crowns. (G. T. Prance.)

undulate throughout the forest in response to changes in leaf biomass and gaps left by fallen trees, and create a complex mosaic of ecological niches. It is in the treefall gaps that the most pronounced differences occur. Depending on its size, a treefall gap will receive considerably more radiation than the adjacent understory (Figure 2.40), which results in a warmer and drier microclimate (Figure 2.41). The quality of light within gaps is also altered significantly (Figure 2.42). The ratio of red: far red light can exceed 1.0 in openings, compared with 0.4 or less under a closed canopy, and this can affect seed germination (Vázquez-Yanes and Smith, 1982).

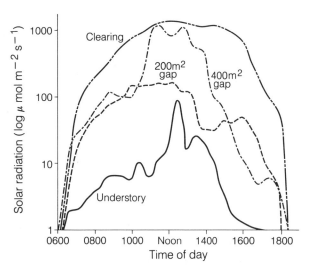

Figure 2.40 The diurnal pattern of photosynthetically active radiation (400–700 nm) on clear sunny days in local environments within lowland tropical rain forest in Costa Rica. (After Chazdon and Fetcher, 1984.) (Reproduced with permission from R. L. Chazdon and N. Fetcher, Photosynthetic light environments in a lowland tropical rain forest in Costa Rica, *Journal of Ecology*, 1984, **72**, 557.)

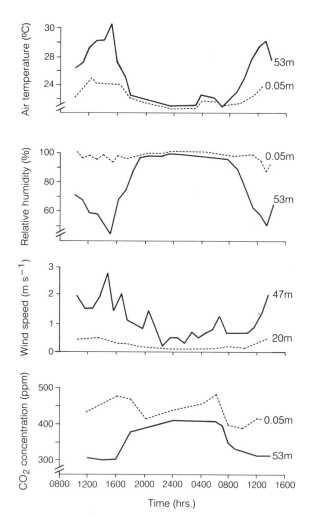

Figure 2.39 Diurnal course of air temperature, relative humidity, wind speed and carbon dioxide concentrations at various levels within the canopy of lowland rain forest in Malaysia. (After Aoki *et al.*, 1975.) (Reproduced with permission from M. Aoki, K. Yabuki and H. Koyama, Micrometeorology and assessment of primary production of a tropical rain forest in west Malaysia, *Journal of Agricultural Meteorology*, 1975, **31**(3), 118.)

2.11 THE FOREST GROWTH CYCLE

Forest regeneration begins once an opening has formed in the canopy. During the gap phase seedlings become established, and many of the light demanding species will rapidly grow into saplings. Advanced growth of secondary species closes the canopy within a few years. The composition of the forest gradually changes as conditions become more favourable for the slower growing forest dominants during the building phase. After many years the forest reaches the relatively stable mature phase which persists until the cycle is repeated by the death of a tree. A small gap normally stimulates the growth of saplings already established in the understory (Whitmore, 1989b). However, in larger openings the young growth usually cannot tolerate the greatly altered

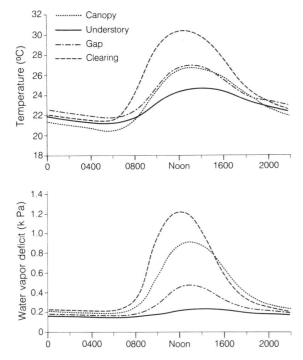

Figure 2.41 Diurnal patterns of temperature and vapour pressure deficit in different environments within lowland tropical rain forest in Costa Rica. (After Fetcher *et al.*, 1985.) (Reproduced with permission from N. Fetcher, S. F. Oberbauer and B. R. Strain, Vegetation effects on microclimate in lowland tropical forest in Costa Rica, *International Journal of Biometeorology*, 1985, **29**, 147, 148.)

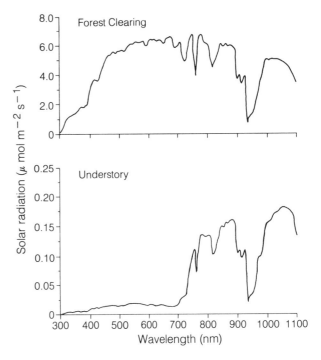

Figure 2.42 Spectral distribution of radiation in full sun, and beneath the rain forest canopy at Barro Colorado Island, Panama. (After Lee, 1987.) (Reproduced with permission from D. W. Lee, The spectral distribution of radiation in two neotropical rainforests, *Biotropica*, 1987, **19**(2), 163.)

microenvironment; in such circumstances the plants which establish first are invasive pioneers. Denslow (1980) recognized three classes of plants according to their dependence on gaps. The large gap specialists are shade intolerant species which require strong light and high temperatures for germination and growth. The small gap specialists germinate in the shade but require gaps for continued growth. Species which can tolerate shade at all stages of development are understory specialists. However, pioneer species are not restricted to the area immediately beneath a canopy opening. The majority establish in the gap perimeter, and the average area colonized by pioneer species in neotropical forests is 3.4 times larger than the projected canopy opening (Popma *et al.*, 1988).

When a large gap arises it is rapidly colonized by species not commonly found in mature forest. These light-demanding species produce numerous small seeds which accumulate in the soil (Swaine and Whitmore, 1988). Tropical forest seed banks are comparatively small, typically with fewer than 1000 seeds m^{-2}; the majority of these are from weedy species and can remain dormant for 2–3 years (Garwood, 1989). Dormancy in these seeds is broken when they are exposed to sunlight, and following germination the seedlings must grow rapidly to avoid shading (Figure 2.43). For example, *Trema orientalis*, a secondary species in West Africa, can grow

to a height of 17 m in 5 years (Swaine and Hall, 1983). The low-density wood of balsa (*Ochroma lagopus*) is produced by similarly rapid stem elongation (Vázquez-Yanes, 1974). Relatively pure secondary stands can establish in large clearings, but in natural treefalls these species must compete with saplings of overstory trees which endure long periods of suppression until such an opening develops. In the lowland forests of Costa Rica 75% of the canopy species are gap dependent (Hartshorn, 1978). Here pioneer species represent less than 1% of the forest cover and survive by leading a nomadic existence. Readily dispersed, lightweight seeds produced in abundance increase the chance that these plants can establish in gaps which open at random throughout the forest.

The floristic composition of small patches of forest changes over time. This cyclical pattern of regeneration occurs because canopy trees are rarely found growing in association with seedlings or saplings of the same species. For some tree species this distribution may be intricately linked to the activity of seed dispersal agents and seed predators. Other species may benefit from the modified microenvironment of a treefall gap; germination is stimulated by the spectral composition of light, fewer seedlings may be lost by damping off (Augspurger, 1983) and subsequent growth rates may increase because of higher light levels. The wide array of microhabitats which occurs in the gaps gradually alters as regrowth proceeds. The variety of niches created by gaps of different sizes on

Figure 2.43 *Cercropia* is a rapidly growing, hollow-trunked secondary tree, but is often outcompeted by hardwood species after about 20 years. (G. A. J. Scott.)

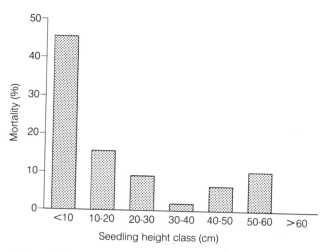

Figure 2.44 Percentage mortality of seedlings of different height classes in lowland rain forest in Malaysia. (After Turner, 1990.) (Reproduced with permission from I. M. Turner, Tree seedling growth and survival in a Malaysian rain forest, *Biotropica*, 1990, **22**(2), 149.)

a recurring basis maintains the high species diversity characteristic of these forests (Denslow, 1987).

2.12 FOREST PRODUCTIVITY

Although many seedlings may become established, the number which survive declines rapidly over time (Fig-

ure 2.44). Many become infected by pathogens, or are consumed by herbivores; some will not survive the deep shade and others will be lost through competition. Once past the seedling stage, individuals are lost at a rate of 1–2% annually as the forest matures (Swaine *et al.*, 1987): this rate of mortality is independent of the size of the trees (Lieberman and Lieberman, 1987; Manokaran and Kochummen, 1987). Few demographic studies have been conducted for tropical species but general inventories, such as that of Jones (1956) in southern Nigeria, show how few large trees ultimately become part of the forest canopy (Figure 2.45). However, it is this relatively small number of individuals which accounts for the bulk of the forest biomass. For example, in a site in central Amazonia 39% of the above-ground biomass was con-

Table 2.10 Height classes and fresh biomass in a central Amazonian forest (after Fittkau and Klinge, 1973)

	Height class (m)						
	< 1.5	1.5–5.0	5–10	10–20	20–30	> 30	Total
Number of individuals							
trees and palms	83650	7450	1525	740	335	80	93780
Biomass (kg ha^{-1})							
leaves	513	3208	1161	3253	6583	3427	18145
stems	936	2993	8864	37832	217685	199727	467101
twigs and branches		1132	3193	14994	111836	68042	199197
Total above ground	1449	7333	13218	56079	336104	267196	685379
large roots	439	1542	1951	6002	23441	15626	49000
fine roots	–	–	–	–	–	–	206040
Total below ground	439	1542	1951	6002	23441	15626	255040
Total above and below ground	1888	8875	15169	62081	359545	282822	940419
Ratios							
leaves:wood	1:1.8	1:1.3	1:10.4	1:16.2	1:50.1	1:77.0	1:36.8
above:below ground	1:0.3	1:0.2	1:0.2	1:0.1	1:0.1	1:0.1	1:0.4

(Source: E. J. Fittkau and H. Klinge, On biomass and trophic structure of the central Amazonian rain forest ecosystem, *Biotropica*, **5**(1), 1990.)

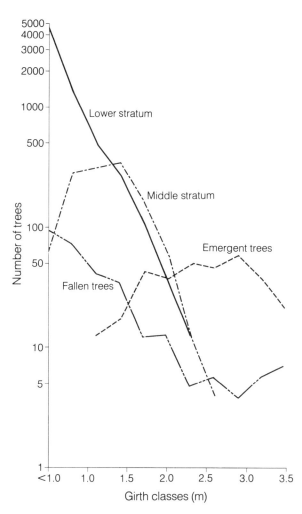

Figure 2.45 Distribution of girths of trees in emergent, middle and lower stories and of fallen individuals in rain forest in southern Nigeria. (After Jones, 1956.) (Reproduced from E. W. Jones, Ecological studies on the rain forest of southern Nigeria. IV (continued). The plateau forest of the Okomu Forest Reserve, *Journal of Ecology*, 1956, **44**, 86, 96.)

Table 2.11 Annual litterfall in tropical forests (after Spain, 1984)

	Litterfall component (t ha⁻¹ yr⁻¹)				
	Total	**Leaf**	**Wood**	**Flowers and Fruit**	**Precipitation (mm)**
Asia					
India	5.57	3.90	1.29	0.32	1219
Malaysia	9.19	6.53	1.97	0.38	2054
Australasia					
Australia	9.68	4.89	1.61	1.29	3609
Africa					
Ghana	9.66	7.41	0.97	0.39	1650
Ivory Coast	9.05	6.25	2.26	0.53	1800
Zaire	9.14	4.69	2.96	1.49	1273
America					
Guatemala	9.33	6.74	1.46	1.00	3747
Panama	11.10	5.83	2.30	1.23	2725
Columbia	8.73	6.64	1.97	0.12	3000
Brazil	6.40	4.29	1.09	0.96	1771

(Source: A. V. Spain, Litterfall and the standing crop of litter in three tropical Australian rainforests, *Journal of Ecology*, 1984, **72**, 956.)

Whereas the composition of the forest is largely determined during the gap phase (Whitmore, 1989b), productivity reaches a maximum during the building phase. Production in natural stands, including the subordinate species, ranges from 16.8 to 32.1 t ha⁻¹ yr⁻¹ for wet lowland sites and decreases to 10.3–24.3 t ha⁻¹ yr⁻¹ in montane forests; in seasonal forests production ranges from 11.4 to 15.5 t ha⁻¹ yr⁻¹ (Murphy, 1975). Such values are necessarily generalized since little growth data are available for the majority of species present in these complex ecosystems. Increases as high as 12 t ha⁻¹ yr⁻¹ have been reported for some commercial species in managed stands (Whitmore, 1984). More accurate measures can be made of the rate at which material is returned to the soil as litterfall: for fine litter this averages about 9 t ha⁻¹ yr⁻¹, and is mostly comprised of leaf material (Table 2.11).

2.13 NUTRIENT DYNAMICS OF TROPICAL FORESTS

Litterfall is the first stage in the return of nutrients to the soil. Although the chemical content of litter varies considerably, certain trends become apparent. In terms of its biomass, the nutrient content of the litter component is proportionally greater than in live above-ground biomass (Table 2.12). Thus litter, which may account for less than 5% of the biomass, can contain as much as 14% of the stored nutrients. The rate at which litter is broken down is important to the overall productivity of the forest, for without an active decomposer population nutrient stocks in the soil would soon be depleted. Decomposition proceeds rapidly under continually warm, moist conditions, and litter supplied throughout the year rarely accumulates on the forest floor. However, in

centrated in only 80 of the more than 90 000 trees that were growing there (Table 2.10). Total above-ground biomass for this site was 685 t ha⁻¹ (fresh weight): this is similar to the value of 664 t ha⁻¹ reported by Kato *et al.* (1978) in lowland forests in Malaysia. In Panamanian forests the equivalent value is 508–560 t ha⁻¹ (Golley *et al.*, 1975), in Puerto Rico 527 t ha⁻¹ (Ovington and Olson, 1970) and in Venezuela 602 t ha⁻¹ (Jordan and Uhl, 1978). The fresh weight of 471 t ha⁻¹ reported by Greenland and Kowal (1960) for forests in Ghana is considerably lower than in the American and Asian lowland forests. A biomass of 477 t ha⁻¹ fresh weight is reported for montane forest growing at 2500 m in New Guinea (Edwards and Grubb, 1977). Biomass declines noticeably in the drier forests: in deciduous forests in Puerto Rico Murphy and Lugo (1986) reported an above-ground biomass of only 82 t ha⁻¹ fresh weight, with roots accounting for a further 69 t ha⁻¹.

Table 2.12 A comparison of the quantities of mineral elements in the above-ground standing crop of vegetation with those returned annually in the litter in various tropical wet forests (after Edwards, 1982)

		Dry weight (t ha^{-1})	Nutrient capital (kg ha^{-1})				
			N	P	K	Ca	Mg
Lower montane forest (New Guinea)	above-ground biomass (ABG)	301	683	37	664	1281	185
	total litter	7.6	91	5.1	28	95	19
	leaf litter	6.4	82	4.7	25	80	17
	total litter: ABG (%)		13.3	13.8	4.2	7.4	10.3
	leaf litter: ABG (%)		12.0	12.7	3.8	6.2	9.2
Lowland forest (Brazil)	above-ground biomass (ABG)	406	2430	59	435	423	201
	litter	7.3	106	2.2	13	18	13
	leaf litter	6.0	86	1.7	10	12	10
	total litter: ABG (%)		4.4	3.7	3.0	4.3	6.5
	leaf litter: ABG (%)		3.5	2.9	2.3	2.8	5.0
Lowland forest (Ghana)	above-ground biomass (ABG)	233	1685	112	753	2370	320
	litter	10.5	199	7.3	68	206	45
	leaf litter	7.0	147	6.1	70	141	38
	total litter: ABG (%)		11.8	6.5	9.0	8.7	14.1
	leaf litter: ABG (%)		8.7	5.4	9.3	5.9	1.2

(Source: P. J. Edwards, Studies of mineral cycling in a montane rain forest in New Guinea: V. Rates of cycling in throughfall and litter fall, *Journal of Ecology*, 1982, **70**.)

Table 2.13 Leaf litterfalls and turnover coefficients for selected tropical forests (after Anderson and Swift, 1983)

Forest locality	Altitude (m)	Litterfall (t ha^{-1} yr^{-1})	Litter standing crop (t ha^{-1})	Turnover coefficient (k_L)	Time (wks)
Malaya, Pasoh	10	6.3	1.7	3.6	14
Malaya, Penang	–	5.4	5.1	1.1	47
New Guinea	100	7.3	5.0	1.5	35
Sarawak	225	5.4	3.2	1.7	31
Ghana	150	7.4	3.0	2.5	21
Nigeria	250	4.7	1.0	2.8	19
Brazil	45	6.1	4.0	1.5	35
Panama	150	7.0	2.8	2.6	20

(Source: J. M. Anderson and M. J. Swift, Decomposition in tropical forests, in *Tropical Rain Forest: Ecology and Management*, Special Publication No. 2 of the British Ecological Society, eds S. L. Sutton, T. C. Whitmore and A. C. Chadwick; published in 1983 by Blackwell Scientific Publications.)

seasonal forests many species drop their leaves at the start of the dry period when decomposer activity is reduced, and a mat of litter may then develop (Hopkins, 1966). Nutrients released during this period can be leached from the soil at the start of the ensuing rainy season (Swift and Anderson, 1989), although most are incorporated in the newly flushing leaves and woody growth. The average length of time required for leaf material to decompose is 24 weeks (Table 2.13) compared with 31 weeks for other fine litter. This is considerably less time than required to decompose the large branches and trunks contributed by treefalls. Anderson and Swift (1983) have suggested that decomposition of woody material over 3 cm diameter takes at least 15 years. However, Lieberman *et al.* (1985b) reported that many dead trees exceeding 22 cm diameter at breast height (dbh) disappeared completely during a 13-year period in Costa Rica, and in Panama as little as 10 years is required for decomposition of dead trees as large as 36 cm dbh (Lang and Knight, 1979).

Although many of the nutrients are stored in the forest biomass where they are resistant to leaching, soil nitrogen reserves often exceed those of the plants and litter, particularly in montane forests (Figure 2.46). Similarly, soil phosphorus levels are high, although only a small fraction may be in a form available to plants. Potassium is generally more abundant in the plant tissue and would be in short supply if not taken up rapidly from the decomposing litter. Thus, Golley *et al.* (1975) estimated that potassium was lost from Panamanian forest soils at a rate of 9.3 kg ha^{-1} yr^{-1}, which is less than 5% of the amount returned annually as litter. Nutrient losses are low because the concentration of roots near the soil surface can effectively absorb nutrients as they become available (Stark and Jordan, 1978). Nutrient uptake is further enhanced by the presence of mycorrhizal infection of the roots (Figure 2.47). Vesicular-arbuscular mycorrhiza formed by over 100 species of zygomycetous fungi are most commonly associated with tropical trees (Janos, 1983). These fungi are not host-specific and infection can spread readily from root to root. Although some of their thread-like hyphae get established inside the root, many others provide an extensive absorptive surface in contact with the soil, thereby enhancing the mineral uptake. In return the host supplies carbon compounds to the fungus. The ectomycorrhizas formed by fungi of the

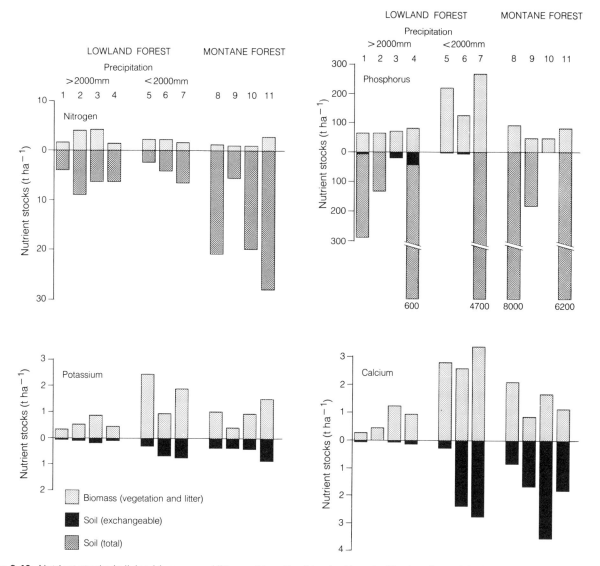

Figure 2.46 Nutrient stocks in living biomass and litter and in soils of tropical forests. The locations of the lowland forest sites are (1) and (7) Venezuela; (2) and (3) Brazil; (4) Ivory Coast; (5) Thailand; (6) Ghana. The montane forest sites are located in (8) Costa Rica; (9) Colombia; (10) New Guinea; (11) Venezuela. (After Jordan, 1985.) (Redrawn with permission from C. F. Jordan, *Nutrient Cycling in Tropical Forest Ecosystems*; published by John Wiley and Sons, 1985.)

class Basidiomycetes are less numerous. They cloak the smaller roots with hyphal strands and develop an extensive network within the intercellular spaces; they are particularly effective in increasing phosphorus uptake. Ectomycorrhizas are often specific to one genus, and rely on wind to disperse their numerous tiny spores. In contrast the large spores produced by vesicular-arbuscular mycorrhizas are usually dispersed by rodents.

Once taken up by the plants, the nutrients are partitioned in varying quantities within the different tissues. The greatest amounts of nutrients are stored in the tree boles, which represent about 80% of the above-ground biomass (Table 2.14). However, on a weight basis, it is the leaves that contain the highest concentrations of elements, and it is this component that must be utilized

efficiently in successive growth and decomposition cycles in order to maintain the high productivity of the forests. Inevitably some nutrients are leached from the soil, but these may not be totally removed from the region, since a high percentage of the drainage water is recycled by evapotranspiration. Brinkmann (1989) estimated that more than 80% of the precipitation falling in Amazonia is retained within the river catchment. This is also the case for the African forest regions, although in Asia a greater amount is lost through run-off. Nutrient losses from leaching are usually offset by inputs in rainfall which, before it reaches the soil, is generally enriched by nutrients leached from the canopy (Table 2.15). Various canopy processes, such as uptake by epiphytes, effectively minimize nutrient losses in undisturbed

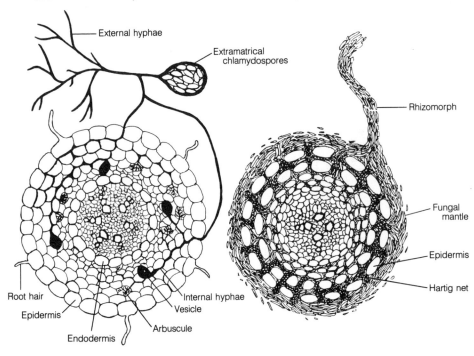

Vesicular arbuscular mycorrhizal root | Ectomycorrhizal root

External hyphae
Extramatrical chlamydospores
Rhizomorph
Fungal mantle
Epidermis
Hartig net
Root hair
Epidermis
Endodermis
Internal hyphae
Vesicle
Arbuscule

Figure 2.47 Diagrammatic representation of the transverse section of roots with vesicular arbuscular mycorrhizal and ectomycorrhizal infections. (After Bagyaraj, 1989.) (Reproduced with permission from D. J. Bagyaraj, Mycorrhizas, in *Ecosystems of the World, Vol. 14B. Tropical Rain Forest Ecosystems: Biogeographical and Ecological Studies*, eds H. Lieth and M. J. A. Werger; published by Elsevier Science Publishers, 1989.)

Table 2.14 Nutrient storage (kg ha^{-1}) and nutrient concentration (% of live weight, in parenthesis) in above-ground plant biomass in rain forest in Puerto Rico (after Ovington and Olson, 1970)

	Leaves	Branches	Boles
nitrogen	118 (1.28)	213 (0.47)	483 (0.25)
phosphorus	7 (0.08)	15 (0.03)	21 (0.01)
potassium	62 (0.67)	111 (0.25)	344 (0.18)
calcium	56 (0.61)	257 (0.57)	581 (0.30)
magnesium	20 (0.22)	51 (0.57)	269 (0.14)

(Reproduced from J. D. Ovington and J. S. Olson, Biomass and chemical content of El Verde Lower Montane Rain Forest Plants, in *A Tropical Rain Forest: A Study of Irradiation and Ecology at El Verde, Puerto Rico*, eds H. T. Odum and R. F. Pigeon; published by Department of Energy, Office of Scientific and Technical Information, Tennessee, 1970.)

forests (Jordan *et al.*, 1980). However, when the forest cover is removed, increased drainage and erosion can result in significant nutrient outflow, even from small plots (Figure 2.48).

The restoration of fertility in disturbed sites depends on rapid turnover of leaves in the regrowth stand. In traditional slash-and-burn agricultural systems an area of forest is cleared and the debris burned. Nutrients derived from the ash improve soil fertility for perhaps 3 years before cropping, leaching or uptake by fast-growing weedy species makes the land unusable. After the land is abandoned nutrients must be accumulated

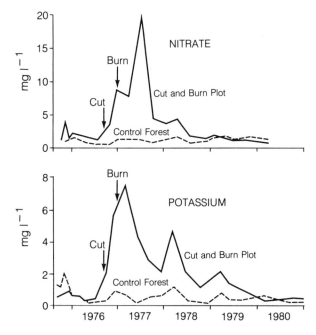

Figure 2.48 Concentrations of NO$_3$-nitrogen and potassium in soil water collected at 40 cm depth over a five-year period in forest and cut-and-burn plots in lowland rain forest in Venezuela. (After Uhl and Jordan, 1984.) (Reprinted with permission from C. Uhl and C. F. Jordan, Succession and nutrient dynamics following forest cutting and burning in Amazonia, *Ecology*, 1984, **65**, 1485.)

Table 2.15 Nutrient content of rainfall and canopy throughfall (after Golley, 1983; Nye and Greenland, 1960; and Edwards, 1982)

	N	P	K	Ca	Mg
	(kg ha^{-1} yr^{-1})				
Venezuela					
Rainfall	21.2	16.7	18.0	16.0	3.1
Throughfall	25.2	5.5	29.5	6.0	4.1
Leaching from the canopy	4.0	− 11.2	11.5	− 10.0	1.0
Litterfall	42.1	2.6	27.3	31.0	8.8
Ghana					
Rainfall	15	0.4	18	12	11
Throughfall	27	4.7	237	41	29
Leaching from the canopy	12	4.3	219	29	18
Litterfall	199	7.3	68	206	45
New Guinea					
Rainfall	6.5	0.5	7.3	3.6	1.3
Throughfall	30	2.5	71	19	11
Leaching from the canopy	32.5	2.0	63.7	15.4	9.7
Litterfall	91	5.1	28	95	19

(Reproduced with permission from P. H. Nye and D. J. Greenland, *The Soil Under Shifting Cultivation*, Technical Communication No. 51; published by CAB International, 1960; also from F. B. Golley, Nutrient cycling and nutrient conservation, in *Ecosystems of the World, Vol. 14A. Tropical Rain Forest Ecosystems: Structure and Function*, ed. F. B. Golley, published by Elsevier Science Publishers, 1983; and from P. J. Edwards, Studies of mineral cycling in a montane rain forest in New Guinea: V. Rates of cycling in throughfall and litterfall, *Journal of Ecology*, 1982, **70**.)

Table 2.16 Nutrients stored in typical tropical fallows at Yangambi, Zaire (after Nye and Greenland, 1960)

	N	P	K	Ca + Mg
	(kg ha^{-1})			
5-year secondary fallow				
Leaves	125	7	80	78
Woody matter	181	13	244	138
Dead wood	4	–	6	6
Litter	80	3	15	71
Roots	179	8	113	129
Total (excluding roots)	391	24	344	292
18-year secondary fallow				
Leaves	149	7	81	77
Woody matter	302	63	307	381
Dead wood	36	1	9	36
Litter	75	2	8	67
Roots	141	35	190	258
Total (excluding roots)	559	73	404	561

(Reproduced with permission from P. H. Nye and D. J. Greenland, *The Soil Under Shifting Cultivation*, Technical Communication No. 51; published by CAB International, 1960.)

once more in the biomass and litter. Nye and Greenland (1960) demonstrated that nutrient accumulation in leaves and fine litter occurs more slowly in older regrowth (Table 2.16) and, apart from phosphorus, most of the nutrients stored in 18-year fallow are present after 5 years. In addition, this younger cover can be cleared and burned more completely, and most of the stored nutrients are returned to the soil. However, nitrogen, carbon and sulphur volatilize during the burn and are lost as gases. Over longer periods the forest biomass becomes increasingly concentrated in the woody trunks, branches and roots. Nutrient stocks in the leaves stabilize about 10

years after abandonment (Figure 2.49). Most of the phosphorus is stored in the foliage, and potassium and calcium are preferentially stored in the stem tissues. In contrast, nitrogen levels steadily decline until countered by inputs from bacterial fixation or rainfall. Nitrogen is fixed biologically by bacteria in root nodules on many species of leguminous trees, by free-living and symbiotic bacteria which are particularly common on the leaves of epiphytes, and by the blue-green algal component of many lichens (Forman, 1975). Additions from rainfall account for less than 7% of total available nitrogen in tropical forests (Edmisten, 1970; Nye and Greenland, 1960).

2.14 THE ROLE OF PLANTS AS A FOOD BASE IN TROPICAL FORESTS

2.14.1 Consumption of plant materials

Comprehensive inventories of forest biomass reveal that leaves account for little more than 1% of the total organic material present in these ecosystems (Table 2.17). The bulk of the organic material is either stored in living wood or incorporated in the soil. However, the relative abundance of each component is by no means proportional to its importance as a food resource for animals. In these regions about 60% of the animal biomass is represented by invertebrates adapted to feed on litter or on the fungi which are decomposing it (Figure 2.50); a further 16% utilize humus deeper in the soil. The large wood fraction is consumed by about 19% of the animal biomass – mostly termites, beetles and insect larvae.

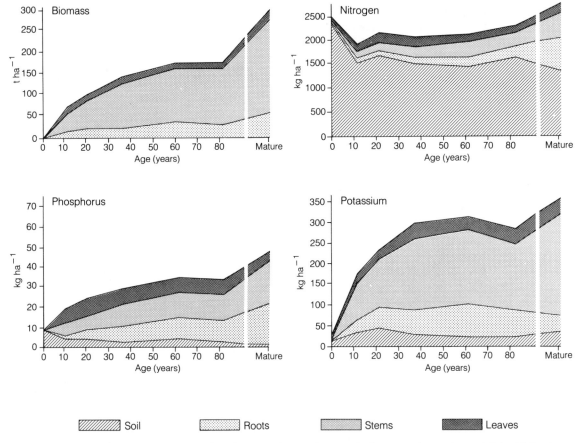

Figure 2.49 Accumulation of biomass, nitrogen, phosphorus and potassium stocks in regrowth stands established following shifting cultivation in lowland rain forest in Venezuela. (After Saldarriaga, 1987.) (Reproduced with permission from J. G. Saldarriaga, Recovery following shifting cultivation, in *Ecological Studies 60 – Amazonian Rain Forests: Ecosystem Disturbance and Recovery*, ed. C. F. Jordan; published by Springer-Verlag N.Y. Inc., 1987.)

The remainder, including many of the more conspicuous herbivores, are mostly dependent on leaves; flowers and fruits, so attractive to birds and bats, are of minor importance.

Although litter is the principal food base for the animal community, the initial breakdown is carried out principally by fungi. During primary decomposition some elements may be released to the soil; others are immobilized within fungal tissues which are subsequently consumed by many soil invertebrates. The termites are the most conspicuous of the decomposer organisms. Although termites consume a relatively small fraction of the litter, they accelerate the decay of large timber, and this is important for the redistribution of nutrients within the ecosystem. As many as 60 species of termite can be found in the humid lowland forests but this number decreases rapidly with elevation. For example, only 10 species of termite have been recorded in the montane forests of Sarawak (Collins, 1989). Some termites feed directly on rotten wood and leaf litter, the cellulose being broken down with the aid of protozoa or bacteria harboured in their hindguts. Others consume the fungi

which they cultivate on the litter. A third group of termites feeds on the organic debris present in the soil. These humivorous termites are particularly abundant in the Asian forests. Their nests are constructed from faeces, soil and saliva and, being unusually rich in nutrients, they provide favourable microsites for the establishment of tree seedlings. Similar soil enrichment occurs in the casts made by humus-feeding earthworms (Cook *et al.*, 1980; Nye, 1955).

2.14.2 Antiherbivore defences

The leaves provide an important source of food for many insects, particularly during their larval stages. In the Neotropics adult leaf-cutting ants can also cause serious defoliation. They harvest live tissue at prodigious rates and transport it to their nests, where it is used to culture fungus. Dirzo (1984) has noted that trees close to a large nest might occasionally be killed by continual defoliation, but reduced growth and less fruit production is the more probable outcome of leaf loss.

Table 2.17 Distribution of organic matter in a central Amazonian forest (after Klinge *et al.*, 1975)

	Dry weight	
	(t ha^{-1})	%
Live biomass above ground		
Leaf matter	8.7	1.1
Branches and twigs	97.2	12.7
Stems	224.7	29.4
Lianas and epiphytes	22.2	2.9
Total	357.1	46.7
Live biomass below ground		
Fine roots	23.5	3.1
Other roots	98.9	12.9
Total	122.4	16.0
Total live biomass	474.7	62.0
Litter biomass		
Standing dead wood	7.6	1.0
Dead wood in litter	18.2	3.4
Fine litter	7.2	0.9
Soil organic matter	250.0	32.7
Total dead biomass	290.6	38.0

(Source: H. Klinge *et al.*, Biomass and structure in a Central Amazonian rain forest, in *Ecological Studies 11 – Tropical Ecological Systems: Trends in Terrestrial and Aquatic Research*, eds F. B. Golley and E. Medina; published by Springer-Verlag N.Y. Inc., 1975.)

However, extensive defoliation can increase seedling mortality. Leaf consumption is of no benefit to the plant, and consequently, various defences have evolved to deter herbivory. Chemical defence is provided by a diverse array of secondary compounds which either are toxic to herbivores, or reduce the digestibility of the intended foodstuff. The nitrogen-containing alkaloids, for example, can affect herbivore enzyme systems; phenols, such as tannin and other organic acids, tend to bind with proteins to form indigestible compounds (Janzen, 1975a). The terpene group, which includes various gums, oils and resins, can physically block the digestive tract, or release toxic products as they are degraded; chicle, used in the manufacture of chewing gum, is one such product. Cyanogenic glycosides which release cyanide when digested are common in many plants including tapioca (*Manihot ultissima*). These compounds are neither uniformly distributed throughout the plant nor ubiquitous throughout the forest ecosystems. The alkaloids are most concentrated in the leaves of many lowland forest species, but they are rarely encountered in montane and cloud forests (Langenheim, 1984). Phenols, which are common in bark and older leaves, are often more concentrated in plants growing on poorer soils (Gartland *et al.*, 1980). Conversely, phenol concentrations are low in fast-growing secondary species which counteract high rates of predation by rapid growth and deployment of new leaf crops (Coley, 1983). Many species have evolved mechanical defensive systems. The incorporation of silica in wood can effectively wear down the chewing parts of insects, while high lignin and fibre contents can reduce digestibility. However, no method can provide complete immunity.

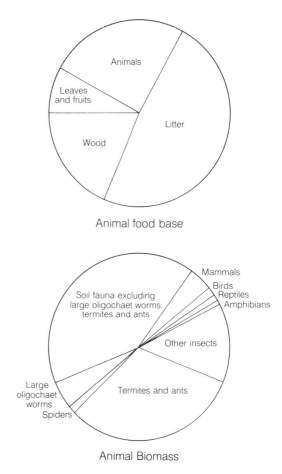

Figure 2.50 General composition and food-base of the animal population in lowland rain forest in Brazil. (After Fittkau and Klinge, 1973.) (Reproduced with permission from E. J. Fittkau and H. Klinge, On biomass and trophic structure of the central Amazonian rain forest ecosystem, *Biotropica*, 1973, **5**(1), 8.)

The defensive strategy exhibited by plants has been linked to their life span and pattern of dispersion (Feeney, 1976; Rhoades and Cates, 1975; Rhoades, 1979). Long-lived species, or those distributed in somewhat clumped patterns, are most likely to suffer predation because they are highly apparent to herbivores. It is assumed that these species would evolve 'quantitative' defences to deter a wide range of potential herbivores; low nutrient content or high fibre content reduces palatibility of tissues in this way. Conversely, species with shorter life spans or those widely dispersed through the forest would develop highly toxic 'qualitative' defences. This defensive strategy would deter the majority of non-adapted generalists, while the feeding opportunities of specialist herbivores which have evolved ways to counter the toxicity are restricted by the phenological, spatial or temporal characteristics of the plant populations. The role of secondary compounds extends beyond the protection of leaves and wood. The bitter taste of many immature fruits is imparted by secondary compounds similar to those found in leaves, and extremely poisonous amino acids, such as canavanine, have been isolated from seed coats. However, these complex defences must be eventually abandoned if the species depend on animals for propagation. The scents and nectars of flowers, and the rich sugars and oils of the mature fruits, are the products of a more solicitous suite of chemicals that are eagerly sought by herbivores.

2.15 THE FOREST ANIMALS

Birds such as the toucans and hornbills are specialized frugivores that often travel in flocks to feed on the widely dispersed fruits in the canopy (Karr, 1989). Smaller species, like the manakins (Pycnonotidae) and bulbuls (Piprinae), feed in the undergrowth; they are common in America and Africa but are poorly represented in Asia. Ground feeders, such as the large cassowaries (*Casuarius casuarius*) of Australia and New Guinea, and the ubiquitous pigeons (Raphidae), make use of fruits which have fallen from the canopy. Apart from parrots, seed-eaters are comparatively rare in tropical forests. Amongst the many insectivorous species are the woodpeckers (Picidae) which feed on bark insects, and the large population of birds that gather insects from the undersides of the leaves (Greenberg and Gradwohl, 1980).

Insects are also consumed by many omnivorous species as a supplement to fruits and seeds. Innumerable cryptic forms and colourations have evolved within the insect population to reduce predatory losses. Others rely on their colour to warn that they are poisonous or distasteful; they are often mimicked by harmless species. A more aggressive form of mimicry occurs between *Lycus loripes*, a distasteful, soft-bodied lycid beetle, and *Elytroleptes ignitus* which eats into it: after absorbing some of the body fluid of its victim, it too tastes repulsive for a period. Despite such precautions, there is a steady depletion of the insect population. Some are lured by the colours and nectar of carnivorous plants like *Nepenthes* which contain protein-digesting fluids in their pitchers. In this way nitrogen and phosphorus is secured by the plants, although some may be lost to spiders which close off part of the pitcher with their webs. The archerfish (*Toxotes jaculatrix*) knocks insects off leaves by a well-aimed squirt of water. However, these are unusual links in the forest food cycle in which the insects are more often consumed by amphibians and reptiles.

The amphibians have adapted to a variety of habitats in the moist forest environment. In addition to the terrestrial species, there is a remarkable diversity of arboreal forms, such as the tree frogs, which have evolved unusual methods of reproduction. Tree frogs of the species *Phyllomedusa* protect their eggs by wrapping them in leaves which overhang a pool; the tadpoles remain there for a while after they hatch and then drop into the water. In *Cerathyla* the female carries her eggs on her back, then deposits them in rainwater which has collected in a leaf cup when they are ready to hatch. The behaviour of *Dendrobates* is more unusual, the eggs being carried by the male until they hatch into tadpoles. In *Pipa pipa*, an aquatic species, metamorphosis progresses beyond the tadpole stage, and young froglets emerge from chambers on the backs of the females. The marsupial tree frog *Gastrothecus* also produces froglets but these are carried in a dorsal pouch. Many tree frogs are green or brown and blend with the foliage and branches, but the bright red colour of *Dendrobates* warns predators that it is highly poisonous: extracts are used by indigenous peoples to tip arrowheads. Equally numerous are the tree-dwelling lizards, most of which are equipped with adhesive lamellae on their feet to help them move on the waxy surfaces of the leaves. The African chameleons are amongst the best adapted of the arboreal lizards. The modified structure of their hands and feet allows them

---

Table 2.18 Bird and mammal communities in the tropical forests of Malaysia and Australia (after Harrison, 1962)

Upper air community
> Birds and bats which hunt above the canopy; mostly insectivorous, but with a large proportion of carnivores.

Canopy community
> Birds, fruit-bats and other mammals; mainly feeding on leaves, fruit, or nectar, but with a few insectivorous and mixed feeders.

Middle-zone flying animals
> Birds and bats; predominantly insectivorous, with a few carnivores.

Middle-zone scansorial animals
> Mammals which range up and down the trunks, entering both the canopy and ground zones; predominantly mixed feeders, with a few carnivores.

Large ground animals
> Large mammals and rarely birds living on the ground, they reach into the cover, or travel widely through the forest; plant feeders, feeding mainly by browsing, rooting for tubers, or occasionally feeding on fallen fruit, with attendant carnivores.

Small ground animals
> Birds and small mammals, capable of some climbing, which search the ground litter and lower parts of tree-trunks; principally insectivores or mixed feeders, but may be solely vegetable feeders; some are carnivores.

(Source: J. L. Harrison, The distribution of feeding habits among animals in a tropical rain forest, *Journal of Animal Ecology*, **31**, 1962.)

to grip the vegetation, and further support is given by the prehensile tail. Once spotted by its independent eyes, the prey of a chameleon rarely escapes the protractile tongue with its adhesive, prehensile tip. The gliding geckos (*Ptycozoon* spp.) and lizards (*Draco volans*) of South East Asia move between trees by extending flaps of skin along their abdomens in much the same way as the flying squirrels (*Aeromys* sp.). The gliding snake *Chrysopelea pelias* launches itself in a similar fashion, but controls its descent by flattening its body. Other snakes are exclusively arboreal, or move through the undergrowth and litter, but some, like the South American anaconda (*Eunectes murinus*), are mostly active in water. Mammals occupy six zones within the forest, according to the level they inhabit in the canopy and the range of foodstuffs available to them (Table 2.18).

2.16 FOREST RESOURCES AND HUMAN ACTIVITY

Until comparatively recently human activity had little impact on the tropical forests. The most primitive cultures engaged in hunting and gathering, and their nomadic life led them to wherever suitable plant and animal products could be found. Living in widely dispersed groups of 15–30 persons, these societies did little to alter the forest ecosystem. Significant disruptions were usually prevented on religous or magical grounds, although dispersal of fruits of preferred species was permitted. Few nomadic tribes still exist. Most evolved more sedentary ways of life, first as gardeners, and later developing shifting agriculture – the typical farming practice in these regions. In traditional systems temporary communities of perhaps 100 persons are established in remote areas, and small patches of forest are cleared by cutting and burning. After 1–2 years the plots are abandoned, and often the village is relocated. The effect on the forest is minimal and plots which have fallen into misuse are not easily distinguished from natural treefall sites. Today the sedentary form of shifting agriculture is more widespread and supports larger populations resident in permanent villages. Intensive land-use allows the plot to be cultivated for 2–3 years and woody, nitrogen-fixing species may be planted with the crops to accelerate soil restoration during the ensuing fallow (Boerboom and Wiersum, 1983). With increasing demands for food, long-practised methods are being gradually altered. Severe soil deterioration resulting from progressively shorter fallows can alter the nature of the regrowth vegetation. In many regions weedy grasses, such as *Imperata cylindrica*, now dominate abandoned land and prevent establishment of the woody species so necessary to the system.

Similar degradation is common in land previously used for plantations and commercial crop production, even though soil deterioration can be minimized by good management (Sanchez *et al.*, 1983). Land clearance

Figure 2.51 (a) Forest clearance in Costa Rica prepares an area for arable crops; (b) a recently cleared forest plot in Brazil. ((a) R. Johnson; (b) G. T. Prance.)

for large-scale agriculture is of a more permanent and extensive nature (Figure 2.51), although crops such as coffee may benefit from shade cast by remnants of the forest cover. Commercial development has resulted in the establishment of the most profitable crops well beyond their natural range. Thus, coffee was introduced to South America from Africa, bananas (*Musa* spp.) now exported from Africa and South America were originally native to Asia, while the extensive rubber plantations of Asia were developed from specimens of *Hevea brasiliensis* reputedly smuggled out of South America. Similarly, many tree species have been introduced to different regions to supplement export of dwindling supplies of local timber species (Figure 2.52). Tropical forestry practices have also undergone dramatic changes in the last few decades. Previously, selected species would be cut by hand, sawn into logs then transported by animals, river or primitive railways. The highly mechanized timber industry of today is much less discriminating, and extensive areas are rapidly devastated. Although an increasing number of species are now being utilized, many more are destroyed by logging.

It is has been estimated that in removing 10% of trees from lowland dipterocarp forest in Malaya an additional

Figure 2.52 A plantation of introduced *Pinus oocarpa* in Honduras. (I. Hyashi.)

Table 2.20 Height–class distribution of trees before and after logging at a site in Malaysia (after Johns, 1988)

Height class (m)	Before logging (%, N = 1138)	After logging (%, N = 559)
< 14.5	37.8	34.6
15–19.5	33.5	34.6
20–24.5	18.5	21.1
25–29.5	4.5	4.8
30–34.5	3.5	3.6
35–39.5	1.5	0.9
> 40	0.7	0.4

(Reproduced with permission from A. D. Johns, Effects of 'selective' timber extraction on rain forest structure and composition and some consequences for frugivores and folivores, *Biotropica*, Vol. 20(1), 1988.)

zon; only 2% of the trees were extracted, but an additional 26% of the tree cover was killed or damaged. Although the actual structure of the forest is altered considerably, its relative composition is not greatly affected (Tables 2.19 and 2.20). Consequently, generalized frugivores and foliovores may not be radically disturbed by such logging practices, although specialist feeders can be disadvantaged, particularly if secondary species are abundant (Johns, 1988). Until recently, there has been little thought given to the future of the land, but this is changing as international concerns are raised about the far-reaching impact of forest clearance.

Brown and Lugo (1984) estimated the area occupied by forests in tropical regions at 19.3 million km^2, of which 62% was in the humid tropics; the remainder was considered wooded savanna. Approximately 68% of the undisturbed productive forest was located in tropical America, 18% occurred in Africa, and Asia accounted for 14%. At the time of the survey only 11% of the productive forest land in tropical America had been logged, compared with 27% in Africa and 49% in Asia. Additional data indicated that forest disturbance by shifting agriculture ranged from 61.6 million ha in Africa to 99.3 million ha in tropical America, and accounted for 15–28% of the total forest region (Lanly, 1982). Land not

55% of the cover is destroyed, leaving only 35% of the forest undamaged (Burgess, 1971). To some extent this results from the structure of the forest. The tall trees crush many smaller individuals when they are felled; others may be pulled down because of intertwined lianas and vines; but in many instances damage can be attributed to unnecessary construction of roads, drag trails and loading sites. Thus, Uhl and Vieira (1989) reported that 9.25 km of roads were constructed for the selective logging of 52 ha of forest in the Brazilian Ama-

Table 2.19 Predominant tree families in logged and unlogged forests in Malaysia (after Johns, 1988)

rank	Before logging family	% sample	After logging family	% sample
1	Euphorbiaceae	27.0	Euphorbiaceae	27.0
2	Dipterocarpaceae	7.7	Dipterocarpaceae	7.7
3	Leguminosae	7.7	Leguminosae	6.8
4	Annonaceae	6.1	Annonaceae	6.8
5	Meliaceae	5.4	Meliaceae	5.4
6	Sapindaceae	5.2	Myristicaceae	5.2
7	Lauraceae	4.8	Sapindaceae	4.8
8	Myristicaceae	4.6	Lauraceae	4.6
9	Burseraceae	4.1	Anacardiaceae	3.4
10	Anacardiaceae	3.2	Myrtaceae	3.4
% of trees in top 10 families	75.8			76.1

(Reproduced with permission from A. D. Johns, Effects of 'selective' timber extraction on rain forest structure and composition and some consequences for frugivores and folivores, *Biotropica*, **20**(1), 1988.)

suitable for logging because of terrain difficulties, or because it had been set aside for parks and reserves, accounted for only 9% of the forested area in Africa, increasing to 29% in tropical America and 52% in Asia. Perhaps for this reason a far greater proportion of the productive forest in Asia is under management, plantations having been established using teak for sawlogs, eucalypts (*Eucalyptus* spp.) for pulpwood and some softwoods such as *Pinus caribaea*, *P. oocarpa* and *P. kesiya* for small timber and firewood (Lanly, 1982). In tropical America, plantations are most widespread in Brazil where fast-growing hardwoods, particularly eucalypts, are grown mainly for pulpwood or fuel; softwoods are also important. Teak is now grown extensively in Africa and accounts for about half of the sawlogs supplied from plantations. Despite increasing interest in plantations, less than 0.1% of the forest region is actively managed for sustainable development (Wood, 1990).

During the period 1981–90 the world's tropical forests were cleared at an estimated rate of 170 000 km^2 yr^{-1} (0.9% yr^{-1}), compared with 113 000 km^2 yr^{-1} (0.65% yr^{-1}) in 1981–85. This change is attributed to understimates of deforestation rates in 1981–5 rather than an increase in the rate of forest clearance in recent years (Whitmore and Sayer, 1992). The rate of clearing varies regionally. In Brazil it is estimated that more than 80 000 km^2 of forest are lost each year at a rate of 2.2% yr^{-1} (Repetto, 1990). About 90% of the Amazonian forests is still intact, but only 12% of the Brazilian Atlantic forests remain (Brown and Brown, 1992). In Central America the forest area is now reduced to about 40% of its original extent and deforestation is continuing at an annual rate of 0.9% (Groombridge, 1992). The rate of deforestation is accelerating in parts of South East Asia (Figure 2.53), where more than 129 000 km^2 have been cleared since 1981 at an average rate of 1.0% yr^{-1}. In central Africa 147 000 km^2 of forest were cleared during the same period (0.6% yr^{-1}), compared with 118 000 km^2 (2.1% yr^{-1}) in West Africa.

Over 80% of the tropical forest is now contained within 14 countries, all of which are faced with rapid population growth, low per-capita incomes and large debt burdens (Wood, 1990). To clear the forest simply to gain much needed revenue is viewed by many as a wasteful use of an irreplaceable resource. Myers (1980) predicted that much of the forest would be lost by the year 2020 unless concerted conservation efforts are undertaken. Not only would this reduce the supply of valuable

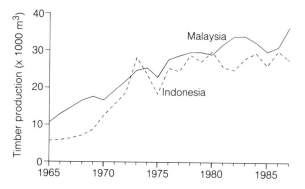

Figure 2.53 Production of tropical hardwoods in Malaysia and Indonesia for the period 1965–88. (After Brookfield and Byron, 1990.)

species, but the serious environmental degradation resulting from such wholesale destruction can only lead to the establishment of a greatly impoverished ecosystem.

The immediate effects of widescale deforestation include the loss of nutrients stored in the timber, deterioration in soil conditions as result of increased leaching and erosion, and a greatly altered microclimate suitable only for a relatively depauperate weedy flora. As plant diversity declines and fewer niches become available, the close links with the animal populations are disrupted. This insidious decay will eventually limit the success of restorative measures. On purely economic grounds the short-term benefits derived from timber sales may be insignificant compared with the potential income that might be derived from the unscreened genetic resources of the forest. Until recently, the Amazon region has been protected from excessive exploitation by its vast size and remoteness. However, the construction of the Trans-Amazon Highway has brought an influx of people ranging from entrepreneurs interested in cattle ranching, timber and mineral resources, to landless peasants seeking a better way of life. Salati and Vose (1984) argued that widespread deforestation could ultimately alter the water balance of the region, and lead to significant changes in the global circulation pattern of the atmosphere. Changes in the energy balance of the Earth have also been postulated as progressively less carbon dioxide is removed from the atmosphere by the dwindling forest cover. Such scenarios have served to focus world attention on the predicament of the tropical forests, and have spurred conservation activities in these deceptively fragile ecosystems.

CHAPTER 3

Tropical savannas

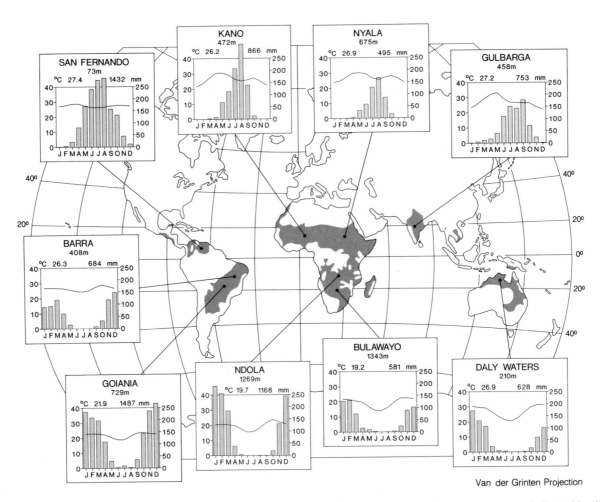

Van der Grinten Projection

Figure 3.1 Distribution of tropical savanna and representative climatic conditions. Mean monthly temperatures are indicated by the line and mean precipitation for each month is shown by the bars. Station elevation, mean annual temperature and mean annual precipitation appear at the top of each climograph.

3.1 INTRODUCTION

The term **savanna** was used originally to describe treeless areas in South America. Now it is more generally applied to a range of vegetation types in the drier tropics and subtropics in which grasses form a conspicuous ground cover. Shrubs and trees are often present resulting in woodlands, open parklands or a dense cover of scrub. The broadest definition of savannas – ecosystems which lie between the forests and deserts of tropical regions (Huntley and Walker, 1982) – attests to the diversity of vegetation that might be included, and numerous regional terms are used to describe distinctive cover types. In South America the more densely wooded areas are referred to as 'cerrado'; fewer trees occur in the 'campos' and 'llanos', and thorn scrub is the dominant cover in the 'caatinga'. In Africa the 'miombo' woodlands can be distinguished from the parklike 'bushveld'. Acacias and scrubby eucalypts occur in the 'mulga' and 'brigalow' of Australia.

Savanna is most widely distributed in tropical Africa where it covers 65% of the continent in a broad belt around the forests of the Congo Basin and West Africa (Figure 3.1). In South and Central America the savanna is more fragmented, and occurs extensively only in eastern Brazil and the Orinoco region. Savanna is also found across northern Australia and throughout the Indian subcontinent, and to a limited extent in South East Asia. Throughout its range, the vegetation is adapted to a pronounced climatic rhythm, actively growing during the rains of the high sun season, and lying dormant in the intervening dry period when fires often burn through the dead litter.

Most classifications of savanna stress the physiognomic characteristics of the vegetation. An early scheme proposed by Shantz and Marbut (1923) for Africa was based on the nature of the grass cover. Three classes were used. The 'high grass–low tree savannas', dominated by grasses such as *Pennisetum* spp. which typically exceed 2 m in height by the end of the growing season, bordered the tropical forests of West Africa. 'Acacia–tall grass savannas', composed of tussock grasses 1–1.5 m in height, occurred over most of the savanna formation,

and 'acacia–desert grass savanna' was present in the drier regions. In later schemes the nature of the tree cover is emphasized. Sarmiento (1983) described four savanna types in South America according to the importance of the tree cover. Trees are absent in the savanna grasslands ('campo limpo'); < 2% tree or shrub cover is found in the tree–shrub savanna ('campo sujo'); wooded savanna ('campo cerrado') has 2–15% tree cover, and 15–40% tree cover occurs in the savanna woodland ('cerrado'). Height and density of tree cover were combined in the classification used by Walker and Gillison (1982) in Australia (Table 3.1). In Cole's (1963) scheme the nature of the grass cover is incorporated with the height and density of the trees (Table 3.2). Although only five categories are proposed, this simple physiognomic approach provides a useful framework for reviewing savanna vegetation.

3.2 REGIONAL PATTERNS

3.2.1 Africa

In Africa, much of the savanna is of the woodland type with tall grasses forming the ground cover beneath a discontinuous canopy of deciduous trees. It occurs extensively in countries such as Angola and Zambia where it is referred to as miombo woodland (Figure 3.2). In central and southern Africa the dominant trees are

Table 3.1 Structural formations in Australia (Walker and Gillison, 1982)

Life form and height of tallest stratum	Tree or shrub cover			
	Dense (70–100%)	Mid-dense (30–70%)	Sparse (10–30%)	Very sparse (< 10%)
Trees > 30 m	Tall closed-forest	Tall open-forest	Tall woodland	Tall open-woodland
Trees 10–30 m	Closed-forest	Open-forest	Woodland	Open-woodland
Trees 5–10 m	Low closed-forest	Low open-forest	Low woodland	Low open-woodland
Shrubs 2–8 m	Closed-shrub	Open-scrub	Tall shrubland	Tall open-shrubland
Shrubs 0–2 m	Closed-heath	Open-heath	Low shrubland	Low open-shrubland

(Reproduced from J. Walker and A. N. Gillison, Australian savannas, in *Ecological Studies 42 – Ecology of Tropical Savannas*, eds B. J. Huntley and B. H. Walker, published by Springer-Verlag GmbH & Co., 1982.)

Table 3.2. Classification of savannas (Cole, 1986)

Savanna woodland
Deciduous and semi-deciduous woodland of tall trees (> 8 m high) and tall mesophytic grasses (> 80 cm high); spacing of the trees more than the diameter of the canopy.

Savanna parkland
Tall mesophytic grassland (grasses 40–80 cm high) with scattered deciduous trees (< 8 m high).

Savanna grassland
Tall tropical grasslands without trees or shrubs.

Low tree and shrub savanna
Communities of widely spaced low-growing perennial grasses (< 80 cm high) with abundant annuals and studded with widely spaced, low-growing trees and shrubs (often < 2 m high).

Thicket and scrub
Communities of trees and shrubs without stratification.

(Reproduced with permission from M. M. Cole, *The Savannas: Biogeography and Geobotany*, published by Academic Press Inc., 1986.)

Figure 3.2 Miombo woodland in Zimbabwe comprised of *Brachystegia spiciformis* and *Julbernardia paniculata*. (B. R. Neal.)

Figure 3.3 (a) Savanna woodland predominantly of *Brachystegia, Isoberlinia* and *Julbernardia* in the Serengeti Plains of East Africa; (b) savanna grassland in Kenya; (c) small acacias such as *A. tortillis* and tree-like *Euphorbia candelabrum* are characteristic species of the drought-deciduous woodlands. ((a) R. T. Coupland; (b) and (c) I. Hyashi.)

species of *Brachystegia, Isoberlinia* and *Julbernardia* which may grow to a height of 20 m; their rather open, flat-topped canopies provide adequate light for the grasses and herbs beneath. Smaller trees, such as *Colophospermum mopane*, or shrub savanna comprised of *Acacia* spp. and woody succulents like *Commiphora pyracanthoides* is found in drier regions or on coarser soils. North of the forested Congo basin the savanna woodlands extend in a broad belt across the continent from Senegal to southern Sudan. In West Africa trees such as *Anogeissus leiocarpus* and *Lophira lanceolata* grow to a height of 12–15 m above a herbaceous layer dominated by tall grasses including *Hyparrhenia, Andropogon* and *Pennisetum* spp. Further east the prominent trees include the baobab (*Adansonia digitata*) and *Sclerocarya birrea*. In East Africa patches of *Albizia* sp. and *Chlorophora excelsa* provide a fragmentary connection with the southern miombo woodlands. Although some species, such as *Burkea* and *Isoberlina*, are found throughout the savanna woodlands, many are more restricted in their distribution. *Monotes kerstingii*, for example, is most commonly found on eroded slopes and *Terminalia macroptera* on poorly drained clays (Cole, 1986).

The physiographic diversity of East Africa supports a variety of plant communities. In upland regions the savanna woodlands may be replaced by evergreen forests which extend up the moist slopes of the volcanic mountains. In contrast, the drier, low-lying regions of the rift valley support extensive tracts of grassland ocassionally interrupted by small acacias or thickets of *Combretum, Euphorbia* or *Commiphora* (Figure 3.3). A similar landscape is found in the bushveld of southern Africa where acacias are scattered across a continuous cover of drought tolerant grasses. However, this simple physiognomy does not reflect the diversity of vegetation in the region (Acocks, 1975). Shrubby species increase in the drier lowveld, where drought tolerant species of *Eragros-*

tis and *Aristida* form an open, tussocky grass cover which eventually merges with the desert vegetation of the Kalahari. In northern Africa the transition to desert is clearly related to latitude; here thorny acacias and coarse wiry grasses form an ecotone along the southern margin of the Sahara Desert.

Figure 3.4 Thorn forests are a distinctive component of the endemic flora of south-eastern Madagascar. (a) Dense stand of *Alluaudia dumosa* (left) and *A. procera* (right); (b) thicket of *Didierea madagascariensis*; (c) columnar form of *Alluaudia adscendens* which grows to a height of 15 m; (d) the leafy stem of *A. adscendens*. (G. E. Schatz.)

The savannas of Madagascar are floristically richer than those of mainland Africa, and the large number of endemic species results in many unique associations. Dense shrub savanna occurs in the most arid parts of south-west Madagascar: here succulent Euphorbiaceae and Didiereaceae form dense thickets above a sparse grass layer (Figure 3.4). Elsewhere, the savanna is largely derived by fire from woodland formations, and

grasses, such as *Aristida, Hyparrhenia* and *Heteropogon*, now form a dense herbaceous layer beneath a scattered overstory of small trees.

3.2.2 South America

The cerrado is the most extensive savanna region in South America; it occurs on the gently rolling plateaus of

Figure 3.5 Open cover of small trees in cerrado near Brasilia. (G. T. Prance.)

Figure 3.6 Widely scattered trees, palms and shrubs in the campo sujo of Brazil. (I. Hyashi.)

east-central Brazil with associated outliers in the states of São Paulo and Pernambuco. In its strictest sense cerrado refers to a dry woodland in which small, rather twisted trees, such as *Curatella americana*, form a continuous cover above a ground layer of xeromorphic herbs (Figure 3.5). This intergrades with: the campo cerrado, characterized by an open cover of short, scrubby trees; the campo sujo, in which taller trees or shrubs provide a widely scattered canopy (Figure 3.6); and the almost pure grasslands of the campo limpo (Figure 3.7) (Eiten, 1972). The dominant grasses are species of *Andropogon*, *Aristida*, *Paspalum* and *Trachypogon*. Throughout this region differences in soils and drainage greatly influence the nature of the vegetation. Gallery forests often occur in damper valley bottoms, and wet campo may be maintained as a distinctive marshy seepage area on some hillsides.

Whereas the cerrado extends across the Brazilian tablelands some 300–1000 m above sea level, the llanos are found at lower elevations (2–300 m) throughout the deeply dissected drainage basin of the upper Orinoco in Venezuela and Colombia. Here treeless grasslands or lightly wooded savannas form the dominant cover. On the higher plains low scattered trees of *Byrsonima crassifolia*, *Curatella americana* or *Bowdichia virgilioides* form an open canopy above the grasses. On steeper slopes this may develop into more densely wooded associations with gallery forests in the valley bottoms. Swamps composed of hydrophilous sedges and grasses and the palm *Mauritia flexuosa* may develop in poorly drained sites. A similar interdigitation of savanna, gallery forest and swamp occurs on the river terraces and alluvial fans throughout the low-lying parts of the llanos. The savanna woodlands of the higher terraces are replaced in turn by semi-deciduous forests and gallery forests on the intermediate and lowest terraces. Much of the bottomlands are occupied by **hyperseasonal savannas** and

Figure 3.7 Predominantly herbaceous cover of the campo limpo near Brasilia. (G. T. Prance.)

'esteros'. The hyperseasonal savannas are subject to months of waterlogging in the wet season, followed by the stress of drought and fire during the dry season. Herbaceous and shrubby plants including *Ipomoea crassicaulis* and *Ludwigia lithospermifolia* are best adapted to these conditions, but some trees, such as *Annona jahnii*, *Pterocarpus podocarpus* and *Spondias mombin*, or the palm *Copernicia tectorum*, may also be present. In contrast, the esteros are flooded in the wet season, but dry out slowly. They are treeless areas dominated by the grasses *Leersia hexandra* and *Hymenachne amplexicaulis*, both of which are highly palatable to the cattle which graze them (Sarmiento, 1983).

Smaller patches of savanna are common further north in the low-lying coastal regions of the Guyanas and Central America, and on many of the islands of the Caribbean. Scattered patches of savanna also occur throughout the Amazon lowlands: their distribution has

Figure 3.8 The palm *Leopoldinia pulcherimma* is conspicuous in this regularly flooded igapó forest. (G. T. Prance.)

Figure 3.9 The undulating terrain of the Gran Pantanal creates a mosaic of swamp, grassland and woodland. (G. T. Prance.)

been related to impervious laterites or seasonal water-logging (Eiten, 1978), although many such areas support denser 'igapó' forest (Figure 3.8). Further south, strad-dling the borders of Paraguay and Brazil, is the Gran Pantanal or 'great swampland' (Figure 3.9). Differences in the degree and duration of waterlogging and drought are reflected in the variety of grassland types which occur across this gently undulating landscape. In the lowest areas species of *Paspalum* and *Panicum* are com-mon; hyperseasonal savannas are dominated by *Sorgha-*

Figure 3.10 Dry caatinga thorn scrub comprised of cacti (*Cereus* sp.) and shrubby *Mimosa hostilis*, *Croton sonderiana* and *Caesalpinia pyramidalis*. (I. Hyashi.)

strum agrostoides; and on the higher land trees, such as *Curatella americana*, become more abundant. The dry woodlands of the 'chaco' lie to the west, where the savanna parklands and shrub savannas are dominated by xerophytic grasses. Thorn scrub and cacti are charac-teristic of the arid western chaco which shares many species with the equally dry caatinga of north-eastern Brazil (Figure 3.10).

3.2.3 Australia

Savanna woodland is found in the northernmost part of Australia and continues into eastern Queensland (Fig-ure 3.11). Stringybark (*Eucalyptus tetradonta*), growing to 15–21 m in height, is common throughout the northerly districts. Beneath its open canopy is an understory of smaller trees and shrubs and a ground cover of bunch grasses. Kangaroo grass (*Themeda australis*) occurs throughout these woodlands. It grows in association with other tall perennial bunch grasses, such as *Sorghum plumosum* and *Aristida pruinosa*, and may be 1.8 m in height at the end of the season. Some of the annual grasses, such as *Sorghum intrans* and *S. australiense*, can exceed 3 m in height, while the shorter species, includ-ing *Aristida hygrometrica* and *Brachyachne convergens*, normally grow to 0.6 m (Shaw and Norman, 1970). In north-eastern Queensland species of ironbark, including *E. crebra* and *E. drepanophylla*, become increasingly com-mon (Figure 3.12), although smaller acacias and mela-leucas (*Melaleuca* spp.) may be locally dominant in heathy sites. Queensland bluegrass (*Dichanthium seri-ceum*) grows abundantly in treeless areas underlain by fine-textured soils which are prone to cracking.

Interspersed with the eucalypt woodlands of eastern Queensland is the brigalow, a transitional woodland formation composed predominantly of *Acacia harpophylla*,

Figure 3.11 Small *Eucalyptus pruinosa* in treed savanna in northern Australia. (O. W. Archibold.)

Figure 3.13 The swollen-stemmed bottle tree (*Brachychiton rupestris*) with its high water content is a conspicuous species in the savannas of eastern Australia. (O. W. Archibold.)

Figure 3.12 Fan palm (*Livistonia decipiens*) in *Eucalyptus crebra* woodland in north-eastern Queensland; tall spear grass (*Heteropogon contortus*) is the dominant species in the herbaceous layer. (G. W. and R. Wilson.)

with the bottle tree (*Brachychiton rupestris*) as a distinctive associate (Figure 3.13). Westwards, the brigalow is replaced by smaller eucalypts, such as poplar box (*E. populnea*), over much of the solodic soils, or by trees like the coolibah (*E. microtheca*) in the river valleys. In these regions the common grasses are *Heteropogon contortus* and species of *Aristida* and *Bothriochloa* (Coaldrake, 1970). The drier climate of central and southern Queensland is marked by the transition to arid shrub and grasslands. Here mulga (*Acacia aneura*), which grows to heights of 3–7 m, is often the sole dominant. The ground cover consists of perennial tussock grasses, such as *Danthonia bipartita* and *Aristida* (Figure 3.14). 'Spinifex', comprised of species of *Triodia* and *Plectrachne*, forms a widely spaced cover of grassy hummocks 1–1.5 m across and 0.5–0.8 m high over much of the semi-arid regions of Western Australia and the Northern Territories (Perry, 1970). Northwards the spinifex merges with the tussocky Mitchell grasses (*Astrebla* sp.), while to the west and south the mulga is once again conspicuous.

3.2.4 India and South East Asia

Savanna is broadly distributed throughout the Indian subcontinent, although much of it has been cleared for

Figure 3.14 Tussock grasslands on the Barkly Tableland in northern Australia dominated by perennial species of *Astrebla*, *Aristida* and *Chrysopogon*. (O. W. Archibold.)

Figure 3.15 Eucalyptus savanna near Port Moresby, Papua New Guinea: the dominant trees are *Eucalyptus alba*, *E. confertiflora*, *E. papuana* and species of *Melaleuca*; the herbaceous layer is dominated by the grass *Themeda australis*. (G. A. J. Scott.)

arable farming. Most of what remains is restricted to rugged, inaccessible areas which are still largely unsettled. Patches of dry-deciduous woodland comprised of small trees, such as *Anogeissus latifolia* and *Diospyros melanoxylon*, still occur in parts of the Deccan Plateau where they provide a much-needed source of firewood; the leafy branches are also cut for livestock feed. *Sehima–Dichanthium* grasslands are found over much of the Indian peninsula, although thorn bushes, such as *Acacia catechu* and *Zizyphus jujuba*, have become increasingly widespread in this heavily grazed association. Similarly, in the semi-arid regions bordering the deserts of Rajasthan, thorny shrubs form a scrubby cover over short grasses. Wooded *Themeda–Arundinella* grasslands derived from humid forests now occur in the Himalayan foothills. Grasslands comprised of *Phragmites*, *Saccharum* and *Imperata* have established on the wetter alluvial soils of the Ganges floodplain (Misra, 1983).

In South East Asia patches of savanna woodland interspersed with tropical forest are found throughout Kampuchea, in parts of Laos and Vietnam, and less extensively in Thailand and Burma. These open woodlands are dominated by deciduous dipterocarp species which may grow to a height of 20 m. Although these woodlands support a rather diverse tree flora with as many as 135 species recorded in some sites, only three species, *Dipterocarpus tuberculatus*, *Pentacme suavis* and *Shorea obtusa*, are common (Blasco, 1983). Differences in stand density and floristic composition are largely controlled by soil conditions, the poorest woodlands being associated with eroded hillslope sites. In flooded or permanently damp areas localized patches of grass savanna may be encountered: *Phragmites karka* and *Saccharum narenga* are common species in such sites.

In addition to mainland sites, savannas occur on many of the islands in the East Indies and the Pacific (Figure 3.15). The comparatively moist, albeit seasonal, climate

of these regions results in mostly wooded forms the specific character of which is influenced by soil conditions and drainage. Drier zones, often lying in topographic rain shadows, typically support shrub communities, while open grasslands are associated with cooler, high elevation sites.

3.3 THE CLIMATE OF SAVANNA REGIONS

The physiognomic diversity of savanna vegetation reflects the different climatic conditions which occur throughout these widely distributed plant formations. In general terms, most savannas are associated with Köppen's Aw climate class. This includes the tropical wet-and-dry climate in which there is a distinct dry season during the low-sun period (Figure 3.16): at least 1 month receives less than 60 mm of rain, and mean monthly temperatures never fall below 18 °C. In many savannas mean annual temperature exceeds 24 °C, but in the higher parts of East Africa and Brazil mean annual temperatures are normally less than 20 °C. Maximum temperatures usually occur at the end of the wet season (Figure 3.1). Although the sun has dropped below its zenith by this time, increased insolation under the cloudless skies can result in mean temperatures above 30 °C. However, during the coldest months mean temperatures as low as 13 °C are not uncommon. In the highland regions mean monthly temperatures can fall below 8 °C, and temperatures below 4 °C have been reported in the acacia shrubland of eastern Australia.

Throughout the tropics the nature of the vegetation cover is more closely determined by the amount and seasonality of the rainfall than by temperature. Annual

Figure 3.16 Wet and dry seasonal aspects of mopane woodland (*Colophospermum mopane*) in Zimbabwe. (B. R. Neal.)

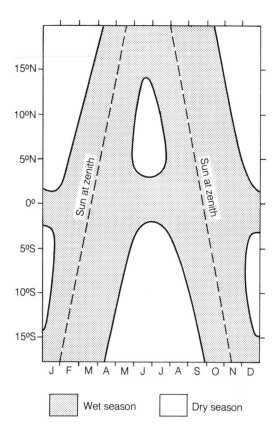

Wet season   Dry season

Figure 3.17 Seasonal distribution of precipitation and duration of the dry season as a function of latitude. (After Richards, 1952.) (Reproduced with permission from P. W. Richards, *The Tropical Rain Forest*; published by Cambridge University Press, 1952.)

precipitation in the savannas varies considerably between regions. The drier areas may receive only 250 mm compared with 1500 mm or more in many of the woodland areas. Irrespective of the total, the rainfall regime is markedly seasonal, and a dry period which can last for several months is a distinctive feature of savannas (Figure 3.17). Nix (1983) has suggested that the severity of the dry season can be approximated by the amount of precipitation received during the driest 3 months of the year. In Africa, Australia and Asia the savannas are generally found in areas which receive less than 25 mm during this period. A similar relationship is evident in much of the cerrado and caatinga of Brazil. However, the savannas of Central America can receive as much as 100 mm of rain during the dry 3-month period, and in the Guianan savannas this value can exceed 200 mm: soil conditions may be important in determining the distribution of forest and savanna in these regions.

3.4 SOILS

Savannas occur extensively on old plateaus which are variously interrupted by escarpments and further dis-

sected by rivers. Such areas are commonly underlain by layers of laterite or gravelly veneers formed from deeply weathered forest soils which developed during the wetter Cretaceous and Tertiary periods (Cole, 1986). The decline of the forests during the dry phases of the Pleistocene resulted in the erosion of the exposed soils and the hardening of the iron-rich clays. Continual weathering has produced nutrient-poor, acidic Oxisols which are subject to drought because of the pronounced seasonality of the rainfall. Much of the clay in these soils is in the form of kaolinite, a mineral whose 1:1 lattice structure affords few sites for exchangeable cations. High concentrations of aluminium produced by clay decomposition can also affect the nutrient balance of the soil: phosphorus is particularly deficient since it becomes immobilized as iron-aluminium phosphate (Eiten, 1972). The cation exchange capacity of savanna Oxisols is low, about 20–40 μeq g^{-1}, and is mostly derived from organic matter (Montgomery and Askew, 1983); fertility is further reduced by low base saturation. Soil analyses for upland sites in cerrado indicate a pH range of 4.8–5.2 with organic matter levels of 1.5–3.0% (Lopes and Cox,

Table 3.3 Nutrient content of savanna soils

		pH	OM (%)	N (%)	P	K	Ca	Mg	Source
					\(\mu\)g g⁻¹				
Venezuela									
llanos	0–10 cm	4.9	2.0	.08	2	123	8	182	Sarmiento, 1984
	10–50 cm	5.1	1.3	–	2	76	1	92	
	50–80 cm	5.1	0.9	–	1	53	1	72	
tree and shrub savanna	0–12 cm	4.4	1.4	.08	0	4	20	2	Eden, 1974
	12–25 cm	4.4	1.1	.06	0	8	148	10	
	25–50 cm	4.8	–	.03	0	4	40	5	
herbaceous savanna	0–5 cm	5.1	1.8	.06	10	23	185	–	Sarmiento and Monasterio, 1969
	5–21 cm	4.8	1.1	.05	12	7	139	–	
	21–52 cm	4.8	0.8	–	15	9	100	–	
Brazil									
cerrado	0–15 cm	4.9	2.3	.07	1	39	66	–	Medina, 1982
campo limpo	0–15 cm	4.9	2.2	.07	1	31	40	–	
savanna woodland	0–15 cm	5.0	2.4	.08	1	43	90	–	
Ghana									
Andropogon savanna	0–5 cm	7.1	0.8	.05	12	59	860	122	Nye and Greenland, 1960
	5–15 cm	6.9	0.7	.04	5	47	780	85	
	15–30 cm	6.4	0.4	.03	2	35	460	65	
Zambia									
wooded *Burkea* savanna	5–30 cm	5.0	–	.04	2	40	83	24	Huntley and Morris, 1982
'miombo' woodland	0–4 cm	5.3	3.3	.14	–	129	460	9	Nye and Greenland, 1960
	4–30 cm	4.9	1.0	.05	–	164	80	4	

(Reproduced with permission from P. H. Nye and D. J. Greenland, *The Soil Under Shifting Cultivation*, Technical Communication No. 51, published by CAB International, 1960; also from G. Sarmiento and M. Monasterio, Studies on the savanna vegetation of the Venezuelan llanos, I. The use of association analysis, *Journal of Ecology*, 1969, **57**; also from M. J. Eden, Palaeoclimatic influences and the development of savanna in southern Venezuela, *Journal of Biogeography*, 1974, **1**, 101; also from B. J. Huntley and J. W. Morris, Structure of the Nylsvley savanna, and E. Medina, Physiological ecology of neotropical savanna plants, in *Ecological Studies 42 – Ecology of Tropical Savannas*, eds B. J. Huntley and B. H. Walker, published by Springer-Verlag GmbH, 1982; also from *The Ecology of Neotropical Savannas* by Guillermo Sarmiento, translated by Otto Solbrig, Cambridge, Mass.: Harvard University Press, copyright © 1984 by the President and Fellows of Harvard College, reprinted by permission of the publishers.)

1977). Similar Oxisols occur in the savannas of central and eastern Africa. Although they have a relatively high clay content, most of these soils are rather porous and tend to dry out quickly. Irrigation, as well as liming and fertilizer treatment, is often necessary for intensive agricultural development (Furley and Ratter, 1988).

Alfisols are more common in the drier savannas. These soils are found throughout northern Africa and in East Africa, across the southern Deccan plateau of India, and in the chaco and caatinga of Brazil. They are mostly derived from crystalline parent materials with a high quartz content which makes them rather light in texture. This, together with the predominance of kaolinite clays, results in a low base exchange capacity. However, base saturation is high (60–90%) and the Alfisols are potentially more fertile than the Oxisols (Table 3.3). Most Alfisols are less than 1 m deep, and continual weathering of the underlying bedrock provides nutrients to the rooting zone (Foth and Schaffer, 1980). Much of the iron is leached from the surface horizons, hence lateritic crusts and plinthites usually do not develop. The productivity of these moderately fertile, shallow soils is limited by the dry seasonal climate. They are also prone to erosion.

The soils of the drier caatinga are mostly derived from sandstones. They are generally quite shallow, although some of the high plateaus are covered with a deeply weathered mantle of kaolinitic clays overlain by sandy ferruginous crusts similar to some Oxisols (Beek and Bramao, 1968). In the chaco of South America comparatively deep soils have developed in sediments eroded from the Andes. Leaching occurs slowly in this dry region and many of the minerals derived from weathering processes are retained in the soil. Calcium carbonate is often present as a layer of hardpan while more soluble salts may accumulate in saline horizons. The resulting soils have a pH near neutral, they are rather low in organic carbon and, because of hardpans or illuvial clay horizons, may be subject to flooding during the wet season (Bucher, 1982).

Entisols are associated with the driest savannas. They are particularly common in southern Africa and northern Australia where they occur as shallow, sandy or loamy soils often containing stones or gravels. Entisols are characteristically low in available phosphorus, and fertility can be further reduced by accumulated salts. In parts of eastern Australia the savannas are found on Vertisols. These deep-cracking soils are composed primarily of montmorillonite clay which expands and shrinks as its moisture content changes. During the dry season, wide cracks can develop to depths of 1 m or more; expansion of the clay as it is rewetted can produce a series of small mounds and hollows. This **'gilgai'** microrelief is common in some parts of the brigalow. Vertisols tend to be acidic, low in phosphorus, and somewhat saline. Their high clay content makes them rather impermeable, but they are distinguished from the hydromorphic soils of more permanently waterlogged savannas by the absence of mottling and gleyed horizons. Vertisols are also common in the savannas of northern India.

3.5 ORIGINS OF THE SAVANNAS

3.5.1 Palynological considerations

Savanna associations have existed for at least 25 million years, but during this time their distribution has been greatly influenced by climatic perturbations. The alternating wet and dry phases of the Pleistocene resulted in periodic expansion of savanna and a consequent fragmentation of the tropical forests (Haffer, 1969). The pollen record in East Africa, for example, indicates that savanna was replaced by forest about 12 000 years BP (Figure 3.18). The Oleaceae, which are today important colonizers in secondary forest sites, began to increase initially and were subsequently replaced by an evergreen forest dominated by the Moraceae, with species of *Musanga* especially abundant. About 7500 years BP this cover gradually changed to semi-deciduous forest with a large component of *Celtis* and eventually to the present cover of grasses, sedges and characteristic savanna plants, such as *Acalypha*. Similarly, the numerous petroglyphs depicting large herbivores, such as giraffe, elephant and rhinoceros, which have been found in the Sahara attest

to a wetter period from 12 000-3000 years BP (Livingstone, 1975).

3.5.2 Climate as a factor in savanna development

Early descriptions of savannas, such as those by Schimper (1903) and Bews (1929), stressed the marked seasonality of rainfall in these regions. More recently, Eiten (1972) used Gaussen's complex xerothermic index based on the length of the dry season to distinguish between forest, cerrado and caatinga regions in Brazil. Nix (1983) has associated tropical vegetation patterns with the severity of the dry season, but Walter (1985) considered that the transition from forest to savanna reflected differences in the water economy of the dominant plants. Perennial grasses, he suggested, are well adapted to these seasonal climates. They grow quickly during the wet season, taking up water with their dense, finely branched root system. Growth continues into the dry season until eventually the leaves die. The shoot meristems, protected by the dried-out leaf sheaths, can survive long periods of drought. Trees will survive as long as they can supply adequate moisture to their meristematic tissues to avoid desiccation and death during the dry season.

3.5.3 The influence of soils and drainage

Factors other than climate have been related to vegetation patterns in areas where savannas and forests are interspersed. Jones (1930), in his account of the Guyanas,

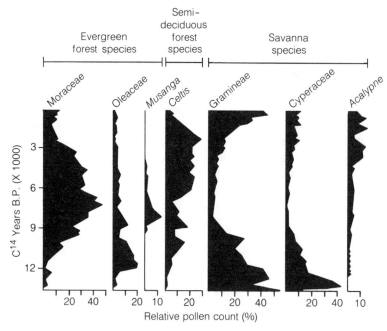

Figure 3.18 Pollen diagram showing reversion to grassland in the Lake Victoria region of East Africa following the establishment of forest during wetter conditions in the early post-glacial period. (After Kendall, 1969.) (Redrawn with permission from R. L. Kendall, An ecological history of the Lake Victoria basin, *Ecological Monographs*, 1969, **36**, 152.)

Figure 3.19 (a) Shrubs and low trees form an open woodland (campiña), often dominated by *Humira balsamifera* on the nutrient poor white-sands areas in central Amazonia; (b) *Cladonia* lichens are prominent in the ground cover in these oligotrophic sites. (G. T. Prance.)

noted that savannas were associated with sandstone districts, whereas soils weathered from igneous rocks, such as granites and gneisses, supported forests. He assumed, like Pulle (1906), Ijzerman (1931) and Hardy (1945), that severe leaching had rendered these sandy savanna soils infertile. Similar relationships between geologic conditions and vegetation cover were described by Waibel (1948) in the Planalto region of Brazil. Here the transition from forest to cerrado and campo was attributed to decreasing soil fertilty and lower soil moisture reserves. Treeless savannas are associated with the dry '**white**

sands' of the Gran Sabana of Venezuela (Sarmiento, 1983). However, comparable white sands, which are essentially devoid of nutrients, support forests in the moist climate of the Amazon (Figure 3.19) (Klinge, 1965; Stark, 1970).

Although forest growth may be restricted by severe moisture stress in the dry season, Bennett and Allison (1928) proposed that seasonal waterlogging could also promote the establishment of savannas. Beard (1953) also considered that natural drainage was the most important characteristic which distinguished the soils of savanna and forested regions in tropical America. He argued that few trees can tolerate the stresses of seasonal waterlogging and seasonal drought imposed by the interaction of climate and topography in savanna regions. The possibility of standing water during the wet season is increased by the lack of organized drainage channels across many of the gently undulating pediplains, and by impervious laterites and iron pans which commonly underlie the shallow savanna soils. Little moisture is retained by these soils in the dry season. Severe moisture budgets are also associated with low-lying areas where impermeable clay-pans are often encountered at depths of 0.5–1 m (Sarmiento and Monasterio, 1975). The clay effectively prevents water from permeating to the deeper horizons, and anaerobic conditions during waterlogging produce pronounced mottling in these soils. Changes in the vegetation cover of southern Africa have been similarly associated with the varying influences of topography, soil depth and the presence of indurated subsurface horizons on soil moisture budgets (Figure 3.20).

Morison *et al.* (1948) emphasized the relationship between soil conditions and topographic setting in their descriptions of the savannas of the Sudan. Higher areas subject to erosion or eluviation were often covered in thin, infertile soils over iron pans. Erosional and depositional processes were in dynamic equilibrium on slopes, and this gave rise to various colluvial soils. Soil formation in low-lying areas was affected by the gradual accumulation of materials and by flooding. (Figure 3.21). In describing the savannas of Brazil, Cole (1960) related soil drainage and fertility to the geomorphological history of the region and thereby provided a convincing explanation of the origin of these enigmatic landscapes. The importance of soil conditions was inferred from the observation that the savannas on the older tablelands are replaced quite abruptly by forest in more recently dissected terrain. The soils on the plateaus are mainly derived from sandstones. They are coarse-textured and infertile and are often overlain by a superficial layer of gravels and pebbles. The residual lateritic iron pans which may also be present are a legacy of the deeply weathered soils which developed in more humid climates under forests. Savanna soils either retain little moisture, or are subject to alternate periods of drought and flood because of perched water tables. Surface

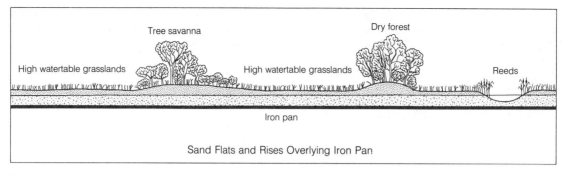

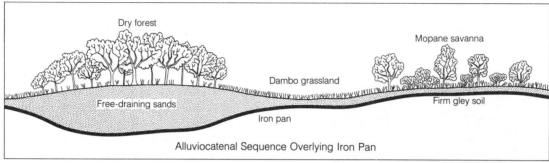

Figure 3.20 Influence of soil conditions and subsurface drainage on ecosystem patterns in the savannas of southern Africa. (After Tinley, 1982.) (Reproduced with permission from K. L. Tinley, The influence of soil moisture balance on ecosystem patterns in southern Africa, in *Ecological Studies 42 – Ecology of Tropical Savannas*, published by Springer-Verlag GmbH, 1982.)

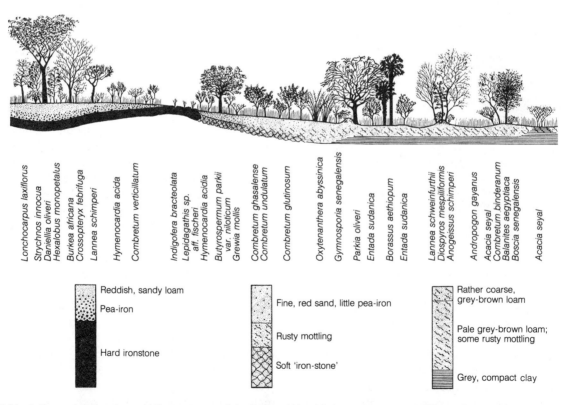

Figure 3.21 Soil–vegetation catenas in the savannas of the Sudan. (After Morison *et al.*, 1948.) (Reproduced with permission from C. G. T. Morison, A. C. Hoyle and J. F. Hope-Simpson, Tropical soil–vegetation catenas and mosaics: a case study in the south-western part of the Anglo-Egyptian Sudan, *Journal of Ecology*, 1948, **36**, 23.)

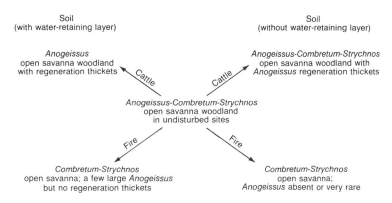

Figure 3.22 The interaction of grazing, fire and soil moisture conditions on savanna vegetation in Nigeria. (After Keay, 1949.) (Reproduced from R. W. J. Keay, An example of Sudan zone vegetation in Nigeria, *Journal of Ecology*, 1949, **37**, 347.)

materials are not removed by erosion in these level sites, and nutrient replenishment is consequently restricted by the slow rate of weathering of the underlying bedrock: Cole (1960) described these soils as 'old and exhausted'. Trees are restricted to the steeper terrain where better drainage and more rapid erosion improves soil conditions.

3.5.4 The impact of fire and grazing

Keay (1949), finding only a general relationship between vegetation, topography and soils in the open savanna woodlands of Nigeria, concluded that the composition of the vegetation was determined by fire and grazing (Figure 3.22). The density of *Anogeissus leiocarpa* increases in grazed areas because the foliage of this tree is distasteful to cattle, whereas fire suppression reduces the abundance of fire-resistant species, such as *Combretum glutinosum*. Bews (1929) had earlier considered that fire was an important factor which maintained the tall grass savannas of Africa in an almost treeless state, and resulted in a boundary with the adjacent forests that may be less than 50 m wide (Clayton, 1958). Fire is now generally regarded as a secondary factor that maintains rather than creates savannas. Rawitscher (1948) suspected that some savannas in Brazil may have been created by felling and burning, since trees became established in protected areas. Kellman (1988) postulated that a burned site will revert to forest only if the closure time is less than the fire recurrence interval, and suggested that slower regrowth on low-fertility soils may be a contributing factor in savanna development in fire-prone environments. However, in wetter climates where savannas often occur as small openings in forest, low fire frequency still does not encourage the growth of colonizing trees (Sarmiento and Monasterio, 1975). The Aripo savannas of Trinidad, which burn infrequently, receive about 2800 mm of rain annually (Beard, 1953).

Fire has become increasingly common in the savannas, and extensive tracts of woodland have been cleared by continual burning; the tree cover is quickly reduced, particularly if the fire is set at the end of the dry season (Trapnell, 1959). Intentional burning has probably occurred in Africa for at least 50 000 years (Rose Innes, 1972) and in Australia for 40 000 years (Nicholson, 1981). In South America this practice became widespread only about 5000 years ago (Batchelder, 1967) and it is unlikely that the small aboriginal population could have created such extensive savannas in this short period. Fire, however, is now used extensively to clear land for domestic livestock. Although grazing and trampling can greatly reduce fire frequency and promote woody regrowth, this is usually prevented by high stock densities and continual burning to stimulate the development of more palatable young shoots. In overgrazed areas the deeper-rooting woody species can increase because more water percolates into the subsoil (Werger, 1983). Grazing is now common throughout the savanna regions, but prior to the widespread introduction of livestock large ungulate populations were found only in Africa. Even here there is no evidence to suggest that this could have resulted in excessive degradation of the woodland. Indeed, it is the diversity of cover types and differences in the dietary requirements of the various animals that maintain these remarkable herds.

3.6 ADAPTATIONS OF PLANTS TO THE SAVANNA ENVIRONMENT

3.6.1 Characteristic life forms

The perennial grasses, herbs and shrubs are well adapted to the seasonal climate and nutrient-poor soils in the fire-prone savanna environment, but although many species are represented in the flora, comparatively few are abundant. These are mainly species of bunch grass.

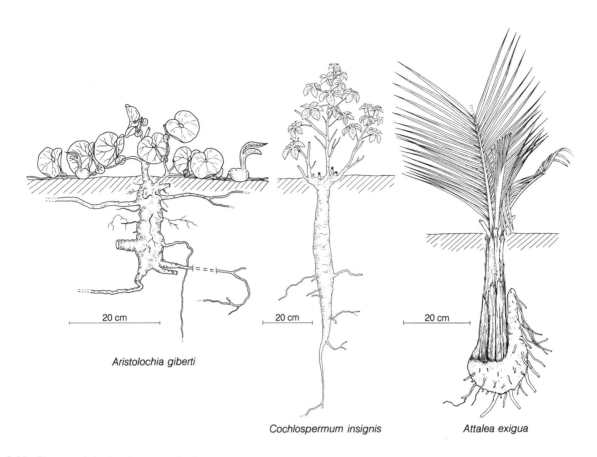

20 cm

Aristolochia giberti

20 cm

Cochlospermum insignis

20 cm

Attalea exigua

Figure 3.23 Characteristic development of subterranean xylopodia in drought-resistant woody species in the savannas of southern Brazil. (After Rawitscher, 1948.) (Reproduced with permission from F. Rawitscher, The water economy of the vegetation of the 'Campos Cerrados' in southern Brazil, *Journal of Ecology*, 1948, **36**, 253, 262.)

In Africa the proportion of the flora comprised of annual species increases from about 20% in more humid areas to over 60% in the arid zones (Menaut, 1983). Similarly, annual herbs are poorly represented in the Brazilian cerrado, but are conspicuous in the drier caatinga (Eiten, 1972). Trees are restricted to sites with more favourable moisture regimes, and are often associated with deeper soils or areas where the hardpan is sufficiently broken for their roots to reach ground water. Most of the trees and shrubs are evergreen with rather large sclerophyllous leaves, but as conditions become drier there is a corresponding decrease in leaf size and deciduous species become more common. In half-woody species, such as *Cochlospermum insignis*, *Craniolaria integrifolia* and the dwarf palm *Attalea exigua*, the aerial parts die back at the end of the growing season and new growth arises from woody underground organs or xylopodia (Figure 3.23). Similarly, most of the perennial grasses and herbs are cryptophytes or hemicryptophytes and a large proportion of their biomass is concentrated underground.

3.6.2 Perennial grasses

The perennial grasses live for 4–8 years and new plants usually arise vegetatively from underground organs.

Most species have coarse, scleromorphic leaves which die back to form a protective layer around the meristematic tissue from which new growth arises the following season. This tunicate form is equally well adapted to drought or fire. Most of the grasses are rhizomatous. In more densely bunched species new tillers develop at the tips of the root-like stems, but in creeping forms they usually arise from axillary buds which are protected by scaly leaves. Others may spread by stolons which grow across the soil surface and take root at intervals. Each clump of tillers develops its own roots, which are generally concentrated in the upper 50 cm of the soil. The densest herbage consists of short, sterile tillers, while flowers and seeds develop on fewer, taller, fertile tillers. In contrast, the tillers of annual species are all potentially fertile.

Most of these species grow vigorously under hot, sunny conditions. In *Paspalum dilatum* net photosynthesis peaks at about 35 °C, and light saturation occurs at about $6000 \, \mu E \, sec^{-1}$ (Cooper and Tainton, 1968). In comparison, the temperate grass *Lolium perenne* becomes light-saturated at $3000 \, \mu E \, sec^{-1}$ and photosynthetic efficiency begins to decline as temperatures rise above 24 °C. In most tropical grasses a ring of chlorenchyma cells surrounds the parenchyma sheath cells of the leaf veins.

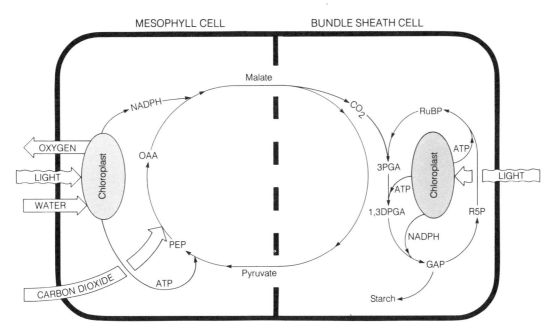

MESOPHYLL CELL   BUNDLE SHEATH CELL

Figure 3.24 Photosynthetic metabolism of C_4 plants in which assimilated compounds are transferred between mesophyll and bundle sheath cells. ATP = adenosine triphosophate; GAP = glyceraldehyde-3-phosphate; NADPH = nicotinamide adenine dinucleotide phosphate; OAA = oxaloacetic acid; PEP = phosphoenol pyruvate; RuBP = ribulose bisphosphate; R5P = ribulose-5-phosphate; 1,3DPGA = 1,3-diphosphoglyceric acid; 3PGA = 3-phosphoglyceric acid. (After Lawlor, 1987.) (Modified and redrawn with permission from D. W. Lawlor, *Photosynthesis: Metabolism, Control, and Physiology*; published by Longman, 1987.)

This is the Krantz anatomy associated with C_4 plants which convert carbon dioxide to malate as an intermediate product during photosynthesis (Figure 3.24). Decarboxylation of the malate in the Krantz cells releases carbon dioxide, which is subsequently incorporated into starches and sugars by the Krebs (C_3) cycle. By maximizing carbon dioxide concentrations in the Kranz cells, the C_4 plants can maintain a high rate of photosynthesis despite high temperatures and strong illumination. However, some tropical grasses do require shade; *Beckeropsis uniseta*, for example, will gradually die out as the associated tree cover is removed (Bogdan, 1977).

3.6.3 Savanna trees

Many of the tree species have also adapted to the stresses of the savanna environment by increased development of underground parts. *Curatella americana* only grows to a height of 6 m, but its shallow roots can be over 20 m in length and secondary roots often penetrate to 6 m (Foldats and Rutkis, 1975). Species such as *Andira humilis* have been aptly described as 'subterranean trees', because of their greatly enlarged roots which can penetrate to depths of 18 m (Rawitscher, 1948). Similarly, the cerrado tree *Caryocar brasiliense* can resemble a very low shrub or herb in frequently burned areas, and much of its woody growth develops underground. Trees such as *Burkea africana* and *Terminalia avicennoides* readily de-

velop a coppiced form in response to burning (Lawson *et al.*, 1968). In contrast to their extensive root growth, above-ground development in savanna trees is generally more conservative. Trees in wooded savannas can grow to 20–25 m in height, but in many regions few trees exceed 12 m, and most are only 2–6 m high, their twisted stems rarely exceeding 50 cm in diameter (Sarmiento and Monasterio, 1983). The thick bark which is characteristic of these trees is considered an adaptation to fire (Figure 3.25). Thus, *Crossopteryx febrifuga*, a thin-barked species, is notably absent from burned sites in Ghana, but thick-barked species are also damaged by fire (Brookman-Amissah *et al.*, 1980).

Although many of the African trees are deciduous, evergreen species are more common in other savanna regions. The old leaves are shed each year after the new crop has developed or shortly before leaf flush, which normally occurs during the middle of the dry season. This is in marked contrast to the grasses and herbs in which all above-ground material dies back. The sclerophyllous character of the leaves, with their thick cuticles, large vascular bundles and deep stomatal pits, appears to be well suited to these droughty environments. However, water loss can be high during the dry season, because in many species transpiration rates are not controlled by stomatal activity. Transpiration in *Byrsonima crassifolia* and *Curatella americana* increases when their leaves are exposed to the sun, although *C. americana* can restrict water loss during the hottest part of the day (Figure

Figure 3.25 A twisted branching pattern and gnarled trunks with deeply furrowed bark are characteristic of many savanna tree species. (G. T. Prance.)

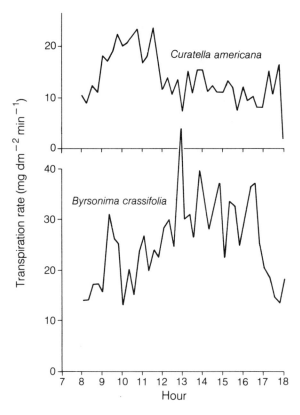

Figure 3.26 Observed maximum diurnal transpiration rates in *Byrsonima crassifolia* and *Curatella americana* in the savannas of Venezuela. (After Foldats and Rutkis, 1975.) (Modified from E. Foldats and E. Rutkis, Ecological studies of chaparro (*Curatella americana* L.) and manteco (*Byrsonima crassifolia* H. B. K.) in Venezuela, *Journal of Biogeography*, 1975, **2**, 173.)

Rawitscher (1948) associated the evergreen habit with deep-rooting species on the assumption that these plants were assured of a supply of water. Indeed, soil water

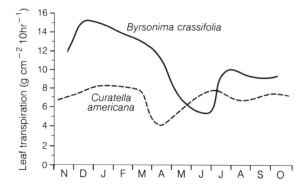

Figure 3.27 Seasonal variation in diurnal transpiration rates in relation to leaf surface area in *Byrsonima crassifolia* and *Curatella americana* in the savannas of Venezuela. (After Foldats and Rutkis, 1975.) (Modified from E. Foldats and E. Rutkis, Ecological studies of chaparro (*Curatella americana* L.) and manteco (*Byrsonima crassifolia* H. B. K.) in Venezuela, *Journal of Biogeography*, 1975, **2**, 172.)

3.26). The pattern of water loss during the year also differs between these species. In *C. americana* transpiration rates decline during March and April because new roots are unable to reach the water table (Figure 3.27). Rapid absorption of rain at the start of the wet season is reflected in increased transpiration beginning in May even though the water table continues to drop. Transpiration rates decline more gradually in *B. crassifolia*: this is attributed to higher field capacity in the rooting zone. The delayed recovery during the wet season results from slower infiltration rates in the heavier soils which support this species.

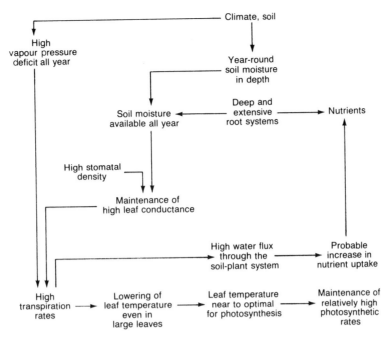

Figure 3.28 Mechanisms contributing to the maintenance of woody species of tropical savannas. (After Sarmiento *et al.*, 1985.) (Slightly modified from G. Sarmiento, G. Goldstein and F. Meinzer, Adaptive strategies of woody species in neotropical savannas, *Biological Reviews of the Cambridge Philosophical Society*, 1985, **60**, 344.)

potentials in the deep rooting zone show little seasonal variation (Meinzer *et al.*, 1983). High rates of transpiration may be necessary to cool the large leaves of the savanna trees in order to maintain optimum temperatures for photosynthesis (Medina, 1982); this is about 24 °C for *C. americana* and 28 °C for *B. crassifolia*. Maximum photosynthetic capacity of these species is nonetheless quite low (3–4 mg CO_2 dm^{-2} hr^{-1}), and further reduction by stomatal regulation of transpiration could seriously restrict root development (Meinzer *et al.*, 1983). High transpiration rates can also concentrate soil nutrients around roots (Nye and Tinker, 1977), and this could benefit plants which are growing in soils noted for their infertility (Figure 3.28). These interrelationships are further enhanced by low surface wettability and reduced leaching from sclerophyllous leaves. In addition, the partitioning of nutrients between the woody tissues and leaves restricts stem elongation to the moist season while leaf replacement occurs during the dry period.

3.7 SEASONAL GROWTH RHYTHMS

The yearly cycle of plant activity in tropical savannas is largely controlled by the markedly seasonal precipitation regime and corresponding changes in available soil moisture. The greatest differences occur in the surface soils, where moisture levels can exceed field capacity during the wet season and later decline to permanent wilting point. Despite this pronounced environmental seasonality, the phenological activity of savanna plants is surprisingly asynchronous and has resulted in many different patterns of leaf activity, shoot development and flowering (Sarmiento and Monasterio, 1983).

Most of the dominant grasses and sedges begin growth at the start of the wet season and come into flower some months later. Vegetative growth continues after seed dispersal, but with declining moisture reserves the shoots begin to wither. A few tillers are produced during the dry season but are short-lived, and by the end of the dry period the above-ground biomass is comprised almost totally of dead straw (Figure 3.29). In other grasses, vegetative growth at the start of the wet season is accompanied by flowering. Seed production is soon completed, but tillering continues until growth is eventually restricted by moisture stress. In perennial species which have a definite resting phase the aerial parts die back completely during the dry season. Shoot development from subterranean perennating organs may commence at the beginning of the wet season. Flowering occurs shortly after, then vegetative growth declines until only the underground organs remain. Alternatively, flowering can occur late in the wet season, followed by a rapid decline in aerial biomass.

Leaf production and new shoot growth in evergreen woody species usually occurs in the dry season. This is also the time for flowering. Leaf growth ceases at the start of the wet season: the foliage becomes progressively senescent and by the dry season it is ready to fall.

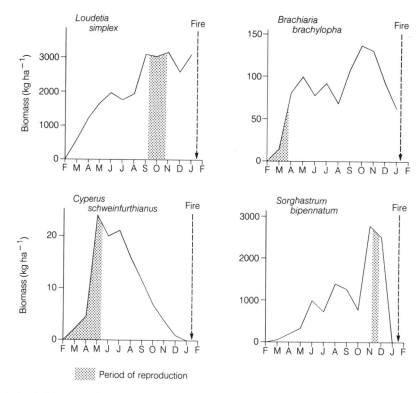

Figure 3.29 Seasonal variation in biomass of selected herbaceous species illustrating different phenological cycles in the burned savannas of West Africa. (After Lamotte, 1975.) (Reproduced with permission from M. Lamotte, The structure and function of a tropical savannah ecosystem, in *Ecological Studies 11 – Tropical Ecological Systems*, eds F. B. Golley and E. Medina; published by Springer-Verlag N.Y. Inc., 1975.)

Conversely, radial growth of stems and branches is restricted to the wet season (Sarmiento *et al.*, 1985). In the deciduous woodlands the trees remain leafless for part or all of the dry season. In Nigeria, for example, leaf development occurs in the wet period from March to June; most leaves fall in the driest part of the year between September and February, and a period of growth extension normally occurs just prior to leaf flush (Hopkins, 1970b). This pre-rain shoot activity might be triggered by seasonally low air temperatures (Jeffers and Boaler, 1966), or by a slight increase in photoperiod at this time of the year (Hopkins, 1970a).

Various phenological strategies are exhibited by the annual plants. Some species germinate soon after the start of the wet season but maximum growth and flowering do not occur till much later. Others, reported only from South America, behave more like desert therophytes and germinate at the beginning or end of the wet season. They complete their life cycle in a few weeks and persist as seeds for the remainder of the year (Sarmiento and Monasterio, 1983). In contrast, the annual plants associated with seasonally waterlogged areas delay germination until the soils dry.

3.8 THE EFFECTS OF FIRE

3.8.1 General character of savanna fires

The annual climatic rhythm is further emphasized by the high probability of fire during the dry season (Figure 3.30). Today, most fires are set deliberately, usually to prepare land for cultivation or to encourage new growth for grazing animals. Consequently, the floristic composition of many savannas has been greatly altered from their original state. Although fires can occur during short rainless periods, most savannas burn at the end of the dry season. Temperatures in the narrow burning zone vary considerably with height above ground. For example, in dry *Cynodon* grassland in Venezuela temperatures can range from 80 °C at ground level to 430 °C at the top of the standing grass cover, and reach 660 °C at the top of the flames (Figure 3.31). Fire spreads rapidly through fine grass fuels when it is fanned by moderate winds, and high temperatures usually persist for only a few seconds. Consequently, temperatures within the soil are not greatly altered. High temperatures are usually restricted to the immediate surface zone, although the

Figure 3.30 Tall tussock grasses provide abundant fuel for the fires that frequently burn through savannas at the end of the growing season. (O. W. Archibold.)

3.8.2 Fire adaptations in savanna plants

The underground organs of the perennial species are rarely damaged by fire but their seeds, like those of the annuals, must be adapted to survive this hazard. The seeds of some grasses and legumes possess hygroscopically active awns which work them into the soil, where they are protected from lethal fire temperatures. The sharply pointed seeds of the Australian spear grasses *Themeda australis* and *Heteropogon contortus* behave in this manner (Tothill, 1969). Significantly warmer temperatures in the blackened soils also stimulate germination in these species. Seed viability in savanna grasses is quickly lost during the wet season and no long-term storage occurs (Mott and Andrew, 1985). In *Themeda triandra* dormancy is broken when the seed is exposed to high temperatures in a fire (Trollope, 1982). Perennial herbs and grasses can avoid the stress of fire if their meristems are encased in tightly packed leaf sheaths. These plants complete their development during the wet season, and by the time of the fire the above-ground biomass is expendable. New growth, stimulated by increased light levels, usually begins quickly after the fire (Blydenstein, 1968).

The results of long-term burning experiments in savanna woodlands, such as those by Trapnell (1959), Charter and Keay (1960) and Ramsay and Rose Innes (1963), indicate that the hotter, late dry-season fires are the most damaging to woody species (Figure 3.33). In the wooded savannas of Africa the dominant canopy trees, species of *Brachystegia*, *Julbernardia* and *Isoberlinia*, are reportedly susceptible to fire (Trapnell, 1959). Small seedlings and suckers are often killed back but they are

depth of heat penetration is affected by soil moisture and the amount of fuel consumed (Figure 3.32).

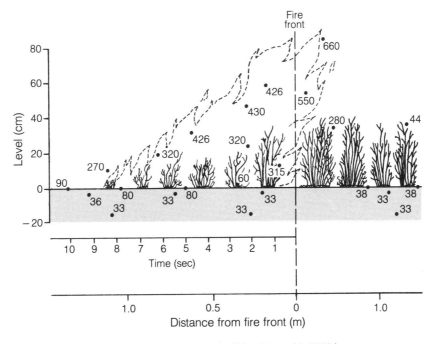

Figure 3.31 Characteristic fire temperatures (°C) in burning savanna. (After Vareschi, 1962.)

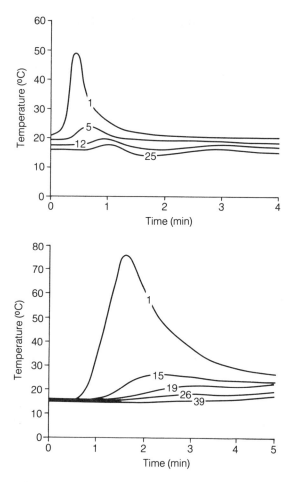

Figure 3.32 Soil temperatures at different depths (cm) during burns in grasslands in New South Wales. (After Norton and McGarity, 1965.) (Reproduced from B. E. Norton and J. W. McGarity, The effect of burning of native pasture on soil temperature in northern New South Wales, *Journal of the British Grasslands Society*, 1965, **20**, 104.)

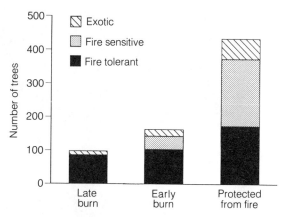

Figure 3.33 The response of the tree cover in the forest–savanna ecotone in Nigeria to different annual burn treatments over a 28-year period. (After Rose Innes, 1972.) (Adapted with permission from R. Rose Innes, 1972, Fire in West African vegetation. Pages 147–73 in E. V. Komarek, Sr., Conf. Chair. *Proceedings, Tall Timbers Fire Ecology Conference: Number 11: Fire in Africa*, 1972, Tall Timbers Research Station, Tallahassee, Florida.)

able to resprout from deep taproots. However, once they are about 1 m tall they are more vulnerable and their survival is largely dependent on an understory of fire-tolerant trees and shrubs which is sufficiently dense to limit the growth of grasses and herbs and thereby reduce the intensity of the fires. Continual burning will eventually lead to the degradation of wooded areas: the surviving trees become coppiced, and new growth is limited to suckers arising from their undamaged root systems (Lawson *et al.*, 1968). Some species can survive annual burning in this manner for at least 40 years (White, 1977). Sprouts can also arise from specialized lignotubers; these woody organs develop from buds in the axils of the cotyledons or young leaves, and gradually become buried as the tree grows.

Shoot buds are occasionally protected by dense scale-like coverings (cataphylls), but in many species they are exposed and escape damage only by growing above the zone of intense heat (Figure 3.34). In species such as the eucalypts, shoots quickly develop from adventitious buds

protected by the bark, the pattern of bud regeneration perhaps being controlled by the variable pattern of fire temperatures around the trunks (Tunstall *et al.*, 1976). Heat transfer to the inner cambial layers is generally reduced by thick bark, and temperature increases are also limited by the density and flammability of the bark. Baobab trees (Figure 3.35) with their thick, moist, porous bark are ideally suited to their fire-prone habitat (Owen, 1974).

3.8.3 Use of fire in savanna management

Fire is used extensively in savanna regions to improve the quantity and quality of forage for wildlife and domestic livestock. Mineral ash, which includes phosphorus, calcium and potassium, accounts for 8–12% of the dry weight of tropical grasses, but usually half of this amount is silica (Bogdan, 1977). Oils and fats make up about 3%, while soluble carbohydrates, sugars and starch, range from 35–55%. Fibre, comprised of lignin and cellulose in cell walls, increases from 25% to 40% dry weight over the growing season, which reduces considerably the nutritive quality of the older forage. Conversely, crude protein, which represents all nitrogenous compounds, can be as high as 20% dry weight in young plants but declines as the season progresses (Table 3.4). However, the crude protein content of grasslands in Malawi reportedly increased from 8% to 16% when they were burned late in the growing season (Lemon, 1968). The mineral content of forage can also increase when growth resumes following burning (Winter, 1987).

Periodic burning to remove dead herbage usually improves the vigour and palatability of grasses. However, long-term studies have demonstrated that the timing

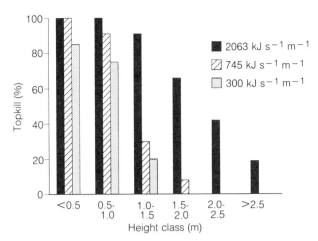

Figure 3.34 Topkill of *Acacia karroo* (sweet thorn) in the thornveld of southern Africa in response to fire intensity. (After Trollope, 1984.) (Reproduced with permission from W. S. Trollope, Fire in savanna, in *Ecological Studies 48 – Ecological Effects of Fire in South African Ecosystems*, eds P. de V. Booysen and N. M. Tainton, published by Springer-Verlag GmbH, 1984.)

Figure 3.35 The baobab tree is well protected from fire, although many are damaged by elephants which tusk the moist, fibrous bark. (B. R. Neal.)

and frequency of fires can affect the density and floristic composition of the vegetation. In the acacia thornveld of Zambia, basal cover was reduced by annual burning but could be increased if fires were set every 3 years (Table 3.5). Different responses occurred in the dominant grasses. *Themeda triandra* decreased considerably with burning at intervals of 1 or 2 years but increased in the 3-year and 5-year treatments; *Cymbopogon plurinoidis* responded best to short-interval burning. In the sandveld, a wooded savanna dominated by *Terminalia sericea* and *Burkea africana*, the grass *Digitaria pentzii* decreased noticeably in the protected plots. It was also sensitive to annual burning. All burning treatments reduced tree and shrub growth. The woody elements in the sandveld appeared to be more susceptible to fire in the infrequently burned sites, perhaps because of greater fuel accumulation and more intense fires (Table 3.6). Tree cover in the thornveld gradually increased if protected from fire for more than 2 years. Here, as in many other regions, frequent fires are considered necessary to control or eradicate encroaching bush.

3.9 PRODUCTIVITY OF SAVANNAS

Annual production of herbaceous material in savannas ranges from 40 g m^{-2} yr^{-1} in drier regions, such as Chad and Senegal, to over 1750 g m^{-2} yr^{-1} in parts of Zaire (Bourlière and Hadley, 1970). Daubenmire (1972a) suggested that shoot production increases by about 100 g m^{-2} for each additional 100 mm of precipitation. This provides only a very general approximation: in Nigeria, for example, grass production in two *Andropogon* savannas with annual rainfalls of 1168 mm and 1300 mm was

Table 3.4 Seasonal variability in the crude protein content (as % of dry matter) of different components of savanna vegetation in Africa (after Owen-Smith, 1982)

	Early growing season	Late growing season	Early dormant season	Late dormant season
Grasses				
Kenya	13.5	10.0	10.0	8.0
Uganda	10.0	8.0	7.0	3.0
Zimbabwe	14.5	5.5	4.5	4.5
Zululand	9.5	6.5	5.5	4.5
Forbs				
Kenya	16.5	15.0	18.0	12.0
Uganda	21.0	15.0	10.0	–
Foliage of woody species				
Kenya	15.0	13.5	18.5	13.0
Uganda	22.5	11.5	16.0	16.0
Zimbabwe	20.0	16.0	13.0	10.0
Transvaal	14.0	12.5	11.5	9.0

(Reproduced with permission from N. Owen-Smith, Factors influencing the consumption of plant products by large herbivores, in *Ecological Studies 42 – Ecology of Tropical Savannas*, eds B. J. Huntley and B. H. Walker, published by Springer-Verlag GmbH, 1982.)

6250 kg and 2310 kg ha^{-1} yr^{-1}, respectively (Ohiagu and Wood, 1979). Considerable variation in biomass occurs during the course of the year in response to climatic and fire-induced growth rhythms. Annual fluctuations are greatest in the aerial growth of herbaceous species which is practically eliminated as fire sweeps through a site. However, new growth in the post-fire season is usually higher than in unburned sites, because it is stimulated by the removal of layers of dead herbage. Thus, San José and Medina (1975) calculated net above-ground production in burned *Trachypogon* savanna in Venezuela at 415 g m^{-2}, compared with 325 g m^{-2} in plots that had not been burned for 4 years. Total biomass in the protected plot exceeded 1100 g m^{-2}. This remained

Table 3.5 The effect of different burning treatments on the importance of common grasses in two savanna sites in Zambia over the period 1949–63 (after Kennan, 1972)

| | Change in % species composition in the ground cover between 1949 and 1963 | | | | | |
| | Thornveld | | | Sandveld | | |
	C. plurinodis	H. contortus	T. triandra	D. pentzii	H. contortus	H. dissoluta
Protected from fire	− 18	− 1	7	− 34	2	22
Burned annually at end of dry season	− 7	6	− 30	− 12	− 10	27
Burned every 2 years at end of dry season	− 12	15	− 13	− 1	− 29	12
Burned every 3 years at end of dry season	20	4	37	− 8	21	16
Burned every 5 years at end of dry season	5	0	7	8	13	− 1

(Source: T. C. D. Kennan, 1972, The effects of fire on two vegetation types at Matapos, Rhodesia. Pages 53–98 in E. V. Komarek, Sr., Conf. Chair. Proceedings, *Tall Timbers Fire Ecology Conference: Number 11: Fire in Africa*, 1972, Tall Timbers Research Station, Tallahassee, Florida.)

Table 3.6 The effect of different burning treatments on tree and shrub cover in two savanna sites in Zambia over the period 1949–1963 (after Kennan, 1972)

| | Number of trees and shrubs | | | |
| | Thornveld | | Sandveld | |
	< 1 m high	> 1 m high	< 1 m high	> 1 m high
Protected from fire	1815	365	1175	420
Burned annually at end of dry season	792	60	765	24
Burned every 2 years at end of dry season	729	19	765	51
Burned every 3 years at end of dry season	1196	103	632	46
Burned every 5 years at end of dry season	1159	120	468	35

(Source: T. C. D. Kennan, 1972, The effects of fire on two vegetation types at Matapos, Rhodesia. Pages 53–98 in E. V. Komarek, Sr., Conf. Chair. Proceedings, *Tall Timbers Fire Ecology Conference: Number 11: Fire in Africa*, 1972, Tall Timbers Research Station, Tallahassee, Florida.)

relatively stable in the absence of fire or grazing, as the growth of new material was equivalent to the amount of dead herbage lost by decomposition. In contrast, underground production was reduced in burned sites, perhaps because of the reallocation of resources to support the vigorous shoot growth (San José *et al.*, 1982).

The amount of material produced each year and the rate at which it accumulates vary considerably according to the floristic composition of the savanna. Menaut and César (1979) noted three patterns of production in adjacent sites in the Lamto savannas of the Ivory Coast (Figure 3.36). Production in marshy sites dominated by *Loudetia simplex* was fairly regular throughout the growing season but declined rapidly once seed production was completed in October (Figure 3.37). In the *Andropogon* grass savanna most species emerged early after the fire, but maximum growth and flowering occurred later in the year. Production was distinctly bimodal in the shady savanna woodland. Maximum live herbaceous biomass at these sites ranged from 6.9 to 9.9 t ha^{-1}. Shrub and tree biomass in the predominantly herbaceous *Loudetia* savanna was less than 40 kg ha^{-1} but exceeded 58 t ha^{-1} in the wooded area. Between 9 and 3850 kg ha^{-1} of the tree and shrub biomass at these sites was leaf material that was renewed annually. Leaf fall usually occurs after fire has removed the herbaceous cover, and so provides consumable material and protection for the soil and litter fauna. Net above-ground productivity for these savannas ranged from 21.5 to 35.8 t ha^{-1} yr^{-1}, but even in the wooded areas, trees and shrubs accounted for only 10–20% of this annual total.

Figure 3.36 Borassus palm (*Borassus aethiopum*) provides an open canopy above a herbaceous layer comprised mainly of *Loudetia simplex* in the Lamto savanna, West Africa. (R. T. Coupland.)

Herbaceous root biomass ranged from 10.5 to 19.0 t ha^{-1}, 80% of which was concentrated in the upper 30 cm of the soil. Tree roots contributed 26 t ha^{-1} to below-ground biomass in the wooded savanna.

In the drier Nylsvley savanna of southern Africa, maximum growth occurs in mid-November, at the start of the wet season, when herbaceous material accumulates at a rate of about 0.5 g m^{-2} day^{-1} (Huntley and Morris, 1982). Peak biomass is not reached until late January or early February, and varies from 450–900 kg ha^{-1} in open areas to 380–600 kg ha^{-1} in wooded sites, depending on rainfall. Approximately half of this biomass is stubble and

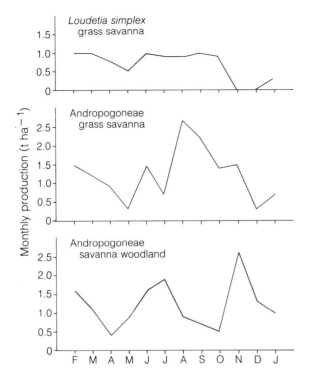

Figure 3.37 Annual cycle of herbaceous production in different types of savanna in West Africa. (After Menaut and César, 1979.) (Reprinted with permission from J. C. Menaut and J. César, Structure and primary productivity of Lamto Savannas, Ivory Coast, *Ecology*, 1979, **60**, 1207.)

cea, account for 78% of the live woody biomass and 41% of the dead wood mass at Nylsvley. Shoot growth in *B. africana* and *O. pulchra* occurs mostly at the start of the wet season at a rate of 40–50 kg ha^{-1} week^{-1} (Cresswell *et al.*, 1982). Production in *T. sericea* is about 10 kg ha^{-1} week^{-1}, but growth continues throughout the wet season from October until April. Much of the shoot extension in these species occurs before the main period of photosynthesis, which suggests that it is dependent on translocation of materials from other tissues. Root growth follows shoot elongation with maximum activity occurring between February and mid-April. Over 95% of the tree roots are concentrated in the upper 50 cm of the soil (Rutherford, 1982). The thickest roots are generally found at depths of 20–40 cm. In *O. pulchra* finer roots can grow upwards, almost to the surface, where they perhaps increase water uptake, while some sinker roots may penetrate more than 2 m to bedrock.

3.10 NUTRIENT STOCKS IN SAVANNA VEGETATION

Total nutrient storage in savanna vegetation is comparatively low and is quite variable over the course of the growing season. Detailed analyses of nutrient content of leaves of savanna trees indicate that evergreen species tend to accumulate lower quantities of all elements than associated deciduous species (Table 3.7). Some of the nutrients sequestered in the leaves are translocated to other tissues prior to leaf fall, although preferential enrichment of soils beneath savanna trees does occur because of long-term litter inputs (Kellman, 1979). In *C. americana*, for example, the nitrogen content of recently fallen leaves is 4.6 mg g^{-1} compared with 9.3 mg g^{-1}

litter. Most of the herbaceous plants are shallow rooting, and more than 80% of their underground biomass is found in the upper 50 cm of the soil: in open sites this amounts to about 3500 kg ha^{-1}, but is less than 2000 kg ha^{-1} in the *Burkea africana* woodland. *Burkea africana*, together with *Ochna pulchra* and *Terminalia seri-*

Table 3.7 Nutrient content of leaves of savanna trees (after Montes and Medina, 1977; Ernst, 1975)

	N	P	K	Ca	Mg
			(mg g^{-1} dry mass)		
Venezuela					
evergreen species					
Curatella americana	9.3	0.8	8.3	4.9	2.7
Byrsonima crassifolia	8.0	0.4	5.4	8.0	3.0
deciduous species					
Genipa caruto	18.0	1.0	19.8	7.3	6.3
Godmania macrocarpa	16.0	0.8	6.5	9.3	3.9
Cochlospermum vitifolium	10.5	1.2	7.8	12.9	3.2
Zambia – miombo woodland					
Brachystegia spiciformis	–	2.1	3.8	14.4	0.3
Julbernardia globiflora	–	1.4	2.7	24.1	0.4
Combretum molle	–	1.1	4.2	21.5	0.4
Terminalia prunioides	–	1.1	3.0	32.7	0.3
Acacia vermicularis	–	1.0	5.8	15.6	0.7

(Reproduced with permission from R. Montes and E. Medina, Seasonal changes in nutrient content of leaves of savanna trees with different ecological behaviour, *Geo-Eco-Trop*, Vol. 4, 1977; and from W. H. O. Ernst, Variation in the mineral contents of leaves of trees in miombo woodland in south central Africa, *Journal of Ecology*, 1975, **63**, 802.)

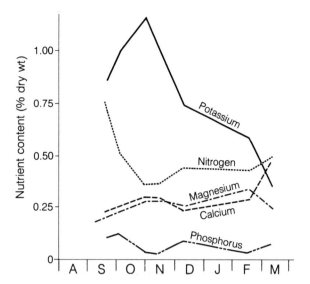

Figure 3.38 Variations in the nutrient content of shoots from the middle of the rainy season until the end of the dry season in hyperseasonal savanna in Venezuela dominated by *Sorghastrum parviflorum*. (After Sarmiento, 1984.) (Reprinted by permission of the publishers from *The Ecology of Neotropical Savannas* by Guillermo Sarmiento, translated by Otto Solbrig, Cambridge, Mass.: Harvard University Press, copyright © 1984 by the President and Fellows of Harvard College.)

Table 3.8 Percentage of nutrients accumulated in the herbaceous cover and in the upper metre of soil at a savanna site in the western llanos of Venezuela (after Sarmiento, 1984)

	Above-ground biomass (%)	Below-ground biomass (%)	Soil (%)
nitrogen	1.6	1.0	97.4
phosphorus	38.9	23.5	37.6
potassium	7.6	2.5	89.9
calcium	5.7	3.3	91.0
magnesium	4.8	1.6	93.6

(From *The Ecology of Neotropical Savannas* by Guillermo Sarmiento, translated by Otto Solbrig, Cambridge, Mass.: Harvard University Press, copyright © 1984 by the President and Fellows of Harvard College, reprinted by permission of the publishers.)

in mature leaves; for phosphorus comparative values are 0.14 and 0.75 mg g^{-1}, and for potassium 4.1 and 8.3 mg g^{-1} (Sarmiento *et al.*, 1985). Translocation is considered a nutrient conserving mechanism for plants growing in poor soils. Similar effects are reported for trees in the miombo woodlands, although concentrations of other elements, such as calcium and manganese, generally increase over the growing season (Ernst, 1975). Seasonal trends in nutrient levels are also evident in herbaceous material (Figure 3.38).

The quantity of nutrients stored in the above-ground biomass accounts for the greatest proportion of the nutrient capital in savanna vegetation. Some of the nutrients in the herbaceous cover are returned to the soil as ash following a fire, but much of the nitrogen is lost as gas and other elements may be lost in particulate form in the smoke. As much as 95% of the nitrogen contained in cerrado vegetation can be lost to the atmosphere in this way, with equivalent losses of 51% for phosphorus and 44% for potassium: this is about triple the amount returned in precipitation (Coutinho, 1990). The heat of the fire can also reduce the population of nitrogen-fixing microbes. However, their numbers are quickly re-established, and soil nitrogen levels often increase because of increased rates of ammonification and nitrification after the thatch is reduced (Nye and Greenland, 1960). Some of the minerals in the ash can be lost in the surface run-off but they are not carried into the deeper soil layers by percolating water. Uptake by vigorous plant growth may limit downward leaching but this does not appear to be an important mechanism for short-term nutrient retention after fire (Kellman *et al.*, 1985). In burned savannas much of the organic matter and nutrient reserves in the soil is contributed by the grass roots. Root growth commences at the start of the wet season (Van Donselaar-Ten Bokkel Huinink, 1966). Nonfunctional root material remains relatively constant during this period, but it is progressively decomposed during the dry season and most of it is reduced during the course of the year (San José *et al.*, 1982).

Studies of nutrient storage in the soil and vegetation of a Venezuelan savanna indicate that the plant cover is an important reserve of available phosphorus (Table 3.8). Approximately 18 kg ha^{-1} of phosphorus is available from the soil (Figure 3.39). Maximum storage in the above-ground vegetation is equivalent to 12 kg ha^{-1}, 75% of which is returned underground at the end of the growing season. The remainder is retained in the dead herbage and is added to the soil together with phosphorus released by root decomposition. Approximately 1 kg ha^{-1} of phosphorus is contributed through rainfall. The low soil reserves and fast recycling rate suggest that productivity is potentially limited by the availability of phosphorus.

Reserves of nitrogen and other elements are much greater in the soil than in the vegetation it supports. The live aerial biomass contains about 60 kg ha^{-1} nitrogen. About 65% of this is eventually transferred to the roots but most of what is left in the dead thatch is lost in the subsequent fire, although some is leached from the straw as throughfall. Root decomposition releases an estimated 10 kg ha^{-1} of nitrogen annually. Nitrogen uptake during the growing season is about 30 kg ha^{-1}. Some of this is available from the mineralization of organic matter, some comes from throughfall and precipitation, and some is added to the system by microbial activity. Losses in drainage amount to only 2 kg ha^{-1} each year.

3.11 SOIL ORGANISMS AND DECOMPOSITION

In the absence of fire or grazing, litter will slowly accumulate until the decreased rate of production is in equili-

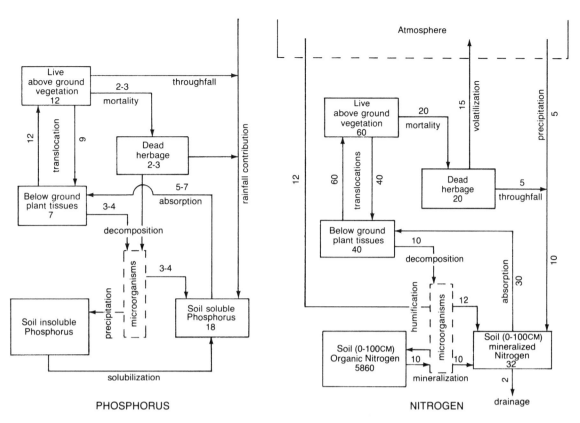

Figure 3.39 The annual cycles of phosphorus and nitrogen in seasonal savanna of *Axonopus purpusii–Leptocoryphium lanatum* in Venezuela; nutrient stocks are given as t ha^{-1} and transfers between compartments as t ha^{-1} yr^{-1}. (After Sarmiento, 1984.) (Reprinted by permission of the publishers from *The Ecology of Neotropical Savannas* by Guillermo Sarmiento, translated by Otto Solbrig, Cambridge, Mass.: Harvard University Press, copyright © 1984 by the President and Fellows of Harvard College.)

brium with the activity of the decomposers (Sarmiento, 1984). Although very few savannas today are not grazed or burned periodically, continual breakdown of dead organic matter over the growing season is essential for maintaining productivity in the nutrient deficient soils. Most leaf litter is decomposed by micro-organisms during the wet season: woody material is largely consumed by termites in the dry season. Earthworms are also important members of the decomposer system in more humid savanna regions. Herbaceous litter decomposes most rapidly. Ohiagu and Wood (1979) have reported that 69% of the grass in Nigerian savannas is broken down annually. Losses of 8–12% each month have been calculated for the Lamto savannas in Guinea (Menaut and César, 1979), although each year over 50% of the dead biomass is ultimately consumed by fire. Similarly, Morris *et al.* (1982) reported losses of over 70% in 8 months for grasses in the Nylsvley savanna of Zambia: comparable losses for tree leaves were less than 10%. Much faster rates have been observed for tree leaves in the miombo woodlands of southern Zaire (Figure 3.40). Here, more than 75% of the leaves of *Pterocarpus angolensis* decomposed in a 6-month period. The lowest rate was 27% for *Parinari curatellifolia* and *Marquesia macroura*. Termites were particularly attracted to the leaves of *Ochna schweinfurthiana*. The termite population, which is estim-

ated at 17.5 million individuals ha^{-1}, accounts for 80% of the biomass of soil organisms in this region.

Termites can be grouped according to their food preferences. The humivorous termites ingest decaying organic matter and mineral soil. They are most abundant in humid areas where their underground activities help maintain soil structure and aeration. Sound dead wood is used by many termites, such as *Nasutitermes*, although this xylophagous diet is regarded as a primitive trait. Foraging termites harvest dead vegetation which is still intact, including twigs and leaves of shrubs and trees, as well as herbaceous materials. All termites harbour flagellate protozoans or bacteria in their digestive tubes to assist with the breakdown of the cellulose in their diets. The fungi that some termites cultivate on debris brought into subterranean gardens assist in the breakdown of lignin, which releases more cellulose (Noirot, 1970). The plant material is used within 5 to 8 weeks of being brought to the nest. These termites are major consumers of dead litter. Termite species, such as *Macrotermes subhyalinus*, *Odontotermes pauperans* and *O. smeathmani*, account for more than 90% of the grass that is decomposed annually in a Nigerian savanna (Ohiagu and Wood, 1979). A density of 57 000 gardens ha^{-1} is reported in the Lamto savannas: these are tended by as many as 4.5 million termites which incorporate 1000 kg of dry litter into the

Table 3.9 Nutrient characteristics of termite mounds and surrounding soil (after Trapnell *et al.*, 1976)

	pH	C (%)	N (%)	K	Ca	Mg
				m-eq 100 g^{-1} dry soil)		
Termite mound						
outer 15 cm of spire	6.4	0.82	0.07	0.83	2.40	1.20
dome at 15 cm depth	8.4	0.50	0.06	1.00	3.45	1.56
dome at 1.4 m depth	7.5	0.46	0.05	1.03	2.70	1.28
dome at 2.9 m depth	5.0	0.50	0.07	0.49	2.52	1.10
Soil						
0–15 cm	5.2	0.86	0.07	0.11	0.15	0.24
15–30 cm	5.0	0.43	0.04	0.06	0.07	0.12
91–122 cm	5.1	0.16	0.02	0.03	0.09	0.17

(Source: C. G. Trapnell *et al.*, The effects of fire and termites on a Zambian woodland soil, *Journal of Ecology*, 1976, **64**, 580, 584.)

soil each year (Lamotte, 1975). Improved soil nutrient status has been observed in large mounds constructed by foraging *Odontotermes* in Zambian woodland soils as a result of this activity (Table 3.9).

Although relatively few genera of termites build above-ground nests, these structures are common features of many savanna landscapes (Figure 3.41). Most are constructed from excreta; in others a mixture of soil and saliva is used. They are invariably closed structures with no permanent openings to the outside. Foraging parties leave the nest by way of holes which lead from subterranean chambers to the soil surface. The largest mounds, such as those built by *Macrotermes bellicosus*, can be 5–6 m high. Their large size reduces diurnal temperature fluctuations, and they are less prone to desiccation in the dry season (Josens, 1983). Species which build

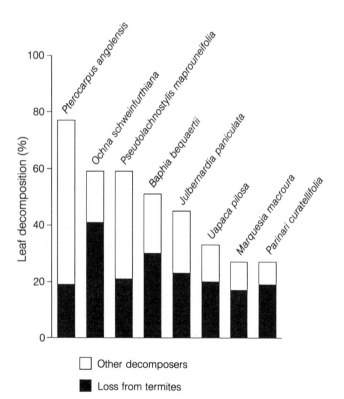

Figure 3.40 Rates of leaf litter decomposition by termites and other activity for various tree species in miombo woodland in East Africa. (After Malaisse *et al.*, 1975.) (Redrawn from F. Malaisse *et al.*, Litter fall and litter breakdown in miombo, in *Ecological Studies 11 – Tropical Ecological Systems*, eds F. B. Golley and E. Medina; published by Springer-Verlag N.Y. Inc., 1975.)

Figure 3.41 (a) Small termite mounds in treed savanna in northern Australia; (b) recolonization of a large termite mound in Tanzania ((a) O. W. Archibold; (b) G. Armstrong.)

small mounds, such as *Trinervitermes geminatus*, may be forced to retreat from some galleries during the hottest part of the day. Although numerous mounds may be present in a region, many are unoccupied. In Nigeria 40% of the 395 mounds ha^{-1} constructed by *T. geminatus* are abandoned (Ohiagu, 1979). Each mound was occupied for an average of 2 years by a population equivalent to 200 termites m^{-2}, and a further 500 m^{-2} were active in adjoining subterranean chambers, some being found at depths of 2 m. It is in such underground nests that the majority of termites are found.

The earthworms also occupy distinctive ecological niches in the savanna ecosystems. At the surface of the soil are the small, short-lived, litter-dwelling species which feed on fragmented grass and leaf debris. The larger humus-feeding worms are confined to the soil, while a third group of earthworms live within the soil but come to the surface to gather litter. The activities of the earthworms are greatly influenced by moisture conditions and at times of drought they remain inactive in mucus-lined cavities in the soil. Consequently, smaller earthworm populations are generally encountered in climatically dry regions. An average density of 2.3 million earthworms ha^{-1} is reported for the Lamto savannas (Lamotte, 1975). The most abundant species is *Millsonia anomala*: it lives close to the soil surface. It is estimated that this species ingests about 500 tons of earth ha^{-1} annually. Like the termites, the principal effect of earthworm activity is to improve the physical structure of the soil. At Lamto, in addition to the soil-dwelling worms, arboreal species are also present, such as *Dichogaster bolaui*, which feeds on debris trapped between the leaf bases of palms (Lavelle, 1983).

Unlike the earthworms and termites, many of the soil-dwelling ants feed on animals, larvae and eggs rather than on plant debris. Although they play no role in litter decomposition, populations can exceed 20 million individuals ha^{-1}, and their activity results in the redistribution and loosening of a considerable amount of soil (Levieux, 1983). Some species, such as *Crematogaster impressa*, live in the grass layers, but apart from building their nests in the hollow stems, no other use is made of the grass. More common are the ants which make their nests in hollow tree branches or build carton nests, either from wood fragments that have been chewed to a paste or from particles of soil. Some of these arboreal ants may consume leaves, seeds and sap but mostly they prey on insects. Some trees, notably the swollen thorn acacias, harbour ants in gall-like swellings. Over 300 swellings up to 8 cm in diameter have been reported on small *Acacia drepanolobium* trees about 2 m in height (Hockings, 1970). Two species of ants commonly inhabit these swellings. The smaller swellings are generally occupied by *Crematogaster mimosae* at an average density of 365 ants per swelling; in the larger swellings *C. nigriceps* is found at an average density of 166 per swelling. Thus a single tree may support over 100 000 ants. In return for sugar from the nectaries and the protection afforded by the woody swellings, the ants provide the tree with a measure of defence against insect predation and browsing by herbivores (Brown, 1960).

3.12 THE INVERTEBRATES OF THE GRASS LAYER

Most of the other invertebrates found in savanna regions live above ground, but are concealed in the grasses. Some, like the crickets and cockroaches, utilize dead material, whereas grasshoppers and the larvae of butterflies and moths feed directly on the herbage. Many of the beetles, spiders and mantids are predators. The invertebrates are the staple diet of many higher animals. Cryptic forms and colorations are therefore common traits in both the predator and the prey species. The grass- and stick-mimics have long, slender bodies and their disguises are further enhanced by deceptive postures (Figure 3.42). Batesian mimicry, the close resemblance of an organism to one which is conspicuous and unpalatable and normally avoided by predators, occurs in several species of butterflies and beetles (Edmunds, 1974). The African monarch butterfly (*Danaus chrysippus*) is mimicked by other species throughout Africa, because the cardenolides which it accumulates from milkweeds and other plants during its larval stage cause vomiting in many of its vertebrate predators. Ants are a common model for mimics since most predators are deterred by their biting and stinging behaviour: the spider *Myrmarachne foenisex* is a common ant-mimic (Edmunds, 1978).

As well as avoiding predators, the invertebrates must survive climatic stresses and fire. Many species enter diapause during the dry season, often resting within the soil where they are also protected from fire. Others, such as the locusts, will migrate to less severe regions (Gillon, 1983). Consequently, there is a considerable fluctuation in invertebrate populations during the course of the year. The intensive studies of Gillon and Gillon (1967) in the Lamto savannas show that the arthropod population in the grass layer is lowest at the end of the dry season: it builds steadily in the wetter period and reaches a maximum in August. Differences from this general trend occur in specific groups. Grasshoppers are most numerous during the early part of the growing season when the new shoots have a high nutrient content. Crickets are particularly abundant later in the year when the amount of dead material is higher. The populations of carnivorous species, such as spiders, fluctuate more irregularly over the year.

Other effects become apparent when the savanna is burned regularly. Not only is the total population of arthropods reduced but also its composition is altered. Grasshoppers are most abundant in burned savannas, whereas cockroaches, mites and the small bugs are more

Figure 3.42 Cryptic forms of the leaf-like katydid and stick-like praying mantis provide camouflage for prey and predatory species. (B. R. Neal.)

Figure 3.43 Fire-blackened surface in recently burned savanna woodland. (O. W. Archibold.)

Table 3.10 Effect of fire on the arthropod population of the Lamto savanna (after Lamotte, 1975)

	Density (individuals 100 m^{-2})	Biomass (g 100 m^{-2})
day before fire	2688	60.1
next day	1710 (64%)	19.5 (32%)
one month later	1044 (38%)	19.8 (33%)

(Source: M. Lamotte, The structure and function of a tropical savannah ecosystem, in *Ecological Studies 11 – Tropical Ecological Systems*, eds F. B. Golley and E. Medina; published by Springer-Verlag N.Y. Inc., 1975.)

numerous in the unburned areas. Many of the smaller arthropods survive the immediate effects of the fire because they are protected by the tightly bunched grass stems or the litter (Table 3.10). The number of indi-

viduals subsequently drops because few species can tolerate the hot, dry conditions or can escape predation on the bare, blackened surface (Figure 3.43). Their numbers soon recover despite heavy consumption by amphibians and lizards (Lamotte, 1975).

3.13 AMPHIBIANS AND REPTILES

Many species of amphibians have adapted to the seasonal dryness of the savanna environment. Some avoid desiccation by living in marshy sites which remain damp all year. Others are active only during the wet season or at night when conditions are more humid. Water is most critical during the breeding season. Eggs which are laid in small ponds usually hatch quickly and the young frogs soon emerge from the water. Mortality rate of the juveniles is very high and many fall prey to snakes and birds. Nonetheless population densities as high as 3000 ha^{-1} have been reported in some of the damper locations in the Lamto savannas (Lamotte, 1983). Numbers are particularly high in wetter years when ponds are more widespread and persistent. Snakes are also more numerous in these wetter locations (Barbault, 1983) and most feed on amphibians. Lizards are less dependent on moisture, and large populations occur even in dry savannas. Like the amphibians, they suffer heavy predation at all stages of their life cycle, and life expectancy is short. The survival curve for the small lizard *Mabuya buettneri* suggests that few adults live more than 3 months (Figure 3.44) and most ultimately perish when the savanna burns. The amphibians and reptiles that survive the fire are further disadvantaged by lack of concealment, and the absence of litter also reduces the invertebrate population on which they feed. Their densities are therefore considerably lower in areas which are burned. Although the amphibians are vulnerable to fire, even in unburned areas several months of wet weather

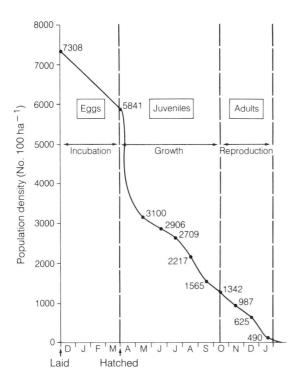

Figure 3.44 Survival curve of the lizard *Mabuya buettneri* in burned savanna in West Africa. (After Lamotte, 1975.) (Reproduced with permission from M. Lamotte, The structure and function of a tropical savannah ecosystem, in *Ecological Studies 11 – Tropical Ecological Systems*, eds F. B. Golley and E. Medina; published by Springer-Verlag N.Y. Inc., 1975.)

are required to re-establish population densities following the dry period. Despite a large increase in numbers, a large proportion of the amphibian population is composed of immature individuals and juveniles (Barbault, 1983) and this is reflected in the slower recovery of snakes and other predators that feed on them.

3.14 THE BIRDS OF THE SAVANNAS

The African savannas support 708 known species of birds. This is about half of the species found in the continent and represents the most varied avian fauna of all of the tropical grasslands (Fry, 1983). In the Neotropics only 521 species of birds are resident in the savannas, and in Australia the reported number of species is 227. The most distinctive species are the large flightless birds – the ostrich (*Struthio camelus*) in Africa, the emu (*Dromaius novaehollandiae*) in Australia, and the rhea (*Rhea americana*) in South America. They are omnivorous like most of the savanna birds, and feed on fruits and seeds and insects. Others have more specialized feeding habits, such as the snake-eating secretary bird (*Sagittarius serpenarius*), and the oxpeckers (*Buphagus* sp.) which remove ticks and other ectoparasites from various large

ungulates (Attwell, 1966), while vultures are particularly abundant in East Africa where there is never any shortage of animal carcasses.

Most of the seed consumed by birds is picked from the ground. In general they are rather unspecialized feeders and their diet is normally determined only by the type of grain available. In response to diminishing food supplies during the dry season, many of the specialized granivores migrate to other regions which receive moisture during alternate seasons. For example, the small weaver-bird *Quelea quelea* migrates across the advancing rain front at the start of the wet season into areas in which new seed crops have ripened. Eventually the seed supply is diminished, and the return journey of the birds coincides with the progressive ripening of seed in areas through which the rains have passed (Ward, 1971). The insectivorous birds usually take prey from the grass layer or from the ground. Large numbers of termites are consumed, and even granivorous species, such as *Q. quelea*, will feed them to their nestlings when they are plentiful (Ward, 1965). During the fire season aerial feeders, such as the kites (*Milvus migrans parasiticus*), are particularly active over the flames, while the marabou stork (*Leptoptilos crumeniferus*) and the pied crow (*Corvus albus*) work through the hot ashes (Gillon, 1983).

Fire is rarely a serious problem to the bird population since most species complete their breeding activities prior to the fire season. However, the eggs and chicks in ground-nesting birds such as the ostrich, guineafowl (Numididae) and francolin (Phasianidae) are especially vulnerable (Brynard, 1964). The insectivores generally breed in the wet period, as do many granivores although their breeding seasons may extend into the dry period when a large amount of seed is still available. The raptors, scavengers and waders normally breed in the dry season but their nests are usually protected in trees or wet sites. Fire is not the only threat to birds and precautions must be taken against predators, particularly during the nesting period. The thorn nests of the weaver birds with their intricately barricaded approaches will usually deter even the most persistent and agile of predators. The red-billed hornbill (*Tockus erythrorhynchus*) is assured even greater protection. The female seals herself and her eggs into a cavity in a tree. She frees herself after the eggs have hatched and then the nest is closed again except for a small feeding slit. Here the young remain until old enough to fly.

In addition to the resident species, the savannas are also important wintering grounds for many birds from colder zones. Nearly 300 species migrate from Eurasia to Africa. Birds from western Europe generally visit West Africa, while those from central Asia tend to congregate in East Africa. About 5% of these migratory species are seed eaters: they tend to winter in the sub-Saharan zone, and arrive at the end of the rainy season when seeds are plentiful (Moreau, 1972). Many insectivores also spend winter here. Those that migrate south of the equator arrive at the beginning of the wet season when the winged termites and ants emerge from their nests in swarms. They are therefore ensured an unusually abundant supply of fat- and protein-rich food during their stay. Fewer species migrate between eastern Asia and Australia, while movements in North America tend to take the birds into the southern United States, the Caribbean and the forested regions of South America (Keast, 1980).

3.15 THE SAVANNA MAMMALS

3.15.1 Rodents

The rodents are the most widely dispersed of the savanna mammals. They are rather inconspicuous, being normally active within the grass cover during the twilight or night-time hours and then retreating to their underground burrows. The principal exception to this is the South American capybara (*Hydrochoerus hydrochaeris*) which grows to a length of 1.25 m and exceeds 0.5 m in height. It is most frequently encountered in wetter parts of the savannas where it feeds on the grasses. Rodents are principally herbivorous, feeding on a variety of plant materials from a range of species as they become available during the year. Often this plant food is supplemented by insects, while some species feed almost exclusively on termites or on other arthropods (Coe, 1972). The decline in the quantity and quality of food during the dry season is reflected in the markedly seasonal pattern of reproduction in most species, and this in turn results in large fluctuations in population densities during the year. Population densities are also influenced by fire and flood. Although rodents can escape direct injury from fire in their burrows, their numbers may decline because of temporary food shortages and lack of cover. Flooding is usually more serious and many rodents drown unless temporary refuge can be found (Sheppe, 1972).

3.15.2 Large ungulates

Unlike the rodents and other small mammals, such as rabbits and hares, the large mammals are less evenly distributed in the savannas. The largest and most diverse communities are found in Africa with over 90 species of ungulates. Nearly all of these are members of the Bovidae (buffalo and antelope). The greatest concentration of animals occurs in East Africa, where species such as wildebeest congregate in herds of several thousand individuals. Elephant, giraffe and rhinoceros are also present, while the carnivores and scavengers further add to the diversity of animal life in this region. Only 21 species of ungulates are native to South America and few of these are resident in the grasslands. Here it is the cervids (deer) that are most common, but even they

tend to browse in wooded areas and seldom leave the protection of the trees (Redford and Fonseca, 1986). In tropical Australia the dominant herbivores are the marsupials. Like their counterparts in South America, they are more numerous in the wooded areas.

The African ungulates are, therefore, a rather unique component of savanna ecosystems. Adaptive radiation in the dominant bovid group is comparatively recent and is attributed to the successive expansion and fragmentation of the forest and savanna regions during the period of climatic instability which culminated in the Pleistocene. Competition between related species may be reduced by geographic isolation, as is the case for many species of antelope (Figure 3.45). In areas where the ranges of the ungulates overlap, ecological separation is achieved through different habitat and food preferences. Species such as Grant's gazelle occupy the plains of Tanzania, whereas impala are usually found in open woodland and the lesser kudu are mostly restricted to dense woodland (Figure 3.46). This effectively separates species with similar food requirements. Other species feed at different levels in the vegetation (Figure 3.47), while the small dikdik feeds amongst the denser shrubs which the larger animals cannot enter.

Alternatively, a site may be occupied by different species in a highly seasonal manner. The successive use of floodplains in the Rukwa valley of Tanzania has been described by Vesey-FitzGerald (1960). These low-lying areas can be flooded for up to 5 months in the wet season. Such conditions are ideal for the hippopotamus: the reedbuck and puku also remain in waterlogged sites. Other ungulates must seek higher ground. The elephants and buffalo move to wooded areas, while the zebras and topi eventually gather in the drier grasslands. Elephants are the first to move back to the floodplain. They enter the dense cover of long grasses and reeds in April or May, followed shortly after by the buffalo. Movement of the zebras and topi occurs in September after the larger animals have opened up the cover by grazing and trampling. The elephants generally retreat to the shade of the woodlands before the end of the dry season, but vigorous sprouting maintains adequate forage for the other animals until the first rains of November.

Similar interactions are reported amongst the grazing animals which inhabit the Serengeti–Mara National Parks of Tanzania and Kenya. Wildebeest account for about half of the estimated 3 million ungulates in this 25 000 km^2 region (McNaughton, 1985). The region also supports 600 000 Thomson's gazelle and 200 000 zebra. During the wet season all these species congregate on the drier knolls where the cover consists of short, actively growing grasses. The nutritive value of the forage is highest at this time of the year. With progressive depletion of the cover, the animals are forced to move into the taller grasses in the lower sites. The zebras are the first to move and are followed in succession by the wildebeest and the gazelles. This sequence of movement reflects

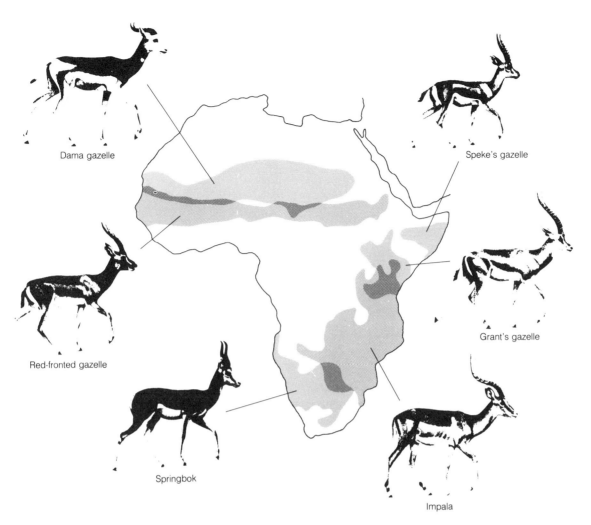

Figure 3.45 Geographic isolation in different species of African antelope. (After Dorst and Dandelot, 1970.) (Reproduced with permission from J. Dorst and M. P. Dandelot, *A Field Guide to the Larger Mammals of Africa*, published by Houghton Mifflin, 1970.)

seasonal changes in the availability of forage and the different dietary tolerances of the animals, and is fundamental to the coexistence of these seemingly competitive species.

3.15.3 Plant–animal interactions in the East African plains

The amount of green biomass available in the wet season exceeds 150 g m^{-2} in parts of the Serengeti, but this can drop to less than 20 g m^{-2} later in the year. Crude protein can be as high as 20% dry weight in young tropical grass plants but declines over the growing season (Table 3.11). Conversely, as the stems elongate more lignin and cellulose is laid down in structural tissues. This fibre is particularly abundant in the uppermost part of the herbaceous cover. The tall grass stems are therefore of little nutritional value to the ungulates. Nonetheless, this obstructive layer must be removed to gain access to

the leaves and herbs below. Zebra are most tolerant of a high-fibre diet. They are non-ruminants and can survive on a low-protein diet because of their fast rate of ingestion. The other species are ruminants: they must chew and regurgitate the food repeatedly before it is broken down sufficiently to pass along the digestive tract. Consumption is therefore restricted when diets are high in fibre (Bell, 1971). Hence, the efficiency with which the animals can utilize the herbage is a major factor governing the order in which they descend into the areas of taller grass, and their subsequent regional migrations (Figure 3.48).

The food requirements of the animals is also related to their size. A zebra with an average body weight of 170 kg consumes 4.8 kg of food daily, whereas the smaller gazelles weighing only 16 kg require 0.7 kg (Sinclair, 1975). The gazelles can therefore remain in the heavily grazed areas for a much longer period. Buffalo will normally precede the zebra into the tall grasses, even though they are ruminants. These heavy, large-mouthed animals

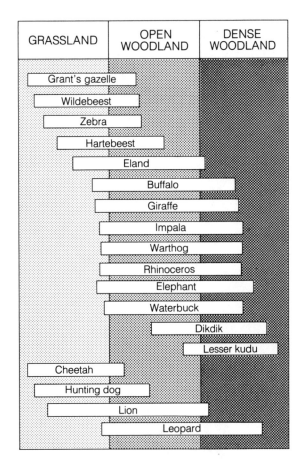

GRASSLAND	OPEN WOODLAND	DENSE WOODLAND

Grant's gazelle
Wildebeest
Zebra
Hartebeest
Eland
Buffalo
Giraffe
Impala
Warthog
Rhinoceros
Elephant
Waterbuck
Dikdik
Lesser kudu
Cheetah
Hunting dog
Lion
Leopard

Figure 3.46 Separation of animal species by habitat preference in the plains of Tanzania. (After Lamprey, 1963.) (Redrawn from H. F. Lamprey, Ecological separation of the large mammal species in the Tarangire Game Reserve, Tanganyika, *East African Wildlife Journal*, 1963, **1**, 65.)

Figure 3.47 Competition between herbivores such as the giraffe and gerenuk is reduced by feeding at different levels in the vegetation. (B. R. Neal.)

Table 3.11 Crude protein content of two common grasses in the Serengeti region of Tanzania at different stages of growth (after Jarman and Sinclair, 1979)

	Green leaf (% dry wt)	Dry leaf (% dry wt)	Stem and sheath (% dry wt)
Early growing season			
Pennisetum mezianum	20.0	–	15.0
Digitaria macroblephora	16.6	–	10.5
Late growing season			
Pennisetum mezianum	9.8	4.3	4.4
Digitaria macroblephora	10.0	4.0	3.8
Dry season			
Pennisetum mezianum	6.8	3.9	3.1
Digitaria macroblephora	11.1	4.9	2.5

(Reproduced with permission from A. R. E. Sinclair and M. Norton-Griffiths, eds, *Serengeti: Dynamics of an Ecosystem*, published by the University of Chicago Press, 1979.)

are necessarily rather unselective in their feeding habits and appear to be more restricted by the quantity of food available rather than its quality (Jarman and Sinclair, 1979). The reduction of the grass cover by the buffalo and zebra not only benefits the wildebeest and gazelles that follow but also opens up the area to non-migratory species, such as the impala and topi, which are equally selective in their food requirements. Consumption of herbage by ungulates is highest in the short grass regions of the Serengeti but even here the large herds remove only 34% of the annual grass production (Table 3.12). Approximately half of the material is reduced

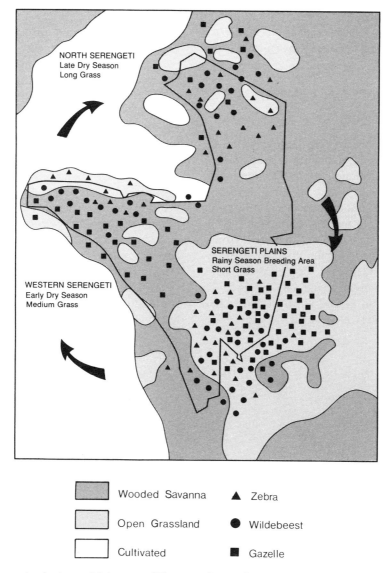

▨ Wooded Savanna	▲ Zebra
▨ Open Grassland	● Wildebeest
☐ Cultivated	■ Gazelle

Figure 3.48 Seasonal movements of zebra, wildebeest and Thompson's gazelle around the Serengeti Plains of East Africa. (After Maddock, 1979.) (Reproduced with permission from A. R. E. Sinclair and M. Norton-Griffiths, eds, *Serengeti*, published by the University of Chicago Press, 1979.)

by detrivores. In the long grassland the ungulates use less than 20% of the grass, and more than 50% of the cover is consumed by fire. Burning similarly removes most of the grass in the wooded kopjes. In these sites ungulates consume only 2% of the grass, which is considerably less than the amount consumed by grasshoppers and rodents.

Table 3.12 Annual production and removal of grasses in the Serengeti region of Tanzania (after Sinclair, 1975)

	Long grassland kg ha^{-1} yr^{-1} (%)	Short grassland kg ha^{-1} yr^{-1} (%)	Kopjes kg ha^{-1} yr^{-1} (%)
grass production	5978	4703	5978
ungulate consumption	1122 (18.8)	1597 (34.0)	122 (2.0)
small mammal consumption	69 (1.2)	4 (0.1)	259 (4.3)
grasshopper consumption	456 (7.6)	194 (4.1)	484 (8.1)
total animal consumption	1647 (27.6)	1795 (38.2)	865 (14.4)
removed by burning	3185 (53.3)	586 (12.5)	3430 (57.4)
removed by detritivores	1146 (19.2)	2322 (49.5)	1683 (28.2)

(Source: A. R. E. Sinclair, The resource limitation of trophic levels in tropical grassland ecosystems, *Journal of Animal Ecology*, 1975, **44**, 516.)

Grass offtake along migration routes ranges from 45% in the wet season to 58% in the dry season, with as much as 89% of the annual production being removed (Onyeanusi, 1989).

3.16 HUMAN IMPACT ON SAVANNA ECOSYTEMS

The native ungulates of Africa are now largely replaced by cattle, sheep and goats. Similar introductions have occurred throughout the savanna regions. Despite high population densities, selective and seasonal feeding habits and long-distance migrations have maintained an ecological balance in the native herds. McNaughton (1985) noted several ways in which a region can benefit from grazing. Defoliation can maintain the plant cover in a state of vigorous growth and, because of the smaller leaf area, soil moisture utilization can be reduced. Productivity can therefore increase in grazed sites. Grazing can also improve the nutritional value of the herbage as a result of rapid mineral recycling in animal wastes. These benefits are lost when introduced herds are improperly managed. All too often the vegetation is depleted by overstocking, or by confining the animals for long periods in overly restricted areas (Werger, 1983). Continual removal of leaf material reduces food reserves. This results in less vigorous regrowth in subsequent seasons and the cover begins to deteriorate. Unpalatable species may increase and bush often spreads into the pastures, either because more water penetrates to the deeper roots, or because fire intensity is reduced as litter production declines. The problem is most serious in the poorly developed regions. These are typically the driest savannas where even woody species are in great demand for fuel and shelter. Continual removal of the protective plant cover soon leads to soil erosion and irreversible destruction of the land. The process has been particularly severe in the Sahelian region of Africa. Here heavy demands on the land, combined with unusually dry conditions, have led to widespread desertification.

CHAPTER 4

The arid regions

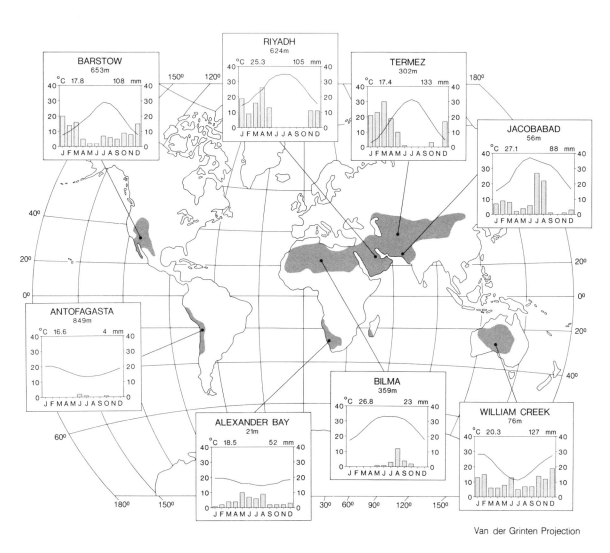

Figure 4.1 Distribution of arid regions and representative climatic conditions. Mean monthly temperatures are indicated by the line and mean precipitation for each month is shown by the bars. Station elevation, mean annual temperature and mean annual precipitation appear at the top of each climograph.

4.1 INTRODUCTION

The arid regions of the world occupy 26–35% of the Earth's land surface. This wide range reflects various definitions based on climatic conditions,

vegetation types, or the potential for food production. Much of this land lies between 15° and 30° latitude (Figure 4.1). Here the air which is carried aloft along the intertropical convergence zone subsides to form the semi-permanent high pressure cells that dominate the climate of the tropical deserts. Adiabatic warming of the air as it descends, coupled with intense radiation under cloudless skies, results in oppressive heat during the high sun season. The tropical deserts are bordered at lower latitudes by semi-arid grasslands which eventually merge with the tropical savannas as summer precipitation increases. Conversely, winter precipitation in the deserts increases towards higher latitudes, and in coastal regions the evergreen trees and shrubs of the Mediterranean climatic regions become more conspicuous. In Asia the tropical deserts form the southern border of the extensive temperate deserts of the continental interior. A similar transition occurs in the high plains of the southwestern United States. The temperate deserts lie in the rainshadow of mountain barriers or are located far inland where moist maritime air rarely penetrates. They experience a continental climate. Temperatures are high in the summer but can drop well below freezing for several months in the winter. Thus, it is aridity rather than continually high temperatures that is the distinctive characteristic of all deserts. Normally this implies a deficiency of precipitation, but the dryness is often intensified by high evaporation rates and by coarse soils which retain little moisture. Life in the desert is conditioned by drought.

4.2 REGIONAL DISTRIBUTION

4.2.1 The deserts of the northern hemisphere

The Sahara, the world's largest desert, covers approximately 9 million km^2 of North Africa. It extends through Egypt to the deserts of the Arabian peninsula which continue eastwards into Iran, Afghanistan and Pakistan, and finally terminate in the Thar Desert of north-western India. The temperate deserts of central Asia lie to the north, separated by the rugged Iranian and Tibetan Plateaus. They present a variety of landscapes. The most westerly of these temperate arid regions is the Turkestan Desert, a low-lying region of internal drainage adjacent to the Caspian Sea. Further east the high mountains of the Tien Shan and Kunlun Shan enclose the sandy wastes of the Takla Makan in north-western China. Beyond lies the Turfan Basin, a block-faulted depression the floor of which is 154 m below sea level, and the desolate plateaus and ranges of the Gobi Desert.

A similar transition to temperate desert occurs in the Basin and Range Province of western North America. Here it is the high Sierras which effectively block the passage of moist air into Nevada and the other states of the interior south-west. Mountain ranges run parallel to the Sierras throughout the northern part of this region, and the intervening desert basins occur at elevations as high as 1500 m. In contrast, the floor of Death Valley lies 86 m below sea level at its lowest point (Figure 4.2). The warmer Mojave and Sonoran Deserts are located further

Figure 4.2 Internal drainage and evaporation has produced extensive salt flats which support little plant life in the floor of Death Valley, California. (O. W. Archibold.)

south. Most of the arid valleys in these deserts lie below 600 m, and extensive areas around the Salton Sea are below sea level. The Chihuahuan Desert of northern Mexico is more rugged but in this region precipitation is mostly restricted by the prevailing anticyclonic conditions rather than topography.

4.2.2 The deserts of the southern hemisphere

Apart from the drier parts of southern Argentina, the deserts in the southern hemisphere all lie within the subtropical high pressure belt. Precipitation is particularly deficient during the southern winter when anticyclonic conditions are most intense. Cold ocean currents also contribute to the stability of the airmasses and even in the Galapagos Islands cactus scrub occurs widely (Figure 4.3). Drought conditions are most severe in the Atacama Desert which stretches as a narrow strip along the coast of Chile from about 18 to 28 °S (Figure 4.4). The outer ranges of the Andes lie less than 100 km inland and further restrict the flow of air from the Pacific. To the north are the 'garua-loma' deserts of Peru. Here coastal fogs (garuas) associated with the Peruvian Current bring some relief to the aridity. Their benefit is lost above 800–1000 m, and arid conditions prevail over much of the western Andes to elevations of 3000 m. The drier parts of Argentina lie in the rainshadow of the Andes. The 'puna' of northern Argentina is a cold, windswept region found at elevations of 3500–4500 m in the high Andes. To the east is the warmer Monte Desert where basins and salt flats are enclosed by mountains, much like the warm deserts of North America. To the south the Monte Desert merges with the dissected plateaus of the cold Patagonian Desert.

The deserts of southern Africa are comprised of three contiguous regions. The Namib Desert occupies a nar-

Figure 4.3 (a) Tree-like *Opuntia echios*, and (b) columnar *Cereus* dominate the cactus scrub on the Galapagos Islands where climate is influenced by the cold Peruvian Current; (c) *Brachycereus nesioticus* is an early colonizer of bare lava in these volcanic islands ((a) and (c) J. T. Price; (b) M. Brown.)

Figure 4.4 Arid salt-pans in the Atacama Desert are essentially devoid of vegetation although species of *Juncus* may establish around the perimeter of some of the saline lakes. (G. E. Wickens.)

row strip of land 80–150 km in width which runs for some 1500 km along the west coast of Africa, from southern Angola to the border of South Africa (Figure 4.5). It is continued south and east across South Africa as the Karoo, which merges with the Kalahari Desert to the north in Botswana. Southern Africa is essentially a high tableland bordered in the east by the rugged Drakensberg Range, which rises to 3000 m, and in the west by an escarpment some 1500 m high. Undulating stony plains, sand dunes, inselbergs and low broken hills interspersed with salt flats and dry river beds occur throughout this region of inland drainage. Although topographic barriers restrict the influx of moist air to the interior, the extreme dryness of the Namib Desert is partially offset by high humidity and fog associated with the cold Benguela Current which flows northwards along the coast.

The most extensive arid region of the southern hemisphere is found in Australia where over 40% of the land is classed as desert: this represents an area of approximately 3.4 million km^2 (Williams, 1979). It is a landscape

Figure 4.5 Sand dunes in the Namib Desert. (B. R. Neal.)

of low relief dominated by broad plateaus and gently depressed basins. Apart from the Macdonnell and Musgrave Ranges which rise to 1500 m, much of central Australia is below 500 m. The lowest sites are found around Lake Eyre, which lies 12 m below sea level. In wet periods the ephemeral streams of the 'Channel Country' may supply sufficient water to fill Lake Eyre, but more commonly extensive salt plains dominate the landscape. To the north is the Simpson Desert, a sandy region across which stabilized dunes run in parallel ridges; dune ridges are also common features of the Gibson and Great Sandy Deserts in the west. The sandy deserts cover some 2 million km^2 of the arid interior.

4.3 CLIMATE

Desert climates are characterized by the scarce and variable nature of their precipitation (Figure 4.1). In hot regions effective precipitation can be further reduced by high evaporation rates, and this parameter is incorporated in the moisture indices derived by Thornthwaite (1948) in his classification of arid climates. Similarly, Meigs (1953) proposed that the temperate and tropical deserts could be classed as extremely arid, arid or semiarid according to the effectiveness of the precipitation. Cold deserts, such as the polar regions and some high elevation sites, are excluded from this scheme if the mean temperature of the warmest month is below 10 °C. This widely accepted classification corresponds generally to natural vegetation zones and agricultural practices in the dry regions of the world. In the most extreme deserts precipitation is so low that all but the hardiest species are eliminated. In the driest parts of the Sahara average annual precipitation is less than 1 mm (Le Houérou, 1986). The coastal deserts are also very dry. Precipitation at Arica in Chile averages only 0.7 mm yr^{-1} (Schwerdtfeger, 1976) and several years can elapse when no measureable precipitation is recorded. However,

coastal fogs which develop over the cold Peruvian Current increase the humidity of the air and provide additional moisture, particularly in central Chile. In the Namib Desert the amount of water which condenses from fogs each year is equivalent to 130 mm of rain (Werger, 1986). Regular dewfalls are also an important source of moisture in many deserts.

Other regions may receive an average of 500 mm of rain annually yet still be classed as deserts. Often the rain is restricted to a short part of the year, or falls in sporadic showers which may cause localized flooding for a brief period. Bell (1979) reported 1-hour rainfall maxima as high as 280 mm in the arid region of Australia. In low-latitude deserts 24-hour totals exceeding 50% of the average annual rainfall have been observed at many locations. Maximum daily rainfalls ranging from 25% to 60% of mean totals are reported in temperate deserts. Such departures from the mean result from unusually intense convectional storms, or because a tropical cyclone penetrated the region. The amount of precipitation received varies considerably from year to year (Table 4.1) but desert organisms have adapted to this unpredictable precipitation regime.

Mean annual cloud cover is generally less than 10% in the Sahara Desert and the Arabian peninsula. This increases to an average of 30% in most other subtropical deserts but can be as high as 50% in the fog bound coastal deserts. In the mid-latitude deserts of Patagonia and central Asia values of 60% are reported (Fitzpatrick, 1979). Well over 3000 hours of sunshine are recorded in most of the hot deserts (Table 4.2) and radiation is intense. During the high sun season, global solar radiation can exceed 700 cal cm^{-2} day^{-1}, although moisture and dust in the atmosphere reduce this by as much as 25%. Further loss of energy can occur because of the relatively high albedo of desert surfaces: on average 28% of the energy is reflected in this way (Fitzpatrick, 1979). Some incoming energy is also converted to latent heat during evaporation of moisture from the soil. Hence, net radiation is often less than half the global radiation received. Despite this, surfaces can become extremely hot, and subsequent heating of the atmosphere often results in air temperatures above 40 °C.

The annual temperature regime varies markedly between different desert regions (Figure 4.1). In the arid parts of Egypt mean monthly temperatures range from 4 to 42 °C (Ayyad and Ghabbour, 1986), and in Australia a range of 11–34 °C is reported (Williams and Calaby, 1985). The mean annual temperature range is smaller in the coastal deserts because of the moderating effect of the ocean: ranges of 13–25 °C are reported for the Namib Desert (Walter, 1986) and 15–21 °C in the Atacama Desert (Schwerdtfeger, 1976). Temperatures in Patagonia are somewhat cooler because of its relatively high latitude but they are also moderated by the maritime location: here the typical annual temperature range is 7–19 °C. Mean summer temperatures as high as 33 °C are

Table 4.1 Extreme precipitation events in desert regions (after Gentilli, 1971; Griffiths, 1972; Lydolph, 1977)

	Max. annual precipitation (mm)	Min. annual precipitation (mm)	Max. 24-hour rainfall (mm)
Africa			
Tamanrasset, Algeria	159	6	48
Cairo, Egypt	63	3	44
Ghadames, Libya	79	6	17
Dongola, Sudan	60	0	36
Alexander Bay, S. Africa	95	22	39
Upington, S. Africa	566	88	119
Windhoek, S. Africa	745	91	86
Swakopmund, SW. Africa	29	0	18
Australia			
Alice Springs	726	60	147
Farina	365	47	127
Oodnadatta	295	29	77
USSR			
Turgay	318	78	93
Kzyl-Orda	187	46	41

(Sources: J. Gentilli (ed.), *World Survey of Climatology Vol. 13. Climates of Australia and New Zealand*, published by Elsevier, 1971; J. F. Griffiths, *World Survey of Climatology Vol. 10. Climates of Africa* (ed. H. E. Landsberg), published by Elsevier, 1972; P. E. Lydolph, *World Survey of Climatology Vol. 7. Climates of the Soviet Union*, published by Elsevier, 1977.)

Table 4.2 Solar radiation for nine arid stations in the northern hemisphere (after Fitzpatrick, 1979)

	Global solar radiation			Net radiation			Total sunshine	
	Daily max. (cal cm^{-2})	Daily min. (cal cm^{-2})	Mean annual (kcal cm^{-2})	Daily max. (cal cm^{-2})	Daily min. (cal cm^{-2})	Mean annual (kcal cm^{-2})	Received (hours)	% of max. (%)
Latitude > 40 °N								
Omsk, USSR	521	48	97	266	−39	36	1900	43
Aralskoe More, USSR	671	104	142	329	−44	48	2450	56
Salt Lake City, USA	667	157	152	319	6	58	3100	71
Latitude 30–32 °N								
El Paso, USA	704	303	192	347	18	72	3600	82
Quetta, Pakistan	715	307	190	344	46	74	3450	79
Giza, Egypt	667	266	176	336	66	77	3500	80
Latitude < 25 °N								
Karachi, Pakistan	576	369	167	267	105	77	3100	71
Dakar, Senegal	579	378	173	280	163	84	2800	64
Wad Medani, Sudan	595	475	171	296	176	91	3800	87

(Source: E. A. Fitzpatrick, Radiation, in *Arid-land Ecosystems: Structure, Functioning and Management*, Vol. 1, eds D. W. Goodall, R. A. Perry and K. M. W. Howes; published by Cambridge University Press, 1979.)

recorded in the warm continental deserts of central Asia, but mean winter temperatures drop to 1 °C. In the cold deserts of Asia mean temperatures can fall to − 22 °C in winter and rise above 25 °C in summer (Walter and Box, 1983d). Similar conditions occur in the temperate deserts of North America: at Reno, Nevada, mean monthly temperatures range from 0 to 20 °C over the course of the year (Bryson and Hare, 1974).

4.4 DESERT LANDSCAPES AND SOILS

Distinctive weathering and erosional processes induced by arid conditions have resulted in a certain uniformity in desert landscapes throughout the world. Despite severe moisture deficits, water is an important geomorphological agent in these regions. Thermal expansion of moisture trapped in rocks may cause disintegration (Winkler, 1977), and even in hot deserts rock shattering can occur when surfaces moistened with dew freeze in the clear night air. Chemical processes, such as the growth of salt crystals, are also effective in rock breakdown (Winkler and Singer, 1972). Rain splash and sheet wash cause pronounced erosion during periods of heavy rain because there is little vegetation to protect the surface. Dry stream channels fill quickly and scour the land, but flow is short-lived, and the heavy sediment loads of these ephemeral streams are deposited on the channel bed or

Figure 4.6 Lag gravels develop when finer materials are removed by erosion. (R. A. Wright.)

Figure 4.7 Aridity and mobility of the substrate limit plant growth on desert dunes. (R. A. Wright.)

as alluvial fans at the base of intricately dissected highlands. These poorly sorted gravels, sands and silts contrast with the rubbly talus which accumulates at the base of escarpments and mesas. Extensive rock pavements covered with a residual layer of gravels are common features of many deserts (Figure 4.6). In Australia these '**gibber plains**' occupy about 1 million km². Similar '**regs**' cover more than 3 million km² of the Sahara Desert where they are distinguished from the bouldery '**hammadas**' and the sandy '**ergs**' comprised of sand sheets and various dune deposits shaped by the wind. In contrast, many internally drained depressions contain ephemeral playa lakes, or are encrusted with residual salt deposits left by evaporation.

Most desert soils are poorly developed **Aridisols** and **Entisols**. Aridisols, the most extensive of the desert soils, are differentiated into **Argids** and **Orthids** according to presence or absence of a subsurface clay horizon. The clayey Argids are associated with older land surfaces where they developed under conditions that were more

humid than at present. Subsequent wind erosion has removed the fine materials to produce the gravelly surfaces so common in the deserts. Orthids include the heterogeneous deposits in younger alluvial fans. With insufficient moisture to leach out soluble salts, concentrations of calcium and magnesium carbonates can be so high that hard layers of caliche form in the profile; while in areas where groundwater is close to the surface, and around playas, sodium salts can form evaporitic crusts. The Entisols are the characteristic soils of the dune areas. These loose, poorly developed soils are readily eroded by the wind (Figure 4.7). Like many Aridisols, they are typically dry, somewhat alkaline and extremely low in organic matter (Table 4.3).

4.5 THE EVOLUTION OF DESERT FLORAS

The geological record suggests that arid conditions have probably existed since Devonian times (Glennie, 1987),

Table 4.3 Surface soil properties of representative desert soils in Australia (after Stace et al., 1968)

	N (%)	OM (%)	pH	P	K	Ca	Mg
				(μ g^{-1})			
desert loam	0.09	1.4	8.3	300	78	240	51
red cracking clay	0.03	0.5	7.1	300	43	540	86
calcareous earth	0.14	2.2	8.3	500	74	680	71
red earth	0.03	–	6.8	200	12	120	18

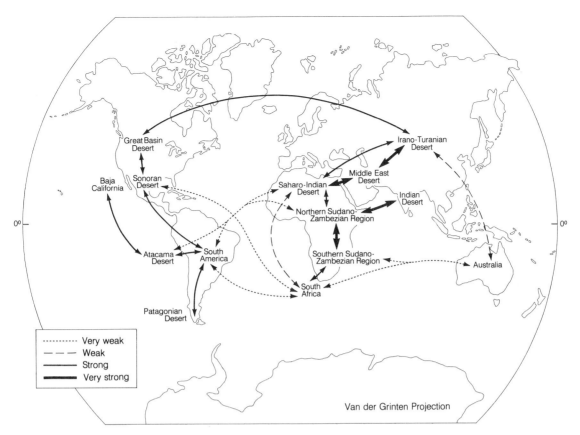

Figure 4.8 Floristic affinities between the major desert regions of the world. (After Shmida, 1985.) (Redrawn with permission from A. Shmida, Biogeography of the desert flora, in *Ecosystems of the World, Vol. 12A. Hot Deserts and Arid Shrublands*, eds M. Evanari, I. Noy-Meir and D. W. Goodall; published by Elsevier Science Publishers, 1985.)

but Axelrod (1958) has postulated that the desert flora originated in the Miocene, expanded during cold, dry phases in the Pliocene, and reached its present distribution only during the Pleistocene. Prior to this the evolutionary development of desert floras would have been restricted by the moist climate of the Cretaceous and Early Tertiary (Axelrod and Raven, 1978). Shmida (1985) has argued that the richness and uniqueness of a desert flora reflects its size, age and period of isolation. Thus, a strong floristic link exists between the interconnected deserts of the Middle East and those of North Africa and Asia (Figure 4.8). Similarities between the floras of the Great Basin of North America and central Asia are explained by the land bridge across Beringia that existed until the end of the Pleistocene. The floristic links between the deserts of North and South America probably arose when the two continents were positioned closer together in early Tertiary times. Whittaker (1977) considered that the variety of life forms which have successfully adapted to the desert environment is indicative of a long evolutionary history. Evolutionary divergence of a single ancient flora is also suggested by the disjunct distribution of similar life forms, such as the Cactaceae of the Americas, the Aizoaceae of southern Africa and the Didiereaceae of Madagascar (Shmida, 1985).

4.6 REGIONAL FLORISTICS

4.6.1 Characteristic plant life forms of the desert

Plant life in deserts is conditioned by available moisture. Some species are entirely dependent on local precipitation; others may grow in depressions and along drainage channels which accumulate water from a wider catchment area (McGinnies, 1968). Only in oases, or where exogenous rivers like the Nile and Colorado carry water across the deserts, is the threat of drought reduced. Therophytes dominate the flora of most deserts (Table 4.4). Germination and development of these short-lived annual herbs is restricted to a brief period following the rains. These ephemeral plants provide a considerable element of seasonality to the landscape depending on the time of the year that growth commences and on their longevity (Table 4.5). The perennial plants, although represented by a variety of life forms, can be grouped into two general classes. **Succulent perennials**, with their remarkable moisture-conserving abilities, are a distinct but rather small component of most desert floras. **Non-succulent perennials**, including grasses, shrubs and trees, are much more common and comprise the bulk of the vegetation cover in all but the most extreme sites.

Table 4.4 The life-form spectra of arid regions. S = stem succulents; P = phanerophytes (trees > 8 m); N = nanophanerophytes (shrubs < 2 m); Ch = chamaephytes; H = hemicryptophytes; C = cryptophytes; Th = therophytes (after McGinnies, 1968)

	Number of species	S	P	N	Ch	H	C	Th
					(%)			
Hot desert climate								
Death Valley, USA	279	3	2	21	7	18	2	42
El Golea, N. Africa	166	0	0	9	13	15	5	56
Ghardaia, N. Africa	300	0	0	3	16	20	3	58
Libya, N. Africa	192	0	3	9	21	20	4	42
Ooldea, Australia	188	4	19	23	14	4	1	35
Hot steppe climate								
Whitehill, S. Africa	428	1	1	8	42	2	18	23
Tomboctou, Mali	134	1	11	12	36	9	3	25
Cool desert climate								
Transcaspia, USSR	730	0	0	11	7	27	9	41

(Reprinted from *Deserts of the World: An Appraisal of Research Into Their Physical and Biological Environments*, edited by W. G. McGinnies *et al.*, by permission of the University of Arizona Press, copyright 1968 Arizona Board of Regents.)

Table 4.5 Life forms of the Sonoran Desert (after Shreve, 1942)

Ephemerals

strictly seasonal:
  winter annuals
  summer annuals
  facultative perennials

Perennials

underground parts perennial:
  perennial roots
  perennial bulbs
perennial shoot reduced to a short leafy caudex:
  leaves succulent
  leaves non-succulent
perennial shoot long but leaves confined to top:
  leaves entire, linear and semi-succulent
  leaves dissected, palmate and non-succulent
leafless stem succulent:
  shoot unbranched
  tall plants with poorly branched shoots
  low or semi-procumbent plants with poorly branched shoots
  richly branched shoots composed of cylindrical segments
  richly branched shoots composed of flattened segments
non-succulent woody plants:
  shoots without leaves, stems green
  low bushes with soft wood
shrubs and trees with hard wood:
  leaves perennial
  leaves drought-deciduous
  leaves winter deciduous

(Reprinted by permission from F. Shreve, Life forms of the North American desert, *The Botanical Review*, Vol. 8, copyright 1942 The New York Botanical Garden.)

4.6.2 Africa

Approximately half of the estimated 3000 species reported from the arid zones of North Africa are found in the Sahara (Le Houérou, 1986). Perennial grasses are commonly associated with dunes and sandy areas where they may grow in association with tall shrubs, such as *Ephedra alata* and *Calligonum comosum*. The stony regs and rocky hammadas are practically barren. In the driest regions perennial species are restricted to drainage depressions and wadis; *Acacia raddiana* and *Tamarix aphylla* are characteristic species in these moister sites. Small shrubs are particularly abundant along drainage channels in areas where precipitation occurs more frequently. More specialized halophytes, including species of *Atriplex, Salsola* and *Suaeda*, are common in wet saline areas. Throughout this region annual plants provide additional variety to the vegetation. They account for 46% of the 755 species reported in the desert flora of Egypt (McGinnies, 1968).

The close relationship between desert species and specific habitats provides a mosaic of plant covers that is essentially repeated throughout the arid lands. Thus, *Calligonum comosum* is again common on the deeper sandy plains of the northern Negev where it grows in association with *Anabasis articulata*, a leafless succulent, and *Retama raetam*, a deeply rooted evergreen shrub (McGinnies, 1968). Grasses, such as *Aristida plumosa* and *Panicum turgidum*, are more common in the sandy parts of the Arabian peninsula (Orshan, 1986), and *Zygophyllum dumosum*, a leaf-shedding shrub, is particularly widespread on gravelly soils. The halophytic species are distributed according to salinity and moisture gradients: *Halocnemum strobilaceum* is common in very saline sites, *Juncus maritimus* is characteristic of slightly saline moist sites, while *Nitraria retusa* is found in deeper, brackish water. Tamarix and acacia usually border the wadis, and the date palm (*Phoenix dactylifera*) is a familiar species of oases.

Unlike the Saharo-Sindian region where similar habitats are occupied by many widely distributed species, the deserts of southern Africa are characterized by a rich and distinctive flora (Figure 4.9). An estimated 5000 species comprise the Karoo–Namib flora, and many are endemic to the region (Leistner, 1979). Werger (1978)

Figure 4.9 (a) *Aloe ferox* and (b) a species of *Euphorbia* in xeric shrubland, Little Karoo, southern Africa. (D. R. Given.)

separates the arid flora of southern Africa into five phytogeographic sudivisions. The northern part is represented by the Namib and Namaland Domains: the Western Cape and Karoo Domains lie to the south, and further inland is the Kalahari Desert. In the dune areas of the Namib Domain perennial grasses form a sparse cover together with the small, deep-rooting, thorny shrub *Acanthosicyos horrida*. Annual grasses grow in association with dwarf shrubs on coarser substrates, although on many gravelly plains only lichens survive. Broad drainage depressions in the northern Namib are commonly occupied by *Welwitschia mirabilis*, a strange and ancient gymnosperm (Figure 4.10). Higher precipitation inland from the coast favours tracts of grassland dominated by species of *Stipagrostis*. Grasses occur extensively in the adjacent Namaland Domain, although shrubs and small trees are more conspicuous in this diverse and often rocky landscape. Species such as *Cochlospermum mopane*, which are commonly associated with savanna areas, may also be present.

The Western Cape Domain is distinguished by its rich flora of succulent plants (Werger, 1986). Some, like species of *Conophytum* and *Lithops*, are inconspicuous plants which simply consist of a pair of swollen leaves lying partially buried in the stony soil. In contrast, the tree-like form of species, such as *Pachypodium namaquensis* and *Aloe pillansii*, may grow 3–4 m in height (Figure 4.11). Many of the succulent species are restricted to the lower elevation sites of the western districts where frosts are uncommon. Winter frosts limit their eastward range, and less succulent shrub-like species, such as *Portula-*

Figure 4.10 *Welwitschia mirabilis* endemic to the Namib Desert produces only two leaves during its centuries-long lifespan; a huge tuberous root projects slightly above the soil and produces reproductive structures most years. (J. A. Aronson.)

caria afra, and the open dwarf shrub formations typical of the Karoo Domain increase in importance. In contrast, the subdued sandy topography of the southern Kalahari supports a very open parklike cover of *Stipagrostis amabalis* in association with *Acacia erioloba* and *A. haemotoxylon*. Taller woodland dominated by *Terminalia sericea* is found in the moister north-eastern parts of this region.

4.6.3 Australia

The flora of central Australia contains about 1200 species, of which 41% are endemic to regions where annual precipitation is less than 250 mm (Perry and Lazarides, 1962). Much of arid Australia supports hummocky **'spinifex'** grasslands dominated by species of *Triodia* (Figure 4.12). The soil is bare between the widely spaced grass hummocks, except for annual plants which appear for a brief period following the rains (Williams, 1979). *Triodia basedowii* provides a cover ranging from 5% to 30% on the sandy soils of the central deserts. To the north

Figure 4.11 Succulent plants of the Western Cape Domain include (a) leaf-succulent flowering stones such as *Conophytum wettsteinii* that grows in rocky crevices; (b) kokerboom (*Aloe dichotoma*); (c) the tapering columnar form of halfmens (*Pachypodium namaquensis*) are crowned with leaves following desert rainstorms. (J. A. Aronson.)

Figure 4.12 Hummocky spinifex grasses (*Triodia*) in central Australia. (O. W. Archibold.)

Figure 4.14 Mitchell grasslands (*Astrebla*) east of Alice Springs, Northern Territories, Australia. (R. T. Coupland.)

and west it is replaced by *T. pungens* which, in association with small trees such as *Eucalyptus brevifolia* and shrubby acacias, forms an arid woodland cover. Elsewhere the hummock grasses gradually merge with open shrublands dominated by mulga (*Acacia aneura*). Mulga is replaced across southern Australia by the arid mallee in which species of *Eucalyptus* form the overstory (Figure 4.13). The mulga shrubland is bordered in the east by *Astrebla* grasslands, in which the dominant grasses form circular tussocks up to 50 cm across and between 0.5 and 2 m apart (Figure 4.14). The cover in all of these communities is structurally similar to the drier savanna formations that surround the arid core. An open cover of low shrubs domi-

nated by species of *Atriplex* and *Maireana* is characteristic of calcareous or saline soils (Figure 4.15). These communities, with their distinctive bluish-grey or greyish-green foliage, are common in the low-lying areas west of Lake Eyre.

4.6.4 South America

The vegetation cover in the northern coastal deserts of Peru and Chile varies from park-like stands of trees and cactus to barren sand dunes (Rauh, 1985). *Prosopis juliflora* is a common tree in the open wooded areas associated with sandy and loamy soils. Grasses, such as *Distichlis spicata* and *Sporobolus virginicus*, are particularly abundant on loose saline soils, while *Salicornia fruticosa* is common in sites where brackish water accumulates. Shrubs, including *Capparis angulata* and *Cryptocarpus pyriformis*, are important colonizers of dunes, although most shrubs are found along river channels where their roots penetrate to the water table. The giant cactus

Figure 4.13 Mallee eucalypts in the western Nullarbor region of Australia. (O. W. Archibold.)

Figure 4.15 Saltbush (*Atriplex*), Nullarbor Plain, South Australia. (O. W. Archibold.)

Figure 4.16 Sparse plant cover in the variable terrain of the Chilean Desert. (a) Pre-puna scrub rising to the high Andes; (b) *Distichylus* sp. stabilizing wind-blown sand; (c) cushion-like form of *Azorella compacta* in its rocky High Andean habitat. (G. E. Wickens.)

Figure 4.17 Grassland of *Stipa speciosa* in sub-Andean Patagonia. (L. Ghermandi.)

Neoraimondia gigantea is a conspicuous plant on coarse stony substrates. Further south is the zone of winter fogs. Here *Tillandsia paleacea* grows on sandy soils: it is entirely dependent on moisture absorbed through hair-like trichomes on its leaf surfaces. The prostrate cactus *Haageocereus repens* is also adapted to these conditions. Other cacti are found on fog-drenched rocky sites, and on the coastal hills tree species, such as *Acacia macracantha*, provide a stunted woodland cover. Numerous annual plants which come into bloom during the foggy months add to the floristic diversity of this region. Although the fogs penetrate many kilometres inland, they bring little

benefit to regions above about 1000 m elevation or to the rain-shadow zones of the interior (Figure 4.16). These intermediate rocky slopes are essentially devoid of plant life. Only at higher elevations, where some orographic precipitation occurs, does the cactus scrub resume. In the Atacama Desert there are no fogs to supplement the meagre rainfall. During rare wet periods a cover of therophytes may develop, or some hardy cryptophytes, such as *Euphorbia copiapina*, may sprout from underground tubers. Often the cover is limited to lichens and algae which depend on dewfall.

Conditions in Patagonia are less extreme. Tussock grasslands form a comparatively narrow zone adjacent to the Andes (Mares *et al.*, 1985). Species of *Stipa* are especially abundant here (Figure 4.17), while in wetter sites rushes (*Juncus* spp.) and sedges (*Carex* spp.) may be locally important. Several shrubs, cacti and perennial herbs are also found throughout this region. In the drier climate further east the small shrub *Chuquiraga avellanedae* and the cushion-like *Nassauvia glomerulosa* are dominant members in the extensive shrub–steppe communities: they form an open cover with grasses, such as *Stipa speciosa* and *S. humilis*. *Atriplex lampa* and other shrubs are common in saline areas.

4.6.5 North America

The deserts of North America can be differentiated by their floristic and structural characteristics (MacMahon and Wagner, 1985). The greatest floristic diversity is found in the warm deserts. Cacti are particularly abundant in the Sonoran Desert (Figure 4.18), including the tall saguaro (*Cereus giganteus*), the multi-stemmed organpipe (*C. thurberi*) and the flat padded beavertails (*Opuntia* spp.). Subtrees which grow 8–10 m in height, such as the paloverde (*Cercidium microphyllum*), mesquite (*Prosopis glandulosa*) and ironwood (*Olneya tesota*), are also com-

Figure 4.18 Characteristic plants of the Sonoran Desert: (a) saguaro (*Cereus giganteus*); (b) ocotillo (*Fouquieria splendens*); (c) barrel cactus (*Ferocactus acanthodes*); (d) teddy bear cholla (*Opuntia bigelovii* var. *bigelovii*); (e) succulent agaves growing beneath the small swollen-stemmed tree *Bursera microphylla*; (f) *Acacia willardiana*, a small tree endemic to northwestern Mexico; (g) Mormon tea (*Ephedra viridis*). ((a) and (g) R. A. Wright; (b)–(d) C. Tracie; (e) and (f) J. A. Aronson.)

Figure 4.19 Joshua trees (*Yucca brevifolia*) in the Mojave Desert, California. (R. T. Coupland.)

Figure 4.20 Transitional arid communities: (a) oak–juniper woodland in Arizona; (b) desert grasslands mainly of Indian ricegrass (*Oryzopsis hymenoides*) in New Mexico. ((a) R. A. Wright; (b) J. T. Romo.)

mon together with numerous smaller shrubs, including bursage (*Ambrosia dumosa*), ocotillo (*Fouquieria splendens*) and the ubiquitous creosote bush (*Larrea tridentata*). Winter annuals are an important component of the Sonoran Desert flora. Much of the Chihuahuan Desert is covered by creosote bush in association with tarbush (*Flourensia cernua*) and white thorn (*Acacia constricta*). Species of *Agave* with their peculiarly long life cycles, and of *Yucca* that flowers only at night, are distinctive plants of this region. Cacti are also common but fewer forms are present than in the Sonoran Desert. Most annuals depend on summer rains in the Chihuahuan Desert. The distinctive species of the Mojave Desert is the Joshua tree (*Yucca brevifolia*), although it is generally restricted to cooler sites at higher elevations (Figure 4.19). Elsewhere in the Mojave Desert, creosote bush commonly grows in association with bursage to form a low shrub cover. Cacti are not conspicuous and subtrees, such as the desert willow (*Chilopsis linearis*), are typically restricted to ephemeral river channels. Winter annuals are especially abundant in the Mojave Desert.

The Great Basin of Nevada and adjacent states lies to the north of the warm deserts. Low growing shrubs, rarely over 1 m high, provide an open cover throughout this region. Big sagebrush (*Artemisia tridentata*) is the dominant species. In the intermontane valleys and plateaus of Oregon, Idaho and Wyoming, sagebrush commonly forms a grass–steppe cover with perennial grasses including species of *Festuca, Agropyron* and *Stipa* (West, 1988). These associations merge with pinyon–juniper (*Pinus monophylla–Juniperus osteosperma*) woodland or stands of ponderosa pine (*Pinus ponderosa*) at higher elevations, and to the south-east with the desert grasslands of Arizona and New Mexico (Figure 4.20). Across Nevada, Utah and northern Arizona, sagebrush often grows in almost pure stands, its grey-green foliage providing a drab monotony to the landscape. Shadscale (*Atriplex confertifolia*) becomes increasingly important in

more southerly parts of the Great Basin, particularly in areas of saline soil or unusually dry sites. Here it forms an open shrub cover with *Artemisia spinescens*, *Grayia spinosa* and *Ceratoides lanata* in association with *Bromus tectorum, Elymus cinereus* and other grasses (Figure 4.21). Distinctive communities dominated by blackbrush (*Coleogyne ramosissima*) are more localized in their distribution: they provide a sparse cover on dry, shallow, calcic soils with perennial grasses, such as *Stipa arida*, *S. speciosa* and *Bouteloua gracilis*.

4.6.6 Asia

The temperate deserts of Asia occupy about 5 million km^2 compared with about 1 million km^2 in North America. Saline soils are common throughout much of the low-lying Turkestan Desert in the western part of this vast region. Shrubs, such as *Artemisia pauciflora* and *Kochia prostrata*, are common dominants in the dry, sodium-impregnated soils of the Caspian Lowlands (Walter and Box, 1983a). With reduced salinity the composition of this desert–steppe cover changes. Grasses such as *Festu-*

Figure 4.21 Characteristic shrubs of the Great Basin Desert: (a) sand sagebrush (*Artemisia filifolia*); (b) shadscale (*Atriplex confertifolia*); (c) winterfat (*Ceratoides lanata*).(J. T. Romo.)

ca sulcata, *Stipa capillata* and *Koeleria cristata* form a dense cover in more favourable salt-free soils. Moister sites support various types of meadows and marshes in which grasses, such as *Agropyron repens* and *Bromus inermis*, are variously associated with sedges, forbs and shrubs depending on salinity and wetness. Further east the cover becomes increasingly dominated by drought-tolerant shrubs, although ephemeral species which bloom in the spring may add to the diversity of the flora.

Conditions become warmer and more arid to the south; ephemeral species and perennials which are active for only part of the summer, such as *Poa bulbosa* and *Carex physodes*, are common amongst the open shrub cover. In sandy areas a dense cover of shrubs and herbs occurs on the stabilized lower slopes: the small tree *Haloxylon persicum* is also characteristic of these sites (Walter and Box, 1983b). Where the sand is still mobile it is first colonized by grasses, followed in succession by perennial herbs and shrubs. The most extreme saline sites may be colonized by *Halocnemum strobilaceum*, but an open shrub cover dominated by *Halostachys caspica* is common elsewhere. Shrubby *Haloxylon ammodendron* may form dense stands in the slightly brackish soils which often surround saline depressions. Areas underlain by clay soils, which flood during wet periods but take in very little moisture, are essentially devoid of vegetation. These sites, known locally as 'takyry', support only algae and lichens or an occasional ephemeral plant.

The Takla Makan of central Asia is an extreme sandy desert in which little vegetation has established (Walter and Box, 1983c). In some areas a sparse cover of herbaceous annuals and the perennial grass *Aristida pennata* grow in association with small shrubs such as *Tamarix ramosissima*. This sandy basin is bordered to the north by the Tarim River and in the south by rivers descending from the snow fields of the Kun Lun Shan. Groves of *Populus diversifolia* and *Ulmus pumila* are common over much of these well-watered floodplains, and species of *Phragmites*, *Typha* and *Scirpus* may form marshlike communities. To the east of the Takla Makan is the Gobi Desert, so named because it is mostly covered with gravels (gobi) and broken stones. Where some sand is intermixed with the coarser materials small shrubs, such as *Anabasis brevifolia* and *Calligonum mongolicum*, provide an open cover. Stands of *Populus diversifolia* may develop where ground water is available, surrounded by shrubs such as *Tamarix ramosissima* and *Haloxylon ammodendron*. To the north the dwarf shrub communities change to desert–steppe and *Stipa* grasslands. Southwards the desert shrubs are replaced at higher elevations by dry montane steppe. This in turn merges with forests of *Picea asperata* on the higher slopes of the Nan Shan Mountains.

4.7 ADAPTATIONS OF PLANTS TO ARID ENVIRONMENTS

The terms drought-escaping, drought-evading, drought-enduring and drought-resisting originally developed by Kearney and Shantz (1912) are still widely used to describe the adaptive strategies of plants in arid regions. The **drought-escaping** plants are the annuals which germinate and grow only when sufficient moisture is available to complete their life cycle. Only their seeds persist during times of drought. The **drought-evading** plants are non-succulent perennials which restrict their growth activity to periods when moisture is available. Typically they are deciduous shrubs which go dormant or die back in dry periods: extreme examples would be lichens and algae which can tolerate extended periods of desiccation. Equally specialized are the fog plants which absorb condensed moisture through their leaves. In many deserts the commonest plants are **drought-enduring** evergreen shrubs. Extensive root systems coupled with various morphological and physiological adaptations of their aerial parts enable these hardy xerophytes to maintain growth even in times of extreme water stress. The **drought-resisting** plants are succulent perennials: the water stored in their swollen leaves or stems is used very sparingly.

4.7.1 Drought-escaping plants

The proportion of annual species in desert floras is inversely related to the amount and reliability of precipitation in a region (Schaffer and Gadgil, 1975). The brief life cycles of these small, shallow-rooted plants commences as soon as moisture is available. Seed set is usually completed within 6 to 8 weeks of germination, but Cloudsley-Thompson (1977) reported that only 8 days elapse between germination and seed fall for *Boerhaavia repens* in the Sahara Desert. Germination is triggered by adequate rainfall. Went (1953) noted that no seedlings of any species emerged in the Sonoran Desert following a 10 mm rainfall: extensive germination occurred only after a rainfall of 25 mm. This was attributed to removal of inhibitors from the seed coats, although the nature of these substances has not been determined (Mayer and Poljakoff-Mayber, 1982). Germination can be reduced when rainfall is excessive, perhaps because diffusable growth-promoting substances are also leached from the seeds (Went, 1953). The annuals develop more vigorously during wetter years, and seed production can increase substantially if the soil remains moist until growth is completed (Beatley, 1967; Mott, 1972). In exceptionally wet years some annuals may persist as facultative perennials (Beatley, 1970). Winter and summer annuals are distinguished on the basis of optimum temperatures for seed germination (Shreve, 1942): this varies from 15–18 °C for winter annuals and 25–30 °C for summer annuals (Beatley, 1974; Went, 1949; Went and Westergaard, 1949). Germination and subsequent development of the annual flora is therefore influenced by temperature conditions prevailing at the time of precipitation. Winter annuals will often grow slowly for some time until stimulated by warmer temperatures (Mulroy and Rundel, 1977).

Continued survival of the annuals requires that adequate seed reserves be maintained through the dry periods until conditions are once again suitable for germination. The brief flush of blossoms attracts many insect visitors. Various characteristics, such as the size, shape and colour of the flowers, and the time of day that they open, have resulted in very specific plant–insect relationships (Orians *et al.*, 1977). In the event that cross-pollination is unsuccessful, many desert annuals will ensure seed production through facultative selfing. Few seeds rot in the desert environment but many are lost to predators, particularly rodents (Brown *et al.*, 1979). Kangaroo rats (*Dipodomys merriami*) consume as much as 95% of the seed of *Erodium cicutarium*, an annual plant of the Mojave Desert (Soholt, 1973). Reduction in the population of annual plants in some years is reported to affect breeding cycles in animals like the kangaroo rat (Beatley, 1969). Birds also consume seeds, although most species supplement their diet with green plant material or insects (Reichman *et al.*, 1979). Seeds are an important food source for many species of ants (Davidson, 1977). Tevis (1958) estimated consumption by harvester ants was

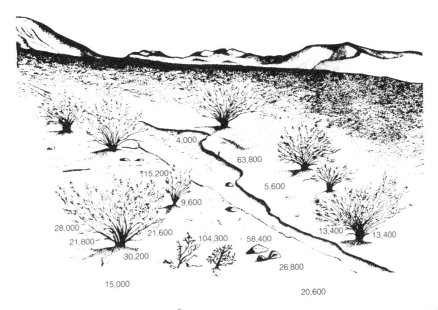

Figure 4.22 Average seed densities (number of seeds m^{-2}) in different microhabitats in Sonoran Desert soils. (After Reichman, 1984.) (Reproduced with permission from O. J. Reichman, Spatial and temporal variation of seed distributions in Sonoran Desert soils, *Journal of Biogeography*, 1984, **11**, 3.)

only about 1% of the seeds produced by the annual flora in the Sonoran Desert, but Whitford (1978) considered that seed foraging by ants in the Chichuahuan Desert was sufficiently intense to affect the abundance of the annual grass *Bouteloua barbata*.

Most of the seed present in desert soils is located in the immediate surface layers but there is considerable variation in the spatial distribution of these reserves. Reichman (1984) reported average seed densities ranging from 4000 to 104 000 seeds m^{-2} in Sonoran Desert soils, although not all of these were from annual species (Figure 4.22). The highest densities are associated with wind-shadows where seeds are protected from further movement by the wind; the lowest values were found in open sites and in dry washes. Seasonal and annual variations in the size of the seed bank are also considerable. Changes in seed densities ranging from 400 to 7700 seeds m^{-2} over a 3-year period have been reported in the Sonoran Desert (Kemp, 1989). Similar fluctuations have been noted in the Mojave Desert (Nelson and Chew, 1977) and in Western Australia (Mott, 1972). Some of this variation can be attributed to growing conditions but the density and distribution of seeds in desert soils is also affected by the foraging activity of animals. Small pocket mice (*Perognathus amplus*) tend to utilize scattered seed resources. Kangaroo rats feed mostly in areas where seeds are more concentrated (Reichman and Oberstein, 1977) and the plants may benefit from the subsequent reduction in seedling densities (Reichman, 1979).

The seeds of most desert annuals have temperature- or moisture-controlled dormancy which may prevent germination, but seed viability is initially high (Mott, 1972).

Some seeds may remain viable for up to 10 years under artificial conditions (Went, 1969), though a period of 2 years is reported by Winkworth (1971) for seeds of buffel grass (*Cenchrus ciliaris*) in the desert. Few viable seeds remain in the soil following the initial flush of germination, and it is speculated that sprouting occurs only if rainfall is sufficient for the continued growth of the seedlings (Juhren *et al.*, 1956; Mott, 1972). This appears to be the case in deserts where the chance of precipitation is relatively high, but more persistent seed banks are predicted for deserts with less predictable rainfall (Brown and Venable, 1986). The seeds of *Chorizanthe rigida*, which are enclosed in the woody involucres of the dead flowers, are released over a period of years as the lignified material is slowly disintegrated by fungi (Went, 1979). A different strategy to delay germination has evolved in *Blepharis persica*. In this species the mature fruits can remain protected by the spiny bracts of the withered parent for several years. Once moistened the fruits open and eject their seeds on to the damp ground, where they germinate immediately: in less than an hour their radicles begin to emerge from the seed (Gutterman, 1972). Similar 'water clock' mechanisms occur in *Plantago coronopus* and *Asteriscus pygmaeus* (Evanari *et al.*, 1982), while seeds attached to the curled stems of the rose of Jericho (*Anastatica hierochuntica*) may be blown considerable distances by the wind before they are released by moistening (Cloudsley-Thompson, 1977).

Mechanisms which prevent all seeds from germinating at the same time can further increase the chance of survival of annual species. Seed **heteroblasty**, in which germination requirements differ for seeds produced by the same plant, has been described for some annuals in the Saharo-Arabian Deserts. The inflorescence of *Aegilops ovata*, for example, forms a compound dispersal unit containing as many as six seeds (Datta *et al.*, 1970). One seed will germinate immediately if water and temperature conditions are favourable. Germination of the second seed is normally delayed till the following year, and several years may elapse before conditions are suitable for germination of the other seeds. Similarly, *Pteranthus dichotomous* may bear as many as seven seeds in its branched pseudocarps, but only a few of these will germinate the first year, thereby ensuring that some seeds remain in the soil (Evanari *et al.*, 1982). **Amphicarpy**, in which the same plant bears both aerial and subterranean fruits, also increases seedling survival (Koller and Roth, 1964). In *Gymnarrhena micrantha*, for example, the larger subterranean fruits are produced first. They are never detached from the parent plant. The aerial fruits, which may not develop in dry years, are dispersed by wind after being released following repeated wetting and drying of the inflorescence, and enable the species to become established in new sites. The seedlings produced by the subterranean fruits are larger and more drought-tolerant than those of the aerial fruits. When conditions are favourable they germinate on the dead parent plant, benefiting from moisture which has penetrated the soil along the channels left by the previous root mass.

4.7.2 Drought-evading plants

Some perennial species are similar to the annuals in that they are active only when moisture is available. The underground organs of these drought-evading plants remain dormant during dry periods but respond quickly when the soil is moistened. New rootlets appear on the short-styled sedge (*Carex pachystylis*) within 12 hours after wetting and quickly spread through the surface soils (Evanari *et al.*, 1982). Shoot development occurs soon after the roots start to form, and active growth of the underground organs commences. Only later do these plants flower and produce fruits but this reproductive phase may be dispensed with in drier years. Water is used freely for as long as it available, but as soon as the soil dries out the leaves and rootlets wither and the plant becomes dormant. The roots of this plant are concentrated in the upper 5–10 cm of the soil. Even for larger cryptophytes, such as the Negev tulip (*Tulipa amplyophylla*) and the desert rhubarb (*Rheum palaetinum*), roots rarely penetrate below 40 cm. These deeper rooting species usually appear only during wetter years. The perennial hemicryptophytes also restrict their activity to the wet season and they die back to ground level in times of moisture stress. Although the root systems of these small plants are usually quite extensive, they possess no specialized mechanisms to conserve water. Transpiration rates are high while water is available, but first the leaves and then the shoots are gradually shed as the soil dries out, and the plants become dormant once more (Evanari *et al.*, 1982).

4.7.3 Drought-enduring plants

Most desert perennials are drought-enduring plants. Photosynthesis may continue for much of the year and so they must replace water lost by transpiration even during periods of extreme drought. These plants usually have extensive root systems which either spread through the surface soils or penetrate several metres below the surface. Depths of 8–10 m are not uncommon, while roots of *Prosopis* sp. have reportedly been found at a depth of 53 m (Phillips, 1963). Branching usually occurs in the capillary fringe above the water table in deep-rooted plants that are able to tap permanent ground-water supplies (Drew, 1979), while short-lived 'rain roots' develop on woody surface roots in response to soil moistening (Kassas, 1966). Root:shoot ratios are comparatively high for most xerophytic shrubs and additional water economy is achieved by various modifications of the leaves. The leaves of paloverde trees (*Cercidium microphylla, C. floridum*) are reduced in size and about 50% their annual photosynthetic output is carried out by the green bark (Adams and Strain, 1968; Adams *et al.*, 1967). The jointed saltwood (*Haloxylon articulatum*) dispenses with leaves altogether. Species such as bean caper (*Zygophyllum dumosum*) and white broom (*Retama raetam*) shed their leaves during dry periods (Evanari *et al.*, 1982). Similarly, the ocotillo (*Fouquieria splendens*) will produce three or four sets of leaves each year, depending on the precipitation regime (Mooney and Strain, 1964). Species such as the thorny saltwort (*Noaea mucronata*) may shed entire branches to restrict moisture loss (Evanari *et al.*, 1982).

Anatomical features, including impregnation of cell walls with waxes and fats, varnish-like coatings over leaf surfaces and thick, hairy cuticles, limit water loss through epidermal tissues; and transpiration rates are reduced when stomata are located in humid pits and grooves, or inside rolled leaves (Kassas, 1966). Mechanical support during times of desiccation is provided by an abundance of lignified tissue. Small leaves or compound leaves, such as those of the acacias and mesquites, allow for efficient heat dissipation and have been selected for in most desert perennials. Leaf temperatures can be further regulated by leaf inclination and surface reflectance. The steeply inclined leaves characteristic of many desert shrubs tend to reduce heat loads during the high sun season, yet intercept far more radiation at other times of the year (Mooney *et al.*, 1977a). The leaves of many perennials are diaheliotropic and follow the sun during the day (Ehleringer and Forseth, 1980). Solar-tracking leaves are also common in annual species: some face the sun to increase photosynthesis, others reduce

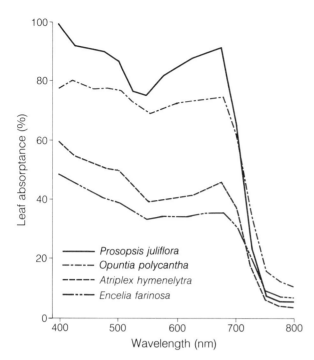

Figure 4.23 Absorptance spectra for wavelengths of photosynthetically active radiation for common species in the Mohave and Sonoran Deserts. (After Ehleringer, 1981.) (Reproduced with permission from J. Ehleringer, Leaf absorptances of Mohave and Sonoran Desert plants, *Oecologia*, 1981, **49**, 367.)

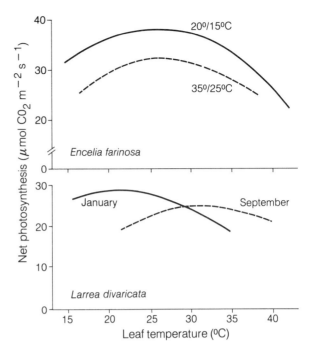

Figure 4.24 Temperature dependence of light-saturated photosynthesis in *Encelia farinosa* compared with the ability of *Larrea divaricata* to acclimatize to seasonal temperature conditions. (After Mooney *et al.*, 1978; Ehleringer and Bjorkman, 1978.) (Reproduced with permission from H. A. Mooney, O. Bjorkman and G. J. Collatz, Photosynthetic acclimation to temperature in the desert shrub, *Larrea divaricata*, *Plant Physiology*, 1978, **61**, 407.)

water stress by avoiding it. Similarly, modification of the leaf surface by hairs, waxes and salt glands can lower energy absorption by 50–80% in some species (Figure 4.23), resulting in reduced tissue temperatures (Gates, 1980) and lower transpiration rates (Smith and Geller, 1980). However, in species such as the drought-deciduous shrub *Encelia farinosa*, transpiration provides significant cooling of the large leaves (Smith and Geller, 1980).

The ability to regulate leaf temperatures enables perennial shrubs to carry out photosynthesis efficiently for extended periods (Figure 4.24). *Encelia farinosa* maintains leaf temperatures near the optimum for photosynthesis throughout the growing season by replacing the larger wet-season leaves with smaller, more pubescent leaves later in the year (Ehleringer and Mooney, 1978). A similar process occurs in *Atriplex hymenelytra* with progressively smaller leaves being produced during the course of the summer. Leaf reflectance increases from 35% to 60% as the leaves age and salt crystallizes on their surfaces from collapsed bladders (Mooney *et al.*, 1977a). Species such as *Larrea divaricata*, which remain active for most of the year, can maintain photosynthetic efficiency by acclimation to seasonal changes in temperature primarily through intrinsic adjustments to the pigment-protein complexes in the chloroplasts (Armond *et al.*, 1978; Mooney *et al.*, 1978). Despite these phenotypic responses, rates of photosynthesis are invariably lower for perennial shrubs than for annual species.

The majority of desert shrubs are C_3 plants, and although they continue to increase production even at very high irradiances, their efficiency is normally reduced by high temperatures, low CO_2 concentrations and increased photorespiration (Ehleringer and Björkman, 1977). Although the desert environment would apparently favour C_4 plants, this method of photosynthesis is common only in halophytic shrubs, perennial grasses in regions of summer rainfall, and summer annuals (Osmond, 1974; Smith and Nobel, 1986).

The ability of C_4 plants to use water efficiently helps to explain the prevalence of this photosynthetic pathway in halophytes. Salinity increases the osmotic potential of the soil solution, and thereby induces physiological drought (Ayyad, 1981). Increased salt concentrations in the cell sap can compensate for this (Jefferies, 1981). Reducing tissue water potential in this way can improve water uptake even in non-halophytes, and the high ion content typical of desert species increases their ability to withstand drought (Osmond, 1979). However, only the halophytes can tolerate high concentrations of sodium salts in the soil. In some of these salt-tolerant species the roots do not take up sodium, but more commonly excess salts are excreted from salt glands, as in species of *Tamarix*, or from salt-accumulating bladders, such as those found on the leaves of *Atriplex*. The sedge *Juncus maritima* sheds entire leaves once they become saturated with salt (Kassas and Batanouny, 1984). The most tolerant halophytes are typically succulent chenopodiaceous shrubs, such as *Halocnemum*, *Suaeda* and *Salicornia*, which are able to dilute the salts using water stored in their fleshy leaves or stems (Jennings, 1968, 1976). However, cacti and other desert succulents are rarely associated with saline soils.

4.7.4 Drought-resisting plants

Succulence is the most obvious characteristic of drought-resisting plants. Most of them flourish in warm, dry climates, yet they are uncommon in the driest desert regions, and some are hardy in areas where freezing temperatures may occur for part of the year (Nobel, 1982a). Desert succulents are generally shallow-rooted, allowing them to respond quickly to light rainfalls. Szarek and Ting (1975) have detected stomatal activity in *Opuntia basilaris* following a rainfall of only 6 mm on dry soil. Water that is surplus to immediate photoysnthesis requirements is stored in the swollen stem or leaf tissues and is slowly used during subsequent dry spells. In the barrel cactus (*Ferocactus acanthodes*) such reserves of water can extend photosynthesis for up to 50 days (Nobel, 1977).

Water loss is reduced in various ways. Species such as *Echinocactus pulchellus* remain underground in the dry season, their shrivelled stems being pulled down by contractile roots: when water is absorbed the cactus swells and re-emerges. In stem succulents, such as cacti, the leaves are reduced to spines and this increases considerably the volume-to-surface ratio. In barrel cactus the ratio is 2.5 cm (i.e. $10\,900\ cm^3 : 4300\ cm^2$), compared with 0.92 cm for a succulent leaf of an agave, and 0.01 cm for many non-succulent leaves (Nobel, 1985). The maximum ratio is achieved by the spherical form of many small cacti. The surface of the stems is generally heavily waxed to reduce cuticular water loss, while the spines not only serve to protect the plant, but also help to dissipate heat. Tissue temperatures beneath the spines of the cholla cactus (*Opuntia bigelovii*) can be reduced by as much as 11 °C in this way.

In *Mammallaria* the spines are drawn together like shutters as water is used up and the plants shrink. The spines open out as moisture becomes available, allowing maximum sunlight to reach the epidermis. Other species dispense with spines and instead depend on irregular surfaces to reduce heat loads: this adaptation is particularly well developed in *Cereus schottii* (Hadley, 1972). Desert succulents are rarely killed by high temperatures, and several species of cacti and agave can withstand temperatures over 60 °C for short periods (Nobel and Smith, 1983; Smith *et al.*, 1984). However, their seedlings are especially sensitive to high-temperature injury, and establishment is often prevented in open areas where soil temperatures can rise to 80 °C. Seedlings of saguaro and other cacti are more likely to survive in the shade of

nurse-plants such as the paloverde (Brum, 1973). Many desert succulents are also sensitive to cold temperatures, and freezing damage can occur to the apex of the stem as a result of radiation heat loss (Nobel, 1980b).

The South American cactus *Neschilenia* grows beneath a layer of translucent quartzite pebbles which protects it from direct radiation as well as desiccating winds and predators (Weisser *et al.*, 1975). Similarly, the small leaf-succulent flowering stones (Mesembryanthemaceae) avoid excessive radiation by growing partly buried in the soil. Only the upper parts of the swollen leaves of the flowering stones are exposed, and these are usually patterned in such a way that they are inconspicuous against the soil. However, the dark dots on the leaf surfaces of *Conophytum* and *Lithops*, or the larger spots on species of *Fenestraria*, are not simply cryptic coloration. They serve as windows which transmit diffuse light on to the chlorophyll-rich layers beneath. Without such adaptations light cannot penetrate far into succulent tissue, so for many species height and orientation can be important. The height of the cactus *Stenocereus gummosus*, for example, varies from 0.7 m in open areas to 4.1 m in denser woodlands where it must compete with other plants for sunlight, while in species of *Opuntia* the cladodes are oriented to intercept maximum radiation during the most favourable part of the growing season (Nobel, 1980a). The leaves of agaves are similarly deployed to maximum advantage (Woodhouse *et al.*, 1980). Even with such arrangements, light intensity may limit photosynthesis, and growth of succulents is generally quite slow.

The water stored by succulents is used slowly during periods of drought, but high tissue water potential limits the ability of roots to extract water from the soil except when it is comparatively moist. Unlike the annual plants and shallow-rooting perennial shrubs with which they compete, the stomata of succulents open in the cool of the night. Transpirational water losses are therefore reduced considerably by restricting CO_2 uptake to the dark hours. Succulent tissue is essential for this nocturnal activity, for it is here that the initial assimilation products are stored before being converted to photosynthates during the daylight period. This unusual two-stage method of photosynthesis is termed **Crassulacean acid metabolism (CAM)**. In CAM plants CO_2 is initially converted to malic acid (Figure 4.25) and tissue acidity levels increase steadily through the night. During daylight the malic acid is decarboxylated and the resulting CO_2 is converted to sucrose and starch by the normal C_3 cycle. Some of the starch is used as the acceptor molecule for additional CO_2 fixation during the next dark period. Maximum CO_2 uptake for most succulents occurs when night-time temperatures are 10–15 °C (Patten and Dinger, 1969) and this greatly increases water use efficiency. Conversely, high daytime temperatures increase the rate at which malic acid is used up and this allows greater CO_2 fixation to occur during the night (Brandon, 1967). Stomatal closure may limit CO_2 uptake

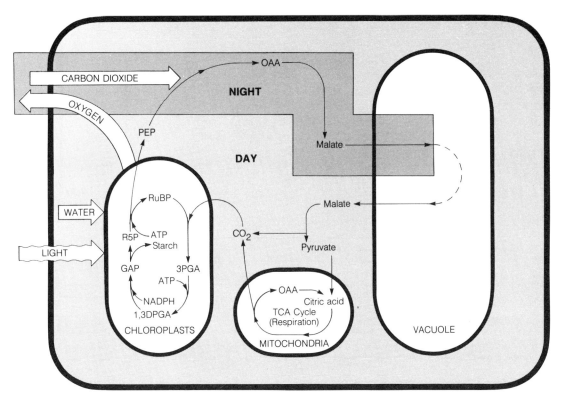

Figure 4.25 Photosynthetic metabolism in CAM plants in which assimilated compounds are transferred diurnally between the cell chloroplasts and vacuoles. ATP = adenosine triphosphate; GAP = glyceraldehyde-3-phosphate; NADPH = nicotinamide adenine dinucleotide phophate; OAA = oxaloacetic acid; PEP = phosphoenol pyruvate; RuBP = ribulose bisphosphate; R5P = ribulose-5-phosphate; 1,3DPGA = 1,3-diphosphoglyceric acid; 3PGA = 3-phosphoglyceric acid. (After Coombs *et al.*, 1983.) (Modified and redrawn from J. Coombs, D. O. Hall and P. Chartier, *Plants as Solar Collectors: Optimizing productivity for energy – an assessment study*; published by D. Reidl Publishing Co., 1983.)

in times of drought but in wet periods the succulents may function as facultative C_3 plants. Such physiological plasticity is most common in regions where rainfall coincides with the long days of the summer (Kemp and Gardetto, 1982). It has also been observed in succulent plants associated with coastal foreshores in temperate regions (Crawford, 1989).

Desert succulents are renowned for their conservative use of water. Measurements by Nobel (1976) indicate that *Agave deserti* transpires only 25 g of water for each gram of CO_2 taken in by its leaves; *Ferocactus acanthodes* loses 70 g H_2O g^{-1} CO_2 (Nobel, 1977). Equivalent values for C_3 and C_4 plants are 450–600 and 250–350 g H_2O g^{-1} CO_2, respectively (Szarek and Ting, 1975). However, succulents are much less efficient than other desert plants in terms of photosynthesis. Not only is CO_2 uptake comparatively low, but also subsequent conversion by the CAM pathway requires more light energy than both the C_3 and C_4 processes (Smith and Nobel, 1986). Growth rates for succulents are consequently slow. Annual growth in *Ferocactus acanthodes* is less than 2 cm (Nobel, 1977) and averages about 4 cm yr^{-1} in *Trichocereus terscheckii*, a cactus which can exceed 5 m in height (Solbrig *et al.*, 1977). Burgess and Shmida (1988) have suggested that succulents are less well adapted to desert conditions than other growth forms in that they are

successful only in comparatively moist deserts with rather predictable rainfall patterns. Thus, even with their capacity to store water, the growth and phenological development of succulents, like all other species, is closely linked to moisture availability.

4.8 THE PHENOLOGY OF DESERT PLANTS

The most obvious seasonal change in the vegetation of arid regions is the flush of ephemeral plants. These appear after adequate rain has fallen, although the germination of winter and summer annuals occurs under quite different temperature conditions. Similarly, leaf and flower development in perennial species are conditioned by increased soil moisture levels. Phenological studies conducted in the Mojave Desert indicate that vegetative growth in some shrubs, such as *Atriplex confertifolia* and *Larrea tridentata*, continues throughout the year (Ackerman and Bamberg, 1974). In *Grayia spinosa* activity is restricted to a few months in the spring and early summer, while species such as *Lycium andersonii* and *L. pallidum* are summer dormant, and resume growth in the autumn (Figure 4.26). Soil moisture increases during the winter, but growth is normally delayed until

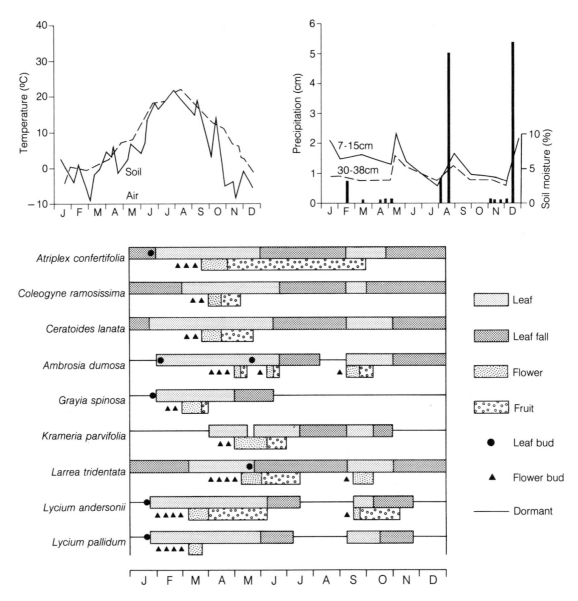

Figure 4.26 Environmental conditions and phenology of common species in the Mojave Desert. (After Ackerman and Bamberg, 1974.) (Redrawn from T. L. Ackerman and S. A. Bamberg, Phenological studies in the Mojave Desert at Rock Valley (Nevada test site), in *Ecological Studies 8 – Phenology and Seasonality Modelling*, ed. H. Lieth; published by Springer-Verlag N.Y. Inc., 1974.)

daytime temperatures are about 20 °C. Flowering is generally completed by May and the deciduous species begin to shed their leaves in June. Further growth occurs in the cooler autumn months if adequate moisture is available. However, *Grayia spinosa* becomes dormant when air temperatures reach about 35 °C and does not resume activity until the following spring. Turner and Randall (1987) subsequently demonstrated that phenological responses of these shrubs can vary by as much as 7 weeks, depending on annual weather conditions (Table 4.6).

Seasonal activity is more varied in the Sonoran Desert where effective rainfall occurs both summer and winter (Solbrig *et al.*, 1977). Here, xerophytic shrubs, such as *Larrea tridentata*, are active throughout the year, while deep-rooting species, such as *Prosopis velutina* and *Cercidium floridum*, remain dormant for up to 5 months in the winter, even though they are supplied with ground

water in all seasons. Conversely, *Encelia farinosa* is dormant only for a short period in the summer, and *Fouqueria splendens* responds by developing new crops of leaves whenever heavy rain falls. Despite these differences in growth activity, flowering occurs in spring or early summer and seed dispersal is completed in time for the summer rains. Comparative data for a summer rainfall site in Argentina indicate that growth for most species starts as soon as temperatures are suitable in the spring and continues till the end of the summer. Here only deep-rooted species, such as *Cercidium praecox*, come into bloom early and will shed their seeds at the beginning of the rainy season. The other perennials flower during the wet season, and produce seeds which do not germinate until the following year.

In the cold deserts low temperatures and frozen soils will naturally limit plant activity to the warmer months

Table 4.6 First dates of leafing, flowering, and fruiting for shrubs at Rock Valley, Nevada from 1972 to 1976 (after Turner and Randall, 1987)

	Ambrosia dumosa	*Grayia spinosa*	*Krameria parviflora*	*Larrea tridentata*	*Lycium andersonii*	*Lycium pallidum*
Leafing						
1972	Feb 17	Feb 18	Mar 14	Mar 1	Feb 16	Feb 18
1973	Feb 28	Feb 15	Apr 18	Mar 12	Feb 26	Feb 20
1974	Feb 25	Feb 25	Apr 2	Mar 18	Feb 15	Feb 4
1975	Feb 27	no data	Apr 29	Mar 18	Jan 24	Jan 24
1976	Feb 18	Mar 4	Apr 9	Mar 26	Mar 4	Mar 4
Flowering						
1972	Apr 16	Mar 4	Apr 13	Apr 19	Mar 9	Mar 7
1973	May 14	Feb 26	May 14	May 14	Apr 3	Mar 13
1974	May 2	Mar 18	May 10	May 10	Apr 2	Mar 26
1975	none	Apr 9	May 21	May 28	Apr 13	Mar 18
1976	May 7	Mar 26	May 21	May 7	Apr 9	Apr 2
Fruiting						
1972	Apr 21	Mar 10	Apr 29	May 2	Mar 14	Mar 15
1973	May 21	Apr 3	May 29	May 21	Apr 18	Mar 22
1974	May 10	Mar 26	May 20	May 20	Apr 13	Apr 2
1975	none	Apr 18	none	June 6	Apr 18	Apr 2
1976	May 14	Apr 9	May 28	May 21	Apr 23	Apr 16

(Source: F. B. Turner and D. C. Randall, The phenology of desert shrubs in southern Nevada, *Journal of Arid Environments*, Vol. 13, 1987.)

but, as in any arid region, growth here is ultimately dependent on available moisture. In the Great Basin region of North America, soil moisture recharge occurs in the spring and plant activity resumes following several months of dormancy. Shoot growth in the dominant evergreen shrubs usually commences in late April: the plants bloom in June and July and the fruits mature in September (Figure 4.27). Root growth normally precedes shoot growth by some weeks and can continue into early winter. Below-ground production can be three times greater than above-ground production in these species because of rapid turnover of the roots (Caldwell and Camp, 1974).

4.9 PRIMARY PRODUCTION IN DESERT REGIONS

Infrequent rainfall, coupled with high evaporation rates and losses through run-off and deep percolation, would appear to leave little water in the soil but the amount available for plant growth varies considerably with depth (Figure 4.28). The surface soils dry out quickly and growth of ephemeral plants is usually limited to a few brief weeks. The deeper rooting plants are ensured water for an extended period but these may go dormant in times of stress. Noy-Meir (1973) considered the minimum precipitation required for any plant growth to be

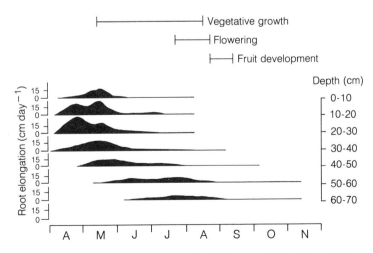

Figure 4.27 Mean daily rate of root elongation in *Artemisia tridentata* in relation to periods of vegetative growth, flowering and fruit development. (After Fernandez and Caldwell, 1975.) (Reproduced with permission from O. A. Fernandez and M. M. Caldwell, Phenology and dynamics of root growth of three cool semi-desert shrubs under field conditions, *Journal of Ecology*, 1975, **63**, 707.)

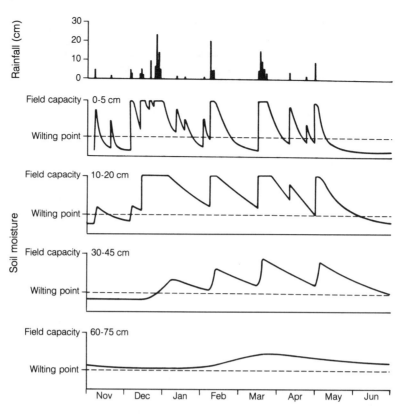

Figure 4.28 Seasonal dynamics of soil moisture at different depths in desert soils in relation to infrequent precipitation events. (After Noy-Meir, 1973.) (Reproduced from I. Noy-Meir, Desert ecosystems: environment and producers, *Annual Review of Ecology and Systematics*, Volume 4, © 1973, with permission by Annual Reviews Inc.)

Table 4.7 Annual productivities of selected desert plants (after Nobel, 1987)

	Desert	Precipitation (mm yr^{-1})	Productivity (kg m^{-2} yr^{-1})
Cryptogams			
Diploschistes calcareus	Negev	50–80	0.01
Annuals			
Amaranthus palmeri	Sonoran	300	0.49
mixed	Negev	163	0.27
Perennial grasses			
Bouteloua eriopoda	Chihuahuan	310	0.19
Hilaria rigida	Sonoran	360	0.23
mixed	Great Basin	270	0.12
Succulents			
Agave deserti	Sonoran	450	0.71
Opuntia ficus-indica	Argentina	300	0.27
Deciduous trees and shrubs			
Artemisia herba-alba	Sahara	300	0.18
Artemisia namanganica	Turkestan	90	0.17
Halocnenum strobilaceum	Arabian	150	0.13
Evergreen shrubs			
Larrea tridentata	Chihuahuan	310	0.25
Deep-rooting phreatophytes			
Prosopis glandulosa	Sonoran	70	0.8–1.3
Prosopis tamarugo	Atacama	1	0.60

(Reproduced with permission from P. S. Nobel, Photosynthesis and productivity of desert plants, in *Progress in Desert Research*, eds L. Berkofsky and M. G. Wurtele; published by Rowman and Littlefield Publishers, copyright 1987.)

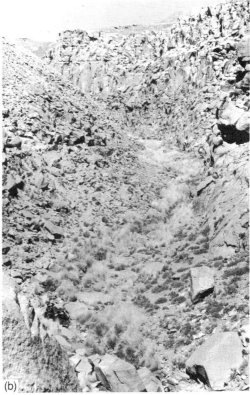

Figure 4.29 The distribution of plants in the Atacama Desert is determined by water availability: (a) *Atriplex glaucesens*, a succulent herb, growing in a runnel; (b) scattered shrubs on the lower scree slopes are replaced by a denser cover of grasses in the dry stream bed. (G. E. Wickens.)

between 25 and 75 mm yr^{-1}. In North American deserts Webb *et al.* (1978) estimated that precipitation of at least 15 mm yr^{-1} is required for growth of annuals and 38 mm yr^{-1} for perennials. Once this threshold is surpassed, estimates of water use efficiency suggest production increases at rates varying from 0.2 to 2 g m^{-2} mm^{-1} (Noy-Meir, 1973; Rutherford, 1980; Szarek, 1979). Annual production in arid regions is generally quite low, but it is also extremely variable according to the nature of the plant cover and the amount of precipitation received in any season (Table 4.7). Le Houérou (1979) reported that average production in North Africa varies from practically nothing to 1200 kg ha^{-1} yr^{-1} depending on rainfall

(Table 4.8). The lower slopes of basins are generally more productive than other topographic sites (Table 4.9) and annual productivity can increase considerably in moist sites, such as drainage depressions and wadis, where water can persist for some time (Figure 4.29). Annual productivity as high as 5700 kg ha^{-1} has been reported for mesquite (*Prosopis glandulosa*) growing in a wash area where mean annual precipitation was only

Table 4.8 Primary production in the semi-arid and arid region of the Sahara (Louw and Seely, 1982)

	Precipitation (mm yr^{-1})	Production (kg ha^{-1} yr^{-1})
Semi arid regions	300–400	800–1200
	200–300	400–600
	100–200	200–400
Arid regions	50–100	100–200
	20–50	~50
	0–20	~0

(Reproduced with permission from G. N. Louw and M. K. Seely, *Ecology of Desert Organisms*; published by Longman, 1982; and H. N. Houérou in *Arid-Land Ecosystems: Structure, Functioning and Management*, Vol. 1, ed. Goodall *et al.*; published by Cambridge University Press.)

Table 4.9 Annual above-ground production in different topographic sites in the Chihuahuan Desert (after Ludwig, 1987)

	Net primary production				
	1971	1972	1973	1974	Mean
	(g m^{-2} yr^{-1})				
bajada alluvial fans	161	292	129	101	171
bajada small washes	37	97	318	179	158
bajada large washes	30	297	456	229	253
basin upper slopes	48	91	179	51	92
basin lower slopes	592	387	292	492	441
basin catchment	52	74	188	191	126
precipitation					
winter(mm)	13	31	75	21	35
spring (mm)	4	14	30	2	12
summer (mm)	77	176	119	158	133
fall (mm)	88	179	4	146	104
year (mm)	182	400	228	327	284

(Source: J. A. Ludwig, Primary productivity in arid lands: myth and realities, *Journal of Arid Environments*, Vol. 13, 1987.)

Table 4.10 Root and shoot biomass for selected desert ecosystems (after Caldwell *et al.*, 1977; Evanari *et al.*, 1975; Rodin and Bazilevich, 1967)

	Root biomass (kg ha⁻¹)	Shoot biomass (kg ha⁻¹)	Root:shoot ratio
Israel			
Artemisia herba-alba	320	350	0.9
Carex pachystylis	1900	440	4.3
Gymnocarpos fruticosum	30	50	0.6
Hammada scoparia	370	350	1.1
Reaumuria negevensis	550	500	1.1
Zygophyllum dumosum	540	630	0.9
Syrian Desert			
Artemisia sieberi	1750	1300	1.3
USSR			
Anabasis salsa	5100	1150	4.4
USA			
Atriplex confertifolia	1886	461	4.1
Ceratoides lanata	1901	281	6.8

(Source: M. Evenari *et al.*, The biomass production of some higher plants in Near-Eastern and American deserts, in *Photosynthesis and Productivity in Different Environments*, ed. J. P. Cooper; published by Cambridge University Press, 1975.)

70 mm (Sharifi *et al.*, 1982). Annual species are particularly responsive in wetter years, and their productivity fluctuates considerably from season to season (Turner and Randall, 1989).

Annual productivity in perennial plants initially accounts for a high proportion of the standing biomass at a site, but because they persist from year to year their annual contribution to the standing crop becomes progressively smaller as the stand ages. Average annual above-ground production is equivalent to 20–40% of standing biomass for deciduous perennials, compared with 10–20% for evergreen species (Noy-Meir, 1973). In cold deserts new shoot growth may be as low as 5% of total plant biomass (Caldwell *et al.*, 1977). Here much

Figure 4.30 The contrasting appearance of (a) a floristically diverse site in the Sonoran Desert supporting various succulents, shrubs and grasses; (b) the widely spaced creosote bushes (*Larrea divaricata*) which dominate a gravelly plain. ((a) R. A. Wright; (b) C. Tracie.)

of the biomass is concentrated underground and root: shoot ratios are comparatively high (Table 4.10). This is not the case with all desert species. Noy-Meir (1973) reported average root: shoot ratios of 0.2–0.5 for desert annuals but values as low as 0.13 are reported by Forseth *et al.* (1984). For many warm-desert perennials the root: shoot ratio is less than 1 (Barbour, 1981). Thus, the assumption by Went (1955) and others that wide spacing between desert shrubs is solely the result of extensive root systems and competition for water has come into question.

4.10 INTERACTIONS BETWEEN DESERT PLANTS

Most deserts are characterized by a very open plant cover. Shreve (1942) reported plant cover in North American deserts ranging from 8% to 15% in simple stands in which species such as *Larrea tridentata* and *Artemisia tridentata* are often the sole dominant. In mixed stands comprised of 4–12 dominant perennials the cover increases to 15–30%, and in floristically diverse stands coverage is as high as 30–60% (Figure 4.30). Shreve (1942) assumed that regular spacing between the dominant plants was necessary to reduce root competition: other species could exist because their rooting habits differed from those of the dominants. Yeaton *et al.* (1977)

have demonstrated such competitive interactions between species in a *Larrea tridentata* community in Arizona. The roots of *Opuntia fulgida* and *Franseria deltoidea* were vertically separated in the soil but they overlapped partially with those of *L. tridentata* (Figure 4.31). Consequently, the structure of the community was determined by the density and distribution of *L. tridentata* which accounted for 69% of the cover at the site.

Measurement of leaf water potentials provided field evidence that regular spacing in *Larrea tridentata* reduces competition for moisture (Fonteyn and Mahall, 1978), but subsequent data compiled by Barbour (1981) indicated that *L. tridentata* is more commonly dispersed in a clumped manner. Phillips and MacMahon (1981) explained this apparent discrepancy in terms of plant size. As growth of the shrubs proceeds, the initial clumped pattern of distribution changes to a random pattern. Uniform distribution occurs only in the biggest plants, which suggests that stand thinning is important in reducing competition in these communities. In addition, several studies have demonstrated that desert perennials become more widely spaced as they grow larger (Nobel, 1981; Phillips and MacMahon, 1981).

Larrea tridentata can occupy a site for many years once it becomes established: one clone in the Mojave Desert may have persisted for over 11 000 years (Vasek, 1980a). Few seedlings arise in undisturbed stands of *L. tridentata*, although they are abundant if the cover is removed

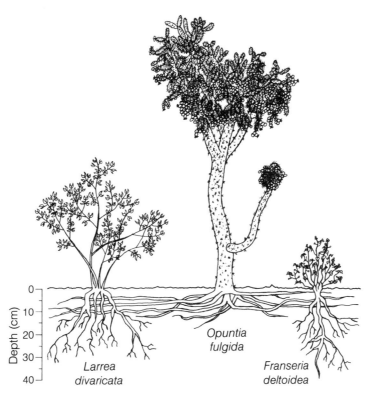

Figure 4.31 Vertical differentiation in the root systems of *Larrea tridentata*, *Franseria deltoidea* and *Opuntia fulgida*. (After Yeaton *et al.*, 1977.) (Reproduced with permission from R. I. Yeaton, J. Travis and E. Gilinsky, Competition and spacing in plant communities: the Arizona upland association, *Journal of Ecology*, 1977, **65**, 592.)

or the soil disturbed (Shreve, 1942). Most seedlings in natural stands are short-lived and little change occurs in the established cover. The floristic composition of most desert communities is remarkably stable and only the relative abundance of each species changes over time (Beatley, 1980; Muller, 1940; Shreve and Hinkley, 1937). However, sequential changes do occur in severely disturbed sites (Lathrop and Archbold, 1980; Vasek, 1980b). Zedler (1981) considered that the nature of such communities was simply an artefact of the survival strategies of the component species, and suggested that subsequent changes of the plant cover were controlled by the frequency and intensity of environmental disturbances rather than by the interaction of the plants on their environment. However, Yeaton (1978) has shown that the distribution of *Opuntia leptocaulis* is related to the presence of *Larrea tridentata*. Initial establishment of the cactus is favoured by the mounds of wind-blown soil which accumulate beneath the shrubs (Figure 4.32). Efficient interception of moisture by the shallow-rooted cactus leads to the demise of the shrub, but the cactus in turn dies because its shallow roots are exposed by erosion. The vacant site is subsequently recolonized by shrub seedlings. McAuliffe (1984) has also reported that saguaros cause a relative increase in stem die-back and increased mortality in paloverde trees which acted as nurse plants during the initial period of establishment of the cacti.

Many species of annuals grow in association with established shrubs (Went, 1942) but some species of shrubs rarely have annuals growing beneath them. Gray and Bonner (1948) concluded that the absence of annuals beneath *Encelia farinosa* was caused by a growth inhibitor produced by the leaves of the shrub, yet a similar substance occurs in the leaves of *Franseria dumosa*, and many annuals grow beneath its canopy (Muller, 1953). Muller and Muller (1956) concluded that allelopathy was of little ecological significance in desert environments, despite finding water-soluble toxins in several desert shrubs. Muller (1953) demonstrated that the associated distribution of annuals with shrubs may be facilitated by the amount of loose soil and litter which accumulates beneath them, rather than being determined by allelopathic affects. The greater amount of material trapped by the intricate branches of *F. dumosa* provides a more favourable habitat for the seeds of annual plants which collect beneath this shrub (Figure 4.33). Toxins from desert plants appear to be quickly inactivated by the soil microflora under natural conditions (Borner, 1960), although Friedman *et al.* (1977) have shown that some

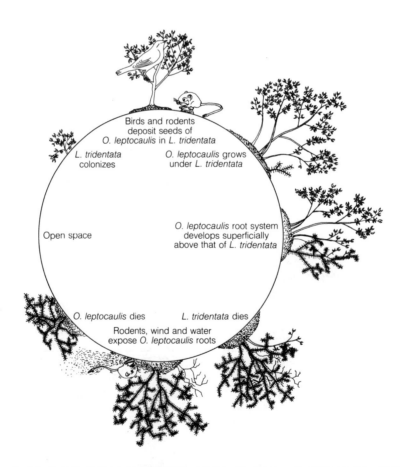

Figure 4.32 Postulated cyclical replacement of *Larrea tridentata* and *Opuntia leptocaulis* in the northern Chihuahuan Desert. (After Yeaton, 1978.) (Reproduced with permission from R. I. Yeaton, A cyclical relationship between *Larrea tridentata* and *Opuntia leptocaulis* in the northern Chihuahuan Desert, *Journal of Ecology*, 1978, **66**, 655.)

| *Franseria dumosa* | *Encelia farinosa* |

Figure 4.33 Comparative growth habits of *Franseria dumosa*, in which each rhizomatous shoot is independently rooted to form an intricate system of branching, and *Encelia farinosa*, in which branches arise above soil level. (After Muller, 1953.) (Reproduced with permission from C. H. Muller, The association of desert annuals with shrubs, *American Journal of Botany*, 1953, **40**, 57.)

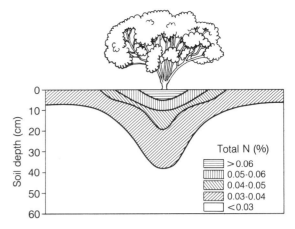

Figure 4.34 Nitrogen concentrations beneath mesquite (*Prosopis juliflora*) in Arizona desert soils as a function of canopy spread and depth. (After West and Klemmedson, 1978.)

annual species are probably suppressed by the strongly aromatic shrub *Artemisia herba-alba* in the Negev. Restricted growth of annuals has also been related to water-repellent soils which have formed around *Larrea tridentata* and other perennials (Adams *et al.*, 1970). Chemical inhibition can also result from salt accumulation around halophytes (Bazilevich *et al.*, 1981; Fireman and Hayward, 1952).

4.11 NUTRIENT CYCLES IN DESERT ECOSYSTEMS

Shreve (1942) considered that the sparse cover in arid areas was insufficient to modify the physical environment. However, conditions beneath established perennials can differ significantly from those of open areas (Charley and West, 1975; Garcia-Moya and McKell, 1970). Nutrient levels are often highest in the surface soils around perennial shrubs (Figure 4.34). These 'fertile islands' develop because of higher litter inputs, more favourable conditions for decomposer organisms and enrichment by animals which live in the protection of the shrubs (West and Klemmedson, 1978). Much of the fine material is quickly broken down by insects or disintegrates during transport by wind or water, although coarse material can persist for many years. Estimates of litter turnover times in desert shrub communities are variable – 20 months in Australia, 2 years in Nevada and 10–14 years in Utah (West, 1981). Chemical analyses presented by Bazilevich *et al.* (1981) indicated that nutrient concentrations are generally highest in the green shoots of desert shrubs (Table 4.11), but this component is a relatively small fraction of the total biomass and as much as 80% of the nutrients are concentrated in the roots. The quantity of mineral nutrients returned to the soil as litter by the roots and shoots of desert shrubs ranges from 57 to 418 kg ha^{-1} each year; calcium, potassium and silicon are particularly abundant. High nitrogen concentrations in the litter result in inputs of 18–136 kg ha^{-1} yr^{-1}. Produc-

tion in desert ecosystems could be severely limited without this continual supply.

Desert soils are generally low in nitrogen and much of it is in unavailable forms. Root nodules are absent from most desert species, and although significant amounts of nitrogen can be fixed by free-living algae and lichen crusts on the soil surface, most of this is quickly lost through denitrification. Some mineralization occurs under clumps of shrubs throughout the year but the availability of nitrogen increases during wet periods, when microbial decomposition is highest. Consequently, the release of nitrogen concides with periods of active plant growth (Kovda *et al.*, 1979). In warm deserts of North America aerial litterfall contributes 6–12 kg N ha^{-1} yr^{-1} (O'Brien, 1978), of which 5–10 kg is added to the soil by decomposition. Input from precipitation and dustfall is estimated at 4–6 kg ha^{-1} yr^{-1} (West, 1978), while losses from wind and water erosion can amount to 6–9 kg ha^{-1} yr^{-1} (Fletcher *et al.*, 1978). New growth of perennials requires 5–20 kg N ha^{-1} each year and, depending on rainfall, a further 1–10 kg N ha^{-1} is needed for the annuals. Over 40 kg N ha^{-1} is available from the soil under the shrubs in the spring, and 5–10 kg N ha^{-1} may be contributed from the adjacent bare areas (Wallace *et al.*, 1978a). The amount of nitrogen contributed from the soil represents approximately 2% of total nitrogen reserves in soil organic matter in desert soils and is restored each year by the activity of the decomposer population.

The leaves of desert perennials contain relatively high concentrations of nitrogen. As much as 60% of this is returned to the stems prior to abscission (Wallace *et al.*, 1978b), but without recycling through litterfall, soil nitrogen would be rapidly depleted. The activity of the detritivores is dependent on adequate levels of soil moisture and the populations of many of these short-lived soil organisms fluctuates considerably during the course of the year (Crawford, 1979). The decomposer organisms are most abundant in the upper soil horizons and show

Table 4.11 The nutrient content of selected desert plants (after Bazilevich *et al.*, 1981)

	N	P	K	Ca	Mg	Total ash
			(% dry weight)			
Artemisia terrae-albae						
green shoots	2.21	0.22	1.96	1.14	0.20	5.76
stems	0.89	0.07	0.86	0.45	0.11	3.22
roots	0.97	0.07	0.50	1.71	0.15	4.26
Anabasis salsa						
green shoots	1.75	0.11	1.02	1.69	0.67	16.28
stems	1.15	0.12	0.69	4.36	0.41	9.01
roots	1.21	0.06	0.35	1.99	0.15	4.08
Haloxylon ammodendron						
green shoots	1.60	0.06	5.31	3.28	2.06	18.14
branches	0.73	0.03	0.60	0.62	0.15	1.79
trunk	0.87	0.01	0.34	0.73	0.05	1.40
roots	1.39	0.05	1.10	1.87	0.95	5.82
Haloxylon persicum						
green shoots	1.62	0.04	2.68	5.39	1.33	11.54
branches	0.69	0.04	0.99	1.03	0.28	2.83
trunks	0.79	0.03	0.40	1.44	0.11	2.36
roots	1.09	0.04	1.42	2.48	1.10	5.92
Carex physodes						
leaves and stems	1.42	0.21	1.63	1.15	0.26	4.35
roots	0.92	0.05	0.48	1.22	0.15	2.85
rhizomes	0.80	0.05	0.32	0.35	0.06	1.11

(Source: N. I. Bazelevich, B. A. Bykov and L. J. Kurochkina, Cycles of mineral elements, in *Arid-land Ecosystems: Structure, Functioning and Management*, Vol. 2, eds D. W. Goodall, R. A. Perry and K. M. W. Howes; published by Cambridge University Press, 1981.)

Table 4.12 The amount of *Kochia sedifolia* forage produced under different grazing treatments (after Trumble and Woodroffe, 1954)

Rate of stocking (sheep km^{-2})		Forge production (kg ha^{-1})					
		1946	1947	1948	1949	1950	1951
nil	(0)	1624	1098	526	818	1187	1333
light	(62)	1490	1109	336	1075	1086	1277
moderate	(124)	2363	1478	818	1176	1501	1904
heavy	(186)	2094	1344	560	974	1411	1590
very heavy	(332)	2621	1579	907	952	1904	1736

(Source: H. C. Trumble and K. Woodroffe, The influence of climatic factors on the reaction of desert shrubs to grazing by sheep, in *Biology of Deserts – The Proceedings of a Symposium on the Biology of Hot and Cold Deserts organised by the Institute of Biology*, ed. J. L. Cloudsley-Thompson; published by the Institute of Biology, 1954.)

a close association with the shrub cover (Freckman and Mankau, 1977). Conversely, the larger litter-feeders such as millipedes and ants remain active between periods of rain (Crawford, 1976; Whitford, 1978; Whitford *et al.*, 1976). Termites are the main detritivores in warm deserts and remain active for much of the year. They account for more than 50% of the annual litter consumption (Johnson and Whitford, 1975) and are the principal consumers of woody material, though each termite species is often very selective in the type of the wood that it attacks (Haverty and Nutting, 1975). Much of this material is eventually incorporated in the soil. The compact, impervious galleries constructed by the termites can remain intact long after they have been abandoned and this can result in a temporary loss of nutrients to the ecosystem (Lee and Wood, 1971; Matthews, 1976).

The possibility that soil nutrients could limit production in arid regions was first reported by Trumble and Woodroffe (1954). They demonstrated that growth in communities of Australian bluebush (*Kochia sedifolia*) was greatly increased by additions of nitrogenous wastes from grazing animals (Table 4.12), as the dung was rapidly broken down by insect larvae, beetles and termites. The role of macro-decomposers is particularly important in arid ecosystems because microbial decomposition is restricted to brief periods when moisture is available (Noy-Meir, 1974). The abundance of detritivorous beetles in the Mojave Desert is also limited by rainfall rather than by food availability, which suggests that detritus can accumulate during dry years (Thomas, 1979). Increased decomposition during wet periods may therefore alleviate nutrient shortages when conditions are most favour-

Table 4.13 The physical and chemical defences of desert plants against herbivores (after Orians *et al.*, 1977)

Life form	Physical defences	Chemical defences	
		Short-lived tissues	Long-lived tissues
ephemerals	leaves easily chewed; no spines	toxins	–
root perennials	leaves easily chewed; no spines	toxins	digestion-reducing substances
deciduous perennials	leaves easily chewed; may have spines	toxins; digestion-reducing substances	toxins; digestion-reducing substances; low nutrient content
evergreen perennials	leaves tough; usually not spinescent	toxins; digestion-reducing substances	toxins; digestion-reducing substances; low nutrient content
succulents	photosynthetic tissue very tough; many spines	–	toxins; digestion-reducing substances; low nutrient content

(Source: G. H. Orians *et al.*, Resource utilization systems, in *Convergent Evolution in Warm Deserts*, eds G. M. Orians and O. T. Solbrig; published by Dowden, Hutchinson and Ross, 1977.)

able for plant growth, and this in turn may determine the success of the consumer organisms.

4.12 HERBIVORY IN ARID ECOSYSTEMS

Herbivores consume only a small fraction of the plant material produced annually in desert ecosystems. In stands of *Larrea tridentata* small mammals consume 2% of net annual above-ground production (Chew and Chew, 1970) and a similar amount may be removed by grasshoppers (Mispagel, 1978). Most herbivores will not consume heavily lignified tissue, preferring leaves, fruits and seeds to woody stems. Soholt (1973) estimated that kangaroo rats removed as much as 11% of these materials produced in a stand of *L. tridentata*. Seeds provide a highly concentrated source of food, and a large proportion is consumed by granivores (Bradley and Mauer, 1971). Soil seed reserves can be reduced considerably through predation by rodents and ants (Brown and Davidson, 1977) and further losses can be attributed to birds and weevils (Orians *et al.*, 1977). Although rodents are the principal seed consumers, most are opportunistic feeders and vary their diets as food resources become available.

Consumption of foliage is deterred by a variety of physical and chemical defence mechanisms (Table 4.13) and the feeding habits of many herbivores are extremely specialized. Eleven of the 15 species of caterpillars that feed on perennial plants in the deserts of Arizona and Argentina confine their activities to a single species (Orians *et al.*, 1977). In contrast, 10 of the 13 caterpillar species that feed on annual plants are generalist feeders. The generalist feeders prefer mature leaves, whereas the specialists mainly consume younger tissue. In addition, the caterpillars which feed on a single species are usually cryptically coloured: this further restricts their activity, although many are active only at night. Grasshoppers are also fairly specialized feeders and their numbers are usually limited by food availability. Most are solitary in their habits, unlike the migratory locusts which periodically swarm in devastating numbers. Locusts are poorly adapted to arid conditions: their eggs hatch only

if they are able to absorb sufficient moisture from the soil, the nymphs are susceptible to desiccation if they do not consume fresh plant tissue and the females will not mature on a diet of dry plant material (Graetz, 1981). However, the flight behaviour of locusts during migration enhances their chances of finding suitably moist breeding areas (Clark, 1971).

4.13 ADAPTATIONS OF ANIMALS IN ARID REGIONS

4.13.1 Insects

Animals must restrict water loss and regulate body temperature through physiological or behavioural adaptations to survive in the inhospitable desert environment. Heat balance can be maintained by evaporative water loss but this strategy is only feasible for larger animals (Figure 4.35). In smaller animals evaporative heat loss would soon lead to desiccation. Many arthropods are therefore equipped with a relatively impermeable, wax-coated cuticle. Several species can alter the structure of

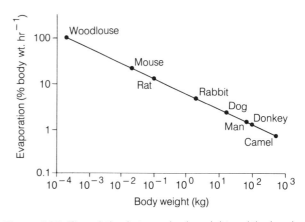

Figure 4.35 The relation between body weight and the hourly rate of evaporation of water required to preserve constant body temperature in desert environments. (After Edney, 1966.) (Reproduced with permission from E. B. Edney, Animals of the desert, in *Arid Lands: A Geographic Appraisal*, ed. E. S. Hills; published by Methuen & Co., 1966.)

the wax layer to allow evaporative cooling to occur under extreme conditions (Loveridge, 1968a). Most arthropods avoid heat stress by being active during cool periods or at night, and retreat underground or to shaded sites during the day. While at rest, water loss from the respiratory surfaces is usually reduced by closing the spiracles through which gas exchange occurs (Edney, 1966). Water loss through respiration can be further reduced if the spiracles are located in pits in much the same way as plant stomata, or if they are placed on the ventral side so that they are concealed when the arthropod rolls into a ball (Cloudsley-Thompson, 1977). Water loss increases during periods of activity. Locusts, for example, lose 65–80% of their moisture through respiration when in flight (Church, 1960; Loveridge, 1968b). Greater water economy is also achieved by reducing the moisture content of excretory products. Arthropods typically eliminate nitrogenous wastes in the form of uric acid or guanine in a dry state, and water in solid waste materials can be reabsorbed in the hindgut.

Arthropods replace water by drinking and by extracting moisture from their food. Water can also be produced by oxidizing hydrogen-containing foodstuffs. Some species can absorb water vapour from humid air through their integument, mouth or rectum (Edney, 1974). The Namib beetle *Onymacris ungulicaris* basks in the fog, and water which condenses on its elytra trickles to its mouthparts (Hamilton and Seely, 1976). A smaller beetle, *Lepidochora discoidalis*, utilizes fog in a different way: it builds trenches to collect the condensed water and then extracts it from the moistened soil (Seely and Hamilton, 1976). Soil moisture is an important source of water for many arthropods (Crawford and Cloudsley-Thompson, 1971).

4.13.2 Reptiles

Many of the adaptations described for arthropods are also found in reptiles. Although lizards and snakes have relatively impermeable skins, evaporation is often the main source of water loss in these animals and can be used for cooling (Mayhew, 1968). Excretory losses are minimized by the production of uric acid which is eliminated in a semi-solid form with other waste products. Herbivorous lizards which take in sodium and potassium from plant materials, or those which imbibe saline water, also excrete these salts from nasal glands (Mayhew, 1968). Some water is lost by this process, although in animals such as the desert iguana (*Dipsosaurus dorsalis*), this may collect in depressions in the nasal passages and reduce moisture loss during exhalation (Schmidt-Nielsen, 1972). Lizards and snakes will drink water when it is available, but the majority are carnivorous and can secure most of their water requirements from their prey. Various species are capable of metabolizing water by oxidation, and some lizards can absorb water through their skin, as reported for Australian

geckos (Warburg, 1966). In a similar way, water on the skin of the spiny agamid (*Moloch horridus*) is conveyed to its mouth by capillary action along a network of fine grooves (Bentley and Blumer, 1962). More acute problems are faced by the amphibians, which are very sensitive to desiccation. These animals are normally active only during the wet season and aestivate as conditions become unfavourable. Similarly, numerous crustaceans which inhabit temporary ponds undergo diapause as the water dries up. Such periods of inactivity also enable reptiles, small mammals and even birds and fish to overcome unfavourable periods in deserts.

Reptiles are ectothermic animals and maintain their body temperatures within narrow limits by various behavioural adaptations. Lizards which are active during the day move in and out of the shade, and sometimes will flatten their bodies against warm surfaces to gain heat. They can lose heat by stiffening their legs to raise their bodies above the hot surface (Mayhew, 1968). Many species are able to change their colour to alter reflectivity of near infra-red wavelengths (Hutchison and Larimer, 1960). For others the period of the day during which they are active changes with the seasons (Figure 4.36). Nocturnal lizards favour cooler conditions but can usually tolerate a broader range of temperatures than their diurnal counterparts. Most remain buried in the soil during the hot part of the day. Desert snakes are also generally intolerant of high temperatures. They spend the day in burrows or in rocky crevices and emerge at night. Some, like the sidewinding rattlesnake (*Crotalus cerastes*) of North America and equivalent forms such as *Bitis peringueyi* of the Namib Desert, remain buried in the sand, leaving the tip of the tail exposed for bait. At night their peculiar method of locomotion enables these species to move quickly over smooth or yielding substrates. Other adaptations common in desert snakes include modified head parts to facilitate burying activity, like the American shovel-nosed snake (*Chionactis occipitalis*), and nasal valves and hornlike scales over the eyes to restrict entry of sand grains, as in the horned viper (*Cerastes cerastes*) of North Africa. The rattlesnakes (*Crotalus* sp.), like other venomous species, are equipped with heat-sensing pits with which they seek out small mammals and other warm-blooded prey.

4.13.3 Mammals

Nearly all of the small mammals found in desert regions are nocturnal and escape the daytime heat by resting in cool burrows. The few species which are active during the day, such as the ground squirrels (*Citellus* spp.), frequently retreat underground and sprawl against the cool soil to disperse body heat (Bartholomew and Dawson, 1968). Rodents and small mammals have no sweat glands and so dispense with evaporative cooling. Very little water is lost through their skin, although in emer-

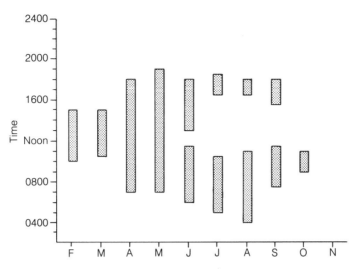

Figure 4.36 Seasonal variation in the time of day during which the lizard *Phrynosoma m'calli* is active on the desert surface. (After Mayhew, 1968.) (Reproduced from W. W. Mayhew, Biology of desert amphibians and reptiles, in *Desert Biology*, **1**, ed. G. W. Brown, published by Academic Press Inc., 1968.)

gency situations water from the mouth can be used to wet the fur (Edney, 1966). Respiratory water losses are reduced because temperatures in the nasal passages are cooler than in the lungs, and this lowers the water vapour content of the air that is exhaled. Water conservation is also achieved by the production of concentrated urine and dry faecal pellets. The ability to concentrate salts in the urine permits species such as the kangaroo rat to consume saline water without adverse effects (Schmidt-Nielsen, 1954). The kangaroo rat can manufacture water by oxidative metabolism and so can exist on a diet of dry seeds. Species that are less well adapted rely on succulent plant tissues for moisture, or feed on insects.

Larger mammals, such as the North American jack rabbits (*Lepus alleni* and *L. californicus*) and carnivores like the kitfox (*Vulpes macrotis*) and bobcat (*Lynx rufus*), reduce heat stress by being active at night and seeking out shady retreats during the day. The largest animals, such as camels, kangaroos and antelopes, have few opportunities to avoid the rigorous environment. However, it is their large size which partly assures their survival. Not only is metabolic heat generation proportionately lower in large animals, but also greater thermal inertia reduces the rate of change of body temperature when exposed to solar radiation (Bartholomew and Dawson, 1968). A comparatively small surface:volume ratio does restrict heat loss by convection and radiation but, unlike humans, body temperatures of these larger animals do not need to be maintained within narrow limits. Adaptive hyperthermia in camels can result in daily fluctuations in body temperature of more than 6 °C (Schmidt-Nielsen *et al.*, 1957) and the stored heat is dissipated during the night. The effect is most pronounced

when animals are dehydrated and water cannot be spared for evaporative cooling (Figure 4.37).

Eland (*Taurotragus oryx*) and oryx (*Oryx gazella*) also regulate body temperatures by hyperthermia but unlike camels, which begin to sweat at temperatures of 40.7 °C, these animals normally reduce heat stress by panting to moisten the nasal sinuses. This lowers the temperature of the blood supplying the brain by 2–3 °C and allows them to tolerate body temperatures as high as 45 °C (Taylor, 1969). Similarly, jack rabbits can allow their body temperature to rise to 44 °C and thermoregulation is achieved by controlling the flow of blood to their large

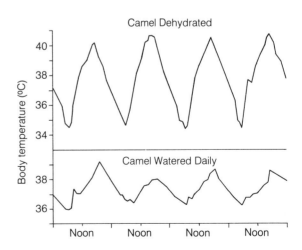

Figure 4.37 Body temperatures of camels in relation to water economy. (After Schmidt-Nielsen, 1964.) (Redrawn from K. Schmidt-Nielsen *et al.*, Body temperature of the camel and its relation to water economy, *American Journal of Physiology*, **188**, 1957, with permission from the American Physiological Society.)

ears (Bartholomew and Dawson, 1968). The ears are devoid of fur, which makes them ideal for heat loss, but conversely, bare skin offers no barrier to heat gain from solar radiation and provides no insulation against cooling during the night. Under these conditions a glossy coat of coarse, short hair provides ideal protection for the larger mammals. However, unless it is well ventilated, evaporation of sweat will only occur from the surface of the coat and any cooling effect is reduced. Thus, the skin of a camel can still be effectively cooled beneath its hairy coat, although sweating normally occurs only if the animal is allowed to drink freely.

The camel, like other ruminants, utilizes some urea for microbial synthesis of protein and this reduces the quantity of water expelled as urine. However, it is the camel's ability to tolerate water losses equal to 25% of its body weight that is most remarkable. The critical limit for water loss for most large mammals is 12–14%; as the blood thickens the circulation rate slows, less heat is dissipated and the animal dies of 'explosive heat death'. In contrast, little water is lost from the blood of a camel and the efficiency of the circulatory system is maintained despite severe dehydration. Although camels can endure long periods without drinking, they very quickly replace the water they have lost, larger animals being capable of consuming as much as 100 litres in about 10 minutes (Schmidt-Nielsen, 1964). Taylor (1969) concluded that eland and oryx could survive indefinitely without drinking, as long as the water content of their forage was higher than 10%. However, as their food supply dwindles they too must seek out water. Hamilton *et al.* (1977) have described how oryx excavate semi-permanent water-holes in the bed of the Kuiseb River in Namibia during the dry season. Like many ungulates, they can travel considerable distances to exploit available resources or to minimize environmental stress.

4.13.4 Birds

Most birds migrate annually to escape the severity of their desert habitat. Species that are permanent residents exhibit characteristics similar to those of the mammals – adaptive hyperthermia, excretion of uric acid and protective plumage. Most rest in the shade during the hottest part of the day but others may soar in the cool air aloft. Moisture is seldom a problem for carnivorous species and others can readily travel to available sources of water. The sandgrouse (*Pterocles namaqua*) of the Namib, for example, returns from its daily excursions with water for its young in its breast feathers (Cade and Maclean, 1967). This species lays its eggs in a slight depression and, like many of the desert birds, must prevent overheating by shading them during the hottest part of the day (Dixon and Louw, 1978). Most birds lay their eggs in a shaded site under a bush or in a rock crevice: the elf owl (*Micrathene whitneyi*) of the Sonoran Desert nests only in the abandoned holes that the Gila woodpecker (*Melanerpes uropygialis*) has cut in saguaro cacti.

4.14 HUMAN IMPACT IN DESERTS

Nomadic pastoralists reduce the stressful impact of the desert environment by moving their herds with the rains. Their domestic animals are well suited to the desert environment. The black Bedouin goat can tolerate a 30% reduction in body weight through dehydration and this enables it to travel for 2–3 days between waterholes. Zebu cattle store fat in their humps in much the same way as camels. This not only provides a source of energy when their food supply is limited but also permits greater heat loss than in animals with subcutaneous fat storage. However, changes in the traditional way of life of the nomads have led to larger, more sedentary herds,

Table 4.14 Changes in the area of the Sahara Desert and its southern boundary from 1980 to 1990 (after Tucker *et al.*, 1991)

Year	Area of Sahara Desert (km²)	Change relative to 1980 (km²)	Change relative to 1980 (%)	Annual precipitation departure from long-term mean (%)	Mean position of southern boundary (latitude)	Annual change in position of southern boundary (km and direction)
1980	8 633 000	–	–	– 13	16.3 °S	–
1981	8 942 000	308 000	3.6	– 19	15.8 °S	55 south
1982	9 260 000	627 000	7.3	– 40	15.1 °S	77 south
1983	9 422 000	789 000	9.1	– 48	15.0 °S	11 south
1984	9 982 000	1 349 000	15.6	– 55	14.1 °S	99 south
1985	9 258 000	625 000	7.2	– 28	15.1 °S	110 north
1986	9 093 000	460 000	5.3	– 21	15.4 °S	33 north
1987	9 411 000	778 000	9.0	– 40	14.9 °S	55 south
1988	8 882 000	248 000	2.9	–	15.8 °S	99 north
1989	9 134 000	501 000	5.8	–	15.4 °S	44 south
1990	9 269 000	635 000	7.4	–	15.1 °S	33 south

(Reproduced with permission from C. J. Tucker, H. E. Dregne and W. W. Newcomb, Expansion and contraction of the Sahara Desert from 1980 to 1990, *Science*, **253**, 300, copyright 1991 by the AAAS.)

and overgrazing has resulted in considerable changes in the plant cover. Perennial grasses are generally replaced by annual species and unpalatable plants may increase in abundance. Total plant cover is reduced, leaving the soil exposed to erosion by wind and water. With declining productivity the land becomes even more heavily used, and in areas where precipitation is at best highly variable, unusually long dry spells can be disastrous. The effect has been particularly severe in the Sahel, which suffered a prolonged drought in 1969–73 and again during the 1980s. Loss of productive land in this way is termed **desertification**.

In addition to overgrazing, there has been increased cultivation in areas that have been used traditionally for raising animals. Although rainfall of 400 mm yr^{-1} is considered necessary for continuous cropping in the Sahel, subsistence agriculture is now attempted in locations where precipitation is less than 250 mm yr^{-1} (Le Houérou, 1977). In these marginal areas the fertility of the soil is quickly exhausted and the abandoned land is left vulnerable to erosion and further deterioration. Desertification can also result from the removal of vegetation for firewood or other products such as waxes and fibres.

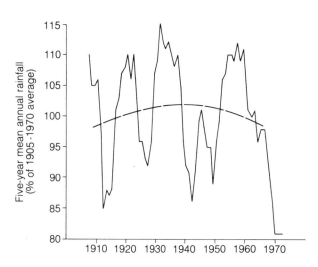

Figure 4.38 Five-year means of annual rainfall in the Sahel region of West Africa expressed as percentages of the 1905–1970 average. The long-term trend is shown by the dashed line. (After Mason, 1976.) (Reproduced with permission from B. J. Mason, Towards the understanding and prediction of climatic variations, *Quarterly Journal of the Royal Meteorological Society*, **102**, 473–98.)

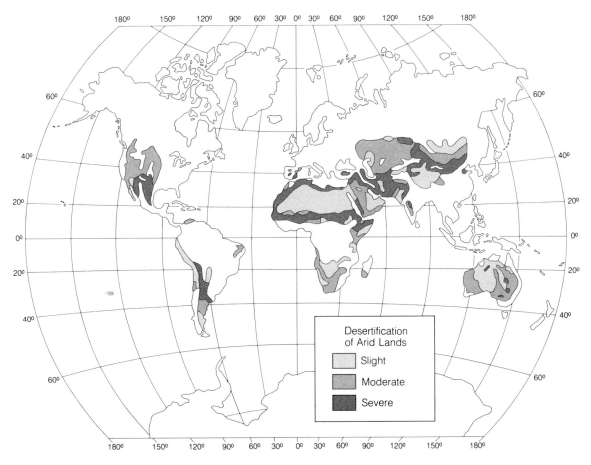

Figure 4.39 World status of desertification of arid lands. (After Dregne, 1983.) (Reproduced with permission from H. E. Dregne, *Desertification of Arid Lands*; published by Harwood Academic Publishers GmbH, 1983.)

Extensive damage from road construction has been noted in countries like Saudi Arabia as a result of oil exploration and associated development (Batanouny, 1979); similar degradation has accompanied mining development. Increased mobility of local inhabitants and improved facilities for tourists have also had some impact on arid ecosystems (Batanouny, 1983).

Equally serious is the deterioration of soil quality resulting from improper irrigation practices. Most commonly salts are brought up by capillary action and subsequently accumulate in the soil as the water evaporates. Dregne (1983) estimated that 21% of irrigated land in arid regions may be adversely affected by salinization, and brackish water is often re-applied to the soil. Fortunately, breeding of salt-resistant varieties has been successful for many crops including cotton, wheat and sugar beets (Schechter and Galai, 1980), while in Egypt rushes (*Juncus* sp.) have been used to reclaim saline areas (Zahran *et al.*, 1979). Changes in the structure of some clays and calcareous soils following irrigation has resulted in the formation of impermeable surface crusts (Elgabaly, 1977).

Arid regions are very sensitive to climatic variations and the extended wet and dry periods which result (Lockwood, 1986). Three major drought periods have occurred in Australia since 1900: conversely, the period 1974–76 was unusually wet and Lake Eyre reached its highest level since European settlement. Similarly, there has been a succession of drought periods in the Sahel (Figure 4.38). The recent droughts are not the most severe on record but their effect has been aggravated by unprecedented demands on the land. Records since 1980 indicate that the area of the Sahara has increased by as much as 1.4 million km^2 because of fluctuations of the southern boundary, as approximated by the 200 mm annual precipitation isohyet (Table 4.14). From 1980 to 1984 southward movement of the desert ranged from 11 to 99 km yr^{-1} but it retreated northwards by 110 km in 1985 and 33 km in 1986. By 1990 the cumulative effect of such fluctuations was to move the southern border of the Sahara about 130 km south of its 1980 position and to increase the area of desert by approximately 7% (Tucker *et al.*, 1991). Cloudsley-Thompson (1977) reported that desert conditions occupied 9% of the Earth's surface in 1882 but had expanded to 23% by 1952. Estimates by Dregne (1983) placed the figure at about 31% of total land area or 47 million km^2. The areas least affected by desertification are the naturally barren deserts which are little used by humans (Figure 4.39), but the adjacent sparsely vegetated grazing lands have been severely affected. Exclusion of livestock from such regions would allow the vegetation to recover slowly (Bedoian, 1978; Hall *et al.*, 1964). However, until alternative economies can be developed even the most stringent agricultural policies will do little to curb the problem in regions like the Sahel, where national policies do not encourage conservation of resources (Thompson, 1977).

CHAPTER 5

Mediterranean ecosystems

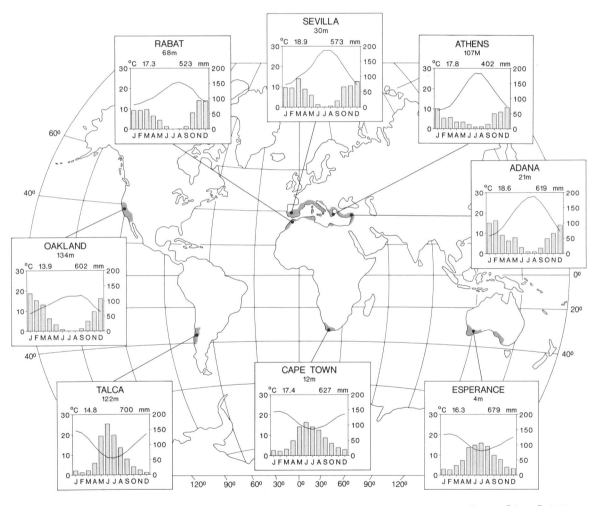

Figure 5.1 Distribution of mediterranean ecosystems and representative climatic conditions. Mean monthly temperatures are indicated by the line and mean precipitation for each month is shown by the bars. Station elevation, mean annual temperature and mean annual precipitation appear at the top of each climograph.

5.1 INTRODUCTION

The mediterranean ecosystems are dominated by evergreen shrubs and sclerophyll trees which have adapted to the distinctive climatic regime of sum-

mer drought and cool moist winters. Such climatic conditions occur in five widely disjunct regions along the subtropical western margins of the continents between 30° and 40° latitude (Figure 5.1). The total land area occupied by the mediterranean plant formations is only about 1.8 million km^2: more than half of this is found in a discontinuous belt around the Mediterranean Sea in southern Europe, the Levant and North Africa, and throughout the Mediterranean islands. Much of the woodland in these heavily populated regions has now been replaced by dense scrub formations known locally as '**maquis**', or by the more open heathy '**garrique**' comprised of shorter aromatic shrubs, such as lavender (*Lavandula pedunculata*) and thyme (*Thymus mastichina*). Intensive grazing by goats has created desert-like steppes in many areas and severe erosion problems have occurred on the rugged hillsides. In southern Africa the corresponding sclerophyllous shrublands and heathlands are restricted to the extreme south-western part of the country where they form an important component of the floristically rich Cape Flora. Here the heathy '**fynbos**' is distinguished from various shrubby '**veld**' types. Similar ecosystems occur in southern and south-western Australia where soils of low fertility support heathy forms of '**mallee**'. On more fertile soils a cover of grasses and herbs develops beneath the low-growing eucalypts. In California it is the '**chaparral**' that dominates the hillsides, with '**matorral**' forming analagous communities in central Chile. The global circulation pattern necessary for the distinctive dry summer climate has precluded any contact between these isolated regions. The similarity of form and functional response of the vegetation to the rigorous mediterranean environment is therefore a striking example of evolutionary convergence, and has resulted in a high degree of endemism within the regional floras.

5.2 CLIMATE

The dry summer climate of mediterranean regions arises from the seasonal change in position of the semi-permanent high pressure zones which are centred over the tropical deserts at about 20 ° latitude north and south of the equator. The persistent flow of stable air out of these high-pressure centres during the summer brings several months of hot, dry weather. Fire is a frequent hazard at this time of the year, particularly when strong winds, such as the Santa Anna of California and the sharav of Israel, begin to blow. Mean summer temperatures are about 25 °C but daytime maxima often exceed 35 °C (Figure 5.1). Cyclonic storms that develop along the polar front pass through the region during the winter, bringing relief from the summer drought. Winter temperatures typically average 10–12 °C, but there is a risk of frost as a result of radiation heat loss at night, cold air drainage from higher terrain, or because of occasional outbreaks of polar air. Frosts may occur on 5–10 nights each year at low elevation sites in the chaparral of California, increasing to 25 nights at higher sites (Hanes, 1981): similar conditions are reported by Kruger (1979) in the fynbos of South Africa. Absolute minimum

temperatures of − 18 °C have been reported for several stations in southern France, and can fall to − 25 °C in more mountainous districts of Turkey (Nahal, 1981). In the Köppen classification system, mediterranean (Cs) climates are restricted to areas where the mean temperature of the coldest month is between 0 and 18 °C, and the mean temperature of the warmest month exceeds 10 °C: a distinction is made between Csa and Csb climates if mean temperatures during the warmest month are above or below 22 °C respectively.

Although mediterranean areas are defined according to temperature conditions, it is the precipitation regime that is most distinctive. Annual precipitation is between 275 and 900 mm with at least 65% falling during the winter months (Aschmann, 1973a). Moisture recharge during the winter months must be sufficient to maintain a continuous vegetation cover despite the summer drought. Moisture effectiveness is determined by temperature and evaporation rates as well as the precipitation regime. A measure of drought severity is provided by the moisture index of Fitzpatrick and Nix (1970). The moisture index is the ratio of actual to potential evapotranspiration and it ranges from 1, when water is non-limiting, to 0 when available soil moisture has been exhausted; a value higher than 0.2 indicafes that there is sufficient soil moisture available for plant growth in mediterranean regions (Figure 5.2). Nahal (1981) similarly distinguished six bioclimates within the mediterranean climate zone based on the seasonal distribution of precipitation (Table 5.1).

5.3 SOILS

Throughout the mediterranean regions there is a recurring physiography of mountainous terrain, dissected tablelands and intervening plains. Alluvial fans and other landforms reminiscent of desert landscapes occur in some of the drier regions. Landslips and slumping are characteristic hazards on steep slopes when the soil layer becomes saturated by heavy rains in the winter. Erosion through rilling and gullying is also common, and river sediment loads are high during periods of heavy rainfall. The continual removal of weathered rock material has resulted in rather thin, infertile soils which are often deficient in nitrogen and phosphorus; profile development is also restricted by the seasonally dry climate and low plant productivity. These shallow, intensely weathered soils can be generally classed as **Xeralfs** (Foth and Schafer, 1980). Xeralfs have a high clay content, which results in poor structure and a tendency to become massive and hard when dry. The clay content increases with depth and often there is a distinct clay layer beneath the A horizon. The pH of the surface soils is near neutral but it becomes increasingly basic in reaction in the deeper horizons (Table 5.2). Base saturation is usually above 50% in the surface horizons and can

Table 5.1 Pluviometric regimes based on ranked seasonal precipitation totals in mediterranean areas (after Nahal, 1981)

	Pluviometric regime	Seasonal precipitation (mm)				
		Autumn(A)	Winter(W)	Spring(P)	Summer(S)	Total
Monaco	AWPS	278	202	182	87	749
Marseille	APWS	232	135	144	61	572
Istanbul	WAPS	170	239	140	58	607
Aleppo	WPAS	43	183	95	4	325
Marrakech	PWAS	64	81	85	12	242
Guercif	PAWS	51	47	77	17	192

(Source: I. Nahal, The mediterranean climate from a biological viewpoint, in *Ecosystems of the World*, *Vol. 11. Mediterranean-type Shrublands*, eds F. di Castri, D. W. Goodall and R. L. Specht; published by Elsevier Science Publishers, 1981.)

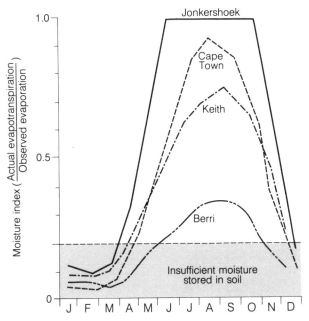

Figure 5.2 Mean monthly moisture indices for mediterranean-climate stations in South Africa and South Australia. (After Specht and Moll, 1983.) (Reproduced with permission from R. L. Specht and E. J. Moll, Mediterranean-type heathlands and sclerophyllous shrublands of the world: an overview, in *Ecological Studies 43 – Mediterranean-type Ecosystems: The Role of Nutrients*, eds F. J. Kruger, D. T. Mitchell and J. U. M. Jarvis, published by Springer- Verlag GmbH, 1983.)

exceed 90% at depth. However, the dominant exchange cations are calcium and magnesium, and soil fertility is generally quite low because the high pH reduces the availability of some nutrients. Phosphorus can be precipitated as insoluble calcium phosphate and many trace elements are unavailable (Specht and Moll, 1983).

The soils in mediterranean regions are greyish-brown or reddish-brown in colour depending on the nature of the parent material, the amount of organic matter present and their degree of development. Calcareous rocks often form '**terra rossa**' as they weather: these iron-stained, stony clay soils are particularly widespread in limestone areas of Yugoslavia (former) and southern Italy. Soil

conditions change consistently with topography, and the reddish soils are replaced by darker lithosols at higher elevations (Zinke, 1973). In California this catenary sequence corresponds to a change in vegetation from annual grasses and herbs at sites below 500 m through shrubland, oak woodland and coniferous forest. Organic carbon content increases in the woodland soils and their colour becomes browner. In some soils there is a similar increase in organic content with depth which suggests that they have developed under more humid conditions in the past (Thrower and Bradbury, 1973). The diverse topography and subsequent geomorphological activity has therefore produced mosaics of older and newer soils throughout the mediterranean regions.

5.4 REGIONAL FLORISTICS

5.4.1 The Mediterranean Basin

The mediterranean plant formations of Europe extend from Portugal and Spain through southern France, Italy, Greece and the Mediterranean islands into Turkey. Much of this area is now covered with scrub, sparse grass or bare rock, but the scattered occurrence of trees suggests

Figure 5.3 Remnant sweet chestnuts (*Castanea sativa*) dominate these wooded slopes in the mediterranean region of southern France, where the species was formerly an important hillside crop. (J. A. Aronson.)

Table 5.2 Nutrient content of surface soils in mediterranean regions (after Miller *et al.*, 1977; Rapp and Lossaint, 1981; Specht and Moll, 1983)

	pH	N (%)	P (%)	K	Ca	Mg
				(meq 100 g^{-1})		
California						
chaparral	6.8	< 0.001	< 0.001	0.2	7.4	1.5
coastal scrub	6.0	0.001	< 0.001	0.3	9.0	6.0
Chile						
matorral	6.3	< 0.001	0.001	0.2	7.6	0.2
coastal matorral	6.0	< 0.001	< 0.001	0.3	2.0	0.3
Southern Africa						
mountain fynbos	4.3	0.070	0.002	0.9	0.7	0.3
coastal fynbos	6.5	0.005	0.060	< 0.1	0.2	0.2
Australia						
heathy woodland	5.7	0.074	0.007	0.3	3.1	1.4
E. diversifolia scrub	8.4	0.078	0.020	1.4	12.3	3.5
E. socialis scrub	8.0	0.061	0.009	0.8	6.9	2.8
France						
Q. coccifera garrigue	7.9	0.34	0.030	0.7	57.6	1.4

(Sources: P.C. Miller *et al.*, Past and present environment, in *Convergent Evolution in Chile and California*, ed. H. A. Mooney, published by Dowden, Hutchinson & Ross, 1977; also R. L. Specht and E. J. Moll, Mediterranean-type heathlands and sclerophyllous shrublands of the world: an overview, in *Ecological Studies 43 – Mediterranean-type Ecosystems: The Role of Nutrients*, published by Springer-Verlag GmbH, 1983, adapted with permission; M. Rapp and P. Lossaint, Some aspects of mineral cycling in the garrigue of southern France, in *Ecosystems of the World Vol. 11. Mediterranean-type Shrublands* (eds F. di Castri, D. W. Goodall and R. L. Specht), published by Elsevier, 1981.)

Figure 5.4 (a) Leaves and fruit of Holm oak (*Quercus ilex*); (b) holm oak woodland artificially cleared of understory vegetation as a fire control measure; (c) through succession holm oak will typically replace Aleppo pine (*Pinus halepensis*) in mixed stands. ((a) R. A. Wright; (b), (c) J. A. Aronson.)

Figure 5.5 Cork oak (*Quercus suber*) woodland in Corsica. ((a) J. T. Price; (b) I. F. Spellerberg.)

that evergreen mixed forests were once dominant in the area (Figure 5.3). The holm oak (*Quercus ilex*), which ranges in size from a shrub to a large tree exceeding 20 m in height, is found throughout the Mediterranean Basin, although it is more abundant in the western districts (Figure 5.4). The cork oak (*Quercus suber*) may be locally dominant on non-calcareous soils (Figure 5.5). Often these two species grow in mixed stands in association with *Arbutus unedo* and various shrubs, but the understory has usually been removed to provide more open park-like stands for grazing. Clearing and overgrazing has reduced much of this cover to maquis scrub dominated by ericaceous shrubs (Debazac, 1983). Similar disturbance on calcareous soils has led to the replacement of holm oak by Kermes oak (*Q. coccifera*) from which the cochineal insect is gathered for its dye. Kermes oak is more tolerant of the arid conditions on these degraded porous soils. Stone pine (*Pinus pinea*), cluster pine (*P. pinaster*) and Aleppo pine (*P. halepensis*) become increasingly common at higher elevations in the western Mediterranean because of efficient seed dispersal and quick growth after fire (Figure 5.6).

The easternmost limit of the mediterranean formation is reached in the coastal areas of Syria, Lebanon and Israel where it merges with the arid lands of the Middle East. *Quercus calliprinos*, an evergreen oak which grows to 10 m, is common throughout this drier region, but deciduous oaks are more abundant here than in other parts of the formation. Conifers, such as the Corsican pine (*P. nigra*) and *P. brutia*, are widespread on burned sites. The cedars (*Cedrus libani*) that were once common in Lebanon and other parts of the eastern Mediterranean have all but disappeared. Desert vegetation extends across North Africa as far west as Tunisia where an arc of mediterranean shrub and woodland extends through northern Algeria and Morocco. The warm, moist conditions again favour *Q. ilex* which grows to elevations of 2000 m on the Atlas Mountains. Aleppo pine and fir (*Abies numidica*), which are also present on the lower slopes, are replaced by cedar (*Cedrus atlantica*) above about 1200 m, although most stands have disappeared because of centuries of exploitation (Thirgood, 1981).

The shrublands of the Mediterranean Basin can be divided into three physiognomic types according to height (Tomaselli, 1981). The maquis is comprised of evergreen shrubs and small trees over 2 m in height; mid-height shrubs (0.6–2 m) comprise the garrique on

Figure 5.6 (a) Aleppo pine (*Pinus halepensis*) growing in typical pasture on calcareous soil in southern France; (b) cluster pine (*Pinus pinaster*) on a rocky hillside in Crete. ((a) J. A. Aronson; (b) R. A. Wright.)

calcareous soils and '**jaral**' on siliceous soils, and the smallest forms (< 0.6 m) make up the '**phrygana**' or '**batha**' communities. The same species may be represented in all of these communities, the differences in size being determined by growing conditions. Wild olive (*Olea oleaster*) and carob (*Ceratonia siliqua*) are common species of the maquis: the densest stands form impenetrable thickets occasionally overtopped by oaks or more commonly by Aleppo pine. Characteristic species in garrique communities include the mastic tree (*Pistachia lentiscus*), the Grecian strawberry tree (*Arbutus andrachne*) and myrtle (*Myrtus communis*). They form a low cover with stunted *Q. coccifera* in association with smaller shrubs, including box (*Buxus sempervirens*), juniper (*Juniperus* sp.) and gorse (*Ulex* sp.), and aromatics such as lavender, thyme and rosemary (*Rosmarinus officinalis*). Species of *Erica* (heath) and *Cistus* (sun rose) are common understory plants on acid soils. Small shrubs such as the brooms (*Cytisus* and *Genista* spp.), provide a low-growing open cover in the phrygana communities (Figure 5.7). Although the shrubs dominate the landscape, there is also a rich herbaceous flora (Figure 5.8) in which annual species are particularly well represented (Table 5.3).

5.4.2 Africa

The mediterranean zone of southern Africa is restricted to the rugged terrain of folded sandstones that form the mountainous rim around the Cape Province. The various sclerophyll communities which comprise the dominant fynbos vegetation of this region can be generally subdivided into Mountain and Coastal types according to their geographic distribution. The **Mountain fynbos** occupies the wetter windward slopes of the mountains (Figure 5.9) but the distinct east–west climatic gradient, together with differences in soil conditions, elevation and aspect, result in a variety of plant associations. In lower sites the vegetation is composed principally of broad-leaved proteoid shrubs, which grow to a height of 1.5–2.5 m. They are found to elevations of 1000 m in more favourable sites (Taylor, 1978). In these communities the taller species of *Protea* form a discontinuous cover above a dense midlayer of shorter proteids and heathy ericoid forms and a ground layer of smaller shrubs and herbs. Most proteids are killed by fire, and frequent burning reduces the structural diversity of this cover. Above 1000 m smaller, ericoid shrubs with narrow rolled leaves become the dominant plant form. Their average height ranges from 0.2 to 1.5 m and many of them are aromatic. Low growing shrubs, 0.2–0.4 m tall, and tussocky hemicryptophytes form a lower stratum in these communities. At higher elevations and on exposed ridges and peaks the cover is comprised mainly of tussocky restioid shrubs which rarely exceed 0.3 m: grasses and some annuals may grow between the clumps of shrubs. Reduced precipitation on the leeward slopes of the

Figure 5.7 (a) Yellow broom (*Genista* sp.) is a common shrub of the eastern Mediterranean region; (b) low-growing spinescent shrub characteristic of phrygana communities. (R. A. Wright.)

Figure 5.8 Spring flowers on the island of Corfu attest to the rich herbaceous flora that is locally represented in some sites. (R. A. Wright.)

Table 5.3 Life-form spectra for various mediterranean shrub communities (Ph = phanerophyte; Ch = chamaephyte; H = hemicryptophyte; C = Cryptophyte; Th = therophyte)

	Number of species	Life forms (%)					Source
		Ph	Ch	H	C	Th	
California							
chaparral	44	41	16	18	11	14	Mooney *et al.*, 1977b
coastal scrub	65	17	19	20	3	41	
Chile							
matorral	108	24	14	20	6	36	
coastal matorral	109	12	15	17	9	46	
Southern Africa							
fynbos	448	34	31	16	15	4	Kruger and Bigalke, 1984
coastal renosterveld	127	12	14	19	45	10	Boucher and Moll, 1981
Australia							
mallee–broombush	288	57	19	9	8	7	Specht, 1981
E. diversifolia heath	274	56	19	10	7	8	
E. behriana–herb alliance	50	48	24	12	2	14	
Israel							
Quercus–Pistachia association	206	47	14	20	8	11	Orshan, 1983
Pinus–Juniperus association	73	8	37	20	15	20	
Arbutus–Helianthemum association	138	20	23	28	17	12	

Figure 5.9 Tall shrub fynbos on lower mountain slopes in southern Africa. (J. A. Aronson.)

tall form an open heath cover with smaller shrubs, grasses and annuals.

5.4.3 Australia

Mediterranean climates occur in the south-western coastal region of Australia and in parts of South Australia and western Victoria. Forests of karri (*Eucalyptus diversicolor*) or jarrah (*E. marginata*) occur in the wetter regions south of Perth in Western Australia (Figure 5.11). Karri grows in areas which receive more than 1100 mm of rain annually but is restricted to acid soils (Rossiter and Ozanne, 1970). This species, which can reach 80–85 m when mature, grows in association with other tall eucalypts, such as the tingle tree (*E. jacksoni*) and marri (*E. calophylla*). *Casuarina decussata* and species of *Banksia* are common in the understory of these wet sclerophyll forests. Jarrah forest occurs on detrital lateritic soils in areas which receive over 625 mm: the understory is similar to the karri forest, although fewer herbaceous species are present. These forests are replaced by wandoo (*E. rudunca*) woodland in regions where annual precipitation decreases to 500–625 mm. Elsewhere the coppiced mallee eucalypts become increasingly dominant, especially on light-textured alkaline soils, with sclerophyllous shrubs providing a heathy cover in less fertile sites (Figure 5.12).

mountains results in a rapid transition to the arid karoo vegetation in which succulent forms are more conspicuous. The area occupied by the **Coastal fynbos** is geologically much younger than the mountain districts. It also receives less precipitation. The sandy soils in the western coastal districts support an open ericoid cover with shrubs growing up to 1 m in height (Figure 5.10). Limestone outcrops along the south coast, and here proteids 1–2 m

Figure 5.10 Coastal fynbos near Cape of Good Hope, southern Africa. (D. R. Given.)

Figure 5.11 Karri (*Eucalyptus diversicolor*) forest in south-western Australia. (O. W. Archibold.)

Figure 5.12 *Banksia* scrub heath in Western Australia. (O. W. Archibold.)

Figure 5.13 The shrub-like form of *Acacia cavens* in espinal established in the subhumid region of Chile. The savanna-like appearance of these communities has resulted from the opening up of the matorral by fire, woodcutting and cattle grazing. The dispersion of nitrogen-fixing *A. cavens* is principally through cattle which consume the fruits. (C. Ovalle.)

The western region is separated from the mediterranean zone of South Australia and Victoria by the acacia shrubland of the arid Nullarbor Plain, although many of the mallee species occur sporadically across the dry southern part of the continent. Mallee is the dominant cover in the south-eastern mediterranean zone. Species such as *E. diversifolia* and *E. incrassata* grow 5–8 m tall on nutrient deficient soils and provide a rather open canopy over *Melaleuca uncinatea* and other mid-height shrubs and shorter sclerophyllous heathland species. In more favourable sites species such as *E. behriana* may grow to heights of 10 m above a ground cover of herbs and grasses with few sclerophyllous shrubs (Specht, 1981). These communities intergrade with sclerophyll forests of stringybark (*E. baxteri*) and messmate (*E. obliqua*) to the south, while woodlands dominated by yellow gum (*E. leucoxylon*) occur on the wetter slopes of the Flinders Range.

5.4.4 Chile

The matorral shrub communities of central Chile occur in the coastal lowlands and on the west-facing slopes of the Andes where they merge with alpine communities at about 2000 m elevation. The majority of matorral species are evergreen shrubs 1–3 m in height with small sclerophyllous leaves, although drought-deciduous shrubs and many spinescent stem-photosynthetic species are also important (Rundel, 1981a). Herbaceous perennials are abundant in the ground cover, including many species of bulbs, ferns and trailing and twining plants. Native annuals become increasingly important in the drier matorral and succulents are also conspicuous in these open stands. In moister sites the shrubs form a dense cover beneath *Salix chilensis, Cryptocarya alba* and other trees which form an overstory 10–15 m in height. Montane matorral replaces the typical form above 1200 m in the

Andes, and is occasionally present on the coastal mountains. Smaller evergreen shrubs form a comparatively open cover with numerous low-growing spinescent forms in these communities. In coastal regions the dominant evergreen shrubs of the matorral, such as *Lithaea caustica* and *Quillaja saponaria*, are replaced at low elevations by drought-deciduous species. *Fluorensia thurifera*, a shrub which grows 1 m in height, occurs widely on drier north-facing sites and, together with succulent species, is an important element in the arid transitional communities in the northern part of the mediterranean zone. At higher elevations the coastal mountains intercept fog and at about 1000 m the matorral is replaced by closed canopy forests of broad-leaved evergreen trees 12–15 m in height. Relict stands of *Nothofagus obliqua* are present in the highest parts. Further inland *Acacia caven* provides a scattered cover above the herbs and grasses of the savanna-like 'espinal' throughout the central valley (Figure 5.13).

5.4.5 North America

The North American chaparral communities occur on the lower slopes of the Sierra Nevada and hilly country throughout much of California (Figure 5.14). Chaparral typically forms a dense cover of evergreen shrubs 1–4 m in height. Chamise (*Adenostoma fasciculatum*) is common over much of the region and is particularly abundant in southern California, where it often forms nearly pure stands on warm south and west facing slopes at elevations of 300–1500 m. Although its small needle-like leaves produce a sparse foliage, few understory plants grow beneath the densely interwoven canopy (Figure 5.15). In coastal areas chamise is replaced by **coastal sage scrub** dominated by California sagebrush (*Artemisia californica*),

Figure 5.14 Oak woodland in the Sierra Nevada, California. (O. W. Archibold.)

an aromatic shrub which grows to about 1.5 m in height. Sagebrush and other subligneous sub-shrubs, such as *Salvia mellifera* and *Eriogonum fasciculatum*, comprise the **'soft chaparral'**. The drought-deciduous foliage on these

species readily distinguishes them from the sclerophyll habit of **'hard chaparral'** species such as chamise. Many of these coastal species are also adapted to the drier parts of the interior mountain ranges where chaparral is eventually replaced by desert communities.

The coastal and inland chaparral associations are separated by the Central Valley of California. In its natural state this low-lying area supported perennial grasslands, but most of this has been eliminated by overgrazing and has been replaced by introduced annual grasses. Chaparral occurs above about 500 m in the Sierra foothills which flank the Central Valley to the east. *Ceanothus cuneatus* (California lilac) may grow in association with chamise throughout its range but mixed stands become increasingly common at higher elevations. In southern California pure stands of *Ceanothus* are considered a fire-successional form, but it replaces chamise as the dominant species in the chaparral of northern California (Hanes, 1981). Similarly, manzanita (*Arctostaphylos* spp.) occurs throughout California but is particularly abundant in higher regions where winter temperatures drop below freezing and snow is common. Scrub oak (*Quercus dumosa*) forms shrubby stands in more mesic sites and mixed woodlands of *Q. agrifolia*, *Q. wilslizenii* and other evergreen oaks may be present on lower hillsides where moisture stress is less severe.

Figure 5.15 The densely branched habit and small leaves are characteristic of mediterranean shrubs: (a) chamise (*Adenostoma fasciculatum*); (b) manzanita (*Arctostaphylos manzanita*) in southern California. (C. Tracie.)

5.5 EVOLUTION OF THE MEDITERRANEAN FLORAS

Despite their limited geographic area, the mediterranean floras are unusually rich. Many widely distributed families are represented but a large number of the species are endemic to each location. Within the Mediterranean Basin local floras range from 1170 species on the island of Cyprus to over 6500 species in the Balkan Peninsula (Taylor, 1978). This compares with about 6000 species in the Cape Province of South Africa, where 212 of the 282 genera are endemic. About 2000 species are reported in central Chile (Rundel, 1981a). The lowest diversity occurs in the chaparral of California with only 900 species. Although the mediterranean floras have evolved in isolation, there is a remarkable physiognomic similarity within these widely separated regions. Global cooling which culminated in the Pleistocene glaciations was accompanied by progressively drier conditions, and it is generally agreed that the distinctive climate of mediterranean regions developed comparatively recently (Axelrod, 1973). Consequently, the present flora must have adapted very rapidly from plants which had evolved during the Tertiary. Many genera of woody plants, such as the oaks, are common to California and the Mediterranean Basin. The similarities between these two regions is attributed to the comparatively late separation of North America from Eurasia in the Late Cretaceous. Raven (1973) and Axelrod (1975) postulated that the northern mediterranean floras evolved from widely distributed ancestral groups in the cool temperate Arcto-Tertiary Geoflora and wet Neotropical Geoflora at least 40 million years ago. In the southern continents the mediterranean floras are mainly derived from tropical elements and were subsequently isolated by the break-up of Gondwanaland (Raven, 1973; Solbrig et al., 1977).

The established flora could have responded to the appearance of the mediterranean conditions at the end of the Pleistocene in several ways. The pollen record clearly indicates that many species were eliminated as the dry summer climate regime intensified. Other species may have evolved evergreen sclerophyllous leaves in response to water stress or low soil nutrient status, and so had been pre-adapted to conditions of aridity (di Castri, 1981). Axelrod and Raven (1978) have suggested that this was the case in southern Africa, where the post-Pleistocene climate of the Cape has favoured adaptive traits that evolved more than 30 million years ago. Species that were less well adapted may have become restricted to more favourable habitats. Centres of species richness in southern Africa are associated with upland areas, and many of the species may persist only because the precipitation regime is more reliable (Deacon, 1983). Thus, the mediterranean flora appear to represent the individualistic response of the component species rather than the migration of whole communities *en masse*

(Livingstone, 1975), and many of the annuals and herbaceous elements have evolved since the Pleistocene.

5.6 THE FORM AND HABIT OF MEDITERRANEAN PLANTS

5.6.1 Evergreen sclerophyllous shrubs

Evergreen sclerophyllous shrubs 1–3 m in height dominate the plant cover in all mediterranean regions. Their size and form is remarkably uniform in similar topographic situations. The evergreen shrubs have small, heavily cutinized leaves with a low surface to volume ratio: this is generally regarded as an adaptation to conditions of high light intensity, nitrogen deficiency or water deficit (Kummerow, 1973). These xeromorphic traits result from thickened cell walls, stronger mechanical tissue and increased venation. The stomata are generally enclosed in pits or grooves on the lower surface, the openings being further reduced by hairs or waxes (Grieve and Hellmuth, 1970). Although the number of stomata is unusually high – 1000 mm^{-2} compared with 100 mm^{-2} in succulents – they are comparatively small (1–2 μm) and stomatal diffusion occurs over less than 0.5% of the leaf area (Larcher, 1980). In some species there is an increased tendency towards vertically oriented leaves at higher elevations. This may reduce heat loads during periods of high sun angle when summer drought restricts evaporative cooling. It also permits more light to penetrate deeper into the canopy (Mooney et al., 1974). The stiff upright branches and smooth bark of the shrubs promote stem flow which tends to concentrate precipitation at the base of the plants (Specht, 1957). Deep-rooting species can draw on moisture reserves throughout the year and are less affected by the summer drought, but the shallow-rooted species must endure periods of severe water stress (Burk, 1978). This is reflected in various morphological and physiological traits. For example, shallow-rooted species of *Cerastes* and *Ceanothus* have thicker leaves and lower rates of transpiration, and close their stomata at lower water potentials than closely related deep-rooted species (Mooney and Miller, 1985).

Species such as chamise and scrub oak, which usually regenerate by sprouting, tend to be more deeply rooted than non-sprouting species (Kummerow and Mangan, 1981), although rooting depths are primarily controlled by soil conditions. The roots of chamise can penetrate 6–8 m in highly fractured substrates (Hellmers et al., 1955) but in shallow soils they grow down less than a metre. Regardless of soil conditions, the fine roots of all species are usually concentrated in the upper 20 cm zone (Kummerow et al., 1977), but in eucalypts and some other species maximum root development will often occur in the deeper, moister subsoil (Specht and Rayson, 1957b). The distribution of roots is often determined by the availability of nitrogen in the soil and they

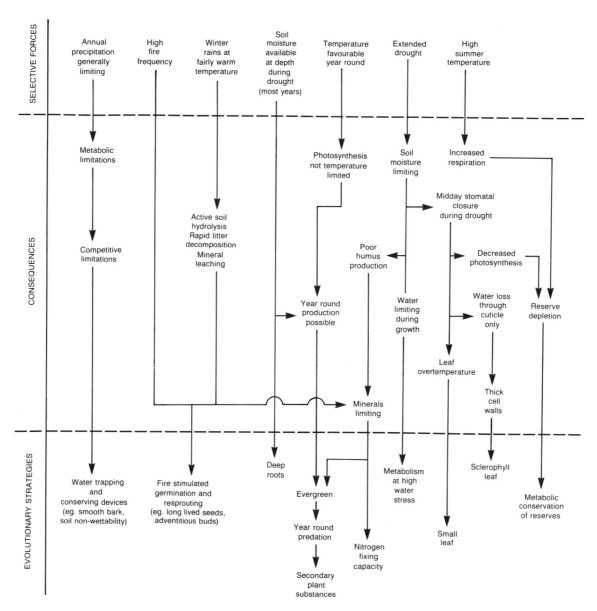

Figure 5.16 An evolutionary model for the mediterranean-climate shrub form. (After Mooney and Dunn, 1970.) (Reproduced with permission from H. A. Mooney and E. L. Dunn, Convergent evolution of mediterranean-climate evergreen sclerophyll shrubs, *Evolution*, 1970, **24**, 296.)

are particularly dense where organic matter has accumulated (Lamont, 1983). The average root:shoot ratio in Californian chaparral species is about 0.5 (Miller and Ng, 1977), compared with 1.5 in Chilean matorral (Hoffman and Kummerow, 1978) and 1.8 in some phrygana stands in Greece (Margaris, 1976). The radial spread of the roots is usually at least 2–3 times greater than the crowns. Consequently, there is considerable root overlap between adjacent plants. However, the bulk of the fine absorptive roots are found under the canopy where they utilize water which enters the soil as throughfall or stemflow, and take advantage of the higher nutrient concentrations which result from decomposition of the plant's own litter (Kummerow *et al.*, 1977). Uptake of

water and nutrients is enhanced by various root modifications, including abundant root hairs, dense clusters of proteid rootlets at the soil surface where they come in contact with water and decomposing litter, nitrogen-fixing symbionts and abundant mycorrhizas (Lamont, 1983).

The selective forces which favoured the evolution of the sclerophyll shrub form are discussed by Mooney and Dunn (1970). They suggest that growth is rarely limited by low winter temperatures or high summer temperatures in mediterranean climates. Consequently, evergreen leaves are potentially less costly than a deciduous strategy which limits photosynthesis to a shorter period (Figure 5.16). However, evergreen leaves must have some mechanism to minimize water stress during the summer drought period. Stomatal closure during the hottest part of the day will reduce water loss, but this also limits CO_2 uptake and the rate of photosynthesis declines. Smaller leaves have lower heat loads (Gates, 1980) and this benefits the shrubs by reducing the amount of water that is needed for transpirational cooling. Organic substances produced by many chaparral shrubs reduce the infiltration capacity of the soil but also conserve soil moisture by restricting evaporation losses. The infiltration capacity of the soil is highest at the base of the stem (Specht, 1957) and the configuration of the branches to promote stem flow therefore ensures that adequate water is available to the roots. The deficiency of nitrogen and other nutrients in many mediterranean soils could account for the large number of nitrogen-fixing species and mycorrhizal infections characteristic of these floras. Losses by leaching are reduced by the slow release of water-soluble nutrients from the evergreen canopy and by the slow rate of decay of the sclerophyll leaves. Finally, the mediterranean flora must be adapted to survive the intense fires which frequently start during the hot, dry summers. The dominant species are therefore those which are able to sprout vigorously after fire or have evolved various fire-dependent seed mechanisms.

5.6.2 Drought-deciduous and dimorphic shrubs

Although the evergreen shrub is the dominant life form of mediterranean regions, other types of plants can also be locally important. Woody deciduous species, perennial herbs and bulbs as well as succulents and annuals are all well represented. These forms are more conspicuous in the open stands which develop in drier habitats or in recently disturbed sites. Some species are drought-deciduous and shed all of the leaves produced during the winter; others reduce leaf activity by wilting or curling (Montenegro *et al.*, 1979) or by abscission of some of the foliage. The chaparral sub-shrub *Salvia mellifera* loses all but its terminal leaves during the drought period but can still maintain positive net photosynthesis except in extremely dry sites. The smaller sub-shrubs are mostly dimorphic chamaephytes which regulate water losses by seasonal modifications of their foliage (Orshan, 1972). Seasonally dimorphic shrubs grow large leaves during the winter and spring and replace them with smaller leaves during the summer. Typically the smaller leaves are produced on short shoots which develop from leaf axils along the main stems (Figure 5.17). The buds form in the spring and the leafy shoots are fully developed by

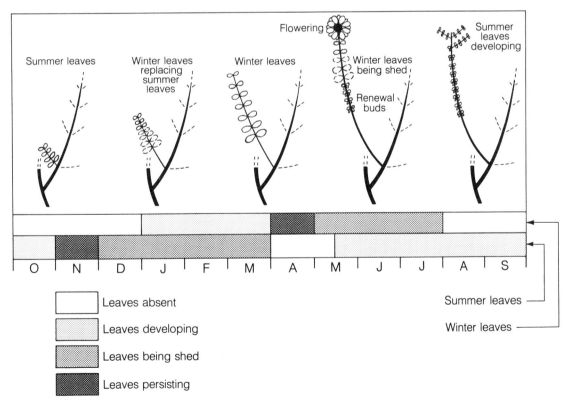

Figure 5.17 Seasonal dimorphism in the mediterranean chamaephyte *Teucrium polium*. (After Orshan, 1963.) (Redrawn and modified from G. Orshan, Seasonal dimorphism of desert and mediterranean chamaephytes and its significance as a factor in their water economy, in *The Water Relations of Plants*, eds A. J. Rutter and F. H. Whitehead; published by Blackwell Scientific Publications Ltd, 1963.)

summer. Normally only the leaves are replaced when the summer shoots eventually elongate into longer stems, but in some species the entire summer shoot is shed. The reduction of leaf area in this way limits transpirational water loss during the summer by 50–76% (Orshan, 1963). The chlorophyll content of the winter leaves is 3–4 times that in summer leaves, and rates of photosynthesis are correspondingly higher. Production is further enhanced because optimum temperatures for photosynthesis are lower in winter foliage (Margaris, 1977). Changes in day length are thought to trigger this exchange of leaves (Berliner and Orshan, 1971; Margaris, 1975). Seasonal dimorphism is characteristic of shrubs with mesophyll leaf morphology but is uncommon amongst sclerophylls (Westman, 1981). Evaporation of aromatic oils during hot weather can also restrict water loss in the sub-shrubs (Meidner and Sheriff, 1976), because they accumulate in and around the leaves and the movement of water vapour must overcome a higher diffusion resistance. Unlike the evergreen shrubs, the sub-shrubs have shallow fibrous root systems which rarely pentrate more than 0.75 m into the soil (Figure 5.18).

5.6.3 Annual and perennial herbs

Other life forms, including annuals and perennial herbs, are usually not conspicuous in mature shrub com- munities, although vines commonly grow into the shrub canopy. These trailing and twining plants may be relatively short-lived herbaceous species or woody forms similar to the grapevine (*Vitis vinifera*), which although widely grown throughout the mediterranean lands is native to north-west India and the Orient. Many of the herbaceous species and annuals are restricted to rocky slopes, or to recently disturbed sites where they appear in the spring for one or two years before being eliminated by the shrubs. Cryptophytes are particularly well adapted to the mediterranean climate. These plants survive the dry summer below ground as bulbs, corms or rhizomes, and produce leaves and flowers during the winter and spring; included in this group are many members of the iris, lily and amaryllis families. Many of the cryptophytes have contractile roots which pull the perennating organs into the ground, and so protect them from excessive heat and drought (Kummerow, 1981). The perennial herbs which persist beneath a shrub canopy normally go dormant in older stands, and are conspicuous only when they come into flower following fire. The post-fire flowering of 'fire lilies' in southern Africa may be a response to increased temperatures (Martin, 1966): in other species it has been attributed to higher light levels (Naveh, 1974; Stone, 1951), or to the release of nutrients in the ash (Erickson, 1965). The duration of a conspicuous understory of herbs and perennial grasses is dependent on the rate at which the shrubs re-establish.

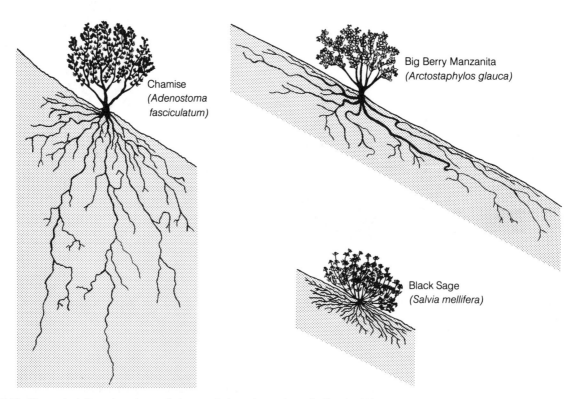

Figure 5.18 Characteristic root systems of chaparral plants in southern California. (After Hellmers *et al.*, 1955.) (Reprinted with permission from H. Hellmers, J. S. Horton, G. Juhren and J. O. O'Keefe, Root systems of some chaparral plants in southern California, *Ecology*, 1955, **36**, 671, 674.)

In chaparral most of the herbaceous plants are usually replaced within 4–5 years (Keeley *et al.*, 1981). Perennial herbs are much more important in the Chilean matorral where they can account for 40% of the ground cover in mature stands (Montenegro *et al.*, 1978): they are also prominent in the fynbos and mallee communities.

Annual plants are particularly well adapted to the seasonal regime of mild moist winters and dry summers, and often account for 40–50% of the species present in mediterranean regions (Raven, 1973). Like the herbaceous perennials, they too are most abundant in disturbed sites. The annuals are especially prominent in the first spring following a fire, but often disappear entirely from the understory within 2–3 seasons. This sequential change in density of annual plants is typical of the chaparral, maquis and more arid fynbos communities (Cody and Mooney, 1978). Some species have been termed **'pyrophyte endemics'** because they are present for only one season and must await another fire before re-establishing. They only survive because they have long-lived seeds, and germination is often stimulated by the presence of charred wood (Keeley *et al.*, 1985). The Chilean matorral is unusual in that there is no post-fire flush of annuals, and even the large component of herbaceous perennials declines significantly following a fire (Mooney *et al.*, 1977b). Annual plants are also conspicuous in gaps which persist amongst the shrubs; such species become increasingly abundant in drier sites and on rocky slopes. Other than being dependent on rainfall, they usually have no special requirements for germination, although subsequent establishment may be controlled by available microsites (Shmida and Whittaker, 1981).

5.7 WATER BALANCE

Plant communities in mediterranean regions are closely associated with moisture gradients. Evergreen forests are restricted to mesic environments and xerophytic scrub becomes predominant in areas where the drought period is more pronounced. In California and Chile the duration and intensity of the drought increases at lower latitudes and the forests are progressively replaced by evergreen shrubs and semi-arid scrub with a large component of drought-deciduous and succulent species (Mooney *et al.*, 1970). In these climates about 60 mm of precipitation must accumulate in the upper soil horizons before deep infiltration occurs, although the amount varies according to the frequency and intensity of precipitation events and the nature of the vegetation cover (Miller *et al.*, 1977). Consequently, deeper rooting chaparral and matorral shrubs are generally restricted to regions where precipitation exceeds 400 mm yr^{-1}, whereas shallow-rooted species, which utilize moisture from depths of 10–30 cm, predominate in regions where annual precipitation is about 250 mm. In southern California chaparral usually occurs above 300 m in response

Table 5.4 Annual water budgets in mixed chaparral on north-facing slopes and chamise chaparral on south-facing slopes in California (after Poole *et al.*, 1981)

	Mixed chaparral	Chamise chaparral
Annual precipitation (mm)	475	475
Interception (mm)	90	57
Soil evaporation (mm)	100	184
Transpiration (mm)	285	154
Subsurface drainage (mm)	0	15
Surface run-off (mm)	0	5

(Source: D. K. Poole, S. W. Roberts and P. C. Miller, Water utilisation, in *Ecological Studies 39 – Resource Use by Chaparral and Matorral*, ed. P. C. Miller; published by Springer-Verlag N.Y. Inc., 1981.)

to increased orographic precipitation (Harrison *et al.*, 1971). Species distributions are also related to topographic exposure. Mixed stands comprised of chamise, scrub oak and species of *Arctostaphylos* and *Ceanothus* occur on north-facing slopes in the chaparral zone. South-facing slopes are vegetated by pure stands of chamise which draw upon deeper water reserves and can also regulate transpirational losses better than comparable species (Miller and Poole, 1979). Despite their southern exposure, the drought conditions in stands of chamise are generally not as severe nor as prolonged as on north-facing slopes (Krause and Kummerow, 1977) because the cover is more open and less water is lost by transpiration (Table 5.4).

In addition to controlling the spatial distribution of the vegetation cover, moisture availability within the soil also regulates seasonal growth activity. As much as 90 mm of water can be retained in the upper 0.3 m of the soil, and several weeks will elapse before it infiltrates to the deeper horizons. The growth cycles of the various plant types are primarily a function of rooting depth and are therefore closely associated with the availability of moisture within the soil. Consequently, growth is usually quite synchronous between similar plant forms. The growth of annuals and shallow-rooted herbs begins in the fall, as soon as the first rains break the drought, and continues until late spring (Figure 5.19). Soil water is lost at a rate of about 2 mm a day in May and June and the surface layers quickly dry out. Growth is progressively later for deciduous shrubs and deep-rooting evergreen species, and continues until all available soil moisture has been used up (Poole and Miller, 1975). In particularly dry years water percolation may be insufficient for the growth of some deep-rooted shrubs (Harvey and Mooney, 1964).

There is no common adaptive strategy to minimize water loss in mediterranean plants, and growth is limited more by moisture conditions than by temperature. Shallow-rooted species, such as *Arctostaphylos glauca* and *Ceanothus greggii*, can reach very low xylem water potentials, but their stomata do little to regulate water loss (Figure 5.20). However, lower leaf conductance in *C. greggii* does reduce annual transpiration to 145–190 mm yr^{-1} m^{-2} leaf compared to 280–290 mm yr^{-1} m^{-2}

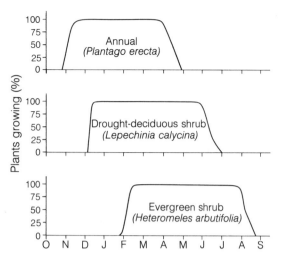

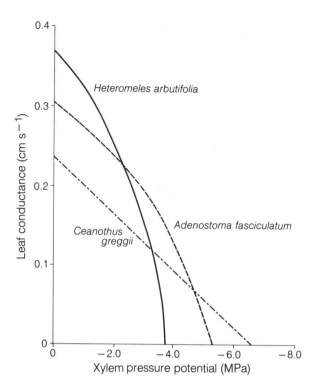

Figure 5.19 Growth cycles of various mediterranean-climate life forms in California as controlled by seasonal water availability at different rooting depths. (After Mooney, 1983.) (Reproduced with permission from H. A. Mooney, Carbon-gaining capacity and allocation patterns of mediterranean-climate plants, in *Ecological Studies 43 – Mediterranean-type Ecosystems: The Role of Nutrients*, eds F. J. Kurger, D. T. Mitchell and J. U. M. Jarvis, published by Springer-Verlag GmbH, 1983.)

Figure 5.20 Relation between leaf conductance and xylem pressure potential in selected chaparral shrubs. (After Miller and Poole, 1979.) (Reproduced by permission from P. C. Miller and D. K. Poole, Patterns of water use by shrubs in southern California, *Forest Science*, **25**(1), 84–98.)

leaf in *A. glauca* (Poole *et al.*, 1981). Xylem water potentials in deep-rooting species, such as *Heteromeles arbutifolia* and *Rhus ovata*, are usually not as low. Water loss is further reduced by stomatal closure at comparatively high potentials, and annual transpiration for *R. ovata* is only 127 mm $yr^{-1} m^{-2}$. Chamise exhibits an intermediate relationship between leaf conductance and xylem pressure potential, but its narrow leaves help to reduce water loss to 150–190 mm $yr^{-1} m^{-2}$.

Young plants are subject to much greater water stress than older individuals until their root systems are fully developed, and this could regulate the population density of some species (Schlesinger and Gill, 1980). In many chaparral plants the leaves and lower branches die and are shed as light levels become limiting beneath the dense canopy (Waisel *et al.*, 1972). The subsequent development of dead stripes along the stems reduces water loss to the healthiest and best illuminated branches, and in older shrubs the proportion of functional stem tissue may be relatively small (Davis, 1973).

5.8 STRATEGIES OF PHOTOSYNTHESIS

The mild mediterranean winters enable many evergreen shrubs to carry out photosynthesis year round, although the rate is periodically reduced by low temperatures or by drought (Mooney *et al.*, 1975). All of these shrubs use the C_3 pathway to fix carbon (Mooney *et al.*, 1974), this method being more efficient in climates where water is limiting during the hot period of the year. The absence of C_4 plants, even among drought-deciduous species,

suggests that this pathway has no adaptive advantage for plants in which most growth occurs in the cooler seasons. Although mediterranean species are C_3 plants, there are noticeable differences in photosynthetic rates between different growth forms. In terms of leaf area, the drought-deciduous shrubs have the highest rates of carbon uptake, averaging 12.2 mg $CO_2 dm^{-2} hr^{-1}$ for Californian and Chilean species (Oechel *et al.*, 1981). Similar values are reported for annual and perennial herbs, but this is approximately double the average rate of 6.5 mg $CO_2 dm^{-2} hr^{-1}$ reported for evergreen shrubs. The higher rate of photosynthesis in drought-deciduous species is accompanied by 2–3 times more water loss by transpiration (Harrison *et al.*, 1971). In these plants photosynthesis decreases significantly when the stomata close in response to drought stress, and eventually ceases when the leaves dry out and are shed. The evergreen species remain active for much longer periods. Economical water use and deep roots enable these species to continue growth for much of the year. Conversely, the drought-evading strategy of the summer-dormant species requires high assimilation rates and rapid growth during the spring.

The optimum temperature for photosynthesis in the evergreen shrubs is generally about 20 °C (Mooney, 1981) but ranges from 15 to 31 °C for common chaparral species (Oechel *et al.*, 1981). The highest value is associated with chamise and this could be an important adaptation which favours its establishment in pure stands on warm south-facing slopes. High photosynthetic rates are maintained over a wide range of temperatures for all species. In chamise, photosynthesis in excess of 85% of the maximum rate occurs over a temperature range of 27 °C. Consequently, the chaparral shrubs exhibit little temperature acclimation, and seasonal differences in photosynthetic rates are largely controlled by the effects of tissue water status on stomatal activity and the subsequent uptake of carbon dioxide (Mooney *et al.*, 1975). Similarly, the temperature optima are little affected by light intensity because the leaves become light-saturated at about 30% full sunlight.

A large fraction of the protein content of leaves is comprised of ribulose bisphosphate carboxylase, the principal carboxylating enzyme of plants (Björkman, 1968), hence the rate of photosynthesis can be directly related to levels of leaf nitrogen. In drought-deciduous species, leaf nitrogen content averages 3.1% compared with 0.9% for evergreen species (Mooney, 1981). Light saturation occurs at lower intensities when plants are grown in low-nitrogen environments (Medina, 1971) and is further reduced when plant growth is limited by water stress (Mooney and Gulmon, 1979). In the Chilean matorral the photosynthetic rate drops to about 15% of maximum during periods of drought, and annual uptake of carbon dioxide for these evergreen shrubs averages $18.7 \, g \, CO_2 \, dm^{-2} \, leaf \, yr^{-1}$ (Oechel *et al.*, 1981). This compares with $20.9 \, g \, CO_2 \, dm^{-2} \, leaf \, yr^{-1}$ for the drought-deciduous species which are active for about 6.5 months, and $14.0 \, g \, CO_2 \, dm^{-2} \, leaf \, yr^{-1}$ for the annual herbs whose growth is limited to only 4.5 months of the year.

5.9 SEASONAL GROWTH PATTERNS IN MEDITERRANEAN VEGETATION

Evergreen shrubs can remain active throughout the year, but there is a distinct annual growth rhythm because photosynthesis is limited by a variety of environmental and physiological constraints. The evergreen chaparral species *Ceanothus greggii* begins growth in early spring (Kummerow, 1983); in this species the optimum temperature for photosynthesis is 15 °C (Oechel *et al.*, 1981). Species with higher optimum temperatures initiate growth later in the spring. Growth can also be affected by diurnal temperature ranges: warm days (23 °C) and cool nights (4–10 °C) are particularly suitable (Hellmers and Ashby, 1958).

Different patterns of shoot development occur depending on whether new buds form at the beginning or end of the growth period (Kummerow, 1983). In the summer-deciduous matorral shrub *Trevoa trinervis* the leaves which develop during the winter rains arise from buds that formed late in the previous growing season. This ensures foliage is rapidly deployed as soon as conditions are favourable (Hoffman, 1972). A similar pattern occurs in shallow-rooted evergreen shrubs (Hoffman and Hoffman, 1976) but those with deeper root systems usually delay growth till later in the spring. In Australian communities the spring growth rhythm is generally restricted to annual cryptophytes and a few shrubs. The majority of the shrubs, trees and perennial grasses remain dormant until mean daily temperatures have risen to 18 °C and the main period of growth is during the summer (Specht and Rayson, 1957a). Here the dominant sclerophyllous shrubs, such as *Banksia ornata* and *B. marginata*, have a dual root system of deep taproots supplemented by an extensive system of shallow lateral roots which become non-functional as the surface soils dry out (Specht and Rayson, 1957b). Seasonal growth patterns in mediterranean regions are further complicated by differences in shoot and leaf longevity exhibited by the seasonally dimorphic species.

Unlike vegetative growth, where there is generally a pronounced seasonal rhythm, flowering and fruiting occur throughout the year with little synchrony between life forms. Comparative studies of chaparral and matorral species have demonstrated a variety of flowering regimes. Species such as *Arctostaphylos glauca* utilize stored carbohydrates to produce flowers and fruits before the period of maximum vegetative growth (Figure 5.21). In *Ceanothus greggii* flowering and fruiting coincide with the period of canopy development, but occur later in *Heteromeles arbutifolia* and chamise. In early-flowering species, flower buds begin to develop soon after leaf initiation and are fully formed by late spring. The buds then enter a period of dormancy until the next growing season. For these species flowering activity is closely linked to growing conditions in the previous year. Flowering is delayed in species such as chamise because the buds develop on the current year's wood after stem elongation is complete. In chaparral, flowering reaches a peak in late spring and fruiting occurs a few weeks later (Kummerow, 1983). Similar spring flowering peaks occur in the floras of the Mediterranean Basin, southern Africa and Australia. However, in the matorral flowering extends over the entire year with only a slight increase in the spring, although 70–75% of the population bears fruit during late summer and autumn.

Mooney *et al.* (1974) have related the phenology of the evergreen chaparral shrub *Heteromeles arbutifolia* to the basic functional demands needed for growth and reproduction. All plants must allocate their limited resources to produce, maintain or protect their growth and reproductive systems. Seasonal patterns of resource allocation develop because fluctuating environmental conditions regulate the cost:benefit ratio of these processes to the plant. In *H. arbutifolia* cambial activity commences in the winter and continues till early summer. This is followed by a period of stem elongation and increasing canopy biomass throughout the summer (Figure 5.22). Flowering and fruiting occur in late summer and fall, and as the above-ground demands decline more resources are

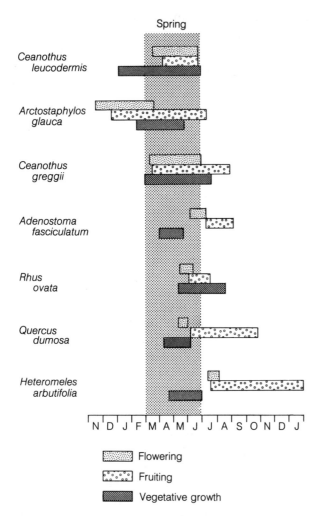

Figure 5.21 Periods of canopy development, flowering and fruiting for the dominant chaparral shrubs in California. (After Mooney *et al.*, 1977b.) (Redrawn from H. A. Mooney *et al.*, The producers – their resources and adaptive responses, in *Convergent Evolution in Chile and California*, ed. H. A. Mooney; published by Dowden, Hutchinson and Ross, 1977.)

diverted into root growth. Chemical analyses of leaves, stems and roots were used to determine the seasonal allocation of carbon for various functions (Mooney and Chu, 1974): phenolics would be used for protection, amino acids and proteins were assigned to metabolism, and starches and sugars were considered as storage compounds. In spring and summer carbon is used primarily in metabolism and structural development, presumably because of competitive demands for sunlight, but growth slows as water becomes limiting, and more carbon is diverted into storage and protection (Figure 5.23). Flowering in different chaparral species may be staggered over a 6-month period and is particularly late in *H. arbutifolia* (Mooney *et al.*, 1974). The mediterranean climate does not confine insect activity to brief periods of the year, and delayed flowering may be an evolutionary response to reduce competition for pollinators. The amount of carbon incorporated into protective com-

pounds in *H. arbutifolia* increases from 8% of total allocation in spring to 25% in winter (Mooney and Chu, 1974), nearly all of which is incorporated into the leaves. New leaves contain tannins and relatively high concentrations of cyanogenic glucosides (Dement and Mooney, 1974): these decrease over the summer then build up again in the fall. The immature fruits are similarly protected, but as they ripen the cyanogenic glucosides are transferred from the pulp to the seeds, and the mature fruits are eagerly consumed by birds.

5.10 BIOMASS AND ANNUAL PRODUCTIVITY

Estimates of above-ground biomass in mature chaparral range from 2.1 kg m^{-2} in a 27-year-old chamise community to 6.3 kg m^{-2} in a 21-year-old stand dominated by *Ceanothus megacarpus* (Rundel, 1983). A comparable value of 2.3 kg m^{-2} is reported for a 17-year-old garrique community comprised of shrubby *Quercus coccifera* in southern France (Mooney, 1981), but above-ground biomasses are lower in mattoral (0.7 kg m^{-2}) and fynbos (0.9 kg m^{-2}). In phryganic ecosystems in Greece above-ground biomass increases from about 0.5 kg m^{-2} in September to 1.2 kg m^{-2} in April. The contribution of leaves to total aerial biomass is about 19% in the spring, but this declines to 12% in the dry season when the smaller summer leaves appear (Margaris, 1976). Similar seasonal changes occur in drought-deciduous Californian coastal sage communities where above-ground biomass is about 0.9 kg m^{-2} when the plants are in leaf: leaves comprises 12% of this total but they are mostly shed in May and June (Gray, 1982). Leaves account for 8–14% of total biomass in chaparral shrubs and as much as 21% in matorral shrubs (Oechel and Lawrence, 1981). About 55% of the total biomass in these plants is comprised of stems, although this is reduced to only 26% in *Lithraea caustica*, and root biomass in this species is correspondingly higher.

Root biomass equivalent to 1.1 kg m^{-2} has been measured in chaparral, with values as high as 2.0 kg m^{-2} in matorral communities (Kummerow *et al.*, 1981). The roots account for 23–74% of the total standing crop of various evergreen shrubs, and consequently there is considerable variation in root:shoot ratios: these range from 0.3 for *Rhus ovata* to 2.9 in *Lithraea caustica* (Table 5.5). The highest root:shoot ratios of 6.3 occur in the Australian healthlands (Specht *et al.*, 1958) and have been attributed to the very infertile soils in this region (Westman, 1983).

The root mass fluctuates considerably during the course of the year as new rootlets develop in response to increases in soil moisture during the spring. In phrygana communities the root mass amounts to about 1 kg m^{-2} at the end of the summer drought but increases to about 1.8 kg m^{-2} at the end of the spring (Margaris, 1976).

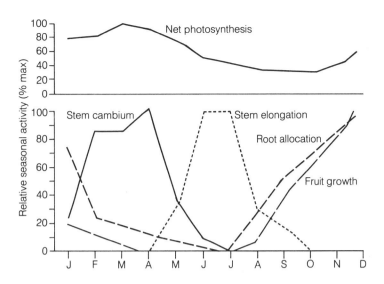

Figure 5.22 Seasonal carbon gain and carbon allocation in the Californian evergreen sclerophyll shrub *Heteromeles arbutifolia*. (After Mooney *et al.*, 1977b.) (Reproduced from H. A. Mooney *et al.*, The producers – their resources and adaptive responses, in *Convergent Evolution in Chile and California*, ed. H. A. Mooney; published by Dowden, Hutchinson and Ross, 1977.)

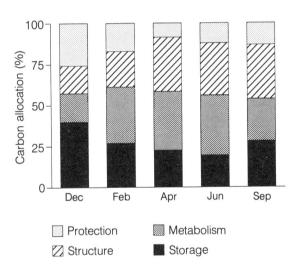

Figure 5.23 Seasonal allocation of carbon to various functions in the Californian evergreen sclerophyll shrub *Heteromeles arbutifolia*. (After Mooney and Chu, 1974.) (Reproduced with permission from H. A. Mooney and C. Chu, Seasonal carbon allocation in *Heteromeles arbutifolia*, a California evergreen shrub, *Oecologia*, 1974, **14**, 301.)

Similar studies in chaparral have shown that most of the seasonal variation in root biomass results from the rapid production and decay of fine roots. Live and dead root biomasses are approximately equal during the winter and spring, with each component contributing less than 0.5 kg m^{-2}. A significant increase in root production occurs as moisture percolates to the rooting zone, but as the soil dries out the live root fraction decreases and a considerable amount of dead material accumulates until decomposition increases with the autumn rains (Figure 5.24). The fine roots – those less than 0.25 cm in diameter – account for about 9% of the total root biomass in mixed chaparral stands (Kummerow *et al.*, 1977) but this is equivalent to a length of 2.7 km m^{-2} (Kummerow *et al.*, 1981).

Typical productivities for chaparral species range from 2046 kg ha^{-1} yr^{-1} for *Ceanothus leucodermis* to 13 946 kg ha^{-1} yr^{-1} for *Arctostaphylos fasciculatum*, the dominant species over most of the formation (Mooney *et al.*, 1977b). Total net primary productivity varies according to the density and composition of the plant cover and its age, and reported values range from 3620 kg ha^{-1} yr^{-1} (Mooney and Rundel, 1979) to 6709 kg ha^{-1} yr^{-1} (Mooney *et al.*, 1977b). Litterfall must be subtracted from these values in order to establish the rate at which the above-ground biomass accumulates. Annual increments ranging from 896 to 3983 kg ha^{-1} yr^{-1} are reported for various mediterranean ecosystems. About one-third of the available carbon assimilated each year by chaparral and matorral shrubs is allocated to new growth, and the remainder is used in respiration (Table 5.6). The new growth is incorporated in different ways. In *Adenostoma fasciculatum* the greatest proportion is added to the stems, whereas *Rhus ovata* incorporates more in foliage and in *C. odorifera* most goes into new root development. The annual carbon budget varies considerably between species and ranges from 556 g dry weight m^{-2} yr^{-1} in *Trevoa trinervis* to 1843 g dry weight m^{-2} yr^{-1} in *C. odorifera* (Oechel and Lawrence, 1981). Although *C. odorifera* allocates the greatest amount of carbon to new growth each year, this species represents only 4.2% of the cover in the sites that were sampled. Consequently, it contributes relatively little to

Table 5.5 Above- and below-ground biomass of dominant species in chaparral and matorral communities (after Shaver, 1983)

	Above-ground biomass (g m^{-2})	Below-ground biomass (g m^{-2})	Root:shoot ratio
chaparral species			
Ceanothus greggii	859.2	251.8	0.3
Arctostaphylos glauca	168.0	151.2	0.9
Adenostoma fasciculatum	901.1	540.7	0.6
matorral species			
Lithraea caustica	57.1	165.6	2.9
Colliguaya odorifera	58.2	34.9	0.6
Satureja gilliesii	60.7	42.5	0.7
Cryptocarya alba	284.8	398.7	1.4

(Reproduced with permission from G. R. Shaver, Mineral nutrient and nonstructural carbon pools in shrubs for mediterraean-type ecosystems of California and Chile, in *Ecological Studies 43 – Mediterranean-type Ecosystems: The Role of Nutrients*, eds F. J. Kruger, D. T. Mitchell and J. V. M. Jarvis, published by Springer-Verlag GmbH, 1983.)

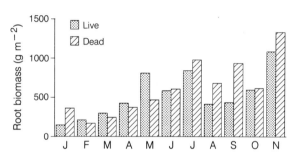

Figure 5.24 Seasonal course of live and dead fine-root dry biomass in a 23-year-old Californian chaparral stand dominated by *Adenostoma fasciculatum*, *Arctostaphylos glauca*, *Ceanothus greggii* and *Rhus ovata*. (After Kummerow, 1983.) (Reproduced with permission from J. Kummerow, Comparative phenology of mediterranean-type plant communities, in *Ecological Studies 43 – Mediterranean-type Ecosystems: The Role of Nutrients*, eds F. J. Kruger, D. T. Mitchell and J. U. M. Jarvis, published by Springer-Verlag GmbH, 1983.)

the annual above-ground productivity of the community (Mooney *et al.*, 1977b). Annual production in mediterranean shrubs can vary greatly between years (Table 5.7); however, the ratio of shrub biomass to shrub production is generally about 3:1. The allocation of new growth between the various terminal shoot components also remains fairly consistent, and approximately 70% of the annual production is incorporated into foliage, 20% into stems and 10% into reproductive structures (Table 5.8).

Productivity is initially quite high in young stands but usually declines as the plants age. In chaparral the mean annual increment of above-ground biomass typically decreases from about 2000 to 1000 kg ha^{-1} yr^{-1} in stands over 5 years old. A similar decline from 1140 to 620 kg ha^{-1} is reported in malee–broombush communities, and in garrique communities from 4200 to 3800 kg ha^{-1} (Specht, 1969). As the growth slows there is a corresponding decrease in the rate of nutrient accumulation (Figure 5.25).

5.11 NUTRIENT UPTAKE AND STORAGE

The soils of mediterranean regions are generally deficient in nutrients. Soil nutrient supply is particularly critical in the impoverished soils of southern Africa and Australia, and many of the plants have specialized mechanisms to maximize nutrient uptake (Lamont, 1983). Some species bear nitrogen-fixing nodules on their roots. Additional nitrogen, fixed by blue-green algae present in epiphytic lichens and in the nectar of some species of *Banksia*, could be washed into the soil by stem flow and throughfall (Lamont, 1980). Parasitic plants are prevalent. Their haustoria spread throughout the soil and attach to the roots of any plants they contact. In *Nytusia floribunda*, a species common to south-western Australia, the haustoria develop on rhizomes and can be more than 100 m in length; the plant itself can reach a height of 12 m which perhaps makes *Nytusia* the world's largest parasite (Lamont, 1983). Carnivorous plants are also abundant; indeed south-western Australia is the centre of distribution for the sundews (*Droseracea*) which trap insects with their sticky glands. However, some of these plants catch but do not digest their prey, and such activity is perhaps simply a way of discouraging herbivory (Lamont, 1983). Development of the fine, proteid root system which penetrates the decomposing litter is stimulated by phosphorus deficiency. Enzyme systems convert the orthophosphates released during decomposition into polyphosphates which are temporarily stored in the roots until they are required for growth later in the year. While none of these mechanisms is unique to mediterranean ecosystems, they are especially prevalent in these regions.

In Australia the widespread distribution of scleromorphic shrubs has been attributed to the low nutrient status of the old and deeply weathered soils, and various communities are differentiated on this basis (Specht and Moll, 1983). On strongly leached soils sclerophylls are present in both the overstory and understory: a ground cover of annual and perennial herbaceous species is found on

Table 5.6 Annual carbon budget for mediterranean plants in California and Chile (after Oechel and Lawrence, 1981)

	Chaparral species				Matorral species		
	New growth	Respiration	Allocation		New growth	Respiration	Allocation
	(g dry wt m^{-2} yr^{-1})		(%)		(g dry wt m^{-2} yr^{-1})		(%)
Arctostaphylos glauca				**Colliguaya odorifera**			
leaves	235	204	31		237	386	34
stems	74	418	34		85	362	24
main roots	22	146	12		38	197	13
absorbing roots	194	146	24		315	223	29
total allocation	525	914			675	1168	
Adenostoma fasciculatum				**Lithraea caustica**			
leaves	72	434	42		267	263	49
stems	190	198	32		48	74	11
main roots	94	108	17		70	154	20
absorbing roots	67	50	10		108	103	19
total allocation	423	790			493	594	
Ceanothus greggii				**Satureja gilliesii**			
leaves	181	151	31		125	161	45
stems	58	265	31		63	145	33
main roots	25	167	18		20	22	7
absorbing roots	108	103	20		48	49	15
total allocation	372	686			256	377	
Rhus ovata				**Trevoa trinervis**			
leaves	109	102	32		87	93	32
stems	64	120	28		61	144	37
main roots	31	72	15		41	59	18
absorbing roots	57	110	25		39	32	13
total allocation	261	404			228	328	

(Source: W. C. Oechel and W. Lawrence, Carbon allocation and utilization, in *Ecological Studies 39 – Resource Use by Chaparral and Matorral*, ed. P. C. Miller; published by Springer-Verlag N.Y. Inc., 1981.)

Table 5.7 Annual shoot production of chaparral and matorral species over three successive years (after Mooney *et al.*, 1977b)

	Maximum shoot weight (g m^{-2} yr^{-1})		
	1971	1972	1973
chaparral species			
Rhus ovata	614.2	77.6	1164.8
Ceanothus leucodermis	369.4	37.1	335.9
Ceanothus greggii	923.0	87.8	1057.0
Heteromeles arbutifolia	646.1	146.4	708.6
Arctostaphylos glauca	1950.1	673.6	1935.6
Adenostoma fasciculatum	273.5	15.4	463.2
Quercus dumosa	379.9	25.3	483.9
matorral species			
Lithraea caustica	403.2	657.1	627.4
Trevoa trinervis	449.2	325.7	443.7
Colliguaya odorifera	410.5	474.0	470.0
Satureja gilliesii	212.7	205.2	217.8
Cryptocarya alba	794.1	592.5	485.0
Quillaja saponaria	276.1	272.4	274.5

(Source: H. A. Mooney *et al.*, The producers – their resources and adaptive responses, in *Convergent Evolution in Chile and California*, ed. H. A. Mooney; published by Dowden, Hutchinson and Ross, 1977.)

moderately leached soils, and a stunted sclerophyllous cover develops on calcium-rich soils. These communities can be distinguished by the nutrient content of their leaves. Foliar nutrient levels are lowest in communities such as the South African fynbos and Australian mallee–broombush, which are associated with strongly leached soils, and increase in chaparral and matorral where the soils are generally more fertile (Figure 5.26). The native flora have evolved several nutrient-conserving mechanisms in response to very infertile soil conditions. The use of leaves and older tissues for nutrient storage during the winter season provides a mechanism for nutrient conservation (Mooney and Rundel, 1979). Without this, nutrients that are released by decomposition during the cool wet season would be leached from the ecosystem.

Tissue analyses of chaparral and matorral species show the highest concentrations of nutrients occur in the leaves, but the bulk of nutrients are associated with stems because they account for proportionately more of the biomass (Table 5.9). Nutrient uptake and storage vary seasonally. In chaparral species, nitrogen is taken up during the winter when growth is minimal: it is stored in old leaves and then translocated to the shoots when they begin their growth in the spring. Conversely, there is a small decline in stem nitrogen levels during the winter. Root nitrogen concentrations increase greatly over the summer. The net effect of these trends is a general increase in nitrogen storage each year (Shaver, 1983). Nitrogen concentrations in *C. greggii* may decline over the year, but this species has the ability to fix nitrogen and so is less dependent on seasonal storage. Changes in phosphorus concentrations over the year are

Table 5.8 Standing biomass and net production of dominant species in chaparral and matorral communities (after Mooney *et al.*, 1977b)

	Above-ground biomass (g m^{-2})	Shoot production (g m^{-2} yr^{-1})			
		Leaves	Stems	Floral parts	Total
chaparral species					
Rhus ovata	2117.1	707.3	135.6	184.9	1027.8
Ceanothus leucodermis	2164.5	142.3	62.3	0	204.6
Ceanothus greggii	3382.5	601.4	204.1	54.4	859.9
Heteromeles arbutifolia	3283.3	395.3	49.5	222.7	667.5
Arctostaphylos glauca	3360.4	1146.9	177.4	79.3	1394.6
Adenostoma fasciculatum	1746.2	126.4	65.1	349.5	541.0
Quercus dumosa	2045.6	369.9	47.9	5.5	423.3
matorral species					
Lithraea caustica	865.0	374.3	67.9	62.0	504.2
Trevoa trinervis	826.5	166.8	185.5	31.5	383.8
Colliguaya odorifera	1385.3	278.2	84.4	75.6	438.2
Satureja gilliesii	768.1	135.3	61.0	12.1	208.4
Cryptocarya alba	1695.3	417.6	66.0	0	483.6
Quillaja saponaria	1278.5	158.0	21.1	1.0	180.1

(Source: H. A. Mooney *et al.*, The producers – their resources and adaptive responses, in *Convergent Evolution in Chile and California*, ed. H. A. Mooney; published by Dowden, Hutchinson and Ross, 1977.)

less variable, but increase in the winter and spring, then decline over the summer. Potassium concentrations in leaves tend to increase over the spring and winter, but there is little change in stem concentrations and in the roots there is a general decline over the spring and summer (Shaver, 1981). Little seasonal variation occurs in the calcium and magnesium levels of chaparral shrubs. In the matorral, calcium levels in the leaves and stems decrease during periods of rapid growth and there is a corresponding decline in root calcium content over winter. Superimposed on these seasonal trends is a general decline in nutrient concentrations in the older tissues. However, calcium is exceptional in that it tends to accumulate with age.

The quantity of nutrients stored in the vegetation and litter of mediterranean ecosystems varies according to the nature of the cover (Table 5.10). The larger evergreen shrubs typically provide a denser cover than drought-deciduous species. For this reason the biomass and nutrient content of chaparral is considerably higher than in stands of coastal sage (Gray and Schlesinger, 1981). There is also considerable variation within similar vegetation types. Nutrient storage is particularly high in communities in which a significant proportion of the biomass is present as foliage. In most mediterranean species the leaves are shed soon after the period of shoot extension is completed, although in some species litterfall occurs throughout the year. In species such as *A. fasciculatum* the leaves can persist for about 2 years before being shed, although normally more than 70% of its foliage is produced each season (Jow *et al.*, 1980). Similarly, leaf longevity is about 1.5 years in *Ceanothus megacarpus* (Gray, 1982). Dead twigs and branches are generally retained, hence leaf fall contributes about 80–90% of the litter in most shrub communities. Exceptions to this in-

clude *A. fasciculatum* in which fruits and flowers may account for 35% of the annual litterfall, and *Quercus agrifolia* which sheds considerable amounts of bark (Mooney *et al.*, 1977b). Annual litterfall in mediterranean ecosystems ranges from 6 to 14% of the aerial biomass (Gray and Schlesinger, 1981). Some elements are reabsorbed from the leaves before they are shed (Table 5.11), but despite this about 10–15% of the nutrients stored above ground are returned to the soil each year (Table 5.12).

5.12 DECOMPOSITION OF ORGANIC MATTER

Litter decomposition in mediterranean ecosystems is limited by low temperatures during the winter and by inadequate soil moisture levels in the summer. Consequently there are two periods of intense mineralization. The first occurs late in the spring in response to increasing soil temperatures; the second is delayed until the start of the rainy period in the fall (Schaefer, 1973). The reabsorption of nitrogen from the leaves prior to abscission greatly increases the carbon:nitrogen ratio, which further limits the rate of microbial decomposition (Schlesinger and Hasey, 1981). Similarly, sclerophyllous leaves contain comparatively high concentrations of phenolics and lignins which are not only resistant to decomposition but may also bind other elements into resistant organic complexes (Gray and Schlesinger, 1981). Although Monk (1966) proposed that these characteristics might reduce leaching losses by slowing down the rate at which nutrients are released, the litter disappears fairly rapidly. Kittredge (1955) reported annual decomposition rates of 13–32% for chaparral litter, equivalent to a litter turnover time of between 3.1 and

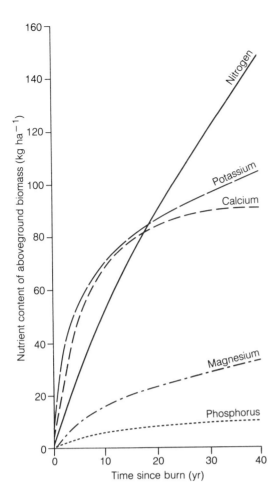

Figure 5.25 Nutrient uptake after fire in chaparral stands dominated by *Adenostoma fasciculatum* and *Ceanothus crassifolius*. (After Specht, 1969.) (Reproduced with permission from R. L. Specht, A comparison of the sclerophyllous vegetation characteristic of mediterranean type climates in France, California, and southern Australia II. Dry matter, energy and nutrient accumulation, *Australian Journal of Botany*, 1969, **17**, 303.)

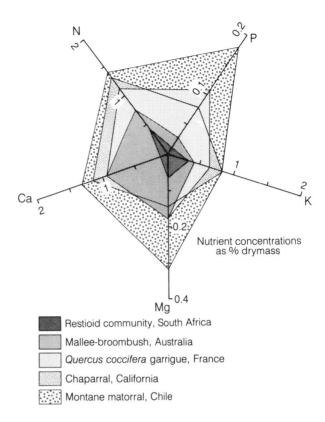

Restioid community, South Africa

Mallee-broombush, Australia

Quercus coccifera garrigue, France

Chaparral, California

Montane matorral, Chile

Figure 5.26 Foliar analysis (% dry mass) of representative mediterranean shrublands. (After Specht and Moll, 1983.) (Reproduced with permission from R. L. Specht and E. J. Moll, Mediterranean-type heathlands and sclerophyllous shrublands of the world: an overview, in *Ecological Studies 43 – Mediterranean-type Ecosystems: The Role of Nutrients*, eds F. J. Kruger, D. T. Mitchell and J. U. M. Jarvis, published by Springer-Verlag GmbH, 1983.)

7.8 years. However, the different elements are not released from the litter at the same rate. Potassium is lost fairly quickly, but calcium and magnesium are released more slowly. Phosphorus is generally lost quite rapidly, although the rate of release is very dependent on the composition of the litter (Schlesinger and Hasey, 1981). Nitrogen is retained in the litter for a relatively long period of time.

Nitrogen mineralization occurs when carbon:nitrogen ratios drop below 30:1, but ratios as high as 80:1 are reported for freshly fallen litter in some chaparral species, decreasing to 44:1 over a 12-month period (Schlesinger and Hasey, 1981). This suggests that nitrogen may not be released during the first year of decomposition, and in fact the nitrogen content of the litter can increase because of additions from precipitation and other sources. Different forms of nitrogen appear in the soil as a result

Table 5.9 Nutrient content of leaves, stems and roots of woody shrubs in chaparral and matorral communities (after Shaver, 1983)

	N	**P**	**K**	**Ca**	**Mg**
	(g m^{-2})				
chaparral species					
leaves	4.44	0.50	2.08	3.82	0.83
stems	13.82	2.08	8.71	9.27	1.89
roots	5.26	1.26	2.06	12.91	0.80
matorral species					
leaves	2.30	0.37	1.80	4.14	0.62
stems	4.15	0.85	4.04	8.29	0.64
roots	5.74	1.24	4.98	12.73	1.34

(Reproduced with permission from G. R. Shaver, Mineral nutrient and nonstructural carbon pools in shrubs from mediterranean-type ecosystems of California and Chile, in *Ecological Studies 43 – Mediterranean-type Ecosystems: The Role of Nutrients*, eds F. J. Kruger, D. T. Mitchell and J. U. M. Jarvis, published by Springer-Verlag GmbH, 1983.)

of the moisture requirements of the various microbial populations. Nitrogen fixing occurs when the soil is relatively moist, but nitrifying bacteria become more active as the soil begins to dry: ammonification occurs during

Table 5.10 Nutrient storage in above-ground biomass (including standing dead wood) and litter in mediterranean ecosystems (after Gray and Schlesinger, 1981)

	Age (yr)	Biomass (kg ha^{-1})	N	P	K	Ca	Mg	Litter (kg ha^{-1})	N	P	K	Ca	Mg
			Nutrients in biomass (kg ha^{-1})						Nutrients in litter (kg ha^{-1})				
California													
chamise chaparral	27	21270	135	19	–	–	–	4170	25	4	–	–	–
chamise chaparral	37	14000	83	17	60	56	6	–	–	–	–	–	–
redshank chaparral	25	30394	134	11	113	234	19	9550	147	22	174	465	172
Ceanothus chaparral	12	41380	264	20	116	225	29	16650	230	12	82	641	45
Ceanothus chaparral	22	75060	472	33	196	389	47	20270	205	6	47	261	67
coastal sage	22	14170	77	12	79	62	19	6167	47	4	18	89	31
France													
Quercus garrigue	13	49693	240	18	250	663	80	–	–	–	–	–	–
Quercus garrigue	17	23500	160	9	85	485	22	–	–	–	–	–	–
Australia													
Banksia scrub	28	50400	305	16	206	279	52	19000	178	7	20	119	31
heath scrub	25	27170	116	6	67	112	38	7410	42	1	3	47	14

(Reproduced with permission from J. T. Gray and W. H. Schlesinger, Nutrient cycling in mediterranean type ecosystems, in *Ecological Studies 39 – Resource Use by Chaparral and Matorral*, ed. P. C. Miller; published by Springer-Verlag N.Y. Inc., 1981.)

Table 5.11 Concentration of inorganic nutrients in fresh and abscissed foliage in chaparral plants (after Schlesinger and Hasey, 1981)

	N	P	K	Ca	Mg
	(% oven dry mass)				
Ceanothus megacarpus					
mature leaves	1.60	0.11	0.76	1.09	0.22
abscissed leaves	0.63	0.03	0.12	1.63	0.24
% change	– 61	– 75	– 84	+ 50	+ 9
Salvia mellifera					
mature leaves	0.83	0.09	1.48	1.42	0.12
abscissed leaves	0.65	0.13	1.35	2.21	0.32
% change	– 22	+ 49	– 9	+ 56	+ 167

(Reprinted with permission from W. H. Schlesinger and M. M. Hasey, Decomposition of chaparral shrub foliage: losses of organic and inorganic constituents from deciduous and evergreen leaves, *Ecology*, 1981, **62**, 764.)

the driest periods (Schaefer, 1973). Although ammonium accumulates in the dry soil, it is generally converted to nitrate and it is in this form that most of the nitrogen is taken up by the plants (Lossaint, 1973).

The bulk of the phosphorus available in the soil is also derived from the breakdown of litter, either through the activity of micro-organisms or through enzymes present on root surfaces (Cosgrove, 1967). All of the common forms of inorganic phosphorus in the soil are highly insoluble, but breakdown of rock phosphate by bacteria has been reported in the calcareous soils around the Mediterranean Basin (Azcon *et al.*, 1976). Mycorrhizal fungi which produce enzymes on their hyphal surfaces have a similar effect. Such activity is inversely proportional to phosphorus availability but widespread occurrence in mediterranean regions may be restricted by dry soil conditions, high concentrations of aluminium and insufficient levels of magnesium and potassium (Lamont, 1983). More commonly the mycorrhizal associations enhance mineral uptake by providing a direct connection between the roots and the site of mineralization, and this reduces the quantity of nutrients lost to leaching. Sclerophyllous shrubs respond vigorously to additions of fertilizer and growth is particularly stimulated by phosphorus. However, the new shoots retain their mesophytic characteristics longer than normal, and on fertilized soils the shrubs are much more susceptible to drought (Specht, 1973).

5.13 NUTRIENT BUDGETS IN MEDITERRANEAN ECOSYSTEMS

In chaparral communities dominated by *A. fasciculatum* about 80% of the nitrogen and 95% of the phosphorus is stored in the soil (Figure 5.27). The annual rate of nitrogen uptake is 3.4 g m^{-2}, of which nearly 90% is translocated to the above-ground tissues. The rate of uptake

Table 5.12 Annual litterfall and nutrient return in mediterranean ecosystems (after Gray and Schlesinger, 1981)

| | Age (yr) | Litterfall (kg ha^{-1} yr^{-1}) | Litter nutrient content | | | | |
| | | | N | P | K | Ca | Mg |
			(kg ha^{-1} yr^{-1})				
Chamise chaparral	27	830	3.8	0.5	–	–	–
Ceanothus chaparral	12	8126	89.3	10.5	34.5	106.3	16.1
Coastal sage	22	1940	15.0	1.4	11.2	25.4	7.8
Banksia scrub	28	5980	63.0	2.5	20.7	38.7	9.3
Quercus garrigue	17	2300	22.2	0.8	9.7	36.5	2.7

(Reproduced with permission from J. T. Gray and W. H. Schlesinger, Nutrient cycling in mediterranean type ecosystems, in *Ecological Studies 39 – Resource Use by Chaparral and Matorral*, ed. P. C. Miller; published by Springer-Verlag N.Y. Inc., 1981.)

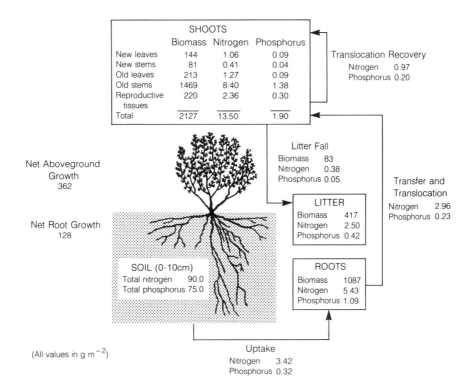

Figure 5.27 Distribution and flux of biomass, nitrogen and phosphorus in the evergreen shrub *Adenostoma fasciculatum* in the California chaparral. (After Mooney and Rundell, 1979.) (Reproduced by permission from H. A. Mooney and P. W. Rundel, Nutrient relations of the evergreen shrub, *Adenostoma fasciculatum*, in the California chaparral, *Botanical Gazette*, **140**, 109–13, published by the University of Chicago Press, 1979.)

of phosphorus is 0.3 g m^{-2} with approximately 70% subsequently incorporated into the aerial biomass. Litterfall returns 0.38 g m^{-2} of nitrogen and 0.05 g m^{-2} of phosphorus to the soil annually, and contributes to nutrient storage in the surface litter. The amount of nitrogen and phosphorus returned in litterfall represents less than 3% of total above-ground storage. This again demonstrates the importance of nutrient translocation out of the leaves prior to abscission. The amount of nitrogen required for new aerial growth each year in this community was estimated at 3.8 g m^{-2}, of which 2.9 g m^{-2} came from the soil and 0.9 g m^{-2} (24%) was supplied from tissue storage. The use of phosphorus stored in plant tissue was

considerably greater, amounting to 46% of the 0.4 g m^{-2} needed (Mooney and Rundel, 1979).

The rate of nitrogen uptake is considerably higher in phryganic communities. Annual uptake is approximately 7.5 g N m^{-2} yr^{-1} of which about 3.9 g m^{-2} is stored in the roots, stems and leaves. The remainder is mostly returned to soil in leaf litter or through root decay and is rapidly mineralized by the microbial population (Margaris, 1976). Nutrient budgets for evergreen oak (*Q. ilex*) woodland and a garrigue (*Q. coccifera*) community in southern France are provided by Lossaint (1973). Nutrient storage in above-ground biomass in the oak woodlands is approximately 570 g m^{-2}. Nitrogen accounts for

Table 5.13 Nutrient storage (g m^{-2}) and annual turnover (g m^{-2} yr^{-1}) in evergreen *Quercus ilex* woodland and garrigue (*Quercus coccifera*) shrub communities (after Lossaint, 1973)

	N	P	K	Ca	Mg
***Quercus ilex* woodland**					
Annual turnover (g m^{-2} yr^{-1})					
Uptake	4.6	0.6	5.1	12.6	0.9
Retention in biomass	1.3	0.3	0.9	4.3	0.2
Return in litterfall	3.3	0.3	1.6	6.4	0.5
Washed from canopy	< 0.1	0.1	2.6	1.9	0.2
Contributed in rainfall	1.5	0.1	0.2	1.0	0.1
Storage (g m^{-2})					
Accumulated in biomass	76.3	22.4	62.6	385.3	15.1
Litter reserves	12.4	0.4	1.0	36.1	0.9
Maximum available in soil	1200.0	40.0	120.0	3000.0	60.0
***Quercus coccifera* shrubland**					
Annual turnover (g m^{-2} yr^{-1})					
Uptake	2.9	0.1	1.3	6.2	0.4
Retention in biomass	0.7	< 0.1	0.3	2.6	0.1
Return in litterfall	2.2	< 0.1	1.0	3.7	0.3
Contributed in rainfall	1.5	0.1	0.2	1.1	0.2
Storage (g m^{-2})					
Accumulated in biomass	16.0	0.9	8.5	48.5	2.2
Maximum available in soil	700.0	6.0	70.0	3410.0	40.0

(Source: P. Lossaint, Water utilisation, in *Ecological Studies 7 – Mediterranean-type Ecosystems: Origin and Structure*, eds F. di Castri and H. A. Mooney; published by Springer-Verlag N.Y. Inc., 1973.)

about 14% of this total and phosphorus 4%, and most of the remainder is calcium (Table 5.13). Underground biomass contributes an additional 155 g m^{-2} of mineral mass. The annual input of nutrients to the soil through litterfall amounts to 12.4 g m^{-2}: in addition 4.9 g m^{-2} is washed out of the canopy and 2.9 g m^{-2} is added directly through precipitation. Most of these inputs are incoporated into the litter layer but 9.4 g m^{-2} is leached away each year. Total nutrient reserves in the litter are equivalent to 51.1 g m^{-2}. Estimates of available mineral reserves in the soil range from 2220 to 4080 g m^{-2}, which is 3–5 times the amount held in the biomass. Total nutrient storage in the garrique community amounts to 77.4 g m^{-2}. Annual nutrient uptake is approximately 10.9 g m^{-2}, of which 3.7 g m^{-2} is retained in the biomass and the remainder is returned through litterfall. Nutrient additions through precipitation are relatively high and this offsets the need for rapid mineralization of the litter. Approximately 80% of the nutrients stored above ground are in woody tissues and are normally released only when fire burns through the vegetation.

5.14 FIRE IN MEDITERRANEAN ECOSYSTEMS

The hot dry weather in the summer months makes mediterranean ecosystems highly susceptible to fire. The average fire frequency in chaparral is about 20–30 years, although stands 80–90 years old are not uncommon (Hanes, 1971; Keeley, 1977). Few fires exceed 400 ha in stands that are less than 20 years old, but as the stands get older the proportion of dead fuel increases and the fires spread rapidly. The largest fires, which can exceed 10 000 ha, usually occur in stands that are more than 30 years old (Wright and Bailey, 1982). Minnich (1983) has suggested that fires of this size arise because modern fire suppression methods allow more fuel to accumulate and that under natural conditions burning would occur more frequently. However, Kruger and Bigalke (1984) reported a natural fire frequency of 40 years in parts of the fynbos, and concluded that fires would have spread more extensively before artificial firebreaks were established. The average fire interval in the woodlands of southern France is 25 years, and in other parts of the Mediterranean Basin the vegetation burns about every 10 years (Le Houérou, 1974). This is similar to the 10–13-year cycle reported for bushfires in Australia (Christensen *et al.*, 1981). Fire occurs much less frequently in the matorral (Rundel, 1981a).

The relatively dense canopy of fine live and dead stems and the volatile oils in the leaves make mediterranean shrub communities very flammable and they burn fiercely. Temperatures of 700 °C have been measured in chaparral fires (Hanes, 1981) and can reach 850–1100 °C in garrique communities (Trabaud, 1981). Fire spreads quickly through these stands. In fynbos the rate of advance ranges from 0.2 to 2.0 m s^{-1}, and high temperatures persist for only a few seconds (Kruger and Bigalke, 1984). Soil temperatures are not greatly affected below depths of 5 cm and there is little damage to subterranean organs. Although material over 20 mm diameter is rarely consumed, the foliage and smaller stems burn readily which results in considerable changes in the distribution of

Table 5.14 Organic matter and nutrient content of plant, litter and soil before and after prescribed burning in chaparral (after DeBano and Conrad, 1978)

	Organic matter		Nitrogen		Phosphorus		Potassium		Calcium		Magnesium	
	before	after	before	after	before	after	before	after	before	after	before	after
					(kg ha^{-1})							
plants	30400	10300	134.3	33.0	10.3	0.8	113.3	23.7	233.7	86.7	18.8	4.4
litter	9550	5200	146.7	138.3	21.8	31.1	173.7	217.4	465.4	601.4	171.7	233.3
soil												
0–1 cm	637	517	181.7	160.0	63.1	56.7	35.1	34.7	676.8	654.3	26.6	26.5
1–2 cm	485	432	155.7	141.0	57.0	51.7	27.7	25.1	702.0	700.6	24.9	29.4

(Reprinted with permission from L. F. DeBano and C. E. Conrad, The effects of fire on nutrients in a chaparral ecosystem, *Ecology*, 1978, **59**, 493.)

nutrients within the ecosystem. DeBano and Conrad (1978) reported that the amount of nitrogen stored in plant tissue was reduced by 75% in burned chaparral, phosphorus by 69% and other elements by 38–79% (Table 5.14). Nitrogen is lost from the litter and surface soils by volatilization. Despite a net decrease in total nitrogen, the amount present as ammonium and nitrates generally increases in burned areas (DeBano *et al.*, 1979) and productivity can increase substantially, particularly when old stands are destroyed. Some potassium might also be carried off in smoke particles when fire temperatures exceed 500 °C (Rundel, 1983) and some of the nutrients in the ash are subsequently lost through erosion. Sediment yields in chaparral during the first year following a fire can be 30 times higher than on vegetated slopes (Rundel, 1981b). Movement of sediment during rainstorms is increased by the presence of water-repellent soil layers. These form when hydrophobic compounds derived from the litter volatilize during a fire and then condense around the soil particles a few centimetres below the surface. The depth and thickness of the water-repellent layer is affected by fire intensity (DeBano *et al.*, 1976). Nutrient losses are offset to some degree by luxury consumption in the post-fire shrub community (Rundel and Parsons, 1980) and by the growth of herbs.

5.15 ADAPTATIONS OF MEDITERRANEAN PLANTS TO FIRE

5.15.1 Vegetative regeneration

Native mediterranean plants are well adapted to fire, and frequent burning is necessary to maintain vigorous growth. When fire burns through the shrub communities it usually kills all aerial shoots, but in some trees new shoots develop from epicormic buds along the trunk and branches (Figure 5.28). Bark protection is common in the eucalypts and oaks. The bark layer is particularly well developed in the cork oak (*Q. suber*) of Spain and Portugal and is stripped off commercially every 8–10 years for the cork (Figure 5.29). The big-cone fir (*Pseudotsuga macrocarpa*) of California can also regenerate by budding (McDonald and Littrell, 1976) but most of the mediterranean conifers are killed by fire. Many of the shrubs regenerate vegetatively, but some are killed by fire and survive only by producing large numbers of seeds which lie dormant for long periods (Biswell, 1974). Chaparral species, such as *Q. dumosa* and *A. glandulosa* which are very resistant to fire, seldom establish through seedlings. Even though mature plants produce large seed crops, the seeds quickly lose their viability (Keeley, 1987a) or are easily killed by fire (Hanes, 1971). These plants survive because their buds are buried in the soil, where they are effectively insulated from lethal temperatures.

Sprouts arise from burls or lignotubers which develop as swellings in the root crown and become progressively larger and more deeply buried as the plant grows. In *Q. dumosa* and other chaparral species, lignotubers which survive for 250 years or more are often 1–2 m across (Hanes, 1965). Lignotubers are common in the flora of Australia (Gardner, 1957) and are particularly large in some mallee eucalypts (Mullette, 1978). New sprouts can also arise from buds on the roots: this is particularly common in the extensively branched root systems of many species in the Mediterranean Basin (Naveh, 1974), and in *Banksia elegans* in south-west Australia (Cowling *et al.*, 1990). The **obligate resprouters** start to regenerate immediately after a fire, even during periods of drought. They grow quickly and recover much of the ground within a few years, providing heavy competition for the dwarf shrubs and herbaceous species as well as for the shrub seedlings which develop.

Facultative resprouters usually do not sprout as vigorously as obligate resprouters, and seedlings will usually supplement vegetative development. Vegetative regeneration in facultative resprouters varies according to the age and vitality of the plants and the intensity and timing of the fire. In stands of *A. fasciculatum* 25% of the plants may fail to regenerate after a fire (Biswell, 1974). The older plants are most susceptible, particularly if burning occurs in the spring (Doman, 1968). Numerous seedlings are produced to compensate for these losses. Many of them die but ultimately there is little difference in the densities of the regrowth and pre-fire stands. The continual replacement of some of the older plants in this way rejuvenates the stand and maintains its vigourous growth (Biswell, 1974).

Although sprouting from the root crown is widespread in mediterranean shrubs, this is considered a

Figure 5.28 (a) Shoot regrowth from epicormic buds in fire-damaged eucalypts in New South Wales; (b) epicormic response in cork oak (*Quercus suber*) growing near Montpellier, France, 8 months after a fire. ((a) O. W. Archibold; (b) L. Trabaud.)

rather primitive and evolutionarily conservative trait. Plants which sprout from long-lived burls may successively occupy a site for a long period but this reduces selection within the population. In non-resprouting species each generation of seedlings is subject to rigorous selection pressures and greater differentiation has occurred in these plants. Of 195 species of evergreen sclerophylls in California, 105 are non-resprouting species of *Arctostaphylos* and *Ceanothus*. These genera

also include 28 species which reproduce vegetatively and have a very different response to fire (Wells, 1969). *A. glandulosa*, a common resprouting species, is rarely killed by fire but produces few seedlings, many of which die (Table 5.15). The non-resprouting species *A. glauca* is killed by fire and is maintained through a large seedling complement. Similarly, seed production in non-resprouting species of *Banksia* is higher than in resprouting species. Seed viability in *Banksia* also varies, ranging from 80–100% in non-resprouting species to 13–50% in sprouters (Cowling *et al.*, 1990).

5.15.2 Seed production and post-fire germination

The Aleppo pine of the Mediterranean regenerates only by seeding: the seeds are released from serotinous cones that are opened by the heat of the fire. Seedfall therefore coincides with the temporary removal of the dense shrub cover and the improved nutrient status of the ash-covered soil. Fire is needed to open the woody fruits which are retained on the branches of *Banksia ornata* in Australia. In other species complete opening is only achieved by repeated wetting and drying following heat treatment (Cowling and Lamont, 1985): this could ensure that seed dispersal ultimately coincides with the autumn rains. Although it is not essential for the release of seeds from the eucalypts and other associated shrubs,

Figure 5.29 Cork oak (*Quercus suber*) in Portugal soon after the thick bark has been harvested from the trunk and lower branches. (J. A. Aronson.)

Table 5.15 Cover of resprouting and non-resprouting chaparral species in burned and unburned stands and seedling density (after Keeley and Zedler, 1978)

Species (and reproductive type)	Stand condition	Cover ($m^2\,ha^{-1}$)	Seedling density (number ha^{-1})	Seedling mortality (% dead after 1 year)
Arctostaphylos glandulosa	unburned	316.0	0	–
(resprouter)	burned	109.0	550	71
Arctostaphylos glauca	unburned	16.4	0	–
(non-resprouter)	burned	17.9	10600	33
Adenostoma fasciculatum	unburned	251.0	0	–
(resprouter)	burned	107.0	14400	39
Ceanothus leucodermis	unburned	42.7	0	–
(resprouter)	burned	27.4	2500	6
Ceanothus greggii	unburned	27.6	0	–
(non-resprouter)	burned	3.9	10600	5
Quercus dumosa	unburned	9.0	0	–
(resprouter)	burned	0.2	0	–

(Source: J. E. Keeley and P. H. Zedler, Reproduction of chaparral shrubs after fire: a comparison of sprouting and seeding strategies, *The American Midland Naturalist*, Vol. 99(1), 1978.)

burning usually promotes mass dispersal in these plants (Gill, 1981). Seeds are rarely stored in the canopy of fynbos shrubs for longer than 5–6 years (Bond, 1985): there is also a progressive decline in seed storage as these stands become older (Bond, 1980). Most of the fynbos species have wind-dispersed seeds, but *Leucadendron platyspermum* is unusual in that fire causes the seeds to germinate in the woody fruits and dispersal is delayed until the seedling is forced out by the growing radicle (Williams, 1972). Germination in the canopy prior to dispersal can also occur in the scrub oak (*Q. dumosa*) of California (Keeley, 1987a).

Seed production begins at an early age in species which are killed by fire. Most chaparral species will bear seeds 3–5 years after a fire (Biswell, 1974). The median age in fynbos is 4 years, but longer periods elapse in slow-growing species that grow in habitats which afford some degree of protection from fire: *Protea magnifica*, which grows in rocky sites, flowers after 6 years, and *P. stokoei* may not flower for 12 years in moist montane sites (Kruger and Bigalke, 1984). Flowering and fruit development in perennial species varies from year to year and this results in considerable fluctuations in the number of seeds produced. Keeley (1977) reported annual seed productions of 0–850 m^{-2} for *A. glauca* and 0–8350 m^{-2} for *C. greggii*. These species produce dormant flower buds and there is a strong correlation between fruit production and precipitation in the previous year. However, there is no relationship between precipitation and seed production in species such as *Banksia*, which do not set dormant buds (Cowling *et al.*, 1987). In most chaparral species the fruits and seeds simply fall close to the parent plant, although in *Ceanothus* the seeds are explosively dispersed over distances of 1–3 m (Parker and Kelly, 1989). Mechanisms which assist in wind dispersal are more common in the proteaceous species of southern Africa, particularly those with serotinous fruits. In non-serotinous species subsequent dispersal is by ants (Slingsby and Bond, 1985). Myrmecochory is

also common in Australian sclerophylls (Berg, 1975), although heat penetration during a fire may be insufficient to stimulate germination if the ants bury the seeds too deeply (Auld, 1986a).

Seed storage in the canopy of species of *Banksia* ranges from 2 to 1340 seeds per plant. The highest values are associated with species, such as *B. leptophylla*, which are killed by fire (Cowling *et al.*, 1987). *B. leptophylla* can store seeds for up to 10 years: each year seed storage increases by about 8%, and over 60% of the seeds within the canopy remain viable. In weakly serotinous species, such as *B. menziesii*, 75% of the seed is replaced annually. Only 3% of the seed remains viable but this species regenerates primarily by resprouting. Canopy storage is common for most of the dominant mallee species (Wellington and Noble, 1985). Seed from woody plants may account for only 5% of the soil seed bank with the remainder coming from herbaceous species (Carroll and Ashton, 1965). A similar situation occurs in the fynbos. However, chaparral plants shed their seed, and long-term storage in the soil rather than in the canopy is the norm. The quantity of buried seed varies markedly between species, and there is little relationship between the size of the soil seed bank and the reproductive strategy of the plants. Amongst species of *Arctostaphylos* the seed banks of obligate seeders ranges from 300 to 28 200 seeds m^{-2} compared with 980–8400 seeds m^{-2} for resprouting species. Seed viability is somewhat lower for resprouting species (Parker and Kelly, 1989). Long-lived perennial species can continue to add seed to the soil until fire disturbs the ecosystem, but species which die out as the stand ages must produce seeds which remain viable for long periods. For many seeds 40 years or more may elapse before they have an opportunity to germinate (Zavitkovski and Newton, 1968).

Some species germinate only after the seeds are scarified by the heat of a fire. This cracks the seed coat and water can then be imbibed by the seed. This is the case for the acacias of southern Africa and Australia (Auld,

1986b; Pieterse and Cairns, 1986). In species of *Ceanothus* scarification by fire must be followed by a period of cold treatment before germination occurs (Quick and Quick, 1961). Numerous seedlings of fire-dependent species develop in the first growing season following a fire but recruitment rarely occurs after this. Factors other than high temperature requirements can also stimulate germination in these fire-prone environments. Many of the species found in the coastal sage scrub of California produce seeds that germinate readily, and hence they are poorly represented in the soil seed bank. Initial post-fire development in these species is usually from sprouts, which quickly come into flower so that seedling recruitment is delayed till the second season (Keeley and Keeley, 1984). Sometimes dormancy is imposed on those seeds which become buried in litter, and they will germinate as soon as the overlying material is burned off (Keeley, 1987a). In other species germination occurs when seeds are exposed to charred wood. The chemical cue is an oligosaccharide which is produced when hemicellulose is heated (Keeley and Pizzorno, 1986). Germination is usually delayed until the water-soluble chemical is leached from the char: the effect is lost if the wood is ashed in an intense fire. Enforced dormancy can also occur because of chemical compounds leached from a mature shrub canopy. The breakdown of these phyto-toxins during a fire probably accounts for the flush of herbaceous plants in recently burned sites (Christensen and Muller, 1975), although growth is also stimulated by increased light levels and improved nutrient conditions in the seedbed.

5.16 COMMUNITY PATTERNS FOLLOWING FIRE

5.16.1 Early post-fire herbaceous cover

The nature and changes in the vegetation cover following fire reflect the regenerative strategies of the various plants. Sprouting begins soon after the fire has burned through the scrub, and many of these plants will persist until the community is again disturbed by fire. Many burned sites are dominated by herbaceous plants which form a temporary cover for 3–4 years before the dense shrub cover re-establishes. This short-lived herbaceous stage is very pronounced in chaparral. Annual plants are particularly abundant in the first year and can account for about 90% of the herb cover at some sites: this declines to about 20–30% by the fourth year as the annuals are gradually replaced by perennial herbs (Keeley *et al.*, 1981). Some of these herbaceous plants may be present in localized sites prior to fire; others are restricted to burned areas. Annual species often persist in rocky sites, in natural openings in the chaparral, or along roadsides or other disturbed places (Horton and Kraebel, 1955). The seeds of these species are easily dispersed and will

usually germinate in the absence of fire (Keeley *et al.*, 1981). Similarly, some perennial herbs remain in a vegetative state in the low-light environment beneath the mature shrub canopy. They grow vigorously once the canopy is removed and most flower in the first year, although they may be rather sparsely distributed. In contrast, some annual and herbaceous species occur only on burned sites. Hanes (1977) refers to these plants as '**pyrophyte endemics**', because their appearance is totally dependent on fire. The '**fire-annuals**' are only abundant in the first post-fire season after seed dormancy has been broken (Keeley and Keeley, 1987). Because the seeds are not readily dispersed, they must endure a long period of dormancy until fire prepares the site once again. The '**fire-perennials**' are also absent beneath mature shrubs, and they too establish in burned sites from seedlings. Although refractory seeds are common in this group, high soil temperatures in exposed sunny sites may be adequate to break dormancy and some species may occasionally appear in cleared but unburned areas (Christensen and Muller, 1975).

5.16.2 Post-fire development of the shrub cover

Unlike the herbaceous cover, the floristic composition of the chaparral shrub cover changes very little after a fire, although the number of individuals of each species in the early post-fire community varies according to the different regeneration strategies. Seedling establishment in obligate resprouters such as *Q. dumosa* is normally successful only in exceptionally wet years. The seedling crop of sprouting species such as *A. glandulosa* is also comparatively small and suffers high mortality. Large numbers of *A. fasciculatum* seedlings appear following fire. This species produces two types of seeds: some require scarification by heat and others germinate as soon as they are mature (Stone and Juhren, 1953). However, seedlings are never present under a mature canopy. Facultative resprouters produce numerous seedlings, but many of them perish because of competition from faster growing sprouts or because of drought and poor root development (Schultz *et al.*, 1955). Species such as *A. fasciculatum*, which resprout vigorously and produce abundant seedlings, are typically best represented in early post-fire sites (Hanes, 1971). The non-resprouting species are necessarily prolific seeders but recruitment is negligible after the first year. This results in even-aged populations which are actively thinned through root competition and increased demands for moisture (Schlesinger and Gill, 1980).

5.16.3 Post-fire succession

Vegetation change in chaparral after fire differs from the normal concept of succession in that the shrubs which

Table 5.16 Survival of chaparral species after fire at Barranca, southern California (after Horton and Kraebel, 1955)

	Years after burning								
	2	3	4	5	7	10	15	20	25
	Plant survival (number 5m^{-2})								
Shrubs and subshrubs									
Adenostoma fasciculatum									
sprouts	2.5	2.5	2.5	2.5	2.5	2.5	2.5	2.5	2.5
seedlings	2.6	2.4	2.0	2.0	1.8	1.6	1.5	1.3	0.9
Ceanothus crassifolius									
seedlings	4.1	3.8	3.6	3.5	3.1	2.5	2.0	1.6	0.9
Erigonum fasciculatum									
seedlings	0.3	0.3	0.3	0.3	0.3	0.3	0.3	0.1	0.1
Salvia mellifera									
seedlings	3.1	3.1	3.0	3.0	2.9	2.8	2.8	2.6	2.0
Perennial herbs (seedlings)									
Convolvulus occidentalis	7.0	2.1	0.3	0.1	0	0	0	0	0
Dicentra chrysantha	0.8	0.8	0.3	0.1	0	0	0	0	0.2
Helianthus gracilentus	0.6	0.6	1.0	1.1	1.0	0.1	0	0	0
Lotus scoparius	3.4	3.0	1.8	1.0	1.6	0.5	0.1	0.1	0
Penstemon spectabilis	6.0	11.9	4.0	1.0	0.1	0	0	0	9.5
Senecio douglasii	0.3	0.4	0.4	1.3	0.5	0.3	0.3	0	0.1
Annual herbs (seedlings)									
Cryptantha intermedia	3.1	13.9	4.6	25.9	0.9	0.8	0.1	0	0.8
Emmenanthe penduliflora	1.0	0.3	0.5	0.1	0	0	0	0	0.6
Erigonum gracile	0.6	1.6	3.5	8.6	0.1	0	0	0	0
Erigonum thurberi	0	3.5	1.1	0.4	0	0	0	0	0
Lessinga glandulifera	0.1	4.0	2.1	9.9	0	0	0	0	0
Oenothera micrantha	0	2.3	3.6	12.4	0.4	1.5	0.4	0	6.6
Phacelia minor	13.5	0.9	0	0	0	0	0	0	0
Salvia columbariae	0.1	1.3	1.5	7.6	3.8	0.4	0.1	0	0
Stephanomeria virgata	23.4	24.6	5.8	6.1	0.1	0	0	0	0

(Source: J. S. Horton and C. J. Kraebel, Development of vegetation after fire in the chamise chaparral of southern California, *Ecology*, 1955, **36**, 251, 254, with permission.)

dominate the mature cover are present in the initial regrowth stands (Table 5.16). Herbaceous species and many short-lived shrubs add to the floristic diversity of recently burned sites, but rarely will new species appear once this initial community has established. Thereafter changes in the composition of the stand are the result of species elimination rather than species replacement (Hanes, 1971). Plant cover increases rapidly and a marked peak occurs 2–5 years after a fire, due mainly to the abundant growth of herbs. Plant cover declines as the herbaceous species die out but steadily increases again to a second peak a few years later. The cover then begins to open up due to the loss of the short-lived shrubs and senescence of the dominants.

McPherson and Muller (1969) suggest that the elimination of the herbs in mature chamise chaparral is caused by allelopathic suppression. Each year water-soluble toxins accumulate on the leaves of *A. fasciculatum* during the dry summer months. With the onset of the rainy period the toxin is washed into the soil, where it prevents germination of most seeds or causes inadequate root growth in more resistant species so that the seedlings ultimately fail. Production of the toxin ceases when the shrubs are consumed by fire, and natural degradation removes it from the soil. The herbs are present only for the few years that the shrub cover is reduced, and are

eventually suppressed as the output of toxins from the canopy increases. Christensen and Muller (1975) subsequently demonstrated that germination was reduced significantly when seeds were watered with rainwater that was collected from beneath mature *A. fasciculatum*. Although several toxic phenolic compounds have been identified in the soil beneath this species (McPherson *et al.*, 1971), Kaminsky (1981) concluded that they did not originate in the canopy but were produced by microbial activity in the litter.

The post-fire development of herbs and sub-shrubs is most striking in the lower elevation sites of southern California. Similar sequences are reported in the Mediterranean Basin (Naveh, 1974), in Australia (Specht *et al.*, 1958) and in southern Africa (Kruger and Bigalke, 1984). Conversely, in Chile the herbs are more prominent in the mature cover, and tend to decrease in abundance in recently burned sites (Table 5.17). The notable absence of 'fire-annuals' in this flora could have resulted from loss of seed viability during the long fire-free periods which occur in this region. In the Mediterranean region, for example, the density and diversity of the annual flora is greatest when fires occur at least every 10 years (Le Houérou, 1981).

It is apparent that fire is necessary for maintaining productivity in all mediterranean ecosystems. There is

Table 5.17 Herb and shrub cover in pre-fire and 1-year post-fire sites in Chile and California (after Keeley and Johnson, 1977)

	Chilean sites		Californian sites	
	pre-fire	1-yr post-fire	pre-fire	1-yr post-fire
number of herbaceous perennial species	19	3	4	14
number of herbaceous annual species	6	3	4	3
herb cover ($m^2\,ha^{-1}$)	6470	698	104	11496
number of shrub species	9	7	4	4
shrub cover ($m^2\,ha^{-1}$)	8838	1807	8270	4856

(Source: S. C. Keeley and A. W. Johnson, A comparison of the pattern of herb and shrub growth in comparable sites in Chile and California, *The American Midland Naturalist*, **97**(1), 1977.)

little growth in old stands and openings begin to develop in the cover as some of the shrubs die. Some sprouts may arise from adjacent plants but there is rarely any seedling development. Hanes (1981) considered chaparral to be a fire-dependent vegetation type, because it cannot replace itself indefinitely without fire and therefore should not be regarded as a true climax vegetation type. Without burning it is assumed that the environment becomes unsuitable not only for the shrubs themselves but also for other species that might succeed them. However, some species that establish from seed may continue to dominate a site for very long periods (Keeley and Zedler, 1978), while in some mesic sites chaparral may be replaced eventually by a low oak woodland (Patric and Hanes, 1964). Establishment of tree species has also been described in senescent fynbos (van Wilgen, 1981) and in coastal heathland in Australia (del Moral *et al.*, 1978). However, the highly combustible nature of the vegetation normally ensures frequent disturbance by fire, and under these conditions the dominant fire-dependent species have a selective advantage (Mutch, 1970).

5.17 PLANT–ANIMAL RELATIONSHIPS IN MEDITERRANEAN ECOSYSTEMS

5.17.1 Seed predation

Chaparral shrubs produce seeds each year but for most species there appears to be no relationship between seed production and the number which remain in the soil (Keeley, 1977). Keeley (1987b) reported no significant change in the size of the seed banks of *Arctostaphylos glauca* and *A. glandulosa* over a 10-year period despite considerable fruiting activity. Little germination occurs during inter-fire periods and losses through senescence and decay are negligible. In these communities seed bank densities are largely controlled by predation and most of the loss occurs within the first year after dispersal (Kelly and Parker, 1990). As much as 80% of the seed placed in artificial caches is consumed within 10 days (Keeley and Hays, 1976). Most of the larger seeds from species such as *A. glauca* are lost to rodents. In mallee eucalypts practically all of the newly fallen seed is removed by ants within a few days (Wellington and Noble, 1985) but

many of the larger seeds are lost to vertebrate predators (Abbott and van Heurck, 1985). Fruits and seeds are also exploited by birds. Schodde (1981) noted five species of frugivorous birds in Australian mallee communities, including the omnivorous emu (*Dromiceius casuarius*) and the highly specialized honey-eater (*Grantiella picta*) and flower-pecker (*Dicaeum hirundinaceum*) which feed on the fruits of mistletoes. Parrots and cockatoos are the most common granivores. They are primarily ground feeders, although some break open fruits in the canopy. Izhaki and Safriel (1990) have shown that germination of seeds of species such as the Palestinian pistachio (*Pistacia palaestina*) is dependent on their consumption by birds.

5.17.2 Herbivory

A large number of seedlings which emerge on recently burned sites are lost to herbivorous mammals (Mills, 1986), and grazing can also limit seedling establishment in openings which occur as the shrubs become senescent (Swank and Oechel, 1991). Young shoots are subject to heavy browsing (Naveh, 1975) but mature sclerophyll leaves are rather tough, have low nutrient and water contents, and contain large concentrations of secondary compounds such as tannins which generally render them unattractive to herbivores. However, during periods of leaf initiation few resources can be spared to deter herbivory, and some of the young foliage will be lost until secondary compounds accumulate (Dement and Mooney, 1974). Herbivory is reduced in some species because leaf initiation is not synchronized with peak insect abundance. In matorral the period of maximum insect activity occurs in the warm wet period from September to December. By this time leaf growth in the drought-deciduous species is coming to an end, and new leaf development has not yet begun in the evergreen shrubs. Consequently, the proportion of leaf area lost to defoliating insects is comparatively small, averaging 2–7% for different matorral communities (Fuentes and Etchegaray, 1983). Foliage losses in some species can be much higher than this, but vigorous vegetative growth will compensate for excessive defoliation (Torres *et al.*, 1980). In Australian eucalypts, as much as 45% of the leaf area is removed (Morrow, 1983). Such a high rate of

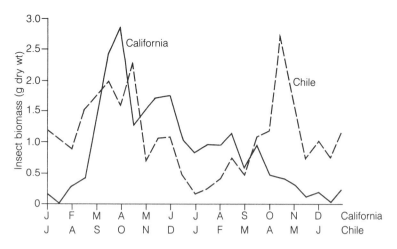

Figure 5.30 Seasonal distribution of insect biomass in mediterranean shrublands in California and Chile. (After Cody *et al.*, 1977.) (Reproduced from M. L. Cody *et al.*, Convergent evolution in the consumer organisms of mediterranean Chile and California, in *Convergent Evolution in Chile and California*, ed. H. A. Mooney; published by Dowden, Hutchinson and Ross, 1977.)

defoliation is contrary to the idea that sclerophyllous leaves are not subject to heavy loss through herbivory. However, the Australian sclerophylls can develop new leaves at any time of the year. This continual replacement of foliage allows several generations of insects to develop each year, hence the unusually high level of leaf consumption.

The activity of the insect defoliators is closely linked to the temporary palatabilty of the young leaves, and so their numbers are low during the dry summer months. Other species will forage for nectar, fruit and seeds as they become available. In California the pronounced spring peak in the insect population coincides with the period of maximum leaf initiation, flowering and fruiting in chaparral, and this is followed by a steady decrease in numbers with little activity occurring in the cool winter months (Figure 5.30). Plant activity is more protracted in Chile and large numbers of insects are also present during the fall when many plants are in flower. The majority of birds in the sclerophyll scrub are insectivores, so their populations fluctuate with that of the insects. Bird densities are higher in the matorral largely because the ground-feeding insectivores are more numerous (Cody *et al.*, 1977). This reflects the greater diversity of foraging sites afforded by the more open and patchy vegetation cover as well as the greater abundance of insects.

In addition to the defoliating insects the plants are visited by various floral herbivores. Small bees and beetles are especially abundant in the chaparral but comparatively few species are effective pollinators. Some are generalist, opportunistic feeders which visit different plant species as they come into flower over the course of the year. Others are very specialized and restrict their activities to one or two closely related species. About 10% of the matorral and chaparral plant species are pollinated by hummingbirds, sphingid moths and large bees. In cooler coastal regions wind pollination becomes more important (Cody *et al.*, 1977).

5.17.3 The effect of fire on animal populations

The fire season coincides with the pupation period for some species of insects in the fynbos (Kruger and Bigalke, 1984) and, being buried in the soil, they escape destruction. Litter-dwelling insects are very susceptible to fire during resting periods but the effects are soon overcome if the fire occurs in the spring, when the population is actively reproducing (Campbell and Tanton, 1981). High mobility and high fecundity are probably the main strategies by which insect populations re-establish after a fire.

Many animals escape bushfires by hiding in burrows or by jumping through the flames (Main, 1981), and they often move away from the burned areas because the habitat is no longer suitable for them. However, the abundance of herbaceous plants in early post-fire communities provides good forage for herbivores. A few small mammals return immediately to the burned fynbos and their populations reach maximum densities 3–4 years after the fire, when the herbaceous cover is most productive. The small mammal population declines as the shrubs re-establish but subsequently rises as the cover becomes senescent (Kruger and Bigalke, 1984). The order in which species appear is determined by habitat preference. Some species feed on the herbaceous ground cover, others consume young shrubby material. The insectivorous shrews must wait until sufficient litter has accumulated to support the invertebrates they feed on. The increased nutritive value and improved accessibility of young fynbos also attract antelopes and other large mammals. In Australia several species of mice (*Pseudomys* spp.) are regarded as fire specialists because they are essentially restricted to heathlands which are regenerating after fire (Braithewaite and Gullan, 1978). In these sites the number of small mammals generally peaks 5–8 years after a fire. The population of some species of kangaroos and wallabies also increases in recently burned sites (Christensen and Kimber, 1975), while others depend on recurring fire to maintain suitable habitat (Catling and Newsome, 1981).

The bird population usually declines after fire. Specialized feeders, such as the sugar birds (*Promerops cafer*) of southern Africa, are absent from burned sites for 4–8 years until their preferred source of nectar is once more available. Nest-site requirements also restrict this species to fairly dense regrowth (Kruger and Bigalke, 1984). The increased abundance of insects which occurs as the plant cover matures is accompanied by the return of the insectivorous birds.

5.18 THE HUMAN IMPACT

Although humans have occupied all of the mediterranean lands for a considerable time, the degree to which they have modified the environment varies considerably. Changes are most widespread in the Mediterranean region where agropastoral activities have been practised for millenia (Figure 5.31). By the end of the 19th century much of the native forest area had been cleared for the cultivation of cereals and legumes or cut for wood and charcoal. Sheep and goat pastures were maintained by repeated burning, and degradation of marginal areas led to the expansion of the scrub cover (Pignatti, 1983). As much as 75% of the agricultural area is used for grazing, although very little would be classed as pasture (Le Houérou, 1981). Centuries of overgrazing have reduced the quality of the forage. Goats are particularly destructive and they are now prohibited from some Mediterranean countries. In more developed regions, such as Spain and Italy, traditional agricultural activities are being replaced by high value fruit and vegetable crops and vineyards on irrigated land, and some of the abandoned farmland is now reverting to mixed oak woodland. However, livestock numbers are still increasing in developing regions like North Africa, and land degradation continues. Later settlement in other mediterranean regions and lower population pressures resulted in less land being utilized for agriculture. In these countries arable farming is restricted to the low-lying valleys, where many specialty crops are now produced under irrigation. Land not suitable for mechanized agriculture is generally maintained for grazing by regular fire management.

Pignatti (1979) has argued that the evergreen oak forests which covered the Mediterranean region before the appearance of humans provided few opportunities for speciation, but this changed as the forest was opened up

Figure 5.31 (a) Olive groves, such as this one in Corfu, are common throughout the eastern Mediterranean Basin; (b) *Quercus ilex* and *Q. suber* form an open canopy above a sown annual legume–grass cover in an agrosilvopastoral system called 'dehesa' practised in southern Spain. ((a) R. A. Wright; (b) R. Joffre.)

and new plant habitats became available. With time this would favour the diversification of the herbaceous annual flora. Naveh and Whittaker (1979) similarly noted that high species diversity occurred when grazing was combined with drought and fire in these physiographically heterogeneous regions. Whereas increased speciation as a result of long and intensive land use is associated with the Mediterranean Basin, this has not been the case in more recently settled regions. Many weedy annuals were introduced to these areas when domesticated livestock were brought from Europe. In Chile and California most of the native grasses and herbs have been replaced by Old World weeds that are better adapted to grazing pressures (Aschmann, 1973b). More recently, commercial timber species, such as Monterey pine (*Pinus radiata*) and eucalyptus, have become widely established. All of these recurring anthropogenic elements enhance the convergent characteristics that are clearly evident in the natural landscape in mediterranean regions.

CHAPTER 6

Temperate forest ecosystems

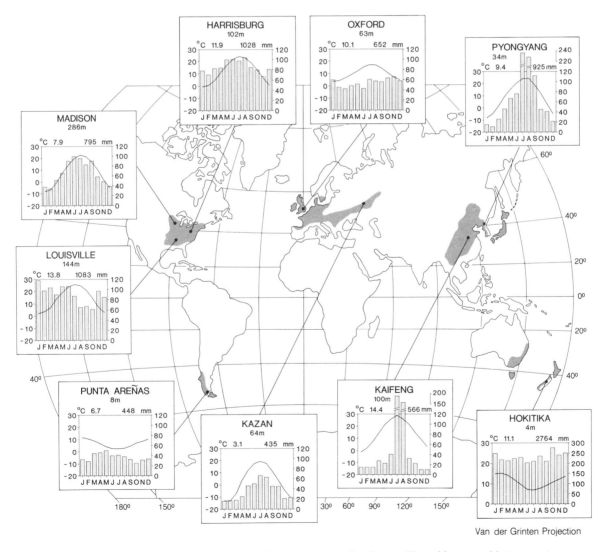

Figure 6.1 Distribution of deciduous forest ecosystems and representative climatic conditions. Mean monthly temperatures are indicated by the line and mean precipitation for each month is shown by the bars. Station elevation, mean annual temperature and mean annual precipitation appear at the top of each climograph.

6.1 INTRODUCTION

Climatic conditions in humid mid-latitude regions alternate between warm moist summers and mild winters. In the northern hemisphere this seasonality has favoured the widespread development of forests dominated by deciduous broadleaf trees. In parts of Asia the deciduous trees are replaced by broad-leaved evergreen species, but it is in the mild, moist climates of the southern hemisphere that the temperate evergreen forests become especially predominant.

The deciduous forests are found in eastern North America, western and central Europe, and in eastern Asia (Figure 6.1). The trees resume growth in the spring in response to increasing temperatures and longer day lengths: many herbaceous species flower at this time, before the canopy casts the forest floor into deep shade through the summer. The end of the growing period is marked by the vivid autumn colours of the foliage, and shortly afterwards the trees enter their leafless winter state. The uniformity which this annual cycle of summer growth and winter dormancy gives to the deciduous forests is enhanced by the wide distribution of many of the genera. Broad-leaved evergreen species become increasingly conspicuous in Japan, South Korea and southern China, and are often associated with conifers in more mountainous districts. Similar mixed evergreen forest occurs in the wet foothills of the Himalayas. In southern Europe the transition to evergreen forest reflects the increased probability of drought in the Mediterranean regions. Evergreen oaks and pines are also widely distributed in the south-eastern United States where they are usually associated with the poorly developed sandy or swampy soils.

In the southern hemisphere temperate deciduous forests are found only in the drier parts of the southern Andes. Elsewhere in southern Chile broad-leaved evergreen rainforests have developed in an oceanic climate that is virtually frost-free. Evergreen forests are also found in New Zealand, Tasmania and parts of south-eastern Australia. Climatic conditions in these regions are similar to those of the Pacific north-west coast of North America, but here the dominant species are conifers such as Douglas fir (*Pseudotsuga menziesii*) and western red cedar (*Thuja plicata*).

6.2 CLIMATE

In the temperate forest regions seasonal variation in climate is determined more by temperature than by precipitation, although conditions are strongly influenced by latitude and continentality (Figure 6.1). In eastern North America the winter climate is dominated by airstreams which bring cold dry air from the Arctic, but cyclonic storms which originate over the oceans frequently bring heavy snowfalls to the region. Winter temperatures

average −5 °C in more northerly deciduous forests but can be as high as 10 °C in southern regions. Spring is a season of thundershower activity caused by the convergence of cold polar air with warm moist air that is brought northwards from the Gulf of Mexico. These maritime air masses dominate the region during the summer. Mean summer temperatures range from 20 to 27 °C across the region but by autumn frost begins to limit plant growth. The length of the frost-free period increases from 120–150 days in the north to more than 250 days in the southern districts. Mean annual precipitation in the deciduous forests ranges from 800 to 1400 mm. Precipitation is generally higher in the south, where as much as 15% of the total comes from tropical storms which effect the region in autumn. Precipitation also increases in the Appalachian Mountains because of orographic effects.

Climatic conditions in Europe become progressively drier eastwards towards the continental interior. In maritime locations, such as the British Isles, mean winter temperatures typically range from 3 to 5 °C, although periods of clear, cold weather occur when polar continental air flows westwards from northern Russia. Summer temperatures average 15–17 °C, and rarely will exceed 30 °C. By contrast, winter temperatures decrease from −4 °C to −15 °C eastwards through the deciduous forest region of central Russia, and average summer temperatures increase to about 22 °C. The easterly flow of air from the Atlantic is intensified during the summer, but in winter the development of a cold anticyclone over the interior of Eurasia tends to block the inflow of maritime air (Wallen, 1970). Summer temperatures inland are moderated by the influx of maritime air but the effect is lost during the winter. Most of lowland Europe receives between 500 and 750 mm of precipitation, much of which comes from the passage of depressions. There is a trend to drier conditions towards the east but the pattern is somewhat obscured by the distribution of highland regions. Snow is rather uncommon in the mild western lowlands but may persist for several months in the forests to the east.

The climate of eastern Asia is characterized by marked differences between the summer and winter seasons due to the monsoonal effect caused by pressure changes over the continent. In winter the general airflow is from the north-west towards the Pacific Ocean. From May till September light onshore winds blow into the region, although late summer is also the typhoon season. Consequently, the precipitation regime has a marked summer maximum: at Beijing, for example, about 90% of the annual rainfall occurs between May and September (Watts, 1969). Annual precipitation decreases towards the north and west, and totals for the region range from about 500 to 1000 mm. The wettest area occurs in the highlands of the Korean peninsula, and between November and April there can be 20–30 days with snow. Elsewhere in temperate Asia fewer than 10 days of snow is the norm. The flow of cold northerly air is very persistent throughout winter, and January temperatures range from about 3 °C in southern coastal districts to −15°C in northen interior locations. By April the area comes under the influence of moist tropical maritime air, temperatures continue to rise and by July mean temperatures range from 22 to 28 °C. The length of the frost-free season towards the northen limits of the deciduous forests is about 225 days, but in the south there are very few days when the average temperature falls below 0 °C.

Southern Chile lies in a zone of persistent westerly winds that blow across the Pacific Ocean and there is little seasonal change in the weather. On the windward slopes of the Andes it usually rains on about 300 days of the year and total annual precipitation can exceed 8500 mm (Miller, 1976). Above 1000 m elevation, precipitation is mainly in the form of snow and permanent ice fields feed glaciers which reach sea level in some fjords. Annual precipitation at the summit of the Andes is about 3000 mm but decreases rapidly to the east: it is on these drier leeward slopes that the deciduous forests occur. Annual temperatures average only 6–10 °C in this region and the annual range is often less than 10 °C. Mean temperatures are similar in the coastal areas but the annual range is smaller, the difference between January and July temperatures being only 4 or 5 °C. It is under these conditions that the sub-antarctic rainforests have developed. A similar climate occurs along the south-west coast of the South Island of New Zealand. Here mean temperatures are about 5 °C in winter and rise to about 14 °C in summer, and annual precipitation as high as 9670 mm has been recorded in some areas (Maunder, 1971). The west coast of Tasmania is noticeably drier, and even in the highlands annual precipitation is usually less than 4000 mm. At these higher elevations mean temperatures range from about 0 °C in winter to 10 °C in summer (Gentilli, 1971). The broad-leaved evergreen temperate forests in these wet regions of Tasmania and New Zealand are floristically and physiognomically similar to those of Chile.

6.3 SOILS

Differences in climate, bedrock and drainage within the various temperate forest regions are reflected in the variety of soil conditions which are represented. The characteristic soils in the deciduous forests of North America are **Alfisols, Inceptisols** and **Ultisols**. The Alfisols are typically associated with glacial materials in more northerly districts. These soils have a thin surface horizon darkened by humus underlain by a paler, leached A_2 horizon and an illuvial clay layer. They have medium to high base saturation and because they contain significant amounts of weatherable material Alfisols are generally quite fertile (Foth and Schafer, 1980). Inceptisols have developed on weathered sandstones and shales in the Appalachian Mountains. Although these soils are strongly leached, large amounts of leaf litter are added to the soil each year. Rapid decomposition of this nutrient-rich material maintains the fertility of these soils. This mull humus produces a deep, darkly stained surface horizon with a crumbly structure which supports a rich herbaceous flora (Braun, 1950). In the hilly Piedmont region which borders the Appalachians to the east and south, the igneous and metamorphic rocks have given rise to comparatively youthful Ultisols. Similar soils are associated with the sands and occasional loess deposits which occur in the coastal plains. They are intensively leached, rather infertile soils with a well-developed clay horizon which is mainly comprised of kaolinite. The predominance of kaolinite together with a low organic matter content reduces the cation-exchange capacity of these soils, and there is a general shortage of exchangeable nutrients. Most of the nutrients are derived from organic matter, hence their fertility is largely dependent on biological recycling. The warm moist climate of the south favours the accumulation of iron and aluminium oxides and the soils in this region are distinctly red or yellow in colour (Foth and Schafer, 1980).

The characteristic soils of the deciduous forest region of Europe are the **brown earths**. They are equivalent to the Alfisols of North America, although in central France and western Germany the soils on calcareous parent materials more closely resemble Inceptisols. The brown earths have a deep, rich humus layer, high nutrient content and abundant biological activity. Distinct horizons are not evident in these loamy soils, but clay content decreases with depth and some nutrients are leached from the surface layers. The removal of calcium in this way results in soils which are slightly acid and pH values typically range from 5.5 to 6.5. The brown earths are so named because of the reddish brown coloration imparted by the uniform distribution of iron oxide within the profile, although the surface layers usually appear dark brown because of their high humus content. In wetter regions more complete leaching produces acid brown earths which tend to be sandy in texture. Under cool, humid conditions **podsols** may develop: they occur in some mountainous areas, in eastern Europe, and in areas of mixed forest where coniferous species are also present. Podsols can also form under deciduous woodland if a thick layer of acid litter accumulates because decomposition is slowed by the shade of a dense canopy (Mackney, 1961).

In east Asia the deciduous forest soils of the mountainous regions are classed as Alfisols, but across the North China Plain **Entisols** have developed from the alluvium deposited by the Yellow and Yangtze Rivers. The North China Plain is a vast, poorly drained floodplain, and the rivers carry heavy sediment loads derived from the badly eroded loess deposits to the west. The granite and gneisses of the highland districts in Manchuria, Korea and Japan form acid brown forest soils some of which are slightly podsolized, especially at higher elevations. Ultisols occur in southern China; they resemble the red and yellow soils in the forests of the south-eastern United States. The Ultisols are distinguished by the amount of weathering and the degree to which oxidation of iron is limited by waterlogging. The red earths are generally drier and less fertile (Foth and Schafer, 1980). Excessive concentrations of chlorides and sulphates can be present in the forest soils of the coastal districts of eastern China (Richardson, 1966).

Most of the soils in the wet temperate regions of the southern hemisphere are highly podsolized. In south-east Australia podsols have developed over a variety of rock types, and because of heavy leaching they all tend to be strongly acid and low in plant nutrients. Typically the soil profiles consist of greyish brown sandy surface horizons overlying yellow clay subsoils (Paton and Hosking, 1970). In parts of Tasmania humus podzols have developed because of impeded drainage caused by impermeable organic layers or iron pans. These soils are often waterlogged during wet periods and this allows plant litter to accumulate in the surface horizons. Similar acidic soils occur in the wetter parts of the South Island of New Zealand, although in the drier eastern areas and throughout the North Island yellow-brown earths of low fertility are associated with most of the forested regions (Wardle *et al.*, 1983). The characteristic soils of Patagonia are either podsols in various stages of development or accumulations of peat (Butland, 1957). Free-draining brown earths can develop on moraine and glacial sands in deciduous woodlands, with acid brown earths under evergreen forests, and in waterlogged depressions peaty gley soils, or very acid peats occur with a pH of 4 or lower (Holdgate, 1961).

6.4 REGIONAL DISTRIBUTION

6.4.1 The North American formation

The deciduous forests of North America extend throughout the eastern United States and parts of Canada from the Great Lakes region south to the Gulf of Mexico. It is a physiographically diverse region. The Atlantic Coast is bordered by a wide level coastal plain, inland from which is the rolling hilly topography of the Piedmont Plateau and the rather rugged terrain of the Appalachian Mountains which rise to about 2000 m. Flanking the mountains to the west is a region of dissected plateaus which eventually merges with the low-lying glaciated terrain around the Great Lakes. The deciduous forests continue further west through the Ozark Plateau and Ouachita Mountains into western Texas, and extend into the Central Lowlands following the river valleys of the Missouri–Mississippi drainage system. The boundaries of this forest region are determined by climatic and edaphic conditions. Coniferous

species are better adapted to the cooler, shorter growing seasons which occur to the north. Higher summer temperatures and more prolonged drought periods favour the prairie grasses in the continental interior. Pines replace deciduous species on the drier sandy soils in the south, and broad-leaved evergreen species also increase in abundance because their growth is not restricted by a period of dormancy in the winter (Braun, 1950).

Tracts of forest now exist only in areas that are not suitable for agriculture and much of this remnant cover is secondary growth. However, it still reflects the varied nature of the original vegetation, and Braun (1950) distinguished several forest types according to the diversity and distribution of the tree species. The greatest number of species is found in the mixed mesophytic forest region of the Appalachians (Figure 6.2). The dominant species in this region include beech (*Fagus grandifolia*), tulip-tree (*Liriodendron tulipifera*), basswood (*Tilia heterophylla*), sugar maple (*Acer saccharum*), sweet buckeye (*Aesculus octandra*), red oak (*Quercus borealis*) and white oak (*Q. alba*). In these moist forests the trees can grow to heights of 35–40 m with trunk diameters of 1–2 m, and their spreading branches provide a completely closed canopy (Figure 6.3). The flora of this region is composed of about 35 species of tall trees but most are only occasionally represented in the canopy. In addition, many smaller trees which may grow to heights of 10–15 m are present in the understory. Some, like the magnolias (*Magnolia tripetala* and *M. macrophylla*) and the dogwood (*Cornus florida*), are noted for their showy blossoms. Many smaller shrubs are common to this region but others, such as the rhododendron (*Rhododendron maximum*), are more localized in their distribution and many are endemic to the southern Appalachians. The deep, humus-darkened soils in this forest type support a rich herbaceous flora and, like the shrubs, many of the herbs are endemic to the region.

The number of tree species decreases towards the west and dominance is shared by progressively fewer species until in the driest regions only the oaks and hickories are common. Eventually, these too are replaced by the prairie grasslands. The most widely distributed trees in the western oak–hickory forests are white oak (*Q. alba*), red oak (*Q. borealis maxima*) and black oak (*Q. velutina*), and the shagbark and bitternut hickories (*Carya cordiformis* and *C. ovata*). However, other species, such as the post oak (*Q. stellata*) and blackjack oak (*Q. marilandica*), are locally present in southern districts and the bur oak (*Q. macrocarpa*) is common in the north. These trees are smaller and the forests are more open than those in the east, but with more light penetrating the canopy there is usually a fairly dense understory of shrubs and herbs. The oak–hickory forests become progressively fragmented and at their western limit are found only in ravines and other moist sites (Figure 6.4)

Beech and sugar maple are the dominant species in the forest region which lies immediately south of the Great Lakes and extends into southern Canada (Figure 6.5).

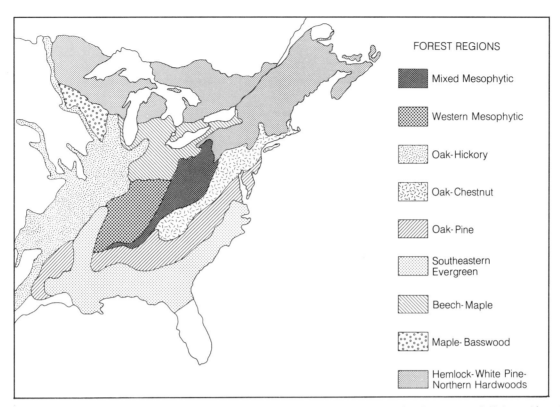

Legend — FOREST REGIONS

- Mixed Mesophytic
- Western Mesophytic
- Oak-Hickory
- Oak-Chestnut
- Oak-Pine
- Southeastern Evergreen
- Beech-Maple
- Maple-Basswood
- Hemlock-White Pine-Northern Hardwoods

Figure 6.2 Major associations within the deciduous forests of the eastern North America. (After Braun, 1964.) (Adapted from E. L. Braun, *Deciduous Forests of Eastern North America*, published by Hafner, 1964, with permission from the Cincinnati Museum of Natural History, Ohio, USA.)

Figure 6.4 Bur oak (*Quercus macrocarpa*) growing in wet draws in the forest–prairie ecotone of the Great Plains. (R. E. Redmann.)

Figure 6.3 Mixed mesophytic cove forest in the Southern Appalachians mainly comprised of sugar maple and basswood. (M. Huston.)

Other species, including elm (*Ulmus americana*), walnut (*Juglans nigra*) and ash (*Fraxinus americana*), may be locally present. Oaks and hickories increase in abundance in transitional areas to the west, while to the south species associated with the mesophytic forest regions become more common. Further to the north-west sugar maple grows in association with basswood (*Tilia americana*), although oaks are very common in the southern part of this region, with ash and elm important elsewhere. The northern limit of the deciduous forests

Figure 6.5 Shrubs and herbs provide an open ground cover beneath beeches, maples and oaks in the Great Lakes region. (M. Huston.)

Figure 6.6 Transitional mixed-wood region of eastern Canada with an admixture of sugar maple (*Acer saccharum*), red oak (*Quercus rubra*) and eastern white pine (*Pinus strobus*). (R. A. Wright.)

occurs in the mixed-wood region which runs as a broad band from northern Minnesota to the maritime provinces of Canada (Figure 6.6). Here, species such as maple, ash and elm which are prominent in the northern deciduous forest are replaced by conifers and by species such as trembling aspen (*Populus tremuloides*) and birch (*Betula papyrifera*), which are distributed throughout the boreal forests.

The eastern slopes of the Appalachians originally supported forests of oak and chestnut (*Castanea dentata*), but with the introduction of chestnut blight the composition of the forest was dramatically altered. The disease was discovered in 1904 and the fungus (*Endothia parasitica*) spread quickly through the region, so that by 1950 all of the chestnuts were dead or seriously affected. Mature chestnut trees had large spreading canopies and their loss created significant gaps in the forest. Chestnut oak (*Q. prinus*), red oak (*Q. rubra*) and red maple (*Acer rubrum*), accounted for about 50% of the gap replacements, all of these species being common associates of the chestnut in earlier stands (Woods and Shanks, 1959). However, the chestnut has not disappeared entirely from this region. The species has re-established as a small understory tree from stump sprouts, and in some stands continues to rank among the 10 most important species (McCormick and Platt, 1980). Although no seedling establishment has occurred, the population is increasing through expansion and division of the the root crowns (Paillet, 1984).

The forests of the Piedmont region of the Atlantic and Gulf States are dominated by oaks and hickories, but the presence of loblolly pine (*Pinus taeda*) and yellow pine (*P. echinata*) on dry, sandy soils and in regrowth forests distinguishes the oak–pine forest region from the oak–hickory region to the west. Although many species of oak occur within this region, they are not uniformly distributed: some, like the post oak, are found mostly on the poorer, thinner soils and others, such as the willow oak (*Q. phellos*), are restricted to the forested bottomlands. Included in the oak–pine region are the 'pine-barrens' of New Jersey and other north-eastern states. Pitch pine (*P. rigida*) is prominent in these sandy, fire-prone environments: it grows to heights of 5–8 m, and small oaks, such as *Q. ilicifolia* and *Q. prinoides*, and shrubs form a well-developed understory beneath the widely spaced pines. The transition to evergreen forests occurs in the southern coastal plains where longleaf pine (*P. palustris*) provides the overstory to scrubby oaks. Longleaf pine is replaced by pitch pine (*P. rigida*) in wetter areas, where it grows in association with a boggy ground cover of grasses and sedges. Stands of bald cypress (*Taxodium distichum*) and water tupelo (*Nyssa aquatica*) occur in the swampy floodplains. Dense stands of sweet gum (*Liquidambar styraciflua*) and other hardwoods are found in bottomlands which are subject to intermit-

Figure 6.8 Ancient oak (*Q. robur*) in southern England. (R. G. Archibold.)

Figure 6.7 (a) Derelict hazel (*Corylus avellana*)–birch (*Betula pendula*) coppice in southern England; (b) pedunculate oak (*Quercus robur*) trees and hornbeam (*Carpinus betulus*) pollards with mixed regeneration in neglected wood–pasture system in Croatia. (G. F. Peterken.)

tent flooding. The transition to evergreen forests is completed by the establishment of live oak (*Q. virginiana*) and slash pine (*P. caribaea*) in the coastal regions, although some deciduous species may occasionally be present.

6.4.2 The European formation

Deciduous trees are widely distributed in Europe, but much of the forest has been cleared for agriculture and what remains has invariably been disturbed by a long period of human occupancy (Figure 6.7). The European formation is essentially divided into two sections by a zone of mixed forests which extends from the Baltic states into the mountainous districts of central and southern Europe. In western Europe the deciduous forest zone extends from northern Portugal to southern Norway, including the British Isles and Ireland. It is replaced by evergreen woodlands in the Mediterranean regions of southern Europe, and by conifers on the slopes of the

Alps and adjacent mountain ranges. The Carpathian Mountains of eastern Europe separate the deciduous forests of the Danube Basin from those in the northern Ukraine and central Russia. Here the forests are interrupted by the southern slopes of the Urals. They continue eastwards as a narrow, discontinuous zone between the steppe grasslands and the northern coniferous forests. The trembling aspen (*Populus tremuloides*) parklands of the Canadian prairies provide a similar extension to the deciduous forest formation in North America.

There are fewer species of trees in the European formation than in the American deciduous forests, but here too different regional associations are recognized. The pedunculate oak (*Q. robur*) is the dominant species in the clay vales of Britain and the heavier lowland soils of north-west Europe (Figure 6.8), and is found as far east as the Caucasus Mountains. It is generally limited to sites below 300 m elevation, usually being replaced in higher sites by sessile oak (*Q. petraea*), birches (*Betula pendula, B. pubescens*) or beech (*Fagus sylvatica*). The abundance of oak has increased because of centuries of planting: it often occurs in monospecific stands but other species such as the elm (*Ulmus procera*), hornbeam (*Carpinus betulus*) and field maple (*Acer campestre*) may be locally present. Elm, which was particularly common in southern England, has been drastically reduced in recent years by Dutch elm disease, the spores of this fungus (*Ceratocystis ulmi*) being transmitted to healthy trees by various bark beetles. Mature oaks typically grow to a height of 30 m. Their broad canopies cast a moderate

Figure 6.9 (a) Oceanic sessile oak (*Quercus petraea*) woodland and (b) associated mossy understory in north-western England. (G. F. Peterken.)

shade within which an open shrubby understory develops. Hazel (*Corylus avellana*) is often the most common species in the shrub layer, although hawthorn (*Crataegus monogyna*), sloe (*Prunus spinosa*) and willows (*Salix* spp.) are frequently present (Tansley, 1953). The evergreen holly (*Ilex aquifolium*) may also occur in the understory, although in unusually favourable conditions it will grow into a tree 18–20 m in height. Many herbaceous species are present in the ground cover and nearly all of these come into flower in the spring. Their distribution is closely related to soil conditions; dog's mercury (*Mercurialis perennis*) and primrose (*Primula vulgaris*) are characteristic of damper clay soils, the wood anemone (*Anemone nemorosa*) and bluebell (*Hyacinthoides non-scriptus*) are common in drier sites, and bracken fern (*Pteridium aquilinum*) is usually associated with well-drained soils.

The sessile oak (*Q. petraea*) is more limited in its distribution than *Q. robur*. It is associated with the maritime climate of western Europe (Figure 6.9) and is further restricted by soil conditions, preferring drier, acid soils to the heavier clays and loams. The diversity of these oak communities is comparatively low because the soils are often deficient in nutrients. Birch, mountain ash (*Sorbus aucuparia*) and holly are commonly found in these oakwoods, with rose (*Rosa canina*) and bramble (*Rubus fruticosus*) in the shrub layer, and grasses and bracken conspicuous in the field layer. The moist climatic conditions which limit the geographic range of the sessile oak

also restricts the beech to north-western Europe. Beech is a tall tree which can grow to a height of 40 m and, because its canopy casts a deep shade, can replace the oaks on well-drained loamy soils. This is normally prevented by clearing and planting and in Britain the beech is mostly limited to chalk and limestone areas which are not suitable for other species. It is more widespread in other parts of Europe where it often grows in association with pedunculate oak, hornbeam and chestnut (*Castanea sativa*). Beech is very common in the mountains of central Europe (Figure 6.10) but further east it is largely replaced by *Q. robur*. Ash (*Fraxinus excelsior*) is also found on calcareous soils: it is not commonly associated with beech but is often present in oakwoods in limestone areas, and also on calcareous clays because it is tolerant of wet soils. However, the usual dominant in very poorly drained sites is alder (*Alnus glutinosa*).

Different species of oak dominate the deciduous forests of southern Europe. *Q. lusitanica* is common in northern Spain and Portugal; in northern Italy *Q. pubescens* is particularly abundant on dry calcareous soils, and further east in the Balkans *Q. trojana* becomes important. All of these species are rather small trees but other larger species, such as the fast-growing Turkey oak (*Q. cerris*), are also present, although the latter prefers cool, humid conditions and so is usually found on hills or mountains (Marcuzzi, 1979). The chestnut is also indigenous to southern Europe and is often found growing with beech at elevations of 500–1000 m throughout this region.

Figure 6.10 (a) Beechwoods (*Fagus sylvatica*) in the Savoy Alps, south-eastern France; (b) floristically poor understory in a young stand of beech in Germany. ((a) G. F. Peterken; (b) R. A. Wright.)

These deciduous forests are replaced by stands of broad-leaved evergreen trees at lower elevations and in more southerly locations, and by conifers on the higher slopes (Eyre, 1968).

6.4.3 The Asian formation

The deciduous forest region of eastern Asia coincides approximately with the lowland areas that border the Yellow Sea from Manchuria to the Yangtze River. Throughout most of this region the forest cover now exists only as groves surrounding temples and shrines. Summergreen forests also occur in the montane region of central and southern Japan and in the lowlands and foothills of northern Honshu and Hokkaido (Numata *et al.*, 1972). In the Shantung Peninsula and similar areas of mainland China the forest cover is composed of several species of oak, including *Quercus variabilis* and *Q. liaotungensis*, elm (*Ulmus macrocarpa* and *U. davidiana*) and maple (*Acer truncatum*), together with lime (*Tilia amurensis*), walnut (*Pterocarya stenoptera*), bay (*Lindera obtusiloba*) and ash

(*Fraxinus chinensis*). Beneath this is a secondary layer of small trees. Numerous flowering shrubs are also present in undisturbed stands and many of these have been introduced to other parts of the world for horticultural use. *Callicarpa japonica*, for example, is a compact shrub with pale pink flowers and violet fruits; *Clerodendron trichotomum* is noted for its white, fragrant flowers; *Styrax obassia* produces pure white bell-shaped flowers carried in drooping racemes; and *Lespedeza bicolor* displays bright rose-purple flowers late in the summer. *Rhododendron micranthum*, which is noted for its unusual flowers, is also native to this region. Where the overstory has been removed dense thickets of hazel (*Corylus heterophylla*), hawthorn (*Crataegus pinnatifida*) and other small trees and shrubs develop (Richardson, 1966).

The lowland plains are almost entirely cultivated but some remnants of forest are found in hilly areas. Here the oaks are again dominant in most communities, with various species of birch (*Betula fruticosa*, *B. dahurica* and *B. platyphylla*), aspen (*Populus tremula*), lime (*Tilia mandshurica*, *T. mongolica*) and maple (*Acer mono*), increasing in importance at higher elevations (Wang, 1961). Willows such as *Salix matsunda* and *S. babylonica*, the renowned 'weeping willow', and balsam poplars (*Populus simonii* and *P. suaveolens*) are still present along water courses.

Coniferous species become more common to the north in areas such as the Lesser Khingan mountains and the Changpaishan massif north of Korea. The flora is more diverse than in comparable forests in Europe and North America. The dominant species throughout this region is Korean pine (*Pinus koraiensis*), which grows in association with Japanese red pine (*P. densiflora*) and Manchurian fir (*Abies holophylla*): at lower elevations species of birch, maple, aspen and oak are important, especially on well-drained soils. In such sites there is a well-developed shrubby understory. At higher elevations species of spruce (*Picea jezoensis* and *P. koyamae*) and larch (*Larix olgensis*) add to the diversity of the conifers. The number of broadleaf species also increases with the addition of rowan (*Sorbus pohuashensis*), lime (*Tilia amurensis*), ash (*Fraxinus mandshurica*), walnut (*Juglans mandshurica*), the Amur cork tree (*Phellodendron amurense*) and others. The shrub layer is equally diverse with spirea (*Spirea salicifolia*), lilac (*Syringa amurensis*), cherry (*Prunus maximowiczii*), hawthorn (*Crataegus maximowiczii*), viburnum (*Viburnum* spp.), currant (*Ribes* spp.) and honeysuckle (*Lonicera* spp.) amongst the species represented (Richardson, 1966).

Broad-leaved evergreen species become more conspicuous in the rugged terrain south of the Yangtze River. The flora is exceptionally diverse with more than 50 broad-leaved and 12 coniferous genera represented in the canopy, totalling several hundred species. Elm (*Ulmus*), lime (*Tilia*) and ash (*Fraxinus*) are each represented by at least eight species, rowan (*Sorbus*) by 32 species and maple (*Acer*) by over 50 species. The coniferous species include the Chinese fir (*Cunninghamia lanceolata*), several

Figure 6.11 Moss and lichen covered trunks of *Fagus crenata* grow above a rich understory of shrubs including *Sorbus*, *Viburnum*, *Hydrangea* and *Acer* in beech woodland in northern Japan. (P. Wardle.)

pines (*Pinus massoniana*, *P. armandi* and *P. tabulaeformis*) and cypresses (*Cupressus duclouxiana*, *C. funebris*), juniper (*Juniperus chinensis*) and hemlock (*Tsuga chinensis*). The dawn redwood (*Metasequoia glyptostroboides*), which was originally widespread in the northern hemisphere and first described from fossil material (see Figure 9.26), also grows in this region together with the maidenhair tree (*Ginkgo biloba*), another relic species that was abundant in the Mesozoic era. Broad-leaved evergreen species, especially oaks (*Quercus* spp.), tanoaks (*Lithocarpus* spp.) and chinquapins (*Castanopsis* spp.), become increasingly dominant towards the south until they are eventually replaced by the tropical monsoon rainforests. Many spices and aromatic species such as the camphor tree (*Cinnamomum camphora*) are associated with this region. Species of Theaceae (tea family), bamboo and ferns are present in the understory.

The forest regions of Japan are analogous to those of the mainland, and here too very little remains except around temples, or in steep ravines and other inaccessible sites. Beech (*Fagus crenata*) is the dominant tree in the deciduous forests of the north (Figure 6.11). It grows in association with various maples, including *Acer mono*, *A. japonicum* and *A. palmatum*, lime (*Tilia japonica*), oak (*Q. mongolica*), ash (*Fraxinus sieboldiana*) and rowan (*Sorbus alnifolia*). Deciduous and evergreen

shrubs are common in the understory including some that are closely related to the bamboos (Numata *et al.*, 1972). Beech is replaced by Japanese wing nut (*Pterocarya rhoifolia*), elm (*Ulmus lacinata*) and the katsura (*Cercidiphyllum japonicum*) in moister sites, with Japanese horse chestnut (*Aesculus turbinata*) and ash (*Fraxinus spaethiana*) more locally distributed. Alder (*Alnus japonica*) is common along watercourses and in other wet areas. Oaks are the characteristic dominants in the temperate evergreen forests of southern Japan, their distribution being influenced by habitat conditions. For example, *Q. gilva* is common on deeper soils, *Q. salicina* occurs on steep slopes and ridges with shallow soils, and *Q. glauca* is associated with limestone areas. Japanese fir (*Abies firma*) and hemlock (*Tsuga sieboldii*) replace the oaks at higher elevations, and in turn give way to hardier conifers in subalpine sites.

6.4.4 The formations of the southern hemisphere

In the southern hemisphere deciduous forests are found only in the drier parts of Patagonia where they are generally confined to the eastern slopes of the Andes south of latitude 45 °S. The cover is of beech and is composed of almost pure stands of lengue (*Nothofagus pumilio*) or nire (*N. antarctica*). Lengue will grow to heights of 25–30 m (Figure 6.12), although at higher elevations it is reduced to shrub-like proportions. Nire is a smaller tree 15–18 m tall but it rarely grows more than a few metres in height, particularly in the eastern limits of the forest region where it forms a scrubby transition to the dry steppes. The undergrowth is comparatively dense and is comprised of small evergreen trees and shrubs including maiten (*Maytenus magellanica*), the holly-leaved mechaia (*Berberis ilicifolia*), smaller barberry bushes (*Berberis buxifolia* and *B. heterophylla*) 1–1.5 m in height, and graceful fuchsias (*Fuchsia magellanica*) which grow to 3 m (Butland, 1957).

The deciduous forest is replaced by temperate rain forest on the wetter western slopes of the Andes. These humid forests are subdivided into two sections: the **Valdivian** rain forests occur in the coastal regions between latitudes 41 and 43 °S, while the **Magellanic** rain forests reach almost to 56 °S. The relatively mild northern region supports a mixed forest cover composed of tall evergreen trees, some of which grow to heights of 50 m. At lower elevations the forests are dominated by species such as *Eucryphia cordifolia*, *Laurelia philippiana* and *Aextoxicon punctatum*, although the deciduous beech *Nothofagus obliqua* may overtop these in some sites. Stands of Patagonian cypress (*Fitzroya cupressoides*) occur in poorly drained lowland sites or in wetter montane zones where they may live to an age of 3000 years. The other conifer commonly associated with this region is the monkey puzzle tree (*Araucaria araucana*) but it is mainly restricted to high elevation sites in the north (Figure 6.13).

Figure 6.12 *Nothofagus pumilio* forest in Andean Patagonia region. (L. Ghermandi.)

Figure 6.13 *Auracaria* forest in southern Chile showing well-spaced trees and lack of undergrowth typical of this vegetation. (D. R. Given.)

The composition of the forests changes above 500 m and evergreen beeches such as *Nothofagus dombeyi* and *N. betuloides* are increasingly common in the canopy, although it is the deciduous species *N. pumilio* which typically forms the abrupt treeline in the western Andes (Veblen *et al.*, 1983).

The floristic diversity of the Magellanic rain forest is reduced because of more severe climatic conditions. *N. betuloides* is the dominant species throughout this region but in coastal locations it is usually accompanied by *N. dombeyi* and *N. nitida*. Deciduous *N. pumilio* may be present at timberline, but southwards the transition to alpine tundra is increasingly through a zone of elfin woodland comprised of stunted and deformed *N. betuloides*. In the extreme south the growth of trees is prevented by wind and spray, and in these cold exposed sites the vegetation alternates between lichen-covered rocks and peat-filled depressions (Butland, 1957). Where tree growth is stunted the branches usually form an impenetrable tangle; elsewhere smaller trees and shrubs provide a dense understory, within which there is a profusion of lianas and epiphytes. The densest growth is often associated with patches of tall bamboos (*Chusquea* spp.) which grow so thickly in disturbed sites that they prevent tree regeneration (Veblen *et al.*, 1983).

Forests dominated by *Nothofagus* are also characteristic of the wetter regions in the South Island of New Zealand

(Figure 6.14). Mountain beech (*N. solandri* var. *cliffortioides*) is the most widespread species in southern districts where it often forms a dense canopy 10–15 m high. Beneath this is a scattered cover of *Coprosoma* shrubs, lichen and mosses, or in wetter sites a dense understory of ferns (Wardle *et al.*, 1983). At higher elevations mountain beech is often accompanied by *N. menziesii* to the timberline. Black beech (*N. solandri* var. *solandri*) replaces mountain beech on warm, dry, lowland sites. It is also found in the North Island where it is an important member of the multi-storied conifer–beech–broadleaf forests. In these communities the tallest trees are conifers; they typically grow to heights of 30–40 m and include species such as matai (*Podocarpus spicatus*) and rimu (*Dacrydium cupressinum*) together with the massive kauri (*Agathis australis*). In swampy areas the dominant species is kahikatea (*Podocarpus dacrydioides*): it grows 50 m or more in height and is New Zealand's tallest indigenous tree. Hardwoods grow to heights of 15–25 m beneath these widely spaced conifers: typical species include kamahi (*Weinmannia racemosa*), and rata (*Metrosideros robusta*) which initially establishes as an epiphyte and, like the tropical figs, subsequently strangles the tree that supported it. Many smaller trees and shrubs are present in the understory as well as tree ferns (*Cyathea dealbata, Dicksonia fibrosa*) and in very mild climates the nikau palm (*Rhopalostylis sapida*). Lianes are also very common in these mixed forests. The ground cover is usually of ferns and mosses.

Figure 6.15 Myrtle beech (*Nothofagus cunninghamii*) growing in association with sassafras (*Atherosperma moschata*) and blackwood acacia (*Acacia melanoxylon*) in the temperate rainforests of Tasmania. (O. W. Archibold.)

Figure 6.14 (a) Lowland mixed beech forest with emergent podocarps and (b) red beech forest (*Nothofagus fusca*), South Island, New Zealand. (A. F. Mark.)

In Australia most of the wet temperate forest occurs in northern and western Tasmania (Figure 6.15). The dominant species is myrtle beech (*Nothofagus cunninghamii*): it is accompanied by sassafras (*Atherosperma moschatum*) on less fertile soils (Paton and Hosking, 1970) and associated species include King William pine (*Athrotaxis selaginoides*), Huon pine (*Dacrydium franklinii*) and celery-topped pine (*Phyllocladus aspleniifolius*). A dense understory of trees, shrubs and tree ferns develops where the canopy is fairly open, but often the canopy is so dense that few plants other than mosses can survive. Mountain ash (*Eucalyptus regnans*) replaces myrtle beech at 750 m elevation, and above 900 m alpine ash (*E. delegatensis*) is the dominant species. Smaller trees and shrubs, such as blackwood (*Acacia melanoxylon*), mimosa (*A. verticillata*) and silver wattle (*A. dealbata*), form a tall dense understory in these forests beneath which is a ground cover of ferns (*Blechnum procerum*) and bracken (*Pteridium aquilinum*). In subalpine areas *Eucalyptus coccifera* and *E. gunnii* are the characteristic dominants.

Similar forests occur in parts of Victoria, but elsewhere on the mainland the characteristic species of the wet sclerophyll forests are eucalypts, such as blackbutt (*E. pilularis*), white mahogany (*E. acmenioides*), tallowwood (*E. microcorys*) and grey gum (*E. propinqua*). In drier areas species such as red gum (*E. blakelyi*), blue gum (*E. leucoxylon*), yellow box (*E. melliodora*) and grey box (*E. moluccana*) are common, with snow gum (*E. niphophylla*) present at high elevations. Equivalent forests occur in the extreme south-west corner of Western Australia state (Figure 6.16): the principal species are karri (*E. diversicolor*), tingle tree (*E. jacksoni*) and marri (*E. calophylla*). Jarrah (*E. marginata*) replaces these species in drier regions and increasing summer drought favours the wandoo (*E. redunca*) woodlands and the mallee eucalypts.

6.5 QUATERNARY HISTORY OF THE TEMPERATE FORESTS

Palaeobotanical evidence suggests that broad-leaved forests have been present in eastern North America since the late Cretaceous, with many genera including *Quercus*, *Acer*, *Fagus* and *Juglans* still represented in the present flora. Gymnosperms were also widespread. Some, like *Pinus*, *Picea* and *Thuja*, have persisted in the region but others, including *Auracaria* and *Podocarpus*, are now found only in the southern hemisphere. Temperate forests were probably confined to high elevation sites during the early Tertiary as tropical and subtropical genera spread northwards, but plants closely related to the modern chestnuts, hickories and limes were associated with this flora. The temperate forest genera were found at latitudes 20–25 ° north of their present limits, with nearly 300 species recorded from Greenland and comparable floras present in Alaska and northern Eurasia. Southward migration occurred as the climate cooled in the mid-Tertiary, but with a change to drier conditions in the late Tertiary the distribution of the forests became more fragmented and resulted in the present disjunct distributions which occur between some species in the forests of the eastern United States and the highlands of Mexico and Guatemala (Braun, 1950).

Successive glacial advances during the Pleistocene caused a further change in the distribution of the temperate forests. The temperate flora was largely displaced by species which are now associated with the northern boreal forests. At the peak of the Wisconsin Glaciation in North America, approximately 18 000 years ago, the edge of the ice sheet was bordered by a treeless tundra zone 60–100 km wide and extending discontinuously along the Appalachians (Figure 6.17). Jack pine (*Pinus banksiana*) forests extended over most of the eastern United States, with spruce (*Picea* spp.) increasing in importance towards the west. The northern hardwood species persisted in a narrow zone south of the conifers, while oaks, hickories and the southern pines covered the coastal plains. Mixed hardwood forests were confined to major river valleys, although spruce was the dominant species

Figure 6.16 Karri (*Eucalyptus diversicolor*) forest in the tall-timber country of south-western Australia. (O. W. Archibold.)

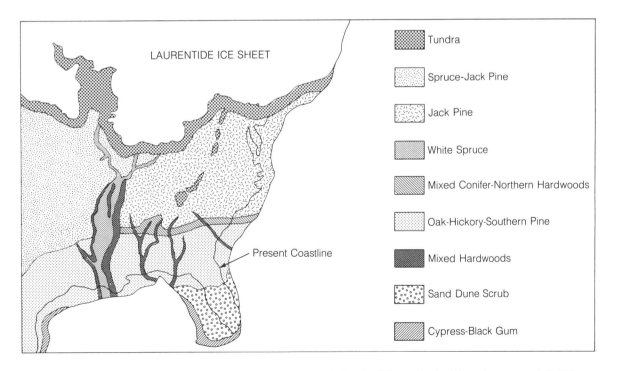

Figure 6.17 Reconstruction of the vegetation of eastern North America during the Wisconsin glacial maximum about 18 000 years BP. (After Delcourt and Delcourt, 1981.) (Reproduced with permission from P. A. and H. R. Delcourt, Vegetation maps for eastern North America: 40 000 yr BP to the present, in *Geobotany II*, ed. R. C. Romans; published by Plenum Publishing Corp., 1981.)

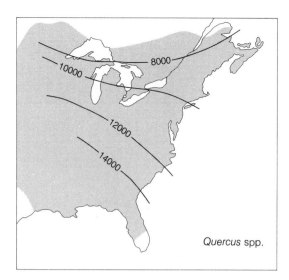

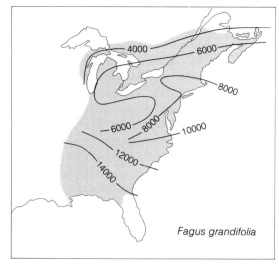

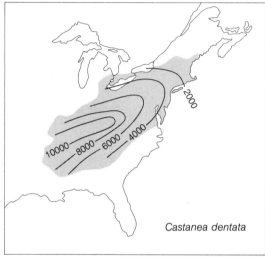

Figure 6.18 Post-glacial migration and present range of selected hardwood species in North America. (After Davis, 1981.) (Reproduced with permission from M. B. Davis, Quaternary history and the stability of forest communities, in *Forest Succession: Concepts and Applications*, eds D. C. West, H. H. Shugart and D. B. Botkin; published by Springer-Verlag N.Y. Inc., 1981.)

in the lower Mississippi. Scrubland covered the Florida peninsula as a result of the drawdown of the water table in the limestone bedrock. This area was formerly assumed to have been a refuge for temperate species during the Pleistocene, but clearly it was too dry (Watts and Stuiver, 1980).

Climatic amelioration began about 16 000 years ago with significant warming at 12 500 years BP. By 10 000 yrs BP there had been a considerable expansion of the mixed hardwood forests, and the oaks and hickories were well established in the western part of their range. Northern genera, such as elm and maple, quickly extended their range and had reached the Great Lakes region 10 000–12 000 years ago (Figure 6.18). The migration of beech was much slower. It moved northwards along the eastern seaboard, then westwards, and only appeared in the pollen record of the Great Lakes region

6000–7000 years ago. Chestnut moved into the region even more slowly. The average rate of range extension for chestnut was 100 m yr^{-1}, compared with 200 m yr^{-1} for beech and maple, 250 m yr^{-1} for elm and hickory and 350 m yr^{-1} for oak (Davis, 1981). It is now apparent that the floristically diverse mesophytic forests of the southern Appalachians have developed in the last 15 000 years, whereas previously this area was considered to have been a refugium during glacial times (Davis, 1983).

A similar vegetation history is known for Europe. The early Tertiary flora in England is represented by palms and tropical species closely related to genera presently found in the islands of the East Indies. This flora was replaced subsequently by woody plants characteristic of the warm-temperate regions analagous to the vegetation in the mountainous regions of southern Asia. Fossil

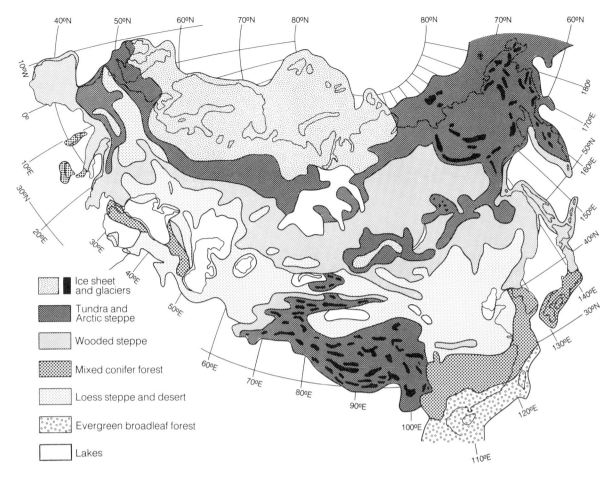

Figure 6.19 Reconstruction of the vegetation of Eurasia during the Wisconsin glacial maximum about 18 000 years BP. (After Frenzel, 1968.) (Redrawn from B. Frenzel, The Pleistocene vegetation of northern Eurasia, *Science*, **161**, 642, copyright 1968 by the AAAS.)

floras from the Netherlands and Poland indicate that many of the modern European species were present by late Tertiary times, and that genera now absent from Europe were eliminated as climatic conditions deteriorated (Pennington, 1969). At the beginning of the Pleistocene species of oak, alder, hemlock, pine and spruce were common throughout western Europe, but were replaced by heathy shrubs during cooler periods. Tundra communities dominated by grasses and sedges accompanied by dwarf birches and willows became established over much of Europe during the last glacial maximum (Figure 6.19). Mean annual temperatures at this time were probably 10–12 °C lower than today, and because of the ice-capped Alps the climate would have been more severe than in eastern North America (Flint, 1971). Wooded tundra occurred in parts of southern Europe and forests were confined to the coastlands of the Mediterranean Basin. The plants responded quickly as conditions changed during the period of deglaciation. In Britain the beginning of the post-glacial period occurred about 10 000 years ago. Oak and elm were widely established in southern Britain by 9000 BP, and within 2000–3000 years had spread throughout the British Isles (Figure 6.20). Alder was present by 8000 BP, lime had arrived about 7500 BP and ash between 6000 and 7000 BP. Beech appeared about 3000 years BP, but unlike other trees its rate of spread did not decrease with time. This suggests that beech was unable to reach the limits of its natural distribution before its range was altered by human activity (Birks, 1989).

Ice sheets originating in Siberia were less extensive than those in Europe. Their southern boundary was approximately 60 °N latitude and in the east they reached only as far as the Laptev Sea. Elsewhere in Asia smaller ice fields developed on mountains, with glacial features reported at about 2000 m in Korea and 2700 m in Japan. Consequently, the forest cover in eastern Asia was less seriously disrupted during the Pleistocene than in Europe or North America, but changes in distribution did occur. In Japan the broad-leaved evergreen forests were restricted to the southernmost parts of Kyushu and deciduous species were found only in the plains (Miyawaki and Sasaki, 1985). The continuation of mainland China into lower latitudes allowed the forest species to retreat southwards as the climate deteriorated, and many ancient genera are still preserved in this flora.

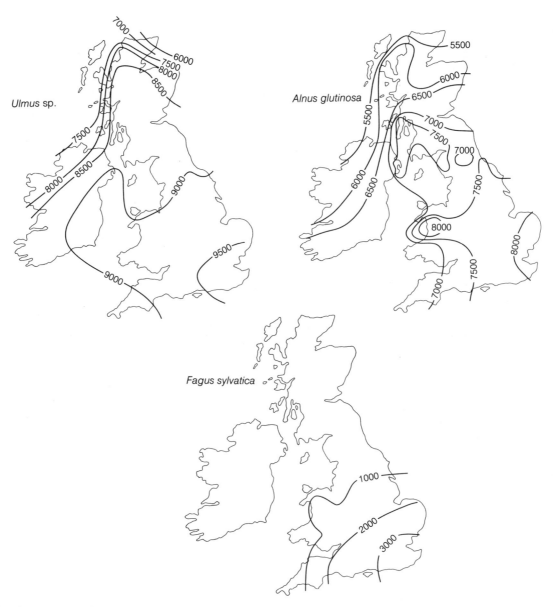

Figure 6.20 The migration of selected hardwood species across the British Isles in post-glacial times. (After Birks, 1989.) (Reproduced with permission from H. J. B. Birks, Holocene isochrone maps and patterns of tree-spreading in the British Isles, *Journal of Biogeography*, 1989, **16**, 510, 516, 520.)

The distribution of temperate forests in the southern hemisphere also corresponds closely with areas that were recently glaciated. Ice covered all of Patagonia from the Pacific coast eastwards across the Andes to the foothills in Argentina. Many species became extinct in this region during the Pleistocene, but *Nothofagus* had re-established in the area by 9000 years BP and had advanced eastwards to its present limit by 5000 years BP (Auer, 1960). The pollen record from northern Patagonia indicates that dense *Nothofagus* forests replaced open

stands of *Weinmannia* as temperatures increased, with *Fitzroya* and *Podocarpus* becoming established about 6500–4500 years BP as conditions subsequently became cooler and more humid (Heusser, 1974). In Australia significant glaciation occurred only in central Tasmania, and it too was accompanied by loss of species from the Tertiary forests that occurred throughout southern Australia. Some of these early types probably survived on low-lying exposed areas in the Bass Straits as sea level dropped. This might explain the present affinities between Tasmania and the mainland (Burbidge, 1960). In New Zealand glaciation was extensive in the mountains of the South Island, and the permanent snow-line was at least 1000 m lower than at present. Over much of the country the vegetation was reduced to subalpine grassland, and forests would have been mainly restricted to the northern peninsula of the North Island and perhaps to the emergent continental shelf (Fleming, 1962).

6.6 THE EVOLUTION OF THE DECIDUOUS HABIT

The deciduous habit is generally regarded as a drought-avoidance adaptation which evolved in dry subtropical regions (Axelrod, 1966). Deciduous trees reduce their demand for water by shedding leaves during the dry season. Conifers have overcome the problem of drought stress by evolving smaller leaves to increase water use efficiency. The wood of conifers is also anatomically simpler than that of deciduous species, being composed almost entirely of **tracheids** which develop in spring and heavy-walled **fibre tracheids** which are laid down in the summer. Tracheids are single cells which can be up to 5 mm in length. Water passes between them by way of 'bordered pits' and moves up the stem at a maximum rate of less than 0.5 m hr^{-1} (Kramer and Kozlowski, 1960). In hardwoods most of the water is conducted along xylem vessels comprised of cells which are interconnected by pores or perforations into tube-like structures that can be more than a metre in length. The wood of trees like birch, beech and maple is described as diffuse-porous because vessels of approximately the same diameter are produced throughout the growing season. Species such as oak and ash have ring-porous wood in which vessels formed in spring are much larger than those formed later in the year. Water is conducted through ring-porous species at rates of $25-60 \text{ m hr}^{-1}$ but this rapid movement is restricted to the youngest tissues. In diffuse-porous species maximum water movement ranges from 1 to 6 m hr^{-1} and, as with the conifers, more than the current year's wood is involved (Kramer and Kozlowski, 1960).

With efficient water conduction hardwood species can maintain high rates of photosynthesis and so shade out competitors, but this advantage is lost in seasonally dry climates and in areas where growth is restricted by long cold winters (Chabot and Hicks, 1982). Average maximum values for net photosynthesis under optimum conditions are $15-35 \text{ mg CO}_2 \text{ dm}^{-2} \text{ hr}^{-1}$ for deciduous trees, $10-18 \text{ mg CO}_2 \text{ dm}^{-2} \text{ hr}^{-1}$ for temperate broad-leaved evergreens and $5-18 \text{ mg CO}_2 \text{ dm}^{-2} \text{ hr}^{-1}$ for conifers (Larcher, 1980). Thus broad-leaved evergreens have the advantage in mild, moist climates where year-round production is possible. In colder climates freezing induces cavitation in large vessels and photosynthesis ceases with the disruption of the water supply to the leaves. The tracheid system of the conifers is less susceptible to cavitation, and appreciable assimilation can continue even though temperatures are below freezing (Jarvis and Sandford, 1986). Despite low photosynthetic rates the conifers can make full use of the short growing season in colder regions, and so gain a competitive advantage over deciduous species.

6.7 LOW TEMPERATURE RESISTANCE IN TEMPERATE PLANTS

Temperate perennial species have evolved resistance to subfreezing temperatures by tolerating the presence of ice in their tissues, or by deep supercooling which prevents ice forming even at temperatures as low as $-40 \,^\circ\text{C}$ (George et al., 1982). In plants with limited frost tolerance ice begins to form in extracellular spaces where there are relatively large amounts of free water. Ice crystals continue to grow as water is drawn out of the cells. This dehydration process increases the concentration of the cell fluids and thereby depresses its freezing point (Gusta et al., 1983). Severe dehydration will eventually kill the cells, and plants with limited hardiness survive because regenerative organs are protected from extreme cold by an insulating layer of snow or are buried in the soil (Burke et al., 1976). Extremely cold-hardy deciduous trees, such as paper birch (*Betula papyrifera*) and trembling aspen (*Populus tremuloides*), and conifers native to boreal regions, also resist low temperature injury through extracellular ice formation. However, temperate forest species survive winter cold by preventing ice formation in the sensitive xylem ray cells by deep supercooling. Freezing is an exothermic process and can be detected in plant tissues by a sudden release of heat. For many temperate species exothermic activity is detected from -41 to $-47 \,^\circ\text{C}$ (George et al., 1974). It is at these temperatures that deep supercooled water freezes in the xylem. Low-temperature exotherms are characteristic of ring-porous species such as red oak and beech and, because the xylem is killed if intracellular freezing occurs, they are restricted to regions with less severe winters (Figure 6.21). This physiological restriction usually does not occur in diffuse-porous species, hence their ranges often extend into colder regions. Winter minimum temperature is therefore one of the principal factors which determine the natural range of many tree species (Sakai and Weiser, 1973).

6.8 PHENOLOGICAL RESPONSES IN DECIDUOUS FOREST ECOSYSTEMS

The seasonal rhythm so characteristic of temperate deciduous forests reflects the physiological response of the plants to periodic changes in their environment. Shorter

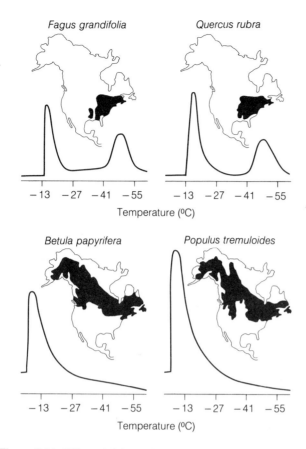

Fagus grandifolia

Quercus rubra

Temperature (°C)

Betula papyrifera

Populus tremuloides

Temperature (°C)

Figure 6.21 Differential thermal analyses and natural geographic ranges of selected hardwood species. In species with more restricted ranges living xylem tissues are generally killed at temperatures corresponding to the second exotherm. (After George *et al.*, 1974.) (Reproduced with permission from M. F. George *et al.*, Low temperature exotherms and woody plant distribution, *HortScience*, 1974, **9**, 521.)

days and cooler temperatures during the winter season are marked by a period of dormancy in which active growth ceases. The onset of dormancy is signalled by the loss of chlorophyll in the leaves; they then turn yellow or red as anthocyanins and other pigments are exposed, and are shed soon afterwards because of physiological changes which occur in the abscission zone at the base of the leaf petioles. The cells in the abscission zone can be distinguished anatomically by their small size and general absence of lignification. In healthy leaves abscission is prevented as long as indole acetic acid (IAA) is transmitted to the abscission zone from healthy leaves. The production of IAA eventually declines, and as the leaves become senescent they begin to produce ethylene, which stimulates enzyme activity in the cells of the separation layer. The production of cellulase and pectinase causes the cell walls to disintegrate; the petiole is weakened and eventually breaks (Goodwin and Mercer, 1983). At the same time cell division occurs adjacent to the stem, and a protective cork-like layer develops. Lateral buds are formed in the leaf axils long before there is a

danger of frost, and terminal buds are present at the tips of the stems, ready to resume growth the following spring.

Hardy woody perennials also undergo various biochemical changes to prepare them for the onset of winter. Cold acclimation occurs in two stages. Cold resistance to temperatures of −15 to −20 °C is initially induced by shortening day lengths in late summer, with hardening to temperatures of −40 °C or lower subsequently achieved when the plants are subjected to light frosts. During the first stages of acclimation starches and sugars begin to accumulate, tissue water content decreases, and hardiness-promoting hormones are synthesized (Weiser, 1970). Abscisic acid (ABA) has been tentatively identified as a hardiness promoter, while gibberellic acid (GA_3) may inhibit this process, although it is the ratio of these hormones rather than their absolute concentration which appears to initiate acclimation (Carter and Brenner, 1985). In the second stage of acclimation enzyme-mediated reactions occur which lead to the reorientation of protein molecules so that they resist severe dehydration, and hence maintain the integrity of the cells and tissues.

The period of deep innate-dormancy which develops in response to short photoperiods and cold nights is ordinarily broken by exposure to low temperatures for periods of one or more months. Normal bud break will not occur in hardened plants without this cold treatment, although requirements differ between species and intraspecifically within geographic races. No chilling requirement is needed to break dormancy in red maples (*Acer rubrum*) originating from the south-eastern United States, whereas normal growth resumes in northern individuals only after one month of chilling (Perry and Wang, 1960). Similar chilling requirements have been reported in sugar maple (*A. saccharum*), the response being conditioned by the length of time the trees are exposed to temperatures below 7 °C (Kriebel and Wang, 1962). After the chilling requirement has been satisfied, warm spring temperatures stimulate the production of gibberellins in the bud scales and leaf primordia, and once this occurs the plants quickly lose their hardiness. In species such as elm (*Ulmus americana*) and red maple, bud break occurs once maximum temperatures reach 15 °C, while temperatures above 21 °C are needed before black ash (*Fraxinus nigra*) and red ash (*F. pennsylvanica*) resume growth (Ahlgren, 1957). The actual date on which this occurs can vary by as much as a month from one year to the next (Table 6.1). This ends a leafless period that lasts for more than 200 days in more northerly parts of the temperate forest.

Carbohydrate reserves in most temperate deciduous trees decrease rapidly in the spring as new leaves and twigs develop and as demands for respiration increase with rising temperatures (Kramer and Kozlowski, 1960). Seasonal fluctuations are less pronounced in evergreen species since they depend more on current photosynthesis for shoot growth in the spring than on stored

Table 6.1 Variation in the timing of phenological events in hardwood species in north-eastern Minnesota, USA. The upper date is the earliest that the response was observed during the 5-year period 1951–56: the lower date is the latest date on which the response commenced (after Ahlgren, 1957).

	Bud break	Leaf flush	Flowering	Pollen shed	Leaf colour change	Leaf fall	Cambial activity starts	Cambial activity ceases
Black ash	27 Apr	20 May	27 Apr	19 May	29 Aug	19 Sep	5 May	20 Jun
(*Fraxinus nigra*)	25 May	8 Jun	25 May	5 Jun	13 Sep	3 Oct	21 May	29 Aug
Red ash	24 Apr	10 May	23 Apr	12 May	6 Sep	19 Sep	6 May	7 Jul
(*Fraxinus pennsylvanica*)	25 May	3 Jun	25 May	4 Jun	20 Sep	3 Oct	20 May	30 Jul
Mountain ash	24 Apr	30 Apr	2 May	13 Jun	20 Aug	29 Aug	15 May	22 Jul
(*Pyrus americana*)	1 May	21 May	1 Jun	16 Jun	3 Sep	8 Oct	5 Jun	5 Aug
American elm	21 Apr	6 May	25 Apr	4 May	16 Sep	3 Oct	14 May	8 Jul
(*Ulmus americana*)	16 May	29 May	8 May	11 May	4 Oct	22 Oct	25 Jun	13 Aug
Bur oak	24 Apr	11 May	12 May	23 May	13 Sep	26 Sep	15 May	16 Jul
(*Quercus macrocarpa*)	21 May	25 May	2 Jun	8 Jun	19 Sep	14 Oct	29 May	23 Aug
Basswood	21 Apr	11 May	24 May	22 Jun	16 Sep	20 Sep	16 May	17 Jul
(*Tilia americana*)	5 May	25 May	28 Jun	28 Jun	20 Sep	10 Oct	25 May	28 Jul
Trembling aspen	2 Apr	27 Apr	14 Apr	29 Apr	26 Sep	4 Oct	28 Apr	4 Aug
(*Populus tremuloides*)	30 Apr	21 May	24 Apr	5 May	12 Jun	14 Oct	23 May	3 Sep
Bigtooth aspen	24 Apr	19 May	23 Apr	5 May	20 Sep	4 Oct	14 May	4 Aug
(*Populus grandidentata*)	11 May	8 Jun	15 May	18 May	4 Oct	14 Oct	22 May	3 Sep
Paper birch	6 Apr	29 Apr	2 Apr	28 Apr	16 Sep	26 Sep	3 May	4 Aug
(*Betula papyrifera*)	5 May	21 May	23 Apr	21 May	26 Sep	10 Oct	21 May	29 Aug
Yellow birch	6 Apr	3 May	2 Apr	13 May	29 Aug	26 Sep	4 May	4 Aug
(*Betula lutea*)	1 May	25 May	16 May	29 May	26 Sep	4 Oct	20 May	11 Aug
Red maple	21 Apr	27 Apr	18 Apr	–	16 Sep	4 Oct	2 May	6 Jul
(*Acer rubrum*)	11 May	18 May	30 Apr	7 May	4 Oct	14 Oct	3 Jun	30 Jul

(Reprinted with permission from C. E. Ahlgren, Phenological observations of nineteen native tree species in northeastern Minnesota, *Ecology*, 1957, **38**, 624.)

Table 6.2 Seasonal changes in the pattern of allocation of photosynthate in the canopy of white oak (after McLaughlin and McConathy, 1979)

Date	Gross photosynthesis (mg CO_2 dm^{-2} h^{-1})	Incorporation leaves (%)	Incorporation branches (%)	Respiration leaves (%)	Respiration branches (%)	Translocation leaves (%)	Translocation branches (%)
April 22–29	5.9	41	3	94	5	–35*	–43
May 17–24	9.0	16	8	26	5	58	45
June 17–24	10.4	22	14	27	6	51	31
August 2–9	9.3	15	14	33	13	52	25
October 4–11	4.0	15	15	14	21	71	35

*negative values indicate a net import of carbon from branches.

(Reproduced with permission from S. B. McLaughlin and R. K. McConathy, Temporal and spatial patterns of carbon allocation in canopy of white oak, *Canadian Journal of Botany*, 1979, **57**, 1411.)

food reserves. In white oak (*Q. alba*) newly expanding leaves utilize about 40% more photosynthate in cell growth and cell-wall thickening than they are producing, although for the remainder of the year the foliage produces considerably more photosynthate than is needed to maintain the canopy (Table 6.2). In the north-eastern United States the leaf buds begin to break about late April and flushing continues for a period of about 2 weeks. Paper birch and trembling aspen are amongst the first species to come into leaf. Beech, sugar maple and oak begin to flush about a week later, but the oak leaves develop more slowly and it can take 4–5 weeks before they are fully expanded (Federer, 1976). The rate at which water vapour diffuses from these newly opened leaves also varies between species because of differences in the size, density and physiological behaviour of the stomates, intercellular air spaces, and characteristics of the boundary layer adjacent to the leaf surface. Diffusion resistance in young birch leaves is less than 5 s cm^{-1}, but is more than 10 s cm^{-1} in oaks: once the leaves are fully developed, resistances drop to 3–5 s cm^{-1} in all species. Late flushing and high diffusive resistance limit transpiration rates in the spring, and this may be an important mechanism to reduce water loss in ring-porous species, such as oak, which must form new conducting tissue at the start of each growing season (Federer, 1976). Diameter growth usually begins early in species like oak and elm, whereas in diffuse-porous trees such as the maples it is delayed until the leaves have expanded (Ladefoged, 1952).

In most species diameter growth continues for several months, and by the end of the season 2–3 mm of wood will normally have been added to the girth of the trunk, although the growth rate is very dependent on environmental conditions. Growth of European oaks increases in years with warm springs or when warm weather occurred in the previous autumn, and in years when the early summer was unusually wet (Fletcher, 1974). Shoot growth usually occurs for a comparatively short period. In the southern United States post oak and white oak complete most of their extension growth in only 11 days (Johnston, 1941): a 30-day period is typical for sugar maple, ash and red oak growing in the north-eastern United States but in birch and trembling aspen shoot growth continues for about 60 days (Kienholz, 1941). Those which grow for most of the frost-free season are characteristically early successional species in which shoot extension commonly exceeds 100 cm each year. Conversely, extension growth in the late successional species is determinate in that the shoot is fully preformed, and subsequent elongation rarely exceeds 50 cm yr^{-1} (Marks, 1975). Birch and trembling aspen are shade-intolerant species which establish in large openings in the forest and occupy the site for 100–150 years. Growth in these species is indeterminate and continues for as long as weather conditions are suitable. Longer-lived species may invest more resources in underground biomass to ensure long-term success under competitive

conditions. However, it is not uncommon for trees like oaks and elms to develop 'lammas shoots' in late summer just before they go dormant.

6.9 THE MICROCLIMATE IN DECIDUOUS FORESTS

Considerable changes occur in the forest microclimate as the trees begin to leaf out in the spring. Between 50 and 70% of the incident radiation penetrates to the forest floor when the trees are in their leafless state but this decreases to less than 10% when the leaves are fully expanded (Figure 6.22). In addition to leaf area, light intensities within a forest depend on the angle of the leaves and the altitude of the sun (Loomis *et al.*, 1967), and considerable spatial and temporal variation arises because of sunflecks and openings in the canopy through which unfiltered light passes (Figure 6.23). Broad deciduous leaves have high absorption in the ultraviolet, visible and the far infrared wavelengths (Gates, 1980), and considerable differences arise in the spectral compo-

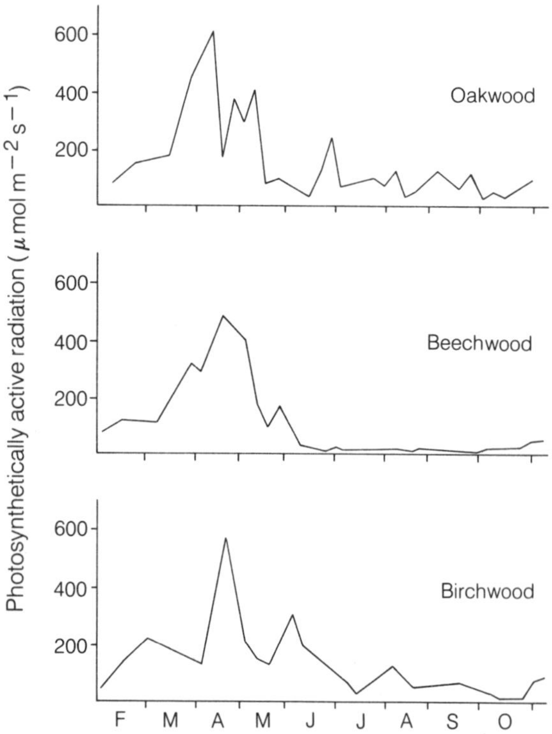

Figure 6.22 Seasonal changes in photosynthetically active radiation (400–800 nm) beneath different hardwood canopies. (After Tasker and Smith, 1977.) (Reprinted from *Photochemistry and Photobiology*, **26**, Tasker and Smith, 1977, with permission from Pergamon Press Ltd., Oxford.)

Figure 6.23 Canopy structure of temperate forests: (a) balsam poplar (*Populus balsamifera*) in the mixed-wood region of Saskatchewan; (b) coigue (*Nothofagus dombeyi*) in the Valdivian rainforest. ((a) R. A. Wright; (b) L. Ghermandi.)

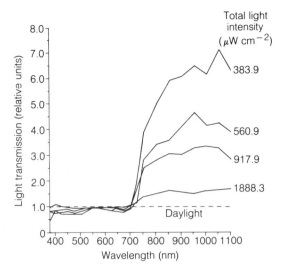

Figure 6.24 Standardized transmission curves of light of differing intensities beneath canopies of European beech (*Fagus sylvatica*) with proportionately more infra-red radiation penetrating the densest canopies. (After Goodfellow and Barkham, 1974.) (Redrawn from S. Goodfellow and J. P. Barkham, Spectral transmission curves for a beech (*Fagus sylvatica*) canopy, *Acta Botanica Neerlandica*, 1974, **23**(3), 228.)

Table 6.3 Estimation of R:FR for shadelight beneath broad-leaf deciduous canopies (after Morgan and Smith, 1981)

	Sky conditions	R:FR
beech	all conditions	0.16–0.97
oak	all conditions	0.12–0.77
birch	all conditions	0.56–0.78
sugar maple	clear	0.8–0.28
	overcast	0.21
sweet chestnut	clear	0.12
daylight		1.15

(Reproduced with permission from D. C. Morgan and H. Smith, Non-photosynthetic responses to light quality, in *Physiological Plant Ecology I: Responses to the Physical Environment*, eds O. L. Lange *et al.*, published by Springer-Verlag GmbH, 1981.)

sition of the light depending on the density of the canopy (Figure 6.24).

The spectral properties of the leaves also change significantly during their life span. The very young leaves of white oak (*Q. alba*) tend to absorb radiation in the visible and short wavelengths, but once they are fully developed there is increased reflectance in the visible green wavelength, and higher absorption in the red and near infrared. Reflectance of green light increases as the chlorophyll begins to disappear, and there is noticeably less reflectance of near infrared radiation just prior to abscission when the leaves are brown (Gates, 1980). Most significant is the seasonal change which occurs in the ratio of red light (600–700 nm) to far-red (700–760 nm). In sunlight the red:far-red (R:FR) ratio is 1.15, but can be less than 0.15 under deciduous canopies (Table 6.3). These wavelengths span the action spectra of phytochrome, a photoreceptor which is converted to its P_r form in far-red light and changes to its P_{fr} form in red light. The balance between these interconvertible forms triggers a variety of plant responses. The high R:FR ratio in spring promotes seed germination, but as the canopy develops R:FR drops and vegetative growth is promoted (Morgan and Smith, 1981). Changes in day length are also sensed by the phytochrome system, and as the nights get longer in the autumn bud growth ceases, the leaves abscise, and the plants prepare for winter dormancy.

Maximum air temperatures are generally lower and minimum temperatures higher within deciduous forests compared with open areas, and the daily temperature

ranges are typically small, especially during the growing season (Lee, 1978). However, energy transfer within the canopy often results in daytime temperatures in the crown layer being a few degrees warmer than in the air above. Energy radiates from the forest during the night, and as the air cools it sinks to the ground, and so creates a temperature inversion. The effect is reversed in winter, and a negative temperature gradient is established between the forest and the atmosphere: this is most pronounced during the daytime.

The size of the crowns and amount of foliage also determine the amount of precipitation that is intercepted by hardwood forests. Over the course of a year interception losses range from 8% under beech to more than 30% in an oak–hornbeam stand (Galoux *et al.*, 1981). These values are very dependent on rainfall intensity and also vary seasonally (Figure 6.25): in mixed hardwood forests in the eastern United States interception losses decrease from 17% in summer to 7% in winter. Similarly, stem flow varies because of differences in the smoothness of the tree bark. The exceptionally high stem flow in beech can lead to acidification of the soil, and pH values below 4 have been reported in Europe because of this (Ellenberg, 1988). In addition, wind speeds are typically reduced by 50–80% within a forest, and this, together with lower temperatures, keeps the air relatively humid. Carbon dioxide produced by plant respiration and by various soil processes usually decreases with height above ground, with noticeably lower concentrations in the canopy during periods of active photosynthesis.

6.10 SEASONALITY AND THE UNDERSTORY

The seasonal changes in microclimate are reflected in changing aspects of the understory during the year (Figure 6.26). Many herbaceous species appear in the early spring before the trees leaf out, but these are replaced by more **shade-tolerant** species as the canopy develops and in autumn and winter evergreen species are most conspicuous. The successive development in the ground flora is linked to the light requirements of the different

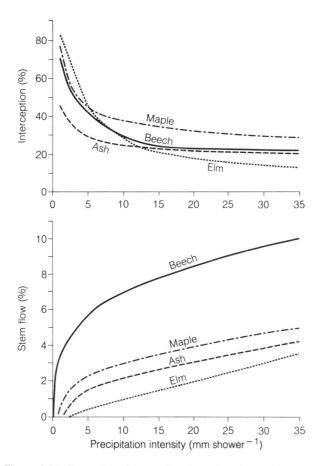

Figure 6.25 Percentage interception (upper) and percentage stem flow (lower) for selected hardwood species in relation to rainfall intensity. (After Kitteredge, 1948.) (Reproduced from J. Kittredge, *Forest Influences*; published by McGraw-Hill Inc., 1948.)

aceous species which appear as the tree canopy develops complete most of their growth before the forest floor is heavily shaded. Other species which are present in the shaded summer flora develop mostly after the canopy is fully expanded. However, they usually persist for some time after the trees lose their leaves, whereas the early summer plants are senescent before leaf fall. For these species maximum assimilation occurs under low light intensities. Many are obligate shade plants and are injured when exposed to higher levels of illumination. In addition, there are evergreen plants in the understory which, although green during the winter, are generally dormant at this time. New leaves are mostly produced in the spring and this is also the usual time for flowering.

The trees also vary in terms of their light requirements: species such as beech and sugar maple are very tolerant of shade but trembling aspen and birch grow best under high light intensities (Table 6.4). Shade tolerance is an indication of a species' ability to withstand competition and is therefore related to its successional status. The ability of trees to survive in sun or shade is determined by their growth rate under differing light intensities (Figure 6.27). Most species show a degree of shade adaptability, and for many seedlings net photosynthesis is often highest under partially shaded conditions. However, higher light compensation points characteristic of shade-intolerant species, combined with high respiration rates during dark periods, result in negative net assimilation in heavy shade (Loach, 1970). In addition, photosynthesis is not uniform within a canopy. Assimilation rates are higher in the brightly illuminated upper crown of white oak (Table 6.5), despite high leaf temperatures and mid-day stomatal closure (Aubuchon *et al.*, 1978). This is generally attributed to the differences in leaf morphology and physiology which result when leaves develop under different light intensities. **Sun leaves** are usually thicker and have more stomata than shade leaves; they are also richer in chlorophyll, contain more soluble protein and have higher levels of RuDP carboxylase (Boardman, 1977). The size and shape of the leaves can also be quite variable in different parts of the crown. In oaks the **shade leaves** are larger and less deeply lobed than the sun leaves in the upper crown (Figure 6.28). Leaves exposed to the sun are inclined at steeper angles

species. Those that complete their development in the early spring are **shade-intolerant** species which are adapted to high light intensities. The light compensation point in these species is rather high, but so too is the saturation light intensity for photosynthesis (Sparling, 1967). Growth may be enhanced by the flush of nutrients that accompanies snowmelt, but these plants are short-lived and there is rapid and synchronous senescence as the canopy closes (Mahall and Bormann, 1978). Herb-

Table 6.4 Shade tolerance of the broad-leaved deciduous forest species of eastern North America (after Baker, 1950)

Very tolerant	Tolerant	Intermediate	Intolerant	Very intolerant
Ostrya virginiana	*Acer rubrum*	*Betula lutea*	*Juglans nigrans*	*Salix* spp.
Carpinus caroliniana	*Acer saccharinum*	*Castanea dentata*	*Betula papyrifera*	*Populus tremuloides*
Fagus grandifolia	*Tilia americana*	*Quercus alba*	*Liquidambar styraciflua*	*Populus deltoides*
Ilex opaca	*Nyssa sylvatica*	*Quercus coccinea*	*Prunus serotina*	*Betula populifolia*
Acer saccharum	*Aesculus glabra*	*Quercus velutina*		*Robinia pseudoacacia*
Cornus florida		*Ulmus americana*		
		Celtis occidentalis		
		Fraxinus americana		
		Fraxinus nigra		

(Reproduced with permission from H. S. Baker, *Principles of Silviculture*; published by McGraw-Hill, Inc., 1950.)

Figure 6.26 Seasonal aspects of deciduous forests: (a) spring carpet of bluebells (*Hyacinthoides non-scriptus*) in an English woodland; (b) mid-summer in beech forest in south-eastern United States; (c) autumn in a stand of aspen (*Populus tremuloides*) in Canada; (d) British oak (*Quercus robur*)–ash (*Fraxinus excelsior*) woodland in winter. ((a), (d) G. F. Peterken; (b) M. Huston; (c) R. A. Wright.)

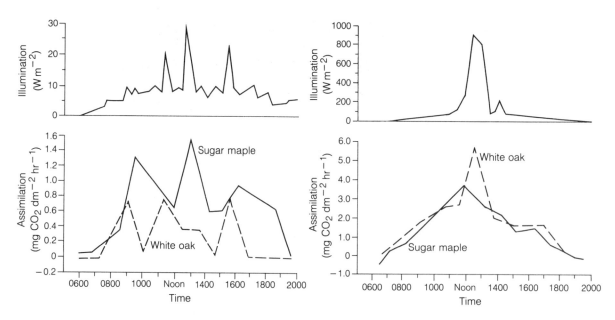

Figure 6.27 Net assimilation rate of CO_2 by sugar maple (*Acer saccharum*) and white oak (*Quercus alba*) in sunny and shaded sites in relation to levels of illumination. (After Geis *et al.*, 1971.) (Reproduced with permission from J. W. Geis, R. L. Tortorelli and W. R. Boggess, Carbon dioxide assimilation of hardwood seedlings in relation to community dynamics in central Illinois I. Field measurements of photosynthesis and respiration, *Oecologia*, 1971, **7**, 281, 284.)

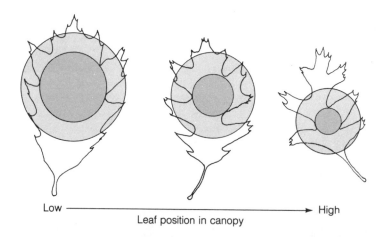

Low ←――――――――――――――――――――→ High

Leaf position in canopy

Figure 6.28 Differences in the shape of leaves of black oak (*Quercus velutina*) growing at various heights within the canopy: the small circle is the largest that can be inscribed on the leaf; the larger circle has the same surface area as the leaf. (After Horn, 1971.) (From H. S. Horn, *The Adaptive Geometry of Trees*, copyright © 1971, reproduced by permission of Princeton University Press.)

Table 6.5 Photosynthetic rates within the crown of white oak (*Quercus alba*) in eastern North America (after Aubuchon *et al.*, 1978)

Crown section	PAR (μ E m^{-2} s^{-1})	Leaf temperature (°C)	Leaf conductance (cm s^{-1})	Rate of photosynthesis (mg CO_2 dm^{-2} h^{-1})
Top	1140	30.8	0.16	7.0
Middle	650	29.0	0.15	5.6
Bottom	390	28.5	0.11	4.6

(Reproduced from R. R. Aubuchon, D. R. Thompson and T. M. Hinckley, Environmental influences on photosynthesis within the crown of a white oak, *Oecologia*, 1978, **35**, 302.)

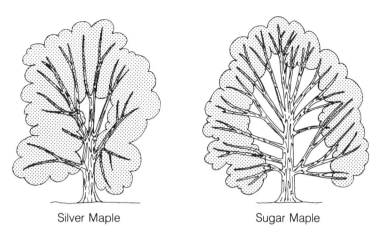

Silver Maple   Sugar Maple

Figure 6.29 Comparative distribution of leaves in the multi-layered canopy of silver maple (*Acer saccharinum*) and the mono-layered canopy of sugar maple (*Acer saccharum*) which has a competitive advantage under shaded conditions. (After Horn, 1975.) (Redrawn from H. S. Horn, Forest succession, *Scientific American*, 1975, **232**, 96.)

Table 6.6 Mean leaf angles of sun and shade leaves in broad-leaved forest species of eastern North America (after McMillen and McClendon, 1979)

	Degrees from horizontal	
	sun leaves	shade leaves
Populus deltoides	75.7	32.3
Prunus americana	33.4	16.4
Fraxinus pennsylvanica	36.8	14.4
Quercus rubra	10.1	11.5
Acer saccharinum	16.9	12.7
Acer saccharum	14.6	7.8

(Reproduced with permission from G. G. McMillen and J. H. McClendon, Leaf angle: an adaptive feature of sun and shade leaves, *Botanical Gazette*, **140**, 437–42, published by the University of Chicago Press, 1979.)

than those in the shade (Table 6.6): the effect is most pronounced in shade-intolerant species. Differences can also be seen in the pattern of branching and the arrangement of leaves in the canopy. Shade-intolerant species have multilayered canopies in which leaves are randomly scattered throughout the branch volume (Figure 6.29). This is quite distinct from the arrangement in shade-tolerant species in which the leaves are concentrated around the periphery of the canopy (Horn, 1975).

6.11 BIOMASS AND PRODUCTION IN TEMPERATE FORESTS

Carbon assimilation in temperate trees is by the C_3 pathway. Photosynthesis occurs over a temperature range of about 5–25 °C, with optimal conditions between 10 and 20 °C (Jarvis and Sandford, 1986). Maximum rates of assimilation under light saturation range from $8.7\,\mu mol\,m^{-2}\,s^{-1}$ in *Populus tremula* to $15.9\,\mu mol\,m^{-2}\,s^{-1}$ in *Liriodendron tulipifera*, with an equivalent value of

$5.1\,\mu mol\,m^{-2}\,s^{-1}$ reported for evergreen black beech (*Nothofagus solandri*). Assimilation rates for leaves vary with their stage of development and age, increasing over summer then falling rapidly as the leaves senesce (Dickmann, 1971; Richardson, 1957). Assimilation rates also vary between sun and shade foliage within the same tree: photosynthetic capacity is generally higher in sun leaves, and assimilation in the strongly illuminated part of the canopy is proportionately higher. The rate of assimilation declines under drought conditions because of stomatal closure (Kozlowski, 1982); nutrient stress similarly limits leaf development and longevity, and lowers leaf chlorophyll content (Linder and Rook, 1984). Consequently, the rate at which plant material accumulates even within similar forest types can vary considerably according to environmental conditions. For example, annual above-ground production in oak–hickory forests can range from $450\,g\,m^2\,yr^{-1}$ on infertile clays to over $2100\,g\,m^2\,yr^{-1}$ on alluvial soils (Peet, 1981). Biomass normally increases steadily over time despite a gradual thinning of the stands but eventually the rate of growth declines. Thus, biomass in oak–pine forests can reach about $400\,t\,ha^{-1}$ after 200 years, with little change in annual production occurring after 45 years (Figure 6.30). The composition of the forest also changes as the shade-intolerant species are replaced by those better adapted to conditions beneath the canopy.

Above-ground biomass in mature deciduous forests typically ranges from 120 to $300\,t\,ha^{-1}$ (Table 6.7) with an additional 30–$80\,t\,ha^{-1}$ present as roots. Gross primary production is comparatively high but much of this is used to meet the respiration demands of the supporting woody tissue; net productivity averages about $10\,t\,ha^{-1}\,yr^{-1}$. In European mixed oak forest, new woody growth accounts for 50% of the total aerial production each year, 26% is used for new foliage and 17%

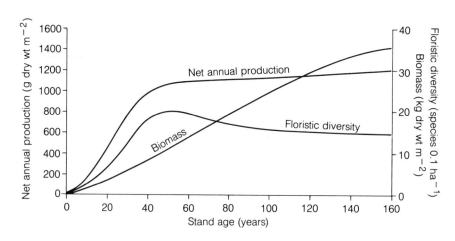

Figure 6.30 Changes in production, biomass and species diversity during post-fire succession in oak–pine forest in eastern North America. (After Whittaker, 1970.) (Reprinted with the permission of Macmillan Publishing Company from *Communities and Ecosystems* by Robert H. Whittaker. Copyright © 1970 by Robert H. Whittaker.)

Table 6.7 Biomass and productivity of temperate deciduous forests (after DeAngelis *et al.*, 1981)

Forest type	Location	Age (yr)	Growing season length (days)	Growing season mean temp. (°C)	Growing season precip. (mm)	Biomass above- (t ha⁻¹)	Biomass below-ground (t ha⁻¹)	Production above- (t ha⁻¹)	Production below-ground (t ha⁻¹)
Mixed oak	Belgium	80	155	13.8	450	121.0	35.0	12.2	2.3
Beech	Denmark	85–90	145	14.1	277	221.3	43.2	15.0	3.8
Mixed oak	UK	80	244	10.0	797	128.4	75.2	9.9	2.7
Oak–hornbeam	Czechoslovakia	50–70	245	13.7	450	161.0	76.1	10.7	–
Oak	Hungary	65	190	17.0	316	221.9	35.6	7.2	–
Beech–maple	USA	110	110	17.5	–	132.8	28.5	9.6	1.9
Tuliptree–oak	USA	50	180	18.1	584	133.8	36.0	7.4	7.1
Oak–hickory	USA	30–80	180	18.1	584	121.6	33.3	7.8	2.0
Chestnut oak	USA	30–80	180	18.1	584	137.9	33.3	9.5	2.0
Beech	Japan	150	244	16.4	1182	292.4	45.2	10.1	1.5

(Source: D. L. DeAngelis, R. H. Gardner and H. H. Shugart, Productivity of forest ecosystems studied during the IBP: the woodlands data set, in *Dynamic Properties of Forest Ecosystems*, ed. D. E. Reichle; published by Cambridge University Press, 1981.)

goes into flowers, fruits and other materials which are returned as litter (Figure 6.31). The herbaceous ground cover contributes about 5% to the total annual production, and the remaining 2% is incorporated into the shrubs. Total above-ground biomass in this oak woodland was 121 t ha⁻¹. Root biomass contributed an additional 35 t ha⁻¹: root material was added at the rate of 2.1 t ha⁻¹ yr⁻¹. Whittaker and Woodwell (1969) provide similar data for 45–60-year oak–pine forest in North America: above-ground tree biomass totalled 64 t ha⁻¹, with shrubs and seedlings contributing a further 15.8 t ha⁻¹. Root biomass was 36.3 t ha⁻¹, and comprised 35% of the total plant biomass at this site (Table 6.8). Production totalled 12.0 t ha⁻¹ yr⁻¹, of which over 70% was above ground; new leaf production was 3.8 t ha⁻¹ yr⁻¹, or 32% of the total. A comparable study in temperate evergreen oak forests in Japan reported a mean total biomass of 445.6 t ha⁻¹ with an annual rate of net production of only 18.3 t ha⁻¹ yr⁻¹ (Kira and Yabuki,

1978). Leaf production at this site averaged 4.1 t ha⁻¹ yr⁻¹, and accounted for 22% of total annual production.

6.12 LITTERFALL IN TEMPERATE FORESTS

Estimates of litterfall in deciduous forests, excluding large trunks, range from 324 g m⁻² yr⁻¹ in Swedish oak–birch forests to 624 g m⁻² yr⁻¹ in Dutch oakwoods (Table 6.9). Leaves contribute 53–88% of the total litterfall (Figure 6.32). Some trunk wood is added to the litter each year: this amounts to an additional 15–33 g m⁻² yr⁻¹ in American deciduous forests and as much as 76 g m⁻² yr⁻¹ in British mixed oak woodland. The amount and type of litter produced varies annually. Litter production in oak–maple forests in Canada ranged from 3.4 to 6.3 t ha⁻¹ yr⁻¹ over a 4-year period, largely because of variations in the woody component (Bray and Gorham,

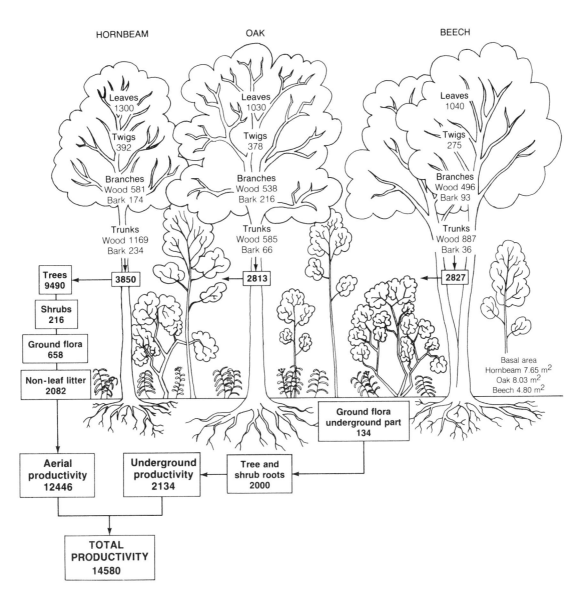

Figure 6.31 Annual primary productivity (kg dry weight ha^{-1}) for different compartments in European mixed-oak forest. (After Duvigneaud and Denaeyer-De Smet, 1970.) (Reproduced with permission from P. Duvigneaud and S. Denaeyer-De Smet, Biological cycling of minerals in temperate deciduous forests, in *Analysis of Temperate Forest Ecosystems*, ed. D. E. Reichle, published by Springer-Verlag GmbH, 1970.)

1964). Similar variations in woody litter input have been reported beneath mixed oak forests in England (Sykes and Bunce, 1970). Increases in woody material usually occur after exceptionally strong winds. Leaf litter varies because of reduced foliage in years with heavy fruiting and seed production, or when insect damage reduces the vigour of the trees.

In deciduous species leaf fall normally occurs over a few weeks at the end of the growing season, although the leaves can be lost at any time during the growing season as a result of injury, disease or unfavourable growing conditions. The leaves of the older trees are usually shed first; however, leaf fall does not commence at precisely the same time each year because the response of plants to the shorter day lengths of autumn is modified by moisture and temperature conditions. Various oaks, including

Q. coccinea and *Q. rubra*, beeches (*F. sylvatica* and *F. grandifolia*) and other species retain **marcescent** (apparently dead) leaves throughout the winter and shed them in the spring, although this is mostly a characterisitic of juvenile trees (Addicott, 1982). Each species has a characteristic pattern of leaf fall. In European oak forests maximum leaf fall in oak occurs in November (Witkamp and van der Drift, 1961): other species shed their leaves earlier, while alder has two main periods of leaf abscission, one in July during which green leaves are shed and the second in October (Figure 6.33). Sykes and Bunce (1970) have also reported loss of green leaves equivalent to 6% of total leaf fall during the spring and summer in British mixed deciduous woodland: this was mainly from birch.

A different pattern of litterfall occurs in temperate broad-leaved evergreen forests. Most evergreen trees in

Table 6.8 Biomass and production in oak–pine forest in the north-eastern United States (after Whittaker and Woodwell, 1969)

	Biomass (dry g m^{-2})		Production (dry g m^{-2} yr^{-1})	
	tree stratum	tree seedlings and shrubs	tree stratum	tree seedlings and shrubs
stem heartwood	1515	–	–	–
stem sapwood	1996	54.3	148.9	4.9
stem bark	812	16.4	26.5	1.8
live branchwood	1410	40.4	193.4	13.3
dead branchwood	180	6.9	–	–
current twigs	55	7.8	53.9	7.8
leaves	412	30.7	351.0	31.0
fruits	21	1.6	20.0	1.6
flowers	2	0.4	2.1	0.4
total above ground	6403	158.5	795.8	60.8
root crowns	1663	234.7		
roots	1662	70.8		
total below ground	3325	305.5	259.7	73.0
total	9728	464.0	1055.5	133.8

(Source: R. H. Whittaker and G. M. Woodwell, Structure, production and diversity of the oak–pine forest at Brookhaven, New York, *Journal of Ecology*, 1969, **57**, 162.)

Table 6.9 Litter production in temperate deciduous forests (after DeAngelis *et al.*, 1981)

Forest type	Location	Age (yr)	Litter components (g m^{-2} yr^{-1})			
			leaves	flower & fruits	branches	total
Beech	Denmark	85–90	269	23	90	382
mixed oak	UK	80	324	21	100	445
Beech	France	mature	345	48	106	499
Oak	Netherlands	140	438	79	107	624
Oak–birch	Sweden	40–200	171	21	132	324
Beech	Sweden	80–100	275	71	76	422
Beech–maple	USA	110	342	27	–	–
Oak–hickory	USA	30–80	410	40	30	480
Chestnut oak	USA	30–80	390	40	15	445
Tuliptree	USA	30–80	370	30	33	433
Beech	Japan	150	349	–	58	–

(Source: D. L. DeAngelis, R. H. Gardner and H. H. Shugart, Productivity of forest ecosystems studied during the IBP: the woodlands data set, in *Dynamic Properties of Forest Ecosystems*, ed. D. E. Reichle; published by Cambridge University Press, 1981.)

Figure 6.32 Leaf litter from deciduous species such as maples (*Acer saccharum*) is rapidly broken down and the nutrients returned to the soil. (R. A. Wright.)

New Zealand shed their leaves in the spring about two months after the peak for non-leaf litter (Bray and Gor-

ham, 1964). A spring peak in leaf fall is also noted in the evergreen forests of Japan (Nishioka and Kirita, 1978) although some leaf litter is supplied to the forest floor throughout the year because of the rather protracted leaf-shedding habit of the dominant species, *Castanopsis cuspidata*. Annual litterfall in these Japanese forests averages 8.44 t ha^{-1} yr^{-1}, of which 50% is contributed as leaf material (Table 6.10). The annual distribution of other fine litter reflects the flowering and fruiting activity of *C. cuspidata*; male flowers are shed in May and June, and the acorns reach maturity in December of the following year. The fall of dead branches occurs mainly in the autumn as a result of strong winds.

6.13 NUTRIENT CYCLING IN TEMPERATE FORESTS

Autumn is the time of maximum litterfall in the mixed hardwood forests at Hubbard Brook in the north-eastern

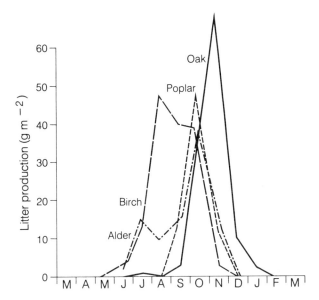

Figure 6.33 Litter production in selected hardwood species in European mixed-oak forest. (After Witkamp and van der Drift, 1961.) (Reproduced with permission from M. Witkamp and J. van der Drift, Breakdown of forest litter in relation to environmental factors, *Plant and Soil*, 1961, **15**, 298.)

Table 6.10 Annual rates of litterfall in temperate broad-leaved forests in Japan (after Nishioka and Kirita, 1978)

| | Litterfall (t ha^{-1} yr^{-1}) | | | | |
	Leaves	**Fine litter**	**Branches**	**Trunks**	**Total**
1968	3.51	1.98	1.80	0.42	7.71
1969	4.37	1.82	0.46	0.53	7.18
1970	4.10	2.29	5.76	1.67	13.82
1971	4.54	1.89	1.39	0.14	7.96
1972	3.18	1.19	0.56	0.60	5.53
Mean	3.94	1.84	1.99	0.67	8.44

(Reproduced with permission from M. Nishioka and H. Kirita, Decomposition cycles, Chapter 4 in *Biological Production in a Warm-Temperate Evergreen Oak Forest of Japan*, eds T. Kira, Y. Ono and T. Hosokawa; published by University of Tokyo Press, 1978.)

Table 6.11 Seasonal variation in input and nutrient content of litter in mature hardwood forest in north-eastern United States (after Gosz et al., 1972)

	Litterfall (dry kg ha^{-1})	**Percentage of litterfall**	**Weight of nutrients (kg ha^{-1})**	**Percentage of nutrients**
Mar–May	383	6.7	5.8	4.1
Jun–Aug	1209	21.2	31.8	22.6
Sep–Nov	2832	49.7	78.6	56.0
Dec–Feb	1278	22.4	24.2	17.2
Total	5702	100.0	140.4	100.0

(Reprinted with permission from J. R. Gosz, G. E. Likens and F. H. Bormann, Nutrient content of litter fall on the Hubbard Brook Experimental Forest, New Hampshire, *Ecology*, 1972, **53**, 782.)

United States. Approximately 50% of the total litter is returned at this time and the rest falls mostly in summer and winter (Table 6.11). Litterfall in these forests aver-

ages 5.7 t ha^{-1} yr^{-1} with a total nutrient content of 140.4 kg ha^{-1}. Practically all the nutrients are contributed by the overstory, but the nutrient content of the litter varies throughout the year depending on the type of tissue being shed. Perennial tissues, such as branches, have lower concentrations of nutrients than the leaves and other deciduous material. The nutrient content also varies according to the time of abscission. Leaves that are lost in the early autumn have a high content of nitrogen, potassium and phosphorus, but the concentration of these elements subsequently declines because they are translocated from senescing tissues. Conversely, elements such as calcium and magnesium are accumulated prior to abscission and concentrations increase as the season progresses (Gosz *et al.*, 1972).

Nitrogen storage in above-ground biomass in temperate deciduous forests is generally less than 400 kg ha^{-1}, although a considerably higher value is reported for beech forest in Sweden (Table 6.12). Phosphorus levels range from 17 to 85 kg ha^{-1}, and potassium from 154 to 456 kg ha^{-1}. Similar variations occur with calcium and magnesium. The distribution of these elements differs between species and between different parts of each species. In European mixed oak forest the amount of nitrogen, phosphorus and potassium stored in woody tissues exceeds that in the leaves by a factor of 4–5, and for calcium and magnesium by a factor of 14–16 (Table 6.13). Leaf biomass in this woodland was 3.5 t ha^{-1} compared with 108.7 t ha^{-1} for woody tissue, hence nutrient concentrations are considerably higher in the foliage. Leaf litter is therefore the principal source of recycled minerals. Average litterfall in temperate deciduous forests is 5.4 t ha^{-1} yr^{-1}, and this accounts for 80% of the nitrogen and phosphorus returned to the soil: approximately 70% of the calcium, 60% of the magnesium and 40% of the potassium is recycled in this way (Cole and Rapp, 1981). Mean turnover time for the organic matter is 4 years: the residence time for nitrogen and phosphorus is generally 5–6 years but other elements, especially potassium, are recycled more rapidly. The average time taken for natural breakdown of fallen leaf litter varies between species. Elm, alder and ash litter is usually mineralized within a year, but decomposition is much slower for red oak and beech litter (Figure 6.34). The amount of litter that accumulates on the forest floor and the quantity of nutrients stored in it therefore depends on the composition of the forest cover (Table 6.14).

Uptake of nitrogen by deciduous trees averages 75 kg ha^{-1} yr^{-1} but the amount of nitrogen required for new growth averages 98 kg ha^{-1} yr^{-1}. This shortfall is supplied by translocation from older tissues. A similar process occurs with phosphorus but other elements, especially calcium, are taken up in greater quantities than needed. The surplus is accumulated in the older tissues and not directly incorporated in the new growth (Cole and Rapp, 1981). Most of the nitrogen is derived from the breakdown of organic matter but some is also added

Table 6.12 Above-ground biomass and nutrient accumulation in temperate deciduous forest ecosystems (after Cole and Rapp, 1981)

Forest type	Location	Age (yr)	Biomass (t ha^{-1})	Nutrient accumulation (kg ha^{-1})				
				N	P	K	Ca	Mg
Mixed oak	Belgium	80	115.1	368	28	200	830	77
Mixed oak	UK	–	113.1	278	17	213	480	26
Beech	W. Germany	80	158.8	404	36	202	303	26
Beech	Sweden	45–130	104.0	1071	85	456	606	105
Beech–maple	USA	110	134.0	367	33	154	402	38
Tuliptree–oak	USA	30–80	109.1	267	21	172	537	59
Oak–hickory	USA	30–80	121.6	369	24	220	856	67
Chestnut oak	USA	30–80	137.3	397	26	242	852	45

(Source: D. W. Cole and M. Rapp, Elemental cycling in forest ecosystems, in *Dynamic Properties of Forest Ecosystems*, ed. D. E. Reichle; published by Cambridge University Press, 1981.)

Table 6.13 Concentration of nutrients in foliage and wood in the mixed oak forest ecosystem at Virelles, Belgium (after Duvigneaud and Denaeyer-De Smet, 1970)

	Nutrient concentration (kg ha^{-1})											
	Quercus robur				*Carpinus betulus*				*Fagus sylvatica*			
	leaves	twigs	wood	bark	leaves	twigs	wood	bark	leaves	twigs	wood	bark
nitrogen	24.0	12.2	63.0	46.0	26.0	18.4	44.0	32.0	21.0	8.3	44.0	13.3
phosphorus	1.5	1.0	6.6	3.1	1.5	1.4	2.8	2.1	1.4	0.7	2.6	0.8
potassium	12.4	3.6	40.0	19.6	10.8	6.6	28.9	13.2	11.4	3.3	38.0	5.7
calcium	12.0	20.4	37.0	243.0	23.0	46.3	59.0	200.0	17.0	12.6	30.0	92.0
magnesium	1.2	2.4	10.9	25.0	1.8	3.3	6.7	6.8	1.4	0.4	14.0	1.0

in rainfall and canopy leaching (Table 6.15). Inputs from precipitation typically account for 10–15% of the nitrogen returned to the soil each year but this can be considerably higher in industrial areas. Most of the phosphorus and calcium is also recycled through litterfall, but canopy leaching is important for potassium and to a lesser degree for magnesium. In undisturbed forests more nitrogen is added by way of precipitation than is lost by leaching. Nitrification is relatively unimportant and very little nitrate is produced in the soil: most of what is lost by leaching is derived from precipitation. Leaching increases following widespread disturbance because excess ammonia produced through decomposition is converted to nitrate: the hydrogen ions which are released subsequently displace potassium, calcium and other cations from exchange sites in the soil (Likens *et al.*, 1970).

Peak nutrient demands occur during the period of canopy expansion, but as the trees get older their requirements decline because growth is mostly through the formation of wood, in which nutrient levels are comparatively low. Similarly, changes in species composition during stand development can result in different nutrient demands: this also results in changes in the composition of the litter. Additional nutrients are released through root decay. The average biomass of fine live roots in temperate forests is 7053 kg ha^{-1}, dead material contributing a further 1780 kg ha^{-1} (Vogt *et al.*, 1986). Turnover of root material averages 5731 kg ha^{-1} yr^{-1} and this contributes 44 kg ha^{-1} of nitrogen and 3.5 kg ha^{-1} of phosphorus to the soil annually. Average nutrient uptake in temperate deciduous forests totals 230 kg ha^{-1} yr^{-1}, with the production of 10.1 tons ha^{-1} of new growth each year (Cole and Rapp, 1981). Nutrient requirements for these forests are moderate and practically all of the demand is met by recycling (Figure 6.35).

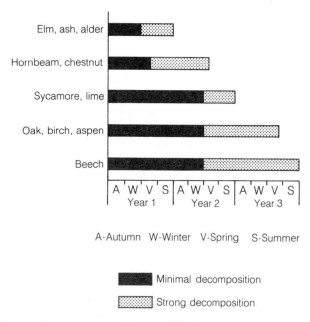

A-Autumn W-Winter V-Spring S-Summer

■ Minimal decomposition

▨ Strong decomposition

Figure 6.34 Mean time for natural breakdown of fallen leaf litter for selected species of European hardwoods. (After Ellenberg, 1988.) (Modified from H. Ellenberg, tr. by G. K. Strutt, *Vegetation Ecology of Central Europe*, 4th edn, published by Cambridge University Press, 1988.)

Table 6.14 Biomass and nutrient accumulation in the organic matter on the floor of temperate deciduous forest ecosystems (after Cole and Rapp, 1981)

Forest type	Location	Age (yr)	Litter biomass (t ha^{-1})	Nutrient accumulation (kg ha^{-1})				
				N	P	K	Ca	Mg
Mixed oak	Belgium	80	4.8	44	2	17	107	5
Mixed oak	UK	–	6.1	74	3	12	101	8
Beech	W. Germany	80	29.7	810	52	83	91	34
Beech	Sweden	45–130	5.2	86	6	104	34	5
Beech–maple	USA	110	48.0	126	8	66	372	38
Tuliptree–oak	USA	30–80	1.5	187	11	14	294	22
Oak–hickory	USA	30–80	27.0	334	22	26	517	32
Chestnut oak	USA	30–80	25.0	298	18	26	318	22

(Source: D. W. Cole and M. Rapp, Elemental cycling in forest ecosystems, in *Dynamic Properties of Forest Ecosystems*, ed. D. E. Reichle; published by Cambridge University Press, 1981.)

Table 6.15 Input of nutrients to deciduous forest ecosystems through litterfall, precipitation and canopy leaching (after Cole and Rapp, 1981)

	Nutrient fluxes (kg ha^{-1} yr^{-1})				
	N	P	K	Ca	Mg
Mixed oak (Belgium)					
litterfall	50.0	2.4	21.0	110.0	5.6
precipitation	13.0	–	5.0	19.0	5.8
canopy leaching	6.9	0.6	16.8	7.1	6.2
Beech (West Germany)					
litterfall	49.4	4.0	15.9	16.2	1.4
precipitation	21.8	0.5	3.7	12.6	2.6
canopy leaching	22.1	0.4	22.5	26.3	3.9
Mixed oak (UK)					
litterfall	63.5	2.6	19.0	83.3	9.7
precipitation	5.8	0.2	3.3	6.9	5.4
canopy leaching	9.2	0.7	35.1	28.0	14.3
Beech (Sweden)					
litterfall	69.0	5.0	14.4	31.7	4.3
precipitation	8.2	0.1	1.9	3.5	0.9
canopy leaching	1.0	0.1	11.2	6.6	2.5
Oak–hickory (USA)					
litterfall	36.5	2.7	19.8	49.1	8.7
precipitation	8.7	0.1	1.0	5.3	1.0
canopy leaching	12.4	0.6	23.9	24.1	3.7
Maple–birch (USA)					
litterfall	54.2	4.0	18.3	40.7	5.9
precipitation	6.5	< 0.1	0.9	2.2	0.6
canopy leaching	5.2	0.7	29.5	5.4	1.6

(Source: D. W. Cole and M. Rapp, Elemental cycling in forest ecosystems, in *Dynamic Properties of Forest Ecosystems*, ed. D. E. Reichle; published by Cambridge University Press, 1981.)

6.14 LITTER DECOMPOSITION

The availability of nutrients in temperate forests is related to the rate at which litter is mineralized. It is usually several weeks before soil organisms begin to attack freshly fallen leaves. Water-soluble sugars and organic acids are leached out during this intervening period and the leaves become more digestible (King and Heath, 1967). As decomposition progresses the C:N ratio decreases, most of the sugars and starches are transformed to carbon

Table 6.16 Rates of litter breakdown by different agents during the first year of decomposition in forest–steppe communities (after Zlotin and Khodashova, 1980)

Decomposition agent	Rate of litter loss (%)		
	Oak	Aspen	Elm
abiotic decomposition	6	6	5
micro-organisms	5	2	4
microfauna	10	9	22
meso and macrofauna	13	26	28
total 1st-year loss	34	43	59

(Source: R. I. Zlotin and K. S. Khodashova, tr. W. Lewis and W. E. Grant, *The Role of Animals in Biological Cycling of Forest–Steppe Ecosystems*; published by Dowden, Hutchinson and Ross, 1980).

dioxide and water, ammonia is released as proteins are broken down, and mineral elements, such as phosphorus and potassium, are leached from the surface. Some of the litter that falls to the forest floor is lost through oxidation, leaching and erosion; the remainder is converted into humus which is continually incorporated into the surface horizons, giving them their dark uniform colour (Edwards *et al.*, 1970).

In temperate regions the initial fragmentation is carried out by earthworms and smaller nematodes, mites (Acari) and springtails (Collembola), but these seldom utilize more than 20% of the litter (Jensen, 1974). Mechanical disintegration provides a more suitable substrate for the growth of fungi and bacteria. Many of these micro-organisms are present on the leaves before they are abscised but their numbers increase rapidly as soon as the litter has been wetted on the ground. During later stages of decomposition the litter is invaded by soil bacteria. Fungal populations develop in a similar way: some species are present prior to leaf fall and persist throughout the winter, others become established the following spring and are active until winter, while a third group appears in late summer almost a year after the leaves are shed (Hogg and Hudson, 1966). The activities of the decomposer population are closely integrated, and litter breakdown proceeds rapidly only when all groups are present (Table 6.16).

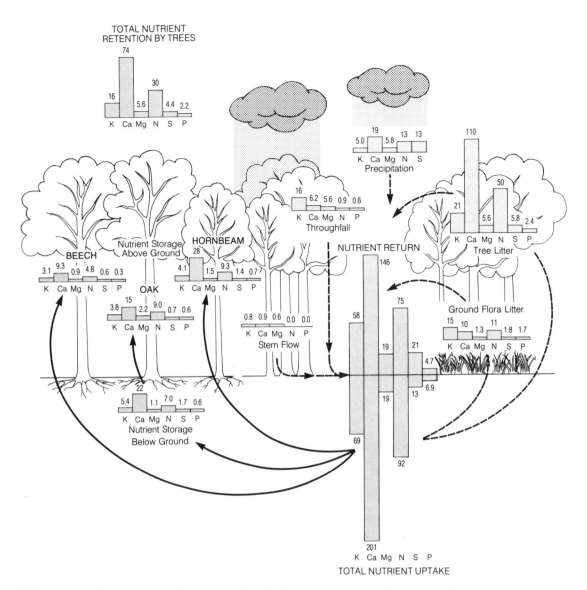

Figure 6.35 Annual cycling of macronutrients (kg ha^{-1} yr^{-1}) in European mixed-oak forest. (After Duvigneaud and Denaeyer-De Smet, 1970.) (Reproduced with permission from P. Duvigneaud and S. Denaeyer-De Smet, Biological cycling of minerals in temperate deciduous forests, in *Analysis of Temperate Forest Ecosystems*, ed. D. E. Reichle, published by Springer-Verlag GmbH, 1970.)

Although the bulk of the nutrients stored in plant tissues are returned in leaf litter in the autumn, other materials are added at various times during the year. In early summer there is a significant addition of nutrients from bud scales and flowers, and this is followed by insect frass (Gosz *et al.*, 1972). The addition of these nutrient-rich materials to the forest floor results in increased saprophyte activity and accelerated decomposition (Zlotin and Khodashova, 1980). Nutrient recycling can also be accelerated by fungi which attack leaf cuticles, as this increases their permeability and favours subsequent nutrient loss through leaching (Jensen, 1974). Similarly, some of the minerals present in canopy leachates come from products deposited on leaf surfaces by aphids and other insects (Carlisle *et al.*, 1966).

6.15 FOREST INSECTS

Insects and their larvae are the primary consumers of leaves and other canopy materials. The insect life cycle in temperate regions is synchronized to environmental conditions and much of the time is spent in obligatory diapause, which ensures that active stages coincide with the most favourable periods for development. Feeding damage on leaves is usually most conspicuous during the larval stages and various symptoms develop as a result of different feeding habits (Figure 6.36). Many caterpillars consume all of the leaf except the largest veins; cankerworms create many small holes in a leaf; and skeletonizing beetles remove only soft material between the veins. Other insects enclose the foliage in silk

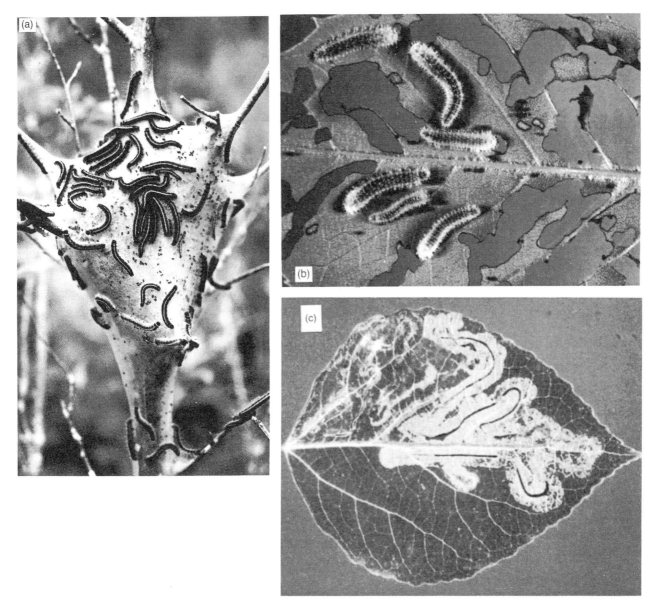

Figure 6.36 Signs of insect damage: (a) prairie tent caterpillar (*Malacosoma californicum lutescens*) larvae resting on nest; (b) spiny ash sawfly (*Eupareophora parca*) larvae feeding on green ash (*Fraxinus pennsylvanica var. subintegerrima*; (c) aspen serpentine leafminer (*Phyllocnistis populiella*) in a trembling aspen (*Populus tremuloides*) leaf. (W. G. H. Ives and H. R. Wong, 1988, with permission of the Northern Forestry Centre, Forestry Canada, Edmonton, Alberta.) (Reproduced with permission from W. G. H. Ives and H. R. Wong, *Tree and Shrub Insects of the Prairie Provinces*, Information Report NOR-X-292; published by Northern Forestry Centre, Northwest Region, Forestry Canada, Edmonton, Alberta, 1988.)

shelters, or protect themselves by folding or rolling the leaves before feeding on them. Browning of the foliage is indicative of leaf-miners which feed inside the leaves. Normally the amount of foliage consumed is comparatively small. For example, the annual loss of leaf tissue in Canadian beech–maple forests is about 6% and in oak forests 11% (Bray, 1964). However, in some years trees can be completely defoliated when species such as the forest tent caterpillar (*Malacosoma disstria*) and gypsy moth (*Lymantria dispar*) reach epidemic population densities.

The effect of defoliation depends on the tree species, and its age and vigour. Trees growing on poor sites or subject to unusual weather conditions and individuals suppressed in the understory are most susceptible to insect attack, but hardwood species are relatively resilient and mortality is unusual unless they have suffered severe insect damage for several consecutive years (Kulman, 1971). Light defoliation will normally have little effect on plant performance, because of increased growth of the remaining leaves in response to higher light levels and improved water and nutrient supply

(Mattson *et al.*, 1988). Significant growth reductions accompany excessive leaf damage and complete defoliation will often result in the death of twigs and branches (Kulman, 1971). Heavy defoliation often weakens a tree and increases its susceptiblity to other insects and disease. Timing, as well as the degree of damage, affects subsequent fitness. If damage occurs early in the growing season the trees will normally refoliate, although growth rates may be reduced. Defoliation later in the season can be more serious because the new leaves may not be cold hardy; consequently, they die before their nutrients are transferred to storage sites.

The populations of most defoliating insects are so regulated by natural predators and weather conditions that they rarely become abundant. Others have cyclical patterns of abundance which culminate every few years in outbreaks of epidemic proportions. The forest tent caterpillar causes extensive damage to a variety of North American hardwood trees when its population peaks every 10 to 12 years. Only one generation of forest tent caterpillars appears each year. Each female moth lays 150–350 eggs in a compact band around small twigs. The young caterpillar is fully formed within 3 weeks but does not hatch till the following spring, when the leaves begin to open. The larvae go through five instar stages over a period of 5–6 weeks, before retreating to the silken webs that they spin over leaves and branches prior to pupation. The adults emerge from the cocoons in 2–3 weeks. The larvae are extremely voracious and during severe outbreaks may completely defoliate all hardwood species, except red maple, over areas as large as 39 000 km^2 (Coulson and Witter, 1984). Trembling aspen is the preferred host, and growth can be reduced by as much as 80% in years when the caterpillar population is at its peak. In birch an 86% reduction has been reported when little more than half of the leaves were removed (Kulman, 1971). Natural outbreaks usually last for only 3–4 years. Unusually cold winters and late spring frosts will kill many larvae, and those which hatch late die of starvation. Some larvae die from disease, others are taken by predators, but most are killed by parasitic flies, especially *Sarcophaga aldrichi*.

Chemical changes have been reported that reduce the palatability of damaged leaves (Wratten *et al.*, 1984) or lower their nutritional quality (Valentine *et al.*, 1983). Baldwin and Schultze (1983) have further suggested that trees with damaged foliage will induce increased levels of phenolics and tannins in nearby undamaged trees, and so influence the activity of phytophagous insects. Haukioja *et al.* (1985) reported a significant increase in pupal development and egg production in autumnal moths (*Epirrita autumnata*) in proportion to distance from birch trees that were artificially defoliated the previous summer. However, the possibility that plants 'communicate' is still uncertain (Fowler and Lawton, 1985). West (1985) demonstrated increased death rates in oak leaf-miners following artificial removal of part of the leaf. Conversely, Edwards *et al.* (1986) noted an increase in palatability in damaged birch leaves. Similarly, Myers and Williams (1984) found that foliage quality in red alder (*Alnus rubra*) declined only after 3 years of heavy defoliation by tent caterpillars.

Less obvious damage is done by other types of phytophagous insects. Sap-sucking aphids and mites feed by piercing the plant tissues with their mouthparts and sucking out the fluids. Associated symptoms include leaf discoloration, curled foliage, deposits of 'honeydew', scales and premature leaf drop. The saliva which is injected into the plant during feeding is usually toxic and causes necrotic areas which interfere with translocation: this can reduce growth and may eventually kill the tree (Coulson and Witter, 1984). Species such as the cicadas appear to have no deleterious effect on the trees during their long nymphal feeding stages but injury results when the adult females cut slits into the twigs for their eggs. This can also induce a defensive response in the form of increased flows of gums and resins (Karban, 1983). Apart from this, damage to woody tissue is mostly carried out by wood borers and bark beetles, and these are typically attracted to dead or dying trees. The intricate galleries that they excavate are therefore an important stage in the breakdown of the wood and the return of nutrients to the soil. However, some of these insects can also spread infectious diseases.

6.16 DISEASES OF TEMPERATE FORESTS

Most diseases of forest trees are caused by fungal infection and result in characteristic symptoms on leaves, stems and roots. Infected leaves may become distorted, covered in moulds and fungal fruiting bodies, or change colour as they wilt and die. Diseased stems and branches are often girdled by cankers, which restrict translocation in the phloem and can result in the deterioration of the roots. Younger shoots are misshapen or killed, while older wood and roots may be susceptible to rots and stains. Dissemination of fungal spores is mostly by wind, but some are spread by splashing water and many rely on insects as specific vectors. Such is the case with Dutch elm disease, which is one of the most serious tree diseases (Figure 6.37).

Dutch elm disease was first recorded in France in 1918 and appeared in Britain in 1927. A severe epidemic in the late 1970s killed most of the elms in southern England (Phillips and Burdekin, 1982). By 1930 it had reached the United States and by 1944 had spread into Canada. The fungus (*Ceratocystis ulmi*) is spread by bark beetles, such as *Scolytus multistriatus*, that emerge from galleries under the bark of infected elm trees. The fruiting bodies develop under dead bark and in insect galleries and are spread throughout the tree in the sap. Some leaves begin to wilt and turn yellow a few weeks after infection, and

Figure 6.38 Regeneration in a gap created in beech forest damaged during an ice storm. (R. A. Wright.)

Figure 6.37 (a) Ivy covered snags of elms in England killed by Dutch elm disease; (b) dieback in jarrah (*Eucalyptus marginata*) in Western Australia caused by the soil-borne fungus *Phytophora cinnamomi*. ((a) O. W. Archibold; (b) Jiri and Marie Lochman, for Department of Conservation and Land Management, Western Australia.)

6.17 FOREST SUCCESSION

Canopy openings account for 9.5% of the area of the hardwood forests of the eastern United States (Runkle, 1982). Approximately 1% of the land is subject to natural disturbance each year (Figure 6.38). The dominant trees, many of which live for 300–500 years, must therefore spend much of their life in the understory. The largest gaps are about 2000 m^2 in area but most are smaller than 150 m^2 (Collins and Pickett, 1987). Closure through lateral growth of branches occurs at a rate of 2% yr^{-1} and saplings take 10–40 years to reach the canopy (Runkle, 1982). Light intensities increase in proportion to the diameter of the canopy opening. Many of the individuals in the understory respond with higher rates of photosynthesis, and become acclimated through changes in leaf orientation and canopy structure (Wallace and Dunn, 1980). A denser cover of understory herbs will characteristically appear in gaps as small as 10 m in diameter (Moore and Vankat, 1986). In addition, the pits and mounds in the soils produced when the trees fall and the decayed logs are readily colonized by herbaceous species (Figure 6.39), many of which establish from seeds brought to these microsites by the ants which nest in the wood (Thompson, 1980). In contrast, the rate of removal of ripe fruit by birds is comparatively high for plants growing in light gaps (Moore and Willson, 1982).

fast-growing twigs may curl into twisted 'shepherd's crooks'. Most trees die within a year. Insects are also responsible for beech bark disease. The beech scale insect (*Cryptococcus fagisuga*) feeds by inserting its long tube-like mouthparts into the bark. The insects themselves have no harmful effect on the tree but the wounds they cause allow entry of the fungus *Nectria coccinea*. Once infected, the bark and outermost sapwood are killed; other fungi invade these areas and the trunk quickly rots and snaps. Some species, such as oaks, are less susceptible to disease but often exhibit a gradual loss of vigour as a result of the combined or sequential action of several deleterious environmental and biotic factors. Symptoms of decline vary with cause and species but typically growth slows, the foliage becomes sparse and there is progressive shoot dieback. The tree is eventually felled by trunk and root rot. The death of the trees opens a temporary gap in the canopy and the process of forest succession begins.

Table 6.17 Seed storage and advanced germination under deciduous forest stands of different ages in the north-eastern United States (after Marquis, 1975)

Species	Basal area of species in the forest overstory (m² ha⁻¹)			Seed storage in the forest floor (×1000 ha⁻¹)			Established seedlings < 1.3 cm dbh (×1000 ha⁻¹)		
	35 yr	65 yr	> 100 yr	35 yr	65 yr	> 100 yr	35 yr	65 yr	> 100 yr
Beech	0.2	4.6	9.0	0.0	42.5	42.5	2.0	3.5	8.8
Birch	0.5	1.4	0.2	142.5	905.0	577.5	0.0	0.0	0.0
Black cherry	6.4	10.4	3.7	337.5	1077.5	380.0	40.5	99.8	6.0
Pin cherry	0.5	< 0.1	0.0	2930.0	4627.5	42.5	0.0	0.0	0.0
Red maple	0.9	1.8	1.6	10.0	22.5	10.0	9.3	20.3	0.5
Sugar maple	9.7	9.7	8.3	< 5.5	0.0	0.0	17.5	1.5	17.5
White ash	0.0	0.5	< 0.1	0.0	0.0	0.0	0.0	0.0	0.0
Yellow poplar	0.5	0.0	0.2	55.0	42.5	55.0	0.0	0.0	0.0

(Reproduced with permission from D. A. Marquis, Seed storage and germination under northern hardwood forests, *Canadian Journal of Forest Research*, 1975, **5**, 480, 481.)

The dominant species in hardwood forests are usually well represented by saplings in the understory but seed reserves in the soil are also comparatively large (Table 6.17). Most of the seeds of sugar maple and beech germinate in the spring of the year following dispersal, and very few remain viable after the first year (Marquis, 1975). Seed germination normally occurs in the first year in birch and red maple but many seedlings also emerge in the second year, whereas ash and yellow poplar usually delay germination for 2–5 years. Conversely, the seeds of pin cherry (*Prunus pensylvanica*) rarely germinate beneath a closed canopy. Pin cherry is a fast-growing, early successional species which establishes quickly in gaps, often in association with other shade-intolerant species such as yellow birch (*Betula alleghaniensis*) and trembling aspen (Figure 6.40). The fugitive strategy of pin cherry is favoured by early and abundant seed production, wide dissemination by birds and, most importantly, a persistent seed bank. Pin cherry seeds can retain their viability for up to 50 years (Marks, 1974). This is a unique characteristic amongst the trees of the hardwood forests of eastern North America and accounts for the immediate appearance of pin cherry in many disturbed sites.

Figure 6.39 Seed germination is often promoted in tip-up mounds and other disturbed microsites. (G. F. Peterken.)

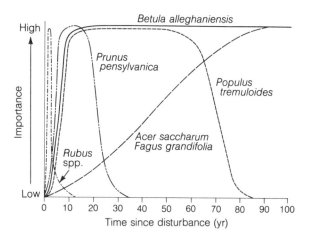

Figure 6.40 Temporal changes in the relative importance of different species following disturbance in North American northern hardwood forests. (After Marks, 1974.) (Reprinted with permission from P. L. Marks, The role of pin cherry (*Prunus pensylvanica* L.) in the maintenance of stability in northern hardwood ecosystems, *Ecological Monographs*, 1974, **44**, 75.)

6.18 HUMAN IMPACT ON TEMPERATE FORESTS

Much of the continuous cover in the temperate forest formations has long since been removed to open up the regions to agriculture: what remains is mostly secondary growth that has re-established in logged areas or on

Table 6.18 Range of forest island characteristics in the Metropolitan Milwaukee Region, Wisconsin, USA (after Levenson, 1981)

	Maximum	Minimum	Mean
size (ha)	39.96	0.03	4.89
canopy stratum			
total number of species	2	18	12
density (number ha^{-1})	929	240	442
basal area (m^2 ha^{-1})	43.30	21.51	29.84
shrub stratum			
total number of species	35	8	19
density (number ha^{-1})	30050	6260	16616
total number of woody species	42	17	25

(Source: J. B. Levenson, Woodlots as biogeographic islands in southeastern Wisconsin, in *Ecological Studies 41 – Forest Island Dynamics in Man-dominated Landscapes*, eds R. L. Burgess and D. M. Sharpe; published by Springer-Verlag N.Y. Inc., 1981.)

Figure 6.41 In many parts of the world, formerly extensive deciduous forests are now reduced to scattered woodlots. (R. G. Archibold.)

abandoned farmland. The remaining areas are variously distributed as forest islands throughout the landscape (Figure 6.41). This island pattern is far from stable and this has important consequences for the plants and animals that are found in these fragmented forest enclaves. Levenson (1981) has described the characteristics of forest islands in a small area of Wisconsin (Table 6.18) that was formerly part of the maple–basswood association. Forty-three sites were represented in the sample, ranging in size from 0.03 to 39.96 ha and including 33 tree species in the canopy with 43 species of saplings and shrubs present in the understory. There was little correlation between floristic diversity and the size of the forest because environmental conditions were rather similar throughout the area. The dominant trees in nearby forested areas include sugar maple, beech and elm. They are well adapted to the moist, shaded conditions within the canopy but are less suited to the drier soils in the isolated forest islands. Thus, there is a critical island size which will no longer provide suitable conditions for their survival and they will be replaced by shade-intolerant pioneer species

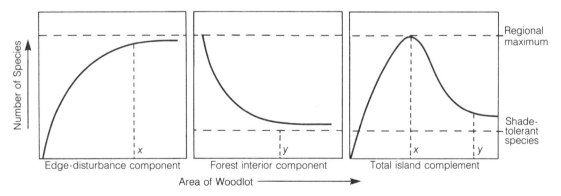

Figure 6.42 The relationship between the length of woodlot perimeters and potential richness of woody species. Given enough area and variety of conditions (left) the number of woody species that could exist becomes asymptotic to the number of species in the region and at some point (x) conditions begin to resemble undisturbed forest. The number of woody species declines in response to changes towards more mesic conditions and low light levels similar to the forest interior (centre): the depletion curve levels off at the number of shade tolerant mesophytic species in the region (y). The combined effect of these functions (right) simulates conditions in forest islands in the eastern United States. (After Levenson, 1981.) (Reproduced with permission from J. B. Levenson, Woodlots as biogeographic islands in southeastern Wisconsin, in *Ecological Studies 41 – Forest Island Dynamics in Man-dominated Landscapes*, eds R. L. Burgess and D. M. Sharpe, published by Springer-Verlag N.Y. Inc., 1981.)

Table 6.19 Change in live tree biomass in northern hardwood forests in Vermont, USA, between 1965 and 1986 attributable to acid rain (after Tomlinson, 1990)

	Biomass (kg ha^{-1})				Net change 1965–1986 (%)	
	1965	1979	1983	1986		
Deciduous species						
Acer saccharum	137600	114900	102900	100000	– 37600	(– 27)
Fagus grandifolia	66700	46600	50000	43600	– 23100	(– 35)
Betula alleghaniensis	39900	51100	45900	41100	+ 1200	(+ 3)
Betula papyrifera	30300	34900	19800	21600	– 8700	(– 29)
Sorbus americana	1800	1500	1100	1300	– 500	(– 28)
Acer spicatum	1670	710	650	300	– 1370	(– 82)
Acer pensylvanicum	980	740	750	600	– 380	(– 39)
Coniferous species						
Abies balsamea	51800	48500	41900	40800	– 11100	(– 21)
Picea rubens	37000	22300	11200	9700	– 27300	(– 74)

(Adapted from G. H. Tomlinson, Table 10.2, p. 217, *Effects of Acid Deposition on the Forests of Europe and North America*; published by CRC Press, 1990.)

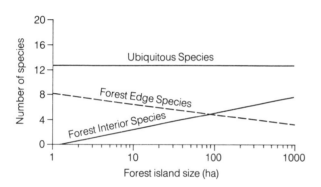

Figure 6.43 The relationship between forest island size and number of forest-interior, edge and ubiquitous bird species in woodlots in the eastern deciduous forests of North America. (After Whitcomb *et al.*, 1981.) (Reproduced with permission from R. F. Whitcomb *et al.*, Effects of forest fragmentation on avifauna of the eastern deciduous forest, in *Ecological Studies 41 – Forest Island Dynamics in Man-dominated Landscapes*, eds R. L. Burgess and D. M. Sharpe; published by Springer-Verlag N.Y. Inc., 1981.)

(Figure 6.42). Species diversity in these communities reaches a maximum in forest islands of 2.3 ha. At least 3.8 ha is required to perpetuate natural forest conditions, although this is very dependent on the frequency and size of gaps created in the canopy. The proportion of edge per unit area of forest increases progressively as the forest is broken up into smaller patches. The edges are richer in species and have higher tree densities than the island interiors, but most are shade-intolerant and are dependent on side-lighting and the drying effect of wind along the perimeters (Ranney *et al.*, 1981). Continual dispersal of propagules from the edges increases the probability of establishment in gaps in the forest interior and the replacement of mesic species.

Changes in the distribution and structure of the forests have important consequences for wildlife. Forest clear-ance has resulted in the extirpation of many animal species, although some have adapted to the new conditions and so have become much more widely distributed. The effects, which have been most extensively documented for regional bird populations, are in many ways analogous to those described for plant species. Bird densities reportedly increase in smaller forest stands, but the species differ markedly in their ability to utilize them because each has specific minimal area requirements (Whitcomb *et al.*, 1981). Consequently, bird populations vary according to the size of the forest islands (Figure 6.43). The number of species associated with the forest interior is positively associated with the size of the forest islands. Edge species decrease in number in larger forest islands, whereas species which utilize either type of habitat occur with equal frequency in all sizes of forest islands. The effect of habitat fragmentation is complicated by a variety of factors. Some species have large territorial ranges and can therefore utilize several isolated forest islands as one large territory, while some non-forest species are now common in smaller forest islands even though they are not dependent on them. Conversely, other species are intolerant of fragmentation and they are subject to regional extirpation as suitable habitat dwindles.

The loss of temperate forests through logging and clearing has been going on for centuries, but since the 1960s indirect losses have also occurred because of increasing damage from **air pollution** and **acid rain**. The initial symptoms of damage from toxic gases usually appear on the leaves. Sulphur dioxide causes bleaching and necrosis of interveinal foliage tissues in broadleaf species. Leaves damaged by fluorides develop marginal necroses, often with a distinct reddish-brown line separating the dead tissue from the healthy parts of the leaf. Small, irregular flecks on the upper surfaces of the leaves is characteristic of ozone damage, whereas peroxyacetyl nitrate (PAN) produces a bronzed or glazed appearance

Table 6.20 Intensity of defoliation in Europe in 1987 attributable to acid rain (after Tomlinson, 1990)

Intensity of defoliation	Country	Damage classes		
		< 10% foliage loss	11–25% foliage loss	> 25% foliage loss
		(%)	(%)	(%)
None	Ireland[a]	95.9	4.1	0
Low	Hungary	85.0	9.0	6.0
	Italy	84.7	12.3	3.0
	Bulgaria	81.7	14.7	3.6
Moderate	Sweden[a]	68.3	26.1	5.6
	France	68.3	22.0	9.7
	Yugoslavia (former)	67.8	21.8	10.4
	Austria	66.5	30.0	3.5
	Belgium	53.5	34.0	12.5
Severe	Czechoslovakia (former)[a]	47.7	36.7	15.6
	West Germany (former)	47.7	35.0	17.3
	Switzerland	44.0	41.0	15.0
	UK	44.0	34.0	22.0
	Netherlands	42.6	36.0	21.4
	Denmark	39.0	38.0	23.0

[a] Conifers only.

(Adapted from G. H. Tomlinson, Table 10.3, p. 218, *Effects of Acid Deposition on the Forests of Europe and North America*; published by CRC Press, 1990.)

Table 6.21 Area of forest in Germany in 1986 and 1987 with greater than 10% foliage loss through acid rain (after Tomlinson, 1990)

	Area per tree species (ha)	Damage (% of area per species)		
		1986	1987	change
Deciduous species				
beech	1256	60.1	65.7	+ 5.6
oak	625	60.7	64.5	+ 3.8
others	626	39.6	41.1	+ 1.5
Coniferous species				
spruce	2884	54.1	48.9	− 5.2
pine	1468	54.0	49.6	− 4.4
fir	172	82.9	79.0	− 3.9
others	357	24.6	28.9	+ 4.3

(Adapted from G. H. Tomlinson, Table 10.6, p. 222, *Effects of Acid Deposition on the Forests of Europe and North America*; published by CRC Press, 1990.)

on the undersurface. Severe damage of this type can usually be traced to specific sources such as smelters and power stations, but with continuing industrial development ambient levels have increased and the problem has become more widespread. Damage on a regional scale is largely attributed to emissions of sulphur and nitrogen oxides which are washed from the atmosphere in precipitation.

The gradual die-back of forests in western Europe and eastern North America has been related to root damage and nutrient imbalance in soils which have become increasingly acidic (Tomlinson, 1990). Damage to broadleaf species in Europe increased by as much as 20% between 1986 and 1987 and is now particularly severe in Denmark, the United Kingdom and the Netherlands (Table 6.19). Sugar maple is particularly vulnerable in North America (Table 6.20) and this has caused concern because of its unique use for production of maple syrup. The young leaves of sugar maple become chlorotic, starting in July or early August, and are shed prematurely. The loss of foliage is initially restricted to the outer ends of the upper branches, but in succeeding years more of the canopy is affected and the trees eventually die: the symptoms are similar for all hardwoods. Conifers are equally susceptible to acid rain (Table 6.21) but damage in some species can be reduced through fertilizer treatment.

Temperate grasslands

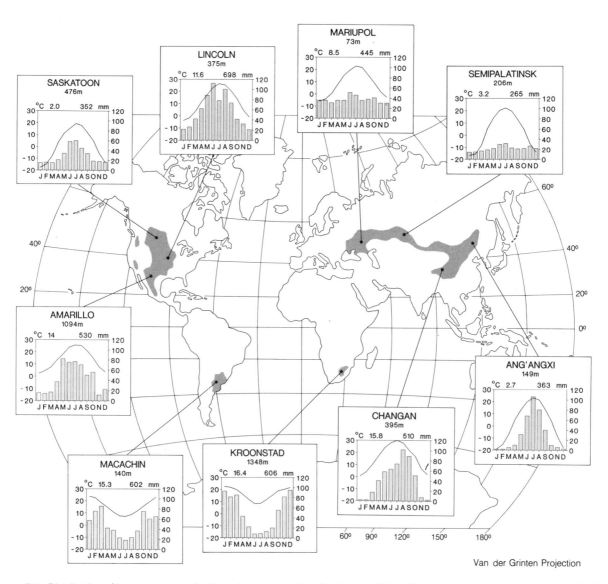

Figure 7.1 Distribution of temperate grasslands and representative climatic conditions. Mean monthly temperatures are indicated by the line and mean precipitation for each month is shown by the bars. Station elevation, mean annual temperature and mean annual precipitation appear at the top of each climograph.

7.1 INTRODUCTION

Temperate grasslands occur in the middle latitudes in regions where the seasonal climate favours the dominance of perennial grasses (Figure 7.1). The most extensive areas are the **prairies** of North America which in their original condition covered more than 350 million ha of the central lowlands. Similar vegetation is found in Eurasia where the **steppes** cover some 250 million ha of rolling plains that extend as a broad belt across the continent from Hungary to Manchuria. In the southern hemisphere grass steppes are represented by the **pampas** of Argentina and the **grassveld** of the high plateaus of southern Africa. Smaller areas of grassland occur in the drier parts of New Zealand with occasional patches also present in south-eastern Australia.

The plant cover in the temperate grasslands appears to be relatively homogeneous but important floristic and structural differences have developed in response to regional and local environmental conditions. The tall sod-forming grasses, which are common in the moister grasslands, are largely replaced by shorter bunch grasses in drier sites where annual species also increase in abundance. Similarly, at higher latitudes, warm season grasses, which have high optimum temperature requirements for growth, are less abundant than cool season species. Sedges and forbs are common throughout the grasslands and in some areas woody plants are conspicuous associates. However, the appearance and composition of the grasslands have been greatly modified by agricultural activities. In many areas the land has been broken for wheat and other crops but, even in areas that are used for grazing, the pastures have often been 'improved' by the introduction of more palatable grasses and legumes. Native species are still present along roadsides and in areas that have not been ploughed but they are subject to different environmental forces than operated in the past. Westhoff (1971) described such areas as subnatural, although relics of the former cover still provide ecologically stable communities throughout the temperate grasslands of the world.

7.2 REGIONAL DISTRIBUTION

7.2.1 The North American formation

The main grassland region of North America extends through the Central Lowlands from southern Canada into northern Mexico through a latitudinal range of more than 30 °. The Central Lowlands is a region of rolling plains which gradually slopes away from the Rocky Mountains. Elevations generally range from 200 to 1200 m, but grasslands extend down to sea level around the Gulf of Mexico and continue as disjunct patches on the high plateaus up to 2400 m. At its eastern limit the grasslands form a long, sinuous and rather indefinite boundary

Figure 7.2 Gallery forest of bur oak (*Quercus macrocarpa*) and chinquapin oak (*Q. muhlenbergii*) associated with a drainage depression in upland tall-grass prairie. (A. K. Knapp.)

Figure 7.3 Scattered groves of trembling aspen (*Populus tremuloides*) are now surrounded by arable land in the aspen parkland region of the Canadian prairies. (R. A. Wright.)

with the oak–hickory forests along the Mississippi Valley and the Ozark Plateau (Figure 7.2). The grasses are increasingly confined to drier, elevated sites in this transitional zone, although the 'Prairie Peninsula' forms a significant extension of the grasslands south of Lake Michigan. Westwards from the Mississippi the land rises gradually and the grasslands are eventually replaced by forests of trembling aspen and pine in the foothills of the Rocky Mountains. Scattered groves of trembling aspen provide a park-like landscape in parts of the Canadian prairies (Figure 7.3). The tree cover increases to the north and species of the boreal forest become more prominent. Far to the south the character of the vegetation changes with the appearance of shrubby acacias and mesquite at lower elevations, but grasslands continue into Mexico along the higher ranges and plateaus.

Many of the grasses are widely distributed throughout the Central Lowlands but floristic diversity increases towards the south, where the growing season is longer,

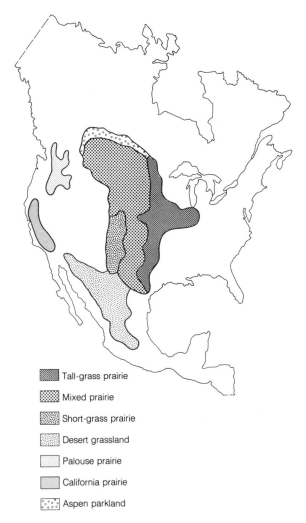

Tall-grass prairie

Mixed prairie

Short-grass prairie

Desert grassland

Palouse prairie

California prairie

Aspen parkland

Figure 7.4 Distribution of temperate grasslands in North America. (After Sims and Coupland, 1979.) (Reproduced with permission from P. L. Sims and R. T. Coupland, Producers, in *Grassland Ecosystems of the World: Analysis of Grasslands and Their Uses*, ed. R. T. Coupland; published by Cambridge University Press, 1979.)

Figure 7.5 Mesic tall-grass prairie in Oklahoma mainly comprised of switchgrass (*Panicum virgatum*), Indiangrass (*Sorghastrum nutans*) and big bluestem (*Andropogon gerardi*). (R. E. Redmann.)

and to the east, where precipitation is higher. The composition of the grasslands changes as different species become more prominent along a latitudinal gradient. For example, thread-leaf sedge (*Carex filifolia*) and needle-leaf sedge (*C. eleocharis*) are largely restricted to the northern prairies, side-oats grama (*Bouteloua curtipendula*) is mostly found in the central and southern prairies, and galleta (*Hilaria jamesii*) is abundant only in the south (Weaver and Albertson, 1956). Grasses typically account for less than 20% of the species present at any site, but will normally contribute more than 90% of the biomass. Relatively few grasses are dominant in a particular habitat and typically most of the cover will be provided by only two or three species (Coupland, 1979). The growth habits of these dominant species provide a convenient physiognomic subdivision of this grassland region into

tall-grass prairie, mixed-grass prairie and short-grass prairie (Figure 7.4).

The dominant species in the eastern tall-grass prairie include big bluestem (*Andropogon gerardi*), Indian grass (*Sorghastrum nutans*) and switch grass (*Panicum virgatum*) (Figure 7.5). These grasses often exceed 2 m in height by late summer when they come into flower and are accompanied by numerous herbaceous species, especially in the spring and fall (Figure 7.6). Mixed-grass prairie occurs throughout central Canada and the northern Great Plains region of the United States; it continues, much restricted in breadth, as far south as Texas. The mixed-grass prairie is comprised of mid-grasses such as little bluestem (*Andropogon scoparius*), needle-and-thread grass (*Stipa comata*) and western wheat grass (*Agropyron smithii*) in association with short-grasses including blue grama (*Bouteloua gracilis*), buffalo grass (*Buchloe dactyloides*) and red three-awn (*Aristida longiseta*) (Figure 7.7). This mixture of species results in a noticeable stratification in the mature grass cover. The taller mid-grasses provide an open cover up to 1.25 m in height, and below this is a denser layer of short-grasses to a height of 30 cm growing in association with various other plants (Figure 7.8). The drought-tolerant short-grasses become progressively more abundant in the drier High Plains to the west. The dominant species in these short-grass prairies

Figure 7.6 Natural tall-grass prairie is noted for its diverse herbaceous flora: (a) coneflower (*Echinacea angustifolium*); (b) common sunflower (*Helianthus annus*); (c) New England aster (*Aster novae-angliae*); (d) blazing star (*Liatris pycnostachya* and silver psoralea (*Psoralea argiphylla*). ((a), (b) A. K. Knapp; (c) C. Kucera; (d) R. E. Redmann.)

are blue grama and buffalo grass: they are 20–50 cm high by the time growth ceases in late summer (Figure 7.9).

Although only a few species are dominant over large parts of the region, differences in topography, soil and drainage result in a mosaic of plant communities. Patches of tall-grass prairie may displace mid- and short-grasses in moister depressions throughout the grasslands, while the short-grasses typically increase on the drier knolls. Smeins and Olsen (1970) have described such a topographic sequence for tall-grass prairie (Table 7.1). Big bluestem is the dominant species on upland and mid-slope sites, but is replaced by prairie cordgrass

(*Spartina pectinata*) at the base of the slopes, with sedge (*Carex lanuginosa*) meadows in areas that are periodically inundated. Differences in soil texture may also influence the composition of the vegetation. For example, in Canadian mixed prairie the *Stipa–Bouteloua* community is characteristic of medium-textured soils developed from undifferentiated glacial till: the *Stipa–Bouteloua* community is commonly associated with drier soils of coarser texture, and the *Agropyron–Koeleria* (June grass) community occurs on the heavy clay soils which have formed on glacial lake deposits (Table 7.2). Variable weather conditions from year to year accentuate

Figure 7.7 *Stipa–Agropyron* mixed prairie in southern Saskatchewan. (R. T. Coupland.)

become established and are especially abundant in disturbed sites, where they usually dominate the early successional communities.

7.2.2 The Eurasian formation

The grasslands of Eurasia form a treeless corridor across the continent in which various regional associations are broadly differentiated according to latitude and altitude. The northernmost component is the '**forest-steppe**' which corresponds with the tall-grass prairie of North America. In the plains of eastern Europe these grasslands are interspersed with groves of oak (*Q. robur*) (Figure 7.11) but further west birch (*Betula verrucosa*) and aspen (*Populus tremula*) are prominent in damper depressions, and in eastern Siberia patches of steppe are surrounded by stands of larch (*Larix sibirica*). Much of this region is now under cultivation and grasslands are resticted to unfavourable sites where agriculture is often limited by poor drainage or salinity. The characteristic grasses are sod-forming species which grow to a height of 1.5 m or more: *Stipa joannis* and *S. stenophylla* are common throughout this region in association with species of *Bromus*, *Poa* and *Agropyron*, various sedges (*Carex* spp.) and numerous forbs. The drier, continental climate of the eastern steppes reduces the diversity of the cover. In particular, the showy spring flowers which appear in the European steppes as the snow cover melts is generally lacking (Figure 7.12). In the European steppes flowering begins in April with the appearance of the mauve blossoms of the anemones (*Anemone patens*) 10–15 cm above the short brown sward, and soon after patches of yellow-flowering spring adonis (*Adonis vernalis*) dot the landscape as the grasses begin to sprout. The most colourful stage is in early summer when the flowering herbs are intermixed with the seed heads of the grasses. By mid-July most of the forbs have begun to wither. The grasses complete their growth in August and the steppes appear dry and monotonous by the time of the first snowfalls (Walter, 1985).

The '**tuft-grass steppes**' are situated to the south of the forest-steppe zone. The dominant plants are densely rooted tussock grasses that are well adapted to the drier conditions, but floristic diversity of these grasslands is considerably lower than in the forest-steppes. In the western parts the cover is largely comprised of *Stipa capillata* and *S. lessingiana*, with *Festuca sulcata* also important. Cryptophytes, such as irises and wild tulips, and winter annuals appear in the spring, and taller herbaceous species including the purple-flowered sages (*Salvia nutans* and *S. superba*) are conspicuous in the summer; shrubby species of *Spirea* and *Caragana* are also common. Further east the grass cover is more open: *F. sulcata* is still present but here it grows in association with other densely rooting, sod-forming species including *Koeleria gracilis* and *Poa botryoides*. Mosses and algae

these patterns through differences in soil moisture availability.

Isolated from the central grasslands is the Palouse prairie, an intermontane bunch grass region occuring in the northwest United States and in the valleys of southern British Columbia. In its original state the dominant species in the Palouse prairie included bluebunch wheat grass (*Agropyron spicatum*), bluebunch fescue (*Festuca idahoensis*) and giant wild rye (*Elymus condensatus*) (Figure 7.10). Species associated with the interior mixed-grass prairies, such as needle-and-thread, western wheat grass and June grass (*Koeleria cristata*), were also present and became increasingly important at higher elevations, but no short-grasses were native to the region. Much of the area now produces arable crops or is used for rangeland. Excessive grazing and fire have favoured the spread of annual grasses, such as downy brome (*Bromus tectorum*), and annual forbs are much more common. Sagebrush (*Artemesia tridentata*) has also become prominent throughout the region.

Some of the grasses native to the Palouse prairie were also important in the grasslands that originally covered the Central Valley of California. The dominant species were perennial bunch grasses such as needlegrass (*Stipa pulchra*) and pine bluegrass (*Poa scabrella*) but many annual species were also present, together with a large and diverse complement of broad-leaved herbs. The character of these grasslands has been completely altered by grazing, fire and other disturbances which have favoured the establishment of introduced annual grasses that are well adapted to the mediterranean climate. Species of *Avena*, *Bromus* and *Hordeum* have now largely replaced the native perennials throughout the valleys and hills (Baker, 1989). The majority of these naturalized grasses originated from Europe and North Africa and were mostly introduced inadvertently in the late 19th century (Table 7.3). A great number of weedy forbs have also

Figure 7.8 Common plants associated with mixed prairie: (a) pasture sage (*Artemisia frigida*); (b) prickly-pear cactus (*Opuntia polycantha*); (c) smooth aster (*Aster laevis*); (d) prairie selaginella (*Selaginella densa*) ((a) J. T. Romo; (b) R. E. Redmann; (c), (d) R. A. Wright.)

are present in the spring but at other times of the year the ground cover is usually quite sparse.

The 'sagebrush-grass steppes' are the southernmost grasslands of Eurasia. They extend discontinuously from the Black Sea into China where they form a wide semi-circle around the Gobi Desert and adjacent arid regions. The sparse grass cover consists mainly of *S. capillata* and *F. sulcata* with drought-tolerant wormwood (*Artemisia maritima*) especially common on the saline soils. A similar plant cover is found on the arid lower slopes in mountainous areas at elevations of 1000–2000 m. This 'mountain-steppe' is replaced at higher elevations by taller grasses and a variety of tall herbs including *Ligularia altaica*, *Delphinium confusum* and *Scabiosa alpestris* (Wang, 1961). In more favourable sites, grasses such as *Alopecurus pratensis*, *Dactylis glomerata* and *Bromus inermis* form luxuriant pastures with many legumes and meadow species adding to the diversity.

Parklands cover much of the area east of the Gobi Desert. Isolated open-grown trees are scattered throughout the grasslands of the central plains of north-eastern China. The most common tree is elm (*Ulmus pumila*) and the region is generally known as the 'elm grasslands'. It is a drought-resistant species and is also present in the 'semi-desert steppe' to the east and south of the Gobi Desert. The dominant species in the plains are tall grasses, such as *Aneurolepidium pseudo-agropyron* and *Stipa baicalensis*, with *Aneurolepidium chinense* and *Artemisia sibirica* important in saline areas and drier upland sites. The coarse steppe grass *Achnatherum splendens* grows in nearly pure stands forming thick clumps 1–2 m tall over much of the westernmost regions. Luxuriant meadows

Figure 7.9 Short-grass prairie at the Pawnee US/IBP site. (R. T. Coupland.)

Figure 7.10 Bluebunch wheat grass (*Agropyron spicatum*) and bluebunch fescue (*Festuca idahoensis*) dominate this Palouse prairie remnant in Idaho. (J. T. Romo.)

Table 7.1 Changes in species composition along a topographic sequence in tall-grass prairie in Minnesota (after Smeins and Olsen, 1970)

Principal species	Species composition (%)		
	level, well drained upland	moderately well drained hillslope	poorly drained lowland
Grasses			
Stipa spartea	37	8	–
Poa pratensis	14	12	–
Andropogon scoparius	12	6	–
Muhlenbergia richardsonis	8	6	–
Andropogon gerardi	–	41	5
Spartina pectinata	–	–	70
Muhlenbergia asperifolia	–	–	10
Calamagrostis inexpansa	–	–	5
Sedges and rushes			
Carex tetanica	4	8	–
Juncus balticus	2	–	7

(Source: F. E. Smeins and D. E. Olsen, Species composition and production of a native northwestern Minnesota tall grass prairie, *The American Midland Naturalist*, Vol. 84(2), 1970.)

Table 7.2 Effect of soil conditions on the composition of Canadian mixed prairie vegetation (after Coupland, 1950)

Principal species	Species composition (%)		
	moist loams	dry loams	clay
Grasses			
Stipa spp.	30.2	17.6	8.7
Bouteloua gracilis	24.7	51.6	–
Agropyron spp.	7.4	2.8	47.6
Koeleria cristata	7.5	2.9	24.6
Sedges			
Carex eleocharis	13.2	7.0	13.8
Carex filifolia	1.5	1.6	–
Forbs			
Artemisia frigida	7.2	6.5	2.3
Phlox hoodii	3.8	1.6	1.3

(Reprinted by permission from R. T. Coupland, Ecology of mixed prairie in Canada, *Ecological Monographs*, 1950, **20**, 283, 287, 292.)

form a transitional zone between the drier grasslands of the central plain and the surrounding forested areas. The flora of this region is extremely diverse and contains both grassland and woodland elements, including seedlings and saplings of birch (*Betula platyphylla*), aspen (*Populus tremula*) and willow (*Salix chingania*), and a large number of herbaceous plants which provide a showy display from spring till late fall (Wang, 1961).

7.2.3 The formations of the southern hemisphere

The largest area of temperate grasslands in the southern hemisphere is the Pampas of eastern Argentina, Uruguay and south-east Brazil: it occupies about 700 000 km^2 of undulating lowland around the Rio de la Plata. Tall and medium bunch grasses occur throughout this

Table 7.3 Origins of naturalized plants in California grasslands (after Baker, 1989)

Origin	Number of species			
	Perennial grasses	Annual grasses	Perennial forbs	Annual forbs
Europe and North Africa	18	28	39	48
Eurasia	3	2	16	7
South Africa	1	1	4	0
Australasia	0	1	4	0
Central and South America	1	1	4	4
North America and Mexico	1	0	4	5
Tropical Africa	1	2	0	0
Old world	1	2	1	0

(Reprinted by permission from H. G. Baker, Sources of naturalized grasses and herbs in California grasslands, in *Grassland Structure and Function: California Annual Grassland*, eds L. F. Huenneke and H. A. Mooney; published by Kluwer Academic Publishers, 1989.)

Figure 7.11 Oak (*Quercus robur*) forest margin in the Kursk district of the former Soviet Union. (R. T. Coupland.)

Figure 7.12 A grassland dominated by *Aneurolepidum chinense* and *Stipa grandis* in the Xilin District of Inner Mongolia, China. (I. Hayashi.)

region, with small patches of trees found in the moister hollows. The dominant grasses are species of *Stipa*, although many different grassland communities can be distinguished according to moisture availability and soil conditions. In the eastern Pampas, species such as *Stipa neesiana, S. hyalina* and *S. papposa* grow in association with other grasses, including *Bothriochloa laguirioides, Panicum bergii* and *Papsalum plicatulum*, with sedges and forbs present between the tussocks. In many communities a single species of grass will dominate the cover (Soriano, 1979). Throughout the Pampas a short-grass community comprised of *Distichlis scoparia, D. spicata* and *Hordeum stenostachys* is associated with alkaline soils, and sedge meadows or dense tussocks of the tall-grass *Papsalum quadrifarium* are common in moister depressions. Drier conditions in the western Pampas favour xerophytic grasses, such as *Stipa tenuissima* and *S. trichotoma*: herbaceous species are almost entirely absent from these communities. With increasing aridity the pampas is replaced towards the Andes by *Prosopis caldenia* woodland and semi-desert communities. To the north the temperate grasses merge with taller subtropical species. The

widespread naturalization of legumes, such as *Trifolium repens* and species of *Melilotus* and *Lotus*, together with the introduction of many European grasses has greatly altered the character of regions which are still in pasture. Elsewhere arable crops have replaced the native grasses of the Pampas.

The temperate grasslands of southern Africa occur on the flat to gently undulating plains and high plateaus in the eastern parts of the country (Figure 7.13). The dry, cold-temperate grasslands are found mainly at elevations of 1200–1500 m (Werger and Coetzee, 1978). *Themeda triandra* and *Eragrostis chloromelas* are the typical dominants with *Digitaria argyrograpta, E. superba* and *Aristida congesta* also important throughout the region, although the composition of the cover varies according to soil conditions. Species such as *Panicum coloratum* and *E. obtusa* are common on droughty, fine-textured soils; *Setaria woodii* and *Pennisetum sphacelatum* are typical species on wet clay soils; *E. lehmanniana* is prominent on sandy soils, and *Sporobolus ludwigii* is found on saline or highly alkaline soils. Many of the dominant species have

Figure 7.13 The veld of southern Africa. (R. T. Coupland.)

Figure 7.14 Various tussock grasslands of South Island, New Zealand: (a) short tussock (*Festuca novae-zealandica*) grasslands at 800 m; (b) silver tussock (*Poa cita*) with *Bulbinella angustifolia* in flower at 1000 m; (c) subalpine snow tussock (*Chionochloa rigida*) grasslands at 1200 m. (A. F. Mark.)

been reduced through overgrazing and they have now been replaced by *Panicum coloratum* and *Eragrostis chloromelas* or short-lived species of *Aristida*. The composition of the grasslands changes towards the east in response to the increased rainfall and cooler temperatures at higher elevations. *T. triandra* and *E. chloromelas* are still widespread but they are less important in these moist, cool-temperate regions and are gradually replaced by species such as *Aristida junciformis*, *Tristachya hispida* and *Eragrostis capensis*.

In New Zealand 'tussock grasslands' occur extensively in the eastern part of the South Island where they lie in the rain shadow of the Southern Alps. They occupy some 57 000 km² at elevations ranging from sea-level to about 1500 m. The dominant grasses are xerophytic bunch grasses, but distinct types are recognized according to their physiognomy and species composition (Coaldrake, 1979) (Figure 7.14). The 'short tussock grasslands' occur up to 1000 m elevation. The dominant species, *Festuca novae-zealandica*, grows to a height of 0.6 m in association with *Poa colensoi* and *Agropyron scabrum* and many herbaceous and woody species. Overgrazing, burning and seeding has favoured the development of a turf-like cover of *Notodanthonia clavata* over much of the short tussock grasslands. 'Tall tussock grasslands' are dominated by species of *Chionochloa* which grow up to 2 m in height. *C. rigida* is the most widespread species, particularly at elevations above 700 m where it grows in association with the common shorter grasses, herbs and woody shrubs. *C. macra* is characteristic of fine-textured soils where it forms a rather open tussocky cover with species of *Celmisia* and *Dracophyllum*. Other species are locally dominant in specific sites. The presence of buried wood and charcoal suggests that much of the grassland in the South Island of New Zealand has been derived from forests which probably grew to elevations of 1200–1400 m: destruction of the forest cover by fire is thought to have begun about 1000 years ago (Connor, 1964).

Continued burning and heavy grazing by sheep have resulted in the subsequent replacement of much of the tall tussock grassland by shorter grasses. Similar grasslands are found in parts of the central volcanic plateau of the North Island: here the dominant species is red tussock (*C. rubra*) but this is replaced by *Poa colensoi* above 1300 m. The grasslands of the North Island are regarded

Figure 7.15 Tussock grasslands in the Kosciusko region of New South Wales, Australia. (O. W. Archibold.)

as a seral cover on pumice soils and are eventually replaced by heath and forest (Coaldrake, 1979).

Small patches of grassland are distributed throughout the temperate woodlands of Australia, where they are associated with small plateaus or cold-air drainage basins (Moore, 1970). The dominant species are tall tussock grasses, such as *Themeda australis, Stipa aristiglumis* and *Poa caespitosa*. These warm-season grasses are now largely replaced by cool-season species of *Danthonia* and *Stipa* and introduced annuals. Shorter 'sod-tussock grasslands' comprised of *Danthonia nudiflora* and a small fine-leaved form of *Poa caespitosa* occur above 1400–1500 m in the south-eastern highlands (Figure 7.15). These are considered as subalpine communities: frosts may occur every month of the year and average minimum temperatures range from − 4 to − 7 °C.

7.3 THE CLIMATE OF THE TEMPERATE GRASSLANDS

7.3.1 North America

The temperate grassland climate is one of recurring drought, and much of the diversity in the vegetation cover reflects differences in precipitation amount and reliability. The Great Plains region of North America has a continental climate in which there is a large annual range of temperature. Precipitation is usually greatest in the early summer when cyclonic storms frequently pass through the region. This is supplemented by thundershowers in the summer and snowfall in the winter. Hail sometimes accompanies thundershower activity: it is most frequently reported from western districts. Precipitation decreases northwards and westwards across the grassland formation. In the south, annual precipitation varies from 500 to 1200 mm, but in the Canadian prairies this decreases to 350–500 mm. As much as 84% of the pre-

cipitation falling in the Great Plains is lost through evaporation (Court, 1974). Crop water balance studies in the Canadian prairies indicate that soil moisture levels in the spring average only 52% of capacity and decrease to 23% at the end of summer: there is little surplus moisture at any time of the year (Hare and Hay, 1974). In addition, precipitation is extremely variable between years, and periods of drought are notorious in the central Great Plains.

Regional temperature patterns are also evident across the grasslands. In summer, air sweeps northwards from the Gulf of Mexico and mean temperatures in July can reach 25–28 °C in central and southern regions, compared with 15–18 °C in the north. The latitudinal temperature gradient is more pronounced in winter when the region comes under the influence of polar air masses. Mean temperatures in January are 5–10 °C across the south, but drop as low as − 25 °C in the north with absolute minimum temperatures near − 40 °C. Consequently, the frost-free period usually exceeds 250 days in the south but ranges from 80 to 100 days in the north. Strong convective turbulence is common in the spring and early summer when temperature and humidity differences in the polar and tropical air masses are at a maximum; this is an important factor in the formation of tornadoes, which are reported more frequently in the Great Plains than in any other part of the world.

7.3.2 Eurasia

The weather patterns of the Eurasian steppes are controlled by the seasonal pressure changes which occur over the large continental land mass. In winter the high pressure system which develops over central Asia diverts the flow of moist air from the Atlantic into the northern forest zones; it also blocks the passage of cyclonic storms from the Mediterranean region. In summer a broad, shallow, low-pressure system develops as the land mass warms under intense and prolonged insolation, and there is increased cyclonic activity. Consequently, precipitation is highest during the warm season, although moisture deficits can occur in spring and early summer. Annual precipitation in the western steppes typically ranges from 300 to 600 mm: the drier parts south of the Ural Mountains receive about half of the water needed to meet potential evaporation demands (Lydolph, 1977). About 25% of the precipitation comes as snow, which may lie on the ground for 100–140 days. The eastward flow of air across the Eurasian plains is blocked by the Ural Mountains. East of the Urals annual precipitation in more northerly grasslands is about 400 mm, but decreases to 200 mm in the south: most of the precipitation in this region comes from thundershowers. Potential evaporation is much higher and there is a continual threat of drought. Winter precipitation is meagre and, because of strong winds, the snow lies unevenly on the ground. The thawing period usually lasts for about 10 days but much of the moisture is lost as run-off because the ground is frozen to depths of 1.5–2 m. Any water that remains is mostly evaporated by the sharp rise in temperature during the spring.

Summers in the steppes are hot and sunny. Mean temperatures in July range from 18 to 24 °C with daytime maxima often above 40 °C. Frequent strong winds increase the rate of moisture loss and promote '**sukhovey**', or desiccating conditions, particularly damaging to the agricultural crops which are now grown throughout the region. The winters are very cold: mean January temperatures are about −20 °C, and minimum temperatures as low as −50 °C are not uncommon. Temperatures rise above freezing in late April and early May across the south-western steppes and about 1 month later in the eastern interior. The first frosts of autumn occur earlier in the eastern steppes: here the average length of the frost-free period is 90 days compared with 165 days in the west.

7.3.3 Southern hemisphere

Climate conditions are less severe in the grasslands of the southern hemisphere because they are located at lower latitudes or are influenced by the adjacent oceans. In the Pampas, mean temperatures in winter range from 6 to 14 °C and in summer rise to 20–26 °C. Although air temperatures rarely fall below freezing, ground frosts are commonly reported between mid-May and mid-September (Prohaska, 1976). Maximum temperatures during summer are usually above 30 °C and can exceed 40 °C in the northern interior. Precipitation decreases from east to west across the Pampas. The more humid areas of Uruguay and southern Brazil receive 1000–1200 mm of rain each year but this decreases to 450 mm in the southwest. Maximum rainfall occurs in the winter in the wetter northern coastal districts, but spring rains become increasingly important in the drier areas and, because they coincide with rising temperatures, conditions are optimal for plant growth. Evaporation demands are comparatively high throughout the Pampas, particularly in the latter part of the year. Annual potential evaporation is about 1000 mm in coastal districts but values as high as 1500 mm are reported for inland sites and most of the region experiences a negative water balance.

The weather patterns in southern Africa are dominated by semi-permanent high-pressure systems that are occasionally displaced by cyclonic disturbances which bring rain as they pass eastwards across the country. Precipitation increases from west to east across the temperate grasslands of southern Africa because of the orographic effect along the high eastern escarpment. Annual rainfall in the drier regions is about 450 mm but most areas receive 600–700 mm, increasing to 800–1000 mm in the eastern limits of the high plateaus. Rainfall occurs most-

ly in the summer months with over 70% of the annual total coming between October and March; much of this is from thundershower activity and, as in other temperate grasslands, it is sometimes accompanied by severe hailstorms. Most of the moisture is lost through evaporation, which averages 1500–2500 mm yr^{-1}. Temperatures decrease with elevation and in summer normally average 20–28 °C with an absolute maximum temperature range of 35–40 °C. Mean winter temperatures range from 10 to 12 °C. The incidence of winter frosts increases in the higher plateaus: here the average duration of the frost period is about 180 days compared with 120 days at lower elevations. During this period the actual incidence of frost varies from 60 to 90 days (Schulze, 1972).

Mean annual rainfall in the grasslands of New Zealand is generally 500–750 mm but is less than 350 mm in some parts of the South Island, where measurable precipitation occurs on fewer than 100 days each year. Periods of drought, in which no measurable precipitation is recorded for at least 15 consecutive days, occur on average three times each year and can persist for up to 6 weeks (Maunder, 1971). Partial droughts, in which less than 0.25 mm of precipitation is recorded over a period of at least 29 days, occur with a similar frequency, and some of these dry spells can last for over 3 months. Most droughts occur in winter. Annual potential evaporation is about 900 mm. The biggest demand is usually recorded in January and most locations have a period in the summer when soil moisture reserves are used up. A winter surplus usually occurs in all but the driest areas. The winters are comparatively mild with mean temperatures of 5–6 °C, although frost may be recorded on more than 100 days each year in some areas, and the snow line usually descends to about 800 m. Mean temperatures in the summer are 15–16 °C, with extremes rarely exceeding 35 °C.

7.4 SOILS

The dominant soils in the temperate grasslands are Mollisols with a relatively thick, dark brown to black surface horizon that is rich in organic matter. These soils are typically well-drained and have good structure so that they remain crumbly when dry: this permits easy penetration of roots and moisture. Regional differences in climate and vegetation have produced a variety of soil types which are generally related to changes in the availability of moisture. Severe droughts rarely occur in the subhumid areas and in these productive grasslands the soil organic matter content ranges from 5 to 10% in the surface horizons. Low soil moisture levels during the warm season slow decomposition and allow humus to accumulate. A large amount of organic matter is supplied by the fibrous roots of the grasses, and significant darkening can occur to depths of 40–60 cm with less intense colour continuing to depths of 1.5–2 m as min-

Figure 7.16 Characteristic soil profile in mixed-grass prairie with top of paler calcic horizon occurring at a depth of 30 cm. (R. E. Redmann.)

Figure 7.17 Saline soil area in Saskatchewan mixed-grass prairie region mostly colonized by *Salicornia rubra*. (R. A. Wright.)

eralized organic materials are carried downwards by percolating water (Foth and Schafer, 1980). Grasses accumulate comparatively large amounts of basic mineral nutrients during the growing season and these are returned to the soil as the plants decay. Base saturation is usually above 50%, although much of this is calcium. These soils are neutral or slightly acid in the upper horizons but become increasingly alkaline with depth because secondary enrichment of calcium carbonate occurs through illuviation.

The grassland soils become progressively thinner and paler in colour in drier regions because less organic material is incorporated into the surface horizons. As potential evaporation demands increase, less water percolates through the profile and salts, particularly calcium carbonate, begin to accumulate as a powdery lime layer or as nodules in the B horizon. There is usually adequate precipitation during the year to leach sodium and other highly soluble salts from the profile. However, in these semi-arid climates the summer rains normally only wet the upper part of the soil, and calcium is deposited at the depth where all available soil moisture is used in evapotranspiration (Figure 7.16). The humus-rich surface horizons in these soils are typically 30–40 cm thick but the organic matter content at the surface is rarely above 3–6%. The high calcium content results in pH values of about 7.0 in the upper part of the profile, increasing to over 8.0 within the calcic horizons. **Solo-**

netzic soils are common in relief depressions where drainage water has carried excess sodium salts (Figure 7.17). These saline soils become increasingly widespread in the arid steppes and eventually merge with Aridisols as the sparse grass cover becomes replaced by xerophytic shrubs and other desert plants.

Although the soils of the Pampas are classed as Mollisols, their properties reflect the warmer climate and higher moisture content of these characteristically ill-drained plains. Most of the area is underlain by dark brown granular soils with a humus layer 35–40 cm thick. Organic matter content decreases from 3–4% at the surface to 1–2% at depths of 60–70 cm: below this the soils are more clayey, and calcareous horizons may be present below 1.5 m. The soils are slightly acid in the upper part of the profile but are weakly alkaline at depth. Base saturation is usually 80–85% but this is mainly due to calcium. Most of these soils have developed in loess deposits derived from volcanic ash. Saline soils with high sodium content occur in closed-relief depressions and on other poorly drained surfaces: they are particularly common in the north and west, where the climate is more arid (Glazovskaya, 1984). Salts are also redeposited in the western Pampas by winds blowing off adjacent desert regions (Foth and Schafer, 1980).

The soils of the southern African grasslands are generally classed as Alfisols, but a variety of types are encountered which reflect the diversity of parent materials, climate and vegetation. In the northern and eastern grasslands the soils are typically neutral red and yellow latosols, but acid sandy soils are found on quartzite bedrock and shallow base-rich soils are associated with outcrops of dolomite (Werger and Coetzee, 1978). Elsewhere weathered shales have given rise to black montmorillonite clay soils, with acid, yellow-grey sandy soils and loams associated with sandstone areas. In poorly drained areas, layers of calcium carbonate and gypsum are present in the soils.

In the grasslands of New Zealand the principal soils are Inceptisols which, in the North Island, are associated with deposits of volcanic ash: in the South Island they have formed on alluvial sediments eroded from the mountains to the west. In the lowlands yellow-grey earths have developed under the cover of tussock grasses; yellow-brown earths are more widespread in the rolling High Country. The humus layer in these soils is only 15–30 cm thick with an organic matter content of about 2%. They are generally light textured and slightly acid, although in flat terrain deep calcareous horizons may be present.

7.5 THE EVOLUTION OF THE GRASSLANDS

The flora of the central Great Plains of North America has few endemic species. This suggests it has developed comparatively recently and attained its present distribution only in post-glacial times (Axelrod, 1985). Although abundant fossil grasses and herbs have been reported in Miocene deposits, associated remains of hardwood species indicate that the region still supported a woodland cover 10–12 million years ago. Drier climatic conditions in the Pliocene (5–6 million years ago) favoured the expansion of the grasslands, but during the Pleistocene spruce forest probably covered most ice-free areas in the northern Great Plains. The forests persisted as late as 12 000–10 000 BP, and grasslands spread into deglaciated areas about 9500 BP. In more southerly regions open woodlands of pine and spruce may have persisted to about 14 000 BP, but a warm steppe-type cover of grasses and herbs intermixed with oak–juniper woodland is indicated by the pollen record of about 10 000 BP. The disjunct distribution of many tree species throughout the Great Plains also suggests that the woodland cover was previously more extensive.

Various forest communities were widespread in Eurasia prior to the Pleistocene and wooded-steppe occupied most of the present grasslands, with patches of steppe grasslands interspersed with desert scrub further south (Frenzel, 1968). Drier conditions during periods of glacial advance favoured the spread of the short-grass and wooded-steppe formations and by the end of the Pleistocene this type of cover formed a broad band from northern France to the plains of Manchuria (see Figure 6.19). Unlike North America, there was no forest belt separating these steppes from the tundra regions to the north, although groves of frost-resistant conifers and hardwoods were present at higher latitudes and gallery forests fringed the major rivers (Flint, 1971). The presence of *Artemisia* in the pollen record indicates that the climate was cool and dry.

In southern Africa it is postulated that the climate of the high plateaus would have been wet and cold during the last glacial maximum, with temperatures 8–10 °C

lower than today (van Zinderen Bakker, 1978). This would have depressed the upper limit of tree growth by as much as 1000 m, and the wooded savanna which had developed in this region by the early Pleiocene (7 million years ago) would have been replaced by alpine grasslands. This alternated with semi-desert vegetation during the warmer, drier interglacials when windblown sand dunes developed in the western plateaus. The temperate grasslands were restricted to the coastal plains of the southern Cape and may not have developed in their present form until 4600 years BP. In New Zealand the pollen stratigraphy suggests that grasslands persisted in the South Island during glacial maxima (Wardle, 1963). Subsequent climatic amelioration favoured the development of forests over much of the eastern lowlands, and the widespread extension of grasslands is attributed to fires set by the moa-hunters (Cumberland, 1962). Paleobotanical records are not available for the Pampas but vertebrate fossils suggest that climatic conditions during the Pleistocene were similar to today (Flint, 1971).

The characteristics of the grasses which enable them to survive the periodic droughts which occur with varying freqency throughout the temperate grasslands also adapt them to the stresses of fire and grazing. Both of these factors probably contributed to the development of grasslands. Although there are no reliable historical records of fire frequency, the risk of fire is well documented in the accounts of explorers and settlers (Pyne, 1986). Forest species readily invade tall-grass prairie when it is left unburned for several decades (Bragg and Hulbert, 1976) and many of the grassy inclusions characteristic of the prairie–deciduous forest ecotone succeeded to forest once frequent wildfires were suppressed following settlement (Schwegman and Anderson, 1986). The conifer forests which covered the Great Plains of North America would have been especially vulnerable to fire during recurrent droughts, and as they were eliminated the grasses were able to spread northwards (Axelrod, 1985). The natural fire frequency for undulating grass prairie is estimated at 5–10 years, increasing to 20–30 years in dissected terrain (Wright and Bailey, 1982), and the persistence of woodlands in more protected coulees and along escarpments attests to the importance of fire as a factor in the development of the grasslands (Wells, 1970). However, many woody shrubs do survive frequent fires and the spread of trembling aspen (*Populus tremuloides*) through suckering is usually reduced only by repeated burning (Svedarsky *et al.*, 1986).

The expansion of the grasslands during the Miocene was accompanied by rapid changes in the mammalian faunas. Grasses and sedges have a high silica content that wears down the grinding teeth of browsing animals, but early horses, camels and other plains-dwelling mammals quickly evolved high-crowned teeth in which growth continually replaces the worn surfaces. Other cursorial specializations, such as speed and endurance, similarly equipped them for life on the open grasslands

(Vaughan, 1986). The large and diverse mammalian population, which in North America, for example, included horses, camels, elephants, rhinoceroses and antelopes and their attendant carnivores (Simpson, 1947), undoubtedly had a considerable impact on the development of the grasslands. Most of the mammals became extinct during the Pleistocene but large numbers of bison (*Bison bison*) and pronghorn (*Antilocapra americana*) still roamed the prairies in historic times. However, their migratory habits would have had quite a different effect on the grasses compared with today's domestic livestock herds.

7.6 GROWTH HABITS OF THE TEMPERATE GRASSES

The dominant species in temperate grasslands are all perennial grasses but they have evolved several different growth habits. In most grasses the main stem is typically very short and barely extends above the soil surface during the vegetative phase. New leaves are constantly produced at the tip of the stem and additional shoots arise in the leaf axils, which in turn produce leafy shoots. This process of **tillering** is characteristic of the tufted, non-creeping bunch grasses such as needle-and-thread (*Stipa comata*), June grass (*Koeleria cristata*) and blue grama (*Bouteloua gracilis*). Tillering is suppressed by a high concentration of auxin in the expanding leaves and is therefore stimulated by the removal of the leaf apex where auxin-synthesis occurs (Beard, 1973). The grasses can be grazed repeatedly during this vegetative stage provided that the short stem is not damaged. The individual tillers normally live for about a year, but once established the bunch grasses are maintained by the continual production of new growth for the 3–4 years that the main stem remains alive. Later in the season reproductive shoots elongate and the flowers are held above the leaves: tillering then ceases and once the fruit is shed the leaves usually begin to die back, so that by winter very little live tissue remains above ground. Occasional seeding is necessary to perpetuate the bunch grasses because the vigour of the clump declines with age. Most grasses are self-pollinated so there is rarely a shortage of seed.

In sod-forming grasses, secondary lateral shoots arise from vegetative buds in the axils of the leaves or in the nodes of older stems and grow as underground rhizomes 10–15 cm below the soil surface (Weaver, 1954). In rhizomatous species, such as western wheat grass (*Agropyron smithii*) and switchgrass (*Panicum virgatum*), shoots and roots develop at intervals along the length of the rhizomes, eventually terminating in a cluster of aboveground shoots. The terminal growing point emerges above ground in response to excessively high temperatures or short day lengths. Linear growth ceases when the tip is exposed to light and chlorophyll begins to form in the leaf scales. At this stage new rhizomes begin to

develop and spread out from the parent plant (Beard, 1973). Rhizomes are the principal method of propagation in the sod-forming grasses and they characteristically form a densely matted layer in the soil. In western wheat grass, for example, some rhizomes can grow over 2 m in length in a single growing season, and within 2 years more than 180 m of rhizomes may permeate 1 m^2 of soil (Mueller, 1941).

Less commonly the secondary lateral shoots grow across the soil surface as stolons; these also develop shoots and roots at their nodes. Species such as buffalo grass (*Buchloe dactyloides*) and creeping bentgrass (*Agrostis stolonifera*) spread in this way. Mueller (1941) reported average growth rates of 3 cm day^{-1} for buffalo grass, and in experimental plots this grass spread over more than 20 m^2 after two years. Stolons and rhizomes are much thicker than the roots of the grasses and are important for carbohydrate storage, which is essential for heat and drought resistance (Julander, 1945).

The root systems of grasses can be divided into three general types according to depth and coarseness, although their development is greatly affected by soil conditions. In short-grass species most of the roots are found within 30–90 cm of the surface, whereas the tall- and mid-grasses often penetrate to 1.5–1.8 m, and some grassland forbs root even deeper. The tillering bunch grasses have dense, fibrous rooting systems that radiate in all directions from the base of the stem. In the drier grasslands the dominant bunch grasses, such as blue grama (*Bouteloua gracilis*), develop a network of fine roots which penetrate to depths of 45–90 cm but radial spread is limited to only 30 cm around the plants (Figure 7.18). All of the roots are less than 0.5 mm in diameter and the laterals are rarely more than 2.5 cm in length (Weaver, 1954). Approximately 90% of the root mass is found in the upper 15 cm of the soil (Weaver and Darland, 1949) so moisture from sporadic precipitation events is absorbed efficiently during the growing season. The rooting depth for mid-grasses, such as western wheat grass (*Agropyron smithii*), is typically about 2 m, but less than 60% of the root mass occurs in the upper 15 cm of the soil. In this species the roots are 0.2–1.0 mm in diameter. The finer roots grow horizontally and rarely exceed 30 cm in length: the coarser roots grow vertically to the maximum depth of water penetration and are less densely branched (Coupland and Johnson, 1965). A similar rooting depth occurs in tall-grasses such as big bluestem (*Andropogon gerardi*) but in these species the larger roots can be up to 3 mm in diameter and branch profusely, with the lateral roots, up to 15 cm in length, spreading densely though the upper soil layers. Species which grow in moist depressions, such as switchgrass (*Panicum virgatum*), have coarse roots 3–5 mm in diameter which may penetrate to depths of 3.5 m: fine branches are usually absent in these species (Weaver, 1954).

A variety of rooting habits also occurs amongst the forbs and small shrubs which are associated with the

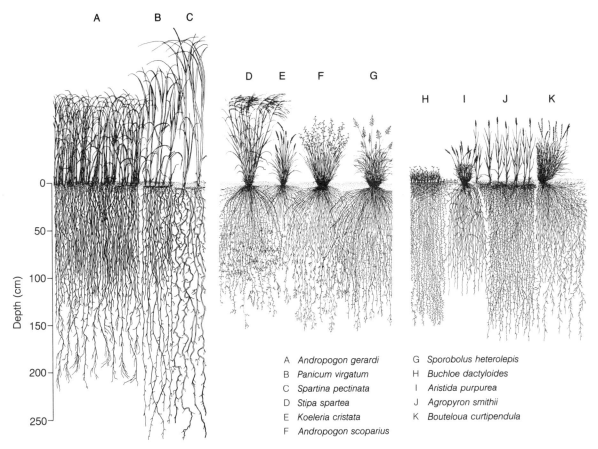

A   Andropogon gerardi   G   Sporobolus heterolepis
B   Panicum virgatum   H   Buchloe dactyloides
C   Spartina pectinata   I   Aristida purpurea
D   Stipa spartea   J   Agropyron smithii
E   Koeleria cristata   K   Bouteloua curtipendula
F   Andropogon scoparius

Figure 7.18 The root systems of common North American prairie grasses. (After Weaver, 1968.) (Reprinted from *Prairie Plants and their Environment: A fifty-year study in the midwest*, by J. E. Weaver, by permission of the University of Nebraska Press. Copyright © 1968 by the University of Nebraska Press.)

grasses (Weaver, 1958). Species such as pasture sage (*Artemisia frigida*) have a fairly shallow system of fine roots which arise from a coarser taproot and spread 20–25 cm laterally through the soil (Figure 7.19). Normally their roots are confined to the upper 90 cm but in drier sites they may penetrate to about 1.5 m. Species such as the smooth goldenrod (*Solidago missouriensis*) and Canada goldenrod (*S. canadensis*) propagate by woody rhizomes which grow just below the soil surface. The clusters of fine roots that are produced at intervals along the rhizomes usually terminate within 30 cm of the surface but coarser roots can penetrate to depths of more than 2 m, with some lateral branching occuring throughout their length. Other forbs have prominent taproots that descend 5–6 m in some soils. In deep-rooting species such as dotted blazing star (*Liatris punctata*), coarse lateral roots arise in the upper part of the soil but fine branching mostly occurs below the main mass of grass roots at depths of about 60 cm. In contrast, the roots of *Lygodesmia juncea* remain unbranched for 1–2 m with some laterals developing at depth. Species with such deep-rooting habits are especially tolerant of lengthy droughts (Weaver and Albertson, 1943).

7.7 XEROMORPHIC CHARACTERISTICS OF THE TEMPERATE GRASSES

Stratification of roots within the soil minimizes competition for moisture, but periods of drought are common and can alter the growth characteristics of the grasses. Morphologically the grasses respond to water deficits by increased root growth and decreased shoot development. Tillering and shoot elongation is reduced, fewer leaves develop, and these are smaller and thinner than normal because the cells are much smaller in size. Younger leaves have a higher diffusion pressure defict than the older leaves and are therefore preferentially supplied with water during periods of water stress. Often the older leaves will die prematurely. This reduces the number of leaves per plant and consequently the root:shoot ratio is increased. Similarly, a period of water deficit causes a general reduction in physiological activity. Internal water stress results in stomatal closure and increases the mesophyll resistance to inward diffusion of carbon dioxide, and so depresses the rate of photosynthesis. Lower tissue water content enhances the hydrolysis of starch: this increases the amount of soluble

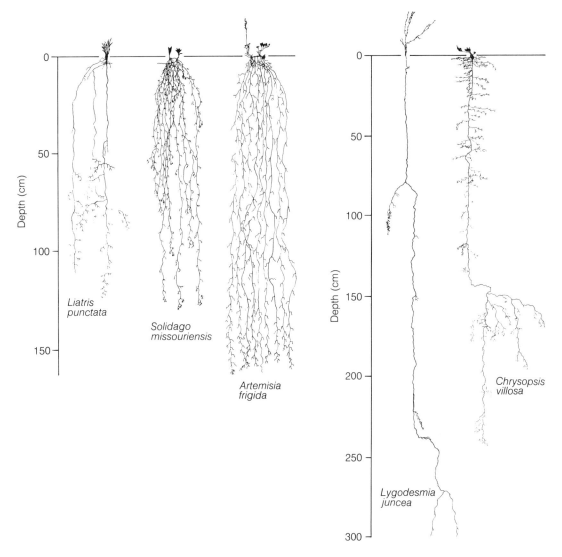

Figure 7.19 Characteristic root systems of herbaceous species in Canadian mixed prairie. (After Coupland and Johnson, 1965.) (Reproduced from R. T. Coupland and R. E. Johnson, Rooting characteristics of native grassland species in Saskatchewan, *Journal of Ecology*, 1965, **53**, 491, 496, 497.)

carbohydrate, but protein synthesis is inhibited and the protein content of the grasses declines (Beard, 1973). Further loss of water from the leaves causes them to droop or roll and often they acquire a blue-green or slaty coloration. Continued water stress will eventually result in mechanical injury of the cells and the death of the tissue. The perennial grasses can resist drought in a state of dormancy. Shoot growth ceases and the above-ground tissues die back and form a mulch which reduces soil moisture losses. New growth is initiated from drought-hardy buds in the crowns, rhizomes or stolons as soon as conditions are favourable.

In addition to a relatively high root:shoot ratio, the drought-resistant grasses tend to have narrow leaves which are less likely to overheat during hot dry weather, sunken stomata and thick, dense cuticles which are impermeable to water. Diffusion resistance is raised by leaf

pubescence as this increases the water vapour boundary layer around the leaf. Water loss can be reduced in some species by folding or rolling the leaves (Redmann, 1985). The shape of the leaves changes because of water loss in the **bulliform** or motor cells. These cells are considerably larger than normal epidermal cells and are arranged in rows that extend along the length of the upper surface of the leaves. Water is rapidly lost through their thin walls during periods of drought, and as they collapse the leaves fold or roll depending on how the cells are arranged in the leaf. This enables the plants to maintain gas exchange with reduced water conductance, although water loss is still regulated by stomatal closure in periods of drought. In drier habitats species with permanently rolled leaves become more abundant. Drought resistance is also achieved through drought hardiness or the ability to survive desiccation. Drought-hardy tissues, such as the buds on rhizomes, are typically comprised of relatively small cells that are devoid of vacuoles. This reduces contraction and mechanical stress as they dry out. High cell-sap concentrations also favour the retention of water, and this further reduces the amount of cellular contraction (Beard, 1973).

7.8 TEMPERATURE AND GROWTH OF TEMPERATE GRASSES

The characteristics which promote drought hardiness also reduce injury from high and low temperatures. High-temperature injury is generally caused by excessive temperatures at the soil surface which damage the stem. Heat hardiness varies according to the age and type of tissue. The lower portion of the crown and the younger leaves are more heat-tolerant than older tissue, and semidormant buds on rhizomes and stolons are more resistant than actively growing tissues. Extended exposure to periods of heat stress causes growth to cease in some cool-season grasses but in most species summer dormancy is largely controlled by water availability, and growth resumes later in the year when conditions are more favourable. Conversely, the distribution of warm-season grasses is mainly related to winter temperatures. The degree of low-temperature hardiness is inversely related to tissue hydration levels, and much of the injury occurs in late winter and early spring as tissue moisture levels increase, or following mid-winter thaws. Direct low-temperature injury results from the explosive growth of ice crystals in tissues with high hydration levels, or from extracellular ice bodies which form when water is drawn out of cells: the protoplasm becomes brittle under extreme dehydration and is subject to mechanical damage. Injury can also occur during thawing if the cell walls expand more rapidly than the protoplasts (Beard, 1973). Most grasses survive the winter because their buds are protected from the cold by an insulating layer of snow and thatch which reduces heat loss from the soil surrounding their stems and buds.

The division of grasses into cool- and warm-season species corresponds with the pattern of distribution of C_3 and C_4 grasses in the temperate grasslands. The C_4 genera mostly originated in tropical or subtropical regions and grow best in warmer climates where photosynthesis can be limited by low intercellular CO_2 concentrations as a result of low stomatal conductance (Ehleringer, 1978). In regions where daytime temperatures are below 25–30 °C the advantage shifts to the C_3 plants, which have comparatively higher rates of photosynthesis at lower temperatures. The effects of seasonal variation in light intensity and temperature on photosynthetic rates within grassland canopies suggest that productivity in C_3 species would increase at higher latitudes and that C_4 species would be superior below 45 °N (Figure 7.20). In North America C_4 species account for

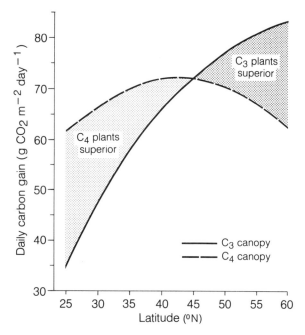

Figure 7.20 Calculated rates of total daily carbon gain for C_3 and C_4 grass canopies at different latitudes within the Great Plains of North America during July. (After Ehleringer, 1978.) (Reproduced with permission from J. R. Ehleringer, Implications of quantum yield differences on the distributions of C_3 and C_4 grasses, *Oecologia*, 1978, **31**, 262.)

68% of the grasses in the southern Great Plains but only 31% in the north (Teeri and Stowe, 1976). This pattern is strongly influenced by temperature minima during the growing season. Most C_4 grasses are sensitive to low temperatures and are injured at 0–10 °C. However, C_4 species such as mat muhly (*Muhlenbergia richardsonis*) and alkali cord grass (*Spartina gracilis*) have been reported between 60 and 65 °N in north-west Canada, where they have adapted to a frost-free period of less than 60 days (Schwarz and Redmann, 1988). The remains of C_4 grasses (species of *Bouteloua*) have been identified in the nests of fossil ground squirrels (*Spheromophilous parryii*) from Pleistocene sediments in Alaska (Guthrie, 1982). The northern C_4 grasses may therefore represent remnants of the 'steppe-tundra' which previously existed in Beringia.

7.9 SEASONAL GROWTH AND PHENOLOGY

The C_4 grasses mainly grow during mid or late summer but C_3 species usually begin growth in early spring. In the short-grass prairie of Colorado new shoots of *Agropyron smithii*, a C_3 species, begin to appear in April, and growth continues during the spring with anthesis occurring in mid-June (Monson *et al.*, 1983). The maximum rate of photosynthesis occurs at 25–30 °C in this species, and CO_2 fixation is irreversibly inhibited at temperatures of 43–46 °C due to heat damage to the leaves (Monson and Williams, 1982). C_4 species such as *Bouteloua gracilis* initiate growth 3–5 weeks later and continue to grow throughout the summer: maximum photosynthesis occurs at temperatures of 40–42 °C, with an upper temperature threshold as high as 53 °C. Flowering in *Bouteloua gracilis* begins in mid-July and continues until early August. Despite these different physiological traits, maximum leaf development occurs during early June in both species with leaf area increasing at a rate of 0.71 cm^2 $tiller^{-1}$ day^{-1} in *A. smithii* and 0.16 cm^2 $tiller^{-1}$ day^{-1} in *B. gracilis* (Monson et al., 1986).

During this period total water loss in *B. gracilis* is 154 mol m^{-2} day^{-1} compared with 145 mol m^{-2} day^{-1} in *A. smithii*. However, by mid-July water loss from *A. smithii* has increased to 182 mol m^{-2} day^{-1} compared with 83 mol m^{-2} day^{-1} in *B. gracilis* for similar levels of CO_2 uptake. The greater use of water by *A. smithii* is offset to some extent by its deeper rooting habit and its restriction to moister depressions. A similar topographic relationship has been reported in mixed prairie where C_4 species such as *Muhlenbergia cuspidata*, *Andropogon scoparius* and *Bouteloua gracilis* dominate the upland communities, and C_3 species such as *Agropyron smithii* and *Stipa virudula* account for nearly all of the cover in lowland sites (Ode et al., 1980). Although C_4 plants have higher water use efficiencies than C_3 plants and are well adapted to the droughty summer climates, temporal differences in growth activity are mainly controlled by the temperature optima of their photosynthetic pathways rather than a response to seasonal moisture gradients (Kemp and Williams, 1980). However, soil moisture levels can affect phenological development: in *B. gracilis*, for example, vegetative growth may continue longer and flowering is delayed in wetter years (Dickinson and Dodd, 1976).

The wide geographic range of many of the grasses in North America has resulted in significant ecotypic variation within species. For example, flowering in *Koeleria cristata*, *Bouteloua gracilis*, *B. curtipendula* and *Panicum virgatum* occurs earliest in populations from the northern and western prairies and becomes progressively later towards the south and east (McMillan, 1959). However, no regional pattern is apparent in species such as *Stipa comata* and *Oryzopsis hymenoides*. They are opportunistic species and flower as soon as conditions are suitable. Resumption of growth in the spring is largely controlled by temperature conditions and the longer period to anthesis in the south results in significantly larger plants: similar ecotypic variation occurs along elevational gradients.

Considerable variation in root production occurs during the course of the growing season because of the availability of water, nutrients and organic matter (Stanton, 1988), and although mean annual root biomass increases in years with higher shoot production, the timing of root growth is usually not consistent from year to year. In blue grama, root growth typically commences shortly before leaf expansion, using food reserves in the crowns or older roots (Ares, 1976). Maximum root growth coincides with the period of greatest leaf expansion. The young roots stop growing when soil water potential reaches − 4 to − 6 MPa, and they die soon after. Late-season root growth is greatest near the crown and as the season progresses root biomass increases in the upper levels of the soil profile. In species such as the bluestems, switchgrass and side-oats grama, root growth is initially quite rapid, declines during the middle part of the growing season and later increases as shoot growth declines (Dalrymple and Dwyer, 1967). Root development is greatly affected by soil moisture availability and increases rapidly during moist periods (Hayes and Seastedt, 1987).

7.10 PRODUCTION AND TURNOVER RATES IN TEMPERATE GRASSLANDS

The leaves of most temperate grasses die back in late autumn and new growth develops from the crowns the following spring. In regions dominated by cool-season grasses, rapid growth occurs in the early part of the growing season as snowmelt percolates into the soil and biomass increases at a rate of approximately 5 g m^{-2} day^{-1}. The period of peak production persists for only 2 or 3 weeks, after which growth declines. A similar rate of production occurs in sites dominated by warm-season grasses, but growth is more erratic and is more dependent on spring and summer rainfall (Sims and Singh, 1978). Peak live biomass for North American grasslands range from 62 to 236 g m^{-2} for short-grass prairie, compared with 101–270 g m^{-2} for mixed prairie and 336 g m^{-2} for ungrazed tall-grass prairie. Peak live biomass in the prairies increases linearly as precipitation during the growing season increases from 100 to 450 mm, but is less dependent on moisture in more humid areas (Figure 7.21). Similarly, production increases from the drier southern steppes northwards across the former USSR, and above-ground biomass in the productive 'meadow-steppes' exceeds 1100 g m^{-2} (Rodin and Bazilevich, 1967).

The length of time that the leaves remain green ranges from a few days to one month (Coupland and Abouguendia, 1974) and transfer of green material to the dead compartment of the canopy begins early in the growing season, particularly in grasslands dominated by cool-season species (Figure 7.22). Consequently, there is a substantial amount of dead material present in the canopy (Table 7.4). As the year progresses the amount of dead material produced during the current growing season gradually increases. Some dead material remains at the start of the next growing season but these older leaves and stems break down rapidly as new growth

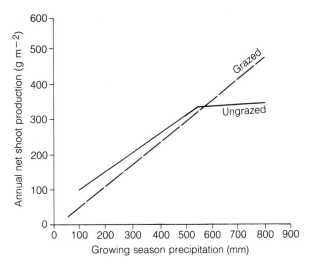

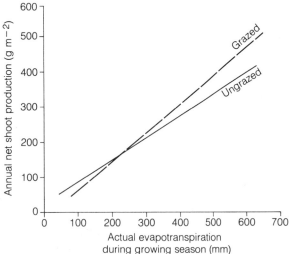

Figure 7.21 The relationship between above-ground net primary production and growing season precipitation and annual evapotranspiration in grazed and ungrazed grasslands in North America. (After Sims and Singh, 1978.) (Reproduced with permission from P. L. Sims and J. S. Singh, The structure and function of ten western North American grasslands III. Net primary production, turnover and efficiencies of energy capture and water use, *Journal of Ecology*, 1978, **66**, 579, 580.)

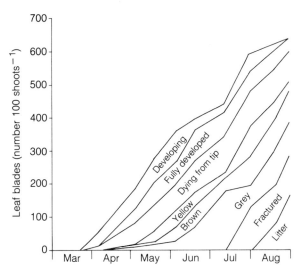

Figure 7.22 Length of time during which leaf blades of northern wheatgrass (*Agropyron dasystachyum*) growing in Canadian mixed prairie remained in various conditions over the growing season. (After Coupland and Abouguendia, 1974.) (Reproduced with permission from R. T. Coupland and Z. Abouguendia, Producers: V. Shoot Development in Grasses and Sedges, *Matador Project – Technical Report No. 51*; published by Department of Plant Ecology, University of Saskatchewan, 1974.)

Table 7.4 Mean biomass of producer compartments in ungrazed prairie in North America (after Sims and Coupland, 1979)

Location	Above-ground biomass (g m^{-2})		
	Green shoots	Dead shoots	Litter
Short-grass prairie			
Pantex, Texas	70	129	331
Pawnee, Colorado	70	65	251
Mixed prairie			
Hays, Kansas	79	117	904
Cottonwood, South Dakota	93	141	496
Dickinson, North Dakota	178	369	457
Matador, Saskatchewan	75	411	238
Tall-grass prairie			
Osage, Oklahoma	152	446	365

(Reproduced with permission from P. L. Sims and R. T. Coupland, Producers, in *Grassland Ecosystems of the World: Analysis of Grasslands and Their Uses*, ed. R. T. Coupland; published by Cambridge University Press, 1979.)

commences. Standing dead material accumulates at an average rate of $0.9 \, \mathrm{g \, m^{-2} \, day^{-1}}$. Subsequently, dead material is incorporated into the soil litter layer at an average rate of $0.8 \, \mathrm{g \, m^{-2} \, day^{-1}}$, and in ungrazed prairies total accumulation ranges from about 100 to 900 g m^{-2}. Although litter accumulates irregularly throughout the growing season, peak litter biomass often occurs early in the spring following the addition of material from the previous year (Sims and Coupland, 1979).

Below-ground production consists mostly of roots and rhizomes, with shoot crowns contributing about 22% of the total underground biomass. Mean below-ground biomass in North American mixed prairie decreases from approximately 2100 g m^{-2} in northern regions to

1150 g m^{-2} in the south. A similar decline from 1350 g m^{-2} to 875 g m^{-2} occurs in short-grass prairie, and a futher decline occurs with increasing aridity towards the west (Sims and Coupland, 1979). Root biomasses of 420–1000 g m^{-2} are reported in various steppe communities in the former USSR (Rodin and Bazilevich, 1967). Between 60 and 80% of the root mass is found in

Table 7.5 Distribution of roots in tall-grass prairie in Missouri (after Dahlman and Kucera, 1965)

Soil depth (cm)	Ovendry root biomass (g m^{-2})			
	April	July	October	January
0–5	766	1107	1025	839
5–10	188	255	238	291
10–15	115	130	151	170
15–25	119	125	161	170
25–36	74	65	79	97
36–46	52	52	70	60
46–56	49	45	65	49
56–66	38	37	45	38
66–76	36	31	44	32
66–86	12	13	23	9
total	1449	1860	1901	1755

(Reproduced with permission from R. C. Dahlman and C. L. Kucera, Root productivity and turnover in native prairie, *Ecology*, 1965, **46**, 86.

Table 7.6 Annual inputs and losses of nitrogen in prairie ecosystems (after Woodmansee, 1978)

	Inputs and losses of nitrogen (g N m^{-2} yr^{-1})		
	Short-grass prairie	Mixed prairie	Tall-grass prairie
Annual inputs			
wet deposition	0.45	0.65	0.75
dry deposition	0.15	0.25	0.25
symbiotic fixation	< 0.05	< 0.05	< 0.05
non-symbiotic fixation	< 0.05	< 0.05	< 0.05
total inputs	0.60	0.90	1.00
Annual losses			
livestock tissue	0.10	0.30	0.40
livestock wastes	0.20	0.60	1.00
ammonification	< 0.05	0.10	0.20
other	0.10	0.20	0.20
total losses	0.40	1.20	1.80
Net nitrogen balance	+ 0.20	− 0.30	− 0.80

(Source: R. G. Woodmansee, Additions and losses of nitrogen in grassland ecosystems, *BioScience*, **28**(7), 449, © 1978 American Institute of Biological Sciences.)

the upper 20 cm of the soil, and all of the shoot crowns are present in the surface layers. The root system is very dynamic and significant changes in biomass occur during the year. For example, in tall-grass prairie the roots and rhizomes are concentrated in the upper 5 cm of the soil (Table 7.5). Root mass in the surface soils increases in the summer and accounts for 60% of total underground biomass in July, then gradually declines over winter. Cooler temperatures may account for the delayed root development in the deeper soil layers. Over the course of the year 200 g m^{-2} of roots and 141 g m^{-2} of rhizomes are lost from the upper 5 cm; approximately 30% of the underground biomass is therefore replaced each year. Similarly, annual turnover of root biomass in short-grass prairie averages 49%, and is estimated at 18% in mixed prairie (Sims and Singh, 1978).

The shoot crowns are important for storage of carbohydrates and their biomass declines as new shoot growth begins. Crown biomass then increases after maximum shoot biomass is attained. Average turnover rates for crowns are higher than for roots, ranging from 82% in tall-grass prairie to 53% in short-grass prairie and 35% in mixed prairie. In Canadian mixed prairies the turnover rate for underground materials is 29% and few roots survive beyond 3 years (Coupland, 1974b). Decay is minimal during the first year, but up to 50% of the dead root mass is mineralized in the second year and most of the material disappears within 3 years.

7.11 NUTRIENT CYCLING

Although plant composition and growth in grasslands is largely controlled by climatic conditions, production is frequently limited by soil nutrient levels, particularly nitrogen and phosphorus (Risser, 1988). Natural input of nitrogen is mainly by precipitation and dry deposition (Table 7.6). The amount of nitrogen contributed by symbiotic associations is comparatively low because le-

gumes and other plants which serve as hosts for nodulating organisms are not abundant (Clark *et al.*, 1980). Additions from non-symbiotic nitrogen-fixing bacteria, such as *Azotobacter* and *Clostridium*, are also low and typically range from 0.1 to 0.2 g N m^{-2} yr^{-1}. Loss of nitrogen is mostly through the volatilization of ammonia in animal wastes and some nitrogen is exported in the tissues of domestic livestock (Woodmansee, 1979). In ungrazed pastures nitrogen losses through leaching, denitrification and volatilization during decomposition are usually minimal. Little nitrogen is lost through leaching because grassland soils are low in nitrates and any soluble nitrogen is quickly taken up by actively growing plants. Similarly, soil moisture and temperature conditions rarely favour denitrification, while ammonia released during decomposition of plant litter is quickly absorbed by the vegetation canopy (Clark *et al.*, 1980). Tillage results in rapid decomposition of organic matter and liberates more nutrients than the crops can use. In Canadian prairies as much as 65% of the nitrogen available for plant growth is lost annually through volatilization, leaching and soil erosion, or is removed in harvested grain (Rowe and Coupland, 1984). Since the 1950s, substantial fertilizer inputs have been required to maintain crop yields.

Above-ground production in short-grass prairie increases following nitrogen applications, especially when irrigated (Dodd and Lauenroth, 1979): increased yields and greater moisture-use efficiency have also been reported in tall-grass prairie (Owensby *et al.*, 1970). Similarly, in undulating short-grass prairie, soil moisture and nitrogen levels increased at the foot of the hill slopes, and above-ground biomass of the grasses is considerably higher than on the ridges (Schimel *et al.*, 1985).

Table 7.7 Efficiency of nitrogen uptake by western wheatgrass (*Agropyron smithii*) and side oats grama (*Bouteloua curtipendula*) grown under various temperature and soil nitrogen conditions (after Christie and Detling, 1982)

Temperature regime (day/night)	Dry weight of plant tissue produced per g nitrogen taken up ($g\ g^{-1}\ N$)	
	Agropyron smithii (C_3 species)	*Bouteloua curtipendula* (C_4 species)
20/12°C		
low soil nitrogen	149	70
high soil nitrogen	73	159
30/15°C		
low soil nitrogen	41	96
high soil nitrogen	33	125

(Reprinted with permission from E. K. Christie and J. K. Detling, Analysis of interference between C_3 and C_4 grasses in relation to temperature and soil nitrogen supply, *Ecology*, 1982, **63**, 1282.

Table 7.8 Average nitrogen content of vegetation and soil in prairie ecosystems (after Bokhari and Singh, 1975)

	Nitrogen content ($g\ m^{-2}$)		
	Short-grass prairie	Mixed prairie	Tall-grass prairie
Green shoots	0.87	1.78	1.35
Standing dead shoots	0.77	1.04	3.36
Litter	1.67	5.52	1.31
Crowns	1.98	1.75	1.26
Roots	4.84	7.45	3.42
Soil (0–20 cm)	129.08	257.73	250.47

(Source: U. G. Bokhari and J. S. Singh, Standing state and cycling of nitrogen in soil–vegetation components of prairie ecosystems, *Annals of Botany*, **39**, 1975.)

Greater nitrogen use efficiency has been reported in C_4 plants compared with C_3 plants (Brown, 1978), and although they are more abundant in low nitrogen soils, this may simply be a consequence of climatic conditions at these sites. Experiments conducted with *Agropyron smithii* (C_3) and *Bouteloua gracilis* (C_4) indicate that the seasonal use of soil nitrogen is related to temperature (Christie and Detling, 1982). When grown under a day/night temperature regime of 20/12 °C, *A. smithii* had a competitive advantage on low nitrogen soils, while *B. gracilis* was favoured by a 30/15 °C regime (Table 7.7).

Nitrogen accumulates in grassland ecosystems mainly in organic forms and most of the annual growth requirement of the plants is met through internal cycling (Clark *et al.*, 1980). Nitrogen is quickly taken up by new plant growth and is mostly stored in crowns and live root mass (Table 7.8). Many of the young roots persist for only 1–2 months, and although nitrogen is rapidly mineralized and recycled, a large proportion is translocated to the green herbage rather than released to the soil. Towards the end of the growing season nitrogen moves from the shoots to the underground tissues, or is stored above ground in the dead foliage. More than half of the nitrogen present in the above-ground dead material is lost by the end of the following growing season: some is transferred to the soil in organic forms bound up in fragments of litter, the remainder is released during decomposition. Considerably more nitrogen is present in root litter and the microbial population associated with it. Amino acids exuded from the roots and sloughed-off root cells, together with some of the microbially immobilized nitrogen, are rapidly mineralized and become available for plant growth, but the remainder is incorporated into the humus and is released very slowly.

Although nitrogen deficiency often limits primary production in temperate grasslands, increased growth has also been reported in tall-grass prairie following application of phosphorus (Reardon and Huss, 1965). The phosphorus contained in plant biomass represents only 2–3% of the total phosphorus present in the soil. Phosphorus is mainly stored in the roots and root litter, but concentration in plant tissues varies with stage of development. Phosphorus is absorbed from the soil solution surrounding the roots, and uptake is maintained by diffusion and mass flow in the root zone and by root growth through the soil. Rapid translocation from the roots occurs during periods of active growth as long as phosphorus levels remain above 0.05% in the exporting tissues (Cole *et al.*, 1977). Phosphorus is withdrawn from senescent tissues but some may be transferred directly to the litter fraction if the plant is damaged; the remainder passes to the decomposer system at the end of the growing season. Readily hydrolysable forms such as nucleic acids and phospholipids are quickly recycled, but some compounds are more resistant to decomposition and the phosphorus remains unavailable in the soil for long periods of time (Clark *et al.*, 1980). Much of the organic litter in grasslands is too low in essential nutrients to sustain bacterial growth without uptake of minerals from the soil. Most of the organic phosphorus present in the soil is therefore of microbial origin. Incorporation of inorganic phosphorus in this way is the most significant pathway in the phosphorus cycle in grasslands. Gains and losses of phosphorus from grassland ecosystems are normally quite small. Low solubility of phosphorus min-

Table 7.9 Percentage of total below-ground food consumption attributed to soil nematodes (after Scott *et al.*, 1979)

	Relative food consumption by nematodes (%)		
	Short-grass prairie	Mixed prairie	Tall-grass prairie
Roots and crowns	67	46	59
Fungi	85	51	23
Arthropods and nematodes	43	88	82

(Source: J. A. Scott, N. R. French and J. W. Leetham, Patterns of consumption in grasslands, in *Ecological Studies 32 – Perspectives in Grassland Ecology*, ed. N. R. French; published by Springer-Verlag N.Y. Inc., 1979.)

erals minimizes leaching even in areas with heavy precipitation but substantial losses can occur through soil erosion (Clark *et al.*, 1980).

Sulphur, like phosphorus and nitrogen, is released from organic matter by decomposition, and deficiencies can arise in low humus soils and in coarse soils where losses occur through leaching. Other minerals such as potassium, calcium and magnesium are largely returned to the soil from organic tissues in aqueous solution. The process is mainly regulated by the physiological age of the leaves and the health and vigour of the plants, with the highest rates of removal associated with mature and senescent tissues. The cations are removed from intercellular spaces in the tissues and combine with dissolved carbon dioxide in the leaching solution to form carbonates. Thus, they are available for plant uptake as soon as they reach the soil unless they are carried beyond the rooting zone in percolating waters (Clark *et al.*, 1980).

7.12 DECOMPOSITION PROCESSES

7.12.1 Soil microbes

The microbial population is concentrated in the rhizosphere, where it is well supplied with dead organic matter. It is estimated that 9 million bacteria per gram of soil are present at depths of 0–5 cm in tall-grass prairie, and some occur to a depth of 50 cm (Risser *et al.*, 1981). Although bacteria are the most numerous micro-organisms in prairie ecosystems, the fungi are the dominant decomposers. Throughout the upper 50 cm of the profile, between 1000 and 2000 m of fungal hyphae permeate each gram of soil. This is equivalent to 635 g m^{-2} dry weight and is at least twice the bacterial biomass. Total microbial biomass in the top 50 cm of the soil in tall-grass prairie is therefore about 950 g m^{-2}: this compares with about 200 g m^{-2} in the upper 30 cm of soil in mixed-grass prairie (Clark and Paul, 1970) and 180 g m^{-2} in the upper 60 cm of soil in short-grass prairie (French, 1979). Considerably fewer micro-organisms are found above-ground. Less than 1% of the microbial biomass is associated with standing live and dead shoots, although their numbers increase considerably as the leaves become senescent and die (Paul *et al.*, 1979).

Under optimal conditions plant litter from tall-grass prairie loses 17% of its weight in 10 days as a result of microbial activity, but bacteria and fungi are especially sensitive to temperature and moisture and normally only 60% of the material is decomposed annually (Risser *et al.*, 1981). In Canadian mixed prairie about 35% of the litter is decomposed by micro-organisms during the growing season (Biederbeck *et al.*, 1974). Here decomposition rates are reduced in the spring because of low temperatures: they increase in early summer when soil moisture and temperature conditions are near optimum and then decline as the soils dry out later in the year. Microbial activity is similarly limited in short-grass prairie, where as little as 15% of the litter is decomposed by micro-organisms (Vossbrinck *et al.*, 1979).

7.12.2 The soil invertebrates

The soil invertebrates are represented by a large and diverse group of organisms which differ markedly in terms of their size, feeding habits and life cycles. The nematodes are the most abundant with average densities in temperate grasslands of 0.5–6 million m^{-2}: this is generally higher than in other terrestrial ecosystems (Petersen, 1982). Many of them feed on bacteria but some of the cells pass undamaged through the intestines and so become widely dispersed within the rhizosphere. Bacterial feeders have high ingestion rates but low growth efficiencies and this accelerates the turnover of nitrogen and other nutrients. Other nematodes pierce the fungal hyphae and suck out the contents. Ectoparasites which feed on the roots of higher plants are also common (Stanton, 1988). In tall-grass prairie it is estimated that 51% of the nematode population is herbivorous, 23% are saprophages and 26% are predators (Risser *et al.*, 1981). Nematodes consume 46–67% of the plant roots and crowns produced in grasslands (Table 7.9). Their populations are closely correlated with live underground biomass, and plant productivity increases significantly when the population is reduced by chemical treatment (Scott *et al.*, 1979). Nematode densities are lowest in the winter and early spring but increase as soil temperatures begin to rise, reaching their maximum in late fall (Willard, 1974).

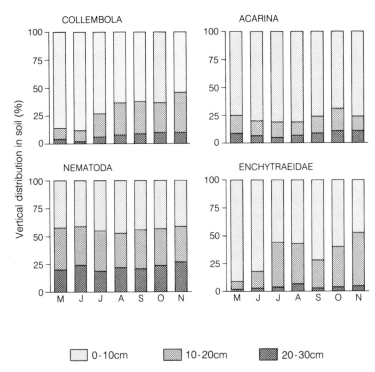

Figure 7.23 Seasonal changes in the vertical distribution of invertebrate groups in Canadian mixed prairie soils. (After Willard, 1974.) (Reproduced with permission from J. R. Willard, Soil Invertebrates: VII. A Summary of Populations and Biomass, *Matador Project – Technical Report No. 56*; published by Department of Plant Ecology, University of Saskatchewan, 1974.)

The most common microarthropods present in the soil and litter are the Collembola and Acari, although they are considerably less numerous than the nematodes, ranging from 140 000 individuals m^{-2} in drier short-grass prairie (Leetham and Milchunas, 1985) to 18 000 individuals m^{-2} in more humid tall-grass prairie (Risser *et al.*, 1981). Like the nematodes, they utilize a variety of food resources: 40% of the soil arthropods in tall-grass prairie are herbivorous and consume the roots and root sap of higher plants as well as algae and microbial tissues, 30% are saprophagous and 30% are predators. Marked fluctuations in the microarthropod population have been reported in tall-grass prairie with maximum densities occurring in the fall. The population is relatively stable in short-grass prairie but here, too, the population peaks at the end of the growing season (Leetham and Milchunas, 1985). The larger invertebrates such as the worms, millipedes and the larvae of some insects mostly break down plant litter by mechanical destruction and the increase in surface area renders them more accessible to microbial and chemical attack (Coleman and Sasson, 1980). The larger soil invertebrates are usually sensitive to desiccation. Population densities are not only higher in more humid regions but their vertical distribution in the soil is usually more variable than other groups. For example, wireworms (Enchytraeidae) are concentrated in the upper 10 cm of soil in Canadian mixed-grass prairies during the spring, when moisture levels are high, but their numbers decline in mid-summer because of differential mortality and migration to deeper layers (Figure 7.23).

7.13 ABOVE-GROUND CONSUMERS

The total biomass of phytophagous animals increases in relation to grassland productivity and is also influenced by the diversity of plant species present. Regional differences in the composition and distribution of the plant cover are therefore reflected in the animal populations which feed on it. Invertebrates occur in the greatest numbers; they feed most intensively on young leaves and stems and their food is often limited to a single species. Small herbivorous mammals and birds mainly consume fresh plant parts and seeds, and large grazing animals are equally selective in their feeding habits (Andrzejewska and Gyllenberg, 1980). Plant production typically varies by 15–50% from year to year. Natural populations of large herbivores are usually not greatly affected because the animals move over vast areas. Small mammal and insect populations often fluctuate dramatically, although these fluctuations are seldom related to food availability.

Whereas the below-ground invertebrates feed mainly on plant sap, the above-ground invertebrates are principally tissue feeders. Grasshoppers are the most important but they tend to assimilate only a small proportion of what they chew. In Canadian grasslands, grasshoppers ingest about 2% of the green shoot production but drop an additional 8%, and nearly all of this material is passed to the decomposer system (Bailey and Riegert, 1973). Grasshoppers have a major impact on litter production in short-grass prairie ecosystems since they can consume as much as 25% of the annual above-ground production (Mitchell and Pfadt, 1974). Damage from grasshoppers is negligible in tall-grass prairie (Risser *et al.*, 1981). Adult weevils and beetles mainly consume green leaves and stems but some feed on pollen or developing seeds and their larvae sometimes cause serious damage to roots. Most caterpillars feed above ground, although cutworms are an important subterranean group. Plant-tissue feeders can increase productivity in such grasses as *Bouteloua gracilis* by delaying senescence of the leaves. Similarly, increased growth of rhizomes has been reported in the bluestems (*Andropogon* spp.) following removal of terminal buds (Risser *et al.*, 1981). Although they are less abundant, the plant-sap feeders such as aphids, thrips and scale insects often have a greater effect on yields than the tissue-feeding insects. Ants are predominantly seed feeders but they do harvest other plant parts and may also transport sap feeders to their nests. The dominant perennial grasses mostly reproduce vegetatively and seed predation is unlikely to affect them adversely, although significant losses have been reported for species such as *Poa annua* and *Agrostis tenuis* because of worms (McRill and Sagar, 1973).

The North American prairie bird populations fluctuate in abundance and composition over the year because a large percentage of the species is migratory. In tall-grass and mixed prairies about 60% of the bird species are seasonal migrants and are present only during the breeding season. Fewer migratory species are recorded in the drier short-grass prairies, probably because there is a greater supply of seeds for the winter residents (Wiens, 1974). The number of breeding species native to the grasslands is comparatively low. Although 19 species are reported from tall-grass prairies, only three species are common; they are the eastern meadowlark (*Sturnella magna*), grasshopper sparrow (*Ammodramus savannarum*) and dickcissel (*Spiza americanum*). In mixed prairie the dominant species include the horned lark (*Eremophila alpestris*), western meadowlark (*Sturnella neglecta*) and Sprague's pipit (*Anthus spragueii*): the ranges of many of the species found here extend to the short-grass prairies. The native species are predominantly omnivorous, feeding on a mixture of seeds and insects during the breeding season, and their relative densities fluctuate from year to year in response to annual differences in above-ground production. For example, the total number of breeding pairs of common species in Canadian mixed prairies ranged from 80 to 145 km^{-2} during a 5-year period: this was attributed to changes in grasshopper production and vegetation development (Maher, 1973).

Several non-passerine species, including the ring-necked pheasant (*Phasianus colchicus*) and sharp-tailed grouse (*Pedioecetes phasianellus*), occur in the grasslands, and hawks and other raptors are conspicuous. In addition, numerous species invaded the grasslands as new habitats were created by changes in land use.

The structure of the small mammal population in the North American prairie grasslands varies as the nature of the vegetation changes in response to regional climatic patterns. Several species of small mammal are widely distributed throughout these grasslands but herbivores, such as the prairie vole (*Microtus ochrogaster*), are predominant only in the productive tall-grass prairies. Granivores such as pocket mice (*Perognathus* spp.) are more common in short-grass prairie, and omnivorous species such as the thirteen-lined ground squirrel (*Spermophilus tridecemlineatus*) are characteristic of the mixed prairies (Grant and Birney, 1979). Many of the species are obligate burrowers that are active in the subsurface litter layers and come above ground only for brief periods to forage or disperse. In drier regions the small mammals spend more time above ground and retreat to their burrows to escape predators, rear their young, or to hibernate. Although many species hibernate during unfavourable periods, some remain active throughout the year, particularly in more southerly tall-grass areas where a well-developed litter layer provides a favourable microclimate.

The annual biomass of small mammals in tall-grass prairies averages 813 g live weight ha^{-1}, compared with 251 g ha^{-1} in shortgrass range and only 76 g ha^{-1} in mixed prairie regions (French *et al.*, 1976). This variation in biomass reflects both the availability of food and the feeding habits of the animals. In the drier grasslands the quantity and quality of herbage is generally not sufficient to support large numbers of herbivores, and litter-dwelling rodents are absent because little plant debris accumulates. In mixed prairie, herbage is marginal and most plants reproduce vegetatively so seed resources are also limited. This restricts both herbivores and granivores; the small mammal fauna is predominantly omnivorous, feeding mainly on beetles, insect larvae and grasshoppers. Only in tall-grass prairie is sufficient herbage produced to maintain a large herbivorous population (Grant and Birney, 1979).

Comparatively little of the above-ground production in temperate grasslands is removed by the natural herbivore populations. In short-grass prairies only 5% of production is eaten or wasted, compared with 12% in mixed prairie and 17% in tall-grass prairie. Mammals account for about 50% of the material removed in the tall-grass prairies, compared with only 5–7% in other types of prairie (Table 7.10). Many large grazing animals, such as the bison and elk, have been extirpated from the prairies, while others like the antelope and deer survive only in greatly reduced numbers (Figure 7.24). It is estimated that some 40 million bison still roamed the plains in 1830

Table 7.10 Production and consumption of plant biomass in prairie ecosystems (after Scott *et al.*, 1979)

	Community energetics (kcal m^{-2} season^{-1})		
	Short-grass prairie	Mixed prairie	Tall-grass prairie
Above-ground parameters			
Net primary production	571.0	1219.0	1292.0
Tissue-feeding mammals	0.5	1.6	20.9
Tissue-feeding arthropods	3.5	6.3	6.9
Plant-sap feeders	3.0	24.7	12.6
Pollen-nectar feeders	0.6	1.0	6.6
Seed-feeding birds	0.6	0.2	0.6
Seed-feeding mammals	0.1	0.2	11.3
Seed-feeding arthropods	0.1	0.2	1.3
Litter-feeding arthropods	1.6	9.9	24.4
Below-ground parameters			
Plant production	3196.0	2372.0	3741.0
Plant-tissue feeders	40.2	244.3	69.0
Plant-sap feeders	192.9	382.4	369.6

(Reproduced with permission from J. A. Scott, N. R. French and J. W. Leetham, Patterns of consumption in grasslands, in *Ecological Studies 32 – Perspectives in Grassland Ecology*, ed. N. R. French; published by Springer-Verlag N.Y. Inc., 1979.)

but this population had been reduced to less than 1000 by the early 1900s (Roe, 1970). Early reports suggested that the large herds consumed so much vegetation that it was difficult to find forage for horses. The feeding habits of bison are quite different from those of the cattle which have replaced them. Bison have a greater preference than cattle for warm-season grasses such as blue grama and buffalo grass at all times of the year, although cool-season grasses are an important part of the bisons' diet in the spring. A much higher proportion of forbs and shrubs is consumed by cattle, especially in heavily grazed areas. The forage selected by bison is lower in crude protein than that eaten by cattle, but they have a higher digestive capacity and their use of the native rangelands is more efficient (Peden *et al.*, 1974).

Figure 7.24 Large ungulates are part of natural prairie ecosystems: (a) bison; (b) mule deer. ((a) A. K. Knapp; (b) J. T. Romo.)

Figure 7.25 Prairie dog town in North Dakota. (M. Brown.)

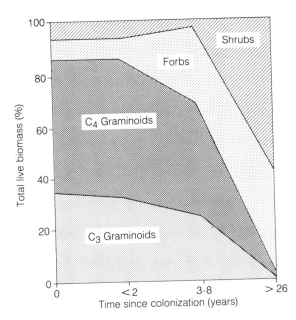

Figure 7.26 Temporal change in per cent contribution of different plant groups to total above-ground live biomass following colonization of a prairie dog town. (After Coppock *et al.*, 1983.) (Reproduced with permission from D. L. Coppock *et al.*, Plant–herbivore interactions in a North American mixed-grass prairie I. Effects of black-tailed prairie dogs on intraseasonal aboveground plant biomass and nutrient dynamics and plant species diversity, *Oecologia*, 1983, **56**, 3.)

Table 7.11 Mean morphological characteristics of greenhouse-grown western wheatgrass (*Agropyron smithii*) plants associated with grazing exclosures and a prairie dog colony in South Dakota (after Detling and Painter, 1983)

	Exclosure	Prairie dog colony
Total number of tillers	12.6	17.0
Number of leaves per tiller	6.7	5.7
Leaf blade length (mm)	230.6	132.6
Leaf blade width (mm)	5.7	3.5
Leaf blade:leaf sheath ratio	1.9	3.8
Height of upper leaf collar (mm)	120.6	25.2
Leaf angle (° from vertical)	27.8	60.1

(Reproduced with permission from J. K. Detling and E. L. Painter, Defoliation responses of western wheatgrass populations with diverse histories of prairie dog grazing, *Oecologia*, 1983, **57**, 67.)

Severe disturbance of the prairie plant cover by native herbivores is now comparatively rare; only around animal burrows, such as those of the prairie-dog (*Cynomys ludovicianus*) are the effects clearly apparent (Figure 7.25). Like the bison, the prairie-dog population has been drastically reduced, but these large sedentary rodents still exist in protected areas in colonies which cover up to 250 ha (Dalsted *et al.*, 1981). The vegetation on the colonies changes markedly over time: the grass cover is reduced and forbs and shrubs increase in importance (Figure 7.26). The dominant grass in this region is western wheatgrass (*Agropyron smithii*), and although it is reduced to about 3% of the cover present in adjacent uncolonized areas, there are significant differences in the morphological characteristics of the two populations (Table 7.11). The increased number of tillers, more compressed leaf angle and the higher leaf blade: leaf sheath ratio in plants from the prairie-dog colonies suggests that they are better adapted to grazing. Native herbivores, such as the prairie-dogs and bison, have probably been a strong selective force in the evolution of the native prairie plants (Detling and Painter, 1983).

In most grasses the meristematic tissues are at the base of the leaf blades so that the tips of the leaves can be removed without stopping growth (Dahl and Hyder, 1977). Sod-forming species of *Stipa* and *Bouteloua* are most resistant because their short crowns are essentially inaccessible to grazing animals. In tall-grass species, elongation of the crown occurs early in the season and this increases the possibility of damage. The greater number of reproductive shoots produced by these taller grasses also renders them more sensitive to grazing, although the palatability of species such as big bluestem (*Andropogon gerardi*) is greatly reduced as the flower stalks mature. Analysis of herbivore diets has also shown that C_4 plants are ingested by insects and small mammals in much smaller quantities than would be expected from their densities in grassland covers: they are often eaten only when preferred species are unavailable (Caswell *et al.*, 1973). The C_4 plants are considered nutritionally inferior because herbivores cannot easily break down the thick-walled bundle sheath cells. Consequently, a large amount of potential food material remains unavailable, and animal populations may be smaller than in comparable stands dominated by C_3 plants (Caswell and Reed, 1976). Regrowth following defoliation is predominantly through tillering (Dahl and Hyder, 1977). Increased rates of photosynthesis in the undamaged foliage and newly produced leaves can stimulate growth and a greater proportion of assimilated materials are allocated to leaf blade development (Table 7.12). However, this is achieved by diverting resources from the roots until sufficient leaf area has developed to supply carbon to the shoots (Bokhari, 1977) and in most plants root growth is markedly reduced following defoliation (Crider, 1955). Such adaptations help to maintain the grasslands now that the migratory herds of native ungulates have been replaced by domestic cattle.

7.14 THE EFFECTS OF GRAZING

The plants available to grazing animals can range from those which are highly palatable to those which are strictly avoided (Daubenmire, 1967). The palatable species are most severely injured, and continual grazing

Table 7.12 Dry matter accumulation in various parts of blue grama (*Bouteloua gracilis*) over a 10-day period following partial defoliation (after Detling *et al.*, 1979)

	Dry matter accumulation			
	Clipped plants		Unclipped plants	
	(mg)	(% of total)	(mg)	(% of total)
New leaf blade growth	20	53	23	33
New leaf sheath growth	8	21	17	25
New crown growth	3	8	9	13
New root growth	7	18	20	29
Total new growth	38	100	69	100

(Reproduced with permission from J. K. Detling, M. I. Dyer and D. T. Winn, Net photosynthesis, root respiration, and regrowth of *Bouteloua gracilis* following simulated grazing, *Oecologia*, 1979, **41**, 130.)

can reduce the quality of the forage as they are replaced by less desirable plants (Figure 7.27). Species which decline in abundance as a result of overgrazing are termed '**decreasers**' (Dyksterhuis, 1949): '**increasers**' become more abundant, but these too may eventually be replaced by unpalatable '**invaders**' in severely overgrazed areas. Decreaser and increaser species are important elements of the natural grassland cover but the invaders are restricted to animal mounds and other locally disturbed sites. The condition of the rangeland is therefore related to its floristic composition rather than to the amount of growth which occurs in a particular growing season (Figure 7.28). The resulting cover is affected by topographic conditions as well as grazing intensity. For example, buffalo grass (*Buchloe dactyloides*) is a conspicuous increaser in mixed prairie in Kansas, especially in heavily grazed upland and hillside sites (Figure 7.29). Although buffalo grass is relished by cattle, it quickly spreads by means of underground growing points and rapidly extending stolons.

Forage production in this mixed-prairie site was generally reduced by grazing (Table 7.13) and similar effects have been reported throughout the western rangelands of North America (Lacey and van Poollen, 1981). Conversely, in tall-grass prairie regions net annual aboveground production can be 25–30% higher under grazing (Sims and Singh, 1978): this is possibly related to faster rates of mineralization and nitrogen uptake under moist conditions (Sims and Coupland, 1979). A considerable amount of dead material accumulates in tall-grass prairie canopies (Table 7.14) and its removal by grazing could raise light levels sufficiently to increase production. Although the amount of dead shoot material in the canopy is generally reduced by grazing, the amount of litter can increase in more productive sites because some of the canopy biomass is added directly to the soil surface by trampling (Quinn and Hervey, 1970).

Photosynthesis is reduced when carbohydrates accumulate in the leaves and might be stimulated if periodic reduction in root reserves induced by defoliation increases the rate at which photosynthates are translocated from the leaves (May, 1960). Increased root exudations caused by rapid translocation of materials following clipping has been observed in western wheatgrass (Bokhari and Singh, 1974), and also in blue grama following tissue removal by grasshoppers (Dyer and Bokhari, 1976). In addition, the application of thiamine, which is present in the saliva of grazing animals, can

Figure 7.27 Although grazing alters the appearance of short-grass (*Bouteloua*) prairie, differences in floristic composition are less pronounced than in other prairie communities. (D. G. Milchunas.)

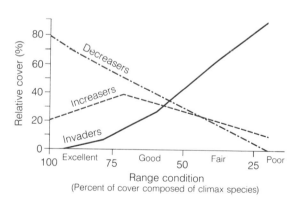

Figure 7.28 Relative contribution of different plant groups as an indicator of rangeland condition following grazing. (After Dyksterhuis, 1949.) (Reproduced with permission from E. J. Dyksterhuis, Condition and management of range land based on quantitative ecology, *Journal of Range Management*, 1949, **2**, 109.)

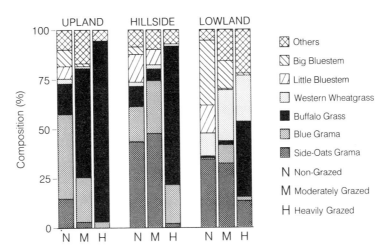

Figure 7.29 Changes in the composition of the vegetation following different grazing intensities on three topographic sites in North American mixed prairie. (After Tomanek and Albertson, 1957.) (Reprinted with permission from G. W. Tomanek and F. W. Albertson, Variations in cover, composition, production, and roots of vegetation on two prairies in western Kansas, *Ecological Monographs*, 1957, **27**, 275.)

Table 7.13 Effects of grazing on forage production in mixed prairie communities (after Tomanek and Albertson, 1957)

	Forage yields (kg ha⁻¹)		
	Ungrazed	Moderately grazed	Heavily grazed
Upland sites			
mid grass	1215	397	12
short grass	1214	1259	1098
forbs	329	495	204
total	2758	2151	1314
Hillside sites			
mid grass	1707	1603	365
short grass	217	398	613
forbs	203	288	177
total	2127	2289	1155
Lowland sites			
mid grass	3549	2056	917
short grass	20	420	528
forbs	97	240	215
total	3666	2716	1660

(Reproduced with permission from G. W. Tomanek and F. W. Albertson, Variations in cover, composition, production, and roots of vegetation on two prairies in western Kansas, *Ecological Monographs*, 1957, **27**, 276.)

enhance growth in side-oats grama (Reardon *et al.*, 1974). Different growth responses have also been noted in blue grama when clipped and treated with bison saliva (Figure 7.30), although total plant production was always higher in plants which had not been defoliated. Loss of vigorous root development under grazing has been reported in many species of grasses: for example, the root biomass of bluestems (*Andropogon* spp.) is reduced by 75% in heavily grazed pastures (Weaver, 1950). However, the effects on total root production are not consistent in all rangelands. In tall-grass prairie, root biomass declines slightly under normal grazing practices. There is little change in root biomass in the warmer short-grass

and mixed prairies but a marked increase can occur in cooler regions if blue grama increases in abundance, because this species has a well-developed root system (Sims *et al.*, 1978).

All rangeland will eventually deteriorate under severe and prolonged overgrazing. Soil moisture is usually reduced in heavily grazed pastures because the soil is compacted by the weight of the animals. The effect is most pronounced in fine-textured soils (Rauzi and Smith, 1973). Infiltration rates are also reduced by the removal of the plant and litter layers (Table 7.15). The impact of raindrops on exposed soil can seal the surface pores, and movement through the soil is restricted because there are fewer root channels and less organic matter present to improve its porosity. Runoff rates can increase markedly under poor grazing management (Dragoun and Kuhlman, 1968) and this can result in significant soil erosion problems. Despite considerable losses of water through interception and transpiration by a grass cover, evaporation rates are lower and reserves of soil moisture are invariably higher in ungrazed pastures (Table 7.16). Reserves of organic matter and nitrogen are decreased by grazing, because less material is added as surface litter and from roots. Nitrogen added to the soil in animal faeces increases microbial activity in grazed sites and this accelerates the breakdown of organic matter. The rate of decomposition can also be increased by higher soil temperatures following removal of the plant cover (Risser *et al.*, 1981).

7.15 FIRE IN THE GRASSLANDS

Fire occurs frequently in temperate grasslands and originally there were few firebreaks to stop its progress

Table 7.14 Average above-ground, litter and root biomasses in ungrazed and grazed grasslands in North America (after Sims *et al.*, 1978)

	Tall-grass prairie		Mixed prairie		Short-grass prairie	
	Ungrazed	Grazed	Ungrazed	Grazed	Ungrazed	Grazed
Above-ground biomass (g m⁻²)						
live shoots	152	166	117	102	70	74
recent dead	118	74	61	43	34	31
old dead	327	77	148	37	68	58
total	597	317	326	182	172	163
Litter biomass (g m⁻²)	256	424	433	120	205	75
Root biomass (g m⁻²)	964	924	1392	1910	885	1027

(Source: P. L. Sims, J. S. Singh and W. K. Lauenroth, The structure and function of ten western North American grasslands, *Journal of Ecology*, 1978, **66**, 285.)

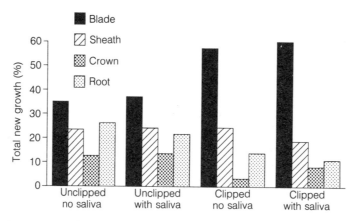

Figure 7.30 Contribution of total new growth of blue grama (*Bouteloua gracilis*) to various plant parts following clipping to 4 cm and foliar application of whole bison saliva. (After Detling *et al.*, 1980.) (Reproduced with permission from J. K. Detling *et al.*, Plant–herbivore interactions: examination of potential effects of bison saliva on regrowth of *Bouteloua gracilis* (H. B. K.) Lag., *Oecologia*, 1980, **45**, 28.)

Table 7.15 Effect of grazing on rate of water intake by silty clay soils in short-grass prairie (after Rauzi and Hanson, 1966)

	Above-ground herbage (kg ha⁻¹)	Surface litter (kg ha⁻¹)	Water intake	
			Dry soil (cm hr⁻¹)	Wet soil (cm hr⁻¹)
heavily grazed	1019	511	2.8	1.3
moderately grazed	1506	447	4.4	1.8
lightly grazed	2093	1232	6.9	3.2

(Source: F. Rauzi, and C. L. Hanson, Water intake and runoff as affected by intensity of grazing, *Journal of Range Management*, **19**, 1966.)

(Figure 7.31). The moisture content of grasses is comparatively low and their erect habit exposes the understory and soils to the drying effect of sun and wind (Vogl, 1974). Fire spreads rapidly through these fine, loosely arranged grassland fuels. Soil surface temperatures in the narrow flame zone are usually below 400 °C, but they can exceed 600 °C in heavy fuel loads. Such high temperatures persist for only a few minutes (Figure 7.32) and although most of the above-ground material is consumed there is little direct impact on soil organic matter, microbial populations or buried seed reserves (Wright and Bailey, 1982). Plant tissue is normally killed at temperatures above 60 °C but the lethal exposure time varies according to moisture content of the material (Figure 7.33). Bunch grasses, such as *Stipa comata*, accumulate large quantities of dead thatch and lethal temperatures can persist for more than 1 hour after the fire has passed. This makes them more susceptible to fire than rhizomatous species, which are less densely bunched and have their growing points buried in the soil (Wright and Bailey, 1982). Frequent and repeated burning is detrimental to all grasses and favours the spread of annual weedy species (Curtis and Partch, 1948).

The absence of fire in North American tall-grass prairie results in an increase in tree cover, particularly in the moister lowland sites, and fire was probably an important

Table 7.16 Effects of grazing on the physical and chemical properties of soil in tall-grass prairie communities (after Beebe and Hoffman, 1968)

	Ungrazed	Moderately grazed	Heavily grazed
Soil moisture (%)			
0–15 cm	27.2	15.8	21.8
15–30 cm	23.4	15.3	16.6
30–45 cm	20.3	14.8	13.4
45–60 cm	15.9	15.3	12.8
Soil temperature (°C)			
0–15 cm	30.8	31.3	32.8
15–30 cm	27.6	28.3	29.7
30–45 cm	26.0	26.0	26.5
Soil organic matter (%)			
0–15 cm	6.0	1.8	4.4
15–30 cm	3.6	0.7	2.3
30–45 cm	2.6	0.4	1.4
45–60 cm	1.3	0.3	0.7
Soil nitrogen (%)			
0–15 cm	0.32	0.12	0.26
15–30 cm	0.22	0.05	0.15
30–45 cm	0.14	0.05	0.09
45–60 cm	0.10	0.03	0.05

(Source: J. D. Beebe and G. R. Hoffman, Effects of grazing on vegetation and soils in southeastern South Dakota, *American Midland Naturalist*, **80**(1), 1968.)

natural factor in maintaining these grasslands (Bragg and Hulbert, 1976). In tall-grass prairies, litter tends to accumulate faster than it decomposes and this reduces the growth and vigour of the plants. Productivity of most native tall-grass species increases following spring fires, although this practice is harmful to local cool-season grasses because they begin growth early in the year (Anderson *et al.*, 1970). Higher yields are maintained for 2–3 years, provided that soil moisture is favourable, and a similar response occurs in below-ground biomass (Table 7.17). Improved growth is attributed to higher light levels, warmer soil temperatures and increases in soil nitrogen (Hulbert, 1988). Soil temperatures can be raised by as much as 9.8 °C on burned areas (Kucera and

Figure 7.31 Fire in tall-grass prairie. (A. K. Knapp.)

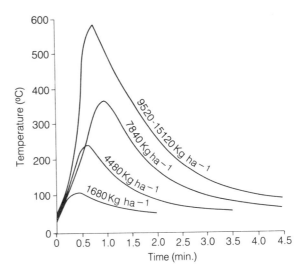

Figure 7.32 Time–temperature curves at the soil surface during prescribed grass fires for varying amounts of fine fuel comprised mainly of tobosagrass (*Hilaria mutica*). Air temperature at the time of the fire was 21–27 °C, relative humidity 20–40% and winds 13–24 km hr^{-1}. (After Wright *et al.*, 1976.) (Redrawn from H. A. Wright, S. C. Bunting and L. F. Neuenschwander, Effects of fire on honey mesquite, *Journal of Range Management*, 1976, **29**, 469.)

Ehrenreich, 1962) but soil moisture levels are invariably lower, particularly in the deeper soil layers (Table 7.18). Burning has little effect on soil chemistry because comparatively small quantities of ash are produced. Release

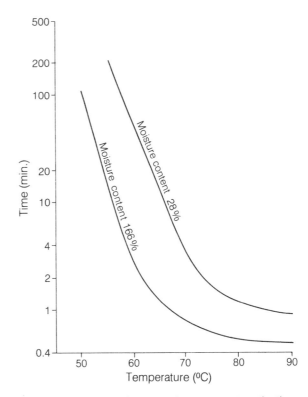

Figure 7.33 The effect of tissue moisture content on the time and temperatures required to kill needle-and-thread (*Stipa comata*) in prairie fires. (After Wright and Bailey, 1982.)

Table 7.17 Effect of different fire regimes on the productivity on two tall-grass prairie communities (after Hadley and Kieckhefer, 1963)

Community type	Years of burn	Green shoot biomass (g m⁻²) 1961	Green shoot biomass (g m⁻²) 1962	Root biomass (g m⁻²) 1961	Root biomass (g m⁻²) 1962	Litter biomass (g m⁻²) 1961	Litter biomass (g m⁻²) 1962	Flowering stalks (number m⁻²) 1961	Flowering stalks (number m⁻²) 1962
Big bluestem	1952, 1959	364	359	1100	1123	1023	1105	16	30
(*Andropogon*	1952, 1959, 1961	1321	591	1212	1262	burned	422	118	53
gerardi)	unburned	302	362	887	900	890	1205	10	28
Indiangrass	1952, 1959	502	531	924	965	790	1107	28	38
(*Sorghastrum*	1952, 1959, 1961	1473	633	950	981	burned	390	132	80
nutans)	unburned	489	476	782	787	1134	1062	17	32

(Reprinted with permission from E. B. Hadley and B. J. Kieckhefer, Productivity of two prairie grasses in relation to fire frequency, *Ecology*, 1963, **44**, 391–2.)

Table 7.18 Average soil moisture levels at various depths in upland tall-grass prairie under different burning treatments (after Anderson, 1965)

Time of burning	Soil moisture (cm H₂O per 30 cm soil) 0–30 cm	30–60 cm	60–90 cm	90–120 cm	120–150 cm	Average
Winter	9.1	8.9	8.6	8.2	7.2	8.4
Early spring	9.1	9.1	8.8	9.3	9.1	9.1
Mid spring	9.1	9.2	8.9	8.9	8.7	9.0
Late spring	9.5	9.5	9.1	9.0	9.0	9.2
Control (unburned)	9.4	9.7	9.3	9.5	9.2	9.4
Average	9.2	9.3	8.9	9.0	8.7	9.0

(Source: K. L. Anderson, Time of burning as it affects soil moisture in an ordinary upland bluestem prairie in the Flint Hills, *Journal of Range Management*, **18**, 1965.)

of calcium, magnesium and potassium can increase pH slightly but the effect is quickly lost because of leaching or removal by wind (Daubenmire, 1968). Nitrogen is volatilized at 200 °C and most of it is lost during a grass fire (Wright and Bailey, 1982). However, soil nitrogen levels will be increased if legumes are more abundant in the post-burn community.

In mixed prairie the grasses are most tolerant of fire in wetter years. However, they suffer considerable losses in years with below normal precipitation, when production in some species can be reduced by as much as 60% (Wright and Bailey, 1982). A similar response is noted in short-grass prairie and it may take 2–3 years before these grassland communities recover (Table 7.19). The reduction in plant growth appears to be related to low soil moisture levels and plant water stress (Redmann, 1978). Fire-tolerant species can therefore benefit from improved moisture conditions when competitors are suppressed: blue grama has reportedly increased after fires in New Mexico following elimination of the scattered overstory of juniper (Dwyer and Pieper, 1967). A similar response has been observed in Texas because of the reduced vigour of competing grasses (Trlica and Schuster, 1969). Production and composition are most affected when fires burn in the spring (Dix, 1960). However, standing dead material is an effective snow trap and spring soil moisture levels can be adversely affected when thatch is burned off late in the year (Trlica and Shuster, 1969). Burning during wetter periods is not particularly harmful to the vegetation: forage quality may be improved because the young shoots are usually more palatable and there is some increase in crude protein levels (Dwyer and Pieper, 1967).

7.16 DROUGHT

Drought is a characteristic feature of temperate grassland climates. Tree-ring data from western Nebraska suggest that 154 drought years occurred between 1539 and 1939 (Weakly, 1943). These dry spells generally persist for several years but become less frequent and less severe towards the northern and eastern parts of the Great Plains (Borchert, 1950). Periods of drought are accompanied by higher temperatures and desiccating winds, so that soil moisture reserves are quickly depleted. Plant production declines sharply in dry years: leaf growth is reduced and stops unusually early, and seed production is negligible. Growth below ground is also less vigorous and roots do not penetrate so deeply into the soil. Drought-hardy species, such as blue grama and buffalo grass, are able to reduce moisture loss by becoming dormant but they will renew growth several times in one season if soil moisture is available (Coupland, 1958). In bunch grasses such as these, the diameter of the clump decreases and some will die if the drought is prolonged. A similar reduction in shoot emergence occurs in rhizomatous species, whereas deep-rooted plants can endure drought conditions by drawing on deep moisture reserves. Rapid percolation in sandy soils can

Table 7.19 Herbage yields in short-grass prairie following burning (after Launchbaugh, 1964; Wright, 1974)

| Time after burn (yrs) | Herbage yields (kg ha^{-1}) | | | |
| | Burned sites | | Unburned sites | |
	Current growth	Litter	Current growth	Litter
Bouchloe dactyloides				
1	1888	–	1673	815
2	2311	343	2159	513
3	1566	1761	1490	1015
Bouteloua gracilis				
1	1882	–	1600	2771
2	1533	783	1397	2894
3	2399	1960	1964	2164
Bouteloua curtipendula				
1	2076	–	3335	–
2	2062	1178	3752	3664
3	1254	2969	1005	6240
Sporobolus cryptandrus				
1	2512	–	2407	–
2	2843	3342	2864	4786
3	3078	3382	2607	4623

(Source: J. L. Launchbaugh, Effects of early spring burning on yields of native vegetation, *Journal of Range Management*, **17**, 1964; and H. A. Wright, Effect of fire on southern mixed prairie grasses, *Journal of Range Management*, **27**, 1974.)

therefore be advantageous to deep-rooting species but conversely the probability of drought stress in shallow-rooted species is increased. In fine-textured soils the downward movement of water may be so slow that only the shallow-rooted species will survive (Coupland, 1974a).

The selective loss of some species and the decreased vigour of others results in a general reduction in plant cover, and the effects of the drought are often intensified by the absence of shade and the protective layer of thatch. Losses are more severe when heavy grazing is continued, and the notorious deterioration of rangeland in the Great Plains from 1933 to 1938 can be attributed to this combination of factors. The greatest damage during this period occurred in the mixed prairies, and as the mid-grasses disappeared several thousand km^2 were converted into short-grass plains (Weaver and Albertson, 1956). Less resistant species such as little bluestem were virtually eliminated, but drought-tolerant species such as side-oats grama and blue grama spread except where the rangeland was overstocked (Albertson and Tomanek, 1965). Some of the forbs also disappeared, although deeper rooting species often increased as the cover of grasses died back. Weedy annuals became much more abundant, especially on land that was covered in wind-blown dust. Germination and growth of species such as Russian thistle (*Salsola pestifer*), green foxtail (*Setaria viridis*) and downy brome (*Bromus tectorum*) were rapid when it rained, and much of the prairie had the appearance of abandoned farmland (Weaver, 1968). By the end of the Great Drought the boundary between the tall-grass and mixed prairies was displaced to the east by as much as 240 km (Weaver, 1968).

Reduction in the cover of tall-grass prairie was less pronounced than in mixed prairie, although in western districts little bluestem range was badly depleted. By 1935 the soil had dried to a depth of almost 2 m but this condition varied with topography and typically little damage was reported in locally moist sites (Robertson, 1939). Elsewhere the tall-grass prairie looked deceptively undamaged, but as the drought continued and the taller grasses died back the destruction of the understory species became apparent. The deterioration of the tall-grass prairies was accompanied by the rapid invasion of western wheatgrass, which spreads by means of long, much-branched rhizomes. Western wheatgrass renews growth early in the spring, and the meagre reserves of moisture that were available during the drought were quickly used up, thereby contributing to the loss of native species (Weaver, 1968). Blue grama, buffalo grass and needlegrass (*Stipa spartea*) similarly extended their range eastwards, and side-oats grama became more prominent in the region.

Recovery of the grasslands began in 1941 when precipitation was well above normal and evenly distributed throughout the growing season. The relict grasses spread quickly by vigorous growth of rhizomes, prolific tillering and an abundance of seedlings. In parts of the tall-grass prairie some of the grasses had remained dormant throughout 7 years of drought, and growth was often delayed until the second wet year. This was characteristic of little bluestem: only a few stems appeared from the old crowns and rhizomes in 1941. Because of its deep-rooting habit, big bluestem was less severely effected by drought and new stems quickly appeared in

Table 7.20 The principal species recorded during the recovery of badly denuded rangeland in the mixed prairie region of North America following the Great Drought of 1933–40 (after Albertson and Weaver, 1944)

First weed stage	
Russian thistle (*Salsola pestifer*)	Common sunflower (*Helianthus annuus*)
Lamb's quarters (*Chenopodium album*)	Poverty weed (*Monolepis nuttalliana*)
Narrow-leaved goosefoot (*Chenopodium leptophyllum*)	Purslane (*Portulaca oleracea*)
Rough pigweed (*Amaranthus retroflexus*)	Spurges (*Euphorbia* spp.)
Tumbleweed (*Amaranthus graecizans*)	Buffalo bur (*Solanum rostratum*)
Prostrate pigweed (*Amaranthus blitoides*)	Horseweed (*Conyza canadensis*)

Second weed stage	
Little barley (*Hordeum pusillum*)	Peppergrass (*Lepidium densiflorum*)
Witchgrass (*Panicum capillare*)	Plantains (*Plantago* spp.)
Stinkgrass (*Eragrostis cilianensis*)	Stickseeds (*Lappula* spp.)
Six-weeks fescue (*Festuca octoflora hirtella*)	Cryptantha (*Cryptantha crassisepala*)
Brome (*Bromus* spp.)	

Early native grass stage	
Sand dropseed (*Sporobolus cryptandrus*)	False buffalo grass (*Munroa squarrosa*)
Western wheatgrass (*Agropyron smithii*)	Windmill grass (*Chloris verticillata*)
Tumblegrass (*Schedonnardus paniculatus*)	

Late grass stage	
Blue grama (*Bouteloua gracilis*)	Red three-awn (*Aristida longiseta*)
Buffalo grass (*Buchloe dactyloides*)	Side-oats grama (*Bouteloua curtipendula*)
Purple three-awn (*Aristida purpurea*)	Squirreltail (*Sitanion hystrix*)

(Reprinted with permission from F. W. Albertson and J. E. Weaver, Nature and degree of recovery of grassland from the Great Drought of 1933 to 1940, *Ecological Monographs*, 1944, **14**, 450.)

areas that were not occupied by wheatgrass and other invasive species. Other species produced large crops of seedlings, which grew well beneath the protective vegetative cover, and few areas of bare soil remained. The vigorous growth of the plants resulted in a rapid thickening of the stands and within 2 years the prairie had assumed a more normal appearance. Increasing amounts of litter accumulated on the soil and many weedy species were excluded by the reduced light levels in the canopy. However, the composition of the tall-grass prairie was quite different from the pre-drought cover. It was several years before the understory vegetation was re-established and native forbs became conspicuous (Weaver, 1954). Recovery was slow in the western parts of the tall-grass prairie where drought-tolerant grasses had become widely established, and mixed prairie still persisted over much of this region 12 years after the drought.

Recovery in mixed prairie was fairly rapid except in areas adversely affected by grazing or by dust deposits. At the end of the drought the vegetation cover in less severely damaged mixed prairie rangeland consisted of tufts of blue grama and buffalo grass less than 10 cm across and rather evenly spaced at intervals of about 30 cm: these were interspersed with peppergrass (*Lepidium densiflorum*) and other weedy species (Weaver and Albertson, 1956). Cover increased quickly through the spread of buffalo grass and sand dropseed (*Sporobolus cryptandrus*) and later by the establishment of blue grama. Although ground cover was quickly restored, as much as 11 years elapsed before its pre-drought composition was attained. On badly denuded rangeland the period of

recovery was prolonged for as much as 17 years. In these areas the initial cover consisted mainly of annual weedy species (Table 7.20) which helped to stabilize the soil. This was succeeded by a second weed stage comprised of species such as little barley (*Hordeum pusillum*), plantain (*Plantago purshii*) and peppergrass, and later by the native grasses. Sand dropseed was particularly abundant in the early native grass stage. It is a prolific seed producer: the tiny seeds are readily dispersed by the wind and seedling development is rapid. Short-grasses such as blue grama and buffalo grass became established later and the weedy species were soon excluded as the dense sod cover became more complete.

7.17 HUMAN IMPACT ON TEMPERATE GRASSLANDS

The effects of drought are often intensified by poor management of the grasslands. Although dry spells have occurred since the 1930s they have generally not caused such disastrous changes to the landscape, largely because the nature and use of the grasslands has changed considerably over the past few decades. Much of the area is now under crop production; the land that is still in pasture has generally been altered by various range management techniques designed to improve its quality and resilience. Most range improvement schemes are designed to increase the quality and quantity of forage available to livestock by increasing the abundance of desirable species through selective plant control. To some extent this can be achieved by the grazing animals themselves. For example, heavy browsing by cattle can effectively control trembling aspen suckers in parts of the Canadian prairies (Fitzgerald and Bailey, 1984). Mechanical methods are commonly used to control growth of woody plants, or more extensively to prepare areas for seeding. Herbicides are typically used to remove infestations of noxious plants or to kill the total plant cover prior to seeding. In other situations herbicides applied in strips have been used to increase forage yields and seed crops in blue grama pastures (McGinnies, 1984). In a similar way prescribed burning is commonly used to increase forage quality and grazing capacity as well as to control brush encroachment. Considerable changes in the composition of the grasslands have also occurred because of the widespread use of introduced species. For example, smooth brome (*Bromus inermis*) and crested wheatgrass (*Agropyron cristatum*), which are both native to Eurasia, are now very common throughout the Great Plains. Use of smooth brome in Canada increased the grazing capacity of some rangeland and extended the grazing season by providing palatable forage in the spring and fall (Looman, 1976). Only a small percentage of native species are present in these pastures. In areas which have been seeded to crested wheatgrass for 30–40 years, native grasses typically account for less than 10% of the vegetation. Although rangeland still occurs extensively across the western plains of North America, over much of this region the vegetation bears only a superficial resemblance to the native prairie that it has replaced.

CHAPTER 8

The coniferous forests

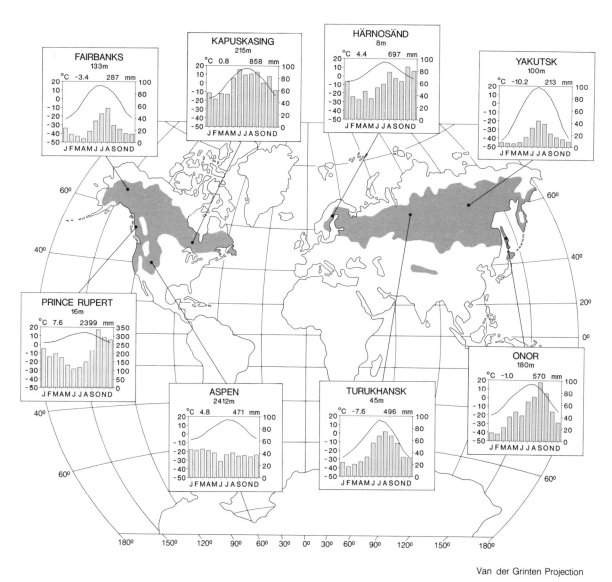

Van der Grinten Projection

Figure 8.1 Distribution of coniferous forest and representative climatic conditions. Mean monthly temperatures are indicated by the line and mean precipitation for each month is shown by the bars. Station elevation, mean annual temperature and mean annual precipitation appear at the top of each climograph.

8.1 INTRODUCTION

Coniferous forests are mainly found in a broad circumpolar belt across the northern hemisphere and on mountain ranges where low temperatures limit the growing season to a few months each year (Figure 8.1). The northern boreal forests cover an estimated 15.8 million km², compared with about 3.3 million km² in temperate mountain areas. The mild, moist climate of the Pacific north-west coast of North America is also favourable for conifers: these forests cover about 0.5 million km² in a narrow coastal strip from Alaska to northern California (Olson, 1975).

Nearly all of the conifers in these regions are evergreen and retain their needle-shaped leaves for several years. Only the larches (*Larix* spp.) shed their leaves annually, although other deciduous conifers, such as the bald cypress (*Taxodium distichum*), occur elsewhere. The geographic position of the southern continents precludes the development of a broad forest zone at higher latitudes and the geographic ranges of the southern conifers are now quite limited. The monkey puzzle tree (*Araucaria araucana*), for example, is native to southern Chile and western Argentina, and the Norfolk Island pine (*A. exelsa*) is endemic to that island. Although fossil *Araucariaceae* have been reported from Europe, the coniferous species of the southern hemisphere are taxonomically unrelated to their northern counterparts. They mostly grow in association with broad-leaved evergreen trees and account for only a small fraction of the total land area under coniferous forest vegetation.

The northern conifers begin photosynthesis as soon as temperatures are warm enough and none of the growing season is lost while new foliage develops. The reduced area of the leaves restricts transpiration and lessens the possibility of physiological drought during winter when water is unavailable from the frozen soil. This xeromorphic characteristic allows conifers to thrive in places where droughts are frequent. The cluster pine (*Pinus pinaster*) and Aleppo pine (*P. halepensis*), for example, are common in the Mediterranean region, and Monterey pine (*P. radiata*), which is native to California, has been planted extensively in Australia and southern Africa. Slash pine (*P. caribaea*) is also an important timber species in the tropics and subtropics, although originally it was found mainly in the savanna lands bordering the Caribbean. Conversely, bald cypress is characteristically associated with swampy sites in the south-eastern United States, and species such as amabalis fir (*Abies amabilis*), western red cedar (*Thuja plicata*) and hemlock (*Tsuga heterophylla*) grow to enormous size in the 'temperate rainforests' of the Pacific Northwest. Similarly, kauri pines (*Agathis* spp.) and yellow woods (*Podocarpus* spp.) are found in the wet tropical forests of Malesia, Fiji and the Philippines.

8.2 CONIFEROUS FOREST CLIMATES

Conifers are the dominant species in cool temperate regions, but the variable structure and composition of the forests reflects the wide range of climatic conditions in which they grow (Figure 8.1). The boreal forests are dominated by cold, dry polar air masses, but temperatures rise quickly during the long days of summer. Mean summer temperatures typically range from 12 to 15 °C but mean winter temperatures often fall to – 30 °C. This large annual temperature range is a distinctive characteristic of the boreal forests. The frost-free season lasts for about 50–100 days, depending on latitude and continentality, but there is always a risk of frost, even in summer, and the growing season usually ends abruptly with a hard freeze. Daylength during the summer increases from about 16 hours in the southern boreal forests to 24 hours north of the Arctic Circle. This tends to compensate for the short growing season, but even where there is continuous daylight daily temperatures usually fluctuate by 10–15 °C because the angle of the sun changes through the day. However, mean summer temperatures in the boreal forests remain above 10 °C for at least 30 days in the northernmost regions and for about 120 days in the south (Walter, 1985). Precipitation is generally less than 500 mm. It falls mainly in the summer, but more than 100 cm of snow can accumulate over the long winter (Figure 8.2). The passage of Atlantic air masses across Scandinavia and the western USSR moderates temperatures and increases precipitation in these regions, although this maritime influence is lost east of the Ural Mountains. Similarly, the movement of Pacific air across North America is restricted by the western Cordillera.

Climatic conditions are most severe in eastern Siberia where mountainous conditions occur at high latitudes. Average temperatures in these extreme sites typically range from – 50 °C in January to 15 °C in July. Temperatures as low as – 71 °C have been reported at Oymyakon,

Figure 8.2 Winter snowpack in a stand of black spruce (*Picea mariana*) in the Yukon Territories, Canada. (P. Marsh.)

a town situated at an elevation of 740 m in the Verkhoyanskii Range: the highest recorded temperature here is 33 °C, giving an absolute range of more than 100 °C (Lydolph, 1977). Temperatures are less severe in the montane forest regions which lie to the south of the main boreal zone. In the Tibetan Plateau winter temperatures near tree-line typically average − 8 °C and increase to 13 °C in summer; absolute temperatures are, respectively, about − 26 °C and 30 °C (Wang, 1961). Although mean temperatures here remain above 10 °C from June to September, the average length of the frost-free season is about 90 days and mean temperatures remain below 0 °C from November to February. Precipitation is more variable and ranges from less than 300 mm in enclosed valleys to more than 2000 mm on exposed slopes. Winters are comparatively clear and dry; most precipitation comes in the summer with the development of the Indian monsoon. In a similar way orographic precipitation occurs as moist Pacific air passes over the western mountain ranges in North America. The windward slopes of the Rockies receive 600–1200 mm but this decreases to less than 400 mm in the eastern foothills. Normal winter temperatures average − 5 to − 10 °C, depending on latitude, and rise to about 15–20 °C in the summer. In the European Alps precipitation varies between 800 and 2600 mm; normal seasonal temperatures average 0 to − 5 °C in winter and 10–18 °C in summer.

In addition to being generally milder and wetter, there are other important differences between the climates of the montane and boreal forest regions. The air is thin at higher elevations and the water vapour content is low. Consequently, incident solar radiation at 3000 m is more than 20% greater than at sea level and there is a proportional increase in ultraviolet wavelengths. The rarefied air causes the ground to warm intensely on sunny days, but the dry air warms more slowly. Significant temperature contrasts between exposed and shaded sites cause windy conditions during the daytime. Heat is lost rapidly at night; temperatures commonly fall by 15 °C and night frosts are common even though daytime temperatures are well above freezing. At night cool air descends into the valleys but updraughts develop during the day as the air in the valley warms. This often causes foggy conditions at higher elevations. Insects are carried into the alpine meadows by these breezes, while pollen brought up from the valleys and deposited on the glaciers in the Alps is sufficient to sustain the ice flea (*Isotoma saltans*) (Marcuzzi, 1979). Warm, dry, gusty winds also develop when air descends on the lee side of mountain ranges. These are the föhn winds which sometimes develop in southern Germany and Austria: in North America they are called chinooks. Often there is a dramatic rise in air temperature and this can result in the loss of foliage in conifers. Lodgepole pine is particularly susceptible, and the discoloured foliage of injured trees usually appears as distinct 'red belts' on the lower mountainsides. The needles are killed because transpired moisture cannot be replaced from the frozen soil, or because the rapid drop in temperature following a warm chinook affects the cold-hardiness of the foliage (Robins and Susut, 1974). Most trees will develop new foliage but some are so weakened that they are eventually killed by secondary agents.

Winters are mild in the coastal forests of North America and average temperatures in January rarely drop below − 5 °C even at higher elevations. In some of the islands and coastal regions of British Columbia the frost-free period exceeds 240 days but it decreases to 120–160 days further inland. Frosts are rare in more southerly regions and average winter temperatures are 5–10 °C. Summer temperatures along the coast range from 10–15 °C in the north to 15–20 °C in the south. Precipitation is heavy throughout the region: over 2000 mm is reported annually in some coastal stations in British Columbia, but further south precipitation decreases to 400–600 mm yr^{-1}. Precipitation is heaviest in autumn and early winter and much of it falls as snow, especially at higher elevations where annual snowfall often exceeds 500 cm (Bryson and Hare, 1974). In Alaska the first significant snowfalls are usually reported in early November and snow persists until late April, but further south the snow season is normally less than 2 months, although this varies greatly with elevation. Sea-fogs are common along the Pacific coast at all times of the year and tend to be densest in areas where the air is further cooled by orographic lifting. In northern California more than 200 mm of moisture is added to the soil as fog drip from the redwood (*Sequoia sempervirens*) canopy, and the distribution of these forests is closely linked to the availability of this supplementary moisture (Azevedo and Morgan, 1974).

8.3 SOILS

The soils of the coniferous forests are relatively young; they are mostly **Spodosols**, but organic soils and thin **regosols** are also common. Spodosols are best developed on sandy parent materials under cool, humid conditions which promote intensive leaching. The plant litter which accumulates on the forest floor is low in nutrients but has a high content of resins, waxes and lignin. Humification is slow and a thick layer of coarse humus characteristically develops. Organic acids are leached out of this surface layer, and water entering the mineral horizons typically has a pH range of 3.5–4.5. Most of the cations are washed through the soil but iron and aluminium are precipitated as sequioxides in the upper B horizon. Some organic matter is also redeposited as an amorphous coating around mineral grains in this part of the soil profile. Although this increases the cation-exchange capacity, base saturation is low and Spodosols are comparatively infertile. Organic soils occur extensively in poorly drained areas throughout the coniferous forests; they consist mainly of relatively undecomposed

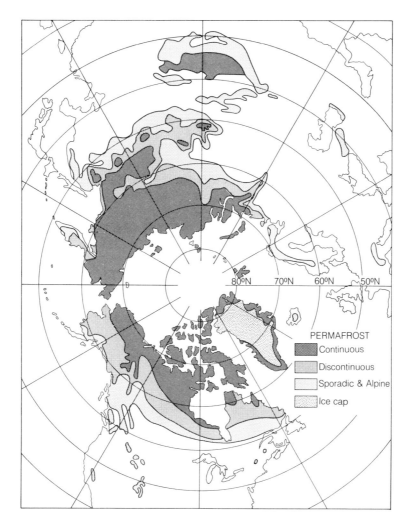

Figure 8.3 Distribution of permafrost in the northern hemisphere. (After Harris, 1986.) (Redrawn and modified from S. A. Harris, *The Permafrost Environment*; published by Croom Helm, 1986.)

sphagnum moss and other organic debris. The surface organic layer is at least 40 cm thick and in some soils as much as 150 cm of debris overlies the mineral subsoil. Most are saturated with water for much of the year and only a few species, such as tamarack and black spruce, can tolerate these conditions. Organic soils warm and cool relatively slowly and many are underlain by permafrost.

Permafrost occurs discontinuously over much of the boreal region (Figure 8.3) and is usually associated with areas that are covered with a deep insulating layer of moss. Thawing may be restricted to depths of less than 1 m in permafrost soils and the trees are necessarily shallow-rooted. Where the soil is not permanently frozen its temperature normally remains above 0 °C, except in exposed areas where little snow accumulates. In these situations frost penetrates to depths of 50 cm or more. Frost activity in the soil causes significant movement of the surface layers and results in hummocks about 50 cm high and 150 cm in diameter. This has a marked effect on

vegetation patterns: lichens grow mainly on the raised portions, and trees, shrubs and mosses are most abundant in the depressions (Zoltai, 1975). Continued freeze–thaw cycles can dislodge the trees and eventually cause them to fall (Figure 8.4).

Although permafrost is found at depths of 1–1.5 m in the mountain ranges of Siberia, it is less commonly encountered in other mountain regions. Heavy snowfalls in the mountains effectively insulate the soil but in exposed sites near tree-line the soil can remain frozen from November until April (Wardle, 1968). In the spring the upper soil layers become saturated with meltwater, while the subsoil remains frozen, and solifluction can occur if the gradient is sufficiently steep. This imperceptible downslope movement of surface materials is a major factor limiting tree growth in some mountain sites (Marcuzzi, 1979). Most mountain forest soils are podsolized but the complex topography and diverse parent materials associated with these regions have produced a variety of soil types. Alfisols are common throughout mountain areas, where they typically develop on glacial till and similar deposits. Most have a fairly high pH and base saturation despite significant leaching: they are principally distinguished from Spodosols by the movement and subsequent accumulation of clay in the B horizon. Inceptisols are also characteristic of many steep, relatively youthful surfaces which support coniferous forests. They have a poorly developed illuvial B horizon, and although physical weathering has broken down the parent material, chemical alteration is usually minimal. These soils are particularly characteristic of volcanic deposits in the Cascade Mountains, Alaska and Kamchatka.

A similar diversity of soil types occurs in the humid coastal forests of North America. Although Spodosols are widespread in this region, the iron content of the podsolic B horizon is usually much lower than in boreal soils. The upper soil horizons are typically darkened by humus and in wetter sites there may be evidence of gleying. In northern regions the soils of the lower valleys are typically derived from alluvial and glacial deposits. At mid elevations the soils are mainly of colluvial origin. The soils are thinner and darkly stained in cooler upper elevation sites and poorly developed Lithosols are typical over much of the steeper terrain. Most of the soils are extremely acid with pH ranges of 3.5–5 in the surface horizons, increasing to about 6 in the subsoil. Plant nutrients are concentrated in the thick surface litter layers, but base saturation is often less than 30% and drops below 15% in the lower mineral horizons. Organic soils are also common in poorly drained sites and several metres of peat can accumulate. Towards the southern limit of the coastal forests, where the climate is drier, the dominant soils are reddish brown Ultisols with loamy or clay loam textures grading to clay in the lower horizons.

8.4 REGIONAL DISTRIBUTION AND FLORISTICS

8.4.1 The northern boreal forest

(a) Structural characteristics

The boreal forests extend as a continuous vegetation belt of varying width across Eurasia and North America. Coniferous forests reach 72 ° 30 'N in the Taimyr Peninsula of Siberia. From this extreme northern limit the forests extend southwards for more than 2300 km but this narrows to less than 600 km in western Siberia (Tseplyaev, 1965). In North America the boreal forest reaches its most northerly position in north-western Canada where the transition to arctic tundra occurs at approximately 69 °N. In western North America the boreal forest merges with the subalpine forests of the Rocky Mountains over a broad ecotone with its southern limit at about 55 °N. Eastwards across the continent the boreal forest region is displaced to lower latitudes. In eastern

Figure 8.4 Black spruce (*Picea mariana*) on hummocky terrain underlain by permafrost. (S. C. Zoltai.)

Canada the northern tree-line occurs at about 58 °N and boreal forests merge, about a 1000 km further south, with mixed forests in southern Quebec, the Maritimes and the New England states. The number of species represented in the boreal forests is comparatively small, and even though many of them are widely distributed there are distinct regional differences in floristic composition. However, the number of species declines as the growing season becomes cooler and shorter at higher latitudes. The hardy species which persist in these depauperate forests are shorter and more widely spaced, and lichens dominate the ground cover. Subdivisions of the boreal forest are defined according to the structural changes which occur as growing conditions become increasingly severe at higher latitudes.

Three structural divisions are generally recognized within the boreal forest. The 'forest–tundra' ecotone is the most northerly division. Further south the forest–tundra is replaced by 'open lichen woodland'; this in turn is replaced by the 'close forest' as the climate becomes less rigorous. The forest–tundra ecotone lies north of the arctic tree-line and is represented by small patches of woodland and individual trees scattered across the tundra (Figure 8.5): it eventually terminates at the northern limit of tree growth. Often the tree species assume a shrubby form, but even in protected sites 100-year-old trees may be less than 2 m tall. In North America the forest–tundra ecotone is up to 300 km wide (Timoney et al., 1992) but is narrower and displaced southwards at higher elevations (Hare, 1950). In Eurasia this zone varies from 20 to 200 km in width. The northern boundary is often quite irregular and trees can extend far into the tundra along valleys and other sites which are sheltered from the cold arctic air (Tseplyaev, 1965). In open lichen woodland the trees are spaced as much as 25 m apart, often in a very regular pattern. Tree growth is slow in wet depressions but denser forest may extend as 'timbered tongues' along moist, well-drained valleys.

The close forests are composed of continuous stands of trees, with mosses, small herbs and shrubs common in the shaded understory. However, the ground flora is less dense and less diverse than in deciduous forests, because the forest floor is in continual shade and is covered by a layer of dry, slowly-decaying resinous needles (Polunin, 1960).

(b) The boreal forests of Eurasia

The Eurasian boreal forests, or 'taiga', extend from eastern Scandinavia across northern Asia to the Pacific Ocean. The approximate latitudinal range of the taiga is 58–68 °N in Europe but this broadens to 49–72 °N in eastern Siberia. Norway spruce (*Picea abies*) is the principal species in northern Europe and forms dense forests on moist, well-drained soils in which dwarf shrubs, such as bilberry (*Vaccinium myrtillus*) and cowberry (*V. vitis-idaea*), form a sparse understory; mosses dominate the ground cover but herbs, such as wood-sorrel (*Oxalis acetosella*) and May lily (*Maianthemum bifolium*), and grasses are also common. Scots pine (*Pinus sylvestris*) is often the sole dominant in the westernmost taiga and the lichen woodlands of the far north but elsewhere it is more restricted to drier sites, poorly drained areas or recent burns. To the south and west the taiga merges with the European mixed forest but some deciduous species, especially birch (*Betula pubescens*), rowan (*Sorbus aucuparia*) and alder (*Alnus incana*) occur throughout Scandinavia and eastern Russia.

In European Russia the northernmost taiga is interrupted by extensive marshlands. The cover is again predominantly Norway spruce and birch with an admixture of Siberian spruce (*Picea obovata*) and larch (*Larix sibirica*). Siberian fir (*Abies sibirica*) and stone pine (*Pinus sibirica*) increase in importance towards the south and east. These species are eventually replaced by stands of aspen (*Populus tremula*), birch (*Betula verrucosa*) and oak (*Quercus robur*) in the transitional woodlands which border the semi-arid steppes. The continuity of the taiga is interrupted by the Ural Mountains. Siberian spruce and larch are the dominant species in the northern parts of western Siberia, but stone pine, fir and Scots pine become more important in the southern districts. Further east monotypic stands of Dahurian larch (*Larix dahurica*) cover vast areas of the harsh Yakut taiga. Other hardy trees such as birch, and less commonly spruce and pine, may be present with floristic diversity increasing in the south where the topography is more varied and montane species begin to appear.

(c) The boreal forests of North America

The boreal forests of North America form a continuous belt across the continent from Alaska to Newfoundland but, like the taiga, they too are composed of relatively few species. White spruce (*Picea glauca*), black spruce (*P. mariana*)

Figure 8.5 Physiognomic character of the boreal forest: (a) isolated stand of black spruce (*Picea mariana*) growing with shrubby tundra species in the protection of a valley in northern Canada; (b) black spruce open lichen woodland in northern Canada; (c) close white spruce (*Picea glauca*) forest near La Ronge, Saskatchewan. ((a) P. Marsh; (b) R. A. Wright; (c) O. W. Archibold.)

and tamarack (*Larix laricina*) are widely distributed, but the ranges of most boreal species are more limited and the composition of the forests change regionally. Black spruce is the dominant species in the open woodlands of the forest–tundra ecotone. It is accompanied by white

Figure 8.6 Treeless area in black spruce (*Picea mariana*) forest on the Canadian Shield. (O. W. Archibold.)

Figure 8.7 Mixed forest region, mainly of paper birch (*Betula papyrifera*), trembling aspen (*Populus tremuloides*) and white spruce (*Picea glauca*) in the southern boreal forest region of Alberta. (R. A. Wright.)

spruce on better-drained sites and tamarack in wetter areas.

The composition of the close forests is more variable, particularly in the complex topography of Alaska and the Yukon. In upland areas the forests are composed principally of white spruce, either in pure stands or mixed with Alaska birch (*Betula neoalaskana*) or trembling aspen (*Populus tremuloides*) in warmer sites: black spruce is common in lowland areas. Tamarack is restricted to boggy sites, and alder (*Alnus crispa*) and willow (*Salix* spp.) form riparian thickets along rivers. Lodgepole pine (*Pinus contorta*) is also important on the lower valley slopes further south.

In central Canada the boreal forest is associated with the shallow soils of the Canadian Shield. Black spruce is the predominant tree over most of this region. It is accompanied by tamarack in poorly drained sites and is often replaced by jack pine (*Pinus banksiana*) on dry, sandy soils and other coarse outwash deposits. Treeless areas are common where underlying rock has been exposed by glaciation (Figure 8.6). The western limit of balsam fir (*Abies balsamea*) is marked by its occasional presence in forest stands in central Canada. Mixed forests of white spruce, trembling aspen, paper birch (*Betula papyrifera*) and balsam poplar (*Populus balsamifera*) occur south of the Canadian Shield where the soils are deeper and the climate is less severe (Figure 8.7).

The character of the boreal forest changes around the Great Lakes as species which are restricted to eastern Canada become more abundant. The appearance of red pine (*Pinus resinosa*), eastern white pine (*P. strobus*), eastern hemlock (*Tsuga canadensis*), eastern white cedar (*Thuja occidentalis*) as well as various hardwood species, such as sugar maple (*Acer saccharum*), yellow birch (*Betula alleghaniensis*) and black ash (*Fraxinus nigra*), add to the diversity of the forest cover. However, white spruce and balsam fir remain the dominant species over much of eastern Canada, although black spruce regains its importance in the rugged windswept lands of Labrador and Newfoundland where the forest resembles the open lichen woodlands of the north.

8.4.2 The montane coniferous forests

Growing conditions at high elevations are quite different from those experienced in the boreal forests and there is little floristic similarity between these regions, except where mountain ranges extend into higher latitudes and elements of both floras become intermixed.

(a) Western Europe

The lower slopes of the Alps are typically clothed in mixed forests of beech (*Fagus sylvatica*), fir (*Abies pectinata*) and Scots pine (*Pinus sylvestris*), with spruce (*Picea excelsa*) and larch (*Larix decidua*) increasing in abundance above 1000–2000 m. Larch is well adapted to dry, sunny aspects and will often form pure stands, but at higher elevations it is progressively replaced by Swiss stone pine (*Pinus cembra*), the common tree-line species in the Alps (Marcuzzi, 1979). The understory in these montane forests is similar to that of the taiga with wide-ranging species such as whortleberry (*V. myrtillus*) and cowberry (*V. vitis-idaea*) dominating the shrub layer and the ground is carpeted in moss. In the Pyrenees and highlands of south-western Europe, Scots pine and silver fir (*Abies excelsa*) are the dominant conifers but drought-tolerant species such as Corsican pine (*Pinus nigra*) are more common in the Mediterranean region where forest cover persists to elevations of 1800–2000 m. In all of these areas the conifers provide a sharp contrast with the vegetation at lower elevations.

Figure 8.8 Upper subalpine forests of birch (*Betula pendula*) and Scots pine (*Pinus sylvestris*) in the Caucasus Mountains. (L. C. Bliss.)

The upper limit of coniferous forest is 800–900 m in southern Sweden but drops to about 400 m in the north. Norway spruce and Scots pine grow to these elevations but birch (*Betula pubescens*) is usually present above this. Similar forests occur on the Urals, although Norway spruce is generally accompanied by Siberian fir and stone pine on the wetter western slopes and Scots pine is the predominant species on the eastern slopes. The tree-line is reached at about 600 m in the northern Urals and rises southwards to over 1000 m. Birch and aspen are more common in the southern Urals and mountain ranges such as the Caucasus (Figure 8.8), and the coniferous forests are gradually replaced by mixed forests in which deciduous species such as oak (*Quercus robur*) and linden (*Tilia cordata*) are important. Larch becomes increasingly abundant further east and grows with Scots pine, birch and aspen in the foothills of the mountainous Altai-Sayan district of southern Siberia. Siberian fir (*Abies sibirica*) is more common at higher elevations and stone pine (*Pinus sibirica*) forms mixed stands with larch at the tree-line. The upper limit of the mountain taiga is about 1800 m in this region (Tseplyaev, 1965).

(b) Continental Asia

Dahurian larch (*Larix dahurica*) is the sole dominant over much of the mountainous region of eastern Siberia. It forms a continuous cover to elevations of 900–1300 m around Lake Baikal and patchy stands persist to 1500 m. A zone of dwarf pine (*Pinus pumila*) occurs at higher elevations, giving way to tundra at about 1900 m. The larch forests are broken by stands of Siberian fir, Siberian pine, birch and aspen in valleys and other protected sites. Larch maintains its dominance throughout the mountains of north-east Asia but spruce, fir and pine become increasingly important towards the south. Broad-leaf species which are common to Manchuria add to the

diversity of these forests, and the region is also noted for species with very restricted geographic ranges. Kamchatka fir (*Abies gracilis*), for example, is unique to the Kamchatka Peninsula, and Sakhalin fir (*Abies sachalinensis*) is a distinctive species of the mountain taiga on the islands of Sakhalin and Hokkaido. Jeddo spruce (*Picea ajanensis*) and Khingan fir (*Abies nephrolepis*) appear around the Sea of Okhotsk, often replacing larch at elevations of 700–1200 m. Above this elevation the forest cover breaks down into thickets of Japanese stone pine and shrubby Erman birch (*Betula ermanii*).

Most of the highland regions of Siberia continue southwards into China and many species are common to both regions. In the mountains of northern Mongolia the principal species are Siberian fir (*Abies sibirica*) and Siberian spruce (*Picea obovata*). Larch (*Larix sibirica*) and Scots pine (*P. sylvestris*) occur in drier habitats and Swiss stone pine (*P. cembra*) is occasionally present in moist places. Birch (*B. pendula*) and aspen (*P. tremula*) are common at lower elevations. Tree-line increases from 1700 to 2600 m in more southerly regions but only *P. cembra* and *L. sibirica* are present in the uppermost forest zone (Wang, 1961): *L. sibirica* is replaced by *L. dahurica* in north-eastern China and Korea and forms the tree-line with dwarf pine (*P. pumila*) at about 1500 m over much of this region. Birch (*Betula platyphylla*) accounts for up to 30% of the tree cover in some of the larch stands and aspen (*Populus tremula*) is occasionally present. Larch casts only a light shade on the forest floor for part of the year, and a dense and varied understory of shrubs develops in these forests. Forests of spruce (*Picea obovata*, *P. jezoensis*) and fir (*Abies nephrolepis*, *A. holophylla*) occur at lower elevations. Numerous broad-leaved deciduous species are present beneath the taller conifers; typical associates include maples (*Acer mandshuricum*, *A. triflorum*), birches (*Betula dahurica*, *B. plataphylla*), limes (*Tilia amurensis*, *T. mandshurica*) and elms (*Ulmus macrocarpa*, *U. pumila*). Smaller hardwood species appear in the undergrowth and beneath this is a rich and varied shrub and herb flora. Larch reappears in the marshy lowland sites, but although the trees are morphologically similar to those at higher elevations, they are considered a different variety. Sedges and grasses are common on the deep peaty soils underlying these monotypic stands.

The simplest forest communities occur in the mountains bordering the steppe-deserts of north-western China. The dominant species is *Picea schrenkiana*. It forms pure stands on the moister north-facing slopes from 1900 m to the tree-line at 3000 m, and grows with aspen and elm down to 1500 m; it is restricted to sheltered ravines on south-facing slopes. The mountains and deeply incised plateaus of the Himalayas lie to the south of the arid steppes, and here the forest flora is exceptionally rich and varied. More than 50 species and varieties of conifers are represented: the principal species are spruces and firs but even these are often limited to specific areas. The greatest diversity occurs in the eastern districts,

Figure 8.9 Forests of *Abies spectabilis* growing at 3700 m in Nepal. (P. Wardle.)

where the dominant species are *Picea likiangensis, Abies squamata* and *A. georgei*. They form tall, dense forests at elevations of 3500–4500 m. *Larix potaninii* is occasionally present at higher elevations. In the north-eastern districts the montane forests extend from 3400 to 3800 m. *Picea likiangensis* is still important here but it is accompanied by *A. faxoniana* and species with more northerly distributions, such as *P. asperata* and *P. noveitchii*. In this district the spruces are restricted to lower elevations and above 3500 m the forest is almost entirely of fir (Figure 8.9). In the south-eastern part of this extensive highland massive the characteristic species is hemlock (*Tsuga yunnanensis*). It occurs abundantly at elevations of 2200–2700 m but is replaced by *Picea brachytyla* at higher elevations with dense forests of *Abies fabri* extending from 3100 m to the tree-line at 4300 m. *Abies delavayi* is the dominant species in south-western districts but further west in Tibet it is replaced by *A. webbia*. *Picea likiangensis* is the most common associate but many other conifers are occasionally present, including the rare dawn redwood (*Metasequoia glyptostroboides*). The spruces grow from 3000 m to the tree-line at 4000 m. Himalayan hemlock (*Tsuga dumosa*) is common at lower elevations; it in turn is replaced by deciduous hardwoods, such as *Betula albo-sinensis* and *Acer tetramerum*. The transition to warm temperate forests in Nepal occurs at about 2500 m with the establishment of evergreen oak (*Quercus semicarpifolia*) (Figure 8.10).

(c) North America

The Western Cordillera of North America forms a continuous mountain chain over the entire length of the continent. Consequently, the composition of these montane forests is less varied than that recorded in the Himalayas and other isolated mountain ranges in Eurasia. Fewer than 40 coniferous species are native to the Rocky Mountains. However, they are not uniformly distributed and four relatively distinct regions are recognized (Daubenmire, 1943). The most northerly region is dominated by boreal species such as white spruce (*P. glauca*) but the presence of subalpine fir (*Abies lasciocarpa*) and lodgepole pine (*Pinus contorta*) distinguishes it from the adjacent boreal forests. Tree-line occurs at about 1000–1200 m in this region but even at lower elevations the cover is generally quite open. Both these cordilleran species are widely distributed; the southern limit of their ranges is approximately 35 °N where they are restricted to elevations of 2300–3700 m (Fowells, 1965). White spruce and black spruce are the only boreal species which persist in the mountains but these are rarely found south of 53 °N. Engelmann spruce (*Picea engelmanii*) becomes increasingly abundant at this latitude but remains a major species of the montane forests only to 45 °N. It grows with subalpine fir above 1200 m to the tree-line at 2200 m (Figure 8.11). Forests of western red cedar (*Thuja plicata*), hemlock (*Tsuga heterophylla*) and Douglas fir (*Pseudotsuga*

Figure 8.10 (a) Stands of evergreen oak (*Quercus semicarpifolia*) growing at 3000 m; (b) birch (*B. utilis*) at 4000 m; and (c) rhododendron at 2900 m in Nepal add to the diverse forest cover of the Himalayas. (P. Wardle.)

Figure 8.11 Engelmann spruce (*Picea engelmannii*) and subalpine fir (*Abies lasiocarpa*) growing near tree-line in the Rocky Mountains of Alberta. (O. W. Archibold.)

Figure 8.12 Ponderosa pine (*Pinus ponderosa*) parkland in the drier eastern valleys of the Rocky Mountains. (M. Brown.)

menziesii) are common at lower elevations in the wetter western districts. The drier valleys to the east support stands of Douglas fir, ponderosa pine (*Pinus ponderosa*) and lodgepole pine, and hardwood species such as trembling aspen and balsam poplar (*Populus balsamifera*) become increasingly abundant.

The composition of the forest changes in the central and southern Rockies. Douglas fir and ponderosa pine are the dominant species in the lower montane forests which, in places such as Colorado, extend from 1600 to 2500 m (Figure 8.12). Lodgepole pine increases in im-

portance in the upper montane zone between 2500 and 2800 m, and in turn is replaced by Engelmann spruce and subalpine fir which form the tree-line at about 3500 m. In addition, there are several species which are more locally distributed, such as the giant sequoia (*Sequoia gigantea*) and the ancient bristlecone pine (*Pinus aristata*). The giant sequoia grows in isolated groves along the western slopes of the Sierra Nevada in central California (Figure 8.13). Most of the trees grow at elevations of 1400–2300 m where they are restricted to moist canyons or areas where underground water is available (Fowells,

Figure 8.13 Giant sequoia (*Sequoia gigantea*) growing in the Sierra Nevada Mountains of central California. (O. W. Archibold.)

Figure 8.14 Coastal rainforest of Sitka spruce (*Picea sitchensis*) in Alaska. (M. Brown.)

1965). They are the bulkiest trees in the world, growing to over 80 m in height and up to 31 m in circumference. Bristlecone pine grows near timberline. It grows extremely slowly and acquires an unusually gnarled and stunted form with many dead branches. Despite severe conditions the bristlecone pines attain great age, with some specimens estimated to be over 6000 years old. Pinyon–juniper woodland is the characteristic cover of the high plains, canyons and lower slopes of Utah, Arizona and New Mexico. Pinyon pine (*Pinus edulis*) is a small, slow-growing tree but its large, thin-coated seeds

are edible and are harvested commercially. Several species of juniper grow with pinyon pine. The most common is Rocky Mountain juniper (*Juniperus scopulorum*), a wide-ranging species which grows up to 15 m tall in this region. Broad-leaved species such as Gambel oak (*Quercus gambelii*) become more common in the mountains of northern Mexico. They alternate with grasslands or deserts at lower elevations but higher up are interspersed with various pines, several of which are distinctive to this region.

8.4.3 The wet coastal forests

(a) The forests of the American Pacific Northwest

The coastal forests of north-western North America extend as a narrow zone from Alaska to northern California. The dominant species are large, long-lived conifers which commonly grow 50–80 m in height with massive trunks 2–3 m in diameter. Approximately 20 species of conifer are native to this region but several distinct associations are recognized. Hemlock (*Tsuga heterophylla*) and sitka spruce (*Picea sitchensis*) are the principal species on the Pacific slopes of the coastal mountains in Alaska and northern British Columbia (Figure 8.14). Sitka spruce is tolerant of salt spray and forms pure stands near the ocean, but is often replaced by amabilis fir (*Abies amabilis*) in well-drained sites and by western red cedar (*Thuja plicata*) in wetter areas. Mixed stands of amabalis fir, mountain hemlock (*Tsuga mertensiana*) and subalpine fir occur above 500–800 m (Rowe, 1977).

Further south sitka spruce becomes confined to the immediate coastal zone; it is a common species in the 'rain forests' of the Olympic Peninsula in Washington state. Western hemlock and western red cedar are the dominant species in the coastal forests of southern British Columbia, Washington and Oregon (Figure 8.15) but the presence of Douglas fir distinguishes these forests from those to the north. Douglas fir occurs extensively in drier sites with western hemlock and western red cedar more conspicuous in moister locations. This type of forest occurs to elevations of 500–1000 m in the coastal mountains. Amabalis fir becomes more abundant at higher elevations but is replaced above 1300–1500 m by mountain hemlock and yellow cypress (*Chamaecyparis nootkatensis*). Above 1700–2000 m the coastal species are replaced by Engelmann spruce and subalpine fir (Franklin and Dyrness, 1969).

The southern limit of the coastal forests is marked by the stands of redwoods (*Sequoia sempervirens*) in California (Figure 8.16). They are the oldest and largest of the coastal species: some are reported to be over 2200 years old and the tallest specimen is recorded at 112 m (Fowells, 1965). Redwoods grow best on deep, moist soil along rivers and on gentle slopes below 300 m, but many stands are dependent on coastal fogs to supplement

Figure 8.15 Flaring buttressed trunk of massive western red cedar (*Thuja plicata*) in British Columbia. (M. Brown.)

precipitation. Consequently, they are restricted to a narrow coastal zone which is often less than 20 km wide. However, they are also sensitive to salt damage and normally do not grow within 2–3 km of the coast. In drier inland sites redwoods are generally replaced by hardwood species such as the evergreen tanoak (*Lithocarpus densiflorus*) and Pacific madrone (*Arbutus menziesii*). Over much of their range the redwoods form mixed stands with Douglas fir and western hemlock.

(b) The moist temperate forests of Asia

The only forests which are comparable to those of the American Pacific Northwest occur in Japan. Although the predominant forest type in the cool temperate regions is Japanese beech (*Fagus crenata*), coniferous species are important in some areas (Kira, 1977). Stands of Japanese red cedar (*Cryptomeria japonica*) are widely distributed throughout the southern islands of Kyushu and Shikoku, where it often grows intermingled with beech, but this type of forest is generally restricted to western Honshu. Over much of Honshu the natural cover is comprised of Nikko fir (*Abies homolepis*) growing in pure stands or mixed with beech. Forests of Japanese fir (*Abies firma*) and Japanese hemlock (*Tsuga sieboldii*) occur at lower elevations along the Pacific coast but they are replaced at higher elevations by endemic species such as Hinoki cypress (*Chamaecyparis obtusata*) and Japanese arbor-vitae (*Thuja standishii*). Japanese hiba (*Thujopsis dolobrata*) becomes more abundant in northern Honshu. Firs

Figure 8.16 Coast redwoods (*Sequoia sempervirens*) in northern California. (O. W. Archibold.)

(*Abies veitchii* and *A. mariesii*), hemlock (*Tsuga diversifolia*), spruce (*Picea jezoensis*) and larch (*Larix leptolepis*)

are represented in the subalpine forests of Honshu. Eastern boreal species dominate the forests on the northern island of Hokkaido. The principal species are Sakhalin fir (*Abies sachalinensis*), Yezo spruce (*Picea jezoensis*) and Sakhalin spruce (*P. glehnii*). These forests descend to sea level in the north-eastern part of the island where summer temperatures are reduced by frequent fogs. The characteristic undergrowth in all of these forests is dense stands of bamboo (*Sasa* spp.). This is similar to the forest understory which occurs in southern Chile, but although the climate here is also cool and moist the Magellanic forests are dominated by broad-leaved evergreen species of *Nothofagus* and conifers are rarely present (Veblen *et al.*, 1983).

8.5 QUATERNARY HISTORY OF THE NORTHERN CONIFER FORESTS

The conifers are an ancient plant group whose fossil history extends back 300 million years to the late Carboniferous period. Many genera have become extinct: others, like the redwoods, now have very restricted geographic ranges but were once widely distributed throughout the northern hemisphere. During Tertiary times vegetation zones were displaced some 20 ° further north than present and places in the Canadian Arctic supported species of spruce, pine, hemlock and fir and several deciduous hardwoods (Larsen, 1980). Climatic cooling began towards the end of the Tertiary and culminated in the Pleistocene epoch. North American boreal species were restricted to areas south of the ice sheets, although some shrubs may have survived in refugia in Alaska and the Yukon (Ritchie, 1982). The conifers spread rapidly northwards as the ice retreated and forests dominated by spruce were established in parts of southern Canada by 12 000 yr BP. Spruce is represented in the pollen record of sites near the present forest–tundra boundary in north-western Canada as early as 9500 yr BP and other boreal species appeared about 6000 yr BP (Figure 8.17).

Climatic conditions in Eurasia during the late Tertiary were warmer and moister than at present, and many thermophilous species were subsequently lost during the Pleistocene. At the height of the Pleistocene the Scandinavian Ice Sheet extended to the Urals where it coalesced with ice which originated in Siberia and reached eastwards to the Laptev Sea and south to 50 ° latitude. Smaller ice centres occurred in the mountains of north-eastern Asia but most of the region was free of ice and was linked to refugia in Alaska via the Bering Land Bridge. Tundra and cold-steppe vegetation covered much of this ice-free land (Young, 1982). Woodland and forest survived mainly in southern Japan and south-east China, with smaller areas adjacent to the Caspian Sea and in some of the mountain ranges of north-western China. Small groves of cold-resistant conifers persisted as a broad zone of cold-steppe vegetation extending from southern Europe to the Pacific coast. These species subsequently spread across the region as the ice retreated. Differentiation of the Siberian taiga therefore began at the beginning of the Pleistocene and is consequently much older than its North American counterpart (Frenzel, 1968).

8.6 ECOPHYSIOLOGY OF CONIFEROUS SPECIES

Rates of photosynthesis in boreal conifers are lower than in the deciduous hardwoods that sometimes accompany them, but their evergreen habit extends the length of the growing season and in some species photosynthesis is maintained during the winter (Krueger and Trappe, 1967). Conifers are C_3 plants with maximum rates of photosynthesis usually recorded between 10 and 20 °C, although assimilation continues over a broad range of temperatures. In black spruce (*P. mariana*), the dominant species over much of northern Canada, the optimum temperature for photosynthesis is 15 °C, but the rate is maintained at 90% of maximum between 9 and 23 °C and is still about 30% of maximum at 0 °C (Vowinckel *et al.*, 1975). The temperature optimum is generally higher in conifers such as ponderosa pine (*Pinus ponderosa*) which grow in warmer climates, but ecotypic differences have been reported in species such as Douglas fir (*Pseudotsuga menziesii*) which grow in a variety of habitats (Sorensen and Ferrell, 1973). The low temperature compensation point for most conifers ranges from -3 to -8 °C, which is the temperature at which their needles freeze (Larcher, 1969), although in European silver fir (*Abies alba*) positive net photosynthesis has been reported at temperatures as low as -18 °C (Jarvis and Sandford, 1986).

Assimilation rates change with needle age and are generally highest in current-year foliage (Freeland, 1952). This decline is attributed to the lower CO_2 conductance of the older needles as a result of reduced permeability of waxes around the stomata (Ludlow and Jarvis, 1971). Assimilation rates also vary on the same tree and are highest in the sun-adapted needles in the upper canopy. In sitka spruce, for example, maximum net photosynthesis rates are about 25% higher in sun needles (Leverenz and Jarvis, 1979). These needles are wider, thicker and have a higher chlorophyll content than those which develop in shade (Leverenz and Jarvis, 1980). The amount of light which penetrates the canopy of closed conifer forests is significantly less than in the open, although sunflecks and larger patches of sunlight filter through to the forest floor (Young and Smith, 1979). Most conifers have low light requirements and photosynthesis is possible under very low light levels. In black spruce, for example, light saturation occurs at about $1000 \mu E \, m^{-2} \, s^{-1}$ during the summer, and the light compensation point is only $35 \mu E \, m^{-2} \, s^{-1}$ (Vowinckel *et al.*, 1975). Shade-intolerant species, such as lodgepole pine and ponderosa pine, have higher light requirements but they outperform shade-tolerant species under optimum conditions. However, some degree of shading occurs within conifer crowns and photosynthetic rates are inevitably reduced below the theoretical maximum (Jarvis and Leverenz, 1983).

Photosynthesis in boreal forests is possible in the weak light of early spring but growth can be limited if the trees are unable to take up moisture from the frozen ground. In black spruce, net photosynthesis is reduced by 50% at soil water potentials of -1.5 MPa, and ceases at -2.5 MPa, although drought stress is rarely this severe (Black and Bliss, 1980). Photosynthesis is reduced because low soil moisture levels cause stomatal closure; this can

occur even in coastal forest species during dry spells. Although the large size of mature coastal conifers provides a reservoir of moisture in times of drought (Waring and Running, 1978), differences in drought resistance affect their geographic distribution (Lassoie *et al.*, 1985). Species such as ponderosa pine and lodgepole pine, which are relatively drought-resistant, are widely distributed in drier southern districts and interior valleys. Amabalis fir and mountain hemlock do not regulate water loss except at very high soil water potentials and so are restricted to very wet regions. However, stomatal activity is controlled by a variety of factors. In species such as lodgepole pine, ponderosa pine and Engelmann

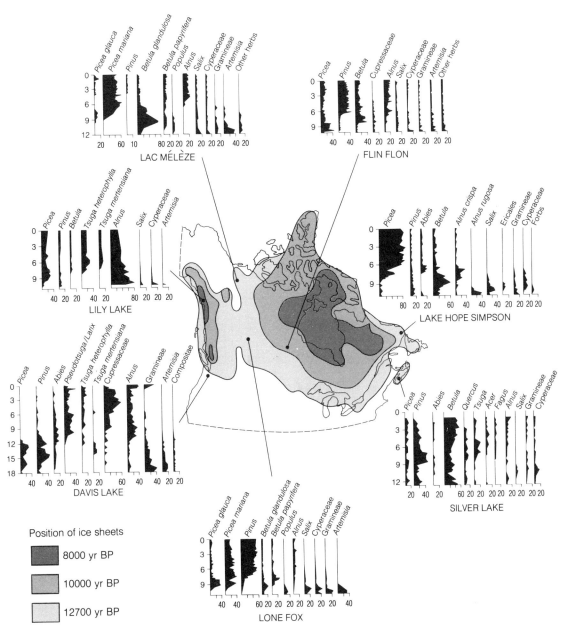

Figure 8.17 Approximate positions of ice sheets at intervals during post-Wisconsin deglaciation with associated pollen diagrams for selected sites in North America. For each pollen diagram the vertical axis is × 1000 years BP and the horizontal axis is percent contribution to the pollen count. (After Ritchie, 1987.) (Redrawn from J. C. Ritchie, *Postglacial Vegetation of Canada*; published by Cambridge University Press, 1987.)

spruce the stomata remain closed if temperatures drop below freezing during the night (Smith *et al.*, 1984). Similarly, improved stomatal control and increased drought resistance are reported in species such as Scots pine and Douglas fir following applications of nitrogen and potassium to the soil (Linder and Rook, 1984).

8.7 ARCTIC AND ALPINE TREE-LINES

Tree growth occurs far beyond the boundary of continuous boreal forest, and various ecological limits are recognized within this broad transitional zone (Hustich, 1979). The contact between the boreal forest and the forest–tundra is often termed the **forest-line** and is located where summer temperatures are rarely high enough for tree seeds to ripen. The northern limit of extensive forest–tundra is represented by the physiognomic **forest-limit** but isolated stands of trees extend beyond this to the tree-line. Arboreal species exhibit a shrubby growth-form north of the tree-line; they survive in sheltered sites within the Arctic tundra and disappear from the landscape at the limit of tree species. A similar transition occurs in mountain areas where a zone of tree-islands and wind-deformed krummholz typically separates the closed forests from the higher alpine communities (Figure 8.18). The arctic tree-line in North America is approximately equivalent to the 10 °C mean summer isotherm, and coincides with the median July position of the Arctic front which separates cool Arctic air masses from milder air originating in the Pacific (Bryson, 1966).

Figure 8.18 Islands of subalpine fir (*Abies lasiocarpa*) near tree-line in the Rocky Mountains. (O. W. Archibold.)

A similar relationship between seasonal air mass dominance and the taiga is reported for Eurasia (Krebs and Barry, 1970), although on both continents a discepancy of 10–14 ° latitude can arise in some years, particularly in mountainous terrain. The zonal divisions within the boreal forest reflect changes in species distribution and growth rates with latitude, and although this is related to airmass frequency (Larsen, 1971), it also correlates with decreased levels of radiation in the north (Hare and Ritchie, 1972).

In Canada global radiation increases from 90 kly yr^{-1} at the arctic tree-line to about 110 kly yr^{-1} at the boundary between open lichen woodland and the closed

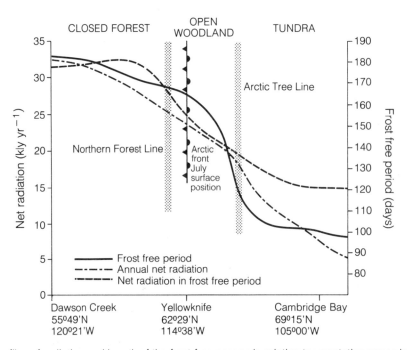

Figure 8.19 Latitudinal profiles of radiation and length of the frost-free season in relation to vegetation zones in northern Canada. (After Hare and Ritchie, 1972.) (From F. K. Hare and J. C. Ritchie, The boreal bioclimates, *The Geographical Review*, **62**, 333–65, reproduced and adapted by permission of the American Geographical Society.)

Figure 8.20 (a) Flagged white spruce (*Picea glauca*) near the arctic tree-line at Churchill Manitoba, and (b) snow protected lower branches of subalpine fir (*Abies lasiocarpa*) at treeline in the Rocky Mountains. (O. W. Archibold.)

forests. Snow remains over northern Canada until late spring, but it lies beneath the canopy of the close coniferous forests and much of the incoming radiation is absorbed so that temperatures rise quickly. In the snow-covered tundra much of the energy is reflected, and consequently the growing season is shorter (Figure 8.19). Net radiation during the thaw season is generally higher than that recorded over the entire year because of considerable energy losses during the winter. However, these climatic differences could result from, rather than cause, the boreal tree-line. Growth is slow in this rigorous climate, and except in the most sheltered sites the trees are flagged or stunted (Figure 8.20). Branches and needles exposed above the snowpack are subject to ice abrasion and desiccation over winter and the trees acquire a shrubby form. Reproduction is usually by layering but occasionally seedlings may become established. The minimum temperature requirement for seed germination in black spruce is 15 °C and this must normally occur by the end of June to avoid seedling mortality from drought (Black and Bliss, 1980). Under present conditions white spruce, black spruce and larch all produce viable seed in tree-line sites in eastern Canada: some are retained in closed cones or are incorporated into the small reserve of seed in the soil and eventually produce seedlings (Elliott-Fisk, 1983). However, seed production is rare at tree-line in the western Arctic; most of the younger trees have established vegetatively and these

isolated stands are possibly relics of more extensive forests which developed in warmer periods (Nichols, 1976). The position of the northern tree-line has changed considerably in post-glacial times. During the warm Hypsithermal period (about 8500–5000 years BP) the tree-line in Canada was located some 250–300 km further north than at present, but had retreated 100 km south of the modern forest limit by 2100 years BP (Nichols, 1975). Mean summer temperatures in northern Canada rose by 1.3 °C between 1880 and 1940 and have subsequently fallen by about 0.7 °C (Kelly *et al.*, 1982). This warming trend is reflected in increased tree regeneration but as yet there has been no advancement of the tree-line (Payette and Filion, 1985).

Vegetation zones are more compressed in mountain regions, but although the alpine tree-line is more abrupt than the arctic tree-line (Figure 8.21), there is generally a transitional '**kampfzone**' in which growth of mature trees becomes increasingly difficult. Seedlings which establish beyond the forest limit are protected by winter snow cover but they become stunted and deformed as exposed branches and buds are killed back. However, some tree-line species, such as dwarf mountain pine (*Pinus mugo*) and dwarf Siberian pine (*P. pumila*), maintain their twisted form even when transplanted to more favourable sites (Wardle, 1974). The elevation of the alpine timberline in North America decreases by 110 m with each degree of latitude (Daubenmire, 1954) and drops from a maximum

Figure 8.21 (a) The alpine tree-line occurs at about 2200 m in the Canadian Rockies; (b) the transition to snow tussock (*Chionochloa rigida*) grassland occurs abruptly at about 1200 m in the Haast Pass, New Zealand. ((a) O. W. Archibold; (b) A. F. Mark.)

elevation of about 4000 m at 20–30 °N to near sea level in extreme northern latitudes. The alpine tree-line, like its northern counterpart, coincides with the altitude of the 10 °C summer isotherm (Tranquillini, 1979). However, summer warmth increases with continentality and the tree-line is invariably higher in mountain ranges that are removed from the moderating influence of the oceans (Figure 8.22). Elevation of the tree-line also varies with local exposure and can drop by more than 150 m on cool, north-facing slopes (Franklin and Dyrness, 1969).

Tree growth declines rapidly at the tree-line and ceases where summer warmth is insufficient to allow new growth to harden before the onset of winter (Wardle, 1974). The start of the growing season is delayed by several weeks at high elevations, and if shoot development continues late into the year the cutinized epidermal layers of the twigs and needles remain thin. This reduces the diffusive resistance of exposed shoots and results in excessive water loss over winter (Baig and Tranquillini, 1976). Cold tolerance is acquired towards the end of the growing season by changes in cell structure and water content. For example, immature needles of *Pinus cembra* will only tolerate a frost of − 2 °C in summer but they withstand winter temperatures of − 43 °C when properly hardened (Tranquillini, 1979). Frost resistance is lost in the spring, and young needles and shoots are often damaged by late frosts, particularly when freezing and thawing occurs rapidly. Conifer needles are seldom damaged

by high temperatures but extreme heating of the soil in the thin mountain air can be lethal for young stems and shallow roots. Surface soil temperatures as high as 80 °C have been recorded in some mountain sites, and can remain above 50 °C for several hours. Stem injury in young conifers begins at temperatures of about 45 °C as the conductive tissues begin to shrink and become constricted. Seedlings are especially sensitive to '**stem girdling**' in this way and are commonly killed after being scorched by the hot soil. Roots can also be damaged

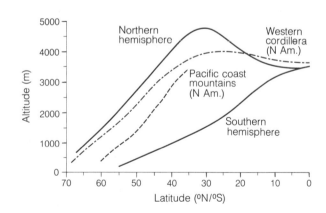

Figure 8.22 The elevation of alpine tree-lines at different latitudes. (After Daubenmire, 1954.) (Redrawn from R. Daubenmire, Alpine timberlines in the Americas and their interpretation, *Butler University Botanical Series*, **11**, 1954.)

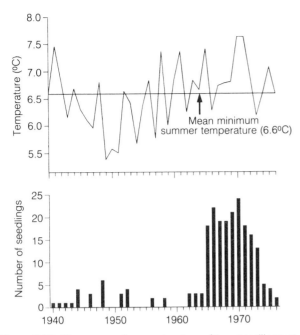

Figure 8.23 The composite age structure of tree seedlings at timber-line sites in the Canadian Rockies in relation to minimum summer temperatures. (After Kearney, 1982.) (Reproduced with permission from M. S. Kearney, Recent seedling establishment at timberline in Jasper National Park, Alta, *Canadian Journal of Botany*, 1982, **60**, 2286.)

by high temperatures but, except in young seedlings, they are generally deep enough to escape injury. Root damage is more common in the spring as a result of frost heaving, when temperatures in the moist soil fluctuate around 0 °C.

The upper limit of tree growth is determined by the elevation at which summer assimilation is balanced by winter losses. This is partly due to the loss of immature tissue but growth at high elevations is also restricted by conditions which reduce net photosynthesis. The dominant tree-line species become light-saturated at comparatively low radiation intensities, but light levels are often suboptimal even in discontinuous stands at tree-line because of cloud or topographic controls (Tranquillini, 1979). During brighter periods air temperatures are usually too low for maximum assimilation, and can be further reduced when light winds cool the foliage or cause stomatal closure. However, leaf respiration is maintained at temperatures as low as −8 °C, and considerable consumption of carbohydrate reserves can occur even in winter. Winter respiration in *Pinus cembra* seedlings at tree-line accounts for 7% of their annual net photosynthetic gain, and assimilation over a period of 20 days in summer is needed to recover this loss. The reduced potential for growth at high elevations is reflected in less shoot extension and radial growth. There is also a change in the shape and arrangement of the foliage; the needles are shorter and narrower, and are more densely crowded about the twig axis. Root growth is limited by

low soil temperatures, and root:shoot ratios decrease rapidly with elevation. All of these factors have an adverse effect on tree growth. Photosynthetic capacity of the trees is reduced by the characteristics of the foliage; small radial growth increments limit the transport and storage of water and assimilates in the stem; and smaller roots restrict water and nutrient uptake (Tranquillini, 1979). Production of seed produces even further demands on food reserves, and viable seed crops become progressively smaller as trees reach their physiological limits. Few seeds are carried above the tree-line by wind, but dispersal by animals can be significant. However, germination is not reliable and seedlings are rarely successful: this ultimately determines the upper limit of the forests. Seedling establishment by subalpine fir and Engelmann spruce at tree-line sites in the Canadian Rockies increased significantly between 1965 and 1973 when minimum summer temperatures were higher than normal (Figure 8.23). However, seed production in these species is intermittent near the tree-line (Franklin *et al.*, 1971), and regeneration may be limited if good seed crops do not coincide with a warm summer (Kearney, 1982).

8.8 PRODUCTION AND BIOMASS IN CONIFEROUS FORESTS

Smaller annual growth increments at higher elevations result in a considerable reduction in timber volume at tree-line. Tranquillini (1979) reported that annual volume increments in 75-year-old stands of *Picea abies* in Europe decreased from $11–12 \text{ m}^3 \text{ha}^{-1}$ to $1.3 \text{ m}^3 \text{ha}^{-1}$ between 1300 m and the tree-line at 1650 m. A similar decrease occurs at the boreal tree-line where annual growth in 150-year-old black spruce averages $3.9 \text{ m}^3 \text{ha}^{-1}$ compared with $62.1 \text{ m}^3 \text{ha}^{-1}$ in more southerly stands (Black and Bliss, 1978). Plant biomass in open lichen woodland in northern Canada ranges from 9.6 to 29.3 t ha^{-1} and increases to $82.3–163.4 \text{ t ha}^{-1}$ in the closed southern boreal forests (Larsen, 1980). Similar values are reported for spruce and pine forests in USSR and Finland (Table 8.1). Production in the boreal forests is extremely variable because of diverse soil and drainage conditions. Growing conditions can be expressed by the '**site index**' which measures the average height of the dominant trees at a specific age. Thus, 100-year-old black spruce growing in different sites in north-eastern British Columbia can range in height from 4.5 to 15 m, and white spruce from 6 to 24 m (Figure 8.24).

Pronounced environmental gradients characteristic of mountain sites also result in considerable local variation in plant performance, but regional trends are often obscured because plant communities change rapidly with elevation (Figure 8.25). In the mountains of Arizona, for example, a biomass of 783 t ha^{-1} is reported for 150-year-old stands of Douglas fir growing on moist north-facing slopes, compared with only 437 t ha^{-1} for similar stands

Table 8.1 Annual production and biomass in selected boreal and montane coniferous stands (after Cannell, 1982)

| | Dominant species | Elevation (m) | Age (yr) | Annual production (t ha⁻¹ yr⁻¹) | Stand biomass | | | |
					Trunks	Branches	Foliage	Total
					(t ha⁻¹)			
Boreal forests								
Alaska	*Picea mariana*	–	130	1.6	86.1	13.0	8.9	108.0
Canada	*Abies balsamea*	500	43	10.5	93.0	16.8	18.3	128.1
	Picea mariana	–	65	–	88.5	10.3	8.3	107.7
	Pinus banksiana	420	44	4.1	73.2	8.8	4.4	86.4
Finland	*Picea abies*	270	260	1.6	67.3	17.1	6.6	91.0
	Pinus sylvestris	140	45	5.1	60.9	7.4	4.4	72.7
USSR (former)	*Picea abies*	140	138	3.2	176.0	17.1	7.4	202.5
Montane forests								
Canada	*Pinus contorta*	1400	100	–	213.3	13.9	12.5	239.7
Japan	*Abies veitchii*	2440	118	8.4	130.3	15.7	14.3	160.3
	Tsuga diversifolia	1790	290	5.8	139.9	51.7	9.9	201.5
Nepal	*Abies spectabalis*	3530	–	–	231.0	28.0	10.7	269.7
	Tsuga dumosa	2760	–	–	429.0	70.0	12.0	511.0
USA	*Abies lasiocarpa*	2720	106	8.6	290.0	50.0	16.2	356.2
	Picea engelmannii	3300	250	–	150.0	28.0	18.0	196.0
	Pinus ponderosa	2500	49	3.9	121.0	20.2	10.6	151.9

(Source: M. G. R. Cannell, *World Forest Biomass and Primary Production Data*; published by Academic Press Inc., 1982.)

in drier sites (Whittaker and Niering, 1975). In nearby stands of 130-year-old alpine fir, total biomass was estimated at 356 t ha⁻¹ compared with 161 t ha⁻¹ for mature ponderosa pine growing on southern exposures. Annual production varied considerably between species, ranging from 5.7 t ha⁻¹ yr⁻¹ for ponderosa pine to 11.1 t ha⁻¹ yr⁻¹ for white fir. On a global scale average biomass in montane and subalpine forests is estimated at 160–320 t ha⁻¹, with net annual production ranging from 5 to 12 t ha⁻¹ yr⁻¹ (Olson, 1975).

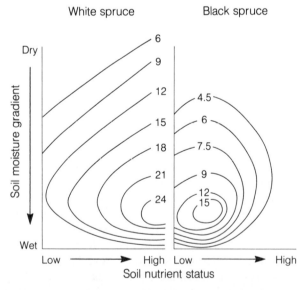

Figure 8.24 Growth of white spruce (*Picea glauca*) and black spruce (*P. mariana*) in northern British Columbia in relation to soil moisture and soil nutrient gradients. Values represent tree height in metres. (After Krajina, 1969.) (Adapted with permission from V. J. Krajina, Ecology of forest trees in British Columbia, in *Ecology of Western North America*, **2** (eds V. J. Krajina and R. C. Brooke), published by the University of British Columbia, Vancouver, 1969.)

Table 8.2 Average maximum age and size of selected forest trees in the Pacific North-west region of North America (after Franklin and Dyrness, 1969)

Species	Age (yr)	Diameter (cm)	Height (m)
Abies amabalis	400+	90–110	45–55
Abies grandis	300+	75–125	40–60
Abies procera	400+	100–150	45–70
Chamaecyperis lawsoniana	500+	120–180	60
Chamaecyperis nootkatensis	1000+	100–150	30–40
Larix occidentalis	700+	140	50
Picea sitchensis	800+	180–230	70–75
Pseudotsuga menziesii	750+	150–220	70–80
Sequoia sempervirens	1000+	150–380	75–100
Thuja plicata	1000+	150–300	60+
Tsuga heterophylla	400+	90–120	50–65

(Source: J. F. Franklin and C. T. Dyrness, *Vegetation of Oregon and Washington*, USDA Forest Service Research Paper PNW-80; published by Pacific Northwest Forest and Range Experiment Station, 1969.)

The forests of the Pacific Northwest are the most productive of any in the world. Typical values for above-ground biomass range from 500 to 1500 t ha⁻¹ but can exceed 3900 t ha⁻¹ in redwood forests (Franklin, 1988). This high biomass results mainly from the great longevity of the dominant conifers (Table 8.2), although annual productivity is also greater than in the boreal and montane forests. Net production in these west-coast forests averages 17.4 t ha⁻¹ yr⁻¹ compared with 9.8 t ha⁻¹ yr⁻¹ in the Rocky Mountains (Grier *et al.*, 1981) but, as in other forest regions, growth is greatly affected by site conditions. Under favourable conditions species such as western hemlock and Douglas fir can exceed 50 m in height within 100 years but may be less than half this on poorer sites (Figure 8.26). Most of the species native to the Pacific Northwest maintain fairly high

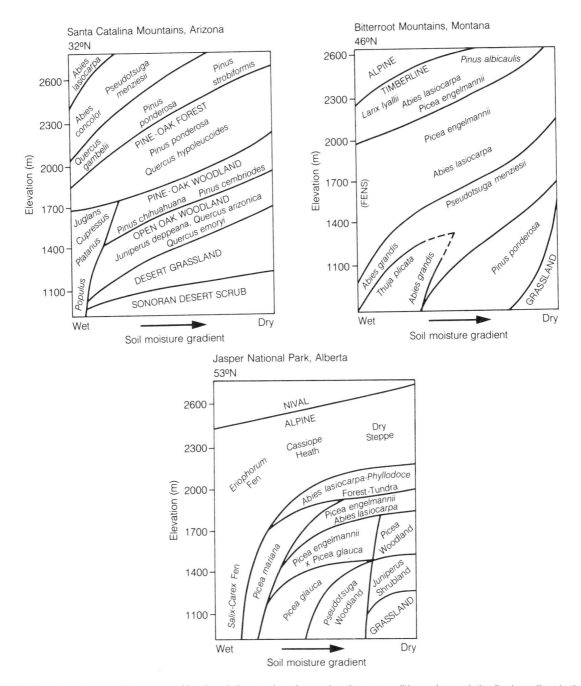

Figure 8.25 Variations in vegetation composition in relation to elevation and moisture conditions along a latitudinal gradient in the Rocky Mountains. (After Peet, 1988.) (Reproduced with permission from R. K. Peet, Forests of the Rocky Mountains, in *North American Terrestrial Vegetation*, eds M. G. Barbour and W. D. Billings; published by Cambridge University Press, 1988.)

growth rates for long periods. In western red cedar, for example, trunk diameter increases by 3–5 mm yr^{-1} and height by 15–30 cm yr^{-1} for at least 200 years (Fowells, 1965): substantial growth therefore continues for much longer than in other forests. Although most biomass is incorporated into woody supporting structures, the foliage component remains important. Projected leaf areas in different forests in the Pacific Northwest range from 9.7 to 13.2 m^2 m^{-2} (Franklin, 1988) compared with

11.4 m^2 m^{-2} in tropical rain forest in Thailand (Kira, 1975) and 6.8 m^2 m^{-2} in temperate deciduous forests in Europe (Duvigneaud and Denaeyer-De Smet, 1970). Gross production is proportional to leaf area and the length of the growing season (Kira, 1975), and although leaf areas of 16.7 m^2 m^{-2} are reported in high elevation Douglas fir forests (Whittaker and Niering, 1975), the climate is not as favourable as on the coast and productivity is consequently lower.

8.9 NUTRIENT CYCLING IN CONIFEROUS FOREST ECOSYSTEMS

The mineral soils beneath mature coniferous forests are comparatively infertile and growth is often limited by the rate at which mineral elements, such as nitrogen, are recycled in the ecosystem through litterfall and decomposition (Cole and Rapp, 1981). Because of their evergreen habit, the nutrient requirement of conifers is usually much lower than for deciduous species. Uptake of nutrients by conifers in Alaska is considerably less than the amount required by stands of birch, trembling aspen and poplar (Table 8.3). Nutrient requirements are lowest for black spruce, although overall productivity for this species averages only 1130 kg ha^{-2} compared with 3660–5650 kg ha^{-2} for white spruce and the hardwoods. Nutrient concentrations are generally highest in the stems and branches of the deciduous species but the nutrient content of conifer needles is comparatively higher than in the broad-leaf foliage (Table 8.4). Although some of the nutrients are translocated from senescing tissues before they are shed, considerable amounts are returned in litterfall and productivity is maintained only

where decomposition is rapid. Coniferous needles decompose more slowly than deciduous litter, because the activity of the microflora is restricted by the high content of waxes, resins and lignin (Millar, 1974). Turnover times for elements in the forest floor ranged from 17 to 84 years in these Alaskan forests (Van Cleve et al., 1983a) but decomposition can take several hundred years when permafrost is present (Cole and Rapp, 1981). Turnover is generally more rapid in the southern boreal forests, although even here nitrogen can be bound up in organic debris for long periods (Gordon, 1983).

As well as being tied up in soil organic matter, additional nutrient stresses arise in boreal forests where a deep carpet of feather mosses, such as *Pleurozium schreberi* and *Hylocomium splendens*, forms the understory. The moss layer insulates the ground and decomposition is reduced because of low soil temperatures (Van Cleve et al., 1983a). Moreover, mosses have no vascular tissues and water and nutrients are absorbed and moved externally in specialized capillary channels formed of large, empty hyaline cells. This 'sponge-like' quality can limit movement of nutrients to the tree roots below (Oechel and Van Cleve, 1986). The moss layer is especially thick

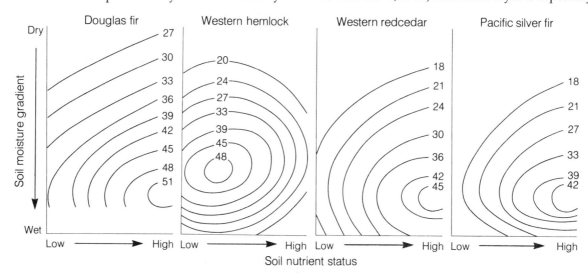

Figure 8.26 Growth of western hemlock (*Tsuga heterophylla*), western red cedar (*Thuja plicata*), douglas fir (*Pseudotsuga menziesii*) and amabalis fir (*Abies amabalis*) in coastal British Columbia in relation to soil moisture and soil nutrient gradients. Values represent tree height in metres. (After Krajina, 1969.) (Adapted with permission from V. J. Krajina, Ecology of forest trees in British Columbia, in *Ecology of Western North America*, **2** (eds V. J. Krajina and R. C. Brooke), published by the University of British Columbia, Vancouver, 1969.)

Table 8.3 Average annual above-ground production and nutrient uptake by forest trees in Alaska (after van Cleve et al., 1983a)

Species	Annual production (kg ha^{-1})	N	P	K	Ca	Mg
		\multicolumn{5}{c}{(g m^{-2})}				
Black spruce	1130	0.51	0.07	0.32	0.26	0.07
White spruce	3660	1.56	0.20	1.41	1.37	0.24
Trembling aspen	5650	6.14	0.74	3.17	3.95	0.77
Paper birch	4700	5.85	0.72	2.24	1.87	0.91
Balsam poplar	5520	5.55	0.41	3.34	5.72	1.17

(Reproduced with permission from van Cleve et al., Productivity and nutrient cycling in taiga forest ecosystems, *Canadian Journal of Forest Research*, 1983, **13**, 751.)

Table 8.4 Distribution of nutrients in above-ground standing crop of forest trees in Alaska (after van Cleve *et al.*, 1983a)

	Biomass	N	P	K	Ca	Mg
			(g m^{-2})			
Black spruce						
leaves	559	4.0	0.3	1.6	8.3	0.7
branches	788	3.2	0.4	1.0	4.5	0.6
trunks	3643	5.1	0.7	3.1	15.7	1.4
White spruce						
leaves	1248	8.5	1.1	5.0	17.7	1.6
branches	1636	8.9	1.3	11.0	20.3	1.7
trunks	13917	10.0	1.1	8.7	33.3	2.2
Paper birch						
leaves	234	5.2	0.6	2.0	1.6	0.8
branches	1630	7.3	0.9	2.6	5.4	1.1
trunk	9357	15.7	1.5	7.3	16.5	3.8
Trembling aspen						
leaves	286	5.9	0.7	2.8	3.4	0.7
branches	1775	5.7	1.0	4.8	20.2	1.4
trunks	10248	15.0	1.4	11.4	44.5	3.3
Balsam poplar						
leaves	269	4.6	0.3	2.8	4.6	1.0
branches	1519	6.1	0.9	6.0	26.4	1.7
trunks	10319	15.6	2.1	24.3	135.8	10.0

(Reproduced with permission from van Cleve *et al.*, Productivity and nutrient cycling in taiga forest ecosystems, *Canadian Journal of Forest Research*, 1983, **13**, 752.)

beneath black spruce and the nutrients it contains are mostly released by fire. Tree growth is therefore enhanced in post-fire sites, but production declines once the canopy closes and more of the nutrient capital is immobilized by mosses (Van Cleve *et al.*, 1983b). Feather mosses are susceptible to desiccation when exposed to sunlight and they are replaced by lichens in the northernmost boreal woodlands (Figure 8.27).

Improved growth of black spruce in boreal woodlands has been related to more favourable soil moisture conditions beneath the lichen mat (Cowles, 1982). However, the high reflectivity and insulating quality of the lichen also lowers soil temperatures (Kershaw, 1985) and a considerable quantity of nutrients remains bound up in the soil organic matter. Nitrogen levels are particularly high in the organic soil layers beneath subarctic lichen woodlands, although other nutrients are mainly stored in the tree biomass (Table 8.5). Few nutrients enter these northern ecosystems in precipitation or through weathering of mineral materials. The nutrients stored in the organic litter are mostly returned to the soil within 2 years but decomposition occurs mainly during the winter period, when the plant cover is inactive, and some elements are subsequently lost through leaching (Moore, 1984). However, nitrogen remains bound up in the litter and concentrations even increase with time, especially in black spruce needles and lichen tissue which have initial C:N ratios greater than 100:1. Nitrogen is therefore considered as the element which is

Figure 8.27 (a) The mossy understory of closed white spruce (*Picea glauca*) forest contrasts with (b) white lichens (*Cladonia* spp.) associated with more northerly stands of open black spruce woodland. ((a) O. W. Archibold; (b) S. C. Zoltai.)

Table 8.5 Biomass and nutrient storage in black spruce-lichen woodland in northern Canada (after Auclair and Rencz, 1982)

Ecosystem component	Biomass (kg ha^{-1})	N	P	K	Ca	Mg
		(kg ha^{-1})				
Tree	28230	98.8	22.6	43.2	64.9	8.8
Shrub	8330	80.4	4.3	10.0	15.3	5.2
Ground cover	9695	61.6	3.9	10.2	5.9	6.5
Arboreal lichen	36	0.2	0.1	0.1	0.0	0.0
Organic soil	45897	394.6	1.1	16.3	1.1	2.5
Mineral soil	820000	777.5	8.6	19.1	14.6	7.2

(Reproduced with permission from A. N. D. Auclair and A. N. Rencz, Concentration, mass, and distribution of nutrients in a subarctic *Picea mariana–Cladonia alpestris* ecosystem, *Canadian Journal of Forest Research*, 1982, **12**, 963.)

most limiting to productivity. Nitrogen-fixing lichens, such as *Stereocaulon paschale*, do provide a source of nitrogen in many northern ecosystems, although activity is limited by dry periods and low temperatures (Kershaw, 1985). Most of the nitrogen remains bound up in the lichen tissues and is ultimately incorporated into the soil organic layers, where it is released slowly by decomposition or more rapidly by fire (Auclair and Rencz, 1982).

Less organic matter accumulates on the forest floor in montane coniferous forests, and the shrubby understory retains fewer nutrients than the mosses and lichens that dominate the ground cover in boreal regions. The bulk of the above-ground biomass and nutrients are contained in the live wood, although the greatest reserves of nitrogen and a considerable proportion of the other elements are held in the needles. In old growth stands of Pacific silver fir, for example, above-ground biomass totals 468.5 t ha^{-1} compared with only 53.5 t ha^{-1} for the litter layer and 244.0 t ha^{-1} of organic matter retained in the soil (Table 8.6). Annual litter production is about

3000 kg ha^{-1} yr^{-1}, and is mainly comprised of needles that are shed between October and March. Large amounts of nitrogen, phosphorus and calcium are returned in litterfall but most of the potassium is transferred as throughfall: nutrient input through precipitation is comparatively small (Table 8.7). The pattern is similar in all coniferous forests, although nutrient inputs from precipitation and throughfall are relatively higher in northern forests because of reduced litterfall. Heavy growths of epiphytic lichens and mosses are common on trees in montane forests with biomasses ranging from 1.9 to 3.3 t ha^{-1} reported in subalpine forests in western North America (Rhoades, 1981). Lichens account for about 1% of the total above-ground biomass in old-growth stands of Pacific silver fir, which is similar to the understory vegetation (Turner and Singer, 1976): lichens account for about 6% of the annual litterfall in these stands. The lichens are mostly dislodged by wind and snow while they are brittle in the winter (Edwards *et al.*, 1960) and some nutrients, especially nitrogen, are recycled in this way (Pike, 1978).

The large trees of the Pacific Northwest accumulate considerable reserves of nutrients in their live biomass, and like other conifers a comparatively high concentration is found in their foliage (Table 8.8). This is the principal pathway whereby nutrients are returned to the forest floor (Table 8.9) and stand productivity is closely related to the rate at which the needles are decomposed. Douglas fir loses about 20% of its needles each year, and mineralization usually occurs within 4–5 years (Johnson *et al.*, 1982). The amount of woody litter increases as the stands mature, and organic matter steadily accumulates on the forest floor because this coarse debris decomposes more slowly. Fallen logs can take more than 250 years to disintegrate, even in these comparatively mild climates.

Table 8.6 Biomass and nutrient distribution in above-ground plant tissue and on the forest floor in old-growth sub-alpine stands of Pacific silver fir (after Turner and Singer, 1976)

Ecosystem component	Biomass (t ha^{-1})	N	P	K	Ca	Mg
		(kg ha^{-1})				
Trees						
Foliage	15.7	173.0	17.0	190.0	259.0	41.0
Branches	17.7	18.0	9.0	120.0	119.0	41.0
Bark	38.7	13.0	8.0	225.0	574.0	20.0
Live wood	265.0	116.0	16.0	284.0	40.0	28.0
Dead wood	127.7	25.0	13.0	137.0	21.0	27.0
Arboreal lichen	1.9	13.0	2.0	8.0	18.0	2.0
Understory	1.8	14.7	1.7	15.8	15.1	0.7
Forest floor						
Wood	10.0	79.0	9.0	10.0	100.0	5.0
Litter	13.0	174.0	12.0	70.0	130.0	14.0
Humus	30.5	397.0	23.0	95.0	317.0	27.0
Soil (0–60 cm)	244.0	–	–	244.0	145.0	88.0

(Source: J. Turner and M. J. Singer, Nutrient distribution and cycling in a sub-alpine coniferous forest ecosystem, *Journal of Applied Ecology*, 1976, **13**, 298.)

Table 8.7 Annual nutrient transfer to the forest floor beneath mature conifer stands

	N	P	K	Ca	Mg	Source
			(kg ha^{-1} yr^{-1})			
Pacific silver fir						
Precipitation	1.3	0.4	0.8	0.6	0.1	Turner and Singer, 1976
Throughfall	1.3	0.1	11.5	5.4	2.1	
Litterfall	16.3	2.0	7.3	39.7	2.3	
Douglas fir						
Precipitation	2.0	0.3	0.9	3.6	1.2	Sollins et al., 1980
Throughfall	3.4	1.2	13.8	7.2	–	
Litterfall	21.8	4.9	8.5	41.9	–	
Hemlock						
Precipitation	1.3	0.2	0.1	2.1	1.3	Abee and Lavender, 1972
Throughfall	3.7	3.0	30.4	2.1	1.5	
Litterfall	32.7	5.6	9.8	63.1	1.1	
Black spruce						
Precipitation	2.3	0.1	1.6	3.9	1.1	van Cleve et al., 1983a
Throughfall	3.5	0.1	2.2	1.3	2.9	
Litterfall	3.6	0.4	0.7	4.1	0.3	
Jack pine						
Precipitation	4.3	0.1	4.1	4.8	0.8	Foster, 1974
Throughfall	3.1	0.1	10.5	5.2	1.1	
Litterfall	19.0	1.4	4.4	12.1	2.1	
Red spruce						
Precipitation	4.8	0.2	0.5	0.1	1.1	Gordon, 1983
Throughfall	8.5	0.2	5.8	7.4	1.8	
Litterfall	20.7	1.8	1.9	19.9	1.3	

(Sources: P. Sollins et al., The internal element cycles of an old-growth Douglas fir ecosystem in western Oregon, *Ecological Monographs*, 1980, **50**, 267, reprinted with permission; N. W. Foster, Annual macroelement transfer from *Pinus banksiana* Lamb. forest to soil, *Canadian Journal of Forest Research*, 1974, **4**, 474, reproduced with permission; van Cleve et al., Productivity and nutrient cycling in taiga forest ecosystems, *Canadian Journal of Forest Research*, 1983, **13**, 756, reproduced with permission.)

Table 8.8 Nutrient storage in forests of the Pacific North-west (after Cole and Rapp, 1981)

	Organic matter (t ha^{-1})	N	P	K	Ca	Mg
		(kg ha^{-1})				
Douglas fir (450 yr)						
foliage	14.1	147	33	81	102	16
branches	54.2	70	17	60	184	15
boles	734.0	349	36	48	401	74
roots	172.8	140	12	50	225	52
understory	5.6	14	2	10	2	–
litter layer	218.5	445	62	80	619	160
soil-rooting zone	120.0	4560	34	660	2040	560
Western hemlock (121 yr)						
foliage	8.0	85	11	31	19	8
branches	50.6	52	17	20	26	5
boles	856.9	584	149	138	433	47
roots	189.2	179	44	61	153	34
understory	4.3	16	3	10	8	2
litter layer	301.1	474	98	118	347	83
soil-rooting zone	776.0	34900	70	720	1600	1600

(Source: D. W. Cole and M. Rapp, Elemental cycling in forest ecosystems, in *Dynamic Properties of Forest Ecosystems*, ed. D. E. Reichle; published by Cambridge University Press, 1981.)

In old stands of Douglas fir, treefalls contribute 0.5–30 t ha^{-1} yr^{-1} and accumulated biomass on the forest floor ranges from 80 to 490 t ha^{-1} (Harmon et al., 1986). However, the concentration of nutrients in fallen logs is fairly low and the amount stored in this debris is often less than in the needles and fine woody litter (Table 8.10). Considerable reserves of nutrients are bound up with the organic matter in the deeper soil horizons and much

Table 8.9 Annual nutrient transfer in various fine litter materials in old-growth hemlock and Douglas fir forests in the Pacific North-west (after Abee and Lavender, 1972; Sollins *et al.*, 1980)

	N	P	K	Ca	Mg
			($kg\ ha^{-1}\ yr^{-1}$)		
Hemlock					
needles	15.6	3.8	5.9	44.1	0.7
fine woody materials (cones, bark, twigs)	13.1	1.3	2.6	14.8	0.3
other (lichens)	2.7	0.3	0.9	0.5	0.0
Douglas fir					
needles (including green foliage)	8.6	3.0	4.1	21.6	1.5
fine woody materials (cones, bark, small branches)	3.8	0.5	1.2	5.6	0.4
other (insect frass, lichens)	2.8	0.3	0.5	1.6	0.2

(Source: P. Sollins *et al.*, The internal element cycles of an old-growth Douglas fir ecosystem in western Oregon, *Ecological Monographs*, 1980, **50**, 269, reprinted with permission.)

Table 8.10 Nutrient concentrations in fine litter and logs in old-growth Douglas fir forests in the Pacific North-west (after Sollins *et al.*, 1980)

	N	P	K	Ca	Mg
			($kg\ ha^{-1}$)		
Fine litter	256.0	50.2	56.3	445.4	133.1
Logs	215.0	6.7	23.7	163.4	16.3

(Source: P. Sollins *et al.*, The internal element cycles of an old-growth Douglas fir ecosystem in western Oregon, *Ecological Monographs*, 1980, **50**, 268–9, reprinted with permission.)

of this is associated with production and turnover of the root system. Root biomass in old-growth stands of Douglas fir is estimated at $153\ t\ ha^{-1}$, or 18% of the total biomass for the community, and about $11\ t\ ha^{-1}$ of this is comprised of fine roots less than 5 mm in diameter (Grier and Logan, 1977). A large proportion of the fine root mass is replaced each year, and because decomposition occurs rapidly this material is important in the overall circulation of nutrients in the ecosystem (Vogt *et al.*, 1986). This is most pronounced at higher elevations where cooler temperatures may inhibit decomposition of the surface litter. In subalpine stands of Pacific silver fir, for example, the mean turnover time of organic matter is reduced from 69 to 16 years by including fine root material in the detrital inputs.

8.10 LITTER DECOMPOSITION

Various fungi may be present in or on coniferous needles prior to shedding but decomposition occurs rapidly only when the waxy cuticle is weathered away. The vascular tissue is broken down quickly once the needles are on the forest floor. The epidermal cells are more resistant, and although they are extensively colonized by fungal hyphae, the needles often remain intact for some time (Millar, 1974). The needles appear greyish and begin to fragment as the cellulose and lignin are consumed by the hyphal webs that spread through the deeper litter layers. Mites which feed on the fungi cause considerable comminution of the needles: after laying

their eggs within the needles, they pass through several larval stages before the adults emerge from a hole cut in the epidermis. Springtails and nematodes are also common in coniferous litter and they too feed mainly on fungi. Coarse woody debris is mainly rotted by fungi but insects play an important role in its decomposition, either by directly consuming the wood or by subsequent introduction of microbes and other invertebrates. Fallen logs can remain intact for several years and fragmentation often begins only after the bark has been lost: this can take up to 30 years in species such as western hemlock (Harmon *et al.*, 1986). Decomposition is also delayed by the initially high moisture content of the sound wood (Grier, 1978). Coniferous wood contains less living tissue than that of deciduous species, and rates of decomposition may be reduced because of lower concentrations of sugars, starches and mineral nutrients. Microbial growth is also restricted by the size and connectivity of the tracheids and other anatomical structures in conifer wood and by various chemicals which inhibit decay (Harmon *et al.*, 1986). Further restrictions on decomposer activity are imposed by cool, moist climatic conditions, and in most coniferous ecosystems organic litter gradually accumulates on the forest floor until it is reduced by fire.

Figure 8.28 Crown fire burning through open lichen woodland in northern Quebec. (J. E. Fitzgibbon.)

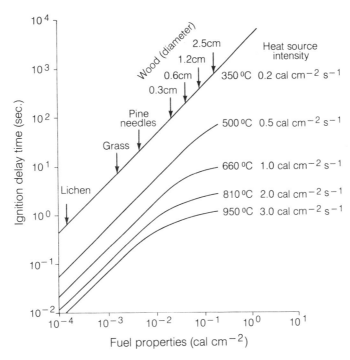

Figure 8.29 Relation of ignition delay time to the physical properties of the fuel and heat source intensity. (After Anderson, 1970.) (Reprinted with permission from H. E. Anderson, Forest fuel ignitibility, *Fire Technology*, **6**, No. 4, copyright © November 1970, National Fire Protection Association, Quincy, MA 02269.)

8.11 FIRE IN CONIFEROUS FORESTS

8.11.1 Fire characteristics

Fire is a natural feature of coniferous forests and evidence of periodic disturbance is preserved in fire scars on tree trunks and in charcoal layers in soils and sediments (Figure 8.28). Differences in fire frequency and intensity influence the structure and composition of the vegetation and abrupt changes occur at fire boundaries. Each year 9000–10 000 fires burn in coniferous forests across Canada: the average annual loss is 2 million hectares, or approximately 0.6% of the forested area. About 35% of the fires are started by lightning and these account for almost 85% of the burned area, because many originate in remote areas and cannot be easily suppressed. Although individual fires larger than 100 000 ha occur in most years, less than 3% are larger than 10 000 ha. However, these few events account for more than 90% of the total area that is burned annually (Higgins and Ramsey, 1992). Fire behaviour is determined by the type and amount of fuel available, and by the rate at which the flames are driven forward. A continuous layer of finely divided fuel must be present for fires to spread. Surface fuels in coniferous forests consist of a mixture of dead needles, dry mosses and lichens and various types of woody litter. A deep organic layer is usually present beneath this surface litter, and live foliage in the crowns is also readily consumed (van Wagner, 1983). Combustion occurs when

the temperature of the fuel reaches about 350 °C and volatile gases are released and mix with oxygen. The time required to reach ignition temperature depends on the moisture content of the fuel, its physical properties and the intensity of the heat source (Figure 8.29). Lichens and other fine fuels burn very quickly but even coarse woody debris ignites readily as fire intensity increases.

Natural fire starts are typically preceded by rainless periods of at least 1–2 weeks accompanied by high temperatures, low humidity and lightning storms: the probability of such conditions determines the 'fire climate'. In Canada the danger of fire is incorporated in the Canadian Forest Fire Weather Index, which is based on noon weather conditions and drying rates of different types of fuel (Figure 8.30). The **Fire Weather Index (FWI)** represents the intensity of the spreading fire in terms of the energy output per unit length of fire front; this is given by:

$$I = HWR$$

where I = fire intensity (kW m^{-1}), H = heat of combustion (kJ kg^{-1}), W = weight of fuel consumed per unit area (kg m^{-2}) and R = rate of spread (m sec^{-1}). Seven forest fire weather zones are defined in Canada on the basis of average FWI values (Figure 8.31). A range of FWI values can occur with varying frequency within each zone: in zone 4, for example, the average FWI is 6–10 but will exceed 30 on 4% of the days during the fire season, and can occassionally exceed 60 (Figure 8.32).

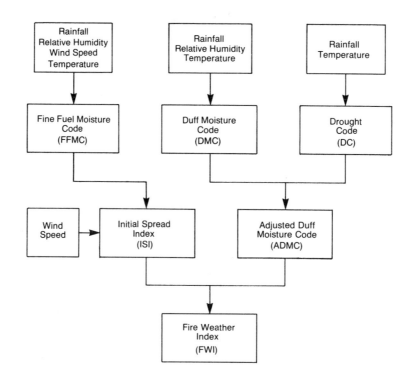

FFMC - moisture content of surface litter and other cured fuels. Nominal fuel load approximately 0.25 kg m^{-2} dry weight.

DMC - moisture content of loosely compacted, decomposing organic matter in upper 5-10cm of forest floor. Nominal fuel load approximately 5 kg m^{-2} dry weight.

DC - deeper layers of compact organic matter. Nominal fuel load approximately 50 kg m^{-2}.

ISI - rate of spread of fire in uniform fuel loads.

ADMC - total fuel available to the spreading fire.

FWI - intensity of the spreading fire.

Figure 8.30 The Canadian forest fire weather index. (After van Wagner, 1974.) (Redrawn from C. E. Van Wagner, *Structure of the Canadian Forest Fire Weather Index*; published by Canada Communications Group Publications, 1974; reproduced with permission of the Minister of Supply and Services, Canada, 1993.)

Fire intensities less than 10 kW m^{-1} are characteristic of smouldering ground fires which burn slowly through deep organic layers as moisture is driven out of the fuel. Surface fires which consume litter on the forest floor as well as shrubs and small trees generate 100–1500 kW m^{-1}. In active crown fires the flames advance through the canopy: they develop as surface fires become more intense during windy periods but fall back to the surface when the winds drop. Fire intensities will normally reach 40 000 kW m^{-1} in crown fires but under extreme conditions this can rise as high as 150 000 kW m^{-1} (van Wagner, 1983). The intensity of a fire not only depends on the amount and condition of the fuel but is also affected by weather conditions and topography. For example, crown fires occur mainly during the afternoon when wind speeds and temperatures increase. Fires spread most readily across level or gently undulating terrain. In dissected terrain, fire activity is affected by complex wind currents and by various temperature and moisture conditions caused by differences in slope and aspect (Wright and Bailey, 1982).

8.11.2 Fire frequency

The fire climate is strongly seasonal at higher latitudes and most fires occur in summer when long days and strong winds cause rapid drying. In the boreal forest the average interval between successive fires is about 60 years, although this varies considerably between different forest types. In central Alaska the average fire cycle ranges from 36 years in stands of black spruce to 113 years for white spruce (Yarie, 1981). Jack pine and lodgepole pine stands are especially prone to fire because they are generally

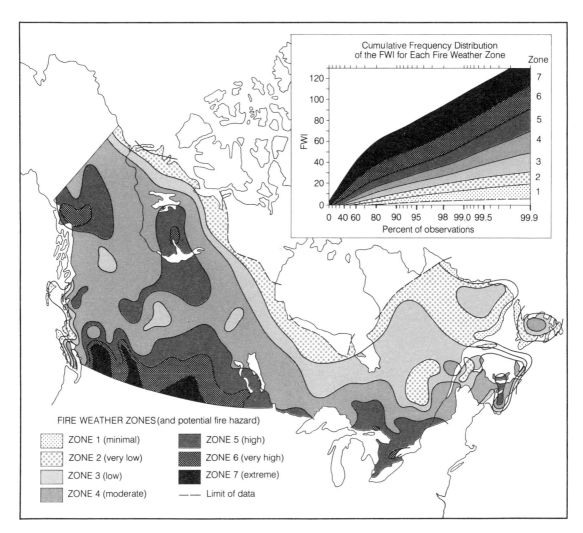

Figure 8.31 Forest fire weather zones of Canada. (After Simard, 1973.) (Redrawn from A. J. Simard, *Forest Fire Weather Zones of Canada*; published by Canada Communications Group Publications, 1973; reproduced with permission of the Minister of Supply and Services, Canada, 1993.)

associated with well-drained sandy soils. A mean fire interval of 28 years has been reported from fire scars on jack pine in central Canada (Delisle and Dubé, 1983). Precipitation increases eastwards across the boreal forests of North America and the fire cycle is considerably longer in these wetter regions; here the average interval between fires is more than 200 years (Wein and Moore, 1977).

Fire frequency is extremely variable in the forests of western North America. Lightning storms are rare in the wet foggy climate of the northern coastal regions and the normal fire interval is 500–800 years. A similar fire record history is reported in the cool, moist forests of Pacific silver fir which grow at higher elevations in the coastal mountains. Fires occur more frequently in the drier regions, and in the coastal Douglas fir zone the fire interval is about 135 years, increasing to about 200 years in areas dominated by western hemlock (Wright and Bailey, 1982). In these regions the most extensive fires coincide with major drought periods but are often intensified by hot dry winds which blow across the mountains. Fire frequency increases towards the south: in the coastal redwood stands of California the average fire interval is only 24 years (Fritz, 1931). Fire intervals as short as 4–8 years are reported in stands of ponderosa pine growing in the interior valleys of the western United States (Weaver, 1974). Forests of Engelmann spruce and subalpine fir grow at higher elevations where moister conditions lessen the chance of ignition, although in drier years fires which start in stands of ponderosa pine often spread into these highly flammable forests.

8.11.3 Fire adaptations in boreal species

Conifers usually survive low intensity surface fires of less than 300 kW m^{-1} but many are killed when fire intensity exceeds 1500 kW m^{-1}. Heat resistance is largely

Figure 8.32 Fire behaviour in jack pine (*Pinus banksiana*) stands in relation to the Canadian Fire Weather Index (FWI) system: (a) slow-spreading ground fire of FWI 14 – flames generally less than 0.6 m high, occasionally reaching the tree crowns when abundant lichen growth provides ladder fuels; (b) taller flames and faster rate of spread at FWI 24; (c) crown fire with flames about 30 m high and rising 10 m above the canopy at FWI 34. (M. E. Alexander and W. J. de Groot, with permission of the Northern Forestry Centre, Canadian Forestry Service, Edmonton.)

dependent on the density, moisture content and thickness of the bark (Figure 8.33). Thick bark is a characteristic of many west coast conifers: the bark of mature redwoods can be 30 cm thick and 15 cm is common in western larch (Fowells, 1965). However, young trees are not well protected and are easily killed by fire. The bark of western red cedar is relatively thin but this species typically grows in moister sites where fires are less frequent. Other characteristic adaptations of conifers which reduce damage from fire include a high, open branching habit, low flammability of the foliage and an absence of the epiphytic lichens that provide ladder fuels into the canopy (Table 8.11). Deep roots also lessen the threat from ground fires.

Jack pine and lodgepole pine have relatively thick bark and they are the only boreal species in North America which can survive repeated scarring from surface fires (Figure 8.34); black spruce, white spruce and larch will occasionally live through a single fire (Rowe, 1983). Adaptation to fire in all of these species is principally

through seed production. The pines produce serotinous cones which remain in the canopy for up to 25 years (Figure 8.35). The resin bonds sealing the cone scales in jack pine melt at temperatures of 60 °C. The heat resistant seeds can withstand temperatures of 150 °C for 30–45 seconds and 370 °C for 10–15 seconds and considerable numbers are released on to the bare mineral soil following a fire (Wright and Bailey, 1982). Black spruce produces semi-serotinous cones which release some seed each year. About 50% of the viable seed is retained in the cones for one year and about 15% persists for 5 years. This seed is released over a period of several months following a fire (Wilton, 1963). Species such as white spruce and balsam fir have no fire-adaptive traits. They are readily killed by fire and re-establish some years later from seed disseminated from unburned stands. Trembling aspen, a common associate in the boreal forests of North America, is maintained almost entirely by root suckering (Figure 8.36) and similar vegetative traits are common in many of the shrubs.

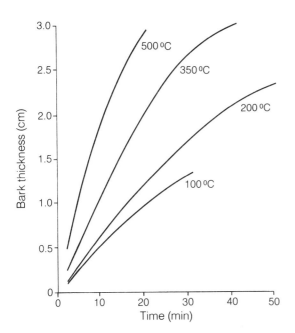

Figure 8.33 Time required to reach 60 °C at the cambium in eastern white pine (*Pinus strobus*). (After Kayll, 1963.) (Redrawn from A. J. Kayll, *A Technique for Studying the Fire Tolerance of Living Tree Trunks*, Department of Forestry Publication No. 1012, published by Canada Communications Group Publications, 1963.)

Forest plants persist in fire-disturbed environments in a variety of ways (Rowe, 1983). Some are classed as 'invaders': they are usually shade-intolerant species, such as fireweed (*Epilobium angustifolium*) and some mosses, whose seeds and spores are readily dispersed by wind.

These fugitive herbs flower and fruit profusely once they are established, but their seeds are short-lived and survival of the species is ensured only by the continual availability of recently disturbed sites. Species such as jack pine which store seed in serotinous cones are termed '**evaders**': they are able to regenerate even if all individuals are killed in a fire. Annuals and biennials, such as pink corydalis (*Corydalis sempervirens*) and bristly sarsaparilla (*Aralia hispida*), are also included in this group. These early post-fire herbs usually disappear within 1–3 years but their seeds are stored in the soil in readiness for the next fire. Similarly, shade-tolerant species such as pin cherry (*Prunus pensylvanica*) and red osier dogwood (*Cornus stolonifera*), which persist into later successional stages, are favoured by fires of low severity that remove little of the surface litter. Seeds buried in the mineral soil are more likely to survive fire than those in the duff layer, but if fires are infrequent deep layers of litter may accumulate and the seeds of the evaders may lose their viability before conditions are suitable for germination. Conifers, such as white spruce and balsam fir, are classed as '**avoiders**'; they are easily killed by fire and only regenerate through seed input from individuals which survived the burn. Shade-tolerant mosses and lichens have also adapted to fire in this way. Few boreal species can successfully resist fire. Cotton grass (*Eriophorum vaginatum*), for example, is classed as a '**resister**': it is protected by its dense tussock habit and will continue to spread vegetatively in the post-fire sites. Species such as trembling aspen, which are killed by fire but subsequently regenerate vegetatively from underground organs, are termed '**endurers**'. The ability to survive in this way is related to the rooting characterisitics of the plants:

Table 8.11 Growth characteristics which affect fire resistance in conifers (after Starker, 1934)

Species	Bark thickness	Rooting depth	Branch habit and canopy	Lichen growth	Flammability of foliage	Typical cause of death
Western larch	very thick	deep	high and very open	light	low	very resistant
Douglas fir	very thick	deep	high and dense	light to heavy	high	crown fires
Ponderosa pine	thick	deep	moderately high and open	light	low	crown fires
Grand fir	moderately thick	shallow	low and dense	light to heavy	medium	root charring and crown fires
Western cedar	thin	shallow	low and dense	light to moderate	high	root charring and crown fires, but often protected by moist habitat
Mountain hemlock	medium	medium	low and dense	light to moderate	high	root charring and crown fires
Lodgepole pine	very thin	deep	moderately low and open	moderate	medium	scorching of cambium
Western hemlock	medium	shallow	moderately low and dense	light to heavy	high	root charring and crown fires
Engelmann spruce	very thin	shallow	low and dense	light to heavy	very high	root charring and crown fires
Sitka spruce	very thin	shallow	moderately high and dense	light to heavy	high	scorching of cambium and root charring
Subalpine fir	very thin	shallow	very low and dense	medium to heavy	high	scorching of cambium and root charring

(Reproduced with permission from T. J. Starker, Fire resistance in the forest, *Journal of Forestry*, **32**, 462–7.)

Figure 8.34 Bark of three common boreal species: (a) white, highly flammable bark of paper birch (*Betula papyrifera*); (b) smooth, thin bark dotted with resin blisters in fire-sensitive balsam fir (*Abies balsamea*); (c) thick, furrowed bark of old jack pine (*Pinus banksiana*) provides some protection from fire. (R. A. Wright.)

the best adapted species are rooted at least 5 cm below the surface of the mineral soil (McLean, 1969).

8.11.4 Seed reserves and seedling establishment

Buried seeds are concentrated in the shallow surface layers of the soil and litter and their survival is dependent on the amount of litter consumed by fire. Seed reserves in coniferous forest soils are comparatively small and the dominant species are rarely represented (Archi-

bold, 1989). Seed production decreases with latitude and altitude and is essentially absent at the tree-line. In these extreme sites species such as black spruce and subalpine fir reproduce mainly by layering, supplemented by occasional seedlings in more favourable years (Payette *et al.*, 1982). Most conifers usually produce seed each year, with heavy crops every few years. Annual seedfalls in white spruce, for example, can range from 5200 to 450 000 seeds ha^{-1}, and in balsam fir from 400 to 550 000 seeds ha^{-1} (Frank and Safford, 1970): seedfalls of 0 to 3 million seeds ha^{-1} yr^{-1} have been reported in stands of

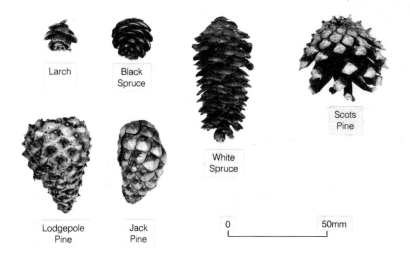

Larch Black Spruce White Spruce Scots Pine Lodgepole Pine Jack Pine

0 50mm

Figure 8.35 The cones of most boreal conifers open when mature, but those of lodgepole pine (*Pinus contorta*) and jack pine (*P. banksiana*) are mostly serotinous and remain sealed until the resin bonds are melted by the heat of a fire. (K. Bigelow.)

Figure 8.36 Regeneration by suckering in trembling aspen (*Populus tremuloides*) gives rise to dense regrowth stands with shrub–herb understory on old fire sites. (R. A. Wright.)

Douglas fir (Reukema, 1982). Seed production is mainly controlled by climatic conditions (Matthews, 1963). Thus, heavy cone crops in Douglas fir depend on a 2-year cycle in which cool, moist summers one year are followed by warm, dry summers the next, with seed dispersal occurring in the third year (Eis, 1973). Seed viability is generally rather low in boreal conifers, averaging 25–27% for species such as white spruce and larch compared with 50–70% in western and montane species (Table 8.12). Most seeds germinate soon after they are shed and quickly lose their viability, although germination is occasionally delayed until the second post-fire season in species such as eastern white pine (*Pinus strobus*), balsam fir (*Abies balsamea*) and jack pine (Thomas and Wein, 1985), and rarely beyond 3 years in black spruce (Viereck, 1983).

Germination and seedling establishment in conifers is generally more successful on burned or mineral seedbeds, because there is a greater chance of reaching the moister soil layers when the overlying litter is reduced (Arnott, 1979). Excessive temperatures caused by the poor thermal conductivity of thick litter layers can also be detrimental to the young plants. In addition, germination can be inhibited by pathogens in the organic debris. For this reason the litter from Engelmann spruce is unsuitable for its own seeds and for those of subalpine fir, Douglas fir and lodgepole pine (Daniel and Schmidt, 1972). Litter of subalpine fir and Douglas fir is also auto-inhibitory, but does not seriously impair germination in other associated species. No detrimental effects are detected with lodgepole pine litter. Similarly, reduced root

Table 8.12 Sylvicultural data for selected coniferous tree species (after Fowells, 1965; US Department of Agriculture, 1948)

Species	Frequency of good seed crops (yr)	Dispersal distance (m)	Viability (%)	Average seeding rates (ha⁻¹)	Seed longevity (yr)
Northern boreal species					
Abies balsamea	2–4	30–45	24–72	2400 (increases to 800 000 after fire)	3
Betula papyrifera	Most years	Close to parent tree	Rapidly lose viability	14 400 000	1.5
Larix laricina	5–14	120	27	2 000 000	3
Picea glauca	2–6	600	25	5 000 000	10
Picea mariana	1	25–50	Good	80 000	5
Pinus banksiana	3–4	60–90	Varies with age	500 000–1 700 000	5
Populus tremuloides	4–5	Several km	Rapidly lose viability	–	0.5
Western montane species					
Abies lasiocarpa	3	–	38	–	1
Picea engelmannii	2–6	200	69	40 000	3
Picea sitchensis	3–4	400	60	Low	–
Pinus contorta	1–3	60	64	40 000	5
Pinus monticola	3–4	120	40	16 000	2
Pinus ponderosa	2	50	Good	100 000	3
Pseudotsuga menziesii	3–7	100–200	Good	3500	4
Sequoia sempervirens	1	60–90	Varies with age	1 200 000	1
Thuja plicata	2–3	120	73	600 000	2
Tsuga heterophylla	3–4	600	56	3 200 000	2

(Source: O. W. Archibold, Seed banks and vegetation processes in coniferous forests, in *Ecology of Soil Seed Banks*, eds M. A. Leck, V. T. Parker and R. L. Simpson; published by Academic Press Inc., 1989.)

growth observed in black spruce seedlings growing in organic soil horizons has been attributed to inhibitory substances produced by microbial decomposition (Mallik and Newton, 1988). Burned seedbeds are also more favourable for black spruce because the seedlings are less likely to be smothered by faster growing mosses (Carleton, 1982). Allelopathic effects have also been suggested between lichens and jack pine and white spruce seedlings in addition to creating unfavourable moisture and temperature conditions on the floor of northern woodlands (Fisher, 1979).

The coastal forest species are mostly shade-tolerant and their seedlings will grow on a variety of seedbeds beneath the mature canopy, provided that there is sufficient moisture. However, even in these wetter regions the thick organic layers are prone to drought, and species such as sitka spruce and western red cedar survive better on mineral soil. Western hemlock seedlings will also tolerate a variety of substrates, but in areas where the forest floor is densely covered with shrubs, herbs and ferns, seedlings are often restricted to fallen trunks and old tree stumps (Fowells, 1965). Seedlings of shade-intolerant species, such as Douglas fir and redwood, grow best in fairly open sites but they are susceptible to drought and are generally most abundant on mineral seedbeds.

8.12 POST-FIRE SUCCESSION IN CONIFEROUS FORESTS OF NORTH AMERICA

The expected age-class distribution of forests which are subject to random fire starts can be described by a negative exponential function (van Wagner, 1978). This model predicts that about 60% of the region will be burned more frequently than the average fire interval, and the remaining stands may escape fire for some considerable time. The interval between fires affects fuel loadings and fire intensities although this is not directly related to stand age (Figure 8.37). Fuel loadings are usually high after fire, especially when dead trees are left standing following a crown fire. Fuel loadings decline as some of the snags collapse and decay but rise again as the stand becomes overmature and some of the trees begin to die. The impact of fire is therefore highly variable in space and time and this maintains the mosaic of canopy and understory vegetation that is so characteristic of the boreal forest. The nature of the regrowth community is determined by the intensity and frequency of fire and the depth of organic litter that is burned off. The inter-relationship of these factors is termed the **fire regime** (van Wagner, 1983).

Light burns will favour fire-tolerant species such as jack pine, but other conifers will usually be killed and subsequent establishment from seedlings is limited when the organic layer is not significantly reduced. Birch and trembling aspen are also sensitive to fire but they

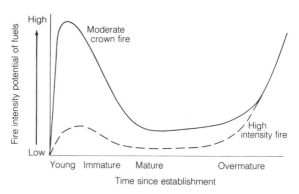

Figure 8.37 Fire intensity cycles in boreal conifer forest as a function of fuel loads remaining after previous fire events. (After Wright and Bailey, 1982.)

are maintained by vigorous vegetative regrowth, which often results in a significant increase in these hardwood species. Hardwoods are also favoured by a short fire interval because they can sprout or sucker before conifers are mature enough to produce large seed crops. Severe burns improve the seedbed and shade-intolerant conifers, such as jack pine, quickly re-establish. The absence of seed trees in extensive burns can delay establishment of fire-susceptible conifers such as white spruce, and shade-tolerant species such as balsam fir may not

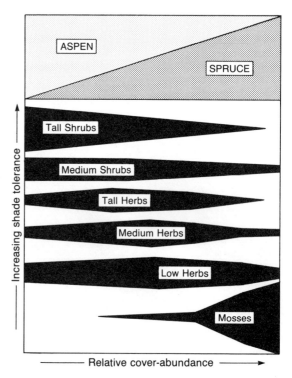

Figure 8.38 The effect of canopy composition on the relative cover-abundance of undergrowth strata in mature mixed boreal forest stands. (After Rowe, 1956.) (Reprinted by permission from J. S. Rowe, Uses of undergrowth plant species in forestry, *Ecology*, 1956, **37**, 462.)

Figure 8.39 A varied ground cover of shade-tolerant shrubs and herbs in the boreal forest of northern Saskatchewan: ferns (*Dryopteris spinulosa*) and horsetails (*Equisetum pratense*) are surrounded by wild sarsaparilla (*Aralia nudicaulis*), palmate-leaved colt's foot (*Petasites palmatus*) and bunchberry (*Cornus canadensis*). (R. A. Wright.)

Figure 8.40 Broad-leaf species such as vine maple (*Acer circinatum*) are characteristically associated with canopy openings in the dense forests of the Pacific Northwest. (O. W. Archibold.)

appear until an adequate ground cover has developed (Thomas and Wein, 1985). The canopy continues to develop with crown closure occurring in about 50 years, and mosses begin to dominate the forest floor. The structure of undergrowth is principally determined by light (Figure 8.38) but species composition is also influenced by moisture and other factors. In drier sites in Canadian boreal forests the tall shrub layer might be composed of wolf willow (*Elaeagnus commutata*) and buffalo berry (*Shepherdia canadensis*) with dogwood (*Cornus stolonifera*) and box elder (*Acer negundo*) present where soil moisture is higher. Similarly, herbaceous species such as purple wild pea vine (*Lathyrus venosus*) and Canada hawkweed (*Hieracium canadense*) are common on drier soils but ferns (*Dryopteris cristata*) and horsetails (*Equisetum arvense*) are characteristically present in damper areas (Figure 8.39). White spruce eventually replaces birch and trembling aspen in most burns, although stands dominated by jack pine and black spruce are commonly perpetuated by fire.

Even-aged stands of black spruce dominate the fire-prone lichen woodland in northern Canada. The trees are invariably killed by fire, and although seedlings establish immediately this may be limited in some sites by low seed viability and poor germination conditions (Black and Bliss, 1980). All of the common subordinate species, such as Labrador tea (*Ledum groenlandicum*), bog whortleberry (*Vaccinium uliginosum*) and dry-ground cranberry (*V. vitis-idaea*), normally regenerate vegetatively. Mosses such as *Polytrichum piliferum* and *Ceratodon purpureus* will also sprout from underground organs (Johnson, 1981). Buried seeds are rarely important in these ecosystems. The lichens are invasive species and rely on transport of sporedia and fragments of thallus from unburned sites,

and even though some are present soon after the fire, more than a century can elapse before species such as *Stereocaulon paschale* achieve their former dominance (Maikawa and Kershaw, 1976). Considerable differences in albedo have been recorded in burned and unburned lichen woodland and temperatures are much higher in the fire-blackened soils (Kershaw, 1977). *S. paschale* is sensitive to high temperature stress and photosynthetic activity virtually ceases at 45 °C (Kershaw, 1985). Consequently, it may not be able to compete with better adapted species until surface temperatures are ameliorated by a mulch of litter and by the shade cast by the slow-growing canopy of black spruce.

Although fires occur less frequently in the forests of the Pacific Northwest, they are often very severe because of high fuel loadings and unusually dry weather. Douglas fir is a fire-resistant species and seedlings are common where older trees act as a source of seed. Weedy herbs appear soon after the fire together with residual species from the original stand. The vegetation in this early stage of succession is characteristically very heterogenous because of differences in the severity of the burn. Species such as ragweed (*Senecio sylvaticus*) and annual willowherb (*Epilobium paniculatum*) are most abundant on heavily burned sites where soil nutrient status is improved by additions of ash: residual herbs

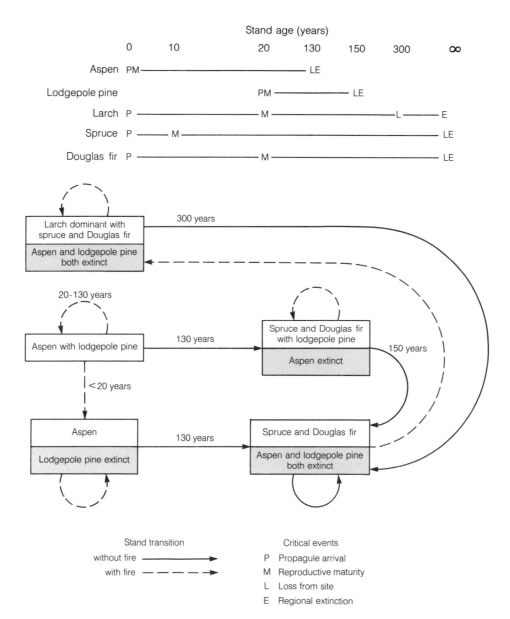

Figure 8.41 Life history characteristics and species replacement sequences in montane conifer forests in Montana. (After Cattelino *et al.*, 1979.) (Reproduced with permission from P. J. Cattelino *et al.*, Predicting the multiple pathways of plant succession, *Environmental Management*, 1979, **3**(1), 45.)

such as star flower (*Trientalis latifolia*) occur more frequently in unburned sites (Franklin and Dyrness, 1969). Similarly, snowbush (*Ceanothus* spp.) is found only in burned areas, whereas red huckleberry (*Vaccinium parviflorum*) persists mainly in patches where the fire did not spread. Herbaceous species, such as bull thistle (*Cirsium vulgare*) and bracken (*Pteridium aquilinum*), increase rapidly during this early weedy stage. This gives way to a shrub-dominated stage after 4–5 years. The shrub stage persists for 15–20 years, by which time the young Douglas firs begin to assume dominance. Shade-intolerant shrubs, such as vine maple (*Acer circinatum*) and salal (*Gaultheria shallon*), will persist for some time in relatively open

stands (Figure 8.40) but the character of the understory changes as the canopy closes over. Western hemlock begins to appear as soil moisture levels increase, accompanied by western red cedar in wetter sites. Seedlings of western red cedar, like those of Douglas fir, grow best in light shade, and although both these species remain as dominant members in many of the oldest stands, they persist because of their longevity, unlike western hemlock which is the shade-tolerant climax species over much of this region (Franklin and Hemstrom, 1981).

The montane forests are composed of several widely distributed tree species whose relative abundance is very dependent on fire. The normal fire cycle in western

Montana, for example, is about 50 years and results in mixed forests of trembling aspen, lodgepole pine, ponderosa pine, Douglas fir, Engelmann spruce and larch. Aspen is favoured by very short fire cycles but dies out under longer fire intervals. The conifers disappear when fires occur very frequently. Lodgepole pine is maintained by occasional burning, but the other conifers become increasingly abundant when fire intervals are very long. These differences are explained by the interaction of fire frequency and the autecological characteristics of the species (Cattelino *et al.*, 1979). Species success in fire-disturbed environments varies according to their method of persistence, their requirements for establishment and the time taken to reach critical stages in their life history. Trembling aspen is a shade-intolerant species which regenerates vegetatively: its life span is about 130 years and once it dies, because there is no other reliable method of regeneration, the species must necessarily disappear from the forest (Figure 8.41). Regeneration of lodgepole pine is dependent on fire to release seed stored in the canopy. Seed production in heavily stocked stands commences when the trees are about 20 years old and continues until they die at about 150 years. The seed source is therefore eliminated if fires

occur prior to maturity or after the trees have died. Seeds of Engelmann spruce, Douglas fir and larch are readily dispersed into burned sites, and will continue to establish for as long as some trees survive in the area. These species, because of their great longevity, are co-dominant in stands that are burned infrequently: once trembling aspen and lodgepole pine are lost from these stands they are unable to re-establish even if the fire regime is subsequently shortened.

8.13 INSECTS AND DISEASES IN THE CONIFEROUS FORESTS OF CANADA

Thousands of insect species and diseases are found in Canadian forests. During the period 1977–81 timber losses from these organisms averaged 63.8 million $m^3 yr^{-1}$ compared with 80 million $m^3 yr^{-1}$ from fire (Honer and Bickerstaff, 1985). Most of the damage is caused by only a few species. The spruce budworm (*Choristoneura fumiferana*), which is indigenous to eastern North America, is the most destructive (Figure 8.42). It is found on many species of conifer but its principal host is balsam fir, although white spruce and red spruce are also seriously affected. Historical evidence indicates that since 1704 outbreaks of spruce budworm in north-eastern North America have occurred at an average interval of 29 years (Blais, 1983). The most recent outbreak peaked in 1978 when over 72 million ha were affected (Hardy *et al.*, 1986). The adult moths lay eggs on the needles in September. Incubation lasts from 7 to 10 days: upon hatching the larvae move towards the interior of the crowns and overwinter, protected by silken webs, in crevices in the bark and other suitable sites. The larvae emerge in spring and move out to the perimeter of the crown, where they begin to feed on staminate flowers and older needles until the buds open. Much of the new foliage is consumed early in the season and radial growth is reduced. Repeated defoliation for 3–4 years will normally result in the death of the trees and losses of weakened trees will continue for some years after the budworm population has subsided (Martineau, 1984). In the absence of spraying or other controls, the outbreaks subside after 7–15 years. This decline appears to be related to unfavourable weather conditions in May and June, together with a reduction in the number of host trees and the absence of new foliage. Spruce budworm is attacked by many predators and parasites but they have little impact when the insect population is high. Artificial control is usually by aerial application of chemical insecticides, although *Bacillus thuringiensis* is now recognized as a safe and effective biological insecticide for spruce budworm (Hulme *et al.*, 1983).

The jack pine budworm and hemlock looper are also important insect defoliators. The jack pine budworm (*C. pinus pinus*) is found throughout the Canadian boreal

Figure 8.42 Stages in the life cycle of spruce budworm (*Choriostoneura fumiferana*): (a) adult spruce budworm; (b) egg mass; (c) second instar larva; (d) young larva in a new shoot: some needles have been removed to reveal larva; (e) mature larva; (f) pupa; (g) close-up of feeding damage on white spruce (*Picea glauca*) shoots. (W. G. H. Ives and H. R. Wong, 1988, with permission of the Northern Forestry Centre, Forestry Canada, Edmonton, Alberta. (Reproduced with permission from W. G. H. Ives and H. R. Wong, *Tree and Shrub Insects of the Prairie Provinces*, Information Report NOR-X-292; published by Northern Forestry Centre, Northwest Region, Forestry Canada, Edmonton, Alberta, 1988.)

forests. Population outbreaks in this species occur at intervals of 10–15 years and generally persist for 3–5 years. Extensive defoliation results in top-kill of the trees, and although stand mortality is less severe than for spruce budworm, over 30% of the trees can be killed when severe outbreaks coincide with drought years (Coulson and Witter, 1984). The hemlock looper (*Lambdina fiscellaria fiscellaria*) is found mainly in eastern Canada where the primary host is balsam fir, although it is also found on white spruce, black spruce and eastern hemlock. The hemlock looper overwinters in the egg stage and the larvae hatch in late spring. The larvae feed on the foliage but they consume only a small part of each needle, which subsequently dies and turns brown. Defoliated trees usually die within a year, although if only part of the foliage is lost growth is reduced but the trees will survive. Hemlock looper is susceptible to fungal infection, which spreads rapidly during cool, moist weather, and spraying trees with water has proven to be an effective method of reducing infestations (Martineau, 1984).

The spruce beetle (*Dendroctonus rufipennis*) is a widespread and destructive pest which attacks the wood of white spruce, Engelmann spruce and sitka spruce. Its life cycle lasts from 1 to 2 years, depending on climatic conditions. In midsummer the females begin to construct egg galleries that run vertically beneath the bark; the males assist by carrying off the wood dust that adheres to resin which exudes from the entry hole. On emerging the larvae construct additional galleries which end in a pupation chamber: here the insect overwinters and the adult emerges in the spring. The bark is loosened by the removal of cambial tissue; the foliage begins to wilt as the flow of sap is interrupted and the tree eventually dies. The mountain pine beetle (*D. ponderosae*), a closely related species, is the most destructive bark beetle in the western forests. Its principal hosts are lodgepole pine, ponderosa pine and western white pine. All of the galleries are constructed in the phloem and the tree is killed by girdling and by blue stain fungi (*Ceratocystis* spp.) which are transmitted by the beetles.

The parasitic dwarf mistletoe (*Arceuthobium americanum*) mainly infects lodgepole and jack pine, but is also found on ponderosa pine. Related species grow on Douglas fir, hemlock, larch and spruce. They cause more damage in the forests of western North America than any other pathogens. The most conspicuous symptom of dwarf mistletoe is the production of witches' brooms. Height growth is suppressed by 30–60%, and this abnormal development reduces the strength of the wood and even its suitability for pulp (Sinclair *et al.*, 1987). Dwarf mistletoe is a dioecious plant which grows systemically in the host tree. It flowers in late spring and early summer and, following pollination, the berries begin to develop on the female plants later in the summer. The berries remain on the plant for about a year before they ripen and forcefully eject a single seed. The seed is surrounded by a hygroscopic, sticky substance that enables

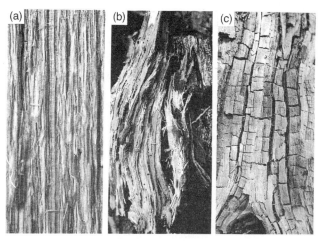

Figure 8.43 Fungal woodrots: (a) white pocket rot in white spruce (*Picea glauca*) caused by *Phellinus pini*; (b) brown stringy rot in alpine fir (*Abies lasiocarpa*) caused by *Echinodontium tinctorium*; (c) brown cubical rot in lodgepole pine (*Pinus contorta*) caused by *Coniophora puteana*. (Y. Hiratsuka, 1987, with permission of the Northern Forestry Centre, Forestry Canada, Edmonton.) (Reproduced with permission from Y. Hiratsuka, *Forest Tree Diseases of the Prairie Provinces*, Information Report NOR-X-286; published by Northern Forestry Centre, Northwest Region, Forestry Canada, Edmonton, Alberta, 1987.)

it to adhere to susceptible host tissues. The seeds travel up to 15 m in this way (Hiratsuka, 1987) but they are also dispersed greater distances by squirrels and birds. The seeds overwinter and germinate in the spring but it takes 3–5 years before aerial shoots develop and another 1–2 years before they flower. The average rate of spread, even in heavily infected stands, is therefore only about 0.5 m yr^{-1} and silvicultural control by clearcutting is effective and economically feasible.

The conifers are susceptible to various root rots and decays. *Armillaria* is the most common root-rotting fungus. This pathogen was originally considered a single species (*A. mellea*) but many distinct species may be involved. The typical symptoms of *Armillaria* root rot are reduced growth, discoloration of the foliage and white mycelial fans under the bark around the base of the trunk. In autumn, honey-coloured mushrooms appear on decayed wood or in the soil surrounding the diseased roots. Infection is mainly by string-like rhizomorphs which grow up to 10 m through the soil. Spores produced by the mushrooms may also be dispersed to dead woody material. Small trees are killed quickly but larger trees will usually survive for several years, although growth declines as the root system deteriorates. The weakened trees are more susceptible to attack from insect pests and other fungi. Prevention of the disease is not possible under normal forest conditions, although use of fungicides and mycophagous nematodes has been investigated. Decay of wood and roots caused by other fungi is

Table 8.13 The effects of site quality on tree species performance in the boreal forests of Saskatchewan (after Kabzems *et al.*, 1986)

Site quality class	Mean annual increment (m³ ha⁻¹)			Rotation age when trees > 9.1 cm DBH (yrs)			Average height at rotation age (m)		
	White spruce	Black spruce	Jack pine	White spruce	Black spruce	Jack pine	White spruce	Black spruce	Jack pine
I+	> 5.0	> 2.4	> 3.2	65	75	60	> 25	> 17	> 22
I	4.3	2.0	2.7	70	90	65	24	15	21
II+	3.8	1.7	2.4	70	95	65	22	14	19
II	3.1	1.4	2.0	75	100	70	20	13	18
II−	2.6	1.1	1.6	75	110	75	17	12	15
III	2.0	0.8	1.3	80	120	80	16	11	13
III−	< 0.7	< 0.5	< 0.8	95	135	90	< 13	< 10	< 12

(Source: A. Kabzems, A. L. Kosowan and W. C. Harris, *Mixedwood Section in an Ecological Perspective*, Technical Bulletin No. 8; published by Saskatchewan Department of Environment and Resource Management, 1986.)

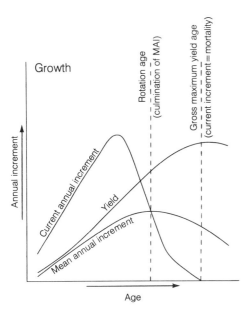

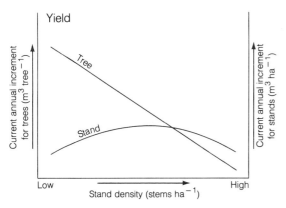

Figure 8.44 The relationship of stand age to yield and growth measured as mean annual increment and current annual increment. (After Bickerstaff *et al.*, 1981.) (From A. Bickerstaff, W. L. Wallace and F. Evert, *Growth of Forests in Canada, Part 2: A Quantitative Description of the Land Base and the Mean Annual Increment*; published by Canada Communications Group Publications, 1981; reproduced with permission of the Minister of Supply and Services, Canada, 1993.)

common in mature and overmature stands. The fungi enter the trees through wounds and fire scars; growth declines and the weakened trees eventually break or are blown down by the wind. Common forms include white pocket rot, brown stringy rot and brown cubical rot (Figure 8.43). Wood with advanced stages of decay cannot be used and allowances for rot are necessary in all timber inventories.

8.14 THE FOREST INDUSTRY IN CANADA

Approximately 453 million ha in Canada is forested, of which about 50% is classed as productive forest land (Forestry Canada, 1992). During 1977–81 the average area harvested totalled 0.76 million ha yr⁻¹ compared with 0.96 million ha yr⁻¹ taken out of production by fire and 0.50 million ha damaged by insects. Losses from fire and insects accounted for about 43% of the average annual growth of the forests, and this is equal to the amount of timber removed by harvest (Honer and Bickerstaff, 1985). Fire and disease affect trees of all ages, whereas logging is more selective: depletion of forest area must therefore be weighted by the age of the trees removed or destroyed. The composition of the forest which grows back after logging often differs from a naturally regenerated stand, and subsequent development is also greatly altered by intensive silvicultural management.

Forest growth varies with environmental conditions, species composition and the age, size and number of trees present. Yield tables are compiled for each species showing the amount of merchantable timber that can be expected under different conditions (Table 8.13). The average amount of growth that has occurred by a given age is expressed by the **mean annual increment (MAI)**. The **current annual increment (CAI)** is the amount of growth that occurs in a specific year, or annually at a specific age (Bickerstaff *et al.*, 1981). MAI changes fairly slowly, especially near rotation age, but net CAI changes

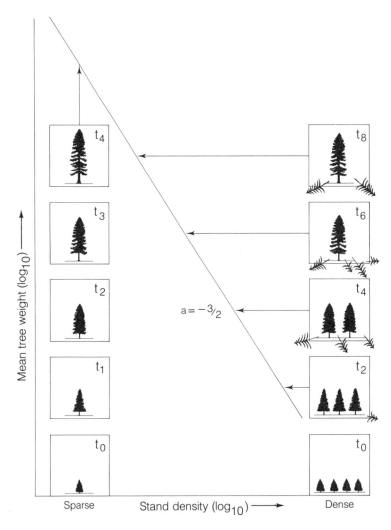

Figure 8.45 The effect of tree density on the self-thinning process in forest stands of increasing age (t_0–t_8). (After Silvertown, 1982.) (Redrawn and modified from J. W. Silvertown, *Introduction to Plant Population Ecology*; published by Longman, 1982.)

rapidly with age (Figure 8.44). Maximum yield occurs when CAI is zero and any volume increase is equalled by mortality: further ageing therefore results in yield reductions caused by the break-up of the overmature stand. The stand reaches its technical rotation age when mean annual volume increment is at a maximum, and should be harvested to optimize wood volume. This occurs before maximum yield is attained.

Tree growth is enhanced in any site when stand density is so low that competition is minimal. For many species the relationship between mean plant weight and stand density is described by the $-3/2$ power law (White, 1980). In lightly stocked stands where competition is minimal the trees will increase in size with no reduction in density until the self-thinning line is reached (Figure 8.45). After this time the loss of suppressed individuals compensates for increased growth, although total stand biomass will continue to increase until the trees reach maturity. Manipulation of stand density is therefore commonly used to optimize tree size and maximize yields.

Without such forest management some stands may never produce merchantable timber. Average stocking of natural stands is usually about 70% of the maximum that is theoretically attainable (Bickerstaff *et al.*, 1981) and harvesting is usually delayed by several years in low-density stands in order to optimize yields (Figure 8.46). Similarly, stands growing on good sites have considerably shorter rotations and higher yields than those associated with poorer sites. Utilization standards also influence silvicultural practices: for example, short rotations are employed when relatively small, low-value trees can be used for pulpwood.

The amount of wood that is removed from a forest to meet management objectives is termed the **allowable cut**. For long-term sustained production the average annual allowable cut must balance the total mean annual increment, but this can be maintained indefinitely only when adequate restocking occurs following cutting. Forest harvesting is mostly by clear-cutting, in which all trees in an area are removed (Figure 8.47), and the unwanted materials are then usually burned to reduce the wildfire hazard. Burning also improves the seedbed and reduces competition from undesirable species. Clear-cutting eliminates the seed source for many species, and natural regeneration is delayed until seed is dispersed from adjacent areas. From a forestry viewpoint the

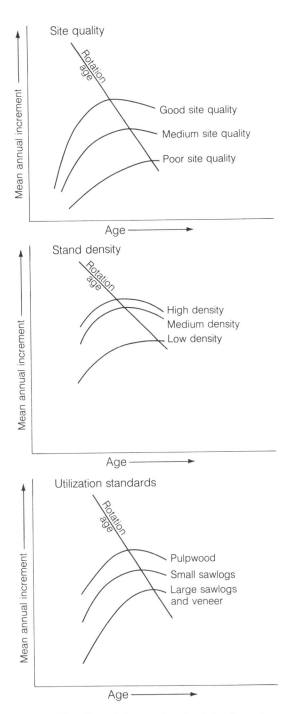

Figure 8.46 The effect of site quality, stand density and utilization standards upon growth and rotation age. (After Bickerstaff *et al.*, 1981.) (From A. Bickerstaff, W. L. Wallace and F. Evert, *Growth of Forests in Canada, Part 2: A Quantitative Description of the Land Base and the Mean Annual Increment*; published by Canada Communications Group Publications, 1981; reproduced with permission of the Minister of Supply and Services, Canada, 1993.)

Figure 8.47 Clear-cut logging in a hemlock (*Tsuga heterophylla*) – Douglas fir (*Pseudotsuga menziesii*) stand in British Columbia. (O. W. Archibold.)

Figure 8.48 (a) Plantation of 15-year-old Douglas fir (*Pseudotsuga menziesii*), Vancouver Island; (b) 6-year-old jack pine (*Pinus banksiana*) plantation in northern Saskatchewan. (O. W. Archibold.)

Table 8.14 Nutrient removal from Douglas fir forest growing in Washington State on high and low fertility soils following conventional harvest and whole tree harvest operations (after Mann *et al.*, 1988)

	Biomass harvested (t ha^{-1})	N	P	K	Ca
		(kg ha^{-1})			
High fertility site					
Conventional harvest	281	478	56	225	23
Whole tree harvest	318	728	96	326	411
Low fertility site					
Conventional harvest	134	161	27	81	–
Whole tree harvest	165	325	56	140	–

(Reproduced with permission from L. K. Mann *et al.*, Effects of whole-tree and stem-only clearcutting on post harvest hydrologic losses, nutrient capital, and regrowth, *Forest Science*, **34**(2), 412–28.)

principal objective of regeneration is to establish a commercial stand as quickly as possible. However, natural regeneration often results in undesirable changes in species composition and inadequate stocking densities. In British Columbia, for example, only 40% of the area which is logged is suitably regenerated by natural means (Pearse *et al.*, 1986). Site preparation and planting occur on about 50% of the harvested areas across Canada (Figure 8.48) but subsequent silvicultural programmes are not very widespread, even though growth rates in managed stands are considerably higher than in comparable natural areas (Brand and Penner, 1991).

The forest industry is becoming increasingly mechanized and intensive; in many areas **whole-tree harvesting (WTH)** is employed, in which all of the tree is removed, including the branches and foliage that would have been left on the forest floor by conventional logging practices. Approximately 60% of the nutrients in merchantable conifers is contained in the trunkwood of conifers. Consequently, nutrient export in biomass is considerably higher in WTH areas (Table 8.14) and this is raising concerns about future productivity. Changes in the nutrient status of the soil have also been reported (Table 8.15). Nutrient concentrations contained in litter are considerably lower in WTH sites compared with uncut stands, whereas nutrient levels are increased by conventional harvest operations. Conversely, soil nutrient concentrations increase with depth in WTH sites and this downward movement suggests that additional losses might be occurring through leaching. Although losses of potassium, calcium and magnesium are generally higher in WTH sites, nitrogen concentrations in leachate are much lower than in conventional harvest sites. This is attributed to the fact that large amounts of soluble nitrogen are leached directly from foliage and other debris left on the forest floor by conventional logging practices. Although WTH increases the risk of nutrient deficiencies, intensive site preparation may be necessary to maintain yields under any logging method (Hendrickson *et al.*, 1989).

8.15 THE CHANGING NATURE OF THE FORESTS

The idea that coniferous forests simply represent a renewable store of merchantable materials is rapidly changing and greater emphasis is now being placed on these areas for non-industrial uses. In the former Soviet Union the forests were officially grouped according to their social, cultural and economic functions (Barr and Braden, 1988). The forests of group 3 were the most extensive and were used for commercial purposes, but in groups 1 and 2 the principal role was resource and environmental conservation. Group 1 forests received the greatest protection and cutting was allowed only to maintain the health of the stands. These forests were important for recreation and were being used increasingly for food production. State Forest reserves provided significant quantities of mushrooms and berries, with other products such as nuts and honey being especially important where hardwood species are abundant. The average annual recreational use of forests in European Russia was estimated at 11 billion hours. This function was often combined with state medico-psychological recuperation services. Group 2 forests were used for various economic and environmental purposes, particularly watershed preservation, but because they were associated with heavily populated regions silvicultural programmes were required to offset over-utilization. Subsistence agriculture was carried on for centuries in the taiga with cattle being raised on hay gathered from small clearings in the forest which were maintained by fire. This practice was later replaced by intensive forage cultivation but most of the farms are now abandoned (Hämet-Ahti, 1983). Only semi-domesticated reindeer herds are still grazed in this marginal forest land. However, wildlife is abundant in these sparsely populated regions and fishing, trapping and guiding provide supplementary income for many inhabitants. Sound ecological management is therefore becoming increasingly important in the commercial development of the coniferous forest resource.

Table 8.15 Nutrient concentrations, organic matter and pH in litter and mineral soil in a Canadian mixed conifer–hardwood stand 3 years after conventional harvest (CH) and whole tree harvest (WTH) operations (after Hendrickson et al., 1989)

	Organic matter (%)	N	K	Ca	Mg	pH
				(mg kg^{-1})		
Litter layer						
Uncut	69.9	6543	848	5302	661	4.19
CH	61.0	6723	873	7353	804	4.58
WTH	54.9	4679	699	5115	624	4.51
Mineral soil (0–5 cm)						
Uncut	6.0	1059	204	359	148	4.11
CH	7.6	1366	246	588	174	4.50
WTH	6.3	1069	224	646	336	4.48
Mineral soil (5–10 cm)						
Uncut	3.8	809	235	352	186	4.48
CH	3.7	916	266	445	216	4.85
WTH	5.4	1463	245	565	369	4.76
Mineral soil (10–20 cm)						
Uncut	3.2	663	254	609	318	5.04
CH	3.1	767	299	612	329	5.18
WTH	3.2	803	254	546	357	5.05

In Canada also, the growing appreciation of the different values of the forests is influencing the way in which this resource is being managed in order to protect environmental quality and yet enhance economic and social benefits. This new initiative is based on the concept of sustainable development in which present needs are met without compromising the ability of future generations to meet their own needs. Thus, forest management expenditure has increased substantially in the past few years. The average area planted and seeded increased from just over 200 000 ha in 1981 to about 500 000 ha in 1991, and more than 800 000 ha of young trees now benefit from forest renewal activities, such as site preparation, stand thinning and pruning, and fertilizer application. In addition, there has been a substantial reduction in the use of pesticides and herbicides. Similarly, significant changes at the processing stage continue to reduce air and water pollution associated with the pulp and paper industry. The introduction of 'minimum content laws' governing the amount of recycled fibres in newsprint has resulted in a greater demand for wastepaper. The projected use of wastepaper in Canadian newsprint production is equivalent to approximately 2% of the country's total softwood harvest.

This new perspective on forest resources in Canada has stimulated the development of various Provincial Forest Ecosystem Classifications. When completed these will provide detailed inventories of forest patterns and recommended harvest practices, based on vegetation types and soil conditions. The use of understory and groundcover species in these classifications provides a more complete assessment of ecosystem attributes. This facilitates the harvesting of alternative forest products such as bark, twigs, mosses and dried flowers for use in floral crafts, the collection of natural foods and herbs, and the supply of botanicals of pharmaceutical value. For example, the Pacific yew (*Taxus brevifolia*) illustrates the benefits of biodiversity in the forests. This species was discarded in conventional forest operations. However, it now supplies taxol for cancer treatment (Joyce, 1993); yew plantations have been established to try to meet the demand that has been jeopardized by previous practices.

CHAPTER 9

The polar and high mountain tundras

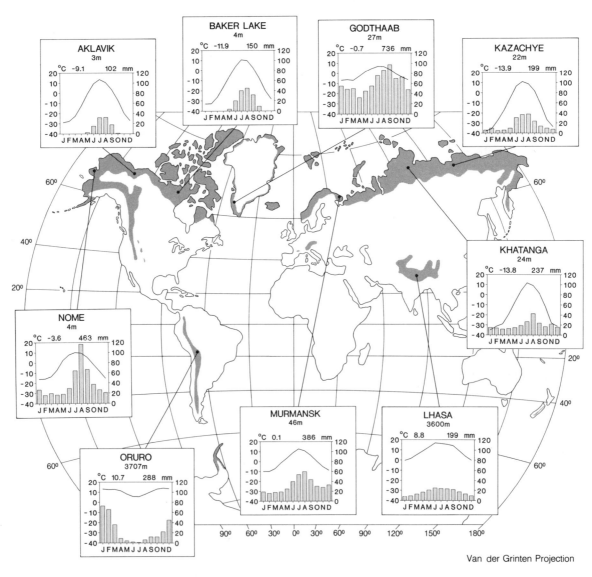

Figure 9.1 Distribution of polar and high mountain tundra ecosystems and representative climatic conditions. Mean monthly temperatures are indicated by the line and mean precipitation for each month is shown by the bars. Station elevation, mean annual temperature and mean annual precipitation appear at the top of each climograph.

9.1 INTRODUCTION

The growth of woody plants is restricted by the lack of summer warmth at high latitudes and high altitudes. The only vascular plants which survive these conditions are highly specialized: they must complete their annual growth quickly during the brief, cool summer and then endure many months of bitter cold. Beyond the tree-line the dominant plants are grasses and sedges: shrubs survive in more favourable sites where snow accumulates in the winter but under more rigorous conditions the vascular plants are mostly replaced by mosses and lichens. In this harsh environment even small differences in microhabitat result in an unexpected diversity of plant communities. These treeless tundra regions cover approximately 25 million km² of the Earth's surface (Figure 9.1).

More than half of the tundra in the world is found in the arctic regions of North America and Eurasia. Only 900 species of flowering plants are represented in this circum-arctic flora. The greatest diversity occurs in Alaska where 600 species are present (Good, 1964), but this decreases with latitude, and less than 100 species are reported in the northernmost islands in the Canadian arctic (Savile, 1972). The latitudinal limit of the tundra is reached at 83.5 °N in the northern tip of Greenland (Billings and Mooney, 1968). Some species are restricted to the arctic, but many others have an arctic–alpine distribution and are found extensively in high mountain areas. Arctic–alpine species are especially abundant in the Rockies and other mountain ranges which extend southwards from the arctic. Alpine tundra covers about 9.5 million km² in the northern hemisphere. These areas are floristically richer than the arctic tundra, with many species endemic to specific mountain ranges.

A few plants, including the crowberries (*Empetrum* spp.) and some grasses and sedges, are bipolar in their distribution and are present throughout the arctic tundra and in parts of Patagonia and the sub-antarctic islands such as South Georgia (Callaghan, 1973). Only two species of vascular plants are known from Antarctica: they are found on the Antarctic Peninsula and some of the adjacent islands. Mosses survive in ice-free sites around the coast of Antarctica and on rocky outcrops and are especially abundant in sites which are manured by penguins (Polunin, 1960). Liverworts, lichens and algae are also found in this cold, dry environment. Less than 0.1% of the world's tundra regions lie south of 60 °S. The total area of tundra in the southern hemisphere is about 1 million km²: it is found mainly in the mountainous regions of South America and New Zealand. Alpine vegetation is also found in tropical areas including the northern Andes, East Africa and Indonesia. The presence of subarborescent species in the tropical mountains is in marked contrast with alpine vegetation elsewhere.

9.2 CLIMATE

The tundra regions at high latitudes are dominated for most of the year by cold, dry continental polar air masses. Moister air enters the region in summer as the sea ice begins to melt. A slight maritime influence occurs in Arctic coastal areas and islands even in winter, and temperatures are less severe than further inland. The effect is especially pronounced in Greenland, Iceland and northern Scandinavia (Figure 9.1). Except in the most favoured locations the average temperature in the warmest month is below 10 °C; temperatures remain above freezing for only 2–6 months of the year. During this brief period of warmth the sun remains above the horizon for most of the day and at extreme latitudes does not set for several weeks. Despite the low sun angle the input of solar energy is considerable; net radiation in the Canadian Arctic is about 225 langleys day^{-1} in July, compared with 275 langleys day^{-1} along the border with the United States (Hare and Hay, 1974). The sun rises only briefly or not at all in winter; for example, at 75 °N the period of total darkness lasts for about 3 months. What little insolation is received is mostly reflected from the snow-covered surface, and energy loss continues during the long, clear arctic nights. Consequently, a negative radiation balance persists for several months over most of the region.

Annual precipitation is generally less than 250 mm in the arctic and comes mostly from cyclonic storms which pass through the region in the warmer months. Snow accounts for about 60% of the precipitation, with light falls usually reported from mid-September until mid-June. Very little melts during this period and a protective snowpack covers the ground to an average depth of 20–40 cm, with drifts 2–3 m deep accumulating in depressions and in sites which are sheltered from the wind. Snowmelt occurs quickly but the presence of permafrost (see Figure 8.3) throughout the Arctic limits infiltration. Where sufficient meltwater collects in depressions and at the foot of slopes the soils remain moist all summer, and differences in soil moisture are reflected by the various plant communities associated with topographic gradients. The climate becomes drier at higher latitudes. Cooler temperatures limit the amount of moisture that is held by the air, and precipitation is less than 100 mm in some locations. In these polar deserts the soils usually dry out soon after the snow has melted and vascular plants are mostly restricted to sites where the snow cover persists until late summer (Figure 9.2).

Climatic conditions in alpine areas are typically more variable than in the arctic. Radiation intensity increases in the clear alpine air and temperatures rise quickly during the day. However, rapid heat loss occurs at night, and although average daily temperatures are low, diurnal fluctuations are more pronounced that in arctic regions: this is particularly apparent in tropical mountains. In the tropics the annual temperature range is small because

Figure 9.2 Late-lying snowpatches in the Canadian arctic. (P. Marsh.)

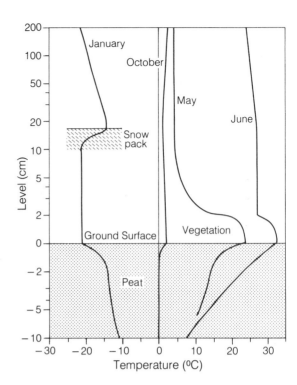

Figure 9.3 Typical temperature gradients in air, snow, vegetation and soil in arctic tundra in northern Sweden. (After Rosswall *et al.*, 1975.) (Adapted with permission from T. Rosswall *et. al.*, Stordalen (Abisko), Sweden, in *Ecological Bulletins 20: Structure and Function of Tundra Ecosystems* (eds T. Rosswall and O. W. Heal), published by the Swedish Natural Science Research Council, Stockholm, 1975.)

there is little change in day length and sun angle throughout the year. Although the range increases with latitude, seasonal temperature differences in alpine tundras are generally less pronounced than in the arctic. Temperatures vary markedly with slope angle and exposure in mountainous terrain, and some valleys may receive direct radiation for only a brief period each day. Precipitation tends to increase with elevation but the highest summits can be relatively dry if most of the moisture is lost on the lower slopes. Precipitation is highest on the windward slopes and a pronounced rain-shadow often occurs on the lee-side of the range. Clouds and fog are common in mountain areas: for example, at 2900 m on the Zugspitze fog is reported on average for 270 days yr^{-1}, compared with only 10 days yr^{-1} at 700 m (Critchfield, 1966). At higher elevations a greater proportion of the precipitation is received as snow. The snowline is encountered at about 5000 m in the tropics and decreases to 3000 m at 40 °N and 600 m at 70 °N. Permanent snow cover occurs at lower elevations in wet coastal mountains and for this reason the snowline is generally lower at equivalent latitudes in the southern hemisphere.

In these harsh environments small differences in microclimate are reflected in the nature of the plant cover. Although the small size of arctic and alpine plants does little to modify climatic conditions, their very stature enables them to take advantage of warmer temperatures close to the ground (Figure 9.3). The effect is most pronounced in dry microsites where less energy is used in evaporation (Weller and Holmgren, 1974). Heat is also dissipated by the persistent winds which blow across the tundra. In the arctic regions of the former USSR the average windspeed is 6–9 m s^{-1} and increases occasionally to 30–35 m s^{-1} (Lydolph, 1977). The strongest winds are usually recorded in winter: this affects the depth and distribution of the snow cover and the protection it affords the plants. Wind speeds are generally higher in alpine regions and commonly exceed 50 m s^{-1} on the open summits in winter. However, the wind patterns are

complex in rugged terrain and conditions can vary considerably over short distances. The combined effects of topography, wind and snow cover are reflected in the characteristic patterns which develop in alpine tundra vegetation (Figure 9.4).

9.3 SOILS AND CRYOGENIC PROCESSES

A variety of soils are found in tundra regions ranging from deep, moist peats to coarse rocky materials which support only mosses and lichens. Many soils in the arctic remain moist during the summer because drainage is impeded by permafrost (Figure 9.5). Only the uppermost layer – the **'active layer'** – thaws out each year. The depth of thawing depends not only on the amount of solar radiation received but also on factors that affect surface thermal properties, such as vegetation cover, snow depth, soil colour and moisture. In coarse materials thawing may occur to depths of 2–3 m, but in most regions the active layer is usually less than 1 m and in extreme sites it may be only a few centimetres thick (Ives, 1974a). Decomposition is slow in these cool, moist environments and over most of the arctic the typical soil profile consists of an organic surface layer of variable thickness beneath which is a strongly gleyed mineral horizon. In Canada these soils are classed as **Cryosols**:

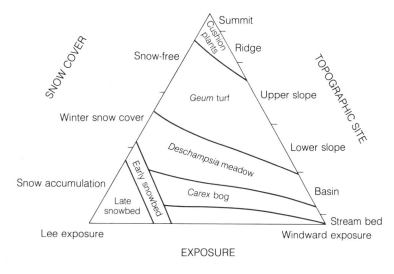

Figure 9.4 Alpine vegetation patterns associated with environmental gradients in the Rocky Mountains. (After Johnson and Billings, 1962.) (Reprinted with permission from P. L. Johnson and W. D. Billings, The alpine vegetation of the Beartooth Plateau in relation to cryopedogenic processes and patterns, *Ecological Monographs*, 1962, **32**, 129.)

they are subdivided according to the depth of the organic layer and degree of disturbance through frost processes which result in various patterned ground features. Sorted and non-sorted circles are common in fine-textured soils, especially in areas where the vegetation and organic layers are thin; they appear as patches of bare ground 1–2 m in diameter (Figure 9.6). Fissures 10–50 cm across, which form when the soil dries, may be

kept open and subsequently enlarged by ice-wedging. The result is a large-scale network of polygonal cracks, which are usually filled with stones (Figure 9.7). Frost heaving assists in the breakdown of organic materials and moves it to deeper parts of the profile, where a distinct band often forms at the base of the active layer. On steeper slopes gravity-induced movement of moist soil can produce tongue-like solifluction lobes 1–5 m high and many metres across. Downslope movement is typically only 5–15 cm yr^{-1} even on slopes of 15 °, but like frost heave features this has a significant effect on the distribution and density of the vegetation (Bird, 1974).

A mat of organic material 5–10 cm thick is present at the surface of most tundra soils. This dark, fibrous layer consists of partially decomposed plant material and is typically very acidic (Table 9.1). Tundra soils are mostly formed from glacial deposits and the underlying mineral

Figure 9.5 Permafrost exposed in a tundra bog. (O. W. Archibold.)

Figure 9.6 High-centred non-sorted circles with arctic avens (*Dryas integrifolia*) in the troughs. (S. C. Zoltai.)

Figure 9.7 Peat-filled tundra polygons in the Canadian arctic. (J. F. Basinger.)

Figure 9.8 Frost hummocks in moist sedge tundra in the Canadian arctic. (J. F. Basinger.)

horizons are of variable texture. Typically the soils are rather silty, but because of the underlying permafrost they remain moist all summer and the lower horizons are invariably mottled. Leaching in tundra soils occurs mainly in the early summer when the soils are supersaturated by snowmelt and water moves laterally through the surface layers. Base saturation is therefore comparatively low in the upper part of the soil and increases with depth (Tedrow, 1977). The bottom of the C horizon is indicated by the permafrost table, although peaty material may be incorporated in the frozen ground beneath. In low-lying areas, where soil drainage is further impeded, the organic layer increases in thickness and commonly exceeds 1 m in bogs. These organic deposits can act as nutrient reservoirs for water-soluble materials moved from the upper slopes by horizontal leaching (Everett *et al.*, 1981).

Well-drained soils develop in coarse materials on ridges and other topographic sites where excess water can move freely through the profile. Except in very gravelly materials, these soils support a complete cover of dwarf heaths, sedges and herbs, and a continuous mat of organic litter forms on the surface. Typically these soils are loose, friable sandy loams. In well-developed soils the profile is stained brown with humus to depths of 50–60 cm. Most of these soils are rather acid but considerable accumulations of carbonates are present in soils which are subject to strong evaporative moisture losses. Usually the carbonates are precipitated as stalactite-like encrustations on the underside of stones in the B and C horizons (Tedrow, 1977). Similar soils are found in the high arctic islands and northern Greenland, where precipitation is low and what little snow that does fall is mostly lost through sublimation or is blown away. In these **polar deserts** the soil is often covered with a thin layer of detritus on which a few low plants and lichens may be established. These coarse-textured, stony soils are so dry that there is very little frost activity. The vegetation cover is more continuous in relief depressions where seepage moisture collects: the surface of these turf-covered soils is usually hummocky because of increased frost action (Figure 9.8). Salt efflorescences may occur on the surface where the soils dry out later in the summer. Organic soils are occasionally present in the wettest sites. Bedrock pavements with little or no loose surficial materials occur throughout the region.

The soils in alpine areas reflect the diverse conditions under which they develop. Most of the parent material is derived by mechanical weathering and the soils are rather coarse-textured and stony. Permafrost occurs in many of the high mountain areas and the soils are typically cold and wet. Frost activity produces patterned ground features similar to those in the arctic (Johnson and Billings, 1962) and profile development is also affected by soil creep on steep slopes. The thinnest soils are the poorly developed Entisols which are characteristic of ridges and other sites where erosion is pronounced. Inceptisols occur on more gently sloping ground: a reasonably thick organic layer is present in these soils and alteration of the subsurface materials is evident in the weakly developed B horizons, in which some mottling may be apparent. Mollisols occur in some alpine meadows where sufficient organic debris has decomposed to form a deep, friable, dark-coloured surface horizon. Deposits of peat occur where decay of organic materials is limited by excessive soil moisture levels. Spodosols are reported locally in well-drained sites on acid parent materials (Retzer, 1974).

9.4 ADAPTIVE CHARACTERISTICS OF ARCTIC AND ALPINE PLANTS

Plant life in arctic–alpine regions is strongly conditioned by the severe environment. The dominant vascular plants are perennial herbs and dwarf shrubs which grow close

Table 9.1 Soil conditions in arctic and alpine tundras

	Depth (cm)	pH	OM (%)	Total N (%)	P	K	Ca	Mg	Source
					(ppm)				
Arctic and Antarctic sites									
King Christian Island 78 °N									
moist meadow	0–5	5.5	–	0.13	6	1			Bell and Bliss, 1978
	5–10	5.3	–	0.31	6	2			
	10–15	6.0	–	0.10	7	1			
dry meadow	0–5	6.1	–	0.18	5	2			
	5–10	6.0	–	0.05	5	1			
	10–20	6.2	–	0.07	6	1			
cushion plant communities	0–5	7.4	–	0.11	1	1			
	5–10	7.7	–	0.08	1	1			
	10–15	7.8	–	0.07	1	1			
lemming nest sites	0–5	5.9	–	0.22	32	3			
	5–10	6.3	–	0.14	20	1			
	10–15	6.5	–	0.15	18	1			
Devon Island (75 °N)									
wet sedge meadow peat soil	0–3	5.9	38.7	1.69	10	393	636	55	Walker and Peters, 1977
	3–6	5.2	40.3	1.97	10	210	796	170	
	6–9	5.0	36.4	2.01	10	108	426	10	
	9–12	4.9	22.6	1.80	5	61	306	< 1	
	12–15	5.0	29.1	1.85	10	78	380	51	
Signy Island (60 °S)									
lichen heath on schist	0–10	5.4	2.3	0.27	4	7	95	28	Collins et al., 1975
moss communities	0–8	4.9	30.9	1.83	27	17	75	35	
	8–25	4.9	9.0	0.61	14	6	25	13	
grass meadows	0–15	5.4	13.8	1.11	8	16	96	58	
penguin rookery	0–5	6.1	10.0	1.80	460	73	106	68	
Alpine sites									
Hardangervidda, Norway (elev. 1200–1320 m)									
Lichen heath	0–2	4.0	17.3	0.76	25	142	430	75	Hinneri et al., 1975
	2–8	4.3	1.2	0.06	4	12	16	5	
	8–18	4.8	1.2	0.05	10	5	11	2	
	18–36	5.2	0.4	0.01	10	3	11	2	
	36–41	5.3	0.4	0.01	7	4	12	2	
Dry meadow	0–9	5.3	42.1	2.60	93	544	7906	227	
	9–15	5.3	1.7	0.14	4	23	1026	23	
	15–45	5.5	0.8	0.07	4	10	426	15	
wet meadow	0–10	5.3	44.4	2.11	22	570	5386	159	
	12–52	5.2	47.5	2.73	4	38	4260	34	
snowbank communities	0–2	4.4	20.9	0.87	45	612	249	129	
	2–35	4.9	1.7	0.07	17	15	23	3	
Mount Washington, USA (elev. 1460–1580 m)									
sedge meadow	0–8	4.1	49.0	1.21	10	55	30		Bliss, 1963
	8–43	4.4	5.9	0.17	18	8	40		
dwarf shrub heath	0–18	4.3	49.0	1.35	7	95	15		
	18–23	4.0	13.0	0.32	5	15	30		
	23–60	4.5	7.6	0.16	21	8	30		
cushion plant communities	0–16	4.2	29.0	0.45	4	43	15		
	16–24	4.4	7.0	0.14	6	15	30		
snowbank communities	0–12	4.0	20.5	0.50	7	30	10		
	12–30	4.4	6.6	0.18	13	13	20		

(Sources: B. D. Walker and T. W. Peters, Soils of Truelove Lowland and Plateau, in *Truelove Lowland, Devon Island: A High Arctic Ecosystem*, ed. L. C. Bliss, published by University of Alberta Press, 1977, adapted with permission; L. C. Bliss, Alpine plant communities of the Presidential Range, New Hampshire, *Ecology*, 1963, **44**, 682; S. Hinneri, M. Sonneson and A. K. Veum, Soils of Fennoscandian IBP tundra ecosystems, in *Ecological Studies 16 – Fennoscandian Tundra Ecosystems, Part 1. Plants and Microorganisms*, ed. F. E. Wielgolaski, published by Springer-Verlag N.Y. Inc., 1975, adapted with permission; K. L. Bell and L. C. Bliss, Root growth in a polar semidesert environment, *Canadian Journal of Botany*, 1978, **56**, 2475, adapted with permission; N. J. Collins, J. H. Baker and P. J. Tilbrook, Signy Island, Maritime Antarctic, in *Structure and Function of Tundra Ecosystems* (eds T. Rosswall and O. W. Heal), published by the Swedish Natural Science Research Council, Stockholm, 1975.)

to the surface of the ground where they can benefit from warmer temperatures during the growing season (Figure 9.9). Most of the grasses and sedges form tussocks in which the dead leaves provide protection from desiccating winds and abrasive ice crystals. The dense mats and compact cushion form of other flowering plants

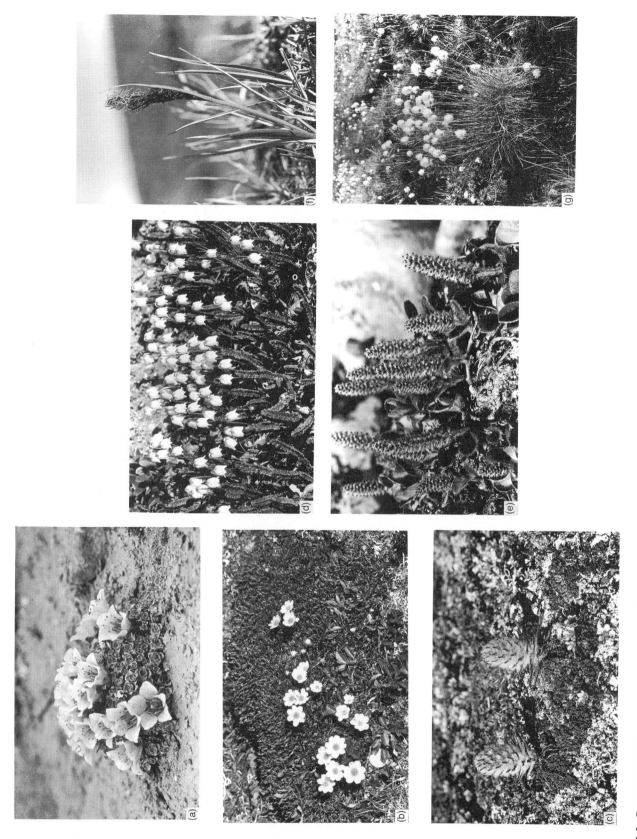

Figure 9.9 Representative tundra plants: (a) purple saxifrage (*Saxifrage oppositifolia*); (b) arctic avens (*Dryas intergrifolia*); (c) woolly lousewort (*Pedicularis lanata*); (d) arctic bell-heather (*Cassiope tetragona*); (e) arctic willow (*Salix arctica*); (f) sedge (*Carex stans*); (g) cotton-grass (*Eriophorum vaginatum*). ((a), (d)–(f) J. F. Basinger; (b), (c), (g) L. C. Bliss.)

Table 9.2 Plant life forms in tundra regions Ph = phanerophytes; Ch = chamaephytes; H = hemicryptophytes; C = cryptophytes; Th = therophytes

Locality	Latitude/ Altitude	Number of species of flowering plants	Ph	Ch	H (%)	C	Th	Sources
Arctic								
Franz Joseph Land	82 °N	25	0	32	60	8	0	Raunkiaer, 1934
Spitzbergen	79 °N	110	1	22	60	15	2	
Vardö, Norway	71 °N	134	2	15	61	13	9	
Iceland	65 °N	329	2	13	54	20	11	
Ellesmere Is.		107	0	24	65	11	0	
Baffin Is.		129	1	30	51	16	2	
Subantarctic Islands	46 °S	151	11	18	58	9	4	Allan, 1937
Alpine								
Swiss Alps	4500 m	8	0	63	37	0	0	Raunkiaer, 1934
Swiss Alps	3000 m	108	0	24	68	4	4	
Norway	2000 m	29	0	45	55	0	0	Dahl, 1975
Mexico	3600 m	38	5	13	76	5	0	Beaman and Andresen, 1966
New Zealand	2000 m	22	0	64	36	0	0	Allan, 1937
New Guinea	3300 m	39	13	23	64	0	0	Costin, 1967
Australia	1800 m	184	1	31	41	18	9	

similarly reduce wind movement amongst the leaves and raise the temperature of the trapped air (Savile, 1972). In some herbaceous species the leaves are arranged as a compressed rosette with a taller flowering stem developing later in the season. Woody species gain protection by growing in depressions which fill with snow early in the winter; any exposed shoots are killed. Low shrubby growth occurs only where there is adequate snow accumulation; in other sites only creeping prostrate forms survive. Few plants other than mosses and lichens can survive in extreme habitats. These plants go dormant in times of severe drought or cold and resume growth when conditions improve – some can endure continual snow cover for several years.

Arctic and alpine plants must not only survive the bitter cold of winter but must also tolerate short, cool summers when temperatures are only a few degrees above the physiological limit for plant growth. The low growth habits of the plants enables them to benefit from the warmer temperatures near the ground but considerable warmth is also gained in deeply pigmented species in which leaves and stems are turned purple by anthocyanins. The degree of pigmentation varies between plant populations and is generally deepest in more exposed sites. Some of these plants can absorb enough energy in the spring to begin growth beneath the snow. Elsewhere favourable microenvironments are created in greenhouse-like spaces which are roofed over with clear ice, or even beneath translucent rock chips (Llano, 1970). Tundra plants are subject to frost at any time during the year: they remain cold-resistant throughout the summer, yet can resume growth immediately conditions are favourable (Billings and Mooney, 1968). Initiation of growth occurs early in the spring but this requires that fully functional leaves are formed quickly. Evergreen shrubs are subject to winter damage and are mostly found in areas where they are protected by snow. Many herbaceous species are semi-evergreen: leaves produced in one summer remain green and functional until they are replaced the following year. In grasses and sedges the basal parts of the leaves remain green and quickly extend as food reserves are transferred to them. This advantage is lost in annual plants, which must develop from seed each year, and few annual species are represented in arctic and alpine floras (Table 9.2). They must set seeds each year in order to survive and this is very dependent on adequate warmth during the brief summer.

Vegetative reproduction becomes increasingly important as growing conditions deteriorate. Various tundra species spread by rhizomes and stolons. Mountain sorrel (*Oxyria digyna*) is capable of indefinite vegetative growth; arctic willow (*Salix arctica*) will continue to root from nodes for up to 100 years (Savile, 1972). However, many species have little vegetative capacity and are maintained by seedling establishment. Most arctic and alpine perennials have pre-formed flower buds, so that flowering activity in one year is largely controlled by growing conditions in the previous season (Billings, 1974b). Pollination is mainly by insects and their activity is strongly regulated by temperature conditions. The shallow, bowl-shaped flowers in species such as arctic avens (*Dryas integrifolia*) and the arctic poppy (*Papaver radicatum*) act as parabolic reflectors: they are also heliotropic and turn towards the sun (Figure 9.10). Flower temperatures can be as much as 10 °C warmer than the surrounding air and so provide a further incentive for pollinating insects (Kevan, 1975). Subsequent seed set is also subject to the vagaries of the climate: it becomes progressively less reliable with latitude and altitude, although heliotropic flower movements also favour seed maturation by increasing the temperature of the pistils (Kjellberg *et al.*, 1982). A warm microenvironment is

Figure 9.10 Heliotropic bowl-shaped flowers in (a) arctic avens (*Dryas integrifolia*) and (b) arctic poppy (*Papaver radicatum*). (J. F. Basinger.)

Figure 9.11 Alpine bistort (*Polygonum viviparum*) with viviparous bulbils developing on lower part of the rachis. (J. F. Basinger.)

similarly created by the densely pubescent inflorescences on species of *Puya* in the Andes (Miller, 1986) and by the down that covers the developing catkins in arctic willow (Krog, 1955). Some species in these rigorous environments are apomictic and form seeds without pollination. Others, such as alpine bistort (*Polygonum viviparum*) and some grasses, are viviparous and produce bulbils which develop into young plants while they are still attached to the inflorescence (Figure 9.11). These subsequently fall to the ground, where they establish a new plant.

Seeds generally ripen shortly before the onset of winter. Most seeds are not dormant but germination is prevented by low temperatures. Optimum temperatures for seed germination in most arctic and alpine species range from 20 to 30 °C and emergence usually occurs shortly after snowmelt. The seedlings have only a brief period to become established and their success is often limited by low temperatures and drought: frost heaving during winter also results in high seedling mortality (Figure 9.12). Growth above ground can be very slow and it may be several years before the plants are big enough to flower. Dormancy is induced by shortening day lengths and cooler temperatures at the end of the growing season, or by drought. Growth in sedges and rushes is usually periodic and development is halted even when there is still an opportunity for growth. Other species are aperiodic and continue to grow until conditions are unfavourable. An extreme example is the low northern

Figure 9.12 Frost hummock and frost-heaved stones in moist arctic meadow. (O. W. Archibold.)

Figure 9.13 Progressive decrease in density of the tundra plant cover occurs with latitude: (a) Low arctic cotton-grass–dwarf shrub heath with *Salix pulchra, S. glauca* and *Betula nana* in northern Alaska; (b) polar semi-desert comprised mainly of *Saxifraga oppositifolia, Luzula confusa* and mosses, Melville Island; (c) polar desert supporting mainly *S. oppositifolia* and cryptogams. (L. C. Bliss.)

rock-cress (*Braya humilis*) which can be in any stage of flowering or fruiting when growth ceases in the autumn; development continues the following year (Sørensen, 1941).

9.5 REGIONAL FLORISTICS

9.5.1 Arctic tundra

Arctic tundra is distributed over a wide range of latitude and the decrease in summer warmth northwards is marked by progressive changes in the structure and diversity of the plant cover (Figure 9.13). In North America three main divisions are recognized (Polunin, 1960). The most southerly is the **low Arctic**, in which a continuous plant cover is present in all but the driest and most exposed sites. Northwards this merges with the tundras of the **middle Arctic**. Fewer species are represented in this region, and the shrubs are smaller and more limited in their distribution. The most northerly zone is the **high Arctic**, where vascular plants are restricted to only the most favourable sites.

(a) North America

A thin cover of grasses and sedges composed of species such as alpine meadow grass (*Poa arctica*) and rigid sedge (*Carex bigelowii*) is characteristic of much of the rolling terrain in the low Arctic tundra of North America. Small shrubs including bog whortleberry (*Vaccinium uglinosum*), crowberry (*Empetrum nigrum*) and narrow-leaf Labrador tea (*Ledum palustre*) are also present. The herbs and shrubs typically grow to heights of 15–30 cm, with an understory of mosses and fruticose lichens. The common mosses include *Hylocomium splendens* and *Polytrichum juniperinum*; lichens are represented by species such as *Cetraria nivalis, Cladonia gracilis* and *Cladina mitis*. The mosses and lichens become more conspicuous in drier upland sites, where they may be accompanied by a scattered cover of xerophilous evergreen shrubs such as white arctic bell-heather (*Cassiope tetragona*) and alpine bearberry (*Arctostaphylos alpina*). In exposed sites, where little snow accumulates, mountain avens (*Dryas octopetala*)

Figure 9.14 Axel Heiberg Island in the Canadian high arctic. (J. F. Basinger.)

and other low-growing cushion plants, such as moss campion (*Silene acaulis*) and purple saxifrage (*Saxifraga oppositifolia*), may accompany the mosses and lichens. The shrub cover is best developed in sheltered sites in the southern tundra where the active layer is unusually deep (Eyre, 1968). Here dwarf birch (*Betula nana*) and willows (*Salix* spp.) grow to heights of 40–60 cm and form open scrub communities with various herbs and smaller shrubs. Locally willows, birches and alder (*Alnus crispa*) grow to heights of 2–5 m in riparian sites where a deep winter snowpack accumulates (Bliss, 1981). Areas of wetland are common in the coastal plains of Alaska and north-western Canada. Sedges, such as water sedge (*Carex aquatilis*) and cotton grass (*Eriophorum angustifolium*), and several species of grass are adapted to these conditions. *Sphagnum* and other mosses are abundant in these wet meadows and aquatic plants, such as variegated horsetail (*Equisetum variegatum*) and mare's-tail (*Hippuris vulgaris*), are present where there is standing water.

Similar communities are found in the middle Arctic, although fewer species are represented and growth and production is reduced. In the high Arctic the vegetation is much sparser. The densest cover is restricted to marshy areas where sedges, cotton-grasses and grasses grow mainly in association with mosses. *Sphagnum* is absent from these sites and is replaced by species such as *Drepanocladus revolvens*, *Distichium capillaceum* and *Mnium riparium*. Woody plants are usually represented only by prostrate willows. Regionally these communities cover 5–40% of the high Arctic. Heath communities are less extensive: they occur sporadically on warmer slopes where dwarf shrubs such as *Empetrum nigrum* and *Vaccinium uliginosum* grow with *Dryas integrifolia*, *Salix arctica* and various sedges. *Cassiope tetragona* is present where snow provides protection in winter. Lichens such as *Cetraria nivalis* are common on the drier uplands, where they are accompanied by *Dryas integrifolia*, various cushion

plants and sedges. *Salix arctica* occasionally grows in the moister depressions associated with patterned ground features (Bliss, 1981). Barren polar deserts cover about 25% of the Canadian high Arctic (Figure 9.14). The vegetation is very sparse in these regions: some mosses are present where moisture is available from late melting snow patches, together with lichens and occasional hardy perennials such as the arctic poppy (*Papaver radicatum*) and snowgrass (*Phippsia algida*).

(b) Asia

Two major latitudinal regions are distinguished in the Soviet Arctic. The tundra zone stretches from the limits of the taiga to the coast and includes Wrangel Island, the New Siberian Island group and most of Novaya Zemlya. North of this is the polar desert zone (Andreev and Aleksandrova, 1981). These regions are floristically and physiognomically similar to those of North America. Shrubby stands of birch and willow are characteristic of the well-drained uplands in the southernmost parts of the tundra. Several boreal species such as bilberry (*Vaccinium myrtillus*) add to the floristic diversity of this region, and small patches of prostrate larch (*Larix gmelinii*) and dwarf stone pine (*Pinus pumila*) may be present in river valleys and other sheltered sites. Mosses, sedges and cotton-grasses grow in poorly drained areas. The shrub cover becomes progressively shorter and more open further north. It is often dominated by low-growing willows, but dwarf shrubs such as *Cassiope tetragona* and species of *Dryas* become increasingly important. The polar desert zone occurs at high latitudes where mean summer temperatures remain below 2 °C. Few species can survive in this region, other than mosses and lichens, and much of the land is barren. The few vascular plants that grow here are mostly found in the polygonal desiccation cracks which develop in the finer substrates that normally only support a lichen crust (Aleksandrova, 1988). Mosses are more abundant in wetter habitats.

(c) Greenland

About 400 species of flowering plants are found in Greenland. Many of these are widely distributed in the Canadian Arctic but some were brought from Europe by early Norse settlers (Good, 1964). There are few endemic species on the island (Böcher, 1963). Many of the species in the moist maritime parts of southern Greenland are of boreal origin but these are progressively replaced by arctic species at higher latitudes. Dwarf shrubs form heaths over much of the available land. In drier sites the cover is mainly composed of crowberry (*Empetrum nigrum* subsp. *hermaphroditum*) and arctic blueberry (*Vaccinium uliginosum* subsp. *microphyllum*). Labrador tea (*Ledum groenlandicum*) is often present in boggy areas and along river channels, where it grows beneath taller willows. Copses of alder (*Alnus crispa*) and willow may be present in fjords and other warm sites, beneath which is a varied ground flora of herbs, grasses, sedges and mosses. Further north the heaths in moister sites are mainly composed of *Ledum palustre* but change to dwarf birch and arctic blueberry as conditions become drier. Arctic marsh willow (*Salix arctophila*) and cranberry (*Oxycoccus quadripetalus microphyllous*) are present in boggy sites. At higher latitudes the heaths are dominated by white arctic bell heather with artic avens: in the more extreme wind-swept places this cover is generally replaced by small patches of xerophilous steppes comprised of rock sedge (*Carex rupestris*), northern wood-rush (*Luzula confusa*) and other graminoids. Alkali grass (*Puccinellia deschampsoides*) is one of the few species that grows where the ground is encrusted with salt.

9.5.2 Antarctic and sub-antarctic tundra

The distribution of antarctic tundra is limited by the availability of ice-free areas in the south polar region. Terrestrial plant habitats in Antarctica are restricted to rocky coastlines, glacial moraines and rugged areas which protrude above the ice sheet. Additional sites are provided by lakes, summer melt-water channels and fumaroles which are locally warmed by volcanic activity (Lamb, 1970). Only two species of flowering plants are native to Antarctica: the grass *Deschampsia antarctica* and a cushion plant (*Colombanthus quitensis*) (Figure 9.15). Both are limited to the Antarctic Peninsula, where they

Figure 9.15 *Colombanthus quitensis* is one of only two flowering plant species that are native to Antarctica. (D. W. H. Walton, British Antarctic Survey.)

grow as far south as 68 °S: they are also found on some of the sub-antarctic islands. The lichens are the most successful and widely distributed plant group in Antarctica (Figure 9.16): they are represented by about 400 species and some have been found on exposed rocks at latitude 86 °S. The most common lichens in rocky sites are species of *Usnea* but others, such as species of *Verrucaria*, grow along seepage channels and around

(a)   (b)

Figure 9.16 (a) Typical Antarctic lichen community; (b) convoluted form of the moss *Bryum argenteum* in coastal regions of North Victoria Land, Ross Dependency. (D. R. Given.)

Figure 9.17 (a) General view of Campbell Island showing sub-antarctic formations of *Dracophyllum* shrubland, *Poa* and *Chionochloa* grassland; (b) the megaherb *Pleurophyllum speciosum* is approximately 1 m in diameter; (c) megaherbfield with *Pleurophyllum speciosum* (flowering), *Anisotome* and *Stilbocarpa*. (D. R. Given.)

melt-water pools, or in coastal habitats where they are exposed to spray and inundation by the tides. About 70 species of moss are reported from Antarctica. They are mostly found in wetter sites, and in some areas on the peninsula deep layers of peat have accumulated beneath them.

A richer flora is present on the cold, wind-swept islands that surround Antarctica (Figure 9.17), but even on the Kerguelen Islands at 49 °S there are only 30 species of vascular plants. Six of these are endemic to this island

group: the Kerguelen cabbage (*Pringlea antiscorbutica*) is the most distinctive and is the only member of this genus. The most common species is *Azorella selago*, which forms small tussocks where it is sheltered from the wind (Polunin, 1960). Meadow-like communities of *Agrostis antarctica, Festuca kerguelensis* and mosses grow in damp depressions: lichens cover the rocky outcrops; and *Acaena adscendens* forms a low shrubby cover in less exposed places. Thirty-five species of vascular plants are reported from Macquarie Island (55 °S). Here too tussocky grasslands form the characteristic vegetation cover at lower elevations. The principal species is *Poa foliosa*, which grows to heights of 1–1.5 m, interspersed with the large-leaved herb *Stilbocarpa polaris*. *Pleurophyllum hookeri* is a distinctive species in sheltered sites at higher elevations: the large rosettes of this silver-leaved herb are conspicuous amongst the grasses, sedges and mosses. In exposed areas the cover is reduced to cushions of *Azorella selago* and mosses such as *Rhacomitrium crispulum*. Similar grasslands of tall, tussocky *Poa flabellata* are found in coastal areas of South Georgia (54 °S) with shrubby *Acaena adscendens* growing in wetter depressions and along stream channels. Much of the drier inland areas support a mixed grassy cover of *Festuca erecta*, *Deschampsia antarctica* and *Phleum alpinum*, which grow in a carpet of mosses and lichens.

9.5.3 The alpine flora

(a) North America

The geographic distribution of the high mountain ranges is reflected in the great diversity of habitats and rich endemic floras of the treeless alpine regions of the world. In North America alpine tundra occurs discontinuously throughout the western mountain ranges from Alaska to Mexico, and less extensively in the northern Appalachians and the mountains of eastern Canada. The dominant species are perennial herbs and dwarf shrubs, many of which are found in the Arctic, and the plant communities are similarly related to local environmental gradients. Meadows and bogs are present in wet, sheltered depressions, heaths cover the drier sites, and lichens and cushion plants are found on exposed ridges.

In the mountains of Alaska there are about 150 species of vascular plants. Over much of this region mountain avens forms a dry meadow community with several grasses (*Arctagrostis latifolia*, *Calamagrostis purpurea* and *Hierochloe alpina*), dryland sedges and rushes (*Carex obtusata*, *Kobresia myosuroides*, *Luzula confusa*), low shrubs (*Empetrum nigrum*, *Rhododendron lapponicum*, *Salix rotundifolia*) and many forbs. Cotton-grasses (*Eriophorum angustifolium* and *E. vaginatum*) and sedges (*Carex aquatilis*, *C. bigelowii*) form tussocky meadows in wetter sites. Several species of saxifrage grow in crevices in the rocky outcrops where they are accompanied by various grasses (*Festuca brachyphylla*, *Hierochloe alpina*), small shrubs and heaths (*Vaccinium uglinosum*, *Arctostaphylos alpina*), ferns (*Cystopteris fragilis*, *Dryopteris fragrans*) and forbs (*Oxytropis mertensiana*, *Minuarta macrocarpa*, *Phlox sibirica*). Arctic bell-heather is common where snow cover provides winter protection and moisture later in the year (Spetzman, 1959).

The alpine flora in the coastal mountains of Washington and British Columbia contains approximately 250 species of vascular plants, although some of these are associated with subalpine krummholtz vegetation (Douglas and Bliss, 1977). The complex topography of this region affects snow depth, the time of melting and soil moisture during the growing season, and this is reflected in the distributional mosaic of plant communities. Snowbank communities are dominated by *Carex nigricans*. Dwarf shrub communities of *Cassiope mertensiana*, *Phyllodoce empetriformis* and *Arctostaphylos uva ursi* are characteristic of better drained sites where the snow melts early, with *Empetrum nigrum* and willows (*Salix nivalis*) present in more exposed sites and northerly exposures. Dense herbaceous covers are found in warm, moist sites and grasses and sedges dominate rich meadow communities on well-drained slopes. Small herbaceous plants such as *Phlox diffusa*, *Oxytropis campestris* and *Solidago multiradiata* are characteristic species of the patchy vegetation cover on the exposed ridgetops.

A total of 210 species are represented in the alpine flora of the central Rocky Mountains (Johnson and Billings, 1962): about half of these also occur in the arctic. Various plant communities are represented. Each is characteristic of a particular habitat, with intermediate cover types developing where they intergrade along environmental gradients. On well-drained slopes at higher elevations, Ross's avens (*Geum rossii*) forms a dense turfy cover with other herbs; sedges and cushion plants become increasingly important in drier, more exposed sites. *Deschampsia caespitosa* forms meadow communities at lower elevations where winter snow cover persists until mid-July, and *Carex scopulorum* is widespread in boggy areas. Dense thickets of flat-leaved willow (*Salix planifolia*) occur along drainage channels in the low alpine zone.

The alpine communities in the southern Rocky Mountains are best developed at elevations of 3500–4100 m (Langenheim, 1962). Dense herbaceous communities dominated by species of groundsel (*Senecio crassulus*), lovage (*Ligusticum porteri*) and lupin (*Lupinus parviflorus*) grow with grasses and sedges at lower elevations: grasslands dominated by *Festuca thurberi* increase in importance on the drier southern and western exposures and steeper sites. High-altitude willow communities are characteristic of sites which are supplied by melting snow during summer. Grasses and sedges, such as *Poa alpina*, *Kobresia bellardii* and *Carex hepburnii*, form meadows at higher elevations: rosette and cushion plants, such as the moss campion and mountain avens, are also present in these communities and in the relatively barren fellfields.

The flora of the northern Appalachians is similar to that of the alpine communities of Scandinavia and central Europe and the Arctic. Relatively few species are shared with the mountains in western North America (Bliss, 1963). Sedge meadows composed mainly of *Carex bigelowii* cover the higher slopes at elevations of 1750–1900 m, integrading lower down with dwarf heath communities composed of *Vaccinium vitis-idaea* var. *minus*, woody *Potentilla tridentata* and three-leaved rush (*Juncus trifidus*). A shrubby cover of *Vaccinium uliginosum* and *Ledum groendlandicum* occurs in the lower alpine zone and in sheltered sites at higher elevations, and where they are protected by winter snowfall. The small evergreen shrub *Diapensia lapponica* is common in wind-swept sites where it grows in association with *Juncus trifidus*, cushion plants, mats of bearberry willow (*Salix uva-ursi*) and other prostrate shrubs. Comparatively rich shrub–herb communities are present where snow patches melt relatively late, including several understory species from the adjacent spruce–fir forests. Streamside habits are equally diverse and several rare species are restricted to these communities.

(b) Eurasia

The distribution of low shrub communities, meadows, heaths and snowbank vegetation is repeated in the mountains of Europe. Crowberry (*Empetrum nigrum* subsp. *hermaphroditum*), which grows up to 50 cm in height, is the characteristic dominant in low heath communities at elevations of 1600–2400 m. It is usually associated with *Vaccinium uliginosum*, *V. myrtillus* and cowberry (*V. vitis idaea*). Alpenrose (*Rhododendron ferrugineum*) is conspicuous at lower elevations on acid soils and trailing azalea (*Loiseleuria procumbens*) becomes important on wind-swept ridges (Polunin and Walters, 1985). The spike-heath (*Bruckenthalia spiculifolia*) is locally dominant in the mountains of south-eastern Europe. In the drier Pyrenees and the Spanish Sierra Nevada spiny shrubs, such as hedgehog broom (*Erinacea anthyllis*), form low cushion heaths, with similar communities occurring in mountain ranges throughout the Mediterranean. Alpine grasslands are present in wetter areas and have long been used for summer grazing. On acid soils they are mainly composed of mat-grass (*Nardus stricta*) and sedges, although numerous flowering plants such as alpine avens (*Geum montanum*), golden cinquefoil (*Potentilla aurea*) and trumpet gentian (*Gentiana acaulis*) add to their variety. Alpine sedge (*Carex curvula*) becomes more common at higher elevations in the Alps where it forms communities with species such as alpine meadow grass (*Poa alpina*) and *Polygonum viviparum* up to 3000 m. *Kobresia mysuroides* is the dominant grass where calcareous soils are present in exposed high elevation sites to 2800 m, but on warmer south-facing slopes it is replaced by moor-grass (*Sesleria albicans*), patches of mountain avens and edelweiss (*Leontopodium alpinum*). Communities of sedge (*Carex nigra*, *Eriophorum angustifolium*) and thread rush (*Juncus filiformis*) are present in boggy hollows but less acid sites and seepage areas are distinguished by a greater number of flowering plants and mosses. Mosses (*Polytrichum sexangulare*) or rushes (*Luzula alpinopilosa*) are common in areas where late-melting snowbanks limit the growing season to only a few weeks, and willows (*Salix herbacea*, *S. reticulata* and *S. retusa*) grow where the snow lies for shorter periods. Cushion plants such as musky saxifrage (*Saxifraga exerata*), Swiss rock jasmine (*Androsace helvetica*) and other small herbs grow on ledges and in crevices on exposed cliffs and scree: a large number of endemic species are restricted to these extreme sites in various mountain ranges.

The alpine flora of Scandinavia is more like that of the Arctic. Mountain birch (*Betula pubescens* subsp. *tortuosa*) forms the tree-line at 700–1000 m in southern Sweden, and immediately above this there is often a scrubby cover of willows (*Salix glauca*, *S. lapponum* and *S. lanata*), or dwarf shrub heaths dominated by *Vaccinium myrtillus*. At higher elevations it is replaced by *V. vitis-idaea* and mats of *Cassiope hypnoides* and willow (*Salix herbacea*), which form grassy heaths with species such as viviparous fescue (*Festuca vivipara*), *Carex bigelowii*, *Luzula confusa* and *Juncus trifidus*. Above 1600 m the vascular plant cover is no longer continuous, and lichens and mosses grow with *Oxyria digyna*, *Silene acaulis* and alpine saxifrage (*Saxifraga nivalis*) on the cliffs and boulder fields near the mountain summits (Kilander, 1965). Further north the distribution of the vegetation is closely related to the depth and duration of the snow cover. The dwarf shrub heaths of *Empetrum nigrum*, bilberry (*V. myrtillus*) and *Dryas octopetala* are found at lower elevations in sites where little snow accumulates. *Cassiope tetragona* becomes more important at higher elevations and grass heaths dominated by *Juncus trifidus* occur extensively. In areas where snow remains for much of the year the heaths are replaced by communities of grasses dominated by *Deschampsia flexuosa*, *Anthoxanthum odoratum* and *Nardus stricta*. Only willows (*Salix herbacea*, *S. polaris*) and mosses are common where the snow melts very late in the year (Gjaerevoll and Bringer, 1965).

The flora of the Urals contains several circum-arctic species, as well as common taiga shrubs such as bilberry and cowberry. Boreal species are less common in the mountain ranges of Siberia where many of the plants are specific to north-eastern Asia (Kuvaev, 1965). In the Verkhoyansk Mountains scattered thickets of dwarf Siberian pine (*Pinus pumila*) mark the transition to the tundra zone at elevations of 900–1300 m, and dwarf birch (*Betula exilis*) becomes increasingly widespread on the drier slopes. *Dryas octopetala* is the dominant species on coarse, stony soils in the upper tundra zone. Berberis-leaved willow (*Salix berberifolia*) and arctic meadow grass (*Poa arctica*) are common in wetter sites where they grow with a variety of herbs including glacier butterbur (*Petasites glacialis*) and alpine bistort (*Polygonum vivaprium*). The vegetation cover is very fragmented on the mountain summits and is largely comprised of lichens. Floristic diversity is much higher in the small patches of alpine tundra which occur discontinuously in the mountains of the southern USSR (former). The rich herbaceous cover shares many species with the Alps and adjacent mountains in south-eastern Europe. The highland vegetation of the Carpathians and Caucasus is mainly represented by alpine meadows comprised of grasses, a great variety of flower-

Figure 9.18 (a) Alpine *Festuca* grasslands with *Ranunculus, Rumex, Pedicularis* and other herbs at 2900 m; (b) *Rhodendron caucasicum* scrub at 3000 m in the Caucasus Mountains. (L. C. Bliss.)

ing plants and dwarf shrubs including species of *Vaccinium, Salix* and *Rhododendron* (Fodor, 1965) (Figure 9.18). On the highest peaks the cover consists of *Juncus trifidus* and sheep's fescue (*Festuca ovina*); in boggy areas *Eriophorum vaginatum, Empetrum nigrum* and swamp cranberry (*Oxycoccus palustris*) grow with *Sphagnum* moss.

The Tien Shan and other mountains of central Asia are distinguished by their diverse and highly endemic floras. For example, a total of 628 species is reported at 3600–4500 m in the Pamir Range (Ikonnikov, 1965). Steppe grasslands and thickets of shrubs cover much of the subalpine zone in this region (Kotov, 1965) (Figure 9.19). Community composition is closely related to soil conditions and exposure. Dense meadows comprised of *Ptiliagrostis purpurea* and other grasses typically cover the south-facing slopes at elevations of 2800–3100 m. The high mountain fescue steppes dominated by *Festuca kryloviana* occur above this to elevations of 3600 m. The north-facing slopes are cooler and moister and several species of *Poa* are present in the tall grass cover. Species of *Kobresia* form a dense turf on the higher slopes up to 4800 m (Figure 9.20).

Figure 9.19 Alpine shrub zone comprised of *Salix anymegensis, Spirea alpina* and *Potentilla fruticosa* at 3800 m in Tibet. (L. C. Bliss.)

Figure 9.20 Alpine tundra, mostly of *Kobresia pygmea* and sedge (*Carex*), at 4760 m in Tibet. (L. C. Bliss.)

Figure 9.21 Alpine tundra on Mt Kosciusko, New South Wales, Australia. (O. W. Archibold.)

(c) New Zealand

Alpine tundra occurs extensively in the South Island of New Zealand and is locally represented by grasslands, herbfields and heath communities above 1800 m in the Snowy Mountains of Australia (Figure 9.21) and above 1200 m in Tasmania (Costin, 1967). In New Zealand the beech forests are replaced rather abruptly by a narrow zone of alpine scrub and extensive tussock grasslands at elevations between 900 and 1500 m, depending on latitude and distance from the sea (Figure 9.22). The scrub zone is composed of tall tussock grasses, large herbs and shrubs: sharp-leaved herbs such as Spanish bayonet (*Acilphylla horrida*) are common in these communities. The alpine grasslands are dominated by species of snow tussock (*Chionochloa* spp.), and various communities can be recognized on the basis of soil conditions, drainage and elevation (Rose *et al.*, 1988) and snowfall regimes (Burrows, 1977). Many flowering plants grow amongst the grasses in this low alpine zone including mountain buttercup (*Ranunculus lyallii*), mountain foxglove (*Ourisia macrophylla*), gentian (*Gentiana montana*) and forget-me-not (*Myosotis australis*). Shrubs are uncommon in the

Figure 9.22 (a) Characteristic subalpine snow tussock–shrub communities in New Zealand; (b) the broad sharp leaves of *Aciphylla scott-thomsonii* and related species found in the alpine grasslands of New Zealand are probably an adpatation to drought rather than protection against animals; (c) mountain buttercup (*Ranunculus lyalli*) in snow tussock grassland, South Island, New Zealand. (A. F. Mark.)

tussock grasslands and are mainly represented by pine-apple scrub (*Dracophyllum menziesii*), snowberry (*Gaultheria depressa*) and various hebes (*Hebe* spp.). The high alpine zone lies beyond the upper limit of the tussock grasslands. Rich herbfield communities are present on deeper soils with little frost activity. The mountain daisy (*Celmisia viscosa*) and blue tussock grass (*Poa colensoi*) are characteristic species in hollows and where rocks provide shelter from the wind (Mark and Bliss, 1970).

On the broad, exposed summits the vegetation cover is reduced to a few species of cushion plants such as *Raoulia youngii*, prostrate shrubs (*Dracophyllum muscoides*), dwarf grasses (*Poa pygmaea*) and lichens. Wind erosion produces a crescentic outline to the vegetation cushions, and these slowly move downwind because regeneration occurs only on the leeside of each clump. Where snow accumulates during the winter the nature of the plant cover varies according to the depth and duration of the snowpack. Few species, other than the moss *Andreaea acutifolia*, survive where the ground is almost permanently covered by snow. Fellfield communities are found in rocky areas where the snowcover is less persistent. Prominent species in the drier fellfields are the 'vegetable sheep' *Raoualia eximia* and *Haastia pulvinaris*: these peculiarly hairy compositae form woolly hummocks up to 2 m across. Small shrubs such as *Hebe buchmanii* and *H. epacridea* are also present together with the speargrass *Aciphylla dobsonii*. The flora is more varied in the western ranges where precipitation is heavier: amongst this group is the unique endemic *Hectorella caespitosa*, the only species in the genus.

(d) Tropical alpine floras

The alpine flora of tropical regions is very different from those at higher latitudes. Tree-line occurs at 3500–4000 m in most of these regions with the upper limit of the alpine zone at about 4800 m: at higher elevations the land is permanently covered with snow (Figure 9.23). The tropical alpine habitat occurs primarily in East Africa and the Andes, and less extensively in the East Indies and Hawaii. These regions are subject to pronounced diurnal temperature fluctuations because of strong daytime insolation and heat loss at night. Frost occurs on most nights, with temperatures often dropping below −5 °C and then rising to 10–15 °C during the day (Smith and Young, 1987). Parallel adaptation to this peculiar environment has resulted in closely convergent life forms. Tropical alpine vegetation consists mainly of tussock grasses and small-leaved shrubs, but the most distinctive species are the giant rosette plants which grow abundantly on the high volcanoes of East Africa and in the Andes.

In East Africa the giant rosette plants are represented by tree-like species of *Senecio* and *Lobelia*. They have thick, unbranched or sparsely branched woody and herbaceous stems, each of which bears a huge rosette of leaves

Figure 9.23 Snow-covered Mt Kilimanjaro at latitude 3 °S rises to 5900 m in the equatorial region of Tanzania. (B. R. Neal.)

and an immense inflorescence (Hedberg, 1964). They have adapted to the peculiar Afro-alpine climate by closing their leaves at night to insulate the apical tissues against frost: shoot temperatures in *Senecio keniodendron*, for example, are maintained at about 2 °C even though air temperatures drop to a few degrees below freezing. In species such as *Lobelia keniensis* additional protection is afforded by water which collects around the base of the inner leaves. Their stems are insulated by dead leaves or thick bark. These plants mainly grow on deep, moist soils in the low alpine zone where they form dense woodlands at elevations of 3800–4300 m. The taller species normally grow to heights of 4–6 m and beneath this canopy there is usually a ground cover of grasses and herbs and a thick carpet of moss. In rocky areas these communities are replaced by scrubby species of *Helichrysum*. However, over much of the drier alpine region the cover is mainly tall, dense tussock grassland comprised of species of *Festuca*, *Agrostis* and *Pentaschistis*. The accumulation of dead stems and leaves in these bunch grasses protects the new growth from extreme temperatures and reduces evaporation of soil moisture. The tussocks also provide insulated nest sites for some birds (Coe, 1969). Disruption of seedlings by frost heaving leaves areas of bare soil in the high alpine zone. Although frost occurs on most nights it does not usually penetrate deeper than 3–5 cm and only the finer substrates are subject to cryoturbation. Spherical balls of moss growing around a central core of soil are found

Figure 9.25 *Draba chionphylla* at 4600 m in the Venezuelan Andes provides a sparse cover near the limits of plant growth. (L. C. Bliss.)

Figure 9.24 (a) Dead leaves provide insulation for *Espeletia timotensis* growing at 4150 m in the Venezuelan Andes; (b) *E. grandifolia* in the Colombian Andes. ((a) L. C. Bliss; (b) G. T. Prance.)

where there is active solifluction and, being unattached, they are rolled by surface movements.

The northern Andes are cool, wet and misty, and on the high tablelands the montane forest cover is replaced by the grasses and evergreen herbs and shrubs of the **páramo**. The páramo extends from western Venezuela to northern Peru at elevations of 3000–4200 m. Further south in Chile it merges with the drier puna. Unlike other alpine communities much of the páramo has been greatly altered by clearing and burning to provide land for cultivation and pasturage (Cuatrecasas, 1968). The lower limit of the alpine zone is often marked by a transition to scrub, and in parts of Peru heavy grazing by llamas and alpacas has lowered the tree-line by nearly 850 m (Crawford *et al.*, 1970). In the grassy páramo the cover is mainly comprised of species of *Calamagrostis*, *Festuca* and *Stipa* but numerous herbs and small shrubs are intermingled with the grasses. Tree-like species of *Espeletia* and *Puya* are also conspicuous with their large rosettes of woolly, evergreen leaves. The taller species can grow

to heights of 14 m and their trunks are normally encased in dead leaves which provide insulation against night frosts (Smith, 1979) (Figure 9.24): others develop a thick woody caudex below ground (Cuatrecasas, 1979). Increased leaf pubescence and changes in the geometry of the rosettes in species such as *Espeletia schultzii* also compensate for lower temperatures and the greater risk of frost at higher elevations (Meinzer *et al.*, 1985). The vegetation cover becomes more open at higher elevations and the bunch grasses are largely replaced by shrubs and various herbs and cushion plants, the most characteristic species being the white woolly *Senecios*. The discontinuous nature of this zone has resulted in a large number of endemic species within this distinctive flora (Cuatrecasas, 1968). Above 4500 m patches of bare ground alternate with a sparse cover of xerophytic grasses, small shrubs, mosses and lichens (Harling, 1979) (Figure 9.25), with permanent snowfields eliminating suitable sites at elevations of 4700–5000 m.

9.6 ORIGIN OF THE TUNDRA FLORAS

Forests grew at high latitudes across North America and Eurasia until the beginning of the Pleistocene epoch. Fossil remains of dawn redwood (*Metasequoia*), swamp cypress (*Glyptostrobus*), maidenhair (*Ginkgo*) and various broad-leaf genera are common throughout the Canadian Arctic (Basinger, 1991) (Figure 9.26) and similar forests are reported in deposits in eastern Siberia (Hills *et al.*, 1974). Tundra species are absent from the early vegetation record, although some may have been present in more northerly mountain areas. Pollen data indicate that much of Alaska was covered by coniferous forests until the late Pliocene (3 million years BP) when they were replaced by herbaceous and shrubby vegetation dominated by sedges, grasses and willows (Wolfe, 1972). In

Figure 9.26 (a) Tertiary (approximately 40 million years BP) fossil forest on Axel Heiberg Island in the Canadian arctic with exposed stumps of dawn redwood (*Metasequoia occidentalis*) and fossilized litter layer; (b) leafy twigs and (c) cones of dawn redwood recovered from the fossilized litter; (d) *M. glyptostroboides*, a close relative of the fossil trees, growing in central China. ((a)–(c) J. F. Basinger; (d) K. Saito.) (Reproduced with permission from J. F. Basinger, *The Tertiary forests of the Buchanan Lake formation (early Tertiary), Axel Heiberg Island, Canadian Arctic archipelago: preliminary floristics and paleoclimate*, Geological Survey of Canada Bulletin 403, 1991.)

these regions the arctic environment developed comparatively rapidly at the end of the Pliocene and has probably existed for little more than 3 million years (Savile, 1972). At this time a circumpolar arctic flora comprised of some 1500 species may have existed at high latitudes but its diversity and distribution was greatly affected by subsequent glacial advances (Löve and Löve, 1974). Some species survived in ice-free refugia, and on islands and along coastlines, or migrated southwards into non-glaciated regions. Those which were unable to adapt to the cold became extinct. Extensive tracts of lowland tundra developed during the glacial maxima in eastern and central North America and across western Europe and Siberia, with smaller areas in Alaska, northern Greenland and the Arctic islands. These expanded and coalesced to form a northern tundra zone as the ice retreated; genetic divergence occurred when the tundra was periodically broken by the northward migration of other types of vegetation during the warmest interglacials (Hoffmann and Taber, 1967). Many arctic plants have apparently migrated from centres in north-eastern Asia, and some were able to reach North America via Beringia (Yurtsev, 1972). Tundra-steppe vegetation persisted in Beringia until about 12 000 years ago, when maritime conditions became established following the rise in sea level.

Although alpine tundra may have originated in earlier geologic times, its distribution would have been severely limited until world climates became progressively cooler in the late Tertiary. It was also during this period that the major mountain ranges achieved their present heights. For example, tectonic uplift in late-Miocene times increased the elevation of the Rocky Mountains by 1500 m and could have lowered temperatures by at least 10 °C. Some alpine species have evolved from lowland elements. This has been important in more isolated mountain ranges, such as those of Central Asia, where many steppe and semi-arid species adapted quickly to the cold, dry mountain environment. Similarly, the endemic species of the Sierra Nevada have close affinities with the desert flora of California and the Great Basin (Packer, 1974). In addition, many arctic plants dispersed to temperate mountain regions during glacial periods along routes established by the altitudinal displacement of vegetation zones. In this way species with bipolar distributions were able to reach the southern hemisphere by way of the Rocky Mountains and the Andes, while many others are widely distributed throughout the length of these mountain systems (Raven, 1963). However, most of the alpine plants in the southern hemisphere appear to be relicts of a tundra flora which evolved in Antarctica before it was covered by ice (Löve and Löve, 1974).

Adjustment to environmental changes is enhanced by the adaptive superiority of polyploid species in which there has been duplication of the chromosomes. The frequency of polyploidy increases in colder climates, and is particularly common in the grasses and sedges which occur abundantly in the Arctic. This increased genetic variability is considered to be a pre-adaptive mechanism which allows species to invade new habitats that are beyond the range of tolerance of the original population. Polyploid species are more tolerant of cooler climates than diploid species but their increased frequency in Arctic environments may reflect the rapid reinvasion of former territory as the ice retreated (Stebbins, 1971).

A number of plant species now exhibit widely disjunct ranges. *Dryas integrifolia*, for example, grows in various parts of the Canadian Arctic and on some mountain ranges, but also occurs locally near Lake Superior (Porsild, 1958). Such a distribution probably originated because this species was previously widely distributed but was subsequently eliminated over most of its former range: the outlying relict population presumably survives because conditions in the area are less suitable in other species. For other species such outliers could have arisen from populations which survived in ice-free refugia. The restricted occurrence of some species in Norway has been explained in this way and, on a larger scale, the phytogeographic similarities between Scandinavia, Iceland, Greenland and Labrador (Ives, 1974b). Originally it was thought that the distinctive alpine flora survived glacial periods on peaks which protruded above the ice sheets. Evidence of glaciation has now been observed on many of these **nunataks**, but even where traces of ice activity are absent these areas may still not have been environmentally suitable.

Ice-free areas existed on many islands in the Canadian Arctic during the Pleistocene, while parts of Alaska and Beringia itself were available for colonization. Other areas also emerged from the sea; several large islands formed along the east coast of Canada, and further south an extensive coastal plain developed. Refugia also occurred along sea cliffs where the land was too steep for a persistent ice cover to develop. At one site in the northwestern United States the pollen record indicates repeated changes in the vegetation over a period of 60 000 years, alternating between tundra species and forests of spruce and hemlock (Heusser, 1972). Freshly exposed glacial moraine is quickly colonized by pioneer species. For example, at Glacier Bay, Alaska, arctic species of willow (*Salix artica*), fireweed (*Epilobium latifolium*) and horsetail (*Equisetum variegatum*) become established within 5 years. A similar herb–dwarf-shrub community is found on hummocky debris which overlies stagnant ice left by the Klutlan Glacier in the Yukon: other species, such as *Dryas integrifolia* and purple reedgrass (*Calamagrostis purpurascens*), begin to appear after 30–40 years (Birks, 1980). Minor climatic oscillations may therefore have presented opportunities for colonization and dispersal for some species during the Pleistocene. Conversely, the persistence of suitable habitats during interglacial periods is equally important for the survival of tundra species. Geographic isolation during these warmer episodes is reflected in the highly endemic alpine floras (Billings, 1974a).

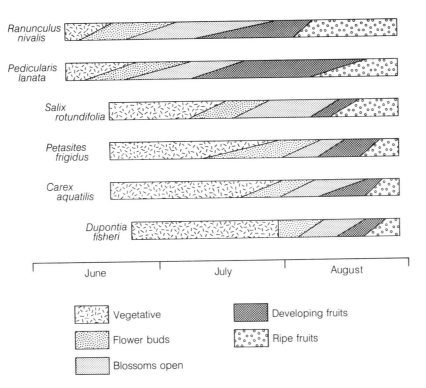

Figure 9.27 Phenology of common tundra species at Barrow, Alaska. (After Webber, 1978.) (Redrawn from P. J. Webber, Spatial and temporal variation of the vegetation and its production, Barrow, Alaska, in *Ecological Studies 29 – Vegetation and Production Ecology of an Alaskan Arctic Tundra*, ed. L. L. Tieszen; published by Springer-Verlag N.Y. Inc., 1978.)

9.7 PHENOLOGY OF TUNDRA PLANTS

The evergreen and semi-evergreen species, which carry some green leaves throughout the winter, resume their growth as soon as conditions become favourable. Evergreen shrubs are more common in the low arctic where they are protected from winter desiccation by higher snowfalls. In the high arctic evergreen species, such as *Cassiope tetragona*, are usually restricted to late-melting snowbanks, which necessarily shortens the growing season. The semi-evergreen habit is characteristic of many rosette plants: leaves formed in late summer, and protected by the older, dead foliage, will resume photosynthesis in the spring as soon as light levels are suitable. In deciduous species new leaves begin to develop when the temperature of the preformed buds rises above 0 °C for a few hours each day (Billings and Mooney, 1968). Leaf development occurs rapidly, and is followed by a relatively long mature phase and a short senescent period (Johnson and Tieszen, 1976). The leaves of grasses and sedges die back during the winter, but growth resumes from basal meristems in the spring. Development continues for as long as conditions are suitable, and unexpanded leaves which remain protected by persistent leaf sheaths will continue their growth the following season (Tieszen, 1978). Annual species are rare in arctic areas but become increasingly abundant in alpine tundras at lower latitudes, where their success depends on germination early in the spring and rapid development during the cool summers. Mosses and lichens are resistant to freezing and have relatively high rates of photosynthesis at low temperatures and low light levels (Oechel and Sveinbjörnsson, 1978). Consequently, their growth is not necessarily restricted to the summer months.

The growing season starts as the snow melts and the soils begin to thaw, and timing is therefore controlled by microtopographic conditions which affect the depth of snowpack. In tundra sites in Alaska vegetative growth begins in early June: the first blossoms appear in late June but most species come into flower in mid or late July (Figure 9.27). Flowering is usually completed by early August and the fruits ripen about one week later. The average period between onset of growth and the appearance of the first ripe fruits is 53 days. The buttercup (*Ranunculus nivalis*) produces mature fruit within 35 days of snowmelt; in woolly lousewort (*Pedicularis lanata*) it can take up to 73 days for the last fruits to mature (Webber, 1978).

The start of the growing season in the high arctic polar deserts is often marked by the development of plants such as *Draba macrocarpa* in 'greenhouses' beneath the ice at the end of May (Alexsandrova, 1988). New leaves start to develop on flowering plants in early or mid June as the snow cover disappears, and the green shoots of the grasses and rushes begin to elongate shortly after. Most species begin to flower in early July, with abundant flowering occurring 2–3 weeks later. Nodding saxifrage (*Saxifraga cernua*) and snowgrass (*Phippsia algida*) come into flower later than other species and most individuals bloom in early August as the fruits of other species are beginning to ripen. Signs of the approach of winter are visible by the end of August as leaves become wilted and close over next year's buds. The flowers of many of the high arctic species are sessile and are buried within the plant cushions as they begin to develop. Only after the fruits are ripe are they raised above the foliage so that the seeds can be dispersed by the wind. The immature flowers and fruits therefore benefit from the extra heat within the plant cushion. Poppy (*Papaver polare*) flowers are borne on long, arched peduncles which are thermotropic and press the flowers towards warmer surfaces (Alexsandrova, 1988). The poppy capsules overwinter and ripen the following year and, like other tundra perennials, seed production is determined by growing conditions in previous years.

In alpine tundra growth and development are also synchronized with snowmelt and therefore can vary by several weeks, depending on site conditions. Bud break is usually earliest in dry heath communities, where there is little snow cover, and becomes progressively later in wet meadow sites: in snowpatch communities species such as tufted hair grass (*Deschampsia caespitosa*) often begin growth beneath the snow cover (Billings and Bliss, 1959). In the Rocky Mountains the snow cover normally remains 2–3 weeks longer in these communities compared with the drier meadows of *Kobresia myosuroides*, but the timing and duration of subsequent phenological stages is generally similar in both communities: flower buds start to open in late June, fruits begin to form in early July and seed set commences in mid July (Fareed and Caldwell, 1975). Growth and development in ubiquitous species varies according to the length of the snow-free season and is further influenced by the effects of slope and exposure on local soil moisture and temperature conditions (Holway and Ward, 1965). Similarly, the start of the growing season can be delayed by several weeks in years with heavy snowfall and when temperatures are lower than normal in the spring (Wielgolaski and Kärenlampi, 1975).

Root development begins as the soils thaw, although maximum activity occurs later in the summer. The optimum temperature for root production in most tundra graminoids is 10 °C and lateral development is further enhanced by better aeration in drier soils (Chapin, 1978). Cotton-grass (*Eriophorum angustifolium*) develops an entire root system each year. The new roots grow down with the thaw and are most abundant at depths where soil temperatures are below 2 °C; some even grow across the top of the permafrost (Shaver and Billings, 1975). The roots remain active even when the surface soil layers begin to freeze and the shoots have ceased to function.

Root elongation decreases as the days become shorter and, in species such as *Eriophorum angustifolium* and *Dupontia fisheri*, ceases at day lengths of 15 hours (Shaver and Billings, 1977). Low soil temperatures in the spring delay the initiation of root growth until shoot growth and even flowering has occurred. Because of this asynchronous activity of shoots and roots the limited resources of the plant are used when conditions are optimal for above- and below-ground production. Photoperiodic control of root growth allows nutrients to be translocated to storage organs even though soil temperatures are still above freezing, and so ensures their availability for rapid early shoot development the following spring. Carbohydrate reserves are also depleted during the period of rapid early-season shoot growth, but most starches and sugars are used for respiration and maintenance of below-ground tissues, and are gradually replaced over the growing season (Chapin and Shaver, 1985).

9.8 LIFE CYCLE AND DEVELOPMENT OF TUNDRA PLANTS

Vegetative reproduction becomes increasingly important in tundra environments because of the high metabolic costs associated with seed production and the uncertainty of seedling recruitment. Seeding occurs intermittently in most populations but the probability is increased by several adaptive traits. In some tundra species flowers develop slowly over two or more growing seasons (Sørensen, 1941): others show an opportunistic response during more favourable years. Flowering is usually promoted by warm temperatures and long photoperiods. In widely distributed species the number of daylight hours required for the production of flower primordia and the opening of the buds the following year are related to a latitudinal gradient. For *Oxyria digyna* this increases from a 15-hour photoperiod in populations at 40 °N to continuous daylight in the Arctic: flowering also occurs more rapidly at higher temperatures (Billings and Mooney, 1968). Similarly, in New Zealand, snow tussock (*Chionochloa rigida*) flowering is irregular because induction occurs only in summers when temperatures are abnormally high (Mark, 1965).

Most tundra plants are pollinated by insects but wind pollination increases in the Arctic as insect activity becomes restricted by lower temperatures. Bumblebees are especially common in the mountains of the northern hemisphere and are active as long as temperatures remain above 10 °C. At higher elevations, where the air is cooler, pollination is mainly by flies and butterflies (Billings and Mooney, 1968). Flies are also the principal pollinators in the Arctic: this is reflected in the predominance of shallow white or yellow flowers ornamented with dark petal spots or purple anthers that function as honey guides and serve to attract the flies (Savile, 1972). Many species are self-fertile, which ensures that some seed is set in most years although seed production generally increases with cross-pollination.

The quantity of seed produced depends on the weather conditions following pollination, and at higher latitudes and altitudes, even in relatively good years, the seed crop is usually quite small. The estimated seed rain in alpine communities in New Zealand ranges from 35 to 8871 seeds m^{-2} (Spence, 1990) compared with 2500 seeds m^{-2} in alpine areas in the eastern United States (Marchand and Roach, 1980) and 300–700 seeds m^{-2} in Norway (Ryvarden, 1971). Few plants complete seed maturation at extreme sites in the high Arctic and many flowers do not even develop as far as anthesis (Bell and Bliss, 1980). Typically the seed is released at the end of the growing season, but in some species the fruits go dormant and seed ripening continues the following summer. Dispersal in arctic species is mainly by wind, and although plumed seeds and fruits are common in many species, other less specialized seeds are blown across the hard, crusted snow during winter. Their journeys often end at the base of a cliff or some other obstruction where snow and debris has accumulated, which makes them favourable sites for plant growth (Savile, 1972). Many seeds are also dispersed by grazing geese and other birds during their spring migrations. In alpine tundra wind is the principal agent of dispersal even in species with no apparent morphological adaptations. The seeds are mostly released before the ground is covered by snow, and relatively few are dispersed during winter (Ryvarden, 1975). The distribution of these wind-dispersed seeds and their subsequent incorporation into the seed bank is related to soil texture. Smaller seeds are mostly trapped by soil particles less than 2 mm in diameter and the majority remain within 1 cm of the surface. The probability of entrapment of larger seeds increases in coarser material but they also become more deeply buried (Chambers et al., 1991).

Optimum temperatures for germination are as high as 20–30 °C in many tundra species. A chilling requirement or a period of after-ripening at higher temperatures is reported in some species (Amen, 1966) but germination is normally prevented by the onset of winter conditions rather than intrinsic seed dormancy mechanisms. Germination usually occurs soon after snowmelt and ceases as the soils dry out. The production of hard-coated, impermeable seeds is therefore a common adaptation to prevent late-season germination in wet meadow species. Although conditions may not be suitable for germination each year, the amount of viable seed that is stored in tundra soils is typically small, ranging from 0 to 1769 m^{-2} in sites in the Canadian Arctic with 80–3367 seeds m^{-2} reported in mountain sites in North America (McGraw and Vavrek, 1989). There is no general relationship between the density of buried viable seed and elevation or latitude, although reserves tend to increase in more productive tundra (Fox, 1983). The absence of abundant seed contradicts the idea that persistence is favoured by low soil temperatures. Long-term viability up to 10 000 years has been claimed for seeds of arctic lupin (*Lupinus arcticus*) retrieved from animal burrows in permafrost (Porsild et al., 1967) but the age of these seeds is disputed.

Seedlings have very little time to become established before the onset of winter. Many seedlings are lost during the first summer because of drought or below-freezing temperatures; seedling mortality in east Greenland, for example, averages 50% each year (Wager, 1938). Seedling growth is very slow and by the end of the first growing season typically only a few small leaves are produced: in some species only the cotyledons appear above ground. Development of the root system proceeds more rapidly. The optimal temperature for root growth in seedlings of high arctic perennials is about 12 °C but weekly growth can be less than 2 mm at normal field temperatures of 1–3 °C (Bell and Bliss, 1978). Root growth is also limited by low soil moisture and lack of nutrients. Poor root development increases susceptibility to frost heaving during the winter and many seedlings are subsequently killed by drought. The common association between seedlings of alpine plants and deep-rooted cushions of moss campion (*Silene acaulis*) may reduce this type of disturbance (Billings and Mooney, 1968). Similar relationships have been reported between seedlings and mature tussock grasses in New Zealand alpine communities (Rose and Platt, 1990). In high arctic tundra seedling survival increases where mats of lichen and moss provide protection from desiccation during the summer (Bell and Bliss, 1980). In contrast, shoot and flower development occurs rapidly in annual plants, such as *Koenigia islandica*, although sizeable seed banks may be necessary to guard against reproductive failure (D. N. Reynolds, 1984).

9.9 PRODUCTIVITY AND BIOMASS

Practically all tundra plants utilize the C$_3$ photosynthetic pathway, although some C$_4$ grasses are found in tropical alpine sites (Tieszen et al., 1979). Optimum temperatures for shoot growth typically range from 15 to 20 °C. In some species assimilation continues at temperatures as low as −5 °C (Tieszen et al., 1981), while upper temperature compensation points can exceed 40 °C (Figure 9.28). Ambient air temperatures are generally below optimum in most tundra regions but this deficiency is minimized by the compact form of the plants. A common characteristic of arctic plants is a reduction in non-photosynthetic supporting tissue and non-chlorophyllous cells in leaf blades at higher latitudes (Tieszen, 1978). Photosynthetic efficiency in these low-temperature environments is also increased by high concentrations of RuBP carboxylase, the enzyme responsible for carbon dioxide fixation (Chabot, 1979).

Maximum rates of photosynthesis are very similar to those reported in temperate species (Limbach et al., 1982)

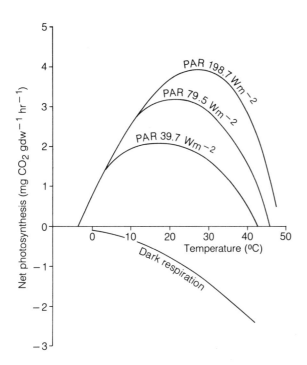

Figure 9.28 The effect of temperature on net photosynthetic rates in the alpine heath *Loiseleuria procumbens*. (After Larcher *et al.*, 1975.) (Adapted with permission from W. Larcher *et al.*, Mt Patscherkofel, Austria, in *Ecological Bulletins 20: Structure and Function of Tundra Ecosystems* (eds T. Rosswall and O. W. Heal), published by the Swedish Natural Science Research Council, Stockholm, 1975.)

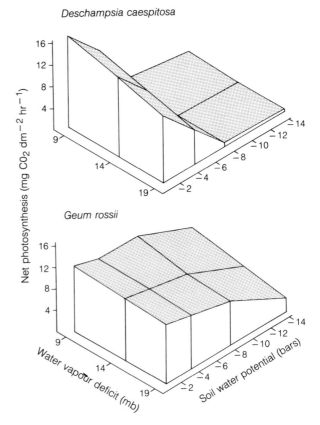

Figure 9.29 Net photosyntheis of *Deschampsia caespitosa* and *Geum rossii* as a function of water vapour deficit and soil moisture stress at a leaf temperature of 20 °C and PAR light intensity of 900 μE m^{-2} sec^{-1}. (After Johnson and Caldwell, 1975.) (Reproduced with permission from D. A. Johnson and M. M. Caldwell, Gas exchange of four arctic and alpine tundra plant species in relation to atmospheric and soil moisture stress, *Oecologia*, 1975, **21**, 96.)

but primary production is reduced because of the short growing season. This is partially offset by the extended periods of daylight at high latitudes. Production in tundra communities is further limited because a significant part of the growing season is used for leaf expansion. In the Arctic, maximum uptake of CO_2 is not attained until late in the summer when leaf area is greatest, even though irradiation and temperatures have begun to decline (Tieszen, 1978). Photosynthesis rates drop as the leaves become senescent, but even though canopy development ceases, a significant proportion of carbon dioxide continues to be assimilated and is presumably transferred into storage below ground.

Water stress in plants occurs because of low soil moisture levels or strong evaporative demands and consequently varies with slope aspect, wind exposure and the drainage characteristics of the site. Tundra species exhibit a general decline in photosynthetic capacity as water stress increases, but in moist sites assimilation is rarely limited by stomatal closure (Johnson and Caldwell, 1975). Not only is water nearly always available to these plants but also their root resistances remain low, even though cold soil temperatures would normally restrict water uptake in temperate species (Stoner and Miller, 1975). However, species such as tufted hair grass (*Deschampsia caespitosa*), which typically grow in wet sites, are much more susceptible to water stress than species such as *Geum rossii* that are more widely distributed

(Figure 9.29). In *Deschampsia caespitosa* the photosynthetic rate quickly drops to less than 10% of maximum as soil and atmospheric water stresses increase, whereas photosynthesis continues at comparatively high rates in *Geum rossii* even when water stress is severe. The high rate of photosynthesis in wet sites is reflected by comparatively greater production and biomass (Figure 9.30). The reduced performance of wet meadow species under stress effectively limits their distribution.

Production is further limited by the low nutrient status of most tundra soils. Low soil temperatures slow the rate of decomposition of organic matter, and mineral weathering is minimal where the soil is underlain by permafrost. Nitrogen is particularly deficient. Most of the nitrogen in tundra ecosystems is fixed by free-living blue-green algae associated with wet mossy sites. Additional supplies are added from species living symbiotically in lichens. Nitrogen-fixing bacteria are found on some leguminous plants, such as species of *Oxytropis*, and also on shrubby avens (*Dryas* spp.). Consequently, differences in ground cover have a considerable impact on

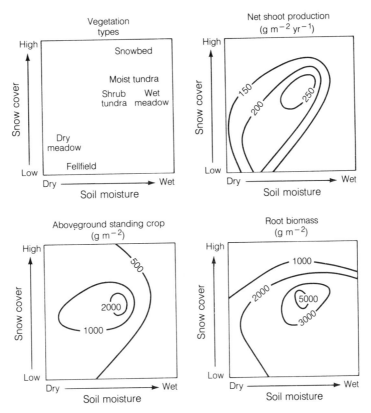

Figure 9.30 The effect of soil moisture and snow cover on net shoot production, above-ground standing crop and root biomass in Rocky Mountain alpine vegetation communities. (After Webber, 1974.) (Reproduced with permission from P. J. Webber, Tundra primary production, in *Arctic and Alpine Environments* (eds J. D. Ives and R. G. Barry), published by Methuen & Co., 1974.)

nitrogen-fixing activity (Figure 9.31). The optimum temperature for nitrogen fixation by blue-green algae and lichens is between 15 and 20 °C (Figure 9.32). Consequently, activity generally increases where dark-coloured mosses or other conditions favour local surface warming (Alexander *et al.*, 1978). The average annual

input of nitrogen by biological fixation in arctic tundra typically ranges from 23 to 380 mg m^{-2} compared with 23–75 mg N m^{-2} yr^{-1} contributed through precipitation (Holding, 1981). The unusually lush plant growth which occurs around nest sites and animal burrows attests to the shortage of nitrogen in tundra soils (Figure 9.33). The

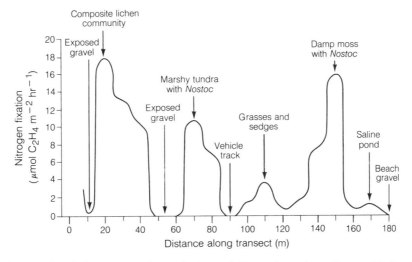

Figure 9.31 Nitrogen fixation rates in relation to vegetation and micro-relief in arctic tundra at Barrow, Alaska. (After Schell and Alexander, 1973.) (From D. M. Schell and V. Alexander, Nitrogen fixation in arctic coastal tundra in relation to vegetation and micro-relief, *Arctic*, 1973, **26**, 134; permission to reprint granted by the Arctic Institute of North America.)

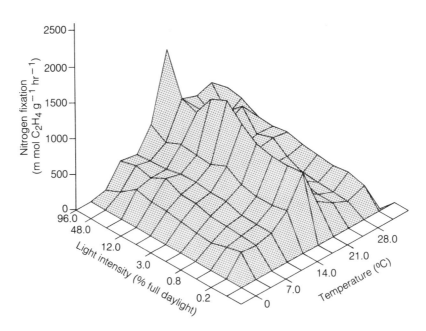

Figure 9.32 Nitrogen fixation rates of the lichen *Peltigera aphthosa* as a function temperature and light in arctic tundra at Barrow, Alaska. (After Alexander *et al.*, 1978.) (Reproduced with permission from V. Alexander, M. Billington and D. M. Schell, Nitrogen fixation in arctic and alpine tundra, in *Ecological Studies 29 – Vegetation and Production Ecology of an Alaskan Arctic Tundra*, ed. L. L. Tieszen; published by Springer-Verlag N.Y. Inc., 1978.)

results of fertilizer trials also suggests that plant growth is limited by low soil nitrogen levels (McCown, 1978).

Plant biomass and annual production decrease with latitude and elevation primarily in response to lower temperatures and shorter growing seasons. These regional trends are most pronounced in the vascular plant cover (Table 9.3). In more extreme polar desert and semi-desert communities of cushion plants and moss the above-ground biomass of vascular plants is usually less than 100 g m^{-2}, and decreases to less than 5 g m^{-2} in sites which support only a sparse cover of herbs and moss. Cryptogam biomass is often much higher and exceeds 1200 g m^{-2} in some sites. Total annual production at these high arctic sites ranges from 7 to 54 g m^{-2}, and is mainly incorporated into the above-ground tissues of the higher plants. In arctic wet sedge–moss communities, above-ground biomass of vascular plants ranges from 85 to 372 g m^{-2} with below-ground biomass of $353–3727 \text{ g m}^{-2}$, and $23–2335 \text{ g m}^{-2}$ for the cryptogams. Annual production in these communities ranges from 71 to 281 g m^{-2}, most of which occurs below ground. Total standing crop and annual production increase progressively through dry sedge meadows, dwarf shrub heaths and low shrub communities. The highest above-ground biomass of 2240 g m^{-2} is reported for low shrub communities in Greenland; the highest annual production is 1605 g m^{-2} recorded in a similar community in South Georgia.

The ratio of above- to below-ground biomass averages 1:0.9 in the polar desert communities, compared with ratios as high as 1:21 in wet sedge meadows (Table 9.4). The shoot:root ratio declines as the proportion of woody species increases. These differences reflect the manner in which stored resources are allocated for seasonal growth. The slow-growing cushion plants retain nutrients in the dead leaves and release them slowly to the fine root system by decomposition (Svoboda, 1977). The slow-growing evergreen shrubs also have comparatively small underground biomass and depend directly on materials assimilated by their persistent leaves rather than large storage reserves. Rapid spring growth in other plants requires translocation of materials from storage. New leaf development in deciduous shrubs and herbs results in a synchronous decrease in biomass of the perennating

Figure 9.33 The productivity of the tundra can be greatly increased by animal wastes deposited around and nests and burrows. (O. W. Archibold.)

Table 9.3 Biomass and net annual plant production in selected tundra ecosystems (after Wielgolaski *et al.*, 1981)

	Vascular plant biomass		Cryptogam biomass	Vascular plant production		Cryptogam production
	above-ground	below-ground		above-ground	below-ground	
	(g m^{-2})			(g m^{-2} yr^{-1})		
Arctic						
King Christian Island, Canada (78 °N)						
moss–herb	3	6	914	3	2	16
Malloch Dome, Canada (78 °N)						
moss–herb	4	17	1207	4	4	25
Truelove Is., Canada (76 °N)						
moss–herb	15	10	232	2	0.5	5
cushion plant–lichen	89	57	64	15	3	3.5
cushion plant–moss	126	50	623	27	5	22
wet sedge–moss meadow	78	691	1097	46	130	105
dwarf shrub heath	145	671	430	18	90	24
Big River, Canada (73 °N)						
cushion plant–lichen	239	203	35	15	6	1
wet sedge–moss meadow	18	3727	23	16	54	1
Barrow, Alaska (71 °N)						
wet sedge–moss meadow	46	996	45	45	–	–
wet sedge meadow	52	1305	30	52	65	27
dwarf shrub	63	471	45	9	16	–
Disko, Greenland (69 °N)						
cushion plant–lichen	66	32	63	32	11	3
herb meadow	396	1030	–	396	520	–
low shrub	2240	200	–	800	40	–
Sub-antarctic						
Signy Island (60 °S)						
Deschampsia turf	327	132	254	260	130	85
moss carpet	0	0	3110	0	0	670
Macquarie Island (54 °S)						
herb meadow	139	670	402	314	550	150
grassland	912	1690	6	1890	3670	21
Alpine						
Hardangervidda, Norway (1230 m)						
dwarf shrub heath	62	191	377	88	100	88
grass–herb meadow	161	545	50	241	245	56
wet sedge	147	1316	175	254	410	173
low *Salix* shrub	804	1305	307	385	290	210
Niwot Ridge, USA (3650 m)						
dry sedge meadow	453	1739	–	164	–	–
wet sedge	141	2927	–	140	–	–
low shrub	1464	4398	–	503	–	–

(Source: F. E. Wielgolaski *et al.*, Primary production of tundra, in *Tundra Ecosystems: A Comparative Analysis*, eds L. C. Bliss, O. W. Heal and J. J. Moore; published by Cambridge University Press, 1981.)

Table 9.4 Average biomass ratios in different types of tundra vegetation (after Wielgolaski *et al.*, 1981)

	Shoot to root (live)	Green to non-green live vascular plants	Live to dead above-ground vascular plants
Polar desert and semi-desert	1:0.9	1:2.3	1:1.9
Wet sedge meadow	1:21.0	1:23.0	1:1.6
Mesic-dry meadow	1:5.0	1:7.7	1:0.8
Dwarf shrub tundra	1:3.1	1:12.0	1:0.6
Low shrub tundra	1:2.0	1:19.0	1:0.2
Forest tundra	1:0.8	1:15.0	1:0.1

(Source: F. E. Wielgolaski *et al.*, Primary production of tundra, in *Tundra Ecosystems: A Comparative Analysis*, eds L. C. Bliss, O. W. Heal and J. J. Moore; published by Cambridge University Press, 1981.)

organs (Tieszen *et al.*, 1981). In deciduous shrubs, carbohydrates and mineral nutrients are mainly translocated from woody stem tissues to the newly expanding leaves in the spring rather than from roots: most of the nutrients are reabsorbed before the foliage is shed (Chapin *et al.*, 1980). In forbs, the early season demand for resources is mainly met by translocation from underground storage and these reserves are slowly replenished by increased nutrient uptake as the soils warm up during the summer. Leaves begin to function very quickly in grasses and sedges and are produced continuously throughout the growing season. Translocation of resources from roots is minimal and the characteristically high shoot:root ratio of these plants may reflect the longevity of the roots rather than their functional storage capacity.

9.10 NUTRIENT STORAGE IN TUNDRA PLANTS

There is considerable variation in tissue nutrient concentrations amongst the different tundra growth forms (Table 9.5). The highest mineral concentrations in above-ground green tissues are typically associated with the forbs and are lowest in evergreen shrubs: this is particularly apparent for nitrogen, potassium and calcium. In most tundra plants the nutrients used to sustain the flush of growth in the spring are acquired before the onset of winter and are rapidly translocated to the newly developing tissues.

Concentrations of nitrogen, phosphorus and potassium in shoots of herbaceous plants increase as growth resumes in the spring but then decline as the growing season progresses (Chapin, 1978). This reduction is partly a consequence of the increased shoot weight but translocation to perennial storage organs reduces aboveground nutrient stocks before peak biomass is reached. As a result there is usually a two- to three-fold change in nutrient concentrations in storage tissues of forbs and graminoids during the course of the growing season (Dowding *et al.*, 1981). In contrast, concentrations of less mobile elements, such as calcium and magnesium, steadily increase during the growing season and remain constant as the leaves become senescent. Seasonal changes in nutrient concentrations are less prononounced in deciduous shrubs because of the greater amount of metabolically inactive tissue. However, as much as 14% of the total nitrogen and phosphorus capital in the stems and large roots is used for leaf development and about half of this amount is removed before the leaves are shed (Chapin *et al.*, 1980). The concentration of nutrients in the older stems and leaves of evergreen shrubs changes very little during the course of the year because there is no need for rapid development of a new crop of leaves in the spring. In these species leaf production continues throughout the growing season and nutrients are mainly supplied simultaneously by withdrawal from older,

Table 9.5 Maximum mineral content in different plant types in tundra communities in Norway (after Wielgolaski *et al.*, 1975)

Plant type	Plant compartment	N	P	K	Ca	Mg
		(% dry weight)				
Willows	green	2.8	0.3	1.0	1.4	0.2
	non-green	1.1	0.2	0.4	1.1	0.1
Ericaceous shrubs						
evergreen species	green	1.1	0.1	0.6	0.5	0.2
	non-green	0.8	0.1	0.2	0.2	0.1
deciduous species	green	1.9	2.1	0.9	0.6	0.2
	non-green	0.7	0.1	0.2	0.2	0.1
Grasses/sedges						
eutrophic species	green	2.5	0.3	2.1	0.4	0.1
	roots	1.1	0.1	0.4	0.6	0.1
oligotrophic species	green	2.3	0.2	1.6	0.3	0.1
	roots	0.8	0.1	0.4	0.3	0.1
Forbs						
N_2-fixing species	green	4.5	0.3	2.4	1.4	0.2
	non-green	2.6	0.1	0.3	1.8	0.1
non N_2-fixing species	green	2.4	0.3	2.2	1.6	0.3
	non-green	1.2	0.2	0.7	1.1	0.2
Mosses	green	1.2	0.2	0.6	0.1	0.1
	non-green	1.1	0.1	0.3	1.0	0.1
Lichens						
N_2-fixing species	shoots	1.5	0.1	0.2	0.2	0.1
non N_2-fixing species	shoots	0.6	0.1	0.4	0.1	0.1

(Source: F. E. Wielgolaski, S. Kjelvik and P. Kallio, Mineral content of tundra and forest tundra plants in Fennoscandia, in *Ecological Studies 16 – Fennoscandian Tundra Ecosystems, Part 1. Plants and Microorganisms*, ed. F. E. Wielgolaksi; published by Springer-Verlag N.Y. Inc., 1975.)

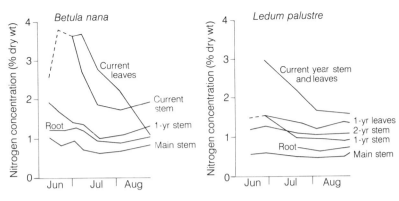

Figure 9.34 Seasonal course of nitrogen concentrations in various parts of dwarf birch (*Betula nana*) and narrow-leaf Labrador tea (*Ledum palustre*) growing in arctic tundra at Barrow, Alaska. (After Chapin *et al.*, 1980.) (Reproduced with permission from F. S. Chapin, D. A. Johnson and J. D. McKendrick, Seasonal movement of nutrients in plants of differing growth form in an Alaskan tundra ecosystem: implications for herbivory, *Journal of Ecology*, 1980, **68**, 194.)

senescing foliage: specialization for winter storage is therefore minimal (Figure 9.34).

Total nutrient storage in tundra vegetation varies with the type of plant cover and the nutrient status of the soil. In alpine tundra in Scandinavia, nutrient reserves are lowest in heath and dry meadow communities growing in oligotrophic sites but increase substantially in more productive wet meadow and low shrub communities (Table 9.6). In Canadian high Arctic tundra communities, estimates of nitrogen storage in hummocky sedge-moss meadow range from 2.7 g N m^{-2} in live shoots of vascular plants to 13.3 g N m^{-2} in roots and rhizomes, with 14.2 g N m^{-2} bound up in dead plant material: an additional 7.1 g m^{-2} of nitrogen were stored in mosses and algae for a total of 37.3 g N m^{-2} (Babb and Whitfield, 1977). Total reserves of phosphorus were similarly estimated at 3.2 g m^{-2}. Nutrient storage in live plants in an adjacent cushion plant–lichen community was estimated

Table 9.6 Mineral accumulation and annual uptake by the vegetation in tundra communities in Norway (after Wielgolaski *et al.*, 1975)

	N	P	K	Ca	Mg
	\multicolumn{5}{c}{(g m^{-2})}				
Alpine lichen heath					
above-ground accumulation	2.4	0.2	0.6	1.0	0.2
below-ground accumulation	1.4	0.1	0.2	0.2	0.1
annual uptake	2.1	0.2	0.5	0.7	0.2
stored in dead plants	2.7	0.2	0.2	0.5	0.1
soil	0.2	0.3	0.5	0.8	0.5
Alpine dry meadow					
above-ground accumulation	1.5	0.1	0.5	1.8	0.1
below-ground accumulation	5.8	0.4	1.3	4.4	0.3
annual uptake	7.4	0.7	3.0	5.8	0.6
stored in dead plants	7.1	0.5	0.9	8.0	0.3
soil	1.2	0.4	0.7	1.4	1.2
Wet meadow					
above-ground accumulation	3.3	0.4	1.0	2.9	0.2
below-ground accumulation	16.6	1.7	2.0	8.5	0.8
annual uptake	13.1	1.3	3.6	6.5	0.8
stored in dead plants	28.1	1.8	2.1	15.5	0.7
soil	1.7	0.2	0.1	0.6	0.1
Willow thicket					
above-ground accumulation	8.5	1.3	2.7	8.4	0.8
below-ground accumulation	10.9	2.1	4.3	8.4	1.5
annual uptake	11.2	1.7	5.2	7.7	1.3
stored in dead plants	10.1	1.4	1.7	9.3	1.2
soil	1.0	0.6	0.4	1.7	1.2

(Source: F. E. Wielgolaski, S. Kjelvik and P. Kallio, Mineral content of tundra and forest tundra plants in Fennoscandia, in *Ecological Studies 16 – Fennoscandian Tundra Ecosystems, Part 1. Plants and Microorganisms*, ed. F. E. Wielgolaski; published by Springer-Verlag N.Y. Inc., 1975.)

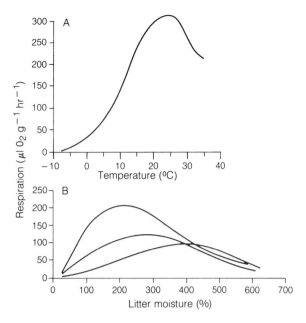

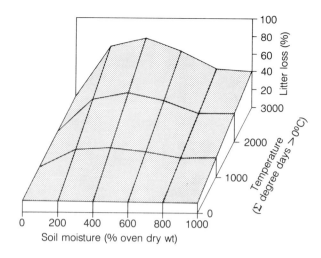

Figure 9.35 The effect of litter temperature and moisture on microbial respiration rates in arctic tundra at Barrow, Alaska. (After Flanagan and Veum, 1974.) (Adapted with permission from P. W. Flanagan and A. K. Veum, Relationships between respiration, weight loss, temperature and moisture in organic residues on tundra, in *Soil Organisms and Decomposition in Tundra* (eds A. J. Holding *et al.*) published by the Tundra Biome Steering Committee, Stockholm, 1974.)

Figure 9.36 The effect of annual soil heat accumulation above 0 °C and soil moisture on first-year litter losses in tundra sites. (After Heal and French, 1974.) (Adapted with permission from O. W. Heal and D. D. French, Decomposition of organic matter in tundra, in *Soil Organisms and Decomposition in Tundra* (eds A. J. Holding *et al.*), published by the Tundra Biome Steering Committee, Stockholm, 1974.)

at 2.3 g N m^{-2} and 0.2 g P m^{-2}, with 3.2 g N m^{-2} and 0.2 g P m^{-2} also present in dead plant tissue.

9.11 DECOMPOSITION

Low temperatures and moist acid soils slow down the rate of decomposition in tundra environments, so that a thick layer of organic debris usually accumulates in the more productive sites. Differences in chemical composition of the various litter types also affect decomposition. Herbaceous litter is low in lignin but has a higher mineral content than woody material. Mosses are low in nutrients but are comparatively high in compounds which inhibit attack by micro-organisms and animals. The high hemicellulose content of lichens similarly limits decomposition to a rather specialized group of organisms (Heal *et al.*, 1981). Some fragmentation occurs through freeze–thaw cycles but relatively little material is broken down in this way. The highest rates of litter decomposition are reported from the warmer, moist oceanic sites such as the sub-antarctic islands where annual weight losses can exceed 50% yr^{-1}, although less than 30% yr^{-1} is characteristic of most tundra sites and the losses are considerably lower in the cold polar deserts. Many leaves remain attached to the plants for a year or more and decomposition of this standing dead material can be retarded by lack of moisture. Reabsorption of nutrients during this period affects decomposition when the leaves are eventually incorporated into the litter mat.

The bacterial and fungal populations are smaller and taxonomically less diverse than in other ecosystems. Several psychrophilic and cold-tolerant forms of bacteria are present and can utilize materials at temperatures near freezing (Widden, 1977). In waterlogged sites, mineralization of organic matter is primarily by anaerobic bacteria and results in the production of methane, but in most sites the initial breakdown of organic matter is carried out by fungi. The growth of most tundra fungi decreases at high substrate moisture levels and is further reduced by acidities above pH 4.5–5.0 (Flanagan and Scarborough, 1974). Maximum growth in fungi generally occurs between 20 and 30 °C and ceases at about 10 °C. However, 10–20% of tundra species are psychrophilic with optimum growth temperatures below 10 °C, and some continue to break down cellulose at temperatures as low as – 7 °C. Little decomposition occurs while the dead material remains standing, but once incorporated into the soil litter layer the population of cellulytic and lignolytic fungi increases and decay is further accelerated by the activity of the soil invertebrates.

The nematodes are the most abundant of the soil invertebrates with densities ranging from 0.1 to 10 million individuals m^{-2}. Differences in density are mainly associated with local drainage conditions and decrease in boggy habitats and with depth (MacLean, 1981). Earthworms are absent from most tundra sites, although they have been reported from herbaceous meadows in northern Russia. However, enchytraeid worms are abundant, particularly in wetter sites, and generally account for 60–75% of the soil fauna biomass. Soil mites (Acarina) are poorly represented and typically number fewer than 50 000 m^{-2}; similar densities are reported for springtails (Collembola). The total biomass of soil invertebrates varies with climatic severity and ranges from 0.2 to 0.4 g dry weight m^{-2} in extreme sites to about 10 g dry weight m^{-2} elsewhere. The soil invertebrates mainly consume the microbial decomposers and their attendant predators, and consequently total invertebrate biomass is inversely related to accumulated organic matter (MacLean, 1974). Soil respiration provides an aggregate measure of decomposition by soil organisms. Some activity occurs at temperatures as low as $-7.5\ °C$, but respiration increases rapidly with temperature and reaches a maximum at about $25\ °C$ (Figure 9.35). Respiration generally ceases at moisture levels below 20% dry weight and is particularly sensitive to substrate moisture at higher temperatures. Variation in rates of decomposition can therefore be attributed to the length and warmth of the active season and differences in moisture conditions (Figure 9.36). The soil microflora and fauna account for 50–100% of the litter decomposed during the first year and any material that remains is mostly lost through leaching of soluble materials: disintegration by physical processes and herbivore activity is comparatively minor (Heal and French, 1974).

9.12 TUNDRA HERBIVORES

9.12.1 Representative groups

Most invertebrates feed on decomposer organisms within the soil but a few larvae feed directly on plant roots. The number of these rhizophagous insects is restricted by severe soil conditions and they are mostly found in subarctic regions, where they are represented by caterpillars of tortricid moths and weevils (Chernov, 1985). Above-ground consumption by invertebrate herbivores is comparatively small and most are very selective leaf feeders (Haukioja, 1981). Aphids, mites and other sucking insects which feed on plant juices are poorly represented, mainly because their development from eggs to adults proceeds without complete metamorphosis and is very dependent on warm temperatures. Most of the common tundra insects pass through various developmental stages and metamorphosis can take several years, depending on weather conditions. For example, the leaf-feeding larval stages of the moths *Gynaephora groendlandica* and *G. rossii* can last up to 10 years in the Canadian high Arctic (Ryan and Hergert, 1977). Some beetles and their larvae feed on the leaves and buds of various flowering herbs, and in shrubby tundra the larvae of sawflies are present on the leaves of willow; damage from leaf-miners is reported in birch. Other insects feed on the floral parts. Bumblebees, large hover-flies, mosquitoes and other insects which drink from the nectaries or consume pollen are essential for propagation and are therefore beneficial to the plants. However, some adult beetles and their larvae devour flowering shoots, buds, flowers and fruits. Weevils

and midge and fly larvae are also found in developing fruits and seeds.

The vertebrate herbivore population in arctic and alpine tundra includes several species of small rodents and hares, larger animals such as caribou, musk oxen and mountain goats, and a variety of birds. The small mammals are represented by over 30 species (Batzli, 1981). In the Alps and other mountain regions the most conspicuous rodents are the marmots (*Marmota* spp.) which hibernate during winter. Other species, such as the pika (*Ochotona* spp.), a small tailless hare, remain active under the snow and store grass in rock clefts and other dry sites for use in the winter. Voles (*Microtus* spp.) are more widely distributed and several North American and Eurasian species are found in both alpine and arctic tundra. In the Arctic the lemmings (*Lemmus* spp. and *Dicrostonyx* spp.) are the most important species in terms of herbivory and subsequent transfer of energy to higher consumers. The food habits of the different species vary according to the season. *Lemmus* eats a variety of plants but feeds mainly on grasses and sedges in summer; mosses may be included in its diet in winter but often the moss is removed simply to expose the young stems and rhizomes of other plants. Moderate grazing early in the season can stimulate new shoot growth in grasses and sedges because the canopy is opened up and the leaves are more fully illuminated. However, production decreases as population densities rise above 50 individuals ha^{-1}, and serious overgrazing occurs during population peaks when densities can exceed 200 individuals ha^{-1}.

9.12.2 Lemming population cycles

The lemming population increases every 3–6 years (Figure 9.37). Similar fluctuations occur in many other rodent populations in the Arctic, with corresponding oscillations in predatory species (Remmert, 1980). A general explanation of these fluctuations is provided by the nutrient recovery hypothesis (Figure 9.38). The tundra ecosystem lacks large available pools of nutrients, and annual inputs through decomposition are small. Following a population crash, much of the nutrient capital is locked up in dead plant and animal tissue and animal waste products. Higher soil temperatures in heavily grazed areas stimulate the decomposer organisms, so that nutrient release from the organic debris accelerates. Grazing intensity is low at this time and the vegetation begins to recover. The high nutrient quality of the new plant growth increases the breeding success of the lemmings and their population grows. Lemmings reproduce 2–4 times a year, with an average of 4–6 young per litter. The growing lemming population soon consumes the available food supply and behavioural interference increases wastage. The quality of the forage declines as more nutrients become tied up in organic tissue: the

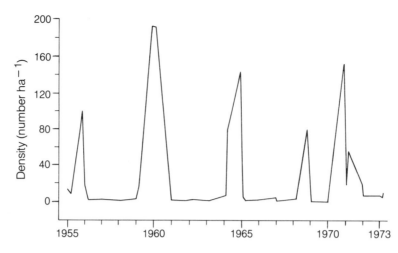

Figure 9.37 Estimates of the density of brown lemmings (*Lemmus* spp.) over a 20-year period near Barrow, Alaska. (After Batzli, 1981.) (Reproduced with permission from G. O. Batzli, Populations and energetics of small mammals in the tundra ecosystem, in *Tundra Ecosystems: A Comparative Analysis*, eds L. C. Bliss, O. W. Heal and J. J. Moore; published by Cambridge University Press, 1981.)

lemmings begin to suffer from malnutrition and starvation and the population quickly declines. The controlling effect of nutrient availability on ecosystem processes is supported by the results of fertilizer and exclosure experiments. The lemming population cycles are eliminated when the plant cover is maintained through fertilizer applications, whereas plant production, litter decomposition rates and the depth of the active thaw layer steadily decline in the absence of grazing, because of shading and the accumulation of an insulating layer of thatch (Schultz, 1969).

Lemmings typically consume less than 0.1% of net annual above-ground plant production in years when population densities are low but this increases to about 25% during peak years (Bunnell *et al.*, 1975). About 70% of the food consumed by lemmings is immediately eliminated in waste products, and because these animals remain active all year there is an accumulation of nutrients

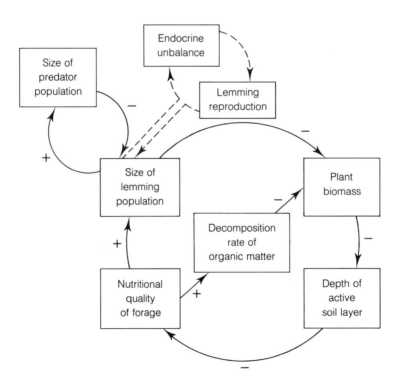

Figure 9.38 Feedback-loop model showing postulated homeostatic controls in the arctic tundra ecosystem. The state of a given compartment either amplifies (+) or counteracts (−) changes in the next compartment. (After Schultz, 1969.) (Reproduced with permission from A. M. Schultz, A study of an ecosystem: the Arctic tundra, in *The Ecosystem Concept in Natural Resource Management*, ed. G. M. Van Dyne; published by Academic Press Inc., 1969.

during winter. Areas which are fed by meltwater can therefore be enriched with phosphorus and other soluble nutrients when they are flushed out in the spring and this flux of nutrients increases considerably in years with high lemming populations (Dowding *et al.*, 1981). The accelerated release of nutrients stored in mosses could account for much of this increase. Winter nests are mainly constructed in low spots where there is greater snow accumulation and this concentration of activity also affects nutrient availability and subsequent plant production. Plant growth in the tundra is especially limited by the availability of phosphorus but uptake begins with the spring thaw, when concentrations are high as a result of decomposition and leaching of animal wastes (Chapin *et al.*, 1978). The nutrient quality of tundra grasses is therefore increased by grazing; this is contrary to the ideas proposed in the nutrient recovery hypothesis and suggests that the population patterns could simply reflect the regenerative capacity of the plants rather than a homeostatic control imposed by the rate at which nutrients are recycled (Chapin, 1978). Other factors unrelated to grazing activity can similarly influence population densities. For example, the lemming population declines when plant production is limited in the previous year because summer temperatures are lower than normal (Batzli, 1981). Conversely, rapid spring melt can eliminate winter nest sites before the ground thaws and summer burrows become available (Bliss, 1977).

9.12.3 Ungulates

Several species of large ungulate are found in the tundra (Figure 9.39). In the Arctic this group is represented by the various races of reindeer or caribou (*Rangifer tarandus*) and the musk ox (*Ovibos moschatus*). The reindeer are migratory species and move into the tundra in the spring where they assemble into large herds, often exceeding 100 000 animals, before dispersing across the tundra to feed on a variety of plants (Table 9.7). Once a suitable area has been reached, the animals normally graze by slowly walking into the wind in order to minimize harassment from mosquitoes (White *et al.*, 1981). The herds reform during the rut as the animals begin their migration southwards to their wintering grounds in open boreal woodlands. Movements during winter are less extensive and are mainly controlled by the amount and type of snow which covers their food: craters 50–60 cm deep are commonly excavated in the hard-packed snow (Pruitt, 1959). The smaller Peary's caribou (*R. tarandus pearyi*) remains in the Arctic islands all year round in habitat which is similar to that preferred by musk oxen. Although these animals do not compete directly for food, the caribou avoid snow which has been disturbed by musk oxen even though the plants are untouched. The musk oxen normally avoid deep snowbanks, preferring to graze on grasses, sedges and

willows in areas which have been blown clear of snow. About 25% of the vegetation cover is removed in heavily grazed areas: this amounts to less than 1% of the forage available over the entire range (Hubert, 1977). However, the food supply can be limited when freezing rain covers the vegetation in a thick crust of ice. The musk ox popula-

Figure 9.39 Large ungulates including: (a) musk oxen on Axel Heiberg Island, Canadian Arctic; (b) mountain goats in the Canadian Rockies; and (c) llamas grazing in the altiplano, Peru, are important members of tundra ecosystems. ((a) A. Soukup; (b) P. M. Sherrington; (c) J. A. Aronson.)

Table 9.7 Forage selection by reindeer and musk oxen. Forage selection is defined as the relative occurrence of a plant in rumen or oesophageal samples divided by the relative occurrence of that plant on the range (after White *et al.*, 1981)

| | Reindeer | | | | Musk oxen | |
| | Norway | | Canada | | Canada | |
	Summer	Winter	Summer	Winter	Summer	Winter
Willows	–	–	4.0	2.9	high	moderate
Shrubs	1.0	4.8	–	–	–	–
Grasses and sedges	0.8	0.5	2.9	5.5	high	high
Herbs	7.0	–	–	–	–	–
Lichens	0.5	1.0	4.0	1.1	low	low
Mosses	0.0	1.1	0.1	0.1	moderate	low

(Source: R. G. White *et al.*, Ungulates on arctic ranges, in *Tundra Ecosystems: A Comparative Analysis*, eds L. C. Bliss, O. W. Heal and J. J. Moore; published by Cambridge University Press, 1981.)

tion has never been very large and the herds were seriously depleted by hunting in the late 19th century. Recently, they have been reintroduced into various locations, but the population is still relatively small and in Canada presently totals about 85 000 animals (Barr, 1991). They live in small groups of 6–12 animals which occassionally intermingle as they wander over their range.

The principal ungulates in alpine areas are species of antelope, goat and sheep. Typical species include the chamois (*Rupricapra rupricapra*) and moufflon (*Ovis aries*) in the Alps, the argali (*Ovis ammon*) and tahr (*Hemitragus jemlahicus*) in the Himalayas, and the bighorn sheep (*Ovis canadensis*) and Rocky Mountain goat (*Oreamnos americanus*) in North America. These sure-footed animals usually roam the rocky outcrops in small groups of 3–10 animals. Population densities are higher for species that live on the mountain steppes, such as the yak (*Bos grunniens*) and Tibetan wild ass (*Equus hemionus kiang*) in the Himalayas, and the llama (*Lama peruana*), alpaca (*L. pacos*) and vicuña (*Vicugna vicugna*) in the Andes. Most ungulates move to lower elevations in winter and graze in the subalpine woodlands or in the meadows which occur below timber-line. However, species such as the Mongolian gazelle (*Procapra gutturosa*) migrate long distances across the plains of Tibet to find areas which are free of snow, while the vicuña remains all year on the grassy puna at elevations of 4000–4800 m. In this peculiar Andean environment snow falls mainly in the summer, but the vegetation protrudes through the soft blanket and the plants grow quickly because of the added moisture and summer warmth (Koford, 1957). Temperatures rarely fall below – 10 °C in the puna, but radiation is intense in the rarefied atmosphere. The fine woolly fleece of the vicuña regulates diurnal gains and losses of body heat, unlike the qiviut of the musk ox which provides insulation against the extreme cold of the arctic winter. Animals which are indigenous to the high mountains must also adapt to the shortage of oxygen through modifications of the cardiovascular system, as in the llama and related camelids, or enlarged nasal cavities as in the chiru or Tibetan antelope (*Procapra picticaudata*).

9.13 SECONDARY CONSUMERS OF THE TUNDRA

Over 100 species of bird breed in the Arctic tundra during the summer but very few are permanent residents. Most species overwinter in a habitat that is very different from the tundra in which they breed. For example, the long-tailed jaeger (*Stercorarius longicaudus*) spends winter far out over the ocean where it steals the prey of other birds. A few species, such as the arctic tern (*Sterna macrura*), occupy similar habitats in summer and winter and embark upon long migrations to Antarctica, where they circle the ice cap before returning to their northern breeding grounds. Many of the smaller birds are insectivores and search for insects on the ground or in the air: others, such as the snow bunting (*Plectrophenax nivalis*), are omnivorous and feed on seeds and insects. The geese, ducks and swans which nest by the ponds and streams mostly consume green foliage and the stems of the grasses and sedges. Predatory species, such as the merlin (*Falco columbarius*) and kestrel (*F. tinnunculus*), follow the small birds as they migrate northwards but others, such as the pomeraine jaeger (*Stercorarius pomarinus*) and long-tailed jaeger, feed mainly on lemmings (Fitzgerald, 1981). Most birds return to warmer regions during the winter but a few species remain in the Arctic all year round. The white winter plumage of the ptarmigan (*Lagopus mutus*) provides some protection from predators and its feathered feet are insulated from the cold. The snowy owl (*Nyctea scandiaca*), which rests on the ground waiting for its prey, is also well camouflaged against the snow. It occasionally feeds on ptarmigan as they break cover after resting in the snow, but mostly it feeds on lemmings which emerge from their winter burrows to forage for food. Lemmings are also taken year round by predatory mammals such as the ermine (*Mustela erminea*), which is small enough to enter the burrows and so eradicates the population in an area before moving on. In summer, the ermine and some predatory birds supplement their diet with eggs and nestlings, especially in years when lemmings are scarce. Animals such as the arctic fox (*Alopex lagopus*) will also eat berries and insects

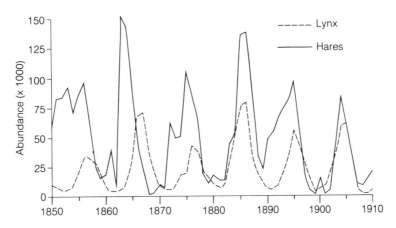

Figure 9.40 Changes in the abundance of lynx (*Lynx canadensis*) and snowshoe hares (*Lepus americanus*) based on records of pelts received by the Hudson Bay Company. (After MacLulich, 1937.) (Reproduced with permission from D. A. MacLulich, *Fluctuations in the numbers of the varying hare* (Lepus americanus), University of Toronto Studies: Biological Series No. 43; published by University of Toronto Press, 1937.)

and the remains of animals taken by wolves and polar bears. In winter the foxes depend on lemmings or dead birds which they find under the snow.

The importance of lemmings in the diets of most predators is reflected in the parallel oscillations in population densities. The effect on the predators varies according to their ability to migrate into or out of an area as the density of the prey changes. Weasels (*Mustela nivalis*) breed during the winter, when there is little alternative prey available, and are considered to have the greatest impact on lemming populations of any predator (MacLean *et al.*, 1974). Consequently, their own populations are regulated by the availability of their prey in much the same way as the classic cycle reported between arctic hares and lynx (Figure 9.40). Snowy owls are also very dependent on lemmings and the number of residents increases in years when prey is abundant. During these periods a pair of owls may rear as many as nine young, but in years when food is scarce it is practically impossible to raise a brood; most of the birds go south when the number of lemmings is low. The snowy owl population frequently increases at intervals of 3–4 years in the southern parts of their range, but these irruptions are not synchronous in all areas and are probably not connected to the lemming cycles, as was previously assumed (Kerlinger *et al.*, 1985).

Small rodents are also the principal food for predators in alpine areas of the northern hemisphere. They are preyed upon by various carnivores including weasels, bears and eagles. Larger animals are taken mainly by the predatory cats such as the lynx (*Lynx canadensis*) and cougar (*Felis concolor*) of the Rockies and the snow leopard (*Panthera uncia*) of the Himalayas. Fewer carnivores are native to the mountains of the southern hemisphere. Consequently, in New Zealand the alpine grasslands were seriously overgrazed by the unchecked growth of

the introduced red deer population: commercial aerial hunting was introduced subsequently as a method of preserving habitat for the endangered takahe (*Notornis mantelli*), a flightless bird which browses amongst the tussock grasses (Rose and Platt, 1987).

The food web in Antarctica is more unusual in that the animals are essentially dependent on the resources of the ocean. The terrestrial fauna is dominated by invertebrates which feed mainly on soil fungi and algae. Unlike the Arctic, lichens do not serve as a food source in Antarctica, although they do provide nesting materials for some sea birds (Lindsay, 1978). Birds such as the skuas (*Catharacta skua*) feed on eggs and hatchlings, but the penguins (*Spheniscidae*) and mammals have become adapted to the sea, although they rest and breed on land. Trampling and nutrient enrichment from their wastes have a considerable effect on the plant cover of Antarctica and the surrounding islands (Figure 9.41). As much as 75% of the vegetation cover has been destroyed over parts of Signy Island and South Georgia in recent years because of the dramatic increase in the fur seal (*Arctocephalus gazella*) population (Lewis Smith, 1988).

9.14 HUMAN IMPACT ON THE TUNDRA

The aboriginal peoples of the Arctic developed a subsistence economy based primarily on the animal resources available along the coast. Seals and whales were the traditional species taken in North America, although fish and seabirds were also important and in some areas the Eskimos were almost totally dependent on caribou (Riewe, 1981). Hunting intensified with the increased demand for provisions by whalers and trappers in the nineteenth century, and animals such as the musk oxen were brought almost to extinction. In Eurasia a nomadic

Figure 9.41 Biotic influences on coastal sub-antarctic vegetation: (a) seal wallows; (b) penguin colonies. ((a) D. R. Given; (b) P. Armstrong.)

way of life developed based on the domestication of the reindeer: these herds now exceed 2.5 million animals, bringing the total population, including wild animals, to about 3.5 million (Chernov, 1985). Management of the herds keeps their numbers within the carrying capacity of the range. In the Hardangervidda region of southern Norway the wild herds are maintained at a population ranging from 10 000 to 13 400 animals, with about 2300 animals being harvested each fall (Østbye *et al.*, 1975).

Opening up the Arctic for oil and mineral exploration in recent years has proved to be the biggest threat to this

Figure 9.42 Eight years after it was used the route of a summer seismic line is still clearly visible with ice wedge thaw occurring where it crosses patterned ground. (L. C. Bliss.)

relatively pristine ecosystem, although such activities are now preceded by comprehensive environmental impact studies. Disturbance of the tundra vegetation cover results in greater absorption of solar energy by the dark surface soils and a subsequent change in the permafrost table (Babb and Bliss, 1974). Thermokarst activity occurs as the ground ice melts, and the land surface may become pitted with thaw lakes and disturbed by slumps and gullying (Figure 9.42). The tundra is particularly sensitive to summer use of tracked vehicles, and even a single pass can result in water-filled ruts which continue to deepen because of the increased thermal conductivity of the saturated peat. The effects are most pronounced in soils with a high ice content. This can result in the formation of deep ponds and a change in the composition of the plant cover which may take hundreds of years to restore (Figure 9.43). Air-cushion vehicles and balloon-tyre vehicles have minimal impact when used in winter, provided that the ground is frozen to a depth of 20 cm and the vegetation is protected by at least 10 cm of snow (S. A. Harris, 1986). Summer use depresses the vegetation mat: this reduces its insulating quality and causes soil temperatures to rise during the summer. Decomposition is accelerated by the warmer conditions and plant growth is stimulated by the increase in available nutrients (Chapin and Shaver, 1981). The effect is comparable to heavy grazing, and recovery time for herbaceous communities is about 2–15 years depending on the nature of the disturbance.

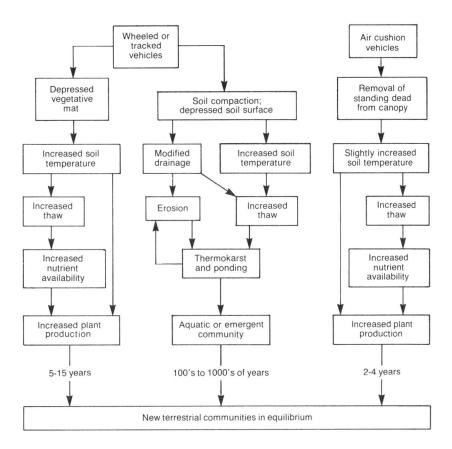

Figure 9.43 Impact of air-cushion vehicles and wheeled and tracked vehicles upon arctic tundra at Barrow, Alaska. (After Webber *et al.*, 1980.) (Reproduced from P. J. Webber *et al.*, The vegetation: pattern and succession, in *An Arctic Ecosystem: The Coastal Tundra at Barrow, Alaska*, eds J. Brown, P. C. Miller, L. L. Tieszen and F. L. Bunnell; published by Dowden, Hutchinson and Ross, 1980.

Permafrost has also been an important factor in the construction and maintenance of pipelines used to transport oil and gas across the tundra. The 1300 km Alyeska pipeline crosses tundra in the northern part of its route from Prudhoe Bay on the Arctic coast to Valdez on the Gulf of Alaska. It is elevated above ground wherever the soils are sensitive to thaw (Figure 9.44) and special crossing points have been established to accommodate the movements of caribou and moose. However, road traffic, work crews and aerial harassment appear to be more disruptive to the animals than the pipeline itself (Bliss, 1983). The other concern is the effect of oil spills on the ecosystem. Crude oil kills plant leaves on contact, but if it does not soak into the soil the effect is minimal because growth of latent buds and stems continues (Figure 9.45). Less energy is reflected by the plant cover under these conditions and this causes a temporary increase in soil temperature, similar to that associated with defoliation by lemmings: the vegetation grows back to its original condition within a few years. However, most plant roots and animals within the upper soil horizons are killed if oil penetrates into the soil before the toxic volatile fraction has evaporated. Damage is usually less severe in wet soils where oil penetration is slower. The addition of carbon-rich, nutrient-poor oil increases microbial demand for soil nutrients, reducing their availability to plants. Reduced water movement within oil-soaked soil is also detrimental to plant growth (Webber *et al.*, 1980).

Figure 9.44 Elevated section of the Alyeska pipeline running across tundra meadow north of the Brooks Range, Alaska. Winter construction from a snowpad minimized environmental disturbance. (W. Barr.)

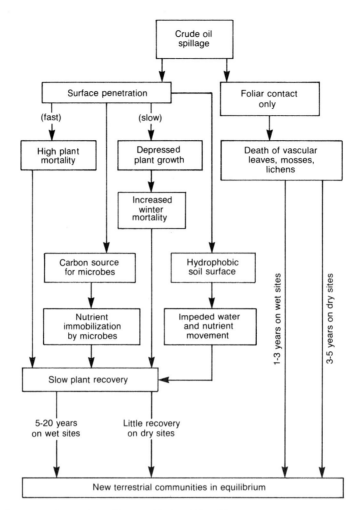

Figure 9.45 Impact of crude oil upon arctic tundra at Barrow, Alaska. (After Webber *et al.*, 1980.) (Reproduced from P. J. Webber *et al.*, The vegetation: pattern and succession, in *An Arctic Ecosystem: The Coastal Tundra at Barrow, Alaska*, eds J. Brown, P. C. Miller, L. L. Tieszen and F. L. Bunnell; published by Dowden, Hutchinson and Ross, 1980.

Human impact in tundra regions is minimized by their inaccessibility. However, wilderness activities have increased considerably in recent years, especially in mountain areas, and serious damage from trampling has been reported in alpine tundra communities (Willard and Mahr, 1970). The human imprint is now evident even in the remotest areas because of global circulation of pollutants. A classic example is the presence of DDT in penguins in Antarctica (Sladen *et al.*, 1966). The transfer of radionuclides through the short Arctic food chain has also been of concern (Osburn, 1974). This problem of biomagnification reflects the inherent simplicity of tundra ecosytems. Ecosystems with low species diversity have traditionally been considered to be very sensitive to disturbance because the loss of a few species will greatly alter community structure and composition. It has also been argued that ecosystem stability is independent of its diversity and is simply a reflection of the ability of the component species to cope with environmental perturbations (May, 1976). Thus, most tundra species have adjusted to their severe and variable environment by diverting a large proportion of their resources into underground storage. However, slow growth rates reduce the chance of recovery following frequent and repeated disturbance. Under these conditions the resilience of the tundra ultimately depends on the geographic scale of the disturbance (Dunbar, 1973). The vast Arctic region is therefore considerably less vulnerable than the small discontinuous areas of alpine tundra.

CHAPTER 10

Terrestrial wetlands

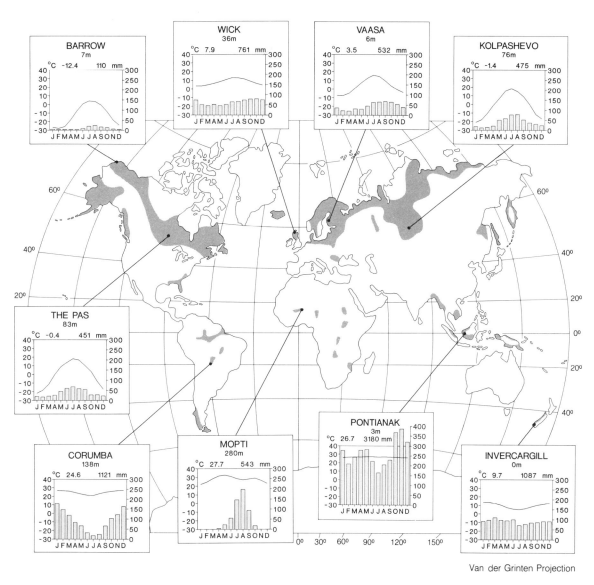

Figure 10.1 Distribution of terrestrial wetlands and representative climatic conditions. Mean monthly temperatures are indicated by the line and mean precipitation for each month is shown by the bars. Station elevation, mean annual temperature and mean annual precipitation appear at the top of each climograph.

10.1 INTRODUCTION

Terrestrial wetlands are specialized plant communities that occur where soil remains saturated for most of the year. The natural diversity of wetland communities reflects the different conditions in which they form. The cool, moist climate at higher latitudes favours the extensive development of bogs and fens, but in temperate and tropical regions swamps and marshes are more widespread. **Bogs** and **fens** are mainly distinguished on the basis of the origin and chemistry of the water that moves through them. Bogs are fed only by the precipitation that falls directly on them and the supply of nutrients is therefore very limited. The peat which accumulates beneath bogs is mainly derived from *Sphagnum* and woody ericaceous shrubs that are well adapted to the acid conditions. Fens are mostly supplied by water from the surrounding area which has picked up nutrients in its movement through the adjacent mineral soils. Sedges and grasses grow abundantly in these minerotrophic conditions. Cool temperatures and anaerobic conditions caused by the high water table favour the accumulation of peat in bogs and fens, but in other wetland communities where organic material breaks down more rapidly various unconsolidated mucks and flocculent mineral soils develop. In **marshes**, decomposition is favoured by seasonal drawdown which periodically exposes the matted vegetation and mudflats. The dominant species in these communities are emergent sedges, rushes and reeds, with grassy meadows and zones of shrubs and trees present where flooding is less severe. Submerged and floating aquatic plants are usually present in deeper channels and pools. Communities in which trees have become established over waterlogged terrain are generally termed **swamps**. The large percentage of aquatic species present in the flora distinguishes swamps from woodlands which grow on river floodplains and other sites which are periodically inundated.

Wetlands develop wherever the interactions of climate, topography and soil conditions cause the water table to remain at or near the surface for much of the year, and so they occur in various climatic regions (Figure 10.1). The water balance of a region is determined climatically by the amount and distribution of precipitation and seasonal evaporation rates. Heavy rainfall does not necessarily induce wetland development. The area of wetlands in the rainy tropics is comparatively small, yet extensive swamps occur around Lake Chad, where total precipitation averages 300 mm yr^{-1}. Most of the world's wetland area is associated with the cool northern climates where evaporation rates are relatively low. However, heavy

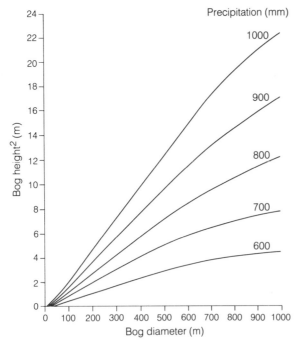

Figure 10.2 The relationship between bog area and height under different total annual precipitation regimes. (After Wickman, 1951.) (Reproduced with permission from F. E. Wickman, The maximum height of raised bogs and a note on the motion of water in soligenous mires, *Geologiska Föreningens i Stockholm Förhandlingar*, 1951, **73**, 419.)

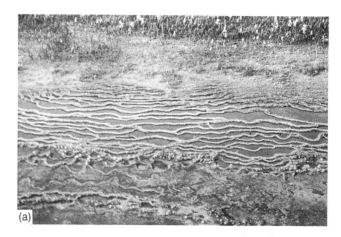

Figure 10.3 (a) String fen in Labrador with shallow pools enclosed by peaty ridges; (b) Stunted larch (*Larix laricina*) trees growing in a ribbed fen in Saskatchewan. ((a) S. C. Zoltai; (b) O. W. Archibold.)

precipitation is usually necessary for the maintenance of deep blanket bogs since they are supplied only by meteoric water (Figure 10.2). The development of fens and swamps in regions with lower precipitation is largely controlled by topographic conditions which affect surface and subsurface drainage.

10.2 REGIONAL DISTRIBUTION

Wetlands cover about 2.5 million km² of the world's land surface and most of this occurs in the boreal and tundra regions of the northern hemisphere. The peatlands of Canada, Scandinavia and the former USSR are the most extensive: together they account for nearly 90% of the total area (Figure 10.1). Most of the remaining area is found in Alaska but peat deposits also occur widely in the British Isles, Germany, Poland and Iceland. In the southern hemisphere the peatlands are mainly confined to New Zealand, southern Chile and the sub-antarctic islands. Swamps and marshes occur extensively in poorly drained areas throughout the world but are especially characteristic of the wet tropical regions of South East Asia, South America and Africa; similar conditions occur in the Everglades of the south-eastern United States. A variety of wetland habitats occur in temperate regions ranging from extensive marshes such as those in the Sanjiang Plain in north-east China to the numerous small waterlogged depressions that form the prairie pothole region of western interior North America.

10.2.1 The wetlands of Canada

Approximately 50% of the world's wetlands are found in Canada, where they are mainly associated with regions of low relief in the boreal forest. Peat accumulations up to 5–6 m thick underlie the bogs and fens which have developed in depressions throughout this region, with thinner organic deposits occurring in the marshes that develop along lake shores and river deltas (Zoltai and Pollett, 1983). The bogs are dominated by *Sphagnum* mosses and ericaceous shrubs such as bog rosemary (*Andromeda polifolia*), Labrador tea (*Ledum groenlandicum*), bog-laurel (*Kalmia polifolia*) and cranberry (*Oxycoccus microcarpus*). Black spruce (*Picea mariana*) may also be present, with larch (*Larix laricina*) and eastern cedar (*Thuja occidentalis*) more common in the richer fens where the tree cover is better developed. Many insectivorous plants are also associated with the oligotrophic bogs of this region including pitcher plants (*Sarracenia purpurea*) and sundews (*Drosera* spp.). The nitrogen derived from digesting animal protein supplements the small amounts which are available from the peaty soils (Larsen, 1982).

Patterned fens which are crossed by low ridges of peat become increasingly common in the northern boreal forests and low arctic regions (Figure 10.3). Where the

Figure 10.4 Black spruce (*Picea mariana*) establishing on small emerging ice-cored palsas in poor fen comprised of *Sphagnum viviparium* and *S. angustifolium*. (S. C. Zoltai.)

ground is almost level the ridges form broad nets around wet hollows and shallow pools, although in gently sloping terrain they are arranged across the gradient in a ladder-like pattern. A scattered cover of stunted larch or black spruce (*Picea mariana*) and ericaceous shrubs is usually present on ridges: this may give way to communities of shrubby birch (*Betula glandulosa*), grass (*Calamagrostis canadensis*) and brown mosses (*Aulacomium palustre, Tomenthypnum falcifolium*) in the driest sites. Plants such as bog-bean (*Menyanthes trifoliata*), *Sphagnum* mosses, sedges and cotton-grasses may be present in the low-lying 'flarks', depending on the depth of water (Vitt *et al.*, 1975). Peat plateaus and smaller ice-cored 'palsas' are also distinctive features in this region (Figure 10.4). They are pushed up above the general level of the wetlands by lenses of ice which form at the base of the peat. These islands usually support black spruce, mosses such as *Hylocomium splendens* and *Pleurozium schreberi*, and other species normally associated with drier sites in the boreal forest.

The wetlands become progressively restricted by decreasing precipitation north of the tree-line, and even though moisture-tolerant species such as the rigid sedge (*Carex bigelowii*) and cotton-grass (*Eriophorum vaginatum*) are common in the region (Figure 10.5), much of the tundra is excluded from wetland classifications because it does not remain waterlogged throughout the year. Here the wetlands are mostly confined to the lowest depressions, or are associated with the patterned ground features that develop in permafrost. The polygonal cracks that develop in the frozen ground are enlarged by ice wedges and the soil that is displaced forms a shoulder around the polygon (Figure 10.6). Small ponds begin to appear which gradually fill with peat derived from sedges, cotton-grasses and mosses such as *Calliergon giganteum* and *Drepanocladus revolvens*. As the centres continue to build they begin to dry out and *Sphagnum*

Figure 10.5 Cotton-grass (*Eriophorum angustifolium*) in boggy tundra at the arctic tree-line. (O. W. Archibold.)

Figure 10.7 Spreading mat of buckbean (*Menyanthes trifoliata*) in a northern Saskatchewan wetland. (R. A. Wright.)

mosses and ericaceous shrubs become established (Zoltai and Pollet, 1983). In the high Arctic the wetlands are mainly restricted to seepage sites or small basins which are fed by meltwater. *Carex stans* is the only abundant vascular species in these sites; most of the cover is provided by *Drepanocladus revolvens* and other mosses.

The distribution of wetlands is more restricted across southern Canada and many of the bogs and fens have evolved through colonization of open water areas. In forested regions, submerged hydrophytes and floating leaf aquatics such as yellow pond-lily (*Nuphar variegatum*) are represented in the pond stage of the hydrarch succession, and sedges and reeds grow in the shallows (Danserau and Segadas-Vianna, 1952). With time cotton-grasses form a floating mat which extends out from the shore, especially where support is provided by logs and other materials. Plants such as buckbean (*Menyanthes trifoliata*) send out buoyant rhizomes through the mat. As it thickens (Figure 10.7), shrubs such as sweet gale

Figure 10.6 Low-centered polygons colonized by various *Sphagnum* species and underlain by permafrost. (S. C. Zoltai.)

Figure 10.8 Black spruce growing in a small basin bog in northern Saskatchewan. (R. A. Wright.)

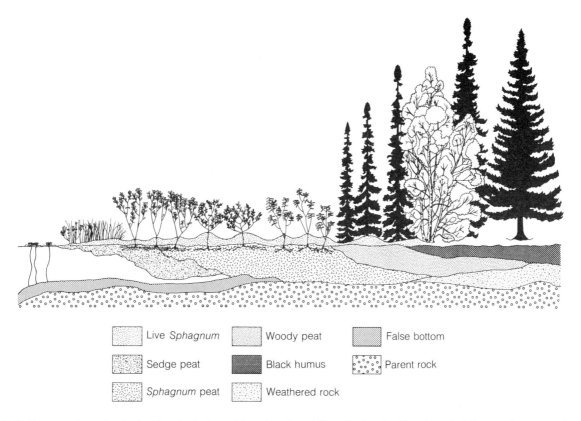

▦	Live *Sphagnum*	▦	Woody peat	▦	False bottom
▦	Sedge peat	▦	Black humus	▦	Parent rock
▦	*Sphagnum* peat	▦	Weathered rock		

Figure 10.9 Transect through a typical bog in the boreal forest region of Canada showing the characteristic zonation generally attributed to successional development. (After Dansereau and Segadas-Vianna, 1952.) (Reproduced with permission from P. Dansereau and F. Segadas-Vianna, Ecological study of the peat bogs of eastern North America I. Structure and evolution of vegetation, *Canadian Journal of Botany*, 1952, **30**, 499.)

(*Myrica gale*), leatherleaf (*Chamaedaphne calyculata*) and Labrador tea become established: the intervening spaces are slowly invaded by *Sphagnum* moss. Black spruce and larch are typically the first trees to grow into the bog as the depression becomes filled with peat (Figure 10.8), but eventually these may be replaced by other conifers or hardwood species such as birch and maple (Figure 10.9). The overgrowth of lakes in this way is referred to as **terrestrialization**, and the rate at which it proceeds in any region is largely determined by the trophic nature of the water and the shape and depth of the lake. In other areas there is evidence that bogs have spread into previously wooded sites by the process of **paludification**. This usually results from a change in climate which causes the land to become oversaturated with water so that peat-forming plants can invade mineral soil. Much of the peatland in the southern boreal forests and around the Great Lakes has developed in this way (Heinselman, 1975).

Peatlands are also characteristic of the mild, wet maritime parts of Canada. Most begin as swamps or fens in moist depressions, but as the surface peats build beyond the influence of the mineral-rich seepage waters there is a gradual succession to bog conditions. In Newfoundland the raised bogs are dominated by *Sphagnum fuscum*, which forms drier hummocks, with *S. magellanicum* and

S. pulchrum more abundant in wetter sites (Figure 10.10). Bog laurel, Labrador tea, small cranberry (*Vaccinium oxycoccus*) and other wide-ranging species grow amongst the mosses. The mosses are mostly replaced by low sedges in the minerotrophic fens which lead out from the bogs, although dense mats of *Sphagnum pylaesii* typically form in the wetter parts of the fens. Bog rosemary (*Andromeda glaucophylla*), sweet gale, leatherleaf and other dwarf shrubs increase in abundance in these

Figure 10.10 Concentric raised bog in Labrador. (S. C. Zoltai.)

Figure 10.11 Prairie potholes surrounded by arable land in Saskatchewan. (P. Grilz.)

Figure 10.12 The prairie wetland region: (a) trembling aspen (*Populus tremuloides*) surrounding zones of sedges and willows in an open water marsh; (b) a shallow marsh in mid-summer with species such as spangletop (*Scolochloa festucacea*), awned sedge (*Carex atherodes*) and creeping spike-rush (*Eleocharis palustris*) occuring within the fringing willows. (R. Zdan.)

nutrient-rich fens (Pollett and Bridgewater, 1973). Open stands of stunted lodgepole pine (*Pinus contorta*) are indicative of bog conditions in the Pacific Northwest. The shrub layer is dominated by bog laurel, Labrador tea and sweet gale with *Sphagnum fuscum*, sedges and cotton-grasses completely covering the ground. Western red cedar (*Thuja plicata*) often grows in swampy areas, where it is usually associated with red alder (*Alnus rubra*). Devil's club (*Oplopanax horridum*), skunk cabbage (*Lysichitum americanum*) and slough sedge (*Carex obnupta*) are the characteristic understory species in the wettest sites where standing water may remain all year. Where the water table is not so high spirea (*Spiraea douglasii*) and salmonberry (*Rubus spectabilis*) are usually present, together with ladyfern (*Athyrium filix-femina*), deerfern (*Struthiopteris spicant*) and various herbs.

Wetlands are an important part of the landscape in the prairies of central Canada and adjacent grasslands in the United States. The northern plains are underlain by fine-textured glacial materials and surface drainage collects in the numerous shallow depressions which characterize the prairie pothole region (Figure 10.11). Precipitation across this 700 000 km² region is comparatively low; most recharge occurs during the spring as the winter snow cover melts. Much of the moisture is subsequently lost through evaporation, and the water level gradually drops over the course of the summer (Figure 10.12). The semi-permanent wetlands are usually bordered by willows such as *Salix bebbiana* and *S. discolor* with various grasses and sedges, including *Poa palustris, Calamagrostis canadensis, Juncus balticus* and other herbs forming a wet meadow community (Millar, 1976). This changes to shallow marsh in wetter areas where the dominant species include reed grass (*Phalaris arundinacea*), awned sedge (*Carex atherodes*) and spangletop (*Scolochloa festucacea*): cattail or reedmace (*Typha latifolia*), bulrush (*Scirpus acutus*) and common reed grass (*Phragmites communis*) may form an emergent deep marsh zone adjacent to the open

water. Precipitation varies considerably in this semi-arid region, and the vegetation is influenced by the salinity of the water as well as the water regime: halophytic species such as Nuttall's salt meadow grass (*Puccinellia nuttalliana*), red samphire (*Salicornia rubra*) and western sea-blite (*Suaeda depressa*) are common in some communities.

10.2.2 The wetlands of the United States

Most of the wetlands in the United States are contiguous with similar communities in Canada but areas such as the Florida Everglades have no northern counterpart (Figure 10.13). The Everglades originally covered about 10 000 km² extending from Lake Okeechobee to the brackish marshes and mangrove swamps around the coast of southern Florida. Throughout this region the dominant species is saw-grass (*Cladium jamaicense*): it often grows in pure stands with plants over 3 m high (Hofstetter, 1983). Maiden-cane (*Panicum hemitomon*) and emergent aquatic plants, such as broad-leaved arrowhead

Figure 10.13 Saw palmetto and dwarf palmetto (*Serenoa repens* and *Sabal minor*) are characteristic understory species in Florida swamp forests. (M. Brown.)

Figure 10.14 Bald cypress (*Taxodium distichum*) swamp forest in the Mississippi Delta. (R. A. Wright.)

(*Sagittaria latifolia*) and southern swamp lily (*Crinum americanum*), add to the diversity of these shallow marsh communities. Most of these areas are flooded during the summer, when precipitation is heaviest, but they dry out in winter and spring and surface water only remains in the deeper depressions and in ponds created by alligators. Pond-lily (*Nuphar luteum*), fragrant water-lily (*Nymphaea odorata*) and other submerged and floating aquatic plants occur in the permanent shallow ponds. Shrubs such as buttonbush (*Cephalanthus occidentalis*) are occasionally present in the swamps, but woody plants are mostly restricted to the '**tree islands**' that form on limestone platforms and other sites which are slightly elevated above the surrounding wetlands. Here the dominant species include holly (*Ilex cassine*), wax-myrtle (*Myrica cerifera*) and red bay (*Persea borbina*), which grow to heights of 10–12 m, but taller trees such as sweet bay (*Magnolia virginiana*) and live oak (*Quercus virginiana*) may also be present. In swampy areas tree islands can develop on peat masses or '**batteries**' which break loose and float to the surface. Buttonbush and bald cypress (*Taxodium distichum*) are usually the first species to become established, quickly followed by leatherwood (*Cyrilla racemiflora*) and holly: fetterbush (*Lyonia lucida*), wax-myrtle and maleberry (*Lyonia ligustrina*) appear later as the surface builds up and conditions become drier (Duever and Riopelle, 1983).

Another type of shrub-dominated wetland is represented by the '**pocosins**' which occur in low-lying areas throughout the south-eastern United States. These communities are mostly ombrotrophic in that paludification has raised their surfaces so much above the mineral soil that all nutrients are supplied by rainfall (Christensen, 1988). *Sphagnum* mosses grow abundantly in these nutrient-limited sites, accompanied by a shrub cover mainly comprised of leatherwood and *Zenobia pulverulenta*, occasionally overtopped by pond pine (*Pinus serotina*). The height and density of the shrub cover increases in more productive sites and a greater number of species is represented. In the absence of fire, white cedar (*Chamaecyparis thyoides*) and pond cypress (*Taxodium ascendens*) become established in some pocosins and eventually a swamp forest community develops.

Swamp forests are characteristic of the river bottomlands and other low relief areas in the south-eastern United States where water tables have risen as a result of post-glacial changes in sea level. Bald cypress is widely distributed throughout these communities and is usually associated with flowing water. The 'knees' which develop from swellings on the upper surface of the horizontal roots are a distinctive feature of this species (Figure 10.14). The smaller pond cypress is more commonly found in areas where water collects and stands for several months during the year: knees do not usually develop in this species. Water tupelo (*Nyssa aquatica*) often grows in association with the cypresses; and smaller trees, such as water elm (*Planera aquatica*) and red bay, and many species of shrub are common in the understory of these swamp forests. Black gum (*Nyssa sylvatica*) and red maple (*Acer rubrum*) grow where the land is flooded less frequently, and in sites which only remain waterlogged during the winter species such as laurel oak (*Q. laurifoia*), elm (*Ulmus americana*) and sweet gum (*Liquidamber styracifolia*) increase in abundance. In some areas open stands of slash pine (*Pinus elliottii*) have developed with their distinctive understories of saw palmetto (*Serenoa repens*) and shrubs such as galberry (*Ilex glabra*) and fetterbush; insectivorous plants such as the sundew (*Drosera capillaris*) and the Venus fly-trap (*Dionaea muscipula*) occur in the herb layer.

10.2.3 The wetlands of western Europe

The most extensive wetlands in western Europe are associated with the boreal forests of Finland. Wetlands are

Figure 10.15 Aapa mire with Scots pine (*Pinus sylvestris*) growing on the strings and hummocks. (S. C. Zoltai.)

amount of runoff which flows into them. In typical sites *Sphagnum fuscum* is the dominant species in the strings. Other mosses such as *Drepanocladus exannulatus* may be present in the flarks but often they remain as open pools. A swampy zone of Scots pine (*Pinus sylvestris*), Norway spruce (*Picea abies*) and birch (*Betula pubescens*) usually surrounds the mires. Narrow-leaved Labrador tea (*Ledum palustre*), dwarf birch (*Betula nana*) and sedges such as *Carex aquatilis* increase in more northerly mires and willows are found in the adjacent swamps. This contrasts with the lawn-like cover of hair's-tail cotton-grass (*Eriophorum vaginatum*) and *Sphagnum papillosum* which is characteristic of the aapa mires in southern Finland. Various types of raised bogs with irregular microrelief are more common in southern Finland. Most of the hummocks support a stunted cover of Scots pine, and beneath this is a shrub layer composed of crowberry (*Empetrum nigrum*), cloudberry (*Rubus chamaemorus*) and *Chamaedaphne calyculata*: heather (*Calluna vulgaris*) is prominent on some of the drier strings. Cotton-grass and *Sphagnum fuscum* are present in the field layer. In the intervening hollows the cover is mainly *Sphagnum balticum*, cotton-grass, white-beak sedge (*Rhynchospora alba*) and Rannoch-rush (*Scheuzeria palustris*). Palsas are present in areas of sporadic permafrost in northern Finland: in these wetlands the cover is mainly *Sphagnum fuscum*.

In Sweden the dominant wetlands are nutrient-poor fens which have developed over the acid soils derived

especially characteristic of the recently emerged lowlands along the northern coast where paludification has occurred continuously for the past 8000 years (Ruuhijärvi, 1983). The principal wetland type in Finland is the gently sloping, minerotrophic '**aapa**' mire (Figure 10.15). It is analogous to the patterned fen of Canada in that a network of dry strings subdivides the treeless centre of the mire into a series of wet flarks. Various types of aapa mire develop, depending on nutrient levels and the

Table 10.1 Distribution of selected species in ombrotrophic and minerotrophic bogs in Sweden (after Sjörs, 1983)

	Ombrotrophic bog	Poor fen	Intermediate fen	Rich fen
Vascular plants				
Eriophorum vaginatum	X	X	X	
Scheuchzeria palustris	X	X	X	
Carex limosa	X	X	X	X
Eriophorum angustifolium		X	X	X
Myrica gale		X	X	X
Menyanthes trifoliata		X	X	X
Carex lasciocarpa		X	X	X
Molinia caerulea		X	X	X
Scirpus hudsonianus			X	X
Carex flava			X	X
Thalictrum alpinum				X
Schoenus ferrugineus				X
Juncus triglumis				X
Carex atrofusca				X
Bryophytes				
Sphagnum balticum	X			
Sphagnum cuspidatum	X			
Sphagnum papillosum		X	X	
Drepanocladus fluitans	X	X		
Drepanocladus purpurascens		X	X	
Calliergon sarmentosum			X	
Paludella squarrosa			X	X
Drepanocladus intermedius				X
Calliergon turgescens				X

(Source: H. Sjörs, Mires of Sweden, in *Ecosystems of the World, Vol. 4B. Mires: Swamp, Bog, Fen and Moor*, ed. A. J. P. Gore; published by Elsevier Science Publishers, 1983.)

Figure 10.16 Borth Bog in Wales, a characteristic European raised mire, is mainly comprised of *Sphagnum* mosses, heath (*Erica tetralix*), cotton-grass (*Eriophorum angustifolium*) and heather (*Calluna vulgaris*.) (J. P. Barkham.)

Figure 10.17 Reeds (*Phragmites communis*) and cattails (*Typha latifolia*) growing beneath alder (*Alnus glutinosa*) is the typical cover in rich fens of East Anglia. (P. Wakely, English Nature.)

from siliceous bedrock. They are dominated by *Sphagnum* mosses but other species such as cotton-grass, bog bean and sweet gale may be present, depending on the microtopography of the surface. The '**poor fen**' is one of several mire types distinguished by differences in species composition associated with nutrient availability (Table 10.1). Floristic composition is similar to that of Finnish mires. Lawn and carpet communities dominated by the sedges and grasses occur in the moister, nutrient-rich fens and will also develop in shallow bogs if the plant roots reach the underlying mineral soil. A dense cover of tall sedges, such as *Carex lasiocarpa* and *C. rostrata*, is characteristic of the most productive sites. Fewer species are found in the ombrotrophic raised bogs which have developed in low-lying areas. Here small shrubs such as crowberry are present on the hummocks: deergrass (*Scirpus caespitosus*) and white-beak sedge are common species in the hollows, and Scots pine usually grows around the bog margins.

Blanket bogs do not occur in Sweden but are increasingly conspicuous in Norway, Ireland and the British Isles, where precipitation is higher. *Sphagnum* mosses are the dominant species in these bogs, the different species being distributed according to moisture levels and nutrient conditions. Faster-growing species, such as *Sphagnum rubellum*, form dry hummocks in highly acidic habitats and are important peat formers, whereas *S. cuspidatum* is prevalent in the wet hollows and does not grow much above the water table. White-beak sedge replaces *S. cuspidatum* in the hollows if the surface is slowly built up. Black bog-rush (*Schoenus nigricans*) is a distinctive species in the bogs of western Ireland. It is usually associated with calcareous swamps or fens; its presence in blanket bogs may be attributed to changes in soil acidity caused by sea spray (Tansley, 1953). Purple moor-grass (*Molinia caerulea*), often accompanied by heather (*Calluna vulgaris*), forms a peaty turf in drier sites, and hare's-tail cotton-grass increases in importance. These species also dominate the various bog communities in the western Highlands of Scotland and other mountainous areas (Figure 10.16).

The floristic composition of the bogs changes where the seepage waters are less acidic. *Sphagnum* mosses are largely replaced by 'brown mosses'; various herbs such as marsh buttercup (*Ranunculus acris*) and marsh willowherb (*Epilobium palustre*) grow amongst the grasses and sedges. The richest fenlands, such as those in East Anglia, occur where mineral-rich waters seep through the peats and silts. Here the dominant plants are sedges and rushes: many species of flowering plants are also present, but there are few grasses (Figure 10.17). Woody plants establish on slightly elevated sites which remain above water during the wetter winter months. Alder (*Alnus glutinosa*) is usually the dominant tree in these mature '**carrs**': hairy white birch (*Betula pubescens*), willow (*Salix atrocinerea*), buckthorn (*Frangula alnus*) and ash (*Fraxinus excelsior*) are common associates. A well-developed shrub layer grows beneath this canopy, together with a great variety of herbs. Reeds (*Phragmites communis*), bulrush (*Schoenoplectus lacustris*) and cattail replace the fen plants around open patches of water. Similar reed-swamps occur throughout the marshlands of Europe.

10.2.4 Wetlands of the former USSR

The wetland ecosystems of the former USSR cover more than 800 000 km². They are similar in form and floristic composition to the wetlands of Canada and Scandinavia. In the arctic regions of central and eastern Asia the typical wetlands are polygonal mires which consist of wet hollows 10–30 m in diameter surrounded by drier ridges (Botch and Masing, 1983). *Arctophila fulva, Dupontia fisheri* and other grasses grow in the hollows with water sedge (*Carex stans*) and hypnoid mosses such as *Drepanocladus* and *Mnium. Sphagnum* mosses are found on the ridges accompanied by grasses and sedges and the diamond-leaf willow (*Salix pulchra*). Cotton-grasses add to the diversity of the wet hollow communities where seepage waters accumulate around pools and river banks. Palsas are more common in the forest–tundra regions of European Russia and Siberia as far east as the Yenisey River. The height of the frozen mounds is about 0.5 m in the northernmost regions but increases to 3.0–4.0 m towards the south: palsas up to 8.0 m in height are reported in some parts of Siberia. The palsas are mostly covered with dwarf birch, narrow-leaved Labrador tea, and bog whortleberry (*Vaccinium uliginosum*) and beneath these is a layer of mosses and lichens; Scots pine, Siberian larch (*Larix sibirica*) and hairy white birch may also be present. Cotton-grasses, sedges and *Sphagnum* mosses are found in the wet hollows. Aapa mires similar to those in Finland occur in the Kola peninsula and around the White Sea. The surface of these treeless wetlands is slightly concave but superimposed on this is a micro-relief of ridges and hollows. Dwarf birch occurs on the ridges where it grows with various shrubs, sedges and deer grass. *Sphagnum fuscum* is common in these drier sites, but is replaced by other mosses in the hollows, where the vegetation is mainly sedges.

The most widely distributed wetland type in northern Asia is the raised string bog. They are especially common in the lowlands of western Siberia, where the cool climate and slow run-off has resulted in extensive paludification. Bogs cover 50–75% of the land in this vast region and account for about half of the total wetland area of the former USSR (Botch and Masing, 1983). The vegetation is mainly determined by the concentric pattern of ridges, hollows and pools which develops on the dome-shaped surface of the bogs. *Sphagnum balticum* grows abundantly in the hollows and marginal channels or 'laggs' which collect the drainage water; *S. fuscum* is the typical moss associated with the dry ridges and it is often accompanied by heather. Scots pine and Siberian pine (*Pinus sibirica*) increase in abundance towards the south, where conditions becomes progressively drier, and the bogs are eventually replaced by sedge fens and marshes. Wetlands occur less frequently in the mountainous eastern districts and are found mainly in the river valleys. Conditions favour the development of sedge fens with a varied cover of trees and shrubs. In Kamchat-ka and other wet coastal regions, blanket bogs of *Sphagnum* have developed in association with sedges, shrubs such as crowberry, and occasionally Dahurian larch (*Larix gmelini*).

10.2.5 The wetlands of the southern hemisphere

The principal wetland regions in the southern hemisphere are associated with the cool, humid climates of Patagonia and New Zealand. *Sphagnum* mosses grow abundantly under these conditions and the presence of cosmopolitan species, such as *S. magellanicum*, provides a floristic link with communities in the northern hemisphere.

In Patagonia raised bogs are best developed in areas which lie south of 52 °S. *Sphagnum magellanicum* is the dominant species at higher elevations and in areas where annual precipitation is 600–1500 mm (Pisano, 1983). It is mostly accompanied by sedges and grasses but evergreen shrubs such as *Empetrum rubrum* become more abundant as the surface builds further above the water table. *Sphagnum* is replaced by other mosses in very wet areas near sea level where heavy precipitation and level terrain keep the vegetation permanently soaked in nutrient-poor water. Where nutrient levels are higher, most of the ground is covered by liverworts and a discontinuous cover of shrubs and grasses usually develops: small trees such as *Nothofagus betuloides* and *N. antarctica* may be present in better drained sites. Cushion-plant bogs occur on sloping terrain where nutrient availability is maintained by abundant precipitation draining across a mineral substratum. The dominant species are *Donatia fascicularis* and *Astelia pumila*, but many other cushion-forming plants and a variety of herbs, grasses, sedges, ferns, shrubs and small trees add to the diversity of these communities. Similar communities dominated by *Bolax gummifera* and *Azorella selago* form thin blanket bogs over bedrock at higher elevations.

Cushion bogs also occur in New Zealand and are characteristic of the high plateaus in the South Island (Figure 10.18). Here the common species are *Donatia novae-zealandiae* and *Phyllachne colensoi* (Campbell, 1983). Bogs and swamps are present in poorly drained sites at lower elevations. *Sphagnum* mosses accompanied by sedges such as *Baumea teretifolia* are the dominant plants in most of the bogs which form in heavy rainfall areas with poor drainage. The 'pakihi bogs' are distinguished by a dense cover of umbrella fern (*Gleichenia dicarpa*): they are associated with the acid, gleyed podsols which overlie impervious iron pans in the wet coastal regions of the South Island. Communities similar to raised bogs have developed in parts of the North Island as a result of accumulations of wet fibrous peat derived from rush-like restiads such as *Sporadanthus traversii* and *Empodisma minus*. Heath-type shrubs are found throughout the wetlands. The most common species is manuka (*Leptos-*

permum scoparium) which grows 4–6 m high and forms the overstory in many communities. It is also common in swamp forests, where it grows beneath kahikatea (*Podocarpus dacrydioides*). Around pools and lakes and in areas which flood periodically, the dominant plants are herbaceous species such as raupo (*Typha orientalis*) or New Zealand flax (*Phormium tenax*), which often forms stands so dense that few other species can complete (Figure 10.19).

10.2.6 Tropical wetlands

The most extensive wetlands in tropical regions are the swamp communities which have developed on river floodplains and around lakes. Swamp forest is the dominant wetland type in wet tropical areas, but where the climate is more seasonal poorly drained sites are mostly colonized by grasses, sedges and reeds. Swamp forests are found throughout the coastal lowlands of Malesia and other parts of South East Asia. They are underlain by coarse woody peat which in some areas is up to 17 m thick. The dome-shaped surface of these deposits is similar to the raised bogs of temperate regions. Similarly, the peat is very low in nutrients and extremely acid, with typical pH values of 3.5–4.5 (Anderson, 1983). A concentric pattern of forest types characteristically develops in these swamps. Around the perimeter of each swamp the vegetation is structurally similar to the multi-storied lowland evergreen forests that occur on mineral soils, and emergent trees grow to heights of 40–50 m. The water table remains close to the surface throughout the year in these communities and many of the species develop stilt roots or buttresses. The composition of the forest changes towards the centre of the swamps and species diversity declines. The trees become progressively smaller and few grow taller than 20 m or exceed 30 cm in diameter. However, stand density increases and the number of pole-sized trees can reach 1300 ha^{-1}. Continued development raises the level of the ground above the water table, and the forest is replaced by open scrubby woodland comprised of relatively few xerophytic species.

In South America most of the forested swamps are associated with the Amazon River and its tributaries. These communities are affected by seasonal variations in water level as well as differences in water chemistry. In the upper Amazon the depth of the river fluctuates by 16–20 m: in the middle reaches the range is 8–14 m and near its mouth 5–7 m (Junk, 1983). Water chemistry is determined by the nature of the bedrock over which it passes and the amount of sediment that is being carried. Three different water types are typically recognized. **White-water** tributaries originate in the Andes and carry heavy sediment loads: nutrient levels are relatively high and pH is about 7. The **black-water** rivers receive drainage from the the bleached sands and podsols of the central lowlands. They contain very few nutrients but

Figure 10.18 Cushion bog of *Thamnolia* with *Dracophyllum longifolium* beyond on Maungatua summit, South Island, New Zealand. (A. F. Mark.)

are stained black by organic materials leached from the raw humus that covers the infertile soils. The black-water rivers are extremely acid with a pH of about 4. **Clear-water** rivers flow from the highlands of central Brazil and the Guyanas: they pick up some nutrients from the crystalline parent rocks but the water remains acidic.

In white-water areas permanent water bodies are usually covered with floating mats of vegetation. In the initial stages of colonization, plants such as water hyacinth (*Eichhornia crassipes*) cover the surface and this provides a substrate for sedges, rushes and other aquatic plants with buoyant root structures. *Montrichardia arborescens*, a tree-like aroid, becomes established once a dense mat has developed, and small trees may also be present. The mats gradually sink as their density increases, and many of the plants are drowned. Gases released during decomposition often cause clumps of dead material to rise to the surface, where the process of colonization begins

Figure 10.19 Flax (*Phormium tenax*) is a characteristic plant of New Zealand bogs. (O. W. Archibold.)

Figure 10.20 (a) Igapó forest bordering the Rio Negro; (b) the depth of the annual flood is clearly marked on the tree trunk. ((a) J. T. Price; (b) G. T. Prance.)

again. Mature mats remain intact as the water level changes during the year; they will conform to the sides of the channels and depression during low water periods, then float up again as the river begins to flood. High temperatures encourage rapid decomposition during periods of low water and little organic material accumulates, even though plant production is considerable. Similar communities are found along the clear-water rivers, although growth is more restricted because of lower nutrient concentrations in the water. The growth of aquatic plants is lowest in areas flooded by black-water rivers. The typical vegetation bordering these rivers is 'igapó', an open forest community which is partially submerged for 6–8 months of the year (Figure 10.20). The flood level is marked by the freshwater sponges that are attached to the trees and shrubs (Figure 10.21). Some of the trees and palms which grow in these sites can withstand flooding for up to 3 years, but there is a distinct zonation of species along the seasonal flood gradient (Keel and Prance, 1979). As the river subsides the drier areas are sparsely colonized by grasses and sedges.

Permanent swamps also occur in poorly drained depressions and bottomlands throughout the savanna regions. The largest area affected in this way is the the Gran Pantanal which is crossed by the Rió Paraguay and its tributaries (Figure 10.22): similar communities are found in the Orinoco llanos of Venezuela. Over much of these regions the land remains flooded for only part of the year, and the soil dries out quickly during the dry season. The dominant plants in these hyperseasonal savannas are tall, coarse grasses and sedges, occasionally accompanied by a few scattered palms. More productive **'esteros'** occur where the soil remains moist throughout the year, and in some areas **gallery forests** occupy the floodplains. These communities usually develop over humic gley soils. In all of these regions the vegetation is affected by slight differences in topography and typically results in a vast mosaic of woodland, savanna and swamp. These diverse lowland communities contrast markedly with the simple wetlands which develop around the mountain lakes in the High Andes. The largest of these is Lake Titicaca at an elevation of 3815 m above sea level. The total flora of Lake Titicaca consists of only eight species of higher plants, most of which are submergent aquatics. *Scirpus tatora*, the reed that is used to construct canoes, is the only emergent species: it grows in sheltered bays around the shoreline. In more exposed areas *Zanniachellia palustris* develops an open turf-like cover in shallow water, but this is eroded away in winter when the lake level drops. Reed-swamps similar to these are more characteristic of the tropical wetlands of Africa.

Tropical wetlands cover more than 400 000 km^2 in Africa and are represented by swamp forests and sedge and reed swamps (Thompson and Hamilton, 1983). The largest areas of swamp forest occur in the central basin of Zaire and West Africa. Several kinds of trees are represented in these communities but the most widespread species are *Raphia farinifera*, a palm, and the tree-like

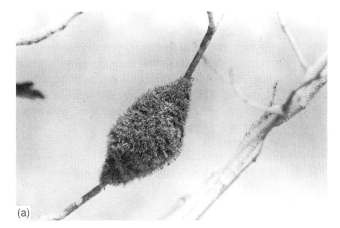

(a)

Figure 10.22 A scattered tree cover established on waterlogged soils in the seasonally flooded Gran Pantanal of Brazil. (G. T. Prance.)

(b)

Figure 10.21 Freshwater sponges growing on trees [(a)] bordering rivers indicate the depth of flooding in the Amazon Basin [(b)]. (G. T. Prance.)

seasonally flooded savannas. Bulrush (*Typha domingensis*), reeds (*Phragmites australis* and *P. mauritianus*) and saw grass (*Cladium mariscus*) are more important at higher elvations. Blanket bogs composed of sedges and occasional patches of *Sphagnum* are present in some wet mountain sites as well as cushion bogs similar to those of New Zealand.

aroid *Cyrtosperma senegalense*. In the drier climate of East Africa the common palm in the swampy valley bottoms is *Phoenix reclinata*, where it grows with small trees such as *Ficus congensis* and *Syzygium cordatum*. At higher altitudes *Syzygium cordatum* forms open stands with shrubs and sedges. The most extensive wetlands in Africa are associated with the complex of shallow lakes in the Chad Basin with similar areas near Lake Victoria and the upper reaches of the Nile. The most widely distributed species in these areas is papyrus (*Cyperus papyrus*) which becomes established in waterlogged soils through the branching growth of its large rhizomes (Figure 10.23). The accumulation of dead material around the plants in lakeside swamps enables them to spread out from the shore as floating mats. This outward expansion is usually limited by winds and currents but detached segments form floating islands over deep water. Partially decomposed materials which fall from the mats form layers of peat which in some valley swamps are 20–30 m deep (Beadle, 1974). Relatively few species grow with papyrus in deeply flooded swamps, but grasses such as *Miscanthidium violaceum* become increasingly common under less severe conditions and form the dominant cover in

Figure 10.23 Mature culms and open umbels of papyrus (*Cyperus papyrus*) in the swamps fringing Lake Naivasha, Kenya. (D. L. DeAngelis.)

10.3 CLIMATE AND WETLAND DEVELOPMENT

The highly absorptive nature of *Sphagnum* moss is an important factor in wetland hydrology. The delicate branching structure of the plants provides a fine spongy mesh which traps large amounts of rainwater. This is absorbed directly by the single layers of hyaline cells which make up the leaves and stems. The ribbed surface of the stems serves as a water transfer pathway through the carpet. Movement along these open channels is entirely dependent on surface tension and under strongly evaporative conditions the supply of water to the upper part of the moss carpet ceases. The shallow-rooting shrubs which develop on the moss hummocks are therefore subject to periods of drought (Ingram, 1983). As the hyaline cells dry out, the mosses become whiter in colour. This increases the albedo of the moss carpet and further evaporation is reduced because a greater amount of solar energy is reflected. Consequently, the seasonal pattern of water loss from *Sphagnum* bogs is quite different from wetland communities dominated by vascular plants.

Decomposition in wetland communities is slowed by anaerobic conditions and the hydrologic regime is progressively altered by the gradual accumulation of peat. Although the deeper layers of peat remain permanently saturated, the water table will normally fluctuate during the year causing the upper layers to dry out to varying degrees, depending on its composition and structure. Water loss occurs in several ways. In addition to losses associated with the plant cover, evaporation from open pools can be considerable and further changes occur because of seepage. The rate of water movement through peat is determined by its botanical composition, the stage of humification and degree of compaction. *Sphagnum* peats are less permeable than sedge peats and the coarser materials derived from *Phragmites* and other reeds. Water movement becomes progressively slower as the peat is broken down and compacted: hydraulic conductivities as low as 6×10^{-8} cm s^{-1} are reported for deep, humified peat layers in blanket bogs, compared with 4×10^{-2} cm s^{-1} for undecomposed surface deposits of *Sphagnum* (Rycroft *et al.*, 1975). The comparatively rapid seepage of water through the upper layers of peat in high rainfall areas minimizes erosion: serious disturbances have been reported in Scottish blanket bogs in which the surface peats have been burned off. The problem is worsened by the development of a gelatinous algal layer of low permeability. Erosion is also caused by water flowing in tunnels beneath the surface of a bog, and on steep ground '**bog bursts**' can occur when the peat becomes saturated during very wet periods. Normally, excess water is rapidly discharged from bogs in surface channels, or is temporarily stored by changes in the size and depth of the pools. The hydrologic regime of bogs is essentially disconnected from regional ground-water flow

and the net effect is to carry water away from the bog. Conversely, fens receive water from the surrounding land and the nutrients acquired by seepage greatly affect the vegetation of these sites. Swamps develop in areas where water collects and are similarly influenced by external conditions. Many swamps are non-accumulative ecosystems because periodic drawdown promotes decomposition of the organic debris.

10.4 WETLAND SOILS AND PEAT ACCUMULATION

10.4.1 Physical characteristics of wetland soils

The majority of wetland soils are organic **Histosols** which develop because decomposition and humification are limited by water-saturated conditions. The suborders of Histosols are determined by the physical structure of the plant litter. In bogs and fens the organic material in the upper layers remains virtually intact, but the botanic origin of the materials is usually more difficult to determine in swamps and other areas where the water table fluctuates. Material classed as peat will normally contain less than 20% inorganic matter with a minimum accumulated depth of 30 cm (Clymo, 1983). Decomposition occurs most rapidly in the uppermost layers or **acrotelm** of peat (Ingram, 1978) and the rate of accumulation is mainly determined by conditions in this layer. In *Sphagnum* communities the acrotelm is comprised of a living moss layer and a litter layer made up of very loose decomposed material (Malmer and Holm, 1984). The plants decay rapidly in the zone where the water table fluctuates and the structural collapse of the mosses coincides with the depth of permanent saturation (Johnson *et al.*, 1990). Beneath this is the **catotelm**, a more compact, waterlogged layer where further decomposition is limited by anoxic conditions. The number and size of the pores is reduced as the macrostructure of the moss changes and the hydraulic conductivity of the peat decreases (Boelter, 1964). This causes the water table to rise and anoxic conditions are maintained even though the depth of peat increases.

Peat accumulates when new material is added to the surface at a faster rate than it is lost by decomposition and compaction. The normal rate of accumulation in temperate bogs and fens is about 6 cm 100 yr^{-1} (Tallis, 1983), although the upper 50 cm of most peat deposits may be less than 200 years old (Ohlson and Dahlberg, 1991). The rate of accumulation varies with the growth rates of peat-forming plants. For example, *S. fuscum* grows 1.4–3.2 mm yr^{-1} in northern Sweden compared with 7–16 mm yr^{-1} in southern Sweden, with rates as high as 30 mm yr^{-1} reported in Germany and Canada (Rochefort *et al.*, 1990). The growth of *S. fuscum* is also affected by climatic conditions; average annual growth rates ranging from 7 to 23 mm have been reported in optimal

Table 10.2 Mass loss from cellulose strips placed in bog hollows and fen over a 12-month period (after Farrish and Grigal, 1988)

Depth (cm)	Bog (% loss from original mass)	Fen (% loss from original mass)
0–5	28	68
5–10	16	62
10–15	9	48
15–20	5	39
20–30	5	29
30–40	6	21
40–50	5	16

(Reproduced with permission from K. W. Farrish and D. F. Grigal, Decomposition in an ombrotrophic bog and a minerotrophic fen in Minnesota, *Soil Science*, 1988, **145**(5), 355, 357.)

Table 10.3 Mass loss of *Sphagnum* in different bog environments over a 12-month period (after Johnson and Damman, 1991)

Location	Species	Mass loss (% original mass)
Blanket bog, UK	*S. papillosum*	8
	S. recurvum	14
	S. cuspidatum	15
	S. rubellum	16
Palsa mire, Sweden	*S. balticum*	4
	S. fuscum	5
	S. lindbergii	7
Oligotrophic bog, Canada	*S. fuscum*	12
	S. magellanicum	18
	S. angustifolium	25
Raised bog, Sweden	*S. fuscum*	14
	S. cuspidum	20

(Source: L. C. Johnson and A. W. H. Damman, Species controlled *Sphagnum* decay on a South Swedish raised bog, *Oikos*, **61**, 1991.)

microtopographic sites at one location in southern Sweden (Wallen *et al.*, 1988). *S. fuscum* decomposes comparatively slowly. The plants remain intact to depths of about 20 cm: they begin to lose their branches at depths of 20–30 cm, and below 30 cm the leaves become detached from the stems (Johnson *et al.*, 1990). With compaction the cumulative length of *S. fuscum* stems in the debris increases from about 50 mm cm^{-3} near the surface to about 250 mm cm^{-3} at depths of 25–30 cm, and the bulk density of the peat increases. Bulk densities normally range from 0.02 to 0.04 g cm^{-3} in the uppermost layers of *Sphagnum* peat and increase to about 0.1 g cm^{-3} at depths of 20–30 cm as the moss loses its structural integrity (Clymo, 1983). This comparative lightness results in a high water-holding capacity which on a weight basis can exceed 3000%.

Although the accumulated depth of peat is affected by compaction, the principal factor controlling the development of deep organic soils is the slow rate of decomposition which occurs below the water table (Clymo, 1965). Measurements of decomposition potential using cellulose strips show that conditions become less favourable with depth and decomposition slows abruptly when the highly reduced waterlogged zone is reached (Table 10.2). Better aeration and warmer temperatures probably account for the greater rate of loss noted in raised hummocks. Direct measurements of decomposition based on mass loss of *Sphagnum* from litter bags placed at the surface of bogs indicate that as much as 25% of the material is lost during the first year of deposition, depending on species and geographic location (Table 10.3): most of this occurs during the first year of incubation. However, the rate of decay is generally faster in species such as *S. cuspidatum*, which usually grows in hollows (Figure 10.24). This intrinsic difference is maintained regardless of environmental conditions and is possibly an important factor in the development of peatland microtopography (Johnson and Damman, 1991). Decomposition occurs more rapidly in minerotrophic sites where larger microbial populations develop in response to less acidic conditions and the higher mineral content in the

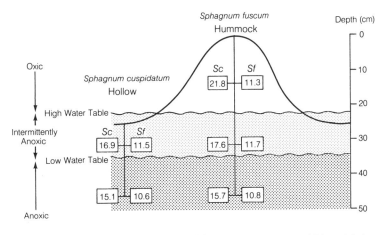

Figure 10.24 Average percentage mass loss over a 22-month period of *Sphagnum cuspidatum* (*Sc*) and *S. fuscum* (*Sf*) from decomposition bags placed in different microenvironments in a raised bog in southern Sweden. (After Johnson and Damman, 1991.) (Redrawn from L. C. Johnson and A. W. H. Damman, Species controlled *Sphagnum* decay on a South Swedish raised bog, *Oikos*, 1991, **61**, 235.)

Table 10.4 Litter mass loss through microbial activity and other soil organisms over a 12-month period in blanket bogs (after Coulson and Butterfield, 1978)

Litter type	Mass loss through microbial activity (% mass loss)	Mass loss through all soil organisms (% mass loss)
Eriophorum vaginatum (leaves)	16	44
Calluna vulgaris (young shoots)	26	48
Sphagnum recurvum (shoots)	16	14

(Source: J. C. Coulson and J. Butterfield, An investigation of the biotic factors determining the rates of plant decomposition on blanket bog, *Journal of Ecology*, 1978, **66**, 634.)

plant debris. For example, litter of *Typha angustifolia* and *Phragmites communis* growing in fenland is completely decomposed in 15–30 months (Mason and Bryant, 1975), whereas only 10–30% of the leaves and shoots of bog plants, such as *Eriophorum vaginatum* and *Calluna vulgaris*, are usually lost over this period (Heal *et al.*, 1978). Although potential loss of material in fens is significantly greater than in bogs, the rate of peat accumulation may not be reduced because plant production is generally higher (Farrish and Grigal, 1988).

10.4.2 Chemical characteristics of wetland soils

Peat is a complex substance containing many organic and inorganic compounds, and its chemical composition depends on the nature and condition of the plant

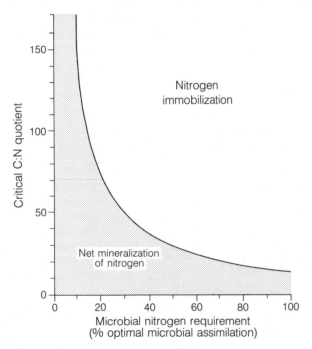

Figure 10.25 Relationship between microbial nitrogen requirement and the critical C:N quotient at which nitrogen immobilization by microbial biomass equals supply by mineralization. (After Damman, 1988.) (Reproduced with permission from A. W. H. Damman, Regulation of nitrogen removal and retention in *Sphagnum* bogs and other peatlands, *Oikos*, 1988, **51**, 295.)

remains. Non-structural carbohydrates such as starches and sugars are rapidly broken down and quickly leached out of the litter. The breakdown of structural materials such as cellulose and hemicellulose by the microbial population is dependent on temperature, redox conditions and the availability of nitrogen in the debris. Lignin and phenols decompose very slowly. These highly refractory compounds form fulvic and humic acids which cause the brown coloration of bog and fen waters (Verhoeven, 1986). Decomposition of *Sphagnum* is inhibited by its high phenolic content and the litter remains intact for a long period of time (Dickinson and Maggs, 1974). Some decomposition occurs through invertebrate activity but this is relatively small compared with losses in other bog plants such as *Rubus chamaemorus* and *Calluna vulgaris* (Table 10.4). Once the material becomes waterlogged it is broken down very slowly by the small number of micro-organisms which are found in the deeper peat layers (Dickinson and Maggs, 1974). The rate of decomposition is related to the concentration of nitrogen in the litter (Coulson and Butterfield, 1978). Nitrogen levels are highest in the uppermost part of the live mosses where this element becomes concentrated through translocation from the older tissues (Malmer, 1988). A further decrease in nitrogen in the upper part of the acrotelm occurs because of uptake by vascular plants which are mostly rooted at this depth (Rosswall and Granhall, 1980). Losses through leaching and denitrification are minimal lower down the profile, and nitrogen concentrations usually increase because of immobilization in microbial biomass.

Nitrogen is immobilized in litter when the C:N ratio is about 15–20, but mineralization occurs at higher values because the amount of nitrogen released by decomposition exceeds requirements for microbial protein synthesis. However, the actual C:N ratio at which mineralization occurs depends on the size and growth rate of the microbial population (Figure 10.25) and on the nature of the plant cover. C:N ratios above 100 are reported in some *Sphagnum* peats (Damman, 1988) compared with 15–20 in fen communities (Wassen *et al.*, 1990). Sedges and reeds growing in minerotrophic sites are higher in nitrogen than *Sphagnum* moss and other bog plants: they decompose fairly quickly and more nitrogen is available through mineralization in fen peats (Verhoeven *et al.*, 1990). However, nitrogen mineralization can occur in

Table 10.5 Loss of nitrogen, phosphorus and potassium (expressed as percentage of original stock remaining) during breakdown of *Sphagnum recurvum* litter (after Brock and Bregman, 1989)

Period of decomposition (days)	Organic matter (%)	nitrogen (%)	phosphorus (%)	potassium (%)
0	100	100	100	100
35	96	91	92	86
70	93	70	70	62
126	93	65	54	5
198	92	61	50	5
288	89	57	46	4
365	82	55	40	4

(Reproduced with permission from T. C. M. Brock and R. Bregman, Periodicity in growth, productivity, nutrient content and decomposition of *Sphagnum recurvum* var. *mucronatum* in a fen woodland, *Oecologia*, 1989, **80**, 48.)

Table 10.6 Chemical properties of rain and surface waters in wetland ecosystems

Wetland type	Location	pH	N	P	K	Ca	Mg	Source
			(mg l^{-1})					
Rain water								
Raised bog	Canada	4.3	–	–	0.0	0.3	0.1	Damman, 1988
Blanket bog	Tasmania	6.0	–	–	0.7	0.7	0.4	
Sedge fen	USA	5.3	0.77	0.08	0.4	1.3	0.2	Richardson and Marshall, 1986
Sawgrass swamp	USA	–	0.54	0.08	1.9	–	–	Steward and Ornes, 1975
Cypress swamp	USA	–	0.26	0.02	0.1	0.2	0.1	Schlesinger, 1978
Surface waters								
Raised bog	USA	3.6	1.34	0.19	1.3	2.4	1.0	Verry, 1975
Sphagnum bog	USA	4.7	2.00	1.51	2.2	2.0	0.5	Schwintzer, 1978
Sphagnum bog	Finland	–	2.25	0.08	0.6	2.2	0.7	Pakarinen, 1978
Sedge fen	UK	7.3	1.21	–	1.4	97.0	4.0	Boyer and Wheeler, 1989
Sedge fen	USA	6.3	0.42	0.05	0.7	16.1	2.2	Richardson and Marshall, 1986
Sedge fen	Poland	7.8	0.46	0.08	1.5	74.0	–	Wassen *et al.*, 1990
Willow carr	UK	7.7	–	–	0.2	13.2	–	Gorham, 1956
Reedswamp	UK	6.4	–	–	0.9	6.7	–	
Cypress swamp	USA	4.5	1.60	0.18	0.3	2.9	1.4	Dierberg and Brezonik, 1984
Papyrus swamp	Uganda	–	1.22	0.03	11.2	7.8	1.6	Gaudet, 1976

Sphagnum bogs if microbial activity is limited by a shortage of readily metabolizable carbon and other nutrients or by toxins (Damman, 1988). For example, nitrogen and other elements are quickly lost from *S. recurvum* during the first year of decomposition but the structural components of the dead cells change very little (Table 10.5).

The nature of the plant cover and its mineral content is primarily determined by the chemistry of the water which supplies a wetland site. The rainwater which nourishes bogs is usually very acidic and low in nutrients (Table 10.6), although the amounts of Na, Mg and Cl typically increase in coastal areas since these ions originate mainly from the sea. Nutrient concentrations in ombrotrophic bogs are highest in the upper part of the moss carpet where materials are deposited from the atmosphere (Malmer, 1988). The mineral content of bog plants is lower than in plants growing in fens and swamps (Table 10.7). Soluble elements such as calcium are washed from the moss canopy and may accumulate in the underlying peat (Figure 10.26). Other elements are lost through litter formation. Potassium is quickly leached from dead moss and concentrations in the underlying peat are typically very low. Organic phosphorus is released slowly during decomposition and tends to accumulate in the hollows, where it is immobilized as iron and aluminium phosphate. Nitrogen is utilized relatively quickly in peat with a high C:N ratio and can also be leached from the drier hummocks into the hollows.

Nutrient concentrations increase in water which has percolated through mineral soil and rock layers, and the greater flow of water through fens further enhances the supply of soluble cations such as potassium, magnesium and calcium (Malmer, 1986). The pH of fen water is normally near neutral because of a higher concentration of calcium (Table 10.6): this increases the availability of other elements and promotes the faster decay and release of nutrients from organic matter (Wassen *et al.*, 1990). However, excessive calcium levels cause immobilization of phosphorus and this element can be deficient in rich fen sites. Comparative studies in German mires indicate that average concentrations of extractable phosphorus increase from less than 10 µg g^{-1} in fens to about

Table 10.7 Tissue nutrient content of plants growing in wetland habitats

	N	P	K	Ca	Mg	Source
			(g m^{-2})			
Bog species						
Sphagnum fuscum	1.36	0.07	0.89	0.36	0.14	Pakarinen, 1978
Sphagnum balticum	2.29	0.14	1.39	0.34	0.18	
Carex lacustris	7.2	0.9	5.00	1.2	0.05	Bernard and Bernard, 1977
Carex rostrata	7.4	1.3	9.5	2.0	1.0	
Eriophorum vaginatum	0.28	0.01	0.18	0.03	0.02	Tamm, 1954
Swamp species						
Typha latifolia	31.52	4.25	37.94	34.84	9.84	Klopatek, 1978
Scirpus fluviatilis	17.26	3.39	15.55	2.60	1.16	
Phragmites communis	43.0	2.0	11.0	2.5	1.1	Mason and Bryant, 1975
Typha angustifolia	13.0	3.2	20.0	7.2	1.6	
Glyceria maxima	44.0	7.1	62.0	3.5	3.4	Dykyjová, 1978
Taxodium distichum	96.7	4.42	21.2	64.6	10.8	Schlesinger, 1978
Cladium mariscus	8.9	0.25	6.5	3.3	0.6	Steward and Ornes, 1975
Cyperus papyrus	103.0	7.8	129.9	19.2	6.0	Gaudet, 1977

$20 \, \mu g \, g^{-1}$ in bogs, even though total phosphorus levels in both vegetation and peat is considerably higher in the fen sites (Figure 10.27). The concentration of extractable nitrogen follows a similar trend, although total nitrogen content of the vegetation and peat changes very little between sites. Extractable nitrogen is predominantly NH_4–N because nitrifying bacteria are inhibited below pH 5, whereas ammonifying bacteria are tolerant of acid conditions (Waughman, 1980). Calcium and magnesium concentrations are highest in fen peats and are taken up by the plants in comparatively large amounts. However, plant-available potassium may be limited in fens because high calcium levels can increase losses through leaching.

10.5 ADAPTATIONS OF WETLAND PLANTS

The *Sphagnum* mosses which grow in bogs and fens are ectohydric: water and dissolved nutrients move to the actively growing capitula through an external capillary process. Water is conducted through the pores in the hyaline cells which surround the stems, branches and leaves, and different rates of flow occur between species depending on cellular structure. The porous hyaline cells and the delicate imbricate structure of the leaves give *Sphagnum* a considerable water-holding capacity which reduces the possibility of desiccation during dry periods, especially in the larger species. However, drought

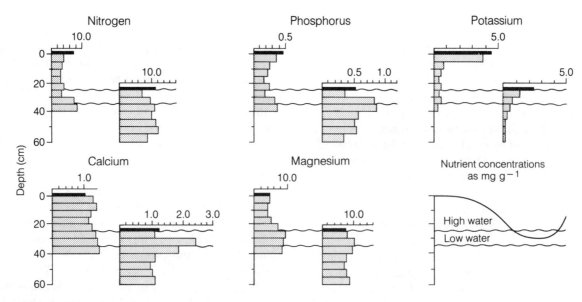

Figure 10.26 Changes in the concentration (mg g^{-1} dry weight) of major cations with depth below *Sphagnum fuscum* hummocks and adjacent *S. magellanicum* hollows in an ombrotrophic bog in southern Sweden; the black line indicates element concentration in the actively growing capitula. (After Damman, 1978.) (Reproduced with permission from A. W. H. Damman, Distribution and movement of elements in ombrotrophic peat bogs, *Oikos*, 1978, **30**, 486.)

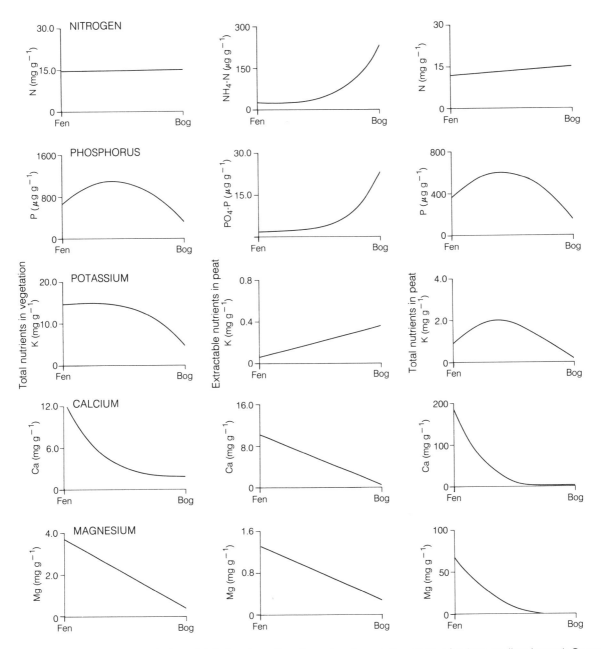

Figure 10.27 Concentrations (per unit dry weight) of major cations in peat and vegetation along a fen-bog gradient in south German mires. (Waughman, 1980.) (Redrawn and simplified from G. J. Waughman, Chemical aspects of the ecology of some south German peatlands, *Journal of Ecology*, 1980, **68**, 1029, 1033, 1034, 1035, 1037.)

tolerance differs between species: some quickly die as they dry out, others develop new shoots lower down the stem if the capitulum is damaged (Figure 10.28). The desiccation resistance of the different species of *Sphagnum* appears to be unrelated to their ecological habitat. For example, *S. auriculatum*, a species which grows in wet hollows, is better able to withstand drought than *S. capillifolium*, which is associated with the drier hummocks (Clymo and Hayward, 1982). The comparatively high rate at which *S. capillifolium* draws water up from the water table during normal conditions is probably the main factor which allows this species and others such as *S. rubellum* to grow on the hummocks (Figure 10.29).

The rate of water flow controls the supply of nutrients to the plants and consequently affects their rate of growth (Clymo, 1973). However, some species of *Sphagnum* cannot tolerate high concentrations of nutrients and most will not survive in alkaline water with a high calcium content. *Sphagnum* has a high cation exchange capacity due to the presence of urionic acid in the cell walls. Soluble ions are held outside the cells until the capacity of the exchange sites is reached: the nutrients are then available for uptake by the living cells. The extracellular binding of ions is a purely physical process and is not controlled by the metabolic activity of the plants (Brown, 1982). The process of cation exchange contributes to the

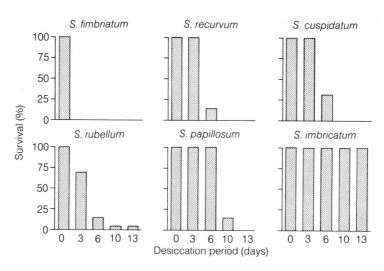

Figure 10.28 The desiccation tolerance of selected *Sphagnum* species. (After Green, 1968.) (Reproduced from B. H. Green, Factors influencing the spatial and temporal distribution of *Sphagnum imbricatum* Hornsch. ex Russ. in the British Isles, *Journal of Ecology*, 1968, **56**, 49.)

acidity of the water in *Sphagnum* bogs, and release of organic acids can also lower pH (Clymo, 1967). Modification of the environment in this way is most pronounced during periods of rapid growth (Clymo and Hayward, 1982).

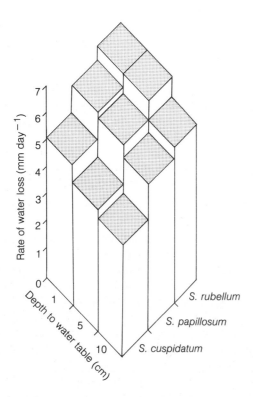

Figure 10.29 Rate of water loss from three species of *Sphagnum* with water tables maintained at different levels under laboratory conditions. (After Clymo, 1973.)

The growth of mosses mainly depends on adequate moisture uptake from the waterlogged peat, but few other plants can survive in the anaerobic conditions which develop in these sites. The redox potential (E_h), which provides an indirect measurement of soil aeration, changes fairly abruptly at the water table (Urquhart and Gore, 1973) and below this sulphides begin to accumulate in the peat. The shortage of oxygen causes problems with root respiration in higher plants and alters the chemistry of the soil so that nutrient uptake is affected. Elements such as iron and mananganese are present in a reduced state in poorly aerated soils, and in this form they are highly soluble and may be taken up in concentrations that are toxic to the plants. In normally drained soils oxygen enters the root by diffusion from the soil atmosphere supplemented by transpirational flow (Armstrong, 1979). The rate of oxygen diffusion in water is about 10^{-4} the rate in air: consequently, saturated soils are poorly aerated and little oxygen enters the root system directly.. The roots of wetland species appear to be no more resistant to anoxia than those of other plants, but adequate ventilation is usually provided by air-space tissue (**aerenchyma**) which extends from the roots to the sub-stomatal cavities of the leaves (Smirnoff and Crawford, 1983). This system of intercellular spaces develops in the mesophyll and palisade tissues by cell separation during maturation or by cell lysis (Sculthorpe, 1967). The effective porosity of aerenchymatous tissues can be as high as 60% compared with about 3% in most terrestrial plants, and sufficient oxygen can be conducted from the atmosphere to the roots to meet respiration requirements (Figure 10.30). Differences in the rate of oxygen diffusion do affect the zonal distribution of marsh plants. For example, wild rice (*Zizania latifolia*) is able to colonize deeper water than *Phragmites australis* because oxygen moves more freely through the tissues of wild rice (Yamasaki, 1984). Oxygen transport to the roots also occurs in trees such as *Nyssa sylvatica* (Keeley, 1979) and more specialized pneumatophores may develop on the upturned roots in species such as swamp cypress.

Aeration of the roots and rhizomes is temporarily suspended in plants such as *Typha latifolia* and *Schoenplectus lacustris* which die back at the end of the growing season. In these species a large reserve of carbohydrate is accumulated in the rhizomes at the end of the growing season and this is subsequently used for new shoot growth in the spring (Steinmann and Bräendle, 1984). The ability to utilize carbohydrate reserves during anaerobic periods varies between species and this may restrict their distribution to specific habitats (Bräendle and Crawford, 1987). In the absence of oxygen the supply of ATP to plant cells ceases, biochemical activity is interrupted and tissue acidification leading to root-tip death can occur within a few hours (Crawford, 1989). Anaerobic metabolism in non-wetland species usually results in an increase in ADH (alcohol dehydrogenase) activity, and ethanol production increases until it reaches toxic levels (McManmon and Crawford, 1971). Wetland species were originally presumed to lack the malic enzyme which catalases the conversion of malic acid to pyruvate and so prevents the accumulation of ethanol. However, active malic enzyme and ethanol production have been detected in many wetland plants (Davies *et al.*, 1974). Ethanol production may act as a compensatory energy source under anaerobic conditions in species such as rice, but this ability to transfer fermentation products from roots to shoots for subsequent aerobic metabolism does not occur in all wetland species (ap Rees *et al.*, 1987). Alternatively, ethanol may be removed via the aerenchyma, or by diffusion into the soil (Keeley, 1979).

The use of oxidized minerals and organic matter as respiratory electron acceptors by anaerobic microorganisms is an important factor in the chemistry of waterlogged soils. In the initial stages of anaerobiosis, oxygen is quickly used up and microbial nitrate reduction occurs. Iron and manganese are reduced to Fe(II) and Mn(II) at lower redox potentials; under extreme conditions sulphate is reduced to sulphide and methane is produced by the dissimilation of organic matter (Ponnamperuma, 1972). Many of these substances are phytotoxic but they are typically transformed into less harmful products by the outward diffusion of oxygen from the root apices of wetland plants with well-developed aerenchyma. Root oxygen diffusion rates of 128–$163 \text{ ng cm}^{-2} \text{ min}^{-1}$ are reported in species such as *Menyanthes trifoliata* and *Eriophorum angustifolium* (Armstrong, 1967a), compared with a rate of only $14 \text{ ng cm}^{-2} \text{ min}^{-1}$ in *Molinia caerulea*, a species which normally grows in sites that are flushed with well-

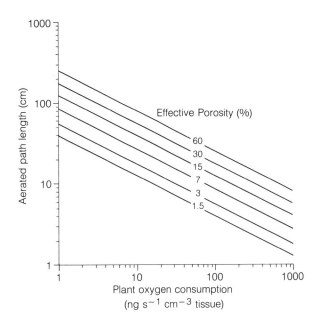

Figure 10.30 Predicted distance to which oxygen will diffuse through plant tissues of different porosity in relation to oxygen consumption. (After Armstrong, 1979.) (Reproduced with permission from W. Armstrong, Aeration in higher plants, *Advances in Botanical Research*, 1979, **7**, 256.)

aerated water (Armstrong and Boatman, 1967). The deposition of iron oxides around the roots is the usual evidence of radial oxygen loss from well-ventilated plants but insoluble iron may also be present in the intercellular gas spaces or even incorporated within the cell walls of the roots (Green and Etherington, 1977). Oxygen diffusion through the aerenchyma accounts for only part of the oxidizing activity in the root cells and rhizosphere; additional oxidation occurs through enzyme-mediated reactions in the root (Armstrong, 1967b).

Immobilization of soluble compounds through oxidation reduces the possibility of iron toxicity and other problems of nutrient imbalance and is typical of many wetland plants. Others regulate the rate of entry of water and solutes into the roots by the xeromorphic nature of their foliage. This restricts transpirational water loss and so provides more time for oxidation of phytotoxins in the rhizosphere. Cross-leaved heather (*E. tetralix*) is able to limit iron uptake in this way (Jones, 1971). The development of xeromorphic foliage may be an adaptative response to the oligotrophic status of most wetland sites, with the small evergreen leaves providing optimum nutrient conservation under these conditions. Alternatively, many of the wetland grasses and sedges renew their root systems annually. This tends to restrict root development to the aerated soil layer which develops when the water table drops during the summer, and eliminates the problem of anaerobic maintenance over winter.

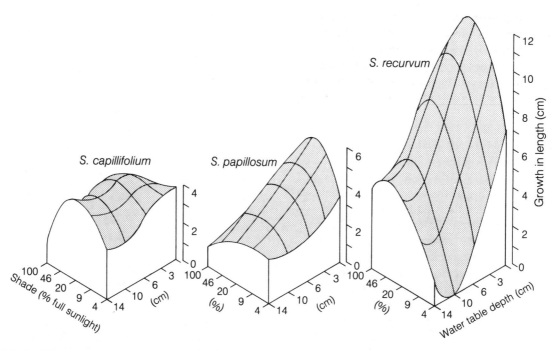

Figure 10.31 Growth in length of three species of *Sphagnum* under experimental conditions in relation to water table depth and shade. (After Clymo and Hayward, 1982.) (Reproduced with permission from A. J. E. Smith, *Bryophyte Ecology*, published by Chapman & Hall, 1982.)

10.6 PRIMARY PRODUCTION AND BIOMASS IN TERRESTRIAL WETLANDS

Comparative estimates of productivity in *Sphagnum* bogs range from 100 to 600 g m^{-2} yr^{-1}. Little light is transmitted deeper than 2–3 cm in a carpet of moss, so that photosynthesis is restricted to this narrow euphotic zone (Clymo and Hayward, 1982). Light attenuation varies according to the structure and density of the mosses. As an example, nearly all the photosynthetic activity in *S. fuscum* occurs in the capitulum compared with only 60% in *S. balticum* with the remainder accounted for in the underlying leaves (Johansson and Linder, 1980). The maximum rate of photosynthesis recorded in *Sphagnum* species is about 1–2 mg CO$_2$ g^{-1} hr^{-1}, and they become light-saturated at about 20–50% full sunlight. Productivity generally increases in wetter sites, although growth can be limited by slower CO$_2$ diffusion rates when the water content of the capitulum is very high. Conversely, the rate of growth declines as the plants dry out during periods of drought (Grace and Marks, 1978): the most sensitive species under dry conditions are those such as *S. recurvum* which normally grow in pools (Figure 10.31).

The majority of wetland vascular species are C$_3$ plants, although papyrus and some other tropical swamp grasses are C$_4$ plants, as is the temperate freshwater sedge *Cyperus longus*. The C$_4$ pathway is characteristic of plants growing in hot dry environments, but the higher rates of photosynthesis may give these plants a compet-

itive advantage in dense wetland communities (Jones, 1986). Seasonal growth patterns are less marked in tropical marshes: in papyrus stands the leaf area index remains at about 8 throughout the year (Jones and Muthuri, 1985). In temperate regions leaf growth in emergent marsh species such as bulrush occurs rapidly in the spring as reserves of assimilates are translocated from the rhizomes. The maximum leaf area index in these stands can exceed 18 (Dykyjová, 1971). The younger leaves are usually inclined at a steep angle, which allows more sunlight to penetrate into the canopy and results in high rates of photosynthesis (Ondok and Gloser, 1978). Assimilation rates decline in older foliage, which are more completely shaded, and is further reduced by the development of leaf aerenchyma (Gloser, 1977). The concentration of chlorophyll in the leaves of bulrushes and reeds decreases towards the leaf base (Szczepanski, 1973). Net production is therefore regulated by the rate of development of the leaf tissue, especially during the early part of the growing season (McNaughton, 1974), and most assimilation occurs in the upper part of the canopy (Ondok and Gloser, 1978). Above-ground production in marshes and swamps typically ranges from 400 to 1500 g m^{-2} yr^{-1} (Table 10.8) but can exceed 3000 g m^{-2} yr^{-1} in some reed-swamps, making them the most productive of any natural ecosystem. Most vascular plants which grow in wetlands are perennial and develop extensive systems of roots and rhizomes which are used to store nutrients during the winter and for vegetative propagation. Below-ground production in reed-swamps

Table 10.8 Maximum annual production by species in selected wetland communities

Dominant species	Location	Above-ground production (g m^{-2} yr^{-1})	Below-ground production (g m^{-2} yr^{-1})	Source
Bog communities				
Blanket bog	Ireland			
Schoenus nigricans		29	66	Doyle, 1973
Molinia caerulea		39	81	
Sphagnum spp.		50	0	
Blanket bog	UK			
Calluna vulgaris		240	183	Forrest and Smith, 1975
Eriophorum vaginatum		176	45	
Eriophorum angustifolium		322	77	
Trichophorum cespitosum		394	581	
Sphagnum spp.		213	0	
Sedge communities				
Carex lacustris	USA	965	208	Bernard and Solsky, 1977
Carex rostrata	USA	780	197	Klopatek and Stearns, 1978
Carex spp.	Canada	820	205	Auclair et al., 1976
Juncus effusus	USA	1592	267	Boyd, 1971
Swamp communities				
Reedswamps	Denmark			
Phragmites australis		781	620	Andersen, 1976
Scirpus lacustris		1325	1224	
Typha angustifolia		807	1800	
Reedswamps	Czechoslovakia			
Acorus calamus		830	550	Hejný et al., 1981
Glyceria maxima		1570	480	
Phragmites australis		3150	2210	
Schoenplectus lacustris		2030	2120	
Sparganium erectum		1500	410	
Typha angustifolia		2050	1420	
Typha latifolia		1550	520	
Freshwater marsh	USA			
Typha latifolia		1585	1092	Klopatek and Stearns, 1978
Scirpus fluviatilis		1116	417	
Carex lacustris		940	134	
Phalaris arundinacea		1353	675	

ranges from about 500 to 1800 g m^{-2} yr^{-1}, compared with about 200 g m^{-2} yr^{-1} in most fens and bogs.

Below-ground biomass in reed-swamps generally ranges from 2000 to 3500 g m^{-2} and is 1–3 times higher than the above-ground biomass. Maximum below-ground standing crops of 410–430 g m^{-2} are reported in stands of *Carex rostrata* and *C. lacustris*, giving a root:shoot ratio of about 0.6 (Bernard and Bernard, 1977). Because mosses have no root systems they contribute nothing to below-ground biomass but ericaceous shrubs direct a considerable proportion of their annual production into their roots. Mean annual above-ground production of *Calluna vulgaris* in blanket bogs in northern England is 130 g m^{-2} yr^{-1}, compared with 221 g m^{-2} yr^{-1} below ground (Smith and Forrest, 1978). Above-ground standing crop at this site averages 740 g m^{-2}, compared with 659 g m^{-2} below ground, giving a root:shoot ratio of

about 0.9. However, a ratio of 16.5:1 is reported for vascular plants in a blanket bog in Ireland where above-ground biomass is only 83 g m^{-2} and live below-ground biomass totals more than 500 g m^{-2}. Mosses account for 209 g m^{-2} of the total standing crop at this site and the algae which grow in the open pools and amongst the mosses contribute a further 360 g m^{-2}, so that the root:shoot ratio for the entire community is about 1:1 (Moore et al., 1975).

10.7 NUTRIENT CYCLING IN WETLAND ECOSYSTEMS

Most of the nutrients present in peat soils are incorporated into organic compounds and are consequently unavailable to the growing plants (Table 10.9). Nitrogen is

Table 10.9 Chemical composition of wetland peats

	Total N (% dw)	Available P (mg 100 g^{-1} dw)	Extractable cations (meq 100g^{-1} dw) K	Ca	Mg	Source
Ombrotrophic bog	1.1	–	1.2	5.0	1.8	Malmer and Sjörs, 1955
Ombrotrophic bog	0.9	–	1.1	7.5	4.7	Rosswall *et al.*, 1975
Blanket bog	–	–	1.0	4.4	4.6	Gore and Allen, 1956
Blanket bog	1.8	16.0	0.5	2.2	–	Walsh and Barry, 1958
Raised bog	1.4	17.0	0.6	2.0	–	
Carex meadow	2.0	9.1	1.6	50.7	9.6	Auclair *et al.*, 1976
Carex fen	2.2	1.3	5.8	812.0	90.6	Richardson and Marshall, 1986
Cypress swamp	1.9	–	0.1	0.1	1.0	Coultas and Duever, 1984

particularly deficient in wetland ecosystems (Morris, 1991). Nitrogen enters bogs mainly through wet and dry atmospheric deposition and microbial activity: in fens and swamps additional nitrogen may be supplied from percolating waters. Biological nitrogen fixation by bacteria and blue-green algae is greatest in neutral or alkaline habitats. The population of nitrogen-fixing bacteria in acid peat bogs is relatively small and mostly restricted to the litter layer (Collins *et al.*, 1978). Epiphytic blue-green algae are generally more abundant and are especially active above pH 6 (Granhall and Lid-Torsvik, 1975). Nitrogen fixation by blue-green algae is strongly light dependent and occurs mainly below the dense capitulum (Basilier *et al.*, 1978). The nitrogen which is fixed and exuded by the algae is subsequently transported along the stem in the upward flow of water (Basilier, 1980).

Biological fixation adds between 0.05 and 3.2 g m^{-2} of nitrogen to blanket bogs in Britain each year compared with 0.8 g N m^{-2} yr^{-1} entering from the atmosphere (Martin and Holding, 1978). Comparative rates of 0.05–1.0 g m^{-2} yr^{-1} for nitrogen fixation and 0.7–1.0 g m^{-2} yr^{-1} for atmospheric deposition are reported for *Sphagnum* bogs in North America (Hemond, 1983; Urban and Eisenreich, 1988); in Swedish bogs nitrogen fixation accounts for 0.20 g N m^{-2} yr^{-1} with 0.22 g N m^{-2} yr^{-1} added from atmospheric sources (Rosswall and Granhall, 1980). Similar rates of nitrogen fixation occur in fens and swamps: for example rates of 0.2–1.3 g N m^{-2} yr^{-1} are reported in fens surrounded by heavily fertilized pastures in the Netherlands, some of this being attributed to micro-organisms associated with the roots of alder (*Alnus glutinosa*) saplings (Koerselman *et al.*, 1989). Symbiotic micro-organisms are also associated with sweet gale (*Myrica gale*), a common bog species, and tissue nitrogen concentrations are greatly increased in these nodulated plants (Dickinson, 1983). Similarly, a high rate of nitrogen fixation (6.0 g N m^{-2} yr^{-1}) occurs in stands of northern manna grass (*Glyceria borealis*) because of micro-organisms in the rhizosphere (Bristow, 1974). There are no symbiotic micro-organisms associated with species of *Typha*: the rate of nitrogen fixation in these communities is about 0.6 g N m^{-2} yr^{-1} and accounts for 10–20% of the community nitrogen requirements. In

cypress swamps about 30% of the nitrogen input comes from microbial fixation and is equivalent to a rate of about 0.4 g m^{-2} yr^{-1} (Dierberg and Brezonik, 1981). A considerably higher contribution from biological fixation is suggested for tropical papyrus swamps (Gaudet, 1976).

Nitrogen mineralization cannot proceed past the ammonium stage in anaerobic environments because the absence of oxygen prevents microbial conversion to nitrate (Patrick and Mahaptra, 1968). Any nitrate that is present in anoxic soils is rapidly converted to nitrous oxide (N$_2$O) or nitrogen by denitrifying bacteria, and considerable losses of nitrogen occur from wetland ecosystems in this way. Fluctuations in the water table may increase losses through denitrification, because nitrate is formed during periods of low water. Any nitrate that remains in the soil is likely to be leached out of the rooting zone and carried away in drainage waters. Displacement of ammonium nitrogen by Fe(II) (ferrous) and Mn(II) (manganous) ions also occurs in acid, anaerobic sediments. Although plant production in bogs is comparatively low, a considerable amount of material accumulates each year as peat because of the slow rate of humification. Consequently, peat bogs act as major nutrient sinks and 65–80% of the annual nitrogen input is retained in the ecosystem (Hemond, 1983; Urban and Eisenrich, 1988). Much of this is transferred to the catotelm, where it is essentially removed from circulation. The remainder is stored in the overlying acrotelm, where turnover time averages 30–50 years (Urban and Eisenreich, 1988), and in the uppermost litter layers, where mineralization may occur in 6–9 years (Rosswall and Granhall, 1980).

Many of the vascular plants which grow in bogs have adapted to these nutrient-deficient habitats by withdrawing nitrogen and other essential elements out of senescing organs before they are shed. Consequently, nitrogen levels in the litter may be too low to support microbial populations. Available nitrogen extracted from the water and incorporated into the microbial tissue may accumulate in the litter during the early stages of decomposition (Polunin, 1984). Nutrient losses from shoot abscission in grasses and sedges are considerably reduced by translocation to roots and rhizomes (Table 10.10). In

Table 10.10 Estimated loss of nutrients from shoots of sedges through translocation to below-ground organs and by leaching (after Chapin, 1978)

	N	P	K	Ca
Nutrient loss through translocation				
(% maximum standing stock)				
Carex aquatilis	53	55	64	0
Dupontia fisheri	21	44	25	0
Eriophorum angustifolium	45	61	48	0
Nutrient leaching				
(% end of season concentration)				
Carex aquatilis	26	54	84	− 45
Dupontia fisheri	27	54	83	− 26
Eriophorum angustifolium	22	38	82	− 115

(Reproduced with permission from F. S. Chapin, Phosphate uptake and nutrient utilization by Barrow tundra vegetation, in *Ecological Studies 29 – Vegetation and Production Ecology of an Arctic Tundra*, ed. L. L. Tieszen; published by Springer-Verlag N.Y. Inc., 1979.)

marsh communities of *Typha glauca* approximately 45% of the nitrogen and phosphorus is similarly lost by translocation by the end of the growing season (Davis and van der Valk, 1983). Potassium is also lost from the shoots, but unlike nitrogen and phosphorus it quickly passes from the rhizomes into the sediment. There is little decrease in shoot calcium levels at the end of the growing season and most of it accumulates in the sediment. Although papyrus grows year round in tropical swamps, a similar pattern of nutrient withdrawal occurs from the older tisues (Gaudet, 1977).

Nutrient translocation is also a characteristic adaptation of the bog shrubs. In deciduous species 45–68% of the nitrogen, 46–80% of the phosphorus and 49–93% of the potassium is withdrawn from the foliage before it is shed, and retention by evergreen species is proportionately higher (Table 10.11). The nutrient content of evergreen leaves declines after the first year and the amounts of nitrogen, phosphorus and potassium in the foliage are generally lower than in deciduous species (Small, 1972a). The combined effect of lower nutrient concentrations and the evergreen habit increases the amount of photosynthate that can be potentially manufactured from each unit of nitrogen that is acquired (Small, 1972b). The widespread occurrence of evergreen shrubs in bogs and other nutrient-deficient habitats is perhaps related to this efficient use of nutrients. Ericaceous species also benefit from the mycorrhizal fungi which infect their lateral roots. The fungal hyphae act as root extensions and absorb nitrogen and phosphorus released from the peat by enzyme action (Dickinson, 1983).

A more unusual response to nitrogen deficiency occurs in carnivorous plants which are commonly associated with bogs. It is mostly insects that are trapped by these plants, but occasionally a small rodent may fall into the larger pitcher plants. A variety of capture mechanisms are used (Fig. 10.32). In pitcher plants, such as species of *Sarracenia* and *Nepenthes*, the prey is guided by nectar, colour or scent into a pitfall trap: escape is prevented by downward-pointing hairs, or by waxes which make the surface too slippery to gain a foothold. The sundews (*Drosera* spp.) and butterworts (*Pinguicula* spp.) secrete adhesive droplets and the leaves then curl over the trapped insects. Suction traps are used by bladderworts (*Utricularia* spp.): aquatic organisms, including protozoa and insect larvae, are swept into these chambers by the inrush of water which occurs when the trigger hairs on the valve are touched. The spring-trap mechanism of the Venus fly trap (*Dionaea muscipula*) is similarly activated by sensitive trigger hairs on the upper surface of the modified leaves. Digestive enzymes are secreted from specialized glands and as the nitrogenous compounds are broken down they are absorbed through the perforated cell walls of the glands and passed into the vascular system (Heslop-Harrison, 1978).

Table 10.11 Mean content and translocation of nitrogen, phosphorus and potassium from the leaves of shrubs growing in bogs in eastern Canada (after Small, 1972b)

	Nitrogen		Phosphorus		Potassium	
	Mean leaf content (%)	Amount translocated (%)	Mean leaf content (%)	Amount translocated (%)	Mean leaf content (%)	Amount translocated (%)
Evergreen species						
Andromeda glaucophylla	1.14	63	0.06	75	0.58	81
Chamaedaphne calyculata	1.36	45	0.08	61	0.39	92
Kalmia polifolia	1.30	49	0.07	56	0.47	79
Ledum groendlandicum	1.41	72	0.09	69	0.64	83
Deciduous species						
Aronia melanocarpa	1.74	54	0.14	73	1.06	86
Gaylussacia baccata	1.66	68	0.08	80	0.68	93
Vaccinium myrtilloides	1.54	45	0.07	47	0.36	73
Virburnum casinoides	2.08	51	0.11	57	0.73	49

(Source: E. Small, Photosynthetic rates in relation to nitrogen recycling as an adaptation to nutrient deficiency in peat bog plants, *Canadian Journal of Botany*, 1972, **50**, 2231.)

Figure 10.32 (a) Insectivorous pitcher plant (*Sarracenia purpurea*) and (b) sundew (*Drosera rotundifolia*) from bogs in northern Saskatchewan (R. A. Wright.)

10.8 THE TROPHIC STRUCTURE OF WETLANDS

The process of nutrient translocation from senescing tissues reduces the mineral content of the plant litter. This creates a poorer substrate for microbial activity and further reduces the population of micro-organisms in acid, waterlogged peats. Consequently, the number of soil organisms which normally feed upon the fungi and bacteria is also restricted. Chemical deficiencies also limit the number of earthworms and isopods in bog peats (Speight and Blackith, 1983). Mites (Acari) and springtails (Collembola) are found only in the surface layers, where they are not affected by waterlogging and anaerobiosis. The density of nematodes is also comparatively low in wet bogs. In these environments most of the soil invertebrate biomass is comprised of enchytraeid worms (Table 10.12).

The majority of the above-ground plant-feeding invertebrates are sap-sucking insects and because the plant litter is largely unchewed it is less readily consumed by the detritivores. Caterpillars and grasshoppers are occasionally present, and slugs and snails are usually found amongst the wet moss and in the pools. *Sphagnum* is relatively inedible and is rarely eaten; most insects are found on the grasses and sedges and amongst the leaf axils of the shrubs. Larger herbivores are poorly represented in bogs, although birds, such as grouse, as well as voles, hares and other small mammals, are occasionally encountered. Consequently, the population of large carnivores is also very limited: foxes, weasels and birds of prey will sometimes move into the bogs from adjacent habitats. The most abundant predators are spiders and beetles: they consume the insects which feed on the plants and may in turn be taken by frogs and birds. A small population of mites, beetles and insect larvae feed on the soil organisms.

Marshes and swamps support a larger and more diverse fauna, and many of these areas are important bird breeding grounds. The primary production of emergent macrophytes is characteristically high in these wetlands and this is further increased by abundant algal growth. The algae are an important food source for a variety of animals including insect larvae, tadpoles and snails, as well as mallards (*Anas platyrhynchos*) and other ducks. Aquatic invertebrates are abundant in the marshes. Some are filter feeders that utilize the bottom sediments, phytoplankton and fine litter particles in the water: others graze and shred the coarser plant fragments. Sap-sucking insects and stem-borers are found on emergent plants but they rarely cause severe damage. However, considerable quantities of plant material are consumed by the larger grazing animals such as rodents and waterfowl.

The muskrat (*Ondatra zibethicus*) is one of the principal herbivores in freshwater wetlands in North America.

Table 10.12 Mean population density and mean standing crop (dry weight) of soil invertebrates in limestone grassland and blanket bog at Moor House, UK (after Coulson and Whittaker, 1978)

	Limestone grassland		Blanket bog	
	Density (number m^{-2})	Standing crop (g m^{-2})	Density (number m^{-2})	Standing crop (g m^{-2})
Lumbricidae	390	23.20	0.04	negligible
Enchytraeidae	80 000	4.10	80 000	2.16
Nematoda	3 300 000	0.14	1 400 000	0.07
Collembola	46 000	0.15	33 000	0.10
Acarina	33 000	0.35	60 000	0.40

(Source: J. C. Coulson and J. B. Whittaker, Ecology of moorland animals, in *Ecological Studies 27 – Production Ecology of British Moors and Montane Grasslands*, eds O. W. Heal and D. F. Perkins, published by Springer-Verlag GmbH, 1978.)

The preferred food of the muskrat are cattails (*Typha* spp.) and bulrushes (*Scirpus* spp.). Muskrat eat green shoots during the growing season and either discard the tougher leaves and stems or use them to build lodges and feeding platforms. In winter they dig out roots and rhizomes. Grazing is usually concentrated around the lodges and much of the plant material is removed for a radius of 2–3 m. The overall impact of muskrats on a marsh is normally quite small and only 1–2% of the total plant production is consumed (Fritzell, 1989), although the continual removal of preferred plant species can eventually affect its floristic composition. In Europe, for example, *Phragmites australis* often colonizes the area around muskrat lodges once *Typha latifolia* has been eradicated (Fiala and Květ, 1971). Similarly, non-dominant marsh species are usually more abundant on abandoned muskrat mounds, which also form attractive sites for other marsh animals (Kangas and Hannan, 1985). Many other rodents are associated with marshes and swamps. The largest of these is the capybara (*Hydrochoerus hydrochaeris*) which inhabits the riverine swamps of the Amazon. Other large herbivores include the moose (*Alces alces*), which grazes in northern marshes during spring and summer. Tropical swamps are similarly utilized by a variety of antelope, such as waterbuck (*Kobus defassa*), lechwe (*Kobus leche*) and sitatunga (*Tragelaphus spekei*), buffalo and even elephants (Sheppe and Osborne, 1971). The most distinctive herbivore is the hippopotamus, which feed mainly on water lettuce (*Pistia stratiotes*). As they move through the swamps they open up tracks for other animals and stir up the bottom sediments, thereby accelerating the release of nutrients.

Waterfowl are generally the most conspicuous inhabitants of the marshes. Some waterfowl, such as geese and swans, feed exclusively on plants, but ducks are mostly omnivorous and more often feed on the invertebrates which live in the water and amongst the plants. Food selection by ducks is influenced by availability and by physiological demands. A high-protein food is required by the egg-laying females and ducklings during the breeding season. For mallards and other dabbling ducks, the normal food includes crustacea, snails and mayflies: the diving ducks, such as lesser scaup (*Aythya*

affinis), will also consume leeches and benthic midge larvae (Swanson and Duebbert, 1989). At this time of the year the ducks depend on the vegetation more for nest sites and protective cover than for food, but a greater amount of pondweed and seeds are consumed later in the summer and especially during the fall migration. Besides the waterfowl, many other species of bird utilize the rich resources of the marshes and swamps. The abundant insect population provides food for species such as the European reed warbler (*Acrocephalus scirpaceus*) and sedge warbler (*A. schoenobaenus*). Wading birds, such as the herons, search for fish, while spoonbills depend on the plankton that they sieve from the water. Several predatory species, such as the marsh hawk (*Circus cyaneus*) and Montagu's harrier (*C. pygargus*), hunt in the marshes. Concealed amongst the plants are various snakes and in some areas alligators, crocodiles and other large reptiles. Other predators, such as foxes and mink, live around the margins of the wetlands but forage in the marshes, where they feed on birds and on voles and other small mammals.

10.9 PATTERNS OF CHANGE IN WETLANDS

Although animal activities do cause changes in the vegetation of the marshes, the effects are normally restricted to the nest sites where the palatable species are continually cropped. However, herbivores such as the muskrat can sometimes have a pronounced impact on their habitat because their populations are highly variable. During favourable conditions the number of muskrats will increase almost exponentially and can reach densities as high as 86 individuals ha^{-1} (Fritzell, 1989). Practically all of the vegetation in a marsh is eaten out when the muskrat population reaches such high densities. The mortality rate then increases as the muskrats are forced into marginal habitats. In prairie marshes, changes in the muskrat population coincide with cyclical changes in the vegetation which appear to be related to precipitation and resultant fluctuations in water level. During periods of high water some of the emergent

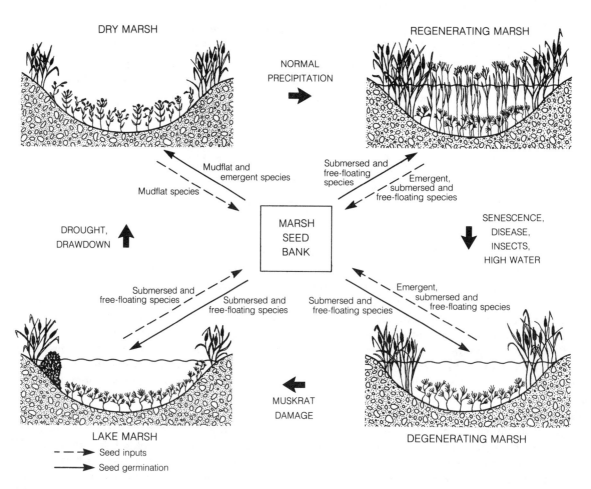

DRY MARSH

NORMAL PRECIPITATION

REGENERATING MARSH

Mudflat and emergent species

Mudflat species

Submersed and free-floating species

Emergent, submersed and free-floating species

MARSH SEED BANK

DROUGHT, DRAWDOWN

SENESCENCE, DISEASE, INSECTS, HIGH WATER

Submersed and free-floating species

Submersed and free-floating species

Submersed and free-floating species

Emergent, submersed and free-floating species

LAKE MARSH

MUSKRAT DAMAGE

DEGENERATING MARSH

- - - → Seed inputs
——→ Seed germination

Figure 10.33 Seed bank dynamics and vegetation cycles in North American prairie marshes. (After van der Valk and Davis, 1978.) (Reprinted with permission from A. G. van der Valk and C. B. Davis, The role of seed banks in the vegetation dynamics of prairie glacial marshes, *Ecology*, 1978, **59**, 333.)

vegetation is killed by prolonged flooding and muskrat activity (Figure 10.33). The degenerating stage of the cycle is characterized by an increase in submersed and free-floating plants, and as the emergent species are replaced the site enters the lake marsh stage. As the water level drops during periods of drought, the exposed dry marsh is colonized by various emergent species and annuals which develop from seeds buried in the mud. The mudflat species are eliminated when the marsh refloods, and during this regenerative stage emergent species spread by vegetative propagation. The establishment of marsh species is therefore dependent on seed germination during periods of low water. Fluctuating water levels are similarly important in the seed bank dynamics of cypress–tupelo swamp forest (Schneider and Sharitz, 1986).

Buried seed reserves in wetland ecosystems are comparatively large. In some prairie marshes over 42 000 viable seeds m^{-2} have been reported in the upper 5 cm of sediment, increasing to 225 000 m^{-2} in surface profiles 35 cm in thickness (van der Valk and Davis, 1979). The dominant plants in the established cover are usually best represented in these seed banks, although seed production by some species is relatively light. For example, flowering shoots of *Phragmites australis* produce fewer than 1000 seeds compared with more than 200 000 in *Typha latifolia* (van der Valk and Davis, 1979). The composition of the seed bank changes with depth, and the presence of large numbers of seeds of mudflat annuals such as nut-grass (*Cyperus odorata*), marsh yellow cress (*Rorippa islandica*) and pale persicaria (*Polygonum lapathifolium*) is presumed to correspond with drier climatic periods and lower water levels. However, germination decreases rapidly with depth of burial and establishment of species such as *Typha glauca* is prevented when the seeds are covered by as little as 1 cm of sediment: other conditions, such as temperature, light and salinity, also affect germination (Galinato and van der Valk, 1986). Seedling establishment can be further limited by the amount of standing or fallen litter that is present (van der Valk, 1986). The windrowing effect, whereby seeds and debris are concentrated along the shoreline of permanent wetlands, can therefore affect subsequent plant recruitment (Pederson and van der Valk, 1984).

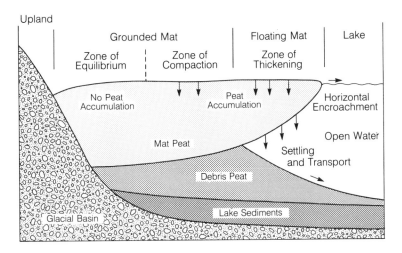

Figure 10.34 Peatland development around temperate lakes in northern latitudes. (After Kratz and deWitt, 1986.) (Reprinted with permission from T. K. Kratz an C. B. DeWitt, Internal factors controlling peatland–lake ecosystem development, *Ecology*, 1986, **67**, 101.)

The gradual accumulation of organic sediment under anaerobic conditions in bogs and fens provides a much longer record of vegetation change and it is mainly these types of wetlands which are used in palynological studies. Most of these peatlands began to form less than 10 000 years ago in the poorly-drained depressions which appeared as the Pleistocene ice sheets receded from North America and Eurasia (Zoltai and Vitt, 1990). Vegetation development in these shallow open-water bodies is recorded by the pollen grains and other plant materials that are preserved in the sediment. The classic successional sequence of plant establishment was assumed to start with free-floating or submerged aquatic plants. Reeds and sedges would colonize the area as more detritus accumulated and these would be replaced by semi-terrestrial communities which were tolerant of seasonal waterlogging as the land continued to build up above the water table. This sequence of terrestrialization would eventually end in forest. The natural climax community in Britain, for example, was considered to be oak (*Quercus robur*) woodland (Tansley, 1953). In wetter situations *Sphagnum* mosses generally form a stable cover which persists for many thousands of years with no evidence of colonization by trees. In these sites the surface layers of peat become progressively more oligotrophic as they emerge above the influence of the percolating groundwater.

Successional development in wetland communities can be evaluated from changes in the underlying sediment (Walker, 1970). In some areas the typical sequence of terrestrialization starts with the establishment of reed-swamps and then passes through a fen-carr stage into an ombrotrophic bog. Sometimes a mat of *Sphagnum* and sedges grows directly over the water. The undersides of these floating mats are usually intermeshed with a network of buoyant roots and stems of species such as leatherleaf. The lake is gradually filled in by accumulat-

ing debris (Figure 10.34). The bottom lake sediments consist of well-decomposed organic material including diatoms and other aquatic organisms. Above this is a layer of moderately decomposed debris peat which has fallen from the floating mat. The surface layer of mat peat consists of the poorly decomposed and interwoven plant roots and stems (Kratz and deWitt, 1986). Lake sediments are always found beneath the peat layers in sites of terrestrialization. However, paludification occurs directly over mineral soils and is usually initiated by a climatic change which causes the water table to rise. Depletion of soil nutrients through leaching might also cause a decline in the forest cover: impermeable iron pans associated with soil podsolization underlie some bogs in upland Britain (Pearsall, 1950). Expansion of the bogs then occurs through seepage and waterlogging of the adjacent downslope terrain, and the lowest peat deposits usually contain woody material derived from the last trees which grew on the underlying mineral soil.

A characteristic **hummock-and-hollow** microtopography develops on *Sphagnum* bogs as they continue to grow. Where the surface is almost level these features form an irregular mosaic, but in sloping sites the ridges and hollows are aligned perpendicular to the gradient. It was originally thought that the development of the hummocks and hollows was a cyclical process. During the building phase the surface of a hollow was slowly raised through the successive establishment of different plant species (Figure 10.35). The vegetation on the former hummocks would be killed by the rising water table, a new hollow would eventually form and the building phase would begin again. However, stratigraphic analyses of peat profiles usually do not reveal the lenticular pattern which would result from this type of regeneration cycle. In some mires the position of the ridges and hollows has remained unchanged for 2500–5000 years; in others an ancient hummock-and-hollow surface now

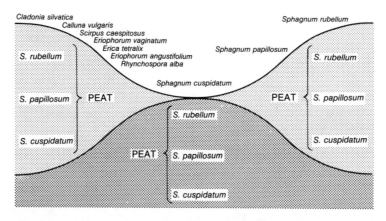

Figure 10.35 Postulated cycle of microtopographic development in *Sphagnum* bogs. (After Tansley, 1953.) (Reproduced with permission from A. G. Tansley, *The British Islands and Their Vegetation*; reprinted by Cambridge University Press, 1953.)

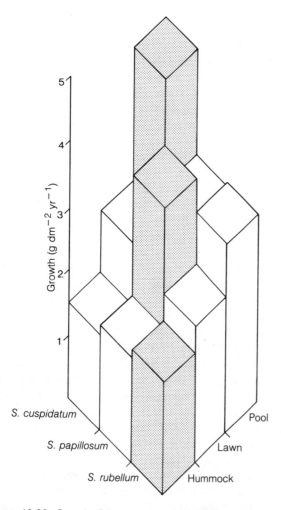

Figure 10.36 Growth of three species of *Sphagnum* transplanted to native (toned) and non-native habitats. (After Clymo and Reddaway, 1971.)

lies buried beneath relatively undifferentiated *Sphagnum* peat (Tallis, 1983). Similarly, comparative studies of bog microtopography over a 60-year period in Sweden shows no evidence of cyclic regeneration (Backéus, 1972).

In some profiles '**recurrence surfaces**' are detected in which there is an abrupt change from strongly humified *Sphagnum* peat to an overlying layer of relatively undecomposed material. These interruptions are sometimes associated with periodic flooding and changes in groundwater flow (Succow and Lange, 1984), fire (Foster and Glaser, 1986), or mass movements in the peat cover on sloping terrain (Pearsall, 1956). However, most of the stratigraphic changes which occur in peat profiles are generally indicative of prevailing climate conditions, with poorly decomposed material only accumulating during wetter periods (Barber, 1981). The upward growth of the *Sphagnum* hummock continues during wet periods as the pools expand, and because the hummock species are better adapted to take up water, they are able to spread into the hollows during drier periods. The distribution of mosses in a bog is therefore more likely to be related to their competitive ability rather than determined by a cyclical pattern of autogenic regeneration (Figure 10.36). Thus *S. rubellum* are most productive on the tops of hummocks where the water table is about 20 cm below the surface, whereas *S. cuspidatum* is the dominant producer in hollows where the depth to the water table is about 3 cm. In the intervening zone the productivity of the mosses varies from year to year so that lawn species will sometimes grow over the pools or collapse into water-filled hollows (Wallen *et al.*, 1988). Changes in species composition apparently related to the overall hydrology of the bog as influenced by climatic conditions rather than an inherent process of cyclical regeneration are evident in the peat stratigraphy of Irish bogs (Walker and Walker, 1961). The effect of weather cycles which produce exceptionally high or low water levels is recorded in bog successions in North America (Schwintzer and Williams, 1974).

10.10 HUMAN IMPACT ON WETLAND ECOSYSTEMS

Peat has been cut and dried for fuel since prehistoric times and reeds have been used for roofing thatch in Europe for centuries. Papyrus is well-known for its use in paper-making in ancient Egypt and the reeds around Lake Titicaca continue to be used to construct canoes. However, most bogs and swamps remained inaccessible to early human cultures except where branches and hurdles were used to form trackways (Casparie, 1972). Consequently, wetland areas have traditionally been regarded as rather unproductive wastelands that could be only be improved through drainage. Extensive drainage of the European wetlands began in the eighteenth century. In places such as the Fens of eastern Britain, a series of canals were cut across the low-lying terrain and pumping was carried out initially by windmills. Subsidence caused by oxidation, shrinkage and compaction occurs when organic soils are drained. Measurements against the Holme Post (Figure 10.37), which was placed in position in 1852, indicate that the surface of Holme Fen has since dropped by about 4 m (Hutchinson, 1980). A rate of subsidence of about 2.5 cm yr^{-1} is indicated by a similar post that was erected in the Florida Everglades in 1924 (Foth and Schafer, 1980). The dry peat is also blown away by the wind. With careful preparation the drained peats can be turned into very productive agricultural land but the biological richness and diversity of these areas is considerably reduced.

Ditching and draining for agriculture continues to be the main wetland management activity. In the prairie pothole region, more than half of the 8 million ha of marshland which existed prior to settlement have now been drained. Those that remain are now surrounded by cultivated fields and receive large inputs of sediment, nutrients and other agro-chemicals. The high nutrient loading of the surface run-off is reflected by increased productivity and higher tissue nutrient concentrations in the emergent plants (Neely and Baker, 1989). Micro-organisms use the litter as a carbon source, and as decomposition proceeds nutrients, such as nitrogen and phosphorus, are normally withdrawn from the water and become immobilized in the sediment. Decomposition occurs more rapidly when the initial nutrient content of the tissue is high and the excess nutrients are released from the litter. Consequently, some prairie marshes which receive agricultural run-off may now export nutrients rather than function as nutrient sinks as is the case with most natural wetlands (Neely and Davis, 1985).

The high productivity and nutrient uptake by emergent marsh plants has resulted in many wetland areas being used in water purification projects. Most of these

Figure 10.37 Holme Post, sunk into the peat in 1848 with its top indicating ground level, has become exposed through fenland drainage. (O. W. Archibold.)

systems are designed to treat domestic sewage, which is especially high in nitrogen, phosphorus and suspended solids. In temperate regions the wastewater is normally pumped into a holding lagoon after primary treatment and is then discharged slowly into the marsh during summer (Kadlec, 1987). Plant growth ceases during winter and the efficiency of the system is reduced, but some purification still occurs providing the water remains unfrozen and still percolates through the peat. Some of the pollutants are removed by sedimentation but mostly they are taken up by plants during the growing season, or utilized by micro-organisms and subsequently released to the atmosphere in gaseous forms. Ultimately, most of the pollutants are incorporated into the sediment. The altered hydrologic regimes and higher nutrient loadings in these treatment systems usually result in changes in the vegetation cover. Algae and floating duckweeds become more abundant and the growth of emergent macrophytes increases: species which are less tolerant of flooding are adversely affected by the continual addition of wastewater to the root zone.

Most of the nitrogen and some of the phosphorus are normally removed as the wastewater passes through the wetland. However, the efficiency of the system depends on the nature of the plant community, the length of time that it has been in use, the type of substrate which

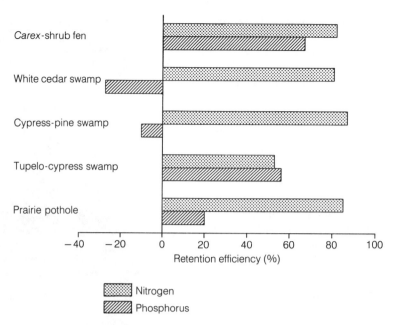

Figure 10.38 Nitrogen and phosphorus retention efficiencies in selected wetland wastewater treatment projects in North America. (After Richardson and Davis, 1987.) (Redrawn with permission from C. J. Richardson and J. A. Davis, *Aquatic Plants for Water Treatment and Resource Recovery*, eds K. R. Reddy and W. H. Smith; published by Magnolia Publishing Inc., 1987.)

underlies the peat and the hydrologic conditions involved. For example, over 80% of the nitrogen and 67% of the phosphorus is removed by a wastewater treatment facility in Michigan that utilizes 700 ha of shrub–sedge fen (Figure 10.38), but this is probably a function of its large size rather than the efficiency of the system. Nitrogen retention has remained at about 80% in a similar project established in a white cedar swamp. Phosphorus retention was also good initially but after 5 years the adsorption capacity of the soil was exceeded and sub-

sequently there has been a net release of phosphorus from this site. Export of phosphorus is similarly attributed to the low adsorption capacity of the sandy substrate which underlies an artificial swamp of cypress pine and bay-trees used to treat secondary effluent in Florida, at the Disney World complex. Retention of phosphorus by soil adsorption is higher in the cypress–tupelo wetland, but without a summer drydown period bacterial nitrification and denitrification are severely restricted and ammonium levels in the wastewater remain unacceptably high (Richardson and Davis, 1987).

Artificial treatment lagoons using aquatic plants now form the basis of many low-cost wastewater treatment processes (Figure 10.39). The efficiency of these systems is generally related to the growth rates and tissue nutrient concentrations of the plants which are used. Floating macrophytes, such as water hyacinth (*Eichhornia crassipes*) and water lettuce (*Pistia stratiotes*), are commonly used in tropical and subtropical regions but emergent marsh plants such as cattails are best suited to temperate conditions. Natural wetlands used for wastewater management are typically unmanaged but the efficiency of artificial systems is usually increased by harvesting the plants (Reddy and DeBusk, 1987). Harvesting floating aquatic plants mainly prevents nutrients from re-entering the water when they die. In emergent species there is a progressive decrease in tissue nutrient concentrations over the growing season (Figure 10.40) and a large increase in nutrient uptake can be achieved if new growth is maintained through frequent harvesting (Table 10.13). Nutrient translocation

Figure 10.39 Water treatment facility at Prudhomme, Saskatchewan, during a period of summer drawdown. Municipal sewage is pumped into the rectangular settling ponds and overflows into marshland reconstructed to provide improved waterfowl habitat. (Duck's Unlimited.)

to less accessible underground storage organs is also reduced, which further increases the demand for nutrients from external sources. The biomass generated from these artifical systems can be used for livestock feed and fertilizer or converted to methane and alcohol fuels. Where artificial wetlands are used to treat industrial effluent, heavy metals and other toxic chemicals can be removed from the harvested plant material. Forested wetland provides long-term storage of nutrients but the comparatively slow growth rate of trees and low nutrient concentrations in the wood make them unsuitable for most artifical treatment projects.

Drainage of natural peatlands to improve tree growth is becoming more widespread as demands for forest products increase (Figure 10.41). For example, height increases of 35–127% are reported for 100-year-old black spruce growing on drained peatlands in Canada (Table 10.14). Tree growth in wet organic soils is usually limited because of poor root development. Lowering the water table improves soil aeration and alters soil temperature regimes so that the surface layers warm more quickly in the spring (Figure 10.42). The fine-root biomass and maximum rooting depth of black spruce and tamarack increase under drier conditions (Table 10.15) and the increased growth that results is attributed to the change in the distribution and structure of the roots and the altered nutrient status of the aerated soils (Lieffers and Rothwell, 1987b). However, growth on drained peat can be limited by nutrient deficiencies and some fertilizer application may be necessary to optimize yields (Mugasha et al., 1991).

Wetlands are clearly more than land reserves that can be converted to forest or cropland through drainage. The legacy of past wetland management is the loss of considerable wetland areas. It is estimated that 30–50% of the wetlands in the United States have now been drained for

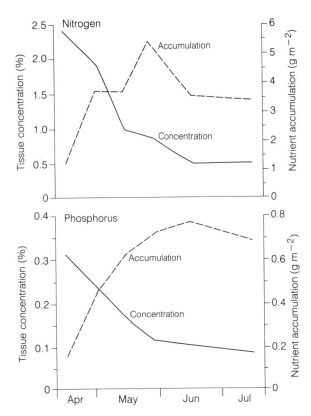

Figure 10.40 Seasonal trends in nitrogen and phosphorus uptake and tissue concentration in cat's-tail (*Typha latifolia*) growing in the south-eastern United States. (After Boyd, 1970.) (Adapted with permission from C. E. Boyd, Production, mineral accumulation and pigment concentrations in *Typha latifolia* and *Scirpus americanus*, *Ecology*, 1970, **51**, 288.)

Table 10.13 Storage and potential uptake of nitrogen and phosphorus by wetland plants species (after Reddy and DeBusk, 1987)

	Nitrogen		Phosphorus	
	Storage (kg ha^{-1})	Uptake (kg ha^{-1} yr^{-1})	Storage (kg ha^{-1})	Uptake (kg ha^{-1} yr^{-1})
Floating macrophytes				
Eichhornia crassipes	300–900	1950–5850	60–180	350–1125
Pistia stratiotes	90–250	1350–5110	20–57	300–1100
Hydrocotyle umbellata	90–300	540–3200	23–75	130–770
Lemna minor	4–50	350–1200	1–16	116–400
Emergent macrophytes				
Typha spp.	250–1560	600–2630	45–375	75–403
Phragmites spp.	140–430	225	14–53	35
Scirpus spp.	175–530	125	40–110	18
Juncus spp.	200–300	800	40	110
Forested wetlands				
Cypress swamp	213	1219	23	86

(Source: K. R. Reddy and W. F. DeBusk, *Aquatic Plants for Water Treatment and Resource Recovery*, eds K. R. Reddy and W. H. Smith; published by Magnolia Publishing Inc., 1987.)

Table 10.14 Calculated height (m) at 100 years of black spruce growing in various wetland habitats in eastern Canada before and after drainage (after Stanek, 1977)

Habitat	Tree height (m)		Relative increase in height (%)
	Before drainage	After drainage	
Open bog	5.2	11.8	127
Treed bog	6.8	12.2	79
Treed fen	11.9	17.2	45
Swamp forest	12.9	17.4	35

(Reproduced with permission from W. Stanek, Ontario clay belt peatlands – are they suitable for forest drainage?, *Canadian Journal of Forest Research*, 1977, **7**, 660.)

agriculture, transportation or urban development (Mitsch and Gosselink, 1986). In some areas the wetlands have been almost totally destroyed: over 90% of the potholes and marshes in Iowa had been drained or filled by the mid 1950s (Leitch, 1989) and 96% of the Mississippi River hardwood swamps had been lost by 1975 (Korte and Fredrickson, 1977). However, as the intrinsic value of wetlands becomes more widely appreciated, earlier government policies which encouraged drainage are being replaced by legislation to ensure that these resources are appropriately managed for a variety of purposes. Present management schemes encourage multiple use of wetlands for environmental protection, recreation and production of renewable resources (Stearns, 1978).

Figure 10.41 (a) Peatland drainage and (b) an established conifer plantation on the Isle of Skye. (O. W. Archibold.)

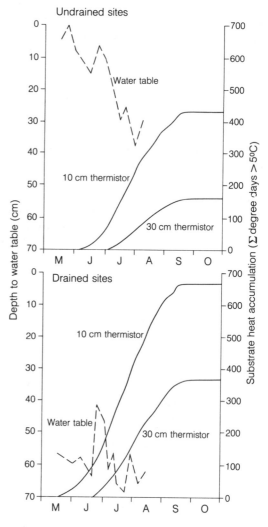

Figure 10.42 Effects of drainage on substrate temperatures in peatlands in western Canada. (After Lieffers and Rothwell, 1987a.) (Reproduced with permission from V. J. Lieffers and R. L. Rothwell, Effects of drainage on substrate temperature and phenology of some trees and shrubs in an Alberta peatland, *Canadian Journal of Forest Research*, 1987, **17**, 99, 100.)

Table 10.15 Mean radial growth (mm yr^{-1}) at 30 cm height and mean biomass (g dry weight m^{-2}) of fine (< 2.0 mm), medium (2.0–10.0 mm), coarse (10.1–50.0 mm) and very coarse (> 50.0 mm) roots of 54-yr-old black spruce and tamarack growing in drained and undrained sites in Canada (after Lieffers and Rothwell, 1987a)

	Mean radial growth at 30 cm height (mm yr^{-1})	
	Drained in 1966	**Wet**
Black spruce		
1962–65	0.64	0.75
1981–84	1.05	0.23
Tamarack		
1962–65	0.13	0.37
1981–84	2.40	0.77

	Mean root biomass (g dry weight m^{-2})							
	Drained in 1966				**Wet**			
Depth (cm)	**fine roots**	**medium roots**	**coarse roots**	**v. coarse roots**	**fine roots**	**medium roots**	**coarse roots**	**v. coarse roots**
Black spruce								
0–10	264.1	521.6	920.0	168.2	41.9	44.1	329.8	0.0
10–20	171.0	67.8	52.2	0.0	31.3	51.4	408.2	1076.0
20–30	54.5	5.7	0.8	0.0	2.9	0.0	0.0	0.0
> 30	4.7	0.8	0.0	0.0	0.0	0.0	0.0	0.0
Tamarack								
0–10	265.3	520.0	1152.7	700.4	47.6	241.6	177.1	0.0
10–20	178.9	155.9	324.9	1578.0	23.1	186.9	275.1	1637.6
20–30	39.3	8.2	0.0	6.5	0.7	0.0	0.0	0.0
> 30	10.4	4.1	0.0	0.0	0.0	0.0	0.0	0.0

(Reproduced with permission from V. J. Lieffers and R. L. Rothwell, Effects of drainage on substrate temperature and phenology of some trees and shrubs in an Alberta peatland, *Canadian Journal of Forest Research*, 1987, **17**, 97–104.)

CHAPTER 11

Freshwater ecosystems

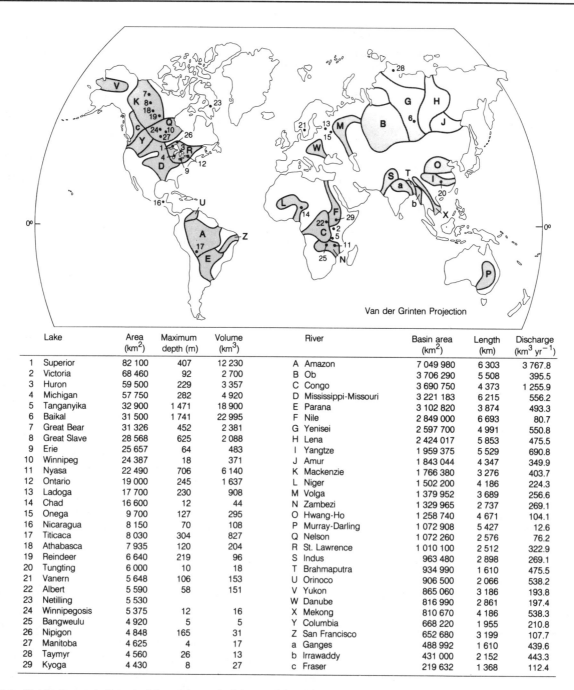

Van der Grinten Projection

	Lake	Area (km²)	Maximum depth (m)	Volume (km³)		River	Basin area (km²)	Length (km)	Discharge (km³ yr⁻¹)
1	Superior	82 100	407	12 230	A	Amazon	7 049 980	6 303	3 767.8
2	Victoria	68 460	92	2 700	B	Ob	3 706 290	5 508	395.5
3	Huron	59 500	229	3 357	C	Congo	3 690 750	4 373	1 255.9
4	Michigan	57 750	282	4 920	D	Mississippi-Missouri	3 221 183	6 215	556.2
5	Tanganyika	32 900	1 471	18 900	E	Parana	3 102 820	3 874	493.3
6	Baikal	31 500	1 741	22 995	F	Nile	2 849 000	6 693	80.7
7	Great Bear	31 326	452	2 381	G	Yenisei	2 597 700	4 991	550.8
8	Great Slave	28 568	625	2 088	H	Lena	2 424 017	5 853	475.5
9	Erie	25 657	64	483	I	Yangtze	1 959 375	5 529	690.8
10	Winnipeg	24 387	18	371	J	Amur	1 843 044	4 347	349.9
11	Nyasa	22 490	706	6 140	K	Mackenzie	1 766 380	3 276	403.7
12	Ontario	19 000	245	1 637	L	Niger	1 502 200	4 186	224.3
13	Ladoga	17 700	230	908	M	Volga	1 379 952	3 689	256.6
14	Chad	16 600	12	44	N	Zambezi	1 329 965	2 737	269.1
15	Onega	9 700	127	295	O	Hwang-Ho	1 258 740	4 671	104.1
16	Nicaragua	8 150	70	108	P	Murray-Darling	1 072 908	5 427	12.6
17	Titicaca	8 030	304	827	Q	Nelson	1 072 260	2 576	76.2
18	Athabasca	7 935	120	204	R	St. Lawrence	1 010 100	2 512	322.9
19	Reindeer	6 640	219	96	S	Indus	963 480	2 898	269.1
20	Tungting	6 000	10	18	T	Brahmaputra	934 990	1 610	475.5
21	Vanern	5 648	106	153	U	Orinoco	906 500	2 066	538.2
22	Albert	5 590	58	151	V	Yukon	865 060	3 186	193.8
23	Netilling	5 530			W	Danube	816 990	2 861	197.4
24	Winnipegosis	5 375	12	16	X	Mekong	810 670	4 186	538.3
25	Bangweulu	4 920	5	5	Y	Columbia	668 220	1 955	210.8
26	Nipigon	4 848	165	31	Z	San Francisco	652 680	3 199	107.7
27	Manitoba	4 625	4	17	a	Ganges	488 992	1 610	439.6
28	Taymyr	4 560	26	13	b	Irrawaddy	431 000	2 152	443.3
29	Kyoga	4 430	8	27	c	Fraser	219 632	1 368	112.4

Figure 11.1 Distribution and characteristics of the major lakes and river basins of the world. (After Herdendorf, 1990; Mielke, 1989). (Reproduced with permission from C. E. Herdendorf, Distribution of the world's large lakes, in *Large Lakes: Ecological Structure and Function*, eds M. M. Tilzer and C. Serruya, published by Springer-Verlag GmbH, 1990.)

11.1 INTRODUCTION

Freshwater ecosystems are associated with the lakes and channels through which precipitation is returned to the ocean as part of the hydrologic cycle. The transition to an aquatic environment occurs where water depth or strong currents prevent the growth and development of terrestrial organisms. Plants and animals have adapted to these conditions in various ways and the structure of aquatic communities reflects the physical and chemical properties of the waterbodies in which they grow. Lakes and ponds function essentially as closed systems, even though water enters and leaves these temporary storage basins in various ways. Water movement is slow and turnover times typically range from 1 to 100 years, depending on volume, depth and rate of discharge (Wetzel, 1975). There is a tendency for materials to accumulate in the still water and consequently ponds and lakes undergo a natural ageing process. The rate of change depends on hydrologic conditions and the size and shape of the basin but it is generally assumed that the open water habitat is eventually lost.

Conditions are quite different in fast-flowing rivers, where much of the fine sediment is transported in suspension. Turnover time in most rivers is less than 20 days and these systems exist in a state of dynamic equilibrium which reflects the balance of gains and losses at any given time. The character of a river changes as it flows away from its source. The clear, turbulent waters of the gravel beds and rocky channels in the upper reaches reflect the erosive power of the river. Sediment loads increase downstream, and in sluggish sections of the river the silts and clays settle out from the turbid water. Sedentary communities similar to those bordering lakes and ponds usually develop in the shallows but in most parts of the river the plants and animals have adapted differently in order to survive in the stronger currents.

11.2 REGIONAL DISTRIBUTION OF FRESHWATER RESERVES

Reserves of freshwater on Earth total some 33.2 million km^3, of which 87% is frozen in polar ice and glaciers, 12% is stored as subsurface groundwater and the remainder is mostly contained in relatively few large lakes. Approximately 125 000 km^3 of freshwater are held in lakes with a combined surface area of about 1.6 million km^2. Freshwater storage in rivers amounts to only 1200 km^3, although 37 000 km^3 is discharged to the oceans each year. The distribution of freshwater lakes and rivers depends on climatic conditions and past geological events, and consequently they are not uniformly distributed throughout the world (Figure 11.1). For example, the concentration of lakes and ponds in Canada and Scandinavia is associated with glacially scoured bedrock surfaces, while those in East Africa are found in the fault depressions of the East African Rift System. Most rivers are located in areas with humid climates although some occur in cooler, drier zones where low evaporation rates compensate for lack of precipitation. In some drainage basins the rivers do not reach the sea. Evaporation exceeds precipitation in these areas of inland drainage and the water which collects in them is invariably saline.

About 20% of the surface reserves of freshwater are contained in the Great Lakes of North America, which have a combined area of 0.25 million km^2. The average depth of water in the Great Lakes ranges from 19 m in Lake Erie to 149 m in Lake Superior, which is 406 m deep in places. Great Bear Lake and Great Slave Lake in north-western Canada are larger than Lake Erie and Lake Ontario in terms of both surface area and volume. These lakes, like most others in North America, were formed as the ice withdrew from the continent at the close of the Pleistocene. The largest lakes in Europe are similarly associated with glacial activity: Lake Vänern and Lake Vättern in Sweden, with surface areas of 5650 km^2 and 1900 km^2, respectively, are the biggest of the European Lakes, although these are considerably smaller than nearby Lake Ladoga (18 000 km^2) and Lake Onega (10 000 km^2) in Russia. Lake Baikal is the largest lake in the former Soviet Union, although its surface area is only 31 500 km^2. The average depth of Lake Baikal is 740 m and parts of the basin descend to 1620 m, giving it a total volume of 23 000 km^3: this is similar to the combined volume of the Great Lakes (24 600 km^3). Lake Tanganyika, located in the Rift Valley of East Africa, is physiographically similar to Lake Baikal. The maximum depth in Lake Tanganyika is 1470 m which, combined with a surface area of 28 000 km^2, makes it second only to Lake Baikal in terms of volume. Lake Victoria has the greatest surface area (70 000 km^2) of the African lakes but it is comparatively shallow with a maximum depth of 80 m, hence its volume is smaller than either Lake Tanganyika or Lake Malawi. Few of the other well-known lakes are larger than 200 km^2 in area and the majority of lakes are much smaller than this. In addition, there is an increasing number of artificial reservoirs, including massive projects such as The Volta Lake (8800 km^2) in Ghana, Lake Nasser (6200 km^2) in Egypt and Lake Kariba (4300 km^2) along the Zambia–Zimbabwe border (Figure 11.2).

11.3 THE FRESHWATER ENVIRONMENT

The physical properties of water are such that it normally remains liquid in most climatic regions. Water has a high specific heat and a considerable amount of energy must be absorbed or lost to bring about a change in temperature. The maximum temperature of a body of water is further regulated by the high latent heat of vaporization (540 cal g^{-1}) which causes energy to be

Figure 11.2 Early flooding stage of Lake Kariba. (G. E. Wicken.)

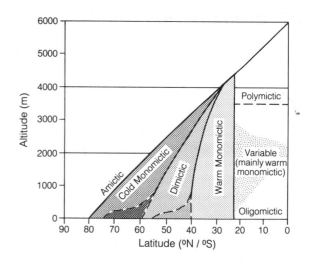

Figure 11.3 Thermal lake types arranged according to altitudal and latitudinal gradients. (After Wetzel, 1975.) (Reproduced from R. G. Wetzel, *Limnology*, published by Saunders College Publishing, 1975.)

dissipated through evaporation; the latent heat of fusion ($80 \, cal \, g^{-1}$) associated with the freezing process similarly reduces heat loss during winter. Consequently, diurnal and seasonal temperature changes occur more gradually and are less extreme than in adjacent terrestrial environments, although this thermal inertia is related to the size of the water body. Water is densest at 4 °C, and because its density decreases until it freezes, ice begins to form at the surface. Only in extreme cases will an entire waterbody freeze. Density is also related to the concentration of dissolved salts: pure water at 4 °C has a density of 1.00 $g \, cm^{-3}$ compared with 1.03 $g \, cm^{-3}$ for sea water. This can affect water circulation in some estuaries and saline lakes. The density of water provides a certain amount of buoyancy to submerged aquatic organisms, while others depend on its viscosity and surface tension to maintain their position.

11.3.1 Temperature

Water is continually mixed as it flows along a river, but in lakes the water remains still unless disturbed by a major inflow or until wind-generated waves and currents develop. Circulation in most lakes is triggered by density changes associated with seasonal temperature regimes. The majority of lakes in the cool temperate regions of the world are **dimictic**, with the deeper water brought to the surface twice each year in spring and autumn. In summer the surface water becomes warmer and less dense as it absorbs solar energy; consequently, it remains at the surface. This warm surface layer forms the **epilimnion** and is separated from the colder water in the **hypolimnion** by the **thermocline** – a zone in which water temperature decreases rapidly with depth. The surface waters are moved by wind and convection currents, but there is no mixing across the thermocline and summer stagnation may occur in the hypolimnion as the

dissolved oxygen is used up. This condition of **thermal stratification** persists until the water in the epilimnion begins to cool in the autumn. The thermocline weakens and sinks as the density of the surface water increases; eventually wind-generated currents reach the hypolimnion and all of the water circulates during the fall overturn. Ice may form on the surface of the lake during winter and an **inverse temperature stratification** may develop as dense water at 4 °C sinks to the bottom. Even though the ice cover prevents wind from moving the water, some circulation occurs because of density currents caused by solar heating or heat transfer from sediments (Wetzel, 1975). The surface waters begin to warm again as winter ends and mixing occurs during the spring overturn period when temperatures reach 4 °C.

The pattern of circulation differs in lakes of other climatic zones (Figure 11.3). In deep polar lakes the temperature of the water at any depth never exceeds 4 °C. In these **cold monomictic** lakes, mixing only occurs in the summer because of the increased density of the slightly warmer surface water. **Warm monomictic** lakes are mainly associated with subtropical regions. Water temperatures in these lakes never drop below 4 °C. These lakes become stratified in the summer but circulation occurs in winter when the surface water is cooler. The lakes of lowland tropical regions rarely circulate because the continual warmth maintains the stability of the water column. Circulation in these **oligomictic** lakes only occurs when thermal stratification is weakened by unusually cold weather. However, circulation is continuous in the **polymictic** lakes of tropical highlands: water temperatures in these lakes are only slightly above 4 °C and mixing occurs because of nocturnal cooling and strong winds.

Water temperatures in rivers are more closely related to atmospheric temperatures and normally reach about

20 °C in temperate regions and 30–35 °C in the tropics. Most rivers have a noticeable diurnal rhythm with temperatures increasing to a maximum in late afternoon. These fluctuations are most pronounced in small, clear streams but diminish as water volume and turbidity increase. Water temperatures in the headwaters of a river are usually several degrees cooler than in the broader lower reaches. This longitudinal gradient reflects the changing hydrologic character of the river and the different environments through which it flows. The turbulent and shallow nature of most rivers prevents the water from becoming thermally stratified, although this can sometimes occur in pools and backwaters, especially if unusually warm or cold water enters from a spring or tributary. Thermal stratification, partly controlled by density differences between sea water and fresh water, does occur in some estuaries where warmer water forms a surface layer.

11.3.2 Light

Water is heated by solar radiation with the amount of warming a function of latitude, altitude and cloud cover. Some energy is reflected from the surface of the water: losses vary with angle of incidence and wave conditions but usually average 5–6% (Wetzel, 1975). The remaining energy is transmitted through the water column where it is either absorbed and changed to heat, scattered by the water molecules and suspended particles and returned to the atmosphere, or used in photosynthesis. The infra-red wavelengths are mostly absorbed within a metre of the water surface: visible light penetrates further into the water but absorption is not uniform for all wavelengths and the spectrum becomes progressively narrower with depth (Figure 11.4). Light transmission in water is greatly affected by the amount of dissolved and suspended materials: whereas some wavelengths can penetrate more than 100 m in deep transparent lakes, this is reduced to less than 1 m in large silt-laden rivers. Light attenuation is important in that it ultimately sets the limit for photosynthesis. Photosynthesis effectively ceases at depths where light intensity is 1% of the incident surface value: this marks the bottom of the **euphotic** zone (Kirk, 1986). The relative depths of the euphotic zone and thermocline in thermally stratified lakes has important consequences for organisms in the hypolimnion. During periods of stagnation, oxygen levels in the deeper water can only be maintained by photosynthesis and therefore depend on adequate light levels in the hypolimnion.

11.3.3 Dissolved gases

Oxygen is a soluble gas and will diffuse into water from the atmosphere. The solubility of oxygen is inversely related to temperature and increases by more than 50% as fresh water cools from 35 °C to freezing (Table 11.1). Similarly, less oxygen dissolves into water at higher elevations because of reduced atmospheric pressures. The amount of gas that remains in solution is also affected by hydrostatic pressure. Consequently, much of the oxygen produced in photosynthesis remains in the water because the higher pressure at depth prevents gas bubbles from forming and rising to the surface until a high degree of suspersaturation is reached. Oxygen diffuses very slowly through water and is mainly carried down from the surface by currents, wave action and turbulence. The levels of oxygen in natural rivers are usually near saturation, especially below rapids and other fast-flowing reaches. However, oxygen profiles in lakes are more variable and reflect the complex interaction of the different physical and biological processes which occur in them. Oxygen levels in **oligotrophic** lakes with little organic production are near saturation at all depths. Oxygen is carried down into the hypolimnion during periods of overturn: very little is used for respiration and decomposition, and some oxygen may be produced by photosynthesis in the clear, deep water below the thermocline (Figure 11.5). In more productive lakes the

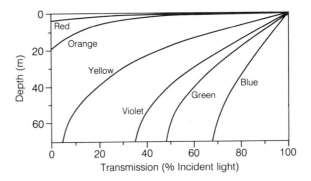

Figure 11.4 Transmission of light of various wavelengths by distilled water. (After Wetzel, 1975.) (Reproduced from R. G. Wetzel, *Limnology*, published by Saunders College Publishing, 1975.)

Table 11.1 Solubility of oxygen in pure water in relation to temperature from saturated air at 760 mm Hg pressure (after Wetzel, 1975)

Temperature (°C)	Oxygen (mg l^{-1})
0	14.2
5	12.4
10	10.9
15	9.8
20	8.8
25	8.1
30	7.5
35	7.0

(Reproduced from R. G. Wetzel, *Limnology*, published by Saunders College Publishing, 1975.)

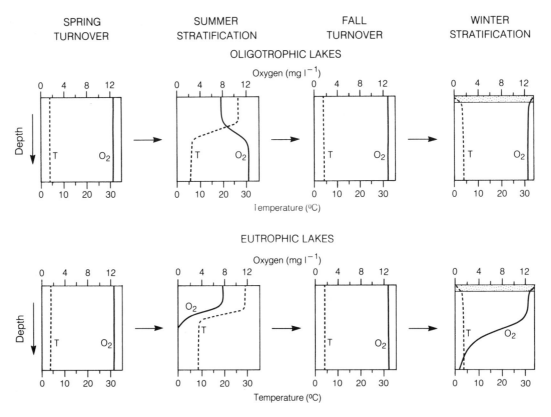

Figure 11.5 Idealized seasonal temperature (T) and oxygen (O$_2$) profiles in temperate dimictic lakes of different trophic status. Winter ice cover is indicated. (After Wetzel, 1975.) (Reproduced from R. G. Wetzel, *Limnology*, published by Saunders College Publishing, 1975.)

demand for oxygen in the hypolimnion is considerably greater. Oxygen reserves are rapidly depleted during periods of stagnation and must be replenished by mixing with the surface waters. Severe oxygen deficits can therefore develop in tropical lakes which mix only rarely. Daily changes in dissolved oxygen concentrations are often superimposed on these seasonal rhythms and are especially noticeable in shallow, productive lakes and in rivers, mainly reflecting gains and losses through photosynthesis and respiration. Photosynthesis and respiration are essentially complementary processes; consequently, the seasonal and diurnal changes in dissolved oxygen levels tend to vary inversely with the concentraton of dissolved CO$_2$.

The amount of atmospheric CO$_2$ dissolved in fresh water ranges from 1.1 mg l^{-1} at 0 °C to 0.4 mg l^{-1} at 30 °C. This is supplemented by CO$_2$ derived from bacterial decomposition and plant and animal respiration. Groundwater seeping into lakes and streams also carries variable amounts of CO$_2$ depending on the nature of the underlying rock. Carbon dioxide is much more soluble than oxygen and it normally reacts with water to form weak carbonic acid (H$_2$CO$_3$), which may then dissociate into bicarbonate (HCO$_3^-$) or carbonate ions (CO$_3^{2-}$), depending on pH (Figure 11.6). In this way the various forms of dissolved CO$_2$ function as a buffer against rapid

shifts in acidity. The process is most effective when calcium is also present in the water. Calcium carbonate present in rock and soil readily dissolves in CO$_2$ enriched water to form calcium bicarbonate, which acts to neutralize the weak carbonic acid. However, calcium bicarbonate only remains in solution in the presence of a certain amount of free CO$_2$ and any process which removes CO$_2$ causes calcium carbonate to precipitate out of solution. In some very productive lakes sufficient CO$_2$ is removed in photosynthesis that the water becomes turbid with suspended calcium carbonate (Strong and Eadie, 1978) and in time a bench of limestone may form in the euphotic zone. Many aquatic plants utilize bicarbonate directly for photosynthesisis (Sand-Jensen, 1987) and this also causes precipitation of calcium carbonate. However, the solubility of CO$_2$ increases in water that contains carbonate and this helps to maintain the equilibrium of the system.

11.3.4 Mineral nutrients

Most of the inorganic CO$_2$ is bound up with calcium but considerable amounts of magnesium carbonate are formed in very alkaline water. These elements, together with sodium and potassium, account for most of the cations

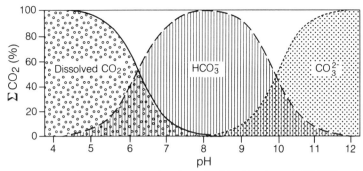

Figure 11.6 Relation between pH and the relative proportions of carbon dioxide, bicarbonate (HCO_3^-) and carbonate (CO_3^-) ions in solution. (After Wetzel, 1975.) (Reproduced from R. G. Wetzel, *Limnology*, published by Saunders College Publishing, 1975.)

Table 11.2 Concentration ($mg\ l^{-1}$) of major soluble nutrients (except nitrogen and phosphorus) in freshwater lakes and rivers

	Ca^{2+}	Mg^{2+}	Na^+	K^+	HCO_3^{2-}	SO_4^{2-}	Cl^-	SiO_2	
					($mg\ l^{-1}$)				**Source**
North American lakes									
Ontario	42.9	6.4	12.2	1.4	115.0	27.1	26.7	0.5	Robertson and Scavia, 1984
Erie	38.0	8.5	–	–	118.0	22.0	15.0	0.3	Goldman and Horne, 1983
Superior	12.4	2.8	1.1	0.6	28.1	3.2	1.9	2.0	
Tahoe	9.4	2.5	6.1	1.7	–	2.5	1.9	–	
European lakes									
Windemere	6.2	0.7	3.8	0.6	11.0	7.6	6.7	–	
Maggiore	22.1	3.8	2.1	1.3	48.0	28.9	2.3	–	DeBernardi *et al.*, 1984
African lakes									
George	17.2	7.4	20.0	4.2	99.0	14.6	8.4	20.0	Goldman and Horne, 1983
Malawi	19.8	4.7	21.0	6.4	70.8	5.5	4.3	1.1	Serruya and Pollingher, 1983
Tanganyika	11.0	38.8	63.1	32.8	206.6	6.3	26.4	13.5	
Victoria	3.9	2.7	9.0	4.1	–	2.5	3.0	4.7	
Chad	10.0	4.8	11.5	5.1	45.0	–	1.0	–	
South and Central American lakes									
Nicaragua	19.0	3.5	17.7	3.9	82.4	9.1	15.9	–	
Titicaca	66.0	32.8	177.0	15.6	–	244.0	248.0	0.8	
Rivers (averaged by continent, and corrected for pollution)									
North America	20.1	4.9	6.5	1.5	71.4	14.9	7.0	7.2	Meybeck, 1979
South America	6.3	1.4	3.3	1.0	24.4	3.5	4.1	10.3	
Europe	24.2	5.2	3.2	1.1	80.1	15.1	4.7	6.8	
Africa	5.3	2.2	3.8	1.4	26.7	3.2	3.4	12.0	
Asia	16.6	4.3	6.6	1.6	66.2	9.7	7.6	9.7	
Oceania	15.0	3.8	7.0	1.1	65.1	6.5	5.9	16.3	
World	13.4	3.4	5.2	1.3	52.0	8.3	5.8	10.4	

(Reproduced with permission from C. Goldman and A. Horne, *Limnology*, published by McGraw-Hill, Inc., 1983; from M. Meybeck, Concentration des eaux fluviales en éléments majeurs et apports en solution aux océans, *Revue de Geologie Dynamique et de Geographie Physique*, 1979, **21**, 227; and from C. Serruya and U. Pollingher, *Lakes of the Warm Belt*, published by Cambridge University Press, 1983.)

in surface waters, although appreciable quantities of reactive silica may also be present (Table 11.2). Most of these ions are leached from the rocks and soils in the drainage basin, with smaller amounts contributed directly from the atmosphere. Granitic rocks are generally deficient in soluble minerals and the water which drains from them is low in carbonates: CO_2 is mainly present as dissolved gas and the water is therefore slightly acidic. Water which has passed through limestone and other soluble rocks is high in bicarbonate and becomes increasingly alkaline as the free CO_2 is bound up in precipitated carbonates. The alkalinity of the water is a general indication of its hardness. **Soft water** draining from igneous rocks contains less than $50\ mg\ l^{-1}$ of dissolved solids. Calcium typically accounts for about 50% of the cations present with progressively lesser amounts of sodium, magnesium and potassium. These elements are mainly present as sulphates, chlorides and

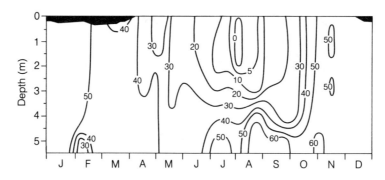

Figure 11.7 Seasonal variation in the distribution of calcium in a hypereutrophic temperate lake in North America. Winter ice cover is indicated. (After Wetzel, 1975.) (Reproduced from R. G. Wetzel, *Limnology*, published by Saunders College Publishing, 1975.)

carbonates. **Hard water** contains proportionately more calcium and magnesium, mostly as carbonate. The total concentration of the ionic components denotes the salinity of the water. Fresh water will normally contain less than $1000 \, \text{mg} \, \text{l}^{-1}$ of dissolved solids, although a value of $3000 \, \text{mg} \, \text{l}^{-1}$ is used in Australia (Williams, 1984). This compares with $35\,000 \, \text{mg} \, \text{l}^{-1}$ for normal sea water (Kennish, 1986) and $300\,000 \, \text{mg} \, \text{l}^{-1}$ in some saline lakes (Williams, 1984). Variation in the salinity of continental surface waters is mainly controlled by climatic conditions and lithology but some differences can be attributed to biological processes (Wetzel, 1975).

Calcium is the most reactive of the major cations present in water, and concentrations in very productive hard-water lakes usually fluctuate seasonally as a result of biogenic decalcification during periods of active growth. This decrease in calcium is directly associated with the utilization of CO_2 in photosynthesis, and thus with seasonal reductions in the concentration of carbonates, but is partly restored through decomposition (Figure 11.7). Biological demands are lower in nutrient deficient soft-water lakes and most rivers, and calcium and carbonate levels in these waterbodies remain fairly constant over the year. Some calcium is utilized directly by plants and animals, principally in structural and skeletal tissue. The amount removed in this way has only a negligible effect on concentration, although the distribution of some organisms is closely related to the amount of calcium present in the water (Macan, 1961). Requirements for magnesium, sodium and potassium are typically lower than for calcium, so that there is very little seasonal change in the concentrations of these elements. Similarly, chloride levels remain fairly constant over time. Sulphate levels are variable because they are affected by the amount of oxygen that is in the water. Seasonal changes in sulphate concentration are most pronounced in **eutrophic** lakes which become deficient in oxygen during the periods of stagnation. Sulphate levels in these lakes usually reach a peak in the spring and then decrease over the growing season, as some of the sulphate is reduced to insoluble sulphide

and lost to the sediments. Sulphates are returned to the water when the sediment is re-oxidized during periods of mixing (Wetzel, 1975).

Of the various essential elements, it is usually only nitrogen and phosphorus that may be sufficiently dilute to cause nutrient deficiencies in natural aquatic communities. Nitrogen is present in water as a dissolved gas but in this form it can only be used by nitrogen-fixing blue-green algae and some bacteria. In temperate regions the rate of nitrogen fixation increases in the spring as light, temperature and nutrient conditions become more favourable for growth. The amount of available nitrogen added in this way is therefore related to the productivity of the waterbodies and typically ranges from 1 to 50% of the annual input (Goldman and Horne, 1983). Nitrogen can also enter waterbodies in precipitation and drainage waters. The concentration of combined forms of nitrogen in natural fresh water normally ranges from 0 to $10 \, \text{mg} \, \text{l}^{-1}$ for nitrates, 0 to $5 \, \text{mg} \, \text{l}^{-1}$ for ammonia (NH_4^+) and 0 to $0.1 \, \text{mg} \, \text{l}^{-1}$ for nitrites. Organic nitrogen, in urea, amino acids and other compounds, typically accounts for more than 50% of the total nitrogen dissolved in lakes and river waters (Wetzel, 1975). The total concentration of phosphorus in natural fresh water ranges from 10 to $50 \, \mu\text{g} \, \text{l}^{-1}$ and is mainly present as organic phosphates and inorganic soluble orthophosphate (PO_4^{4-}). The concentration of exchangeable phosphorus is mainly determined by the characteristics of the drainage basin and rate of sedimentation of organic debris. Because of reduced biological demand, inorganic phosphate levels in river water are usually higher than in lakes, and are less subject to seasonal variation. Phosphorus is rapidly taken up by aquatic plants and productivity is eventually limited by low availability. Some phosphorus is directly regenerated in the surface waters through zooplankton grazing, and rapid recycling of phosphorus in this way can extend the period of high productivity. However, most phosphorus is lost in organic debris and is released through bacterial decomposition or detritivore activity. Phosphorus levels in temperate lakes are therefore highest in the spring

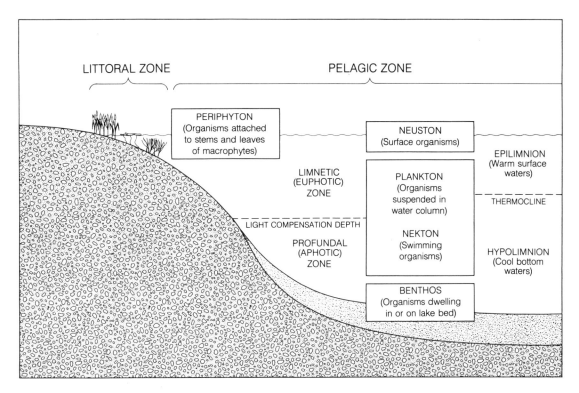

Figure 11.8 Zonation in temperate freshwater lakes and the characteristic classes of organisms that are present. (After Odum, 1971.) (Adapted from E. P. Odum, *Fundamentals of Ecology*, 3rd edn, published by Saunders College Publishing, 1971.)

when regenerated nutrients are returned to the euphotic zone through mixing.

11.4 LIFE FORMS AND AQUATIC HABITATS

The dominant plants in freshwater ecosystems are the algae. These grow attached to the substrate or live as free-floating phytoplankton which drift with the currents. Higher plants, such as duckweed (*Lemna minor*), float on the surface of the water and occasionally mats of reeds become detached and are carried into the deeper water of the **pelagic** zone. However, most macrophytes root in the sediment or grow on rocky substrates. In lakes the habitat conditions change as water depth increases away from the shore, and a markedly concentric pattern of vegetation zones may sometimes develop. Light penetrates to the bottom of the shallow-water **littoral zone** (Figure 11.8). The distance this zone extends from the shoreline depends on the morphometry of the lake basin and the transparency of the water. This is the only zone present in shallower lakes and ponds in which rooted plants grow throughout the basin (Figure 11.9). In deeper lakes the illuminated surface waters form the **limnetic zone**. The transition to the deep-water **profundal zone** occurs at the limit of photosynthesis by green plants. In rivers, zonation is predominantly longitudinal and reflects their changing hydrologic characters as they

flow away from their sources (Figure 11.10). Plants similar to those of lakes are found in rivers but their distribution and relative abundance are greatly influenced by the strength of the current and the mosaic of conditions that it produces. Pools and rapids are the distinctive habitats in these lotic environments. Water flow is reduced in the deeper pools and sediment settles to the river bed. The water is shallower and flows with greater turbulence in the rapids. The fine sediment is swept out of the channel and the plants which grow here are firmly attached to bedrock or boulders.

Reeds, bulrushes and other emergent macrophytes are restricted to the shallowest water of the littoral zone and typically grow to depths of about 1 m. A zone of floating-leaved macrophytes such as the water-lilies (*Nuphar* spp. and *Nymphaea* spp.) grow further out from the shore in water depths of 1–3 m (Wetzel, 1975). These plants are anchored in the mud by stout rhizomes but their buoyant leaves are supported on long, flexible petioles that extend to the surface. The submerged aquatic macrophytes tolerate deeper water although the growth of these plants is ultimately limited by low light intensities – even in very clear lakes they rarely establish below 10 m. Quillwort (*Isoetes* spp.) and charophyte algae such as *Chara* spp. and *Nitella* spp. are usually the only plants which survive at these depths. Most submergent angiosperms grow in shallow water, and although their vegetative parts remain below the surface, some species do produce aerial flowers. Free-floating macrophytes,

Figure 11.9 A characteristic zonal pattern of sedges and other hydrophytes develops around the shores of larger lakes and shallow ponds. ((a) P. Marsh; (b) R. Zdan.)

such as pondweed (*Lemna* spp.) and water hyacinth (*Eichhornia crassipes*), may grow in the limnetic zone which extends across the open water, but it is usually the microscopic algae that are most abundant in this zone (Figure 11.11).

Most of the plants which grow in streams and rivers resist the energy of the flowing water by attachment to various substrates. **Epilithic algae** colonize hard surfaces that are swept free of loose sediment. Others are **epipelic** and form a film over mud and silt or grow as **epiphytes** attached to other plants. Many species only grow where there is continuous water movement. Their distribution is affected by the speed of the current. Thus some species only grow in rapids where turbulent flow maintains high concentrations of dissolved gases (Hynes, 1970). The high CO_2 requirement of aquatic mosses also tends to restrict them to the rocky sections of a river channel. A few very specialized angiosperms grow submerged in rapids and waterfalls: they are attached to the rock by means of modified roots which either spread like lichens or form a fine mat which permeates every surface irregularity. Most aquatic angio-

sperms root into soft sediments, and although some are more common in rivers than in lakes, none grow exclusively in lotic habitats. They are mainly found in slow-moving stretches of a river where silt has settled out of the water; nevertheless some species are able to root in coarser gravels where the current is faster. Free-floating macrophytes are rarely abundant in temperate rivers but some may occur in sluggish backwaters or become entangled in the stems of other plants. However, rapidly growing mats of water hyacinth and other floating aquatics are common in tropical rivers where they become so dense that navigation and fishing are virtually impossible. Phytoplankton are also adversely affected by strong currents: they mostly enter rivers from adjacent lakes but generally perish in the turbulent flow. Phytoplankton do survive in the pools and shallows, where the current is slower, and they can persist in the boundary layer where the water in contact with the channel bed is almost stationary. Conditions in the sluggish sections of wide, meandering rivers are more suitable for phytoplankton and their density and diversity typically increase downstream (Greenberg, 1964).

11.5 MORPHOLOGICAL ADAPTATIONS IN AQUATIC PLANTS

The presence of stomata, thin cuticles and other relic features suggest that aquatic angiosperms are highly specialized plants that have returned to the water in fairly recent evolutionary times: there is little fossil evidence of aquatic plants before the Eocene (Sculthorpe, 1967). The emergent macrophytes resemble terrestrial plants more closely than other aquatic plants and are mostly equipped with air-space tissue to ensure adequate exchange of gases with the atmosphere. They are basically adapted to waterlogged soils, with many species also found in marshes and other wetland habitats.

11.5.1 Morphological diversity in angiosperms

(a) Submerged angiosperms

The morphological adaptations of truly aquatic plants are most pronounced in species that grow totally submersed in water. The need for structural tissue is greatly reduced. The stems, petioles and leaves of most submersed angiosperms contain little or no lignin. Thus they remain very flexible and readily bend into a shape that offers least resistance to the current. The cellular structure is also greatly modified, and because there is no cambium, secondary growth does not occur. Consequently, the vascular system is greatly reduced compared with terrestrial plants, and in some species the xylem vessels are entirely replaced by a cavity. Similar cavities in the leaves and roots permit rapid diffusion of

Figure 11.10 Habitat conditions vary along the course of a river as size and flow regime change. ((a)–(c) R. A. Wright; (d) O. W. Archibold.)

Figure 11.11 Characteristic life forms of aquatic plants: (a) cat's-tails (*Typha latifolia*) and (b) water plantain (*Alisma plantago-aquatica*) are common emergent plants; (c) floating-leaved yellow pond-lily (*Nuphar variegatum*); (d) water hyacinth (*Eichhornia crassipes*), a free-floating aquatic macrophyte; (e) duckweed (*Lemna minor*), a small free-floating plant, has covered the water surface with its dense scum-like growth; (f) large-leaved water crowfoot (*Ranunculus aquatilis*) with its white flowers above the water surface. ((a) and (e) O. W. Archibold; (b) and (f) R. A. Wright; (c) and (d) M. Brown.)

gases throughout the plant. The leaves of most submergent plants are narrow and elongated, and in some species, such as water milfoil (*Myriophyllum verticillatum*), are finely dissected into thread-like strands. This general form reduces mechanical damage from water movements. The ratio of surface area to volume is also increased and so improves the efficiency of gas exchange, light absorption and nutrient uptake, which is also an important function of the foliage.

Leaf shape in submersed angiosperms usually varies with growing conditions and considerable morphological plasticity can occur even on a single stem. Such **heterophylly** is a conspicuous characteristic of amphibious species of *Ranunculus* and *Potomageton* which can grow either fully submersed or partially exposed to the atmosphere. Typically the aerial leaves are morphologically similar to those of terrestrial plants but those growing in deeper water are more dissected. In *Ranunculus flabellaris* the submersed leaf form is induced by low temperature conditions (Johnson, 1967) or high CO_2 concentrations (Bristow, 1969), and in *Hippuris vulgaris* by the attenuation of far-red light in water (Bodkin *et al.*, 1980). Nearly all submersed angiosperms root in loose sediment and spread by means of rhizomes or stolons. However, hornwort (*Ceratophyllum* spp.) has no roots and floats freely below the surface of the water, although colourless root-like branches will sometimes anchor the plant to the substrate. Conversely, the leaves and stem of *Tristicha* are very reduced and the strap-like roots which anchor it in its rocky habitat are green and photosynthetic (Hynes, 1970).

(b) Floating-leaved angiosperms

The floating-leaved angiosperms must tolerate the mechanical force of wind and waves and consequently grow most abundantly in relatively sheltered sites. The cosmopolitan water-lilies are familiar representatives of this group (Figure 11.12). They are perennial plants which are anchored in the mud by large rhizomes. The leaves are borne on long, flexible petioles which are about 20 cm longer than the depth of the water and so compensate for wave action and fluctuations in water level. Columnar air-spaces add to the buoyancy of the petioles, which are strengthened by specialized parenchymatous cells (Cutter, 1978). The strong, thick leaves are typically circular in shape with a waxy, water-repellent cuticle. They float horizontally on the water surface and the shade they create will often limit the distribution of submerged aquatic species. Larger leaves, such as those of the Amazon water lily (*Victoria amazonica*) which grow up to 2 m in diameter (Figure 11.13), are reinforced underneath by a network of stiffened veins, but normally venation is greatly reduced in all aquatic species (Wetzel, 1975). Flotation is provided by gas-filled lacunate tissue in the under part of the leaf and by localized masses of spongy tissue. The upper leaf surface, which is

Figure 11.12 Red water lily (*Nelumbo nucifera*) in Kakadu National Park, a World Heritage Site in Northern Territory, Australia. (O. W. Archibold.)

in contact with the air, is equipped with functional stomata, but unlike most terrestrial species they are absent or non-functional on the lower surface.

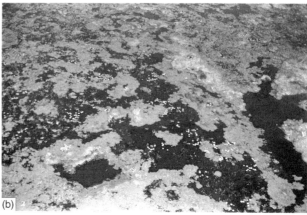

Figure 11.13 Amazon water lily (*Victoria amazonica*); its large floating leaves appear as the paler spots in the aerial view of a lake in Brazil. (G. T. Prance.)

(c) Free-floating angiosperms

Plants which are free-floating are more variable in structure. The larger free-floating plants may consist of a stemless rosette as in water lettuce (*Pistia stratiotes*), or develop as clusters of upward-thrusting leaves as is the case with water hyacinth (*Eichhornia crassipes*). Both of these plants are kept afloat by the spongy tissue in the petioles. Water-repellent hairs cover the surface of the leaves of *Pistia* and prevent the plant from becoming waterlogged. Similar hairs are found in the aquatic fern *Salvinia auriculata* but in this species they provide buoyancy by trapping air. Small plants such as duckweed (*Lemna* spp.) are not differentiated into stems and leaves but consist of plate-like thalli usually less than 1 cm in diameter beneath which there is a short, simple root. It is kept afloat by numerous lacunae which in some species are so elongated that the plant develops a cylindrical form. *Wolffia arrhiza* dispenses with leaves, stems and roots: it is the smallest known flowering plant. Although it is barely visible to the naked eye, this tiny globular plant will at times cover the surface of a sheltered lake with a green scum much like the microscopic phytoplankton.

11.5.2 Morphological diversity in the algae

The algae are an extremely diverse group of non-vascular plants that lack roots, stems and leaves. They may exist as single cells, as loosely organized colonies or as filaments. Most are photosynthetic and so produce oxygen as a by-product of their metabolic activities. Some of the unicellular algae possess hair-like flagella and, apart from a rigid cellulose cell wall, are difficult to distinguish from the flagellated protozoa. The various divisions of the algae are based primarily on the biochemical composition of their pigments and cell walls, the type of energy reserves they manufacture and the presence or absence of locomotory flagella (Table 11.3). The Rhodophyta and Phaeophyta are almost exclusively marine.

The **blue-green algae** (Cyanophyta) are a conspicuous and widespread group in freshwater habitats (Scagel *et al.*, 1965). They occur as smears or cushions on all types of substrates or as free-floating individuals in the plankton. The cell walls of blue-green algae are composed of an inner cellulose layer encased in a mucilaginous pectic sheath which may appear yellow, brown, red or blue depending on the type of pigment present and the pH of the water or substrate. The simplest blue-green algae are unicellular free-floating organisms but colonial and filamentous forms also occur. Few of the non-filamentous forms are able to fix nitrogen, although this is an important characteristic of *Nostoc*, *Anabaena* and other filamentous blue-green algae that possess **heterocysts**. These are enlarged, thick-walled cells that are connected by protoplasmic threads to the rest of the filament. Because the heterocysts lack photosystem II and do not generate oxygen, they help to maintain an anaerobic condition in the cytoplasm. This is essential for nitrogen fixation because nitrogenase, the enzyme involved, cannot function in the presence of oxygen (Goldman and Horne, 1983). Seasonal variations in nitrogen fixation are closely related to heterocyst development (Horne and Goldman, 1972): it is repressed by high nitrogen concentrations in the water and if iron and molybdenum are unavailable for nitrogenase synthesis.

The **dinoflagellates** are members of the Dinophyceae: their characteristic golden or greenish-brown colour results from the presence of the pigment **peridinin**, a xanthophyll that is unique to this group (Scagel *et al.*, 1965). Most dinoflagellates are unicellular and are encased in a cellulose wall composed of discrete sculptured plates that are tightly cemented together in a pattern unique to each species. In *Peridinium*, as an example, the plates are perforated with pores whereas in *Ceratium* the plates extend as long horns. The development of horns, spines and other wing-like structures in the photosynthetic dinoflagellates reduces the rate of sinking and so increases their ability to remain in the photic zone. Most possess two morphologically dissimilar flagella which emerge through a small pore in the cellulose wall. One flagellum trails behind the cell and by its whiplash action propels the organism forward at rate of about 1 cm s^{-1}. The other flagellum encircles the cell in a transverse groove and provides rotational movement.

The golden or **yellow-brown algae** (Chrysophyceae) are members of the division Chrysophyta which also includes the diatoms (Bacillariophyceae). Species of golden algae, such as *Ochromonas*, are generally unicellular and encased in a pectic sheath consisting of two overlapping shells (Scagel *et al.*, 1965). One or two flagella are present in motile species. The golden colour of the Chrysophyceae results from the presence of carotenes and xanthophylls in the chloroplasts which effectively mask the chlorophyll. The Chrysophyceae are particularly common in lake plankton during the cold season, and are usually the dominant algae in cold polar lakes (Hobbie, 1984). The closely related **yellow-green algae** (Xanthophyceae) are similarly abundant in alpine lakes or in the cool water issuing from springs. Silicon deposited within the cell wall of the **diatoms** makes them more ornate than the Chrysophyceae, although their structure is very similar and again consists of two overlapping shells or frustules. Most diatoms are unicellular, but some may be loosely aggregated into filaments and star-shaped colonies by mucilaginous pads or long spine-like processes. However, the vegetative cells are not flagellated. The predominant freshwater diatoms belong to the order Pennales which are distinguished by their rather elongate shape. An unsilicified groove or raphe forms the central axis in most of these diatoms and this allows them to move, albeit extremely slowly ($0.2–2.5 \text{ μm s}^{-1}$), through cytoplasmic streaming. This is not sufficient to

Table 11.3 Characteristics of the common aquatic algae (after Scagel *et al.*, 1965)

	Common habitat	Common morphology	Flagellation	Pigments	Cell wall	Storage products
Cyanophyceae (blue-greens)	lakes and oceans	filamentous	none	chlorophyll *a*, β-carotene, xanthophylls	cellulose and pectin	starch and proteins
Dinophyceae (dinoflagellates)	oceans, lakes and estuaries	unicellular or colonial	biflagellate	chlorophyll *a + c*, β-carotene, xanthophylls	cellulose and pectin	starch, fats and oils
Chrysophyceae (yellow-browns)	streams, lakes and oceans	unicellular or colonial	uni- or biflagellate	chlorophyll *a + c*, α-carotene, xanthophylls	cellulose and pectin	fats, oils and sugars
Bacillariophyceae (diatoms)	oceans, lakes, rivers and estuaries	unicellular	generally none	chlorophyll *a + c*, α-carotene, fucoxanthin	silicon and pectin	fats, oils and sugars
Xanthophyceae (yellow-greens)	lakes	unicellular or filamentous	biflagellate	chlorophyll *a*, β-carotene, xanthophylls	cellulose and pectin	sugars
Euglenophyceae (euglenoids)	mainly polluted lakes	unicellular	usually uniflagellate	chlorophyll *a + b*, β-carotene, xanthophylls	generally absent	sugars
Chlorophyceae (greens)	rivers, lakes and estuaries	unicellular, colonial or filamentous	generally biflagellate if present	chlorophyll *a + b*, β-carotene	cellulose and pectin	sugars
Phaeophyceae (browns)	oceans and estuaries	long fronds	reproductive cells only	chlorophyll *a + b*, α + β-carotene, fucoxanthin	cellulose, pectin and phycocolloids such as algin	sugars and derivatives such as mannitol
Rhodophyceae (reds)	oceans and estuaries	unicellular, colonial or filamentous	none	chlorophyll *a*, α + β-carotene, xanthophylls, phycobilin	cellulose, pectin and mucilages such as carrageenin	starch and fats

(Source: R. F. Scagel *et al.*, *An Evolutionary Survey of the Plant Kingdom*; published by Wadsworth Publishing Co. Inc., 1965.)

keep them from sinking and buoyancy is mainly provided by oil reserves in the cytoplasm.

The **Euglenophyta** are considered to be one of the most primitive groups of flagellates. They are mainly unicellular freshwater organisms and have both plant- and animal-like features. *Euglena* lacks a firm cell wall and is enclosed in a flexible membrane or pellicle so that its shape changes. The anterior end of the cell is drawn in to form a gullet through which food is ingested. From one to three flagella protrude from the gullet and either propel the organisms along a spiral path through the water or simply act as a rudder. The Euglenophyta are particularly abundant in water that is rich in organic matter. Phagotrophic forms such as *Peranema* are common where fine debris or small plankton are available, while *Astasia* is very tolerant of polluted conditions and normally occurs where putrefaction is present. Other non-photosynthetic forms live within nematodes, flatworms and other invertebrates, and one is known only from the intestinal tract of some frog tadpoles (Scagel *et al.*, 1965).

The **green algae** (Chlorophyta) are also predominantly freshwater plants. They are widely distributed as free-floating or motile forms in the plankton or as larger attached plants on various substrates. Most have a rigid cell wall composed of an inner cellulose layer and an outer pectic layer which, in the stoneworts (Charo-

phyceae) and some other species, is impregnated with calcium carbonate (Scagel *et al.*, 1965). The characteristic grass-green colour of this group is due to the presence of chlorophyll *a* and *b* in well-organized chloroplasts. As in higher plants, carbohydrate reserves accumulated through photosynthesis are mainly stored as starch. The great morphological diversity in the green algae is indicative of their evolutionary development and three evolutionary lines are recognized. The simplest green algae of the **volvocine** line are unicellular forms, such as *Chlamydomonas*, which may be motile or non-motile and exist in isolation or fastened in a mucilaginous matrix as colonies. In colonial volvocine algae, such as *Gonium*, 4–32 identical cells join together, but in the more advanced forms, such as *Volvox*, the size of the colony increases to 500–50 000 cells and there is a tendency towards some division of labour. Members of the **siphonous** evolutionary line are distinguished by their multinucleate cells. The most primitive forms consist of a single vesicle or, when occasional septation and nuclear division occur, as a colonial thallus, as in *Pediatrum*. More specialized branching forms include *Cladophora glomerata*: this species can grow to over a metre in length and is attached to the substrate by rhizoid-like structures which develop at the base of the filaments (Hynes, 1970). Repeated division in one or more planes occurs in members of the

tetrasporine evolutionary line and in the most advanced forms, such as *Stigeoclonium*, prostrate and erect structures develop from parenchymatous-like tissues similar to that of higher plants.

11.6 ECOPHYSIOLOGY OF AQUATIC MACROPHYTES

Atmospheric CO_2 is the dominant source of carbon for emergent and floating-leaf macrophytes but submersed plants depend on aqueous carbon sources, either as dissolved CO_2 or as bicarbonate (HCO_3^-). The slow diffusion rate of CO_2 in water imposes a major limitation on photosynthesis in submersed plants and gas exchange is increased through the modified structure of the leaves and stems. However, diffusion resistance at the boundary layer remains comparatively high (Black *et al.*, 1981), especially when there is insufficient water movement to disrupt the stagnant microzone that surrounds the leaf (Westlake, 1967). For this reason many submersed plants will supplement CO_2 with HCO_3^- which is taken in by active transport (Maberly and Spence, 1983). Photosynthesis is therefore indirectly affected by the pH of the water since this influences the equilibrium between the different forms of dissolved inorganic carbon. The ability to use HCO_3^- is a particular advantage in alkaline hardwater habitats. Freshwater macrophytes with no apparent affinity for HCO_3^-, such as the aquatic mosses, usually have very low CO_2 demands or grow in habitats where CO_2 is oversaturated. Other species utilize alternate sources of CO_2. For example, the leaves and roots of the aquatic fern *Isoetes lacustris* are permeated by extensive air channels which transport CO_2 from the sediment to the chloroplasts lining the leaf lacunae, and return oxygen to the rhizosphere (Richardson *et al.*, 1984).

The submersed macrophytes are adapted to the low light levels which occur in water. Most species function as shade plants and become light saturated at an irradiance lower than 50% full sunlight. Consequently, observed net photosynthetic rates are characteristically low (Van *et al.*, 1976). Photosynthesis occurs mainly in the upper part of the submersed canopy, but the structure and arrangement of the foliage and shade adaptations in the lower leaves makes some photosynthesis possible even in dense vegetation (Westlake, 1980). Some loss of photosynthetic capacity can also result from encrustations of epiphytic algae on the older, lower leaves. Net photosynthetic rates are generally highest in the morning, but decline later in the day because the oxygen which is produced diffuses through the plant and stimlates photorespiratory carbon oxidation which ultimately produces available CO_2 (Hough, 1974). In terrestrial C_4 plants the rate at which photorespired CO_2 is lost is greatly reduced because the enzyme phosphoenolpyruvate carboxylase, which fixes CO_2 in photosynthesis, is not inhibited by the presence of oxygen. This is

also an important mechanism for conserving CO_2 in some aquatic species which grow in an environment in which gas levels are suboptimal. However, the C_4-type photosynthetic pathway in aquatic plants such as *Hydrilla verticillata* differs from the terrestrial counterpart in that Krantz anatomy is absent. A similar C_4 pathway occurs in species such a *Littorella uniflora* which derive their CO_2 from the soil (Bowes, 1987). A CAM-like process in which CO_2 is captured at night is reported in species of *Isoetes* (Keeley, 1987c).

Unlike terrestrial species, the CO_2 compensation point in submersed aquatics varies with growing conditions. This is reflected by changes in the photosynthesis: photorespiration ratio (Holaday *et al.*, 1983). High temperatures and long day lengths during summer decrease the CO_2 compensation point; as it drops, photorespiration and oxygen inhibition decline and there is a corresponding increase in net photosynthesis under C_4-like conditions. This persists until the temperatures begin to cool and the days get shorter, and net photosynthesis declines as carbon assimilation occurs through a C_3-like process. Optimal water temperatures for photosynthesis in different species range from 20 to 35 °C, further obscuring the relationship with terrestrial C_3 and C_4 plants. Some species remain photosynthetically active under winter ice cover at temperatures as low as 2 °C (Boylen and Sheldon, 1976). However, in most temperate waterbodies primary production of submersed macrophytes is minimal during the winter. Growth begins in the spring, as temperature and light intensity increase, and reaches a maximum in early summer. The rate of accumulation declines in late summer and autumn as growing conditions deteriorate and the plants become senescent.

11.7 GROWTH CHARACTERISTICS OF PHYTOPLANKTON

The phytoplankton population consists of perennial species – the **holoplankton** – which are always present, even though their numbers may be extremely low at certain times of the year, and intermittent species – the **meroplankton** – which are dormant for part of their life cycle. Both types are subject to strong seasonal fluctuations in population density which are primarily controlled by the interaction of light, temperature and nutrient availability. The optimal combination of factors varies between species and consequently the composition of the phytoplankton population changes considerably during the year. In typical temperate waterbodies phytoplankton densities are at a maximum in the spring when there is a large increase in the number of diatoms. This spring **bloom** is followed in the summer by smaller irregular peaks of various flagellates, and later during the smaller autumnal bloom by an increase in blue-green algae, diatoms and dinoflagellates. At higher latitudes there is usually only a single summer peak in phytoplankton biomass, compared with a single, and often less dramatic, winter peak in the tropics where conditions are less variable (Wetzel, 1975). Seasonal changes of this type are less pronounced in natural river systems but an increase in the plankton population generally occurs during the summer in large temperate rivers, and in tropical rivers during the dry season when less water is flowing (Hynes, 1970).

Phytoplankton are C_3 plants (Ganf, 1980) and depend on adequate light conditions for growth, but in these organisms growth is essentially analogous to cell division, and replication is closely regulated by external temperatures (Reynolds, 1984). Thus the combination of increasing light, rising temperatures and the return of nutrients to the photic zone in the spring provides ideal conditions for growth. Theoretically the number of phytoplankton increases exponentially: the rate is related to the surface-to-volume ratio of the individual cells or colonies, and average generation times range from 3 to 180 hours under optimum conditions (Tilzer *et al.*, 1980). However, the actual rate of population growth is reduced by predation, parasitism and other factors (Figure 11.14). Holoplanktonic diatoms such as *Asterionella* often dominate the spring pulse in temperate lakes but their growth is ultimately limited by low concentrations of silica. Silica is not readily recycled and mostly comes from river inflow during winter. It is incorporated in the cell walls of the diatoms and upon death is lost through sedimentation. The diatoms are succeeded by green algae and later by increasing numbers of blue-green algae during the autumn bloom. The increase in the blue-green algae depends on ammonia released by decomposer organisms in the sediment and brought to the surface during the fall overturn. Meroplanktonic diatoms such as *Melosira* are also an important component of the autumn bloom, because organisms which were dormant in the sediment become resuspended in the water.

In addition to the seasonal changes in illumination intensity, there is a regular diurnal rhythm which may be further altered by sky conditions. Consequently, the depth in the water column where light conditions are optimal for a given species of phytoplankton varies continually during the growing season. The phytoplankton compensate for this by altering their position in the water. Vertical migrations in flagellated species are a phototactic response to variations in light levels, and the direction of movement can be positive or negative in any given species. In general these phytoplankton descend as irradiance increases during the morning and move towards the surface later in the day (Tilzer, 1973). In blue-green algae vertical migrations are controlled by changes in density caused by the formation and collapse of gas-filled vacuoles. Vacuole formation occurs under low light intensities and the increased buoyancy causes the cells to rise in the water. The size of the vacuole is regulated by the turgor pressure within the cells: this is

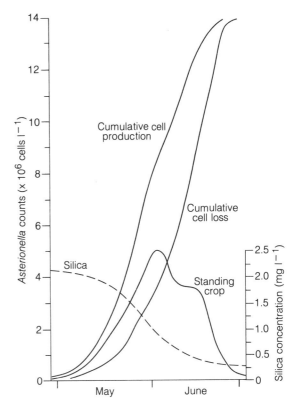

Figure 11.14 Spring and early summer production and loss of the diatom *Asterionella formosa* and associated silicate concentration in the epilimnion of Lake Windemere, UK. (After Lund *et al.*, 1963.) (Reproduced with permission from J. W. G. Lund, F. J. H. Mackereth and C. H. Mortimer, Changes in depth and time of certain chemical and physical conditions and of the standing crop of *Asterionella formosa* Hass. in the North Basin of Lake Windermere in 1947, *Philosophical Transactions* B, 1963, **246**, 275.)

controlled by the rate at which sugars accumulate in the cell as a result of photosynthesis and so enables the blue-green algae to remain at an optimum depth in the water. Planktonic diatoms have no method of controlling their position within the water column and are moved passively by the currents. However, benthic diatoms can move vertically in the sediment by secreting mucilage. The cells move towards the surface of the sediment before dawn and begin to bury themselves during the afternoon. This pattern of activity may optimize nutrient uptake from the sediment (Kirk, 1983).

Excessive light levels inhibit photosynthesis in phytoplankton and assimilation generally decreases when radiation intensities exceed 60% of maximum surface values (Tilzer *et al.*, 1980). Vertical movements in the water column compensate for this to some degree and the usual effect is that maximum photosynthesis occurs at some depth below the surface of the water (Figure 11.15). However, this reduction in photoassimilation in surface waters is not entirely due to low plankton densities. Additional adaptation is achieved by altering the amount of pigment in the cells and in most species the amount of chlorophyll decreases at higher light intensities (Tilzer and Goldman, 1978). Light intensity becomes a limiting factor once it falls below 15% of maximum surface values and the photosynthetic rate declines exponentially below the zone of light saturation. Daily changes in light intensity can therefore produce considerable short-term variability in the photoassimilation profile (Sephton and Harris, 1984). The zone of photoinhibition is directly related to the clarity of the water and optimal light intensities can occur at considerable depth in some lakes, although photosynthesis in deep water may be limited by cold temperatures. Conversely, light intensities are greatly affected by the growth of plankton themselves and high biogenic turbidity in productive lakes can at times eliminate the photoinhibitory zone. Light attenuation within stands of macrophytes can similarly reduce phytoplankton photosynthesis in the littoral zone. The plankton may also be adversely affected by the increase in pH which occurs as available CO_2 is withdrawn from the water, and by the excretion of organic compounds by macrophytes (Wetzel, 1975). Nutrient limitations to phytoplankton production, particularly phosphorus, are reported in many lakes (Hecky and Kilham, 1988).

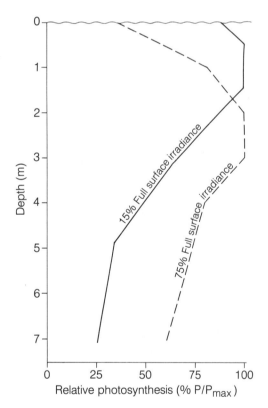

Figure 11.15 Depth profiles of relative photosynthesis (% P/P$_{max}$) in Lake Windemere, UK, under different intensities of surface irradiance by wavelengths of 400–700 nm. (After Belay, 1981.) (Reproduced with permission from A. Belay, An experimental investigation of inhibition of phytoplankton photosynthesis at lake surfaces, *The New Phytologist*, 1981, **89**, 62.)

11.8 PRIMARY PRODUCTION AND BIOMASS IN FRESHWATER ECOSYSTEMS

Productivity and biomass in the illuminated surface waters are greatly affected by the relative abundance of the different plant types. Thus, primary production in small lakes and ponds is disproportionately high compared with the volume of water they contain, especially if aquatic macrophytes have become established in the shallows. Estimates of above-ground productivity for submersed aquatic macrophytes in temperate inland waters range from 10 to 500 g m^{-2} yr^{-1} compared with more than 1000 g m^{-2} yr^{-1} in some tropical lakes (Kirk, 1986). Values for submersed macrophytes are generally lower than for floating-leaf plants, especially in tropical regions where production of species such as water hyacinth can exceed 4000 g m^{-2} yr^{-1} (Gopal, 1987). Underground production is comparatively low in most submersed plants but in floating-leaf species, such as water-lilies, the root systems account for as much as 80% of the plant biomass. Losses from senescence and abrasion from water movement during the active growing period typically account for 2–10% of the annual production: similar losses may occur through grazing by ducks, mammals and fish (Wetzel, 1975). Consequently, maximum seasonal biomass can be considerably less than total annual production (Rich *et al.*, 1971).

In a similar way phytoplankton populations rarely increase at the rates predicted for the constituent species. As an example, in Lake Constance, grazing losses in the spring account for as much as 82% of production and are a major factor controlling the seasonal succession of species in the photic zone (Tilzer, 1984). Species with high reproductive rates that appear early in the season are usually small and are readily consumed by the filter-feeding zooplankton (Sommer, 1981). The larger diatoms, dinoflagellates and colonial phytoplankton are more resistant to grazing and this transition to less edible forms is a common phenomenon in many lakes. Losses in these species is mainly through sedimentation, which in Lake Constance increases from about 15% of production during the summer to over 50% in autumn. Residual losses also occur through respiration, death and lysis of cell materials in the water column (Jassby and Goldman, 1974). The net effect is that measurements of phytoplankton biomass (B) greatly underestimate annual production (P), and P:B ratios exceeding 100 are reported in some lakes (Winberg, 1980).

Variations in annual net productivity of phytoplankton are mainly controlled by the length of the growing season and nutrient availability. The short growing season at higher latitudes reduces phytoplankton production to less than 15 g C m^{-2} yr^{-1}: in temperate latitudes this increases to 100–600 g C m^{-2} yr^{-1} (Hammer, 1980), with values of 300–640 g C m^{-2} yr^{-1} reported for many tropical lakes (Serruya and Pollingher, 1983). The higher productivity in the tropical zone may reflect the longer growing season but the photosynthetic efficiency in these lakes is also comparatively high, averaging 2–3% compared with < 1% in temperate lakes (Wetzel, 1975). However, strong seasonal trends in productivity are reported in many tropical lakes (Vincent *et al.*, 1986) which suggests that growing conditions are more variable than traditionally thought, perhaps because of periodic differences in mixing depth (Robarts, 1984). Similarly, phytoplankton production decreases with altitude, but although values as low as $2.5\,g\,C\,m^{-2}\,yr^{-1}$ have been measured in lakes in the Canadian Rockies, mean annual production in Lake Titicaca is $511\,g\,C\,m^{-2}\,yr^{-1}$. Considerable variation in phytoplankton production occurs within this basic geographic pattern because of differences in nutrient availability, and a distinction is traditionally made between nutrient-poor oligotrophic lakes and more fertile eutrophic lakes: in this classification production in temperate **oligotrophic** lakes is below $25\,g\,C\,m^{-2}\,yr^{-1}$ and exceeds $75\,g\,C\,m^{-2}\,yr^{-1}$ in **eutrophic** lakes (Rodhe, 1969), although these values are considered too low for tropical lakes. Differences in nutrient availability and length of growing season give an extreme range of phytoplankton production in freshwater lakes from about $4\,g\,C\,m^{-2}\,yr^{-1}$ in Char Lake in the Canadian Arctic (Kalff and Welch, 1974) to about 1400 $g\,C\,m^{-2}\,yr^{-1}$ in Lake George, Uganda (Ganf and Horne, 1975).

11.9 PHYTOPLANKTON NUTRIENT DEMAND

The mineral requirements of the phytoplankton are similar to those of other plants, although in addition to mineral elements many species require small quantities of organic micronutrients. The most common of these water-soluble organic growth compounds is vitamin B_{12} (cobalamine) but vitamin B_1 (thiamine) and vitamin H (biotin) are also required by some of these **auxotrophic** species (Table 11.4). Some vitamins enter the water in precipitation but mostly they are supplied by bacterial synthesis and are derived through decomposition of organic matter. Photosynthesis can proceed for some time when only water and CO_2 are present but growth is eventually prevented by the lack of essential nutrients. The water around phytoplankton is rapidly depleted of nutrients and, under stationary conditions, nutrient uptake is eventually limited by the diffusion gradient which becomes established between the cells and the water. Continued nutrient uptake therefore requires that phytoplankton maintain steep concentration gradients by swimming or sinking through the water (Lund, 1965). Consequently, the phytoplankton are rarely distributed evenly throughout a lake but instead take advantage of spatial differences in nutrient concentrations (Lehman and Scavia, 1982). The ability of organisms to exploit this

Table 11.4 General requirements (– none known; + rare; ++ few species; +++ many species) for vitamins among freshwater algae (after Wetzel, 1975)

	Biotin	Thiamine	Vitamin B_{12}
Cyanophyceae	–	–	+++
Dinophyceae	++	–	+++
Cryptophyceae	+	+	++
Chrysophyceae	+	+++	++
Bacillariophyceae	–	++	+++
Xanthophyceae	–	–	–
Euglenophyceae	+	+	++
Chlorophyceae	–	++	+++

(Reproduced from R. G. Wetzel, *Limnology*, published by Saunders College Publishing, 1975.)

chemical heterogeneity through diel migration has been demonstrated in motile flagellates and consequently they are less prone to nutrient limitations than other forms. Nutrients are taken up and stored during the diurnal migrations and later assimilated in the euphotic zone. In this way the phytoplankton can utilize the relatively high nutrient concentrations that occur in the hypolimnion (Salonen *et al.*, 1984). Luxury consumption of nutrients has been reported in some species. For example, the freshwater diatom *Asterionella formosa* can store phosphorus in concentrations about 100 times greater than the amount needed for normal functioning, which is sufficient for the population to continue to double for about seven generations even though external phosphate levels appear to be limiting (Fogg, 1980).

The relative growth of phytoplankton increases with nutrient concentration until a maximum rate is reached, and thereafter becomes independent of concentration. However, the total amount of organic matter produced under light-saturated conditions is eventually limited by nutrient availability. In freshwater ecosystems production is usually limited by phosphorus and less commonly by nitrogen or silica. Phosphorus requirements are comparatively high for green algae. Blue-green algae have lower phosphorus requirements and because they can fix nitrogen they are especially abundant in lakes with low nitrogen:phosphorus ratios. The growth of diatoms is often limited by low concentrations of silica but their phosphorus requirements are generally lower than for green or blue-green algae and they are the dominant phytoplankton in lakes with high silica:phosphorus ratios (Tilman *et al.*, 1982). Small, motile algae are less susceptible to nutrient limitations because they have a greater surface area to cell volume and can move away from chemically deficient water. Phytoplankton with these characteristics are proportionally more abundant in oligotrophic lakes. In extremely oligotrophic lakes productivity can be directly limited by low nutrient availability but in most waterbodies nutrients are sufficiently concentrated to sustain maximum growth as long as they remain in solution. Production will, however, decline as nutrients are progressively accumulated in the biomass. Thus seasonal changes in abundance and composition of the phytoplankton population may be attributed to competition for limited nutrient resources (Sommer, 1989), although the systematic changes which occur under different trophic conditions appear to compensate for nutrient limitations (G. P. Harris, 1986).

The ratio of total nitrogen to total phosphorus (TN:TP) exceeds 200 in oligotrophic lakes and the growth of phytoplankton is therefore mainly controlled by the availability of phosphorus. Less than 10% of the total phosphorus is present as readily available dissolved inorganic phosphorus and sustained production is dependent on rapid recycling in the photic zone. As much as 70% of the annual production in oligotrophic lakes is carried out by algal picoplankton (0.2–2.0 μm diameter) which include chroococcoid cyanobacteria and tiny eukaryotic microalgae (Stockner and Shortreed, 1989). Specific growth rates, which are analagous to productivity measurements, of 0.8–1.5 day^{-1} are reported for picoplankton in Lake Superior (Fahnenstiel et al., 1986) and the short regeneration time combined with negligible sinking rates (Bienfang and Takahashi, 1983) help to maintain an adequate supply of nutrients in the illuminated surface water. In these lakes the phosphorus may be recycled at least 20 times during the growing season, so that despite high specific growth rates the standing crop remains low (G. P. Harris, 1986). Specific growth rates are comparatively slow for the larger phytoplankton which grow in eutrophic lakes but eventually a considerably greater standing crop accumulates. The TP:TN ratio in eutrophic lakes is below 15, and because productivity is less dependent on rapid phosphorus recycling, more of the nutrients are lost through sedimentation. Production in these lakes is therefore dependent on internal nutrient regeneration through microbial activity in the sediments and subsequent transfer to the euphotic zone during periods of circulation. Conversely, production in oligotrophic lakes is mainly regulated by the geochemical nature of the drainage basin and continual nutrient influx from rivers.

11.10 DECOMPOSITION AND NUTRIENT CYCLING

11.10.1 Physical processes

Dissolved (nominally < 0.5 μm diameter) and particulate organic matter which accumulate in aquatic ecosystems provide energy for various benthic detritivores and microbial decomposers. Most of the organic carbon is present as soluble non-humic forms which are more rapidly assimilated than particulate matter. The dissolved organic carbon present in eutrophic lakes is mainly derived from the photosynthetic activity of the phytoplankton. Organic carbon is released to the water as glycollate during periods of active growth and under some conditions as much as 95% of the total carbon fixed in photosynthesis is excreted in this way (Fogg, 1975). Similarly, soluble carbon compounds such as sugars are rapidly lysed from dead phytoplankton cells under aerobic conditions, although losses occur more slowly in anaerobic environments and a greater amount remains as refractory compounds, which are highly resistant to further degradation (Wetzel, 1975). Grazing by zooplankton also assists in the breakdown of organic material, although not all of the ingested phytoplankton is necessarily digested and assimilated (Table 11.5): some of the phytoplankton pass back into the water unharmed; others may be dead or only partly digested. Comparatively little dissolved organic carbon is added to the water directly by grazing, but excretion of fine particulate matter and rapid lysis of materials from dead zooplankton are important processes in decomposition (Saunders, 1980).

Proportionally more carbon originates from terrestrial sources in rivers and oligotrophic lakes. Dissolved organic matter enters as leachates from the plant canopy or from material in various stages of decay. Particulate matter falls directly into a river from the overhanging canopy or is carried in the run-off which drains the land. Leaching losses in particulate leaf material vary between

Table 11.5 Assimilation efficiency of three zooplankton species feeding on different algae (after Schindler, 1971)

| Food | Algal type | Assimilation efficiency (assimilation rate / ingestion rate × 100) | | |
		Daphnia longispina	Diaptomus gracilis	Cyclops strenuus
Elakatothrix	green, unicell	100.0	31.3	19.0
Oocystis	green, colony	10.5	13.7	8.0
Gloeocystis	green, colony	13.6	44.2	18.2
Coelastrum	green, colony	20.8	29.1	6.2
Microcystis	blue-green, colony	17.9	45.3	8.9
Anabaena	blue-green, filament	50.8	73.5	25.9
Oscillatoria	blue-green, filament	25.6	29.7	3.7
Asterionella	diatom colony	100.0	20.1	38.0
Cryptomonas	unicellular flagellate	91.6	100.0	18.5

(Source: J. E. Schindler, Food quality and zooplankton nutrition, *Journal of Animal Ecology*, 1971, **40**, 591, 593.)

Table 11.6 Rate of weight loss in leaves and needles by leaching in stream water (after Saunders, 1980)

Substrate	Leaching rate (% day^{-1})
Fagus sylvatica	8
Fraxinus americana	11
Pinus sylvestris	11
Alnus glutinosa	13
Picea abies	14
Quercus robur	14
Betula verrucosa	16
Populus tremuloides	19
Salix lucida	23
Fraxinus excelsior	25
Cornus amomum	27

(Source: G. W. Saunders, Organic matter and decomposers, in *The Functioning of Freshwater Ecosystems*, eds E. D. Le Cren and R. H. Lowe-McConnell; published by Cambridge University Press, 1980.)

Table 11.7 Average decomposition rates of particulate leaf organic matter in stream water during winter (after Petersen and Cummins, 1974)

Substrate	Decomposition rate (% day^{-1})
Slow rate of decomposition	
Fagus grandifolia	0.25
Populus grandidentata	0.38
Populus tremuloides	0.46
Quercus alba	0.52
Quercus borealis	0.27
Medium rate of decomposition	
Acer platanoides	0.76
Acer rubrum	0.62
Alnus glutinosa	0.75
Carpinus caroliniana	0.83
Carya glabra	0.89
Juglans nigra	0.70
Salix lucida	0.78
Fast rate of decomposition	
Acer saccharum	1.07
Fraxinus americana	1.20
Tilia americana	1.75

(Source: R. C. Petersen and K. W. Cummins, Leaf processing in a woodland stream, *Freshwater Biology*, **4**, 1974.)

species (Table 11.6), ranging from 8% day^{-1} for ash (*Fagus sylvatica*) to 27% day^{-1} for dogwood (*Cornus amomum*). The residual leaf biomass decomposes more slowly at rates of 0.1–1.8% day^{-1} (Table 11.7). Consequently, the leaves remain in the water from 50 to 400 days and so provide a stable substrate for the micro-organisms and invertebrates that feed on them (Saunders, 1980).

11.10.2 Microbial processes

The amount of material that is eventually incorporated into the sediment depends not only on the type and amount of debris but also on the physical and chemical characteristics of the waterbody. The settling rate of dead phytoplankton varies according to their size and

Table 11.8 Maximum sinking rates of dead phytoplankton (after Reynolds, 1984)

Phytoplankton species	Sinking rate (m day^{-1})
Asterionella formosa	0.93
Cyclotella praeterissima	0.82
Fragilaria crotonensis	0.97
Melosira italica	0.98
Stephanodiscus astraea	2.39
Synedra acus	0.63
Tabellaria flocculosa	0.89

(Source: C. S. Reynolds, *The Ecology of Freshwater Phytoplankton*; published by Cambridge University Press, 1984.)

shape but even for larger colonial species it is typically less than 1.0 m day^{-1} (Table 11.8). Most of the organic matter can therefore be decomposed as it sinks slowly to the bottom of deep, well-oxygenated oligotrophic lakes: under such conditions as much as 99% of the particulate organic matter produced in a lake decomposes in the water column (Ohle, 1956). Greater amounts of organic debris settle to the bottom of productive eutrophic lakes and so decomposition is more dependent on microbial activity in the anaerobic sediments. The density and distribution of bacteria in lakes therefore varies with trophic status. In oligotrophic lakes bacterial concentrations are highest within the photic zone: a noticeable decrease in density occurs at the bottom of the photic zone but thereafter it remains relatively uniform throughout the water column. Bacterial densities are higher in mesotrophic lakes but change relatively little with depth, whereas in eutrophic lakes bacterial densities normally increase with depth. Bacterial populations are greater in lake sediments than in the water column, with the highest densities recorded at the surface of the sediments, but decrease rapidly beneath this thin, oxygenated transitional zone (Saunders, 1980). The distribution of bacteria within the sediment varies spatially according to the amount of organic debris that is produced, with highest densities occurring in the littoral sediments beneath stands of macrophytes; smaller populations occur in deep water areas and they are least abundant along wave-swept shorelines and other unproductive sites (Wetzel, 1975). Bacterial populations fluctuate as a result of seasonal and annual variations in organic production, nutrient status of the water and grazing by zooplankton, although rapid changes are also reported for which there is no obvious cause.

The bacterioplankton suspended in the water column are mostly heterotrophic organisms with diverse metabolic requirements (Pedrós-Alió, 1989). The majority are chemoheterotrophs and utilize dissolved or particulate organic matter as a source of carbon and energy. Species of *Pseudomonas* and *Cytophaga* are aerobic; they take in dissolved oxygen and release CO_2 to the water. Anaerobic forms such as *Desulfovibrio* utilize sulphate (SO_4^{2-}) as an electron acceptor to oxidize dissolved organic matter and in so doing produce hydrogen sulphide (H_2S). The CO_2 and H_2S released by these bacteria are subsequently used in other trophic pathways. *Chromatium*, for example, is an anaerobic photoautotroph which requires light energy to process CO_2 and H_2S by photosynthesis. Although the C_3 pathway is ultimately used to manufacture carbohydrates, photosystem II is lacking and consequently no oxygen is produced. Chlorophyll *a* is not present in these cells but is replaced by bacteriochlorophylls and other pigments which are especially sensitive to the blue, green and ultraviolet wavelengths which usually penetrate deep into the water. **Chemoautotrophic** bacteria oxidize reduced inorganic compounds to obtain energy and use CO_2 as their carbon source. For example, *Nitrosomonas* converts the ammonia generated by bacterial decomposition of protein and other nitrogenous organic compounds into nitrite. Similarly, bacteria of the genus *Thiobacillus* oxidize H_2S and other reduced sulphur compounds to sulphates.

The aerobic bacteria are the principal decomposers in the epilimnion and it is assumed that different species are adapted to utilize the various resources available (Pedrós-Alió, 1989). Some may depend on products excreted by the phytoplankton or else colonize dead cells. The free-living bacteria utilize dissolved organic matter: they are most abundant in the spring when the population of small phytoplankton increases rapidly. Sinking rates are typically only 1–2 mm day^{-1} for free-living bacteria (Jassby, 1975), but heavy grazing by zooplankton can reduce the bacterial population to a minimum during the distinctive clear-water phase which occurs in early summer in some lakes (Lampert *et al.*, 1986). Particulate material is decomposed by attached bacteria: this group reaches its maximum density in early summer when the supply of debris increases because the phytoplankton are larger and subject to lower grazing losses (Pedrós-Alió and Brock, 1983). Attached bacteria grow more slowly than free-living forms but they absorb carbon at a higher rate and store this excess in extracellular mucilaginous material. However, sinking rates for attached bacteria can be as high as 0.9 m day^{-1} and as much as 67% of the total daily bacterial production may settle out of the water column (Ducklow *et al.*, 1982).

The basic metabolic requirements of the bacteria determine their distribution within a lake. Some exist in a completely anaerobic environment but others require the presence of both reduced and oxidized compounds: their relative densities reflect the degree of stagnation occurring within the water column. In oligotrophic lakes the concentration of dissolved oxygen tends to increase with depth and the transition to anaerobic conditions usually occurs within the sediment. Oxygen demands are higher in eutrophic lakes and water in the hypolimnion becomes anoxic during periods of stratification. Under these conditions bacteria such as *Thiobacillus* and *Nitrosomonas* are concentrated in the transitional metalimnion, where some oxygen diffuses down from the overlying epilimnion and reduced compounds such as H_2S and NH_4 are available from the hypolimnion

(Pedrós-Alió, 1989). Similarly, the photosynthetic sulphur bacteria, such as *Chromatium*, can only exist where there is sufficient light to oxidize H_2S through the reduction of CO_2. The distribution of these types of bacteria in holomictic lakes changes when the thermocline breaks and the water begins to circulate: during these periods they become more evenly distributed within the water column. The highest bacterial densities occur in the upper few millimetres of sediment. These bacteria are mainly saprophagic and metabolize the organic detritus by absorbing the oxygen which diffuses into the sediment from the overlying water. Only anaerobic bacteria, such as sulphate-reducing *Desulfovibrio* and methane-producing *Methanobacterium*, exist below this oxidized micro-zone. Other bacteria exhibit mutualistic relationships with heterocystous blue-green algae *Anabaena* (Paerl and Kellar, 1978) or, like *Vampirococcus* and *Daptobacter*, are predators of phototrophic bacteria (Pedrós-Alió, 1989) and so are only indirectly affected by oxygen and carbon sources.

11.10.3 The role of zooplankton and detritivores

Microbial transformation of organic matter within the water column and sediment is an integral part of the nutrient regeneration process in aquatic ecosystems. However, zooplankton and benthic detritivores also function as decomposer organisms in that they ingest dead organic matter and excrete ammonia, CO_2 and other substances that are necessary for primary production. The detritus-feeding zooplankton are mainly rotifers and small non-predatory cladocerans such as *Daphnia* which filter particles as large as 20–25 µm in size (Nauwerck *et al.*, 1980). They graze on planktonic algae in oligotrophic lakes but detritus may account for a large proportion of their diet in eutrophic lakes, even though it may not be readily assimilable (Saunders, 1969). Thus particulate organic matter is continually reprocessed in the epilimnion by non-selective feeders. A detritus turnover time of only 5 days has been calculated for Lake Michigan during the summer when the *Daphnia* population is at its highest density and this may increase its subsequent utility as a microbial substrate (Scavia and Fahnensteil, 1988). The dominant invertebrate detritivores are chironomid larvae and oligochaete worms, especially in the poorly oxygenated sediments of the profundal zone. Deposit-feeders gather the particles of food as they settle on to the lake bed. Others filter the food from water which is pumped through their U-shaped burrows or is passed through inhalant and exhalant siphons. Those that live within the sediment feed on low-quality refractory debris or the bacteria which they scrape from it.

Bacteria can sequester dissolved nitrogen and phosphorus at extremely low concentrations (Currie and Kalff, 1984), and although some is subsequently released to the water, bacterial biomass constitutes a potentially large nutrient sink. In Lake Ontario, for example, free-living chroococcoid cyanobacteria can account for nearly 40% of total primary production during times of peak abundance (Caron *et al.*, 1985). However, these free-living bacteria are consumed by a variety of organisms within the water column and this, rather than detrital processes, is now considered the principal method of nutrient regeneration in aquatic ecosystems (Stockner and Antia, 1986). These bacteria not only serve as a food source for heterotrophic micro-organisms, but a considerable proportion are also ingested by planktonic algae (Bird and Kalff, 1986). Recycling nutrients through the food web depends on the degree of microbial activity in the epilimnion and on the presence of organisms such as *Daphnia*, which are an essential link between the picoplankton and higher organisms (Stockner and Shortreed, 1989). Community structure and production is therefore influenced by both producers and consumers. Nutrient availability determines the potential biomass in each trophic level and so exerts a 'bottom-up' control on freshwater pelagic ecosystems (Bartell *et al.*, 1988). Conversely, the phytoplankton community may be regulated by 'top-down' processes or grazing intensity. 'Bottom-up' control is most pronounced at the producer level and becomes progressively weaker further up the food web as predator-mediated effects intensify. The relative importance of these controls also varies with the trophic status of the lake. In eutrophic lakes 'top-down' effects are strong between piscivores and zooplankton but have little impact at the zooplankton–phytoplankton level, whereas the grazing by zooplankton has the greatest impact on community structure in oligotrophic lakes (McQueen *et al.*, 1986).

The interactions between phytoplankton and zooplankton are complex. The density of some phytoplankton is directly reduced by grazing but production can also be stimulated by the rapid turnover of nutrients (Sterner, 1986). Nutrient regeneration by zooplankton grazing is most important in large, eutrophic lakes where the extensive pelagic zone is unaffected by nutrient inflows from streams and other processes in the littoral zone (Carney and Elser, 1990). Under these conditions the process of nutrient regeneration is ultimately tied to the feeding rates of the zooplankton and this is often sufficient to supply most of the phytoplankton demands for nitrogen and phophorus. These nutrients are mostly excreted as NH_3 and PO_4, in which forms they are immediately available for uptake. Thus ambient concentrations will normally remain low during periods of high productivity (Lehman, 1984). Conversely, silica is not readily digested by zooplankton and is usually concentrated in faecal pellets which sink fairly quickly from the euphotic zone. Thus the progressive reduction in silica through grazing can also be a factor in the seasonal phytoplankton succession. However, grazing by zooplankton can be highly selective and the preferred species may simply be removed while the others benefit from the regenerated nutrients.

11.11 FUNCTIONAL ROLE OF ZOOPLANKTON

11.11.1 Feeding habits of zooplankton

The zooplankton are a relatively passive group of organisms and so are mostly associated with lentic habitats where they are not swept away by strong currents. In most lakes the dominant zooplankton are rotifers and crustaceans; some protozoa and planktonic larval stages of insects and other organisms may also be present. Most rotifers are non-predatory organisms that feed on bacteria, phytoplankton and dead detrital material that they sweep into their mouth cavities by means of the surrounding cilia. The food particles are typically quite small (3–15 µm), although larger cells are sometimes taken, ruptured and the contents ingested. The crustacean zooplankton are mainly represented by species of cladocera and copepods. The cladocerans, or water fleas, which include species of *Daphnia*, range in size from 0.2 to 3.0 mm. Food consisting of bacteria, small phytoplankton and detritus is filtered from water that is continually moved by complex movements of the legs (Wetzel, 1975). The food collects in a ventral groove between the base of the legs and is moved forward to the mouth, where it is ground up by the large, chitinized mandibles. The maximum size of the food particles ranges from 10 to 80 µm and is determined by the morphology of the setae on the legs that initially entrap and filter the material. The planktonic copepods are mostly non-predatory organisms that feed on algae up to 50 µm in size. Primarily filter feeders, they consume small organisms and detritus which are collected in currents set up by the antennae. They are nonetheless rather selective feeders and various species coexist because of differences in the size of the food particles they consume. Other copepods, such as the larger *Cyclops*, are raptorial feeders and seize their food directly in their mouthparts then force it into the mandibles for maceration prior to swallowing. The food is often used uneconomically and a considerable amount of detritus is added to the water in this way (Nauwerck *et al.*, 1980). The larger predatory species are mostly organisms of the littoral benthic zone. Other non-planktonic copepods are adapted to seize and scrape particles from the sediments and macrovegetation (Wetzel, 1975).

11.11.2 Reproduction and population dynamics

The zooplankton are well adapted to utilize the short-lived increases in phytoplankton density. The rotifers reproduce parthenogenically for most of the year. Generation times typically range from 1 to 3 days under optimum conditions but can take 2–3 weeks depending on temperature and quality of the food supply (Nauwerck *et al.*, 1980). Reproduction in cladocerans is also mainly parthenogenic. The eggs develop in a dorsal brood pouch so that the majority of species do not produce free-living larval forms. Parthenogensis continues until

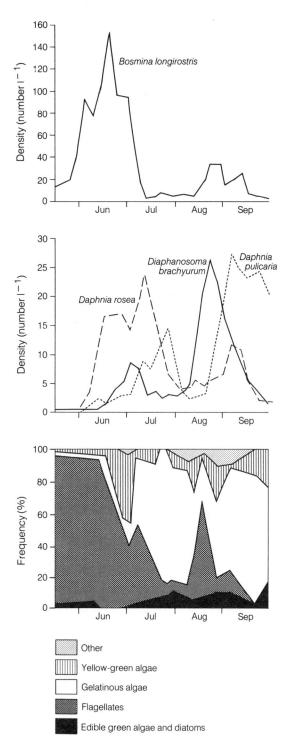

Figure 11.16 Seasonal succession of zooplankton and algal groups in a temperate North American lake. (After DeMott, 1983; Kerfoot et al., 1988.) (Reprinted with permission from W. R. DeMott, Seasonal succession in a natural *Daphnia* assemblage, *Ecological Monographs*, 1983, **53**, 328; and from W. C. Kerfoot, C. Levitan and W. R. DeMott, *Daphnia*–phytoplankton interactions: density-dependent shifts in resource quality, *Ecology*, 1988, **69**, 1810.)

conditions become unfavourable, often as a result of cooler temperatures, changes in photoperiod or a decline in the quantity and quality of available food. Some of the eggs then develop into males but others are haploid and require fertilization (Hutchinson, 1967). Copepods reproduce sexually and the fertilized eggs are carried in one or two sacs by the female. Some of the eggs may be dropped into the sediment, where they undergo a period of diapause. Others hatch into free-swimming larval nauplii which are released to the water and develop into adults through a series of moults and instar stages. The rate of development in copepods varies between species. Some may produce several generations in a year but in others development is interrupted by long periods of aestivation. Mortality is high during the juvenile stages because of predation and cannibalism. Adult and immature copepods make up most of the overwintering population of zooplankton in temperate lakes, although large numbers of cladocerans in diapause and rotifer eggs are also usually present. Zooplankton activity increases in the spring as the lake temperatures rise and more food becomes available, with maximum biomass occurring in late summer. Zooplankton biomass can fluctuate considerably from year to year in temperate lakes. This has been attributed to climate conditions which affect the development of phytoplankton in some years. For example, the diatom population in Lake Windemere in England begins to grow earlier in years with warmer summers and may be declining by the time the *Daphnia* begin to reproduce (George and Harris, 1985).

The composition of the zooplankton population is similarly linked to seasonal changes in phytoplankton abundance and to the ensuing competition for this variable food resource. Studies in Vermont have demonstrated that *Daphnia rosea* has a competitive advantage over similar cladocerans in the early spring: microflagellates such as *Chlamydomonas* are the dominant phyoplankton at this time of the year and this sustains the rapid increase in the population of *D. rosea* as it hatches from resting eggs (Figure 11.16). The typical brood size in *D. rosea* is 5–8 eggs per adult female initially but declines steadily as algae which are resistant to digestion become more abundant (Kerfoot *et al.*, 1988). *D. rosea* is later replaced by *D. pulicaria*, which is better able to assimilate the gelatinous green algae that grow vigorously in midsummer and autumn. It has been generally assumed that the rate of food collection and assimilation by zooplankton increases with body size, so that the larger species are considered to be competitively superior when food resources are limited (Brooks and Dodson, 1965). Thus, the small rotifer *Bosmina longirostris* is replaced early in the year by *D. rosea* as soon as the flagellate algae become scarce. The summer decline in *D. rosea* is accompanied by a secondary peak in the rotifer population and an increase in the small cladoceran *Diaphanosoma brachyurum*.

The growth and reproductive response of zooplankton is affected by food availability. Large species of *Daphnia* are unable to reach reproductive maturity at low food concentrations, whereas smaller species can mature in 5–14 days, depending on food supply (Tillmann and Lampert, 1984). Although threshold food concentrations are lower for smaller species, they are less resistant to starvation. Juvenile stages are also more vulnerable to starvation (DeMott, 1989) and some of the seasonal changes which occur in zooplankton species can be accounted for in this way. Some zooplankton are intrinsically adapted to food availability in lakes of different trophic status. Thus *Diaptomus dorsalis*, a copepod associated with eutrophic lakes, grows comparatively faster when food is plentiful, whereas *D. floridanus* and *D. mississippiensis*, which co-occur in oligotrophic lakes, perform best when food is limited (Elmore, 1983). Variation in food availability is one of several factors which affect the fecundity of the zooplankton and similar changes can arise from seasonal temperature rhythms and predation. Thus *Diaphanosoma* is a warm-water organism: its eggs hatch in late spring and maximum population densities in temperate lakes are consequently delayed compared with other cladocerans (Kratz *et al.*, 1987). Unlike *Daphnia* and other large cladocerans, *Diaphanosoma* is unaffected by the increase in cyanobacteria which occurs during midsummer in many temperate lakes, and consequently dominance shifts to the smaller species (Threlkeld, 1986). *Diaphanosoma* is directly limited by competition and the rapid increase in abundance coincides with the midsummer decline in *Daphnia* (Kerfoot *et al.*, 1988).

11.11.3 Predation by higher organisms

The removal of *Daphnia* and other grazing species through predation is an important factor in the seasonal succession of zooplankton in lakes. Plankton are consumed by a variety of animals including fish, waterfowl and insects as well as by the predatory zooplankton such as rotifers and copepods. The majority of the planktivores are selective feeders and the prey on which they feed is mainly determined by size, visibility and motility (Zaret, 1980). The predators can be classified according to the methods they use to locate and capture their prey. The invertebrate predators are raptorial feeders which mostly feed on smaller (< 1 mm) zooplankton (Gliwicz and Pijanowska, 1989). The vertebrate predators consume larger prey and are either non-visual filter feeders that select their food at random or are visual particulate feeders that remove the most conspicuous prey (Greene, 1985). The impact of predation is determined by seasonal changes in predator abundance and their feeding rates. In most aquatic ecosystems the fish are the dominant predators. All fishes are planktivores in their juvenile stages and will consume most species of zooplankton. Fish normally hatch in large numbers in the spring and early summer but mortality is high and less than 1% will

usually survive to the end of the first season (Gliwicz and Pijanowska, 1989). Predation is most pronounced in late spring and early summer, when the density and feeding intensity of juvenile fish is highest: hatching is presumably timed to coincide with peak zooplankton abundance. Planktivorous fish typically gather their prey by taking in water through the mouth and releasing it via the opercular cavity. The ability to escape from the resulting current of water varies between potential prey species. Attacks directed at calanoid copepods are the least successful and the characteristic evasion rate is about 50%. Cyclopoid copepods are captured more frequently and most cladocerans will be drawn in by the suction (Wright and O'Brien, 1984).

The prey that is taken into the mouth must be retained on the gill rakers, and the fineness of these filtering structures to some extent determines the ultimate fate of the zooplankton. However, feeding selectivity varies with food availability and smaller prey are actively pursued when resources are scarce (Hall *et al.*, 1976). The high predation rates have a considerable effect on the size distribution of the zooplankton community (Figure 11.17). The introduction of planktivorous fish into lakes is generally accompanied by an increase in rotifers and other small zooplankton, whereas the larger cladocerans and copepods have a competitive advantage when predation is minimal. Consequently, differences in the zooplankton population occur from year to year in response to varying fish densities (Cryer *et al.*, 1986). Selective removal of the most abundant prey species facilitates the coexistence of potential competitors and the effect of predation is usually to increase diversity of the zooplankton population, as the available resources are no longer monopolized by a few fast-growing species (Gliwicz and Pijanowska, 1989). Zooplankton diversity is lowest in the spring before most of the planktivorous fish hatch and the invertebrate predators begin to multiply. Later in the year the dominance shifts to those species which are less vulnerable to predation because they are either very small or highly transparent, or are better able to avoid capture. The subsequent recovery of the prey species occurs when predation pressure is relaxed because the predators begin to die off, are themselves consumed by higher carnivores or turn their attention to other sources of food. However, the predators will normally consume a very large percentage of the available prey and occasionally this has resulted in the elimination of certain species of zooplankton from some lakes (Goldman *et al.*, 1979).

11.12 FOOD-WEB INTERACTIONS IN LAKES

In highly productive lakes the detrital food chain provides an indirect flow of energy from large algae to small grazers. The smaller phytoplankton which are

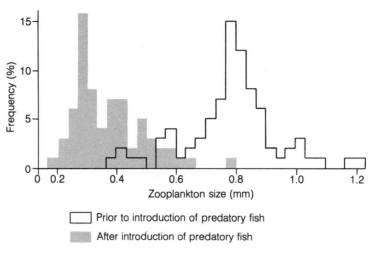

Figure 11.17 Size distribution of crustacean zooplankton before and 20 years after the introduction of the predatory fish *Alosa aestivalis* into a temperate lake in North America. (After Brooks and Dodson, 1965.) (Redrawn from J. L. Brooks and S. I. Dodson, Predation, body size, and composition of plankton, *Science*, **150**, 33, copyright 1965 by the AAAS.)

characteristic of lakes of low productivity are mostly grazed directly by the zooplankton. The trophic status of a lake is therefore reflected in the distribution of plankton biomass within different size classes and is ultimately related to the way in which the organisms interact (Stein *et al.*, 1988). The total biomass in a lake is determined by resource availability but its distribution within size classes is controlled by energy transfer efficiencies. The smaller species of phytoplankton generally exhibit faster growth, higher nutrient uptake efficiencies and more rapid nutrient turnover rates than larger species but this is counteracted by higher grazing losses. Similarly, body size, through its effect on threshold food levels and food selection, can affect life history strategies and the structure of the zooplankton population (Lynch, 1980). The size distribution of the phytoplankton and zooplankton in a lake is functionally related through grazing: changes in the size structure of the zooplankton are reflected by a corresponding adjustment in the phytoplankton population (Bergquist *et al.*, 1985) and conversely the grazing population is affected by the size of phytoplankton that is available. The feeding activity of fish is similarly regulated by size. In planktivorous species the structure of the gill rakers determines the range of zooplankton that is ingested, while other species are gape-limited and thus restricted in the size of prey that they can consume.

Large organisms must grow through juvenile stages and so are subject to size-related advantages and disadvantages at each stage of their life history. For example, adult *Daphnia* will normally outcompete smaller species of zooplankton (Vanni, 1986) but their competitive ability is poor during juvenile stages (Neill, 1975). Similar size-dependent interactions occur between piscivorous fishes. Spawning time and subsequent growth of fishes vary with water temperature and in some species this will determine their vulnerability to predation. Largemouth bass (*Micropterus salmoides*) spawn earlier in years when water temperatures rise quickly in the spring. This gives them a size advantage over the later-hatching shad (*Dorosoma petenense* and *D. cepedianum*) on which they prey (Adams and DeAngelis, 1987). The predator grows quickly under these conditions and the prey remains vulnerable throughout the season. The largemouth bass hatch later in years when the water temperature is cooler and so they are too small to feed on the young shad. Growth rates are slower on the alternative diet of zooplankton and macroinvertebrates and the shad therefore retain their size advantage. Size-dependent trophic interactions of this type can have a profound influence on the structure of a lake community (Carpenter *et al.*, 1985).

The variable role of the zooplankton is a key determinant of community structure in that they serve either as food for planktivores or are themselves consumers of phytoplankton. The zooplankton population is reduced in years when spawning times and growth rates favour the planktivores and consequently grazing pressure on the phytoplankton is relaxed. For example, in Oneida Lake, New York, the relative abundance of *Daphnia pulex* is controlled by the number of yellow perch (*Perca flavescens*) which hatch each year (Mills and Forney, 1988). In years when young fish are abundant *Daphnia* is generally replaced by smaller zooplankton. However, the size of the young perch is subsequently related to the biomass of *Daphnia* that remains available, because growth is slower when competition for food is more intense. The smaller perch are then more vulnerable to predation from walleye (*Stizostedion vitreum vitreum*) and fewer individuals are recruited into the population. The biomass of *Daphnia* is inversely related to the biomass of nanoplankton on which it feeds, but as food becomes scarce brood size decreases and *Daphnia* productivity declines. *Daphnia* are also unable to utilize the blue-green algae which become abundant later in the summer: this produces a bottleneck in the food web and productivity at higher trophic levels necessarily declines.

11.13 TROPHIC RELATIONSHIPS IN STREAMS

Phytoplankton are much less abundant in lotic environments and primary production is mainly carried out by attached algae and the macrophytes that grow along the channels. Some of the higher plants are eaten directly (Lodge *et al.*, 1988) but more commonly they harbour a diverse population of attached and free-living organisms which serve as part of the food base for these communities. Animals such as mayflies, caddis-flies and amphipods feed on filamentous algae and this, rather than phytoplankton, is the principal food for aquatic herbivores in streams. Some invertebrates also feed on the mosses which grow in stream channels but these plants, like the macrophytes, mostly enter the food chain as detritus (Gregory, 1983). Unlike lakes, which essentially function as closed systems, much of the energy available to stream communities is derived from allochthonous materials that enter the drainage network from terrestrial sources and are passed downstream in the flowing water. Fewer trophic levels are usually involved in the lotic foodweb: the organic matter is initially processed by benthic invertebrates, which in turn are consumed by fish and other vertebrate predators.

The direct influence of the riparian zone on stream ecology is greatest in the small headwater channels (Figure 11.18). These smaller streams are generally heterotrophic in that community respiration exceeds gross primary production and the organisms depend mainly on organic matter of riparian origin (Cummins, 1988). These communities are dominated by invertebrates which are specialized in collecting or shredding coarse particulate organic matter. The organic drift is collected in various ways. Mayfly nymphs, caddis-worms and similar insects trap the food particles in fringes of hairs on their mouthparts or forelegs. The larvae of some insects construct tubes of silt and silk surmounted by strings of saliva to catch suspended organic matter or, like *Hydropsyche*, spin silken nets. Small animals are also caught in this way and are consumed by carnivorous larvae, and some nets are so fine that they trap bacteria. The shredders mostly consume the micro-organisms and fungi which grow on leaf tissue or on dead wood (Cummins and Klug, 1979). Animals such as snails and stoneflies specialize in scraping algae from rocks and other surfaces, while detritivores consume the fine particulate matter that lodges in the stream bed. Reduced shading and higher nutrient concentrations in mid-sized rivers favours the growth of aquatic plants. Production exceeds respiration in these autotrophic communities and the excess organic matter is transported downstream.

Figure 11.18 Large, moss-covered logs contribute to the food base of this small stream in Tennessee. (M. Huston.)

The quantity of fine particulate organic matter increases in these rivers: the density and diversity of collectors and scrapers increase accordingly, although the shredders are less abundant than in the headwaters. Primary production is reduced by the high turbidity of the largest rivers, and the invertebrate population is dominated by collectors that feed on the fine particulate organic matter which is suspended in the water (Cummins, 1988).

Although predatory species such as insect larvae and fish are present to varying densities in all streams and rivers, their effect on community structure is usually small (Allan, 1983). Many of the invertebrate species feed on chironomids during early stages of development but will consume a more varied diet as they grow larger. The range of organisms consumed is largely determined by size and availability rather than selectivity. Fishes are also general predators, although they differ in their feeding habits. The bottom feeders mostly consume the larvae of aquatic insects, whereas surface and water column feeders depend on terrestrial insects and aquatic invertebrates. Fish and invertebrate predators are voracious feeders and can consume most of the available prey. Heavy predation can be sustained by the continual drift of organisms in the water (Mundie, 1974) but some of the prey may escape detection through adaptive behaviours. Thus habitat selection in crayfish (*Orconectes propinquus*) varies with predator density and the smaller, more vulnerable individuals are less active (Stein and Magnuson, 1976). Similarly, some mayflies can detect and avoid invertebrate predators through chemical stimuli (Peckarsky, 1980). **Drift**, which represents excess production that is carried downstream in the current, is an important source of food for many predators. Various adaptations such a hooks and suckers, sticky secretions and heavy stone cases reduce the likelihood of animals being swept away (Hynes, 1970). Many animals remain hidden under stones during the day and emerge to feed at night. This reduces the chance of detection by fish and other predators that must first sight their prey: the number of organisms in the drift generally peaks just after sunset. Many of the insect larvae in the drift will fly upstream to lay their eggs after they emerge, and so perpetuate the cycle. Although many fish are opportunistic feeders, the smaller prey are generally consumed less frequently than expected from their density in the drift (Allan, 1983) and the size of organisms again becomes a factor in the structure of the food webs.

11.14 LAKE SUCCESSION AND CULTURAL EUTROPHICATION

Although the structure of freshwater aquatic ecosystems is influenced by the complex interactions of producers and consumers, the potential productivity of each trophic level is ultimately regulated by nutrient availability. The nutrient-poor oligotrophic lakes are distinguished from the nutrient-rich eutrophic lakes in a given geographic region on the basis of differences in productivity. In temperate regions the productivity of oligotrophic lakes is limited by the low input of inorganic nutrients from external sources: these lakes are typically large and deep, and because there is little organic production, decomposition and nutrient recycling is minimal in the oxygen-rich hypolimnion (Wetzel, 1975). All lakes are ephemeral in a geological sense in that they are gradually filled in with sediments. The traditional assumption has been that lakes become more fertile over time because of the continual input of nutrients into a progressively smaller volume of water (Hutchinson, 1973). The water in oligotrophic lakes remains well oxygenated throughout the year, but as the lake bed is steadily raised the ability of the hypolimnion to supply oxygen to the decomposer organisms begins to decline. Thus the characteristic oxygen deficit of shallow eutrophic lakes in temperate regions could result from the reduced volume of the hypolimnion rather than a change in nutrient availability. The natural eutrophication of lakes through sedimentation may take many thousands of years but during most of this ageing process they will retain good water quality and exhibit a diverse biological structure. However, conditions are different in tropical regions: oxygen depletion in the hypolimnion is characteristic of these amictic lakes and is therefore not indicative of their trophic status (Thornton, 1987).

Although the trophic status of a lake is affected by various natural factors, in many parts of the world these are now less significant than the human activities which occur within the drainage basin (Figure 11.19). Phytoplankton production increases rapidly as more phosphorus and nitrogen becomes available, and the undesirable growths of algae associated with polluted nutrient-rich lakes are now synonymous with eutrophication. The discharge of organic wastes is the most common disturbance to aquatic ecosystems. The main source of organic enrichment is domestic sewage, but other materials are added through various industries associated with food-processing, tanning and textiles, paper-making and petrochemicals (Hellawell, 1986). The principal effect of organic discharge is the depletion of dissolved oxygen (Table 11.9). Oxygen is used up by the heterotrophic micro-organisms which decompose organic matter, and the degree of pollution is typically reported in terms of this **biological oxygen demand** (BOD). Toxic substances which rarely if ever occur in natural waterbodies are added through industrial processes. They are not readily biodegradable, and although initial concentrations may be relatively low, they tend to accumulate in the food chain to the detriment of the higher organisms. Inert solids from coal washeries and similar operations increase turbidity of the water until they eventually settle out and in doing so smother most of the benthic organisms. The discharge of heated water used for industrial cooling purposes also affects the distribution and activities of aquatic organisms.

Cultural eutrophication in temperate lakes is usually associated with an increase in available phosphorus and the dense growth of blue-green algae that result. Large amounts of phosphorus are released from municipal and industrial sources, with smaller amounts derived from various diffuse sources within drainage basins. The phosphorus discharged in municipal sewage is primarily in a soluble form that is easily assimilated by aquatic plants (Young et al., 1982): it can also be removed from wastewater relatively easily and this is the principal activity in most eutrophication control programmes. For example, the restoration of Lake Washington in the United States was achieved by lowering phosphorus levels through sewage diversion (Edmondson and Lehman, 1981). Annual phosphorus input into this lake exceeded 200 000 kg in some years and as much as much 70% of this originated from sewage. The high phosphorus levels resulted in dense growths of the filamentous blue-green algae *Oscillatoria* and water clarity was greatly reduced (Figure 11.20). Algal growth declined as excess phosphorus was gradually incorporated into the sediments and by 1975 the lake was considered to have recovered from eutrophication. However, nutrient diversion does not eliminate the problem of eutrophication but simply transfers it to another location. Phosphorus can be removed from sewage by chemical precipitation using aluminium and iron salts: in this way effluent concentrations can be reduced to 0.3–1 mg P l^{-1} (Ryding and Rast, 1989). Regulatory procedures which limit the use of phosphate detergents have also proved effective. Similarly, restrictions on the type of land-use activities permitted within a drainage basin have been adopted in some countries in order to reduce nutrient inputs through run-off.

Although phosphorus limitations are reported for some tropical lakes, algal growth in these regions is more commonly limited by low nitrogen levels. A

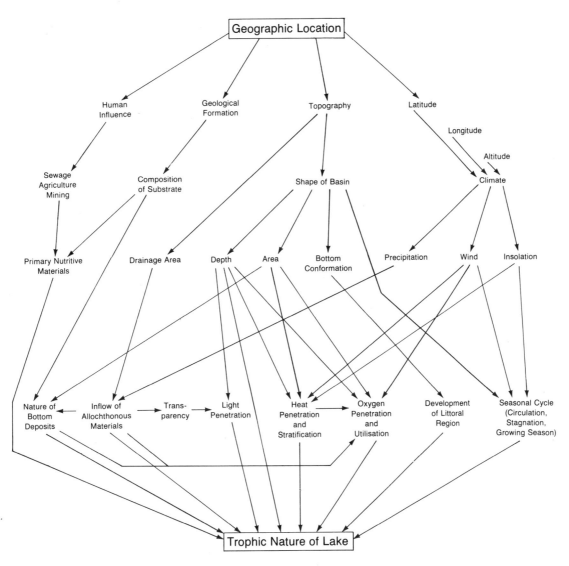

Figure 11.19 Physical and chemical factors affecting the trophic status of lakes. (After Rawson, 1939.) (Adapted from D. S. Rawson, *Some physical and chemical factors in the metabolism of lakes*, American Association for the Advancement of Science Publication 10, copyright 1939 by the AAAS.)

phosphorus concentration of 25–30 µg l^{-1} is considered as the lower limit of eutrophy in temperate lakes but 50–60 µg l^{-1} is suggested as a more realistic index in tropical regions. Nitrogen concentrations in eutrophic tropical systems are similar to those recorded in temperate lakes and typically range from 0.2 to 1.0 mg l^{-1}. Consequently N:P ratios as low as 1:1 commonly develop in tropical lakes. Nitrogen-fixing blue-green algae are naturally abundant under these conditions and increased growth of aquatic macrophytes in eutrophic lakes is usually the biggest problem (Thornton, 1987). Eutrophication is comparatively rare in natural tropical lakes, although unusual nutrient enrichment can occur through hippopotamus and other wildlife species. Nutrient input to waterbodies is increasing because of soil erosion following forest clearance but most enrichment

problems in the tropics are associated with effluent entering artificial lakes constructed near urban centres.

Various methods can be used to treat eutrophication directly within a lake (Cooke *et al.*, 1986). Primary production can be reduced by adding alum (aluminium sulphate) or sodium aluminate to precipitate the phosphorus. An alternative procedure requires water to be diverted to increase the flushing rate. This method is particularly useful in reservoirs where algae can be flushed out or where nutrient-rich bottom waters can be drawn off before they mix with the epilimnion. The amount of phosphorus released into the water from the sediment can be reduced by maintaining oxygenated conditions in the hypolimnion, as this causes greater binding of phosphorus and ferric hydroxide. Vigorous circulation to break the thermocline will similarly help to aerate the

Table 11.9 Environmental effects of sewage discharge and industrial effluents (after Hellawell, 1986)

	Principal environmental effect	Potential ecological consequence
Organic enrichment		
1. High biochemical oxygen demand caused by bacterial degradation	Reduction in dissolved oxygen	Elimination of sensitive oxygen-dependent species; increase in some tolerant species; change in community structure
2. Partial biodegradation of proteins and other nitrogenous materials	Elevated ammonia concentrations; increased nitrite and nitrate levels	Elimination of intolerant species through ammonia toxicity; reduction in sensitive species; potential for increased plant growth in nutrient-poor waters
3. Release of suspended solids	Increased turbidity and reduced light penetration	Reduced photosynthesis by submerged plants; abrasion of gills; interference with normal feeding activity
4. Deposition of organic sludge	Release of methane and H_2S under anaerobic conditions; blanketing of substratum	Elimination of normal benthic community
Inert solids		
1. Particles in suspension	Increased turbidity and reduced light intensity abrasive action	Reduced photosynthesis in plants; feeding impaired by reduced visibility or interference with collecing mechanisms
2. Deposition of material	Blanketing of substratum; loss of interstices; substrate instability	Change in benthic community; biological diversity reduced.
Toxic wastes		
1. Presence of poisonous substances	Adverse affect on water quality	Community composition changes through elimination of sensitive species; sub-lethal effects such as reduced reproductive capacity and changes in behaviour

(Source: J. M. Hellawell, *Biological Indicators of Freshwater Pollution and Environmental Management*, published by Elsevier Applied Science Publishers Ltd, 1986.)

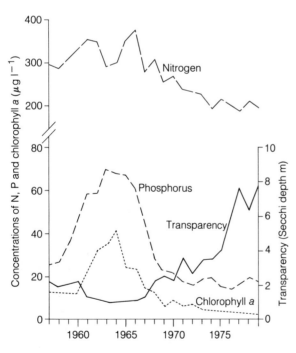

Figure 11.20 Changes in the properties of Lake Washington, USA, following elimination of secondary sewage effluent in 1968. (After Edmonson and Lehman, 1981.) (Adapted from W. T. Edmondson and J. T. Lehman, The effects of changes in the nutrient income on the condition of Lake Washington, *Limnology and Oceanography*, 1981, **26**, 24.)

deeper water. Other processes to limit nutrient exchange from the sediment include injection of calcium nitrate in solution to stimulate microbial denitrification, covering the lake bed with particulate materials such as fly ash, reducing lake levels to expose the sediments to the atmosphere, or simply removing nutrient-rich sediment by dredging. Algicides, such as copper sulphate, and herbicides provide effective short-term control of heavy plant growth, and mechanical harvesters can be used when immediate removal is necessary for recreation and other activities. Algal growth can also be reduced through biomanipulation; for example, carnivorous fish are commonly introduced into a waterbody to prey on the smaller planktivores and thereby increase the zooplankton grazing rate. Although aquatic plants can be effectively used to treat sewage-laden waters in specially constructed lagoons, the harvesting of rooted macrophytes from the littoral zone has less effect on water quality in lakes because most of the nutrients taken up by these plants are derived from the sediment (Carignan and Kalff, 1980). However, nutrients are released into the water when the plants decompose and eutrophication may increase if the plants are not removed.

11.15 HUMAN IMPACT ON RIVERS

Most rivers have been affected in some way by different forms of pollution or changes in flow regime as a result of engineering projects and various land-use activities within the drainage basin. As with lakes, the principal source of pollution in rivers is organic wastes derived from municipal and industrial sources. Rivers polluted with organic effluent show a characteristic change in physical and biological properties downstream from the point of discharge, which reflects the degree of dilution and dispersion that occurs in the flowing water (Figure 11.21). Typically, organic sludge is deposited on the river bed and the dissolved oxygen content of the water falls sharply because of the increase in saprophytic microorganisms, which form slimy growths of '**sewage fungus**' wherever there is a copious supply of biodegradable materials. 'Sewage fungus' is not a single organism but colonies of slime-forming bacteria such as *Sphaerotilus natans* growing in association with fungi and filamentous algae such as *Stigeoclonium tenue*: sedentary protozoa are also important constituents of the 'sewage fungus' community (Hellawell, 1986). Anaerobic bacteria are also common in the sludge deposits and break down the organic matter into acetic acid and methane. The distribution of algal species indicates the degree of pollution. For example, microscopic *Oscillatoria* usually grows abundantly near sewer outfalls, where it is often accompanied by larger species such as *Cladophora glomerata*. Most macrophytes are severely limited by the high turbidity and sedimentation which occurs with the discharge of sewage. Epiphytic growths of 'sewage fungus' are also detrimental. However, *Potamogeton pectinatus* is comparably tolerant of pollution, and species such as *Nuphar lutea* and *Lemna minor* will also survive some organic enrichment (Haslam, 1978).

The response of macroinvertebrates to differing degrees of pollution is so distinctive that it is often used as a standard measure of water quality. In clean rivers the typical macroinvertebrate community is comprised of numerous species, each represented by relatively few individuals. Few species can tolerate severe pollution. The distinctive species in badly polluted rivers are tubificid worms which feed on the bacteria in the sludge. They may be accompanied by chironomid larvae or 'bloodworms', but these organisms cannot endure prolonged periods of anoxia and so are more commonly found in less polluted areas. The appearance of *Asellus aquaticus* in European rivers is indicative of gradually improving conditions, and the presence of *Gammarus pulex*, caddisflies and mayflies further downstream are signs that the river has reached the first stage of recovery (Hellawell, 1986). Fish are equally sensitive to organic pollution, mainly because of the reduced concentration of dissolved oxygen. The natural distribution of fish in rivers is partly controlled by different oxygen levels in the water: salmonids are associated with cool, oxygen-saturated headwater streams, whereas 'coarse' fish are found in the

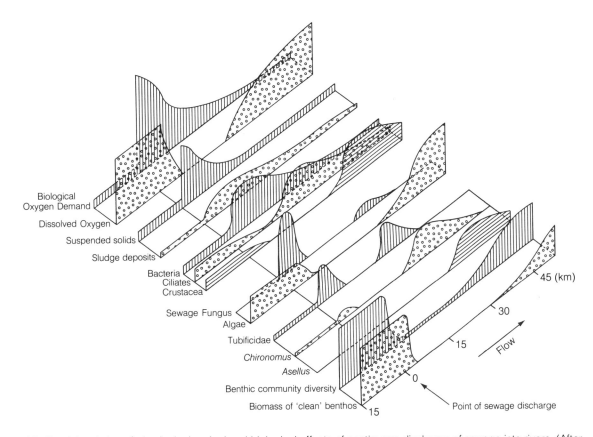

Figure 11.21 Spatial variation of physical, chemical and biological effects of continuous discharge of sewage into rivers. (After Bartsch, 1948.) (Modified and redrawn from A. F. Bartsch, Biological aspects of stream pollution, *Sewage Works Journal*, 1948, **20**(2), 292–302.)

Table 11.10 Change in water quality in non-tidal rivers and canals in England and Wales during the period 1958–1980 (after Toms, 1985)

Water quality class	1958 (km)	1970 (km)	1975 (km)	1980 (km)
Unpolluted	24950	28500	28810	28810
Doubtful	5220	6270	6730	7110
Poor	2270	1940	1770	2000
Grossly polluted	2250	1700	1270	810
Survey total	34690	38400	38590	38740

(Source: R. G. Toms, River pollution control since 1974, *Water Pollution Control*, **84**; published by the Institute of Water Pollution Control, 1985.)

lower reaches where oxygen levels are lower. Although some species can survive periods of oxygen deficiency, they mostly avoid heavily polluted waters. High concentrations of ammonia in organic wastes are also toxic to fish. Similarly, suspended solids interfere with feeding efficiency and smother the spawning gravels as they settle out of the water. Rivers are essentially self-purifying systems but this is rarely achieved because of the varied and heavy usage they receive throughout their length. However, even severely polluted rivers can be restored with adequate conservation measures; such is the case with the River Thames (Gameson and Wheeler, 1977) and other rivers in the United Kingdom (Table 11.10).

A more permanent effect on the ecology of a river occurs when the the hydrologic regime is altered with the construction of reservoirs and similar projects (Table 11.11). The most significant change following the construction of a dam is the creation of an artificial lake, and the upstream lotic environment is gradually eliminated as the water floods the river valley. Important changes in the remaining parts of the upstream ecosystem have been reported where migratory fish are unable to pass the dam. Decomposition of the uncleared vegetation results in temporary anaerobic conditions and heavy silting may occur until the new shoreline is formed. The effect of the artificial lake on the downstream ecology depends on the design of the dam, as this determines the quality of water exported from the lake. If water is released from the bottom, as is the case with many hydroelectric installations, it is cold and low in dissolved oxygen. However, it may contain large amounts of bacteria and plankton and carry nutrients out of the reservoir along with hydrogen sulphide and other reduced compounds which can be toxic to aquatic organisms. Conditions are reversed when water is discharged over the top of a dam: the results are similar to a natural lake with nutrients being trapped in the reservoir and heat exported (Odum, 1971). The effect of a dam also varies according to the way in which the flow is regulated: rapid fluctuations in discharge, as is characteristic of power generation, and unusually low water conditions, such as occur with irrigation schemes, are the most disruptive for the biota. Dredging and straightening of channels also tend to have an adverse effect on aquatic communities in that they create unnaturally uniform

Table 11.11 Environmental effects of reservoir construction and channel modification (after Hellawell, 1986)

	Principal environmental effect	Potential ecological consequence
Reservoir construction and operation	Barrier to normal river flow	Prevention of upstream migration of fishes and some invertebrates; interference with downstream drift of plankton and invertebrates.
	Increased water depth	Loss of normal riverine community; inundation of adjacent land.
	Hydrologic regime altered	Sudden disharge to meet demands for power generation, reduced summer flow from irrigation dams and shoreline fluctuation during discharge and recharge of reservoirs can sweep away organisms or leave them stranded.
	Changes in the physical and chemical properties of the stored water (for dams discharging from bottom)	Reduced primary productivity downstream because nutrients held in reservoir; low dissolved oxygen levels can eliminate sensitive species; reduced pollution tolerance.
	Changes in the biological character of the stored water	Reservoir supports higher algal and zooplankton populations; potential for increased fish production; downstream drift supports filter feeders and collectors.
Channel modifications to increase flow	Changed physical dimensions; straightening; reduced roughness	Loss of microhabitats and reduced community diversity.
Bank modifications	Removal of trees; raising banks to reduce flooding	Loss of riparian habitat; increased light encourages growth of aquatic plants; loss of detrital input.
Channel maintenance	Removal of substrate by dredging, control of vegetation.	Habitat modification; removal of benthic fauna and flora; invasion of more resistant species.

(Source: J. M. Hellawell, *Biological Indicators of Freshwater Pollution and Environmental Management*; published by Elsevier Applied Science Publishers Ltd, 1986.)

Table 11.12 Nitrogen and phosphorus input to waterbodies from non-point sources (after Loehr *et al.*, 1989)

Non-point source	Total nitrogen load (kg ha^{-1} yr^{-1})	Total phosphorus load (kg ha^{-1} yr^{-1})
Agricultural regions		
Rural cropland	0.1–2.9	2.1–79.6
Pasture	0.1–0.6	3.2–14.0
Land receiving manure	0.8–2.9	4.0–13.0
Idle land	0.1–0.3	0.5–6.0
Feedlots	10.0–620.0	100.0–1600.0
Urban areas		
Residential	0.8–2.2	5.0–7.3
Commercial	0.1–7.6	1.9–11.0
Industrial	0.4–4.1	1.9–14.0
Forest	< 0.1–0.9	1.0–6.3
Atmosphere		
Forested regions	0.1–0.5	1.0–11.3
Rural–agricultural	0.1–1.0	10.5–38.0
Urban–industrial	0.3–3.7	4.7–25.0

(Source: S.-O. Ryding and W. Rast, eds, *Man and the Biosphere Series, Vol. 1 – The Control of Eutrophication of Lakes and Reservoirs*; published by UNESCO Publishing, 1989. Reproduced with permission from UNESCO Publishing, Paris.)

conditions (Hynes, 1970). Similarly, exotic species may extend their range along canals and other waterways: the introduction of the sea lamprey (*Petromyzon marinus*) and alewife (*Alosa pseudoharengus*) into the Great Lakes via the Erie and Welland Canals had a devastating effect on lake trout and other indigenous species. This resulted in an almost total collapse of the commercial fishery (Hartman, 1988). In tropical regions the creation of new lakes and waterways is notorious for spreading water-borne diseases such as schistosomiasis.

Most of the sediment formerly carried in the river is deposited in the reservoir, and turbidity decreases downstream. For example, the sediment load of the lower Nile averaged 134 million t yr^{-1} prior to the closure of the Aswan High Dam, but peak turbidity levels declined from about 3000 mg l^{-1} to less than 40 mg l^{-1} after 1964 (Latif, 1984). The silt that was carried in the turbid water during the flood season was an important source of nutrients in the delta and provided a source of detritus for the inshore sardine fishery of the Mediterranean Sea. The fishery has subsequently declined and fertilizers are now required to maintain agricultural production on the delta (Moss, 1980). However, fish production in Lake Nasser has steadily increased, but as with all artificial lakes the representative species are very different from the original riverine population. Riverine species are generally active swimmers and can maintain their position against the current, although they must also rest in slack water or behind obstacles to dissipate the lactic acid which rapidly accumulates in their tissues. Some are equipped with suckers to adhere to rocks or are flattened dorso-ventrally so that they can remain close to the river bed. Most species are associated with well-aerated water (Hynes, 1970). Tropical fish are also adapted to the marked seasonal changes in habitat conditions which occur in natural watercourses. In the dry

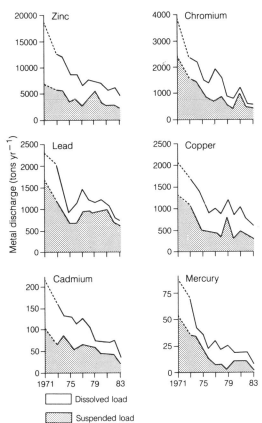

Figure 11.22 Decrease in average annual metal loads in the River Rhine at the German–Dutch border during the period 1971–83 (after Malle, 1985). Chemical analysis of suspended sediment ceased in 1983, but continued monitoring of total metal loads indicate that by 1991 concentrations of zinc had dropped to 1659 t yr^{-1}, chromium 277 t yr^{-1}, lead 277 t yr^{-1}, copper 360 t yr^{-1}, cadmium 5.5 t yr^{-1} and mercury 2.8 t yr^{-1}. (K.-G. Malle, personal communication.) (Redrawn with permission from K.-G. Malle, Metallgehalt und schwebstoffgehalt im Rhein II, *Zeitschrift fur Wasser und Abwasser Forschung*, 1985, **18**, 208.)

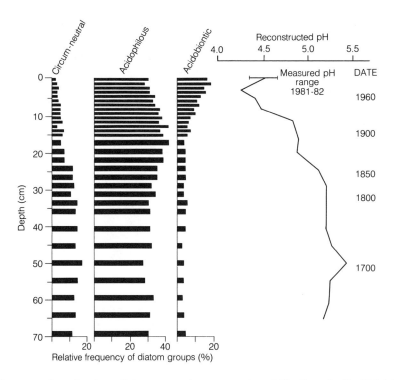

Figure 11.23 Changes in the proportion of various diatom groups in response to acidification of a Scottish lake. (After Battarbee *et al.*, 1985.) (Reprinted from R. W. Battarbee *et al.*, Lake acidification in Galloway: a palaeoecological test of competing hypotheses, with permission from *Nature*, **314**, 351, copyright 1985 Macmillan Magazines Limited.)

season they are confined to narrow channels and isolated pools and, because food is scarce, the condition of the fish is generally poor. As the river level rises the fish migrate upstream and out on to the floodplains where they spawn. The eggs hatch quickly and the fry feed on the increased supply of plankton and invertebrates that come with the flood. The fish are forced back to the main channel as the water level drops. Some survive the dry period by using up stores of fat. The lungfish (*Protopterus annectens*) aestivate in mucous coccoons, and the catfish (*Melapterurus* spp.), which is also a facultative air-breathing species, in burrows in the mud: other species depend on drought-resistant eggs (Moss, 1980).

Rivers are affected by all of the natural and anthropogenic processes that occur within their drainage basins. Chemicals emitted to the atmosphere through urban and industrial activities enter the water in ionic form or as dry fall-out. Soil erosion is a major source of phosphorus in agricultural areas and the volatilization of ammonia from feedlots adds nitrogen to the atmosphere. Farming can also contribute considerable nutrient pollution through wastewater drainage and leaching of fertilizers and other agricultural chemicals. Apart from feedlots, nutrient additions from non-point sources are comparatively low, especially in land which is not used for crop production (Table 11.12). Logging can similarly affect water quality and streamflow characteristics. At Hubbard Brook, New Hampshire, water temperature, stream flow, nutrient concentrations and sediment loads all in-

creased following deforestation (Borman and Likens, 1979). The addition of toxic pollutants to rivers through industrial processes or as pesticides and herbicide residues from agriculture and forestry is widely documented (Hellawell, 1986). The nature and character of these materials has changed over the years. Until quite recently most toxic substances, such as ammonia, phenols, tars and various metals, were associated with heavy industries. Evidence of the reduction of toxic emissions through better effluent treatment is seen in the water quality records of many rivers (Figure 11.22). However, the use of organic chemicals has risen considerably in recent years; many of these become concentrated in the food chain and the chronic long-term effects of these substances are receiving increased attention (Meybeck *et al.*, 1989).

Similarly, the wet and dry deposition of acidic substances such as SO_2 and NO_x which enter the atmosphere through combustion of fossil fuels has caused significant changes in many aquatic ecosystems. The diversity of the algal community typically declines in acidified waters and acid-tolerant macrophytes such as *Sphagnum* and *Juncus bulbosus* may increase in abundance (Stokes, 1986). Microbial activity falls sharply below pH 5.0: the decomposer bacteria are replaced by fungi and litter decays more slowly. Similarly, the plankton is increasingly dominated by acid-tolerant copepods and cladocerans such as *Bosmina* and *Diaptomus*, which replace *Daphnia* and other sensitive species (Mierle *et al.*, 1986). The invertebrate population also provides a good indication of pH conditions. Most snails and small mussels cannot survive below pH 6.0; freshwater shrimp (*Gammarus pulex*) are not found below pH 5.5; and mayflies and the freshwater louse (*Asellus aquaticus*) disappear from streams at pH 5.0. A threshold pH of 5.0 is also characteristic for most species of freshwater fish. Young migratory fish are especially vulnerable to the surge in acidity that generally accompanies snowmelt (Hesthagen, 1986): non-migratory species decline because high egg and fry mortalities result in recruitment failure, and their populations are dominated by older fish (Rosseland, 1986). A long-term record of acidification is provided by the diatom microfossils preserved in lake sediments: the method is based on the variation in pH tolerance between species (Battarbee, 1984). In Scottish lakes the reconstructed pH profile indicates an increase in acidity in the mid nineteenth century (Figure 11.23) and a period of rapid acidification from 1950 to 1970, followed more recently by a reduction in acid deposition (Battarbee *et al.*, 1988).

CHAPTER 12

Coastal and marine ecosystems

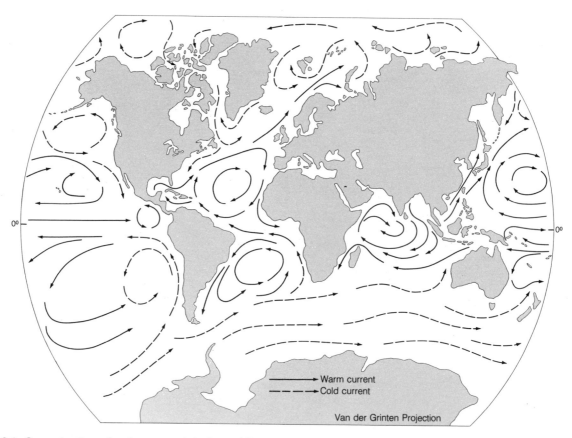

Figure 12.1 General pattern of surface currents in the world's oceans.

12.1 INTRODUCTION

The seas and oceans cover approximately 360 million km² or about 70% of the Earth's surface. The ocean basins are interconnected and currents generated by the wind keep the surface waters in continuous circulation (Figure 12.1). Deep water is brought to the surface in zones of upwelling where winds move the surface water away from the continents and further mixing occurs because of differences in water temperature and salinity. Although conditions are less variable than on land, the limits of tolerance of most marine organisms are comparatively narrow and their distribu-

tion is determined primarily by the interrelated effects of water depth, latitude and distance from shore (Barnes and Hughes, 1988). The average depth of the oceans is 3700 m, but in the deeper trenches the ocean floor is more than 10 000 m below the surface. Some organisms occur at these extreme depths but most species are found in the shallower water that surround the continents: these shallow waters represent about 8% of the total area of the oceans (Tait, 1972). Several zones are recognized in the open oceans on the basis of depth (Figure 12.2). The **epipelagic** zone extends from the surface to a depth of 200 m. The transition to the deep ocean environment occurs at the edge of the continental shelf and is marked

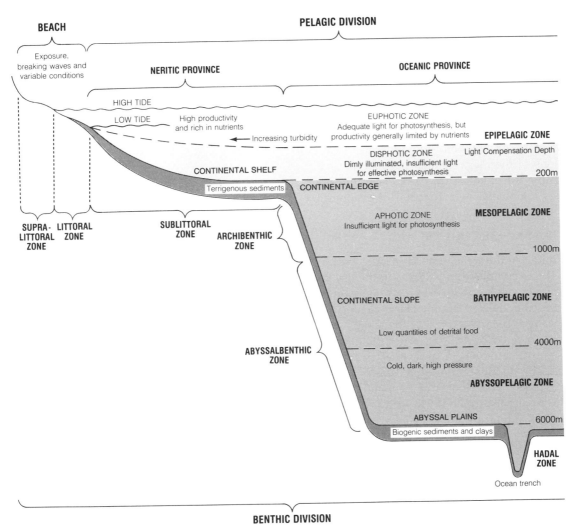

Figure 12.2 Principal divisions of the marine environment. (After Tait, 1972.) (Modified from R. V. Tait, *Elements of Marine Biology: An Introductory Course*; published by Butterworth and Co., 1972.)

by the 200 m depth contour. Diurnal and seasonal changes in illumination distinguish the epipelagic zone from the **mesopelagic** zone from 200 to 1000 m and the **bathypelagic** zone from 1000 to 4000 m, and the **abyssopelagic** zone below 4000 m. The deep ocean trenches form the **hadal** zone. In these deeper regions of the ocean the only source of light is bioluminescence generated by some of the organisms. Water temperatures become less variable with depth and turbulent mixing induced by wave action on the surface ceases at the bottom of the epipelagic zone. The open ocean is inhabited by planktonic organisms which float or drift in the water and by the more powerful swimming animals which comprise the **nekton**. The sedentary organisms that live on or in the seashore and seabed are known collectively as the **benthos**.

Differences in such factors as the duration and depth of tidal inundation and the nature of the substrate result in more varied conditions along the seashore. Attached algae able to resist the force of breaking waves dominate the littoral vegetation along rocky coastlines in temperate regions. Similar conditions in the tropics favour the development of coral reefs and atolls. Salt marshes develop in protected intertidal areas where mud and silts accumulate along temperate shorelines: in tropical regions such habitats support various species of mangrove. The most abundant plants on sandy foreshores are microscopic diatoms. Vascular plants such as marram grass establish on the dunes which form behind the beach. Equally distinctive communities are associated with the continually changing conditions associated with estuaries where fresh and saline waters mix at the mouths of rivers. The influence of the sea decreases with distance inland from the shoreline in all of these habitats and results in a pronounced zonation of organisms. Along slowly eroding rocky shorelines this zonation is essentially permanent but in salt marshes and other habitats where accretion is relatively rapid, such zonation represents successional change and is indicative of the dynamic character of the marine environment.

12.2 CHEMICAL AND PHYSICAL NATURE OF THE OPEN OCEANS

12.2.1 Salinity

Sea water contains dissolved solids and small amounts of gases in solution. Sodium, magnesium, calcium, potassium, chlorine and sulphur account for more than 90% of the total salts in solution. Although the relative concentrations of the major elements remain constant, the salinity (S, ‰) of the surface waters in the open ocean varies regionally, as a function of evaporation (E, cm yr^{-1}) and precipitation (P, cm yr^{-1}):

$$S‰ = 34.6 + 0.0175(E - P).$$

Surface salinities in the wet equatorial regions typically average < 34.5‰ compared to > 35.5‰ in the dry subtropical high pressure belt, with extremely high salinities of 40‰ recorded in the Red Sea. In polar regions salinity increases locally as sea water freezes and the salts are left behind, or is lowered by melting of ice. Similar variations in salinity occur where freshwater rivers drain into the sea. The distribution of organisms is often affected by the freshwater 'plumes' at the mouths of large rivers. For example, anadromous fish such as Arctic char (*Salvelinus alpinus*) migrate along the coast of the Beaufort Sea in the relatively warm (5–10 °C), slightly brackish (10–25‰) water which forms during the summer melt season (Craig, 1984). Fresh water is less dense than salt water and remains at the surface under these conditions. The density of sea water is also affected by temperature and increases until it reaches its freezing point at about – 2 °C. Denser water produced by evaporation or cooling at the surface therefore has a tendency to sink.

These convection-like currents are important for the circulation of the deep ocean waters which are beyond the influence of waves and winds. This downward flow originates mainly in the polar seas where the water is continually cooled by energy transfer to the atmosphere (Figure 12.3). Water pressure increases by about 1 atmosphere for each 10 m of depth: this also increases the density of the water, so that the cold, deep ocean waters are effectively isolated from the warmer surface layers.

12.2.2 Temperature

Sea surface temperatures vary with latitude and season and range from – 2 °C in polar regions to 28 °C in some parts of the tropics. Surface temperatures remain relatively constant at high and low latitudes but a seasonal range of 8–10 °C is not unusual in mid-latitude oceans. In tropical oceans the warm surface layer is separated from the cold, dense bottom waters by a permanent thermocline at depths of 100–500 m, below which temperatures slowly decrease towards the bottom (Figure 12.4). In temperate oceans a temporary thermocline develops during summer, usually at depths of 15–40 m, but disappears as the water cools in winter and convectional mixing occurs to depths of several hundred metres (Tait, 1972). A permanent thermocline is usually present at depths of 500–1500 m, although the temperature gradient is less pronounced than in tropical oceans. Temperatures in the polar seas vary only slightly with depth because the water is continually mixed as the surface layers cool and sink, although irregularities of 2–3 °C occur where warmer water flows polewards from the temperate oceans.

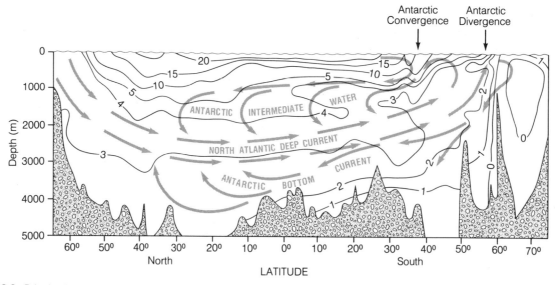

Figure 12.3 Principal subsurface currents and temperatures (isotherms in °C) in the western Atlantic Ocean. (After Tait 1972; Open University, 1989b.) (Sources: modified from R. V. Tait, *Elements of Marine Biology: An Introductory Course*; published by Butterworth and Co., 1972; also from Open University Course Team, *Seawater: Its Composition, Properties and Behaviour*, 1989, with permission from Pergamon Press Ltd., Headington Hill Hall, Oxford, OX3 OBW, UK.)

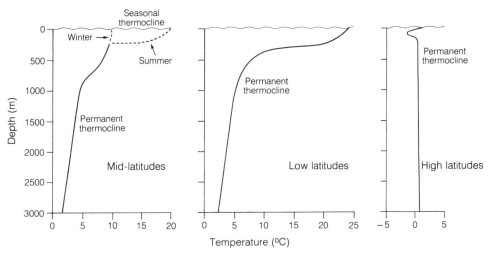

Figure 12.4 Temperature profiles in the deep oceans. (After Tait, 1972.) (Reproduced with permission from R. V. Tait, *Elements of Marine Biology: An Introductory Course*; published by Butterworth and Co., 1972.)

12.2.3 Dissolved gases

Sea water contains several dissolved gases including nitrogen, oxygen and carbon dioxide. Nitrogen is the most abundant with concentrations varying from 8 to 15 ml l^{-1}. In gaseous form nitrogen is utilized only by nitrogen-fixing bacteria such as *Clostridium* and blue-green algae such as *Trichodesmium*, which are especially abundant in tropical and subtropical waters (Goering *et al.*, 1966). Most of the nitrogen needed by marine organisms is derived from nitrates and ammonium absorbed by the phytoplankton in the illuminated surface waters. However, the average concentration of nitrate is only about 0.5 ppm, and ammonium is even less abundant. Nitrates are rapidly depleted and almost undetectable in many surface waters (McCarthy, 1980), and although they are brought back to the surface in areas of upwelling, it is estimated that the amount returned is less than losses through sedimentation (Fogg, 1982). Production in the sea is often limited by this shortage of nitrogen, unlike fresh-water ecosystems where phosphorus is usually in short supply. Nitrogen is especially limiting in tropical oceans where a permanent thermocline develops. In coastal waters the concentration of dissolved inorganic nitrogen is supplemented by river and groundwater inputs.

The oxygen content of sea water typically varies from 2 to 8 ml l^{-1}, depending on water temperature and biological activity. The concentration of dissolved oxygen increases at higher latitudes because the gas is more soluble in cold water. The density currents which flow out from the polar seas therefore carry oxygen-rich water into the deep ocean basins. Oxygen concentrations are also high at the ocean surface, where photosynthesis supplements oxygen dissolved from the atmosphere. Conversely, an oxygen minimum occurs at intermediate depths where the gas is removed by animal respiration and by bacterial oxidation of organic debris. Oxygen concentrations as low as 0.1 ml l^{-1} are characteristically recorded in this zone. The water may become completely anoxic in poorly ventilated basins such as the Black Sea. Hydrogen sulphide is formed under these conditions through the activity of sulphate-reducing bacteria, and the water is unusually rich in carbon dioxide (Riley and Chester, 1971).

Carbon dioxide is present in sea water at considerably higher concentrations than in the atmosphere but it is mostly present in chemically bound forms, mainly as bicarbonates of alkaline metals such as sodium, calcium and potassium. As in fresh water, the relative proportions of the different forms of inorganic CO_2 vary with pH (see Figure 11.6). Large quantities of CO_2 are utilized in photosynthesis, but as the gas is removed from sea water by plant growth, the equilibrium of the system is maintained by the dissociation of carbonic acid and subsequent conversion of carbonates to bicarbonates. Consequently, CO_2 is never in short supply under natural conditions. Some CO_2 is bound up as calcium carbonate in the shells and skeletons of marine organisms and may eventually be incorporated into sedimentary limestones. However, carbonates are redissolved at great depth because the acidity of the water increases under pressure (Riley and Chester, 1971). Consequently, carbonates are not conspicuous in sediments below about 6000 m.

12.2.4 Light

The radiant energy which enters the sea is absorbed by water, plant pigments or dissolved dye-like substances known as '**Gelbstoff**'. Most of the incident energy is converted into heat because of the strong absorption of infra-red wavelengths (Figure 12.5). Primary production

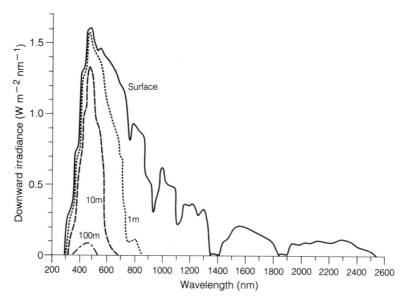

Figure 12.5 The spectrum of downward irradiance in the sea. (After Jerlov, 1976.) (Reproduced with permission from N. G. Jerlov, *Marine Optics*; published by Elsevier Science Publishers, 1976.)

is confined to the **euphotic zone** where light intensity exceeds 1% of maximum daylight. In the clearest oceans the euphotic zone extends to depths of about 100 m. Although there is insufficient light for plant growth below this depth, radiation in the blue wavelengths can be detected at depths of 800–900 m in the waters of the Mediterranean and Caribbean (Clarke and Denton, 1962). In coastal waters the depth of the euphotic zone usually ranges from 10 to 30 m and may fall below 3 m in turbid estuaries. Light attenuation in pure sea water is lowest in the blue-green wavelengths (425–450 nm) which corresponds to the absorption and action spectra of the chlorophylls and carotenoids that are present in all groups of marine algae (Halldal, 1974). Wavelengths of 525–575 nm are absorbed by fucoxanthin and peridinin found in diatoms and dinoflagellates, and by the phycoerythrins in red algae. Blue-green algal pigments, such as phycocyanin, absorb wavelengths of 535–650 nm, and chlorophyll is again sensitive to the longer 665–680 nm wavelengths. Gelbstoff mostly absorbs ultraviolet radiation and the shorter visible wavelengths. These substances are released from brown seaweeds and phytoplankton and are also contributed by rivers discharging into the sea. Gelbstoff is gradually removed from the oceans by photo-oxidation and microbial degradation, or is absorbed by particulates as they sink to the depths (Yentsch, 1980).

Many marine organisms are known to emit luminescent light at night or at depths where surface illumination is weak or absent. Phosphorescence is a common feature of tropical oceans: it is especially intense in small bays in the Caribbean where it is caused by nearly continuous blooms of the dinoflagellate *Pyrodinium bahamense* in the agitated water (Smayda, 1980). The luminous flash of blue light in dinoflagellates appears to deter predators. A similar escape mechanism is found in some crustaceans and squid, and the downward-directed photophores in the fish of the upper mesopelagic zone (200–700 m) make them less conspicuous against the faint background of light when viewed from below (Tait, 1972). In deep ocean species luminescence may serve as a mating signal or function as a lure to attract prey. Although bioluminescence occurs at all depths in the oceans, its intensity, frequency and duration vary as the populations change. For example, in the North Atlantic Ocean maximum bioluminescence during the night is recorded at depths of 100 m with over 160 flashes min^{-1}; a secondary peak of 90 flashes min^{-1} occurs at 900 m, decreasing to about 1 flash min^{-1} at 3750 m (Clarke and Hubbard, 1959). Several bioluminescent systems have been identified, all of which require some form of oxygen. In most animals bioluminescence is produced in specialized light organs or photopores, but in some fish symbiotic bacteria provide a continuous source of light that is regulated by screening pigments or opaque membranes. Typically the light is produced by direct oxidation through simple substrate–enzyme reactions, of the luciferin–luciferase type described for fireflies (Smith, 1989); this system is characteristic of the dinoflagellates and species such as the boring clams (*Pholas*) and crustaceans (*Cypridina*).

12.3 COASTAL WATERS AND LITTORAL HABITATS

The coastal waters extend from the edge of the land to the limits of the continental shelf. This is the **neritic** province, and includes the **littoral** zone along the shoreline and the shallow waters of the **sublittoral** zone which extend from the low tide-mark to a maximum depth of about 200 m (Figure 12.2). Conditions in the littoral zone are much more variable than in the open oceans; temperature, salinity and turbidity change rapidly because of the action of waves and tides. The nature of this very dynamic environment is determined by the properties of the parent rocks and the type of geomorphic processes which are acting on them. Rocky shorelines develop where resistant materials are exposed to the the powerful force of breaking waves. The energy of the waves is concentrated on the headlands by refraction once they reach shallow water (Figure 12.6). Most of the energy is expended as the waves break offshore, but the water still collapses with considerable force and most of the loosened rock is swept away. Only those organisms which are firmly attached to the substrate remain in these environments.

Sandy beaches usually form in sheltered coves between the rocky headlands and at other points along a coast where waves and currents allow materials to accumulate. Here the typical shoreline consists of the beach, a surf zone and the shallow waters in which bottom sediments are kept in motion by longshore transport processes: wind-driven sand often accumulates as dunes on the landward side of the beach. The form of the beach reflects the size and frequency of the waves. Beaches tend to build up during periods of relative calm as sand is pushed onshore, whereas larger storm waves carry material out to sea where it is deposited as offshore bars. Because the underwater slope is usually fairly gentle, most of the wave energy is dissipated in the shallow water. The spilling breakers advance up the beach as a turbulent sheet of water, then drain back to the sea. Considerable amounts of sand are moved down the beach in this way. The surface of a sandy beach is continually disturbed by wind and waves and, except in unusually sheltered areas, is not a suitable habitat for attached organisms. Plants are therefore rare on sandy beaches, although many species have adapted to the peculiar conditions in the adjacent dunes. Conversely, many animals live buried in the sand where conditions are relatively stable.

Mud flats from where there is negligible wave action and currents are weak. Mostly they occur in estuaries where the fine silts and clays carried by rivers settle out in areas with little tidal flow. The largest estuaries are associated with low-relief coastal areas and represent regions where fresh water mixes with sea water. Pronounced longitudinal gradients occur within estuaries. Salinities are highest at the seaward end, although fresh water tends to flow over salt water because it is less dense. Consequently, there is a net outflow of water at the surface: this is compensated by a net inflow of salt water near the bottom, although the degree of stratification varies according to the size and shape of the estuary

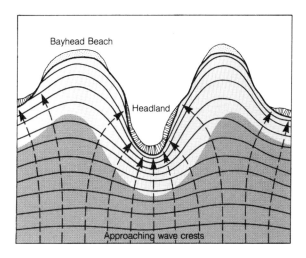

Figure 12.6 Wave refraction along an embayed coastline. (After Strahler and Strahler, 1992.) (Reproduced with permission from A. H. and A. N. Strahler, *Modern Physical Geography*, 4th edn, published by John Wiley and Sons Inc., copyright © 1992.)

and the discharge of the river (Pritchard, 1955). Little light penetrates below 1–2 m in these turbid waters and primary production is often limited to a fraction of the water column. Salt marshes are the characteristic communities in the intertidal reaches of temperate estuaries, and in the tropics mangroves are the dominant plants. Species distributions are related to the microtopography of the mud flats, and vertical zonation is just as pronounced as on the rocky shores and sandy beaches.

The sublittoral zone comprises the shallow sea floor associated with the continental shelf, and extends from the low tide-mark to depths of about 200 m. The average width of the continental shelf is 75 km and accounts for about 8% of the area of the world's oceans (Figure 12.7). However, the present shelf ecosystems are of recent origin since most areas were periodically exposed by changes in sea level during the Pleistocene. The continental shelf is for the most part covered with terrigenous deposits eroded from the land and carried to the sea by rivers. The sands, silts and clays are irregularly distributed. Some areas show a progressive decrease in grain size away from the shore but more commonly the coarsest materials are found on the outer shelf, where they were deposited by rivers during glacial periods (Guilcher, 1963). Following the retreat of the shoreline, deposition on the outer shelf is presently very low. Deposition rates of $1–100 \text{ cm } 100 \text{ yr}^{-1}$ are reported nearer to shore depending on the nature and intensity of terrestrial geomorphic processes. The highest rates occur where large rivers carry heavy loads of sediment to the sea. Terrigenous accumulation is slow in drier areas and the deposits are more commonly composed of organic calcareous materials and fine sediments contributed by the wind. In colder regions much of the debris is carried by glaciers

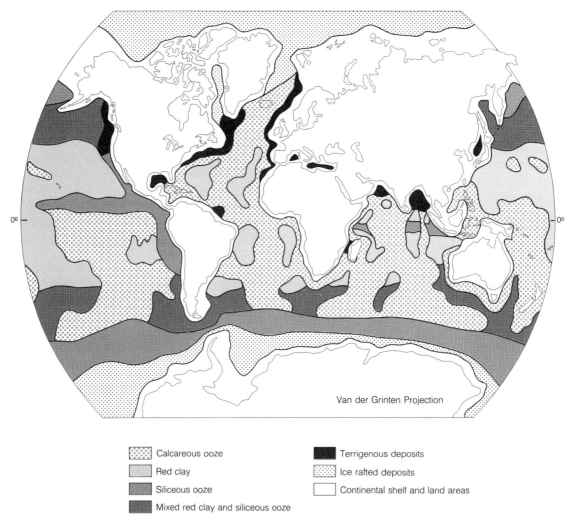

Figure 12.7 Distribution of the principal types of pelagic sediments. (After Open University, 1989a.) (Reprinted from Open University Course Team, *Ocean Chemistry and Deep-sea Sediments*, 1989, with permission from Pergamon Press Ltd., Headington Hill Hall, Oxford, OX3 OBW, UK.)

Legend:
- Calcareous ooze
- Red clay
- Siliceous ooze
- Mixed red clay and siliceous ooze
- Terrigenous deposits
- Ice rafted deposits
- Continental shelf and land areas

and is typically a poorly sorted mixture of stony materials, sands and muds. Some areas are kept free of sediment by strong tidal currents: in various places in the English Channel, for example, the seabed is formed of exposed bedrock.

12.4 PELAGIC SEDIMENTS

The terrigenous shelf deposits contrast with the pelagic sediments of the deep ocean. The pelagic sediments are mostly fine textured organic materials composed of the skeletal fragments of planktonic organisms (Figure 12.7). Calcareous oozes derived from the shells of foraminifera and pteropods are the most widespread on the abyssal plains to depths of about 4500 m. Below this depth hydrostatic pressure increases the solubility of calcium carbonate and the deposits contain a higher proportion of silica. The characteristic deposits in tropical

oceans at depths of 4500–8000 m are siliceous oozes comprised of radiolarian skeletons. Siliceous deposits also occur around Antarctica and in the North Pacific; they are associated with areas of high diatom productivity. The sediments in the deepest parts of the ocean are referred to as inorganic 'red clays'. They are composed of silica and aluminium oxide with iron, manganese and other minerals often present in the form of nodules (Arrhenius, 1963). The inorganic deposits accumulate at a rate of 1–2 mm $1000 \, yr^{-1}$. Their distribution is restricted to the deep ocean areas where organic production is low and most of the biogenic material is dissolved as it sinks through the water. A sedimentation rate of 1–5 cm $1000 \, yr^{-1}$ is typical for biogenic sediments, depending on the productivity of the ocean. These sediments also provide a record of environmental change. The relative dominance of warm and cold water species of pelagic foraminifera in ^{14}C-dated cores have been used to determine temperature conditions during the Pleistocene

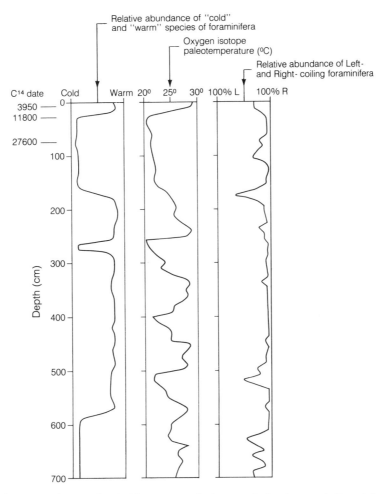

Figure 12.8 Pleistocene paleotemperatures estimated from oxygen isotope analysis, changes in the relative abundance of warm and cold water planktonic Foraminifera and the ratio of left- and right-coiling shells of *Globorotalia truncatulinoides* in Atlantic deep-sea sediment cores. (After Ericson *et al.*, 1961.) (Reproduced from D. B. Ericson *et al.*, Atlantic deep-sea sediment cores, *Geological Society of America Bulletin*, 1961, **72**, 267, 277.)

(Emiliani and Flint, 1963). Similar analyses use the ^{18}O:^{16}O ratio in the shells (Emialini, 1955). The ratio depends on the temperature of the sea water during the period of shell formation, with ^{16}O increasing during colder glacial periods and ^{18}O increasing during the warm interglacials (Figure 12.8).

12.5 PLANT LIFE IN MARINE ENVIRONMENTS

Plant life is confined to the illuminated surface waters of the ocean. In shallow coastal waters the dominant marine plants are attached algae: aquatic vascular macrophytes increase in importance in salt marshes and other shoreline environments which are periodically flooded by the tide. Attached plants are restricted by light requirements to a maximum depth of about 120 m. The brown algae (Phaeophyceae) are the most conspicuous. They are almost entirely marine in distribution and are mainly

associated with rocky shorelines. Included in this group are the large kelps, such as *Macrocystis* and *Sargassum*, which grow to lengths of 50 m and form dense subtidal forests in tropical and subtropical regions. The red algae (Rhodophyceae) are the most widely distributed of the larger marine plants. They occur most abundantly in the tropical oceans where some species grow to depths of 120 m. Only the deep-water species are bright red in colour. Others, such as *Porphyra*, grow high in the intertidal region where they become dry and brittle when they are exposed to the atmosphere for several days. These plants are usually dark green or black in colour. Heavy calcification of the outer cell wall occurs in some genera of red algae. They are important in the formation of coral reefs, and the cementing action of the algae binds together the calcareous remains of the organisms (Scagel *et al.*, 1965). Most of the large algae are classed as haptophytes in that they are attached to bedrock or large-calibre substrate materials, although sea lettuce (*Ulva* sp.) and other green algae (Chlorophyceae) are rhizophytes and typically root in fine muds and silts. Submerged vascular plants are represented by monocotyledenous species such as eel grass (*Zostera marina*). They also grow in fine sediments which are less disturbed by water movements. Most angiosperms are restricted to salt marshes and estuaries or the specialized dune and cliff environments.

Some macrophytes, such as *Sargassum*, remain metabolically active even after they become detached from the substrate, and seedlings of *Rhizophora* and other mangroves are readily dispersed by ocean currents. Plants such as these are relatively uncommon and the characteristic species in offshore waters are diatoms, dinoflagellates and other microscopic algae. Diatoms are widely distributed in all marine environments but are especially abundant in colder, nutrient-rich coastal waters. Most diatoms are free-floating planktonic species but some are also present on the surface of mud flats, in salt marshes and attached to other algae and aquatic plants. They occur either as individual cells, ranging in size from 0.01 to 0.2 mm, or are loosely connected by mucilaginous pads into filamentous chains. The diatoms generally exhibit two peaks of abundance. One occurs in the late spring as water temperatures begin to rise and the amount of sunlight increases; a second less intense bloom occurs in early autumn as nutrients are returned to the surface.

The dinoflagellates are second only to diatoms as primary producers in the sea: species of *Peridinium* and *Ceratium* are most abundant. They are also present in brackish water and in beach sand. Some occur as parasites in fish and copepods or as symbionts in the tentacles of sea anemones (Scagel *et al.*, 1965). Reproduction in dinoflagellates is by cell division, and under optimum conditions the concentration of dinoflagellates may become so dense that the water is discoloured. High densities of *Gymnodinium* and *Gonyaulax* are responsible for the '**red tides**' that are reported in some coastal waters and which can result in the death of many marine animals (Brongersma-Sanders, 1957). In a related phenomenon known as '**mussel poisoning**', a nitrogenous toxin accumulates in the tissues of filter-feeding invertebrates (Schantz, 1960). The toxin, which is gradually eliminated over a period of months, can cause paralysis and occasionally death when the shellfish are eaten by humans or other mammals. Extensive fish kills in the Indian Ocean are caused by similar blooms of the blue-green algae *Oscillatoria*. In some cases the fish may be suffocated because their gills become clogged with the filamentous algae. Alternatively, the nocturnal respiratory demands of the algae in the absence of photosynthesis may cause a critical reduction of oxygen in the water. Other blue-green algae, such as the red-pigmented species *Trichodesmium erythreum*, can also discolour the water: the Red Sea is so named because of periodic blooms of this species.

Other planktonic producers include the algae-like **coccolithophorids** and **silicoflagellates**: they are closely re-lated to the diatoms. The coccolithophorids are distinguished from other golden algae by the presence of heavily calcified discs and plates embedded in the cell walls; an elaborate internal siliceous skeleton is characteristic of the silicoflagellates (Scagel *et al.*, 1965). Motile coccolithophorids are abundant in the nanoplankton of the North Sea, and some like *Emiliani huxleyi* form massive blooms in Norwegian coastal waters from mid to late summer (Smayda, 1980). They also occur extensively in warm tropical waters. Species which lack photosynthetic pigments are reported at depths of 2500 m in the Mediterranean Sea. These colourless species ingest particulate organic matter and generally grow more rapidly than those capable of photosynthesis (Provasoli, 1963). The silicoflagellates are also small, single-celled organisms. They are mostly associated with cold, nutrient-rich waters such as the zone of continuous upwelling in the Antarctic.

12.6 FACTORS CONTROLLING MARINE PRIMARY PRODUCTION

12.6.1 Light requirements of marine phytoplankton

Net photosynthesis in the deep sea ceases at depths where light levels fall to about 1% of incident surface illumination, although viable phytoplankton may descend deeper than this for part of the day. Changing weather patterns cause significant variations in daily rates of production and seasonal differences become more pronounced with latitude (Figure 12.9). The amount of light available to the plants varies with the transparency of the water. The highest rates of production occur where all incident radiation is absorbed by phytoplankton, but as productivity increases a greater proportion of the incident light energy is intercepted by dead organisms and other suspended debris. In tropical oceans most of the light is absorbed by the water and the clear blue colour is indicative of low productivity. Here the light compensation depth occurs at about 100–120 m, compared with 25–50 m in more turbid temperate and high-latitude oceans. The distribution of phytoplankton varies with light conditions. The pigment systems of dinoflagellates are sensitive to low light intensities and these organisms are usually most abundant near the bottom of the euphotic zone. Similarly, maximum photosynthesis in marine cyanobacteria is reported at depths where little light penetrates, or when light intensities are reduced at the beginning and end of the day. Planktonic forms of green algae are also most common in deeper water. The diatoms tolerate a wider range of light conditions but are generally most abundant in the surface waters (Raven and Richardson, 1986).

Full sunlight inhibits photosynthesis in most marine phytoplankton: maximum primary production usually occurs some metres below the surface. However, the

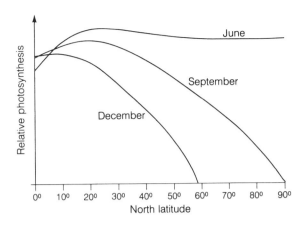

Figure 12.9 Seasonal differences in relative primary production in northern hemisphere oceans as a function of latitude. (After Ryther, 1963.)

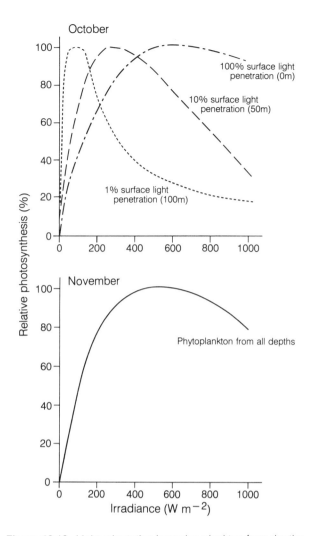

Figure 12.10 Light adaptation in marine plankton from depths of 0, 50 and 100 m in the Sargasso Sea. Light intensities at these depths are equivalent to 100%, 10% and 1% of surface irradiance. In October (upper) the water column is stratified, and deep-water populations from below the thermocline function as 'shade' plants and become light saturated below 400 μE PAR m^{-2} sec^{-1}. In November (lower) when the thermocline breaks and the water column mixes the phytoplankton from all depths function as 'sun' plants. (After Ryther and Menzel, 1959.) (Reproduced with permission from J. H. Ryther and D. W. Menzel, Light adaptation by marine phytoplankton, *Limnology and Oceanography*, 1959, **4**, 495.)

relationship between photosynthesis and light intensity is not constant over time. Phytoplankton adapt physiologically to ambient light levels by varying the amount of pigmentation or the ratio of the different pigments present in each cell. For example, in some marine diatoms the chlorophyll *a* content increases at lower light intensities (Chan, 1978), and in some dinoflagellates there is a substantial increase in the light-harvesting pigment peridinin (Prézelin, 1976). Similarly, the spectral quality of light in the underwater environment can also affect pigment content and chloroplast structure in marine phytoplankton. In diatoms such responses are triggered by exposure to weak blue-green light, and so increases the viability of non-motile species which sink below the surface of the oceans (Vesk and Jeffrey, 1977). Such depth-related adaptations are especially characteristic of thermally stratified oceans: under these conditions the phytoplankton from deeper parts of the euphotic zone become light-saturated under strong irradiance and their photosynthetic rate declines (Figure 12.10).

12.6.2 Nutrient requirements of marine phytoplankton

All of the chemical elements required by marine phytoplankton occur naturally in sea water. Productivity is never limited by elements such as potassium, calcium and magnesium, which are present in relatively high concentrations. Similarly, there is an adequate supply of trace elements. However, even in naturally productive areas of the ocean the concentration of nitrogen is considerably lower than on land, and productivity in marine ecosystems is generally limited because of this. Not only is nitrogen progressively removed by nutrient uptake, but the depth of the euphotic zone is also reduced as the density of phytoplankton increases and any nutrients

remaining in the water can no longer be utilized (Ryther, 1963). The principal form of available nitrogen in sea water is nitrate and concentrations are typically highest in deeper water, where it is released from organic detritus by bacterial oxidation (Figure 12.11). Phytoplankton production is very dependent on rapidly regenerated nitrogen, with nitrates released from the sediment contributing only 10–20% of the requirement (Eppley and Peterson, 1979). Nitrite is present at much lower concentrations and is usually concentrated near the base of the euphotic zone, where it is produced by phytoplankton growing under low light intensities. Ammonium, derived

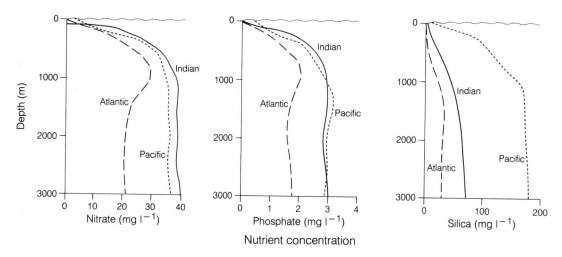

Figure 12.11 Vertical distribution of nitrate, phosphate and silicate in the Atlantic, Pacific and Indian Oceans. (After Sverdrup *et al.*, 1942.) (From H. U. Sverdrup, M. W. Johnson and R. H. Fleming, *The Oceans: Their Physics, Chemistry and General Biology*, © 1942, renewed 1970, pp. 241, 242, 245; reprinted by permission of Prentice Hall, Englewood Cliffs, New Jersey.)

from decomposition of organic material, urea from animal excretion and other dissolved organic nitrogen compounds, is also utilized as a source of nitrogen in varying amounts by phytoplankton (McCarthy, 1980). In addition, dissolved nitrogen gas is fixed by some species of blue-green algae but only about 0.2% of the nitrogen used in primary production enters the oceans in this way (Fogg, 1982).

Inorganic phosphorus is less abundant in sea water than nitrogen. The ratio between these elements remains essentially constant and is proportional to the rates at which they are taken up by phytoplankton. Although there is a general correspondence between phosphorus concentrations and phytoplankton productivity, phosphorus is usually not considered to be a growth-limiting factor in marine ecosystems because the supply of nitrogen is exhausted more rapidly (Riley and Chester, 1971). Some phosphate is regenerated in the surface waters through zooplankton excretion. Turnover times of 10–40 days are reported for phosphorus in the North Pacific compared with 3–5 days for ammonium and 10–15 days for urea-N (Eppley *et al.*, 1973). The growth of diatoms and other phytoplankton having silicified skeletons may be limited by the concentration of dissolved silicon. Thus the blooms of diatoms which occur in zones of upwelling end as silicon is depleted: the seasonal succession from thick-walled to thin-walled species of diatoms in Antarctic waters is similarly attributed to reduced silicon concentrations (Paasche, 1980). Vitamins are also required by many diatoms, dinoflagellates and other marine algae. Vitamin B_{12} is the most important of these substances and is mostly synthesized by bacteria. It is present in the euphotic zone at concentrations of 0.2–2 ng l^{-1}, but like other nutrients exhibits seasonal and regional variations. For example, in the Sargasso Sea concentrations increase during winter, reaching a max-

imum in April at the start of the spring diatom bloom. Primary production drops rapidly as the supply of vitamin B_{12} is depleted. The diatoms release vitamin B_1 (thiamine) and so condition the water for *Emiliani huxleyi*, a coccolithophore which requires only thiamine (Menzel and Spaeth, 1962).

12.7 NUTRIENT REGENERATION IN THE DEEP OCEANS

The supply of nutrients in the surface waters is depleted when the phytoplankton die and sink from the euphotic zone or are passed along the food chain. The amount of organic debris that is incorporated into the bottom sediments is relatively small. In the Sargasso Sea the average rate of deposition is only 46 mg m^{-2} day^{-1} on the abyssal plains at a depth of 5360 m, and is comprised mainly of fine-grained skeletal remains. Organic carbon accounts for 3% of the detritus: this is equivalent to 0.5% of the total carbon production in the region (Honjo, 1978). Considerably more particulate organic matter is remineralized by the microbial population in the water column and returned to the surface by vertical mixing. Microbial activity is normally greatest at the upper boundary of the thermocline and generally coincides with the zone of maximum bioluminescence (Sorokin, 1981). A similar layer exists within the immediate surface waters, and at greater depths where detritus is carried down by deep currents such as the Antarctic Intermediate Water. This layered vertical distribution is a permanent feature of tropical oceans but develops seasonally in temperate waters. Although the microbes are individually very small (0.3–0.8 µm), 20–30% of the marine population normally form aggregates greater than 5 µm in size, and some may develop into visible flakes of 'marine snow' 2–7 mm in size. In this way they become available to the coarse filter-feeders that are normally unable to ingest single microbial cells. The average standing stock of detritus in the open ocean is 300–400 g m^{-2}: it is mainly composed of the remains of organisms and faecal pellets, although colonising bacteria and protozoa account for about 5% of the detrital biomass (Sorokin, 1981).

The downward flux of particles through the thermocline depends on their size and density and the viscosity and stability of the medium. Sinking rates are further affected by the feeding activity of organisms which migrate vertically in the water column. A sinking rate of 43–95 m day^{-1} is reported for marine snow in surface waters (Shanks and Trent, 1980), compared with rates of 20–2700 m day^{-1} for faecal pellets from various marine organisms (Angel, 1984). The detritus is rapidly colonized by bacteria and these are subsequently consumed by ciliate protozoans. The material begins to fragment within a few days. This alters the rate of sinking and the downward flux of fine particles typically drops from about 30% of primary production through the thermocline to less than 5% below 400 m.

Bacteria account for a large proportion of the microheterotroph production in the oceans, equivalent to 10–30% of primary production (Fenchel, 1984). It is estimated that they consume from 20 to 40% of total fixed carbon in the form of phytoplankton exudates and leachates from dead cells as well as from detrital material. Bacterial densities in oceanic waters remain fairly constant over time, which suggests that production and consumption rates are roughly equivalent. Some are recycled through lysis of cells but the majority are lost through the grazing activity of nanoplanktonic flagellates. These in turn are consumed by ciliates, rotifers and small crustaceans.

Detritus that sinks through the water column and is incorporated into the bottom sediments provides the food source for many benthic organisms and ultimately passes to the bacteria. In this way nitrogen, which is generally considered to be limiting in marine ecosytems, is regenerated and may be returned to the pelagic zone by turbulent mixing. Nitrate returned from the depths, together with nitrogen added by fixation, has been termed '**new nitrogen**' (Dugdale and Goering, 1967), as opposed to '**regenerated nitrogen**' that is recycled within the euphotic zone itself. The productivity of ocean waters is related to the ratio of new to total nitrogen and values range from 0.06 in oligotrophic waters of the North Pacific central gyre to 0.66 for productive upwelling waters off Peru (Olson, 1980).

Intense bacterial activity occurs around the deep-sea hydrothermal vents where unusual ecosystems are supported by chemosynthetic primary production (Grassle, 1986). Hydrothermal vents are associated with volcanic activity along the mid-oceanic ridges and are especially characteristic of the Galapagos Rift and other regions in the eastern Pacific where sea floor spreading is occurring at rates of 6–18 cm yr^{-1}. The fluids emanating from the vents are rich in biologically important compounds with greatly elevated concentrations of calcium, silicates and large amounts of hydrogen sulphide. The reduced inorganic substances in the hydrothermal fluids are the source of energy for primary production. Most bacteria isolated from these environments oxidize sulphur but others oxidize managanese, hydrogen and methane or use CO_2 as a carbon source. The most common microorganisms are *Thiothrix*-type or *Leucothrix*-type bacteria which grow as microbial mats on the glassy basalts or on the shells of bivalves. *Methanococcus jannaschii*, a recently discovered species of methane-producing bacteria, has an optimum growth temperature of 85 °C and other species may remain active at temperatures as high as 250 °C. Productivity of micro-organisms in the vent fluids ranges from 9.5 to 29.4 µg C g^{-1} hr^{-1}, depending on temperature. In the Galapagos region productivity in the flourishing communities is 2–3 times that in the surface waters.

12.8 PRODUCTIVITY IN MARINE PELAGIC ENVIRONMENTS

12.8.1 Regional productivity of the deep oceans

The rate at which nutrients are returned to the surface is controlled by physical processes and these are partly reflected in regional differences in marine productivity (Figure 12.12). Productivity is low in most tropical oceans because of the slow upward diffusion of nutrients through the thermocline: in these regions the growth of phytoplankton is essentially controlled by turnover rates within the euphotic zone. Production rates are generally 50–150 mg C m^{-2} and remain more or less constant throughout the year. The highest production is recorded where nutrient-rich water is brought to the surface in equatorial regions of semi-permanent upwelling (Figure 12.13). This effect becomes more pronounced towards the eastern side of the ocean basins where nutrients are entrained from coastal waters. Similarly, productivity in the Sargasso Sea is periodically enhanced by the influx of nutrient-rich water which originates as pinched-off meanders in the Gulf Stream (Blackburn, 1981). Deep mixing during severe storms may also bring up nutrients from below the thermocline.

Productivity is also low in Arctic waters, mainly because of light and nutrient limitations (Dunbar, 1970). Annual production ranges from 10 to 25 g C m^{-2} yr^{-1} north of 70 °N, and decreases to as little as 1–5 g C m^{-2} yr^{-1} under the ice cap (Nemoto and Harrison, 1981). The increase in day length at high latitudes results in prolonged periods of illumination during the summer, followed by several months of total darkness. A considerable amount of energy is lost through reflection

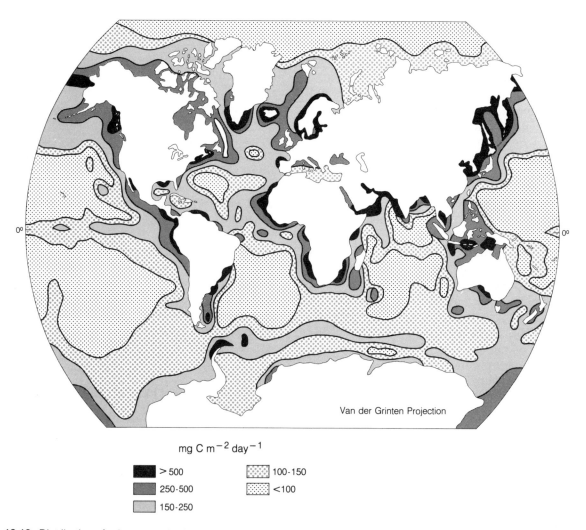

mg C m^{-2} day^{-1}

■ > 500   ▨ 100-150

▨ 250-500   ▫ <100

▨ 150-250

Figure 12.12 Distribution of primary production in the world oceans. (After Koblentz-Mishke *et al.*, 1970.) (Adapted with permission from *Scientific Exploration of the South Pacific*, ed. W. S. Wooster, National Academy of Sciences, 1970. Courtesy of the National Academy Press, Washington, DC.)

because of the low solar angle, or is absorbed by the snow-covered sea ice which covers as much as 60% of the Arctic Ocean during summer. Phytoplankton begin growth under the ice with pronounced algal blooms occuring in open-water areas as the ice recedes (Digby, 1953). Most of the production occurs within the upper 10 m of the water column (Grainger, 1979), but considerable numbers of microalgae are also present in the **epontic** flora at the base of the sea ice. The dominant Arctic ice algae are pennate diatoms such as *Nitzschia* and *Navicula*. They are completely frozen into the ice during winter and lose their pigmentation but become active again in the spring. The ice algae are often present at much greater densities than in the water column (Meguro *et al.*, 1967). They provide a source of food for organisms such as polychaete worms and copepods that also live in the ice as well as those that feed below the surface as the ice disintegrates. The salinity of Arctic surface waters is comparatively low because of seasonal ice-melt and river

discharge into this relatively enclosed ocean basin. The stability of the water column is maintained by this low-density surface layer, but some circulation occurs because of intrusions of warm water from the Atlantic Ocean (Sverdrup *et al.*, 1942). Nutrients returned to the surface waters in this way accumulate during winter but are rapidly depleted over the brief summer (Figure 12.14).

In contrast, the Antarctic is traditionally noted for its high productivity as a result of the continuous upwelling of nutrient-rich water around this continent (Walsh, 1971). Even during periods of peak phytoplankton growth there is an adequate supply of nitrates, phosphates and silicates, although limiting concentrations of vitamin B$_{12}$ are reported in some areas (Carlucci and Cuhel, 1977). Estimates of primary production in the region vary from 16 to 100 g C m^{-2} yr^{-1} (El-Sayad, 1978). Productivity is highest in the coastal waters and around the sub-Antarctic islands, and can exceed 3.6 g C m^{-2} day^{-1} (Mandelli and

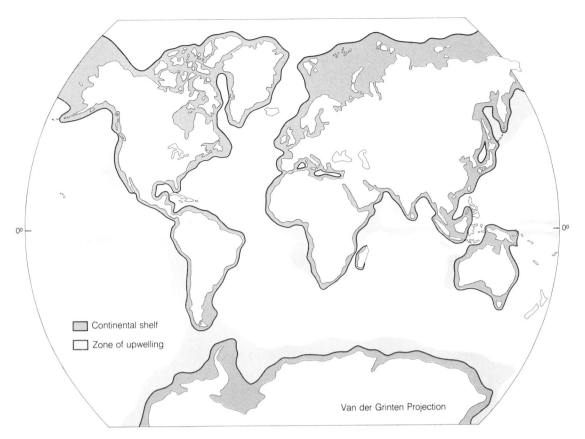

Figure 12.13 Limits of the continental shelf and main areas of upwelling in the world oceans. (After Wright, 1977.) (Zones of upwelling reprinted from J. B. Wright (ed.), *Oceanography*, copyright 1977, with kind permission from Elsevier Science Ltd., The Boulevard, Langford Lane, Kidlington, OX5 IGB, UK.)

Burkeholder, 1966). However, these environments represent less than 6% of the area of the southern oceans; average productivity over most of the region is only $0.13 \text{ g C m}^{-2} \text{ day}^{-1}$. The growing season is limited by the short austral summer to a period of 3–4 months and during this time the average depth of the euphotic zone is 78 m. Assimilation rates are generally highest at depths of 10–20 m where light intensities are 25–50% of maximum surface values. This is attributed to photoinhibition, and production within the water column is correspondingly very sensitive to fluctuating light conditions (Holm-Hansen *et al.*, 1977). In some species photosynthesis continues until light intensity falls to 0.01% of maximum incident levels: these organisms are mostly associated with the bottom layers of sea ice but can also account for some of the production measured in very deep water. Light is especially important in areas of instability where currents carry the surface water below the light compensation depth. The removal of phytoplankton from the zone of optimum illumination before they attain maximum growth may account for the unusually low productivity reported in some Antarctic waters (Hasle, 1956).

Phytoplankton production in temperate oceans is strongly related to seasonal variations in nutrient supply (Figure 12.15). Nutrients are brought up by convectional

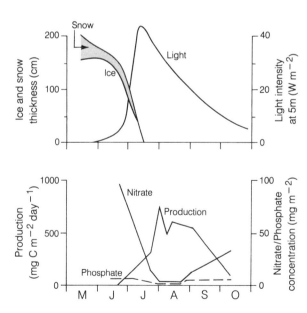

Figure 12.14 The characteristic cycle of phytoplankton production and associated environmental parameters in the waters of Frobisher Bay in the Canadian Arctic. (After Grainger, 1979.) (Reproduced with permission from E. H. Grainger, Primary production in Frobisher Bay, arctic Canada, in *Marine Production Mechanisms*, ed. M. J. Dunbar; published by Cambridge University Press, 1979.)

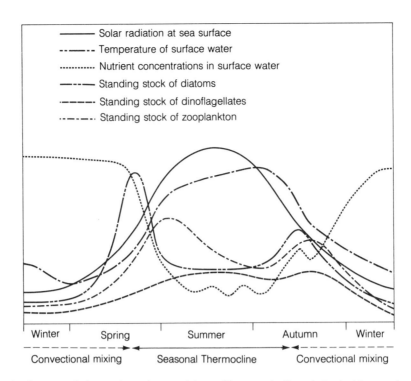

Figure 12.15 General cycle of seasonal change in environmental conditions controlling phytoplankton production. (After Tait, 1972.) (Reproduced with permission from R. V. Tait, *Elements of Marine Biology: An Introductory Course*; published by Butterworth and Co., 1972.)

mixing during the winter and gradually accumulate in the surface water because growing conditions are unfavourable for phytoplankton. At this time of the year growth is mainly limited by low light intensity and most phytoplankton are below the critical depth for effective photosynthesis. Production increases quickly in the spring as water temperatures rise and radiation becomes more intense, but later the water column becomes stabilized with the development of the seasonal thermocline. Mixing during the summer is restricted to the action of winds and waves. The phytoplankton remain in the well-illuminated surface waters and the available nutrients are soon depleted. Consequently, production declines over the summer until the thermocline breaks in the autumn and nutrients are returned to the surface by convectional mixing. The effect is short-lived as growing conditions deteriorate with the approach of winter. The seasonal cycle of phytoplankton growth in temperate oceans varies regionally with hydrographic conditions (Colebrook, 1982). The start of the growing season is closely correlated with the difference in water temperature between spring and summer. This is a measure of the stability of the water column and is therefore affected by depth and latitude. Thus the growing season in the Atlantic Ocean begins later and becomes progressively shorter in the deeper northern waters (Figure 12.16).

The density of zooplankton increases as more food becomes available and typically peaks in early summer. The delay following the spring phytoplankton bloom

occurs because zooplankton reproduction and growth is initially limited by low temperatures: this time-lag consequently increases with latitude (Heinrich, 1962). As growth of phytoplankton becomes progressively limited by nutrient availability, a greater proportion is lost to grazing. For example, in Swedish fjords, consumption of phytoplankton increases from 3% to 60% of daily production as the season progresses (Båmstedt, 1981).

12.8.2 Productivity in coastal waters

Productivity is higher over the continental shelves than in the deep ocean and ranges from 200 to 1000 g C m^{-2} yr^{-1} (Walsh, 1981). This high productivity is attributed to greater nutrient recycling in the shallow water as a result of wind-induced mixing. Upward movement of water occurs where near-shore winds produce surface currents that move away from the coast. Nutrient-rich water is continually brought up to the euphotic zone and optimum conditions are maintained for a longer period. The principal areas of coastal upwelling occur along the eastern margins of the Pacific and Atlantic Oceans: here strong currents are generated by the winds that blow out of the subtropical high-pressure cells towards the equator. The Peru Current causes upwelling along the western coast of Peru and the California Current affects the coastal waters of the south-western United States. Similar upwellings off north-

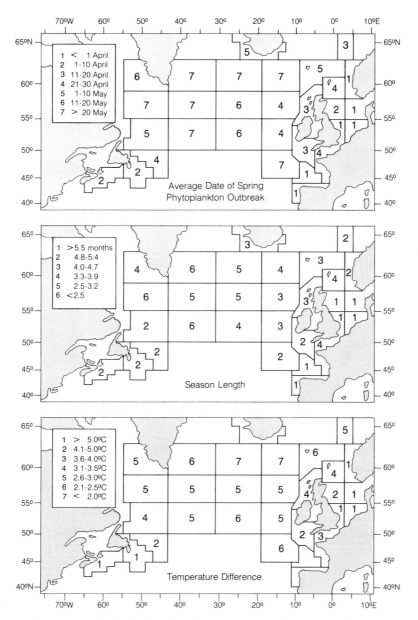

Figure 12.16 Average time of the spring outbreak of phytoplankton, length of growing season and difference in surface water temperatures in late winter and early spring in the North Atlantic Ocean. (After Robinson, 1970.) (Reproduced with permission from G. A. Robinson, Continuous plankton records: variations in the seasonal cycle of phytoplankton in the North Atlantic, *Bulletins of Marine Ecology*, 1970, **6**, 341, 343; with acknowledgement to the Natural Environment Research Council and the Sir Alister Hardy Foundation for Ocean Science.)

west Africa are associated with the Canaries Current, and off south-west Africa with the Benguela Current. Although water containing high concentrations of nutrients is brought up from depths of 50–200 m, mixing usually only occurs within the upper 10–20 m. Consequently, most phytoplankton remain within the euphotic zone and are carried away from the coast (Barber and Smith, 1981). The strong counterflow which is present below the surface carries plankton and debris back towards the coast. Nutrient regeneration by heterotrophic organisms in the deeper water helps to maintain high concentrations of dissolved elements.

As in the deep ocean, nitrogen is considered to be the primary limiting nutrient of phytoplankton production in coastal waters. Here 30–70% of the nitrogen assimilated by phytoplankton is derived from the sediments through microbial degradation of organic matter (Billen and Lancelot, 1988). Soluble and particulate nitrogen compounds are rapidly broken down to ammonia. It is in this form that nitrogen is most readily assimilated by phytoplankton, and conversion to NO_2^- and NO_3^- through nitrification does not occur in shallow seas. The seasonal changes which occur in the concentration of **dissolved inorganic nitrogen (DIN)** are most pronounced in mid and high latitudes. The renewed growth of the phytoplankton in the spring results in a marked decrease in DIN. The nitrogen is returned to the water at each trophic level as excreta and dead organic matter which must be decomposed by bacteria before it is again available. Nitrogen, in the form of ammonium, is released from coastal sediments at rates of 10 to 78 mg N m^{-2}. Further offshore, increasing amounts of ammonium are regenerated within the water column by bacteria (Harrison, 1978). Production may also be regulated by the influx of nitrate from the land or brought from the deep waters beyond the continental edge (Paasche, 1988).

Production is enhanced by the continual availability of nutrients in the surface waters. In some shallow coastal areas, tidal mixing causes vertical transport of nutrients. Tidal currents are produced when the water that accumulates along a coast during high tide moves seawards. The movement of the water is controlled by several factors and although surface currents are extremely variable, their velocities decrease only gradually with depth. For example, tidal currents in 165 m of water in the Bay of Biscay range from 20 to 26 cm sec^{-1} compared with 30–42 cm sec^{-1} at the surface (Cartwright and Woods, 1963). Tidal currents can be sufficiently strong and turbulent to carry nutrients to the surface in deep coastal waters. Regions of high productivity also occur along thermal fronts where warm water overlying the thermocline is forced to the surface by the inflow of nutrient-rich, tidally-mixed water (Savidge and Foster, 1978). The process is enhanced where surface density differences are maintained by the continual seaward flow of brackish water, as occurs in Puget Sound in the Pacific Northwest region of the United States (Winter *et al.*, 1975). The high productivity of marine frontal systems is reflected in local concentrations of fish, birds and marine mammals, and is often associated with conditions leading to the development of 'red tides' (Le Fèvre, 1986). Other factors also contribute to variable productivity in coastal waters. Even though nutrient concentrations are generally high in estuaries, phytoplankton production can be light-limited because of high turbidity (Walsh, 1983). Areas of unusually low productivity can also develop where freshwater discharge is exceptionally high. Thus lenses of nutrient deficient water are characteristically present near the mouth of the Amazon (Gibbs, 1970) and seasonal '**estuarization**' causes the development of saline fronts in the Bay of Bengal during the wet monsoon (Amos *et al.*, 1972).

12.8.3 Productivity of benthic plants

The water is too deep or turbid for light to penetrate to the seabed over most continental shelves and plant production is mainly by phytoplankton. Attached algae begin to appear in shallower subtidal waters. Species of *Laminaria*, *Macrocystis* and other brown algae, which are collectively known as **kelps**, are most abundant. Their growth continues until irradiance is reduced to about 1% of full sunlight: this is equivalent to a depth of 8 m in the turbid waters of the North Sea but increases to about 95 m in the Mediterranean (Lüning and Dring, 1979). For some crustose red algae, limiting light conditions occurs at depths where irradiance falls to 0.05% of surface values. The maximum depth for benthic plants therefore ranges from 10 to 175 m, depending on water clarity. The kelp communities are very productive and under optimum conditions production can exceed 1300 g C m^{-2} yr^{-1} in beds of *Laminaria* and 1000 g C m^{-2} yr^{-1} in *Macrocystis* (Mann, 1982). As with phytoplankton, light and nutrient supply are the principal factors which affect the growth of kelp. Juvenile stages of *Laminaria* will grow under very low light intensities but mature plants become light-saturated only at extremely high irradiance. Consequently, productivity in dense stands is often limited by self-shading, although this is reduced somewhat by the movement of the fronds and the high rate of attrition at their tips (Kain, 1979). Substantial amounts of nitrogen are taken up and stored during winter by temperate species of *Laminaria* when growth is reduced by poor light conditions. These reserves are typically exhausted before the end of the growing season and growth is limited, except where constant agitation of the water ensures resuspension of nitrates in the sediments (Johnston *et al.*, 1977).

Kelp beds are restricted by the requirements of the gametophyte stages to temperate and subpolar regions where sea temperatures range from 0 to 20 °C (Figure 12.17). Species of *Laminaria* are common in the North Atlantic and north-west Pacific: *Macrocystis* and *Nereocystis* mainly grow in the north-east Pacific (Figure 12.18), and *Ecklonia* is dominant in the waters around southern Africa and Australia. Within each region the distribution of the species reflects the competitive advantages of their different growth forms. For example, *Laminaria digitata* is the dominant species in the sublittoral kelp beds of Europe. It has a short, flexible stipe and buoyant, strap-like blades which bend without damage when exposed during low spring tides. *L. digitata* is very productive throughout the growing season and outcompetes *L. saccharina* and *L. hyperborea* in shallow water. Unlike these other species, *L. digitata* stores very little photosynthate and cannot draw on reserves in winter, when light levels in deeper water frequently fall below the compensation point. Under these conditions the slower growing species *L. saccharina* has a competitive advantage. In still deeper water *L. saccharina* is overtopped by *L. hyperborea*, which holds its blades in well-illuminated water by means of a long, stiff stipe (Lüning, 1979). The distribution of benthic seaweeds is not always determined by interspecific competition. In some areas the lower limit of the seaweeds is controlled by the activities of herbivorous animals, especially sea-urchins. Grazing by sea-urchins (*Strogylocentrotus droebachiensis*)

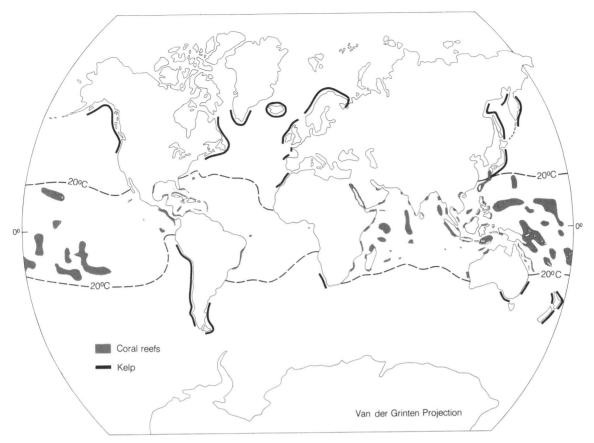

Figure 12.17 The general distribution of coral reefs and kelp beds in relation to the 20 °C mean annual surface water temperature isotherm. (After Mann, 1973; Wells, 1957.) (Redrawn with permission from J. W. Wells, Coral reefs, *Geological Society of America Memoir 67*, 1957, **1**, 630; also from K. H. Mann, Seaweeds: their productivity and strategy for growth, *Science*, **182**, 976, copyright 1973 by the AAAS.)

in North American waters is periodically relaxed in unusually warm years when an amoeboid pathogen causes mass mortality (Scheibling and Stephenson, 1984). The extinction of Stellar's sea-cow (*Hydrodamalis gigas*) in the late eighteenth century has been linked to the eradication of kelp by sea-urchins following the extermination of the predatory sea otter (*Enhydra lutris*) by hunters (Barnes and Hughes, 1988).

12.9 CORAL REEFS

Kelps are absent from tropical regions, except where they are associated with cold currents, and the characteristic ecosystems in the warm coastal waters are the coral reefs. The growth of coral is restricted to regions where mean annual water temperatures are above 18–20 °C (Figure 12.17). The present distribution of coral reefs covers an area of about 2 million km² (Achituv and Dubinsky, 1990). The richest assemblage of reef-building corals is found in the western Pacific with about 500 species, compared with only 65 species in the Caribbean. Most coral grows in clear, shallow water less

than 15 m deep (Buddemeier and Kinzie, 1976). The reef-building hermatypic corals are anthozoan coelenterates which produce external skeletons of calcium carbonate. Their growth and calcification is dependent on the primary productivity of symbiotic unicellular algae which live on the surface tissues of the coral organisms (Goreau, 1961). Active coral growth is therefore limited by insufficient light for photosynthesis to depths of 25–30 m (Figure 12.19).

Three major types of coral reef are distinguished on the basis of morphology and relation to adjacent land. **Fringing reefs** are mostly associated with rocky shore-lines: they are particularly well developed along the shores of the Red Sea where there is minimal input of freshwater and terrigenous sediment (Ladd, 1977). **Barrier reefs** lie offshore and are separated from the land by a shallow lagoon. The Great Barrier Reef of Australia is of this type. It consists of more than 2100 individual reefs lying in water that is rarely more than 60 m deep and covers an area of about 230 000 km². The third type of coral reef is the **atoll**, which typically encloses a circular lagoon. Atolls are mostly associated with volcanic cones that have slowly subsided below the surface of the ocean.

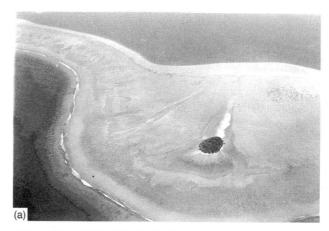

Figure 12.18 (a) Bull kelp (*Nereocystis*) along the west coast of British Columbia at low tide; (b) underwater view of bull kelp showing the buoyancy of the stipes. (R. DeVreede.)

Figure 12.19 (a) The submerged reef flat and transition to deep water along the reef front are clearly visible around an island reef at Truk, Micronesia; (b) conditions in the shallow water on the reef flats are very different from the wave and surge zone. (R. DeVreede.)

The action of wind, waves and currents causes conspicuous zonation in coral reefs. Active reef-building corals are only present on the windward edge of the reef below the mean low-water level of spring tides. The steep seaward reef front is usually sculptured into a series of grooves and spurs and in deeper water may be faced by talus. The upper part is essentially barren but at depths of 10–15 m, where the waves have little effect, the delicate foliate corals characteristic of sheltered lagoons increase in abundance (Wells, 1957). A ridge of crustose coralline algae up to 1 m high commonly forms on top of the coral. Some low-growing, spreading forms of coral survive in this turbulent environment where they are kept moist by waves and surge. Behind the algal ridge is the **reef flat** where pools and channels cut in the rock provide a variety of microhabitats at low tide (Figure 12.20). Elsewhere patches of coral sand and gravel may accumulate on the reef flat, and boulders of reef rubble are conspicuous along its outer edge. Coralline algae interspersed with a variety of smaller corals form the predominant cover of the outer reef flat, with low coral terraces growing where waves are refracted across the

Figure 12.20 Exposed reef flat, Palau, Micronesia. (R. DeVreede.)

surface. Patches of green algae and seagrasses are common on the sandy inner reef flats. The growth of corals increases where the water is deeper and is most prolific in the enclosed lagoons. **Patch reefs** or **microatolls** formed by larger corals in water 1–2 m deep provide shelter for many delicately branching forms. In most lagoons the water is 25–50 m deep: the slow circulation is ideal for coral growth, although they are restricted to areas where little sediment accumulates. Continual deposition of sediment can result in the formation of coral cays and perhaps to the eventual establishment of shrubs and trees (Figure 12.21).

Net primary production for coral reefs typically ranges from 200 to 1000 $g\,C\,m^{-2}\,yr^{-1}$ (Lewis, 1981). This high productivity partly reflects the increased density of phytoplankton compared with the surrounding ocean, but is mainly due to the growth of benthic plants on the reef flats. However, the reef flats comprise only a small part of most coral reefs, and average net productivity for entire reef systems is estimated at only 47 $g\,C\,m^{-2}\,yr^{-1}$, which is only slightly higher than in the adjacent oceans (Smith, 1988). Dense mats of filamentous green, red and blue-green algae are common on reef rocks and coral sands, with net production ranging from about 200 $g\,C\,m^{-2}\,yr^{-1}$ for patches of blue-green *Oscillatoria* to more than 1300 $g\,C\,m^{-2}\,yr^{-1}$ for non-calcareous red algae, such as *Wurdemannnia* (Sournia, 1976). Where present, the 'algal turfs' are heavily grazed by fish and invertebrates, and although they may never achieve high biomass they appear to be a very important part of the reef ecosystem (Vine, 1974). The coralline red algae grow abundantly on coral reefs with net production ranging from 240 $g\,C\,m^{-2}\,yr^{-1}$ in Pacific atolls (Marsh, 1970) to 370 $g\,C\,m^{-2}\,yr^{-1}$ in Caribbean reefs (Wanders, 1976a). This is generally higher than the productivity of fleshy algae, but the major role of the corallines is reef-building rather than as a source of food for consumers. Macrophytic algae are generally not abundant on coral reefs, although production as high as 2550 $g\,C\,m^{-2}\,yr^{-1}$ is reported for beds of *Sargassum* on shallow reefs in Curaçao in the Caribbean (Wanders, 1976b). Seagrasses such as turtle-grass (*Thalassia testudinum*) may also root into the soft sediments which collect on coral reefs. Net annual productivity of the seagrasses ranges from 370 to 800 $g\,C\,m^{-2}\,yr^{-1}$ (Lewis, 1981) and they are an important source of food for invertebrate herbivores and fish (Ogden and Lobel, 1978). The symbiotic zooxanthellae account for 5% of the tissue biomass in reef-building corals (Odum and Odum, 1955). However, net productivity of these intracellular dinoflagellates is relatively high, with values ranging from 530 to 770 $g\,C\,m^{-2}\,yr^{-1}$ reported for coral reefs in the Caribbean (Lewis, 1981). Conversely, the free-living phytoplankton normally contribute very little to the total productivity of coral reefs (Scott and Jitts, 1977).

Phytoplankton production in coral reefs is limited by the low concentration of nutrients in tropical waters,

Figure 12.21 Coast cottonwood (*Hibiscus tiliaceous*) typically establishes along island shorelines on the Great Barrier Reef, Australia. (G. W. and R. Wilson.)

whereas benthic producers benefit from nutrients supplied by various internal mechanisms. In tropical oceans plankton production is limited by the availability of nitrogen (Smith *et al.*, 1986), but in coral reefs the supply of nitrogen is maintained by biological fixation. The seagrasses often support heavy growths of epiphytic blue-green algae (Goering and Parker, 1972), and nitrogen-fixing bacteria in the rhizosphere also are an important source of nitrogen in coral reefs (Patriquin and Knowles, 1972). The rate of nitrogen fixation on reef flats reaches 25 $g\,m^{-2}\,yr^{-1}$ and is higher than in any other marine ecosystem (Capone and Carpenter, 1982). Fish may contribute to these high rates of nitrogen fixation by grazing the benthic algae that compete for space with the cyanobacteria (Wilkinson and Sammarco, 1983). Additional nitrate may be supplied through upwelling and advection of nutrient-rich water, as is the case for the Great Barrier Reef (Andrews and Gentien, 1982). Similarly, fringing reefs around islands may receive some nitrate from groundwater discharge and run-off (Lewis, 1987).

The nitrogen fixed by coral reef organisms passes into the ecosystem by way of the grazing and detritus food chains and as dissolved organic materials released by the algae (Mann, 1982). Ammonium is excreted by fish and other reef-grazing animals and in this form can be utilized directly by coral zooxanthellae (Muscatine and D'Elia, 1978). Some ammonium also accumulates in reef sediments through the feeding activities of protozoa and larger detritivores, but this is then oxidized to nitrite and nitrate by bacteria. Most of the nitrate is assimilated into plant and bacterial tissue or accumulated within the coral (D'Elia, 1988), although some may be exported in the water that flows across the reef (Webb *et al.*, 1975) or lost from detrital sediments through denitrification (Smith, 1984). Net primary production in coral reefs is considerably lower than gross primary production and this has been explained traditionally by *in situ* nutrient cycling. However, production is limited by nutrients, and even the excreta from schools of migratory fish may be sufficient to enhance the growth of coral (Meyer and Schultz, 1985). Abnormal nutrient enrichment can alter the structure of coral reef ecosystems through the overgrowth of algae and shade from increased phytoplankton production (Smith *et al.*, 1981). Thus coral reefs may function as disclimax communities that are maintained by exporting nutrients in the water that continually flows across them (D'Elia, 1988).

Secondary production on coral reefs is mainly supported by detrital matter in the form of algal fragments, faecal pellets and coral mucus which, together with the colonizing bacteria, is initially consumed by filter feeders and suspension feeders (Sorokin, 1978). The corals themselves are sessile suspension feeders. They are most prolific in strong currents along the seaward fringes of the reef, where they capture particulate material and zoooplankton in their tentacles or by means of mucus filaments. The corals provide a substrate and shelter for a great diversity of animals. As much as 60% of the coral rock may be removed by the activity of boring animals such as sponges, polychaete worms and molluscs (MacGeachy and Stearn, 1976) but most animals live externally, protected by their cryptic colours and forms (Patton, 1976). Typical herbivores that consume the benthic algae include sea urchins and molluscs as well as many species of fish. Most of the reef fish are opportunistic carnivores that feed on zooplankton, sessile organisms or other fish, and some are coral predators (Goldman and Talbot, 1976). The most notorious coral predator is the crown-of-thorns starfish (*Acanthaster planci*) that periodically causes mass mortality on some reefs (Endean, 1973). However, colonization of denuded areas can occur quite rapidly (Wallace *et al.*, 1986) and the effects on the reefs may not be as catastrophic as was previously believed (Rowe and Vail, 1984). Periodic disturbance of coral ecosystems has also been reported as a result of hurricanes (Highsmith *et al.*, 1980), unusually cold sea water temperatures (Porter *et al.*, 1982) and heavy rainfall during periods of tidal emersion (Stoddart, 1969), although most damage in these sensitive ecosystems can now be attributed to increased recreational use and pollution.

12.10 COASTAL VEGETATION

12.10.1 Rocky shorelines

Rocky shorelines are the product of marine erosion and the organisms which survive in these habitats are adapted to the physical stresses of waves and tides. The intertidal zone is especially severe because of the changing conditions which occur with the constant ebb and flow of the tide. The plants are subjected to varying periods of ex-

Figure 12.22 Air bladders provide buoyancy to the fronds of the brown seaweed *Ascophyllum nodosum*. (O. W. Archibold.)

posure and must tolerate periods of turbulence and desiccation as well as changes in temperature and salinity. In temperate regions the most conspicuous plants are species of *Fucus* and *Ascophyllum* and other large brown seaweeds (Figure 12.22). They are also present at higher latitudes, although their distribution is limited by freezing temperatures and ice abrasion (Stephenson and Stephenson, 1954): here the brown seaweeds are mostly represented by the subtidal species which grow beneath the ice. Thus *Laminaria* and *Alaria* are found throughout the Arctic and the range of *Macrocystis* extends to Antarctica. In tropical regions light intensities and temperatures in the intertidal zone are too high for most macroscopic algae and they are consequently restricted to areas below the level of mean low tides. Smaller species of brown algae such as *Sargassum* and *Turbinaria* grow in the warm subtidal waters throughout the tropics. The intertidal zone along rocky tropical shores is mainly colonized by blue-green and filamentous green algae which form a dense, turf-like cover, with encrusting lichens present in the spray zone.

Differing physiological requirements result in a pronounced vertical zonation of the seaweeds along rocky shorelines that is best developed in temperate regions. Species of kelp grow at the lowest tidal levels on moderately sheltered coasts but are replaced by *Alaria esculente* in more exposed areas (Figure 12.23). The greatest diversity occurs within the main intertidal area where species

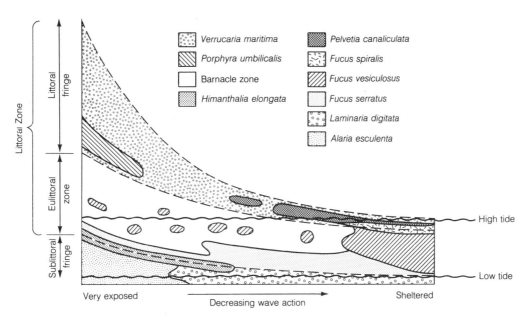

Figure 12.23 The distribution of common European seaweeds in relation to tidal range and exposure to wave action. (After Tait, 1972.) (Reproduced with permission from R. V. Tait, *Elements of Marine Biology: An Introductory Course*; published by Butterworth and Co., 1972.)

such as *Fucus serratus* and *Himanthalia elongata* are replaced by *Ascophyllum nodosum*, *F. vesiculosus* and *Pelvetia caniculata* on the upper shore. Black lichen (*Verrucaria maritima*) occurs near the high tide-mark and is especially prominent on exposed wave-swept headlands, where it is often accompanied by tar-like smears of blue-green algae. Green algae, such as *Enteromorpha compressa* and *Ulva lactuca*, are also characteristic species of the lower spray zone and shallow intertidal water: *Ulva* is especially abundant in areas polluted with sewage.

The vertical zonation of the seaweeds is related to their tolerance of desiccation during periods of low water

Figure 12.24 Zonation of seaweeds in a sheltered inlet on the coast of British Columbia. The upper limit of the seaweeds is sharply defined by the desiccation tolerance of *Fucus*; seagrasses occur at the bottom of the exposed rocks. (R. DeVreede.)

(Figure 12.24); nevertheless there is little evidence that desiccation avoidance is a primary adaptive mechanism in seaweeds (Schonbeck and Norton, 1979a). The rate of water loss in marine algae is largely dependent on the ratio of surface area-to-volume (Dromgoole, 1980), so that species which lose water quickly will also rehydrate quickly and can resume metabolic activities soon after re-immersion. The rate of photosynthesis in many seaweeds increases to a maximum at 10–20% water loss, perhaps because of an initial increase in the rate of CO_2 diffusion as the plant surface dries out (Johnson *et al.*, 1974). Photosynthesis in fucoids increases to a steady rate within 15–30 minutes of re-immersion after moderate drying, but interspecific recovery under more severe conditions is clearly related to the natural pattern of zonation (Dring, 1982). The rate of water loss depends on weather conditions, and a high rate of photosynthesis is maintained in exposed seaweeds during cloudy periods (Figure 12.25). Although the rate of photosynthesis can remain high during periods of exposure, a decline in growth has been reported in some species because of nutrient stress (Schonbeck and Norton, 1979b). Seaweeds do not have roots and so nutrient uptake occurs only during periods of immersion. Consequently, the zonation of the seaweeds may also reflect their ability to sequester nutrients (Table 12.1).

The lower zonal limits of the seaweeds are affected by the competitive ability of each species. Competition for light is considered the most important factor affecting species distributions on the lower shore and so the taller seaweeds, such as the kelps, are at an advantage (Dayton, 1975). Understory species – mainly red algae – are

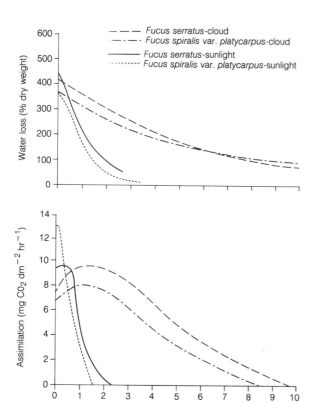

Figure 12.25 Water loss (upper) and assimilation rates (lower) of European *Fucus serratus* and *F. spiralis* var. *platycarpus* in relation to emersion during sunny and cloudy days. (After Stocker and Holdheide, 1938.) (Adapted with permission from O. Stocker and W. Holdheide, Die Assimilation Helgolander Gezeitenalgen Wahrend der Ebbezeit. *Zeitschrift fur Botanik*, **32**, 1938, published by Gustal Fischer Verlag Jena, Germany.)

Figure 12.26 Sea palms (*Postelsia palmaeformis*) growing on a mussel bed. (R. DeVreede.)

also present and will increase in abundance in the absence of the taller plants. Similarly, *Fucus spiralis* and *Pelvetia caniculata* grow below their normal range in areas cleared of *Ascophyllum nodosum* and *F. vesiculosus* (Schonbeck and Norton, 1980). *Pelvetia* grows relatively slowly and it is soon overtopped by the other fucoids. *Ascophyllum* is more tolerant of shade and will persist beneath the cover of *Fucus*; being a relatively long-lived species, *Ascophyllum* achieves dominance after 3–4 years when the *Fucus* dies. This replacement is further affected by the selective grazing of limpets and littorinid snails which feed preferentially on *Fucus*. Rapid colonization

by short-lived 'fugitive species', such as *Enteromorpha* and *Ulva*, occurs following disturbance of the established plants (Southward and Southward, 1978). In a similar way some algae grow attached to barnacles and mussels which are eventually washed away by the increased drag of the plants. The distinctive zone of sea palm (*Postelsia palmaeformis*) along parts of the Pacific Coast of North America is maintained in this way (Figure 12.26) (Dayton, 1973). Seasonal changes in distribution associated with tidal behaviour can also affect the stability of the algal zones and some species form distinct bands only at certain times of the year (Umamaheswararo and Sreeramulu, 1964).

In temperate regions, the most productive species along rocky coastlines are the subtidal kelps with potential net annual production typically exceeding 1000 $g\,C\,m^{-2}$: corresponding production by intertidal seaweeds ranges from 500 to 1000 $g\,C\,m^{-2}$ (Mann and Chapman, 1975). Kelp growth occurs mainly during winter and spring and is sustained by stored photosynthates. The primary storage products are laminarin and mannitol, and by late summer these account for over 40% of the dry weight of the plants (Boney, 1965). This assimilatory surplus develops because respiration rates decline in the summer when photosynthesis is at a max-

Table 12.1 Mean linear growth (mm month^{-1}) of two species of brown algae growing in normal sea water and in sea water enriched with different concentrations of nutrients under two artificial tidal regimes (after Schonbeck and Norton, 1979b)

	Tidal regime (hr submerged 12 hr^{-1} cycle)			
	Pelvetia canaliculata		*Fucus spiralis*	
Growth medium	**2 hr**	**10 hr**	**2 hr**	**10 hr**
Normal sea water	1.7	5.3	1.5	10.9
+ 5% nutrient enrichment	2.8	7.2	4.5	21.6
+ 20% nutrient enrichment	5.0	5.1	4.5	20.7
+ 100% nutrient enrichment	5.6	7.8	15.1	12.2

(Source: M. Schonbeck and T. A. Norton, The effects of brief periodic submergence on intertidal fucoid algae, *Estuarine and Coastal Marine Science*, **8**, 1979.)

Figure 12.27 Sea pink (*Armeria maritima*) and grasses provide a sparse cover on a rubbly sandstone cliff. (O. W. Archibold.)

Figure 12.28 Deformed trees and sparse ground cover attest to the severe environment along the sea cliffs of Newfoundland. (O. W. Archibold.)

imum. The ability to mobilize storage products during periods of low light intensity is especially advantageous for kelps growing at high latitudes. Growth in fucoids occurs mainly during the summer: the need for storage products is less important and levels of mannitol and laminarin typically fluctuate by less than 5% during the year. Although potential growth is high in *Laminaria*, losses of dissolved organic material account for as much as 40% of net production and are especially high following periods of desiccation (Sieburth, 1969) or damage (Moebus and Johnson, 1974). Productivity is further reduced where space is limited by the presence of barnacles and other sessile animals. Consequently, average productivity throughout the variable rocky coastal habitat is estimated to be about 100 g C m^{-2} yr^{-1} (Mann, 1982).

Terrestrial plants begin to appear on the cliffs beyond the influence of breaking waves, but only very hardy species can tolerate the conditions which are peculiar to this maritime environment. Comparatively few species have adapted to the wind and shallow droughty soils in these spray-drenched sites. The fragmentary cover which develops in crevices is dominated by annuals and small perennials. Their seeds are resistant to high salinities (Okusanya, 1979a), and growth in some species is stimulated by sea spray (Okusanya, 1979b). In Britain, for example, the sea pink (*Armeria maritima*) usually grows closest to the sea (Figure 12.27). It controls salt levels in

its tissues by active excretion through specialized glands on the leaf surface (Crawford, 1989). Sea plantain (*Plantago maritima*) is also extremely tolerant and occurs widely in very exposed places. Salt-tolerant grasses, for example common salt marsh grass (*Puccinellia maritima*) and crested hair-grass (*Koeleria gracilis*), and plants such as sea lavender (*Limonium vulgare*) which are commonly associated with salt marshes may also be present, together with the maritime lichens *Verrucaria* and *Lichina* (Chapman, 1976). Short-turf communities, dominated by grasses such as red fescue (*Festuca rubra*), Yorkshire fog (*Holcus lanatus*) and common-bent grass (*Agrostis tenuis*), become increasingly widespread along the cliff tops, where relative abundance is controlled by grazing and trampling as well as exposure (Figure 12.28). Seabird colonies also affect the nature of the plant cover: species such as sea campion (*Silene maritima*) may become locally abundant but even the most resistant plants are eliminated around heavily utilized sites (Figure 12.29).

12.10.2 Sandy beaches and dunes

A progressive reduction in cover of macrophytic algae occurs as the calibre of beach materials changes from coarse boulders and pebbles to sand. The growth of

Figure 12.29 Plant cover is limited around the nesting sites of albatrosses and boobies in the Galapagos Islands. (M. Brown.)

seaweeds and of rooted plants is precluded by the absence of firm surfaces and by the abrasive action of the sand as it is moved by the waves (Figure 12.30). Thus the flora of sandy beaches is restricted to benthic microalgae and the phytoplankton of the surf zone. The benthic microflora is mainly comprised of diatoms that are attached to the sand grains and are relatively immobile. Light for photosynthesis rarely penetrates more than 5 mm below the surface of the sand (Brown and McLachlan, 1990) but living micro-organisms are often mixed to depths of 20 cm by wave activity: they can survive several months of burial and resume photosynthesis when brought back to the surface (McIntyre, 1977). Unattached microalgae are common in more sheltered areas where the sand is finer and firmer: these species exhibit rhythmic vertical movements in response to diurnal or tidal stimuli and form mats on the surface of the sediment. Similar movements are reported for the diatoms in the

Figure 12.30 Continual movement by wave action restricts macrophytes to the upper beach along exposed sandy shorelines. (O. W. Archibold.)

surf zone. These accumulate at the water surface during the day and will often become stranded on the beach as deposits of foam. Once cell division is completed the diatoms produce a mucous coat which allows them to adhere to grains of sand during the night. Primary production in the surf zone is estimated at 20–200 $g C m^{-3} yr^{-1}$ and represents an important food source for benthic filter feeders and bacteria: the highest values are associated with more exposed beaches. Conversely, production by the benthic microflora, which ranges from 0 to $50 g C m^{-2} yr^{-1}$, is highest in sheltered bays (Brown and McLachlan, 1990).

Although *in situ* production is comparatively low, additional organic matter is brought in with each tide and concentrated in the permeable sand. The amount which is trapped increases in finer sands, but because of increased residence time it is also more susceptible to microbial attack (Pugh *et al.*, 1974). The interstitial food web is dependent on dissolved and particulate matter that is flushed into the pore spaces by wave action and filtering. Mineralization occurs mainly through bacterial activity. Some assimilable nutrients are excreted by detritivores such as lugworms (Arenicolidae) that also live in the sand, but the principal role of these organisms is mechanical fragmentation as well as improved oxygen diffusion through sediment disturbance (Brown and McLachlan, 1990). The regenerated nutrients are flushed out of the sand quite rapidly. Most of the organic nitrogen is returned as nitrate and is immediately available to phytoplankton, although ammonia may accumulate in poorly oxygenated sands. The bacteria are consumed by protozoa and by the nematodes, copepods and oligochaete worms which comprise the meiobenthos: these in turn are captured by larger predatory worms. Many species of mollusc can be found within the sand. They are filter feeders that draw in water and suspended particles of food by means of inhalant siphons that extend to the surface. Their distribution is influenced by the texture of the beach deposits and tidal activity. In Europe the razor shells (*Ensis* spp.) are characteristically restricted to more exposed sandy beaches, whereas clams (*Mya arenaria*) and furrow shells (*Scrobicularia plana*) are adapted to finer, anaerobic muds. Coarse debris that accumulates along the strandline may be utilized by scavenging amphipods such as sand hoppers (*Talitrus* spp.), seaweed flies (*Coelopa* spp.) and rove beetles (*Bledius* spp.) which are subsequently preyed on by birds. Birds such as oystercatchers (*Haematopus* spp.) also probe the sand for molluscs at low tide. A considerable proportion of the intertidal invertebrate production may be consumed by shore-birds but relatively little is returned to the beach ecosystem in waste products, feathers and carcasses (Hockey *et al.*, 1983).

In areas with strong and persistent onshore winds, sand from the beach is carried inland and deposited as dunes (Figure 12.31). Entrainment and transport of beach sand are directly related to wind speed. The minimum

Figure 12.31 Persistent onshore winds have created extensive sand dunes along the Pacific coast of North America. (O. W. Archibold.)

Figure 12.32 *Desmoschoenis spiralis* colonizing coastal sand dunes in South Island, New Zealand. (A. F. Mark.)

threshold for sand transport is 4.5 m sec^{-1}, and above this the rate of movement is proportional to the cube of the wind speed. Differences in the size, shape and density of the grains affect the ease of entrainment and it is mostly sands of fine and medium texture (125–300 μm) which eventually accumulate as dunes. Surface adhesion increases in wet sands and so good dune development is limited to areas where extensive sand flats are able to dry out between tides (Boorman, 1977). The form of the dune is initially determined by the strength of the wind and the amount of sand available. A series of ridges typically form in less exposed situations where there is an abundant supply of sand, but this changes to hummocky and parabolic forms where a limited amount of sand is moulded by strong winds. The growth and form of the dunes are further influenced by the establishment of vegetation, with pronounced linear ridges usually occurring where the sand is almost completely covered by plants.

The initial formation of embryo dunes above the strandline is caused by the deposition of sand around pioneering plants. Dune-initiating plants are usually small, salt-tolerant annual forbs, such as sea rocket (*Cakile maritima*) and saltwort (*Salsola kali*). Their seeds and

fruits are buoyant and can withstand long periods of immersion in sea water (Ignaciuk and Lee, 1980), but although they remain viable in the sand for several months, few germinate if buried below 8 cm (Lee and Ignaciuk, 1985). Continued development of the dune depends on the establishment of dune-building species, which are mainly grasses (Figure 12.32). The dominant species in Europe is marram grass (*Ammophila arenaria*), with sea lyme grass (*Elymus arenarius*) more prominent at higher latitudes: *A. breviligulata* and *E. mollis* occupy similar niches in North America. In tropical areas woody species typically accompany grasses such as *Sporobolus* and *Aristida*. Similarly, *Spinifex* grasses and shrubby *Acacia* are common dune-building species in Australia (Doing, 1985). All of these species are salt-tolerant to some degree, but more importantly they thrive under conditions of sand accumulation. For example, growth of *A. breviligulata* is stimulated by burial (Table 12.2) and will survive up to 0.6 m of deposition in a year (Disraeli, 1984), although continual burial does not favour seedling establishment (Laing, 1958). Seedlings of dune-building species are also sensitive to desiccation, and most spread vegetatively by tillering from rhizomes (Huiskes, 1979). This, together with extensive root systems, helps

Table 12.2 The effect of burial by sand on the growth and development of *Ammophila breviligulata* (after Disraeli, 1984)

Depth of burial (cm)	Plant height (cm)	Number of tillers plant^{-1}	Number of buds tiller^{-1}	Live biomass (g m^{-2} dry weight)			
				Aerial shoots	Rhizomes	Buried culms	Roots
2.5	34.8	2.3	1.3	161	101	0	134
7.5	37.3	3.0	3.0	292	187	3	205
12.5	46.5	3.7	5.0	423	272	100	274
17.5	52.9	4.5	7.2	553	357	318	343
22.5	55.1	5.3	9.6	684	442	703	411
27.5	57.2	6.0	12.3	814	527	1300	479
32.5	59.8	6.8	15.3	945	611	2160	548

(Source: D. J. Disraeli, The effect of sand deposits on the growth and morphology of *Ammophila breviligulata*, *Journal of Ecology*, 1984, **72**, 150.)

Figure 12.33 (a) Sea holly (*Eryngium maritimum*) is one of the first plants to establish as coastal dunes in Europe become stabilized by (b) pioneering marram grass (*Ammophila arenaria*); (c) a rich herbaceous community develops on the stabilized dunes over time. (O. W. Archibold.)

to bind the sand, but plant vigour declines as the dunes become stabilized. In *Corynephorus canescens* this has been attributed to mineral deficiencies caused by the absence of young absorptive roots which normally develop when the stem is buried (Marshall, 1965). For example, a decline in above-ground biomass from 398 to 125 g m^{-2} is reported in 1-year- and 10-year-old stands of *A. arenaria* in Sweden: root biomass had increased from 45 to 210 g m^{-2} during this period (Wallen, 1980).

Sand dunes become progressively older away from the shore and the plant communities which they support reflect the change in growing conditions which occurs over time (Figure 12.33). Embryonic dune soils are mainly composed of grains of quartz. They have a low cation exchange capacity and are consequently deficient in nutrients. Most have a high content of calcium carbonate because of the incorporated fragments of shell which, together with the salt spray, makes them very alkaline. The slow accumulation of organic matter increases the

acidity of the soil and accelerates the removal of the calcium. Initially the principal source of nutrients is tidal litter but this is superseded by atmospheric deposition as the dunes get bigger (Boorman, 1977). Nitrogen is considered a critical factor for the establishment of *Ammophila arenaria* (Willis, 1965) and the success of this species has been attributed to the presence of nitrogen-fixing bacteria in the rhizosphere (Abdel Wahab, 1975). This is also a characteristic of sea buckthorn (*Hippophäe rhamnoides*) and consequently nitrogen steadily accumulates in the soil beneath these shrubby plants (Silvester, 1977). Conversely, a progressive decrease in the amount of calcium-bound plant-available phosphorus has been demonstrated in dune sequences in New Zealand. However, this is replaced by organically bound materials so that the supply of phosphorus is regulated by the amount of litter produced and subsequent rates of mineralization (Syers and Walker, 1969a; 1969b). A similar decline in potassium has been reported in older dune soils in Australia (Thompson, 1981), although this element is usually not considered to be critically deficient in calcareous sands as it is continually supplied in sea spray (Etherington, 1967). The contribution of nutrients by sea spray drops rapidly away from the beach (Barbour, 1978) and most are lost by rapid leaching through the coarse sand (van der Valk, 1974). The small amount that is intercepted by the plants is generally absorbed directly by the leaves (Boyce, 1954) or is carried into the soil by stem flow (van der Valk, 1977). Nutrient input from sea spray is supplemented by the breakdown of organic material produced *in situ* or brought to the dunes by birds that feed on the adjacent beaches. Insects consume some of the litter that accumulates in the lee of the vegetation hummocks and at the base of the dunes, but most of the organic matter is decomposed below ground by the interstitial biota (Brown and McLachlan, 1990).

The foredunes are often referred to as **yellow dunes** because much of the surface is not covered by vegetation. Here the plants are vulnerable to drought and heat stress. The limited rooting depths of the plants effectively isolates them from the water table once the height of the dunes reaches 3–4 m. Comparatively little moisture is retained in porous sands but the problem of drought is reduced by subsurface condensation of dew during the night: fog is also an important source of moisture for many of the plants in coastal dunes. However, water deficits and wilting are still reported in many species during dry periods (Willis and Jefferies, 1963). Potential cooling by transpiration is reduced during dry spells and this may cause heat injury in an environment where surface temperatures can exceed 60 °C (Boerboom, 1964). Winter annuals which appear in the spring and complete their growth before the beginning of the hot dry summer are therefore very characteristic of sand dune floras (Pemadasa and Lovell, 1974). The appearance of mosses and lichens in the ground cover characteristically marks the beginning of the **grey dune** phase (Chapman, 1976). The older stable dunes support a variety of plant covers but are typically dominated by grasses, heaths or shrubs with numerous non-maritime species adding to the diversity of their floras. The varied topography of the stabilized dunes often results in higher moisture levels in the depressions. Such areas may support a distinctive cover of rushes (*Juncus* spp.), sedges (*Carex* spp.) and other plants, such as willows (*Salix* spp.), that are commonly associated with damp ground. Pondweeds and algae are normally present in permanently flooded sites with some halophytic species occurring in areas that are periodically inundated by the sea (Crawford and Wishart, 1966).

12.10.3 Salt marshes

Salt marshes occur where fine sediments accumulate along sheltered coastlines in the temperate and high-latitude regions of the world (Figure 12.34). They are often associated with river estuaries but also occur where the coastline is protected by islands, sand bars and spits. The dominant plants are rooted macrophytes which are variously adapted to the environmental stresses associated with tidal inundation (Figure 12.35). The sublittoral vegetation along muddy shores consists of species such as eelgrass (*Zostera marina*) which are only exposed for brief periods during extreme low-water spring tides. Continued deposition of sediment brought in by the tide raises the level of the land, and the gradual change is reflected in the establishment of species which are increasingly tolerant of prolonged exposure to the atmosphere. Colonization by vascular plants begins as soon as a stable substrate is available and this is initially enhanced by mucus secreted by diatoms (Coles, 1979). The rate of deposition increases as the plant cover develops and the sediment becomes more firmly bound by the roots (van Eerdt, 1985); the coarser materials are mostly deposited at lower elevations on the salt marsh and along drainage creeks (Phleger, 1977). The rate of accretion is typically less than 1 cm yr^{-1} and mainly occurs in the lowest parts of a marsh which are frequently covered by the tide (Steers, 1977). A spatial pattern begins to develop in the plant cover as the duration and frequency of tidal flooding is altered by sediment accumulation. Competitive replacement occurs in response to these environmental changes but long-term successional development is often interrupted by processes that alter the physiography of the salt marsh.

All salt marsh plants must endure periodic inundation by sea water, and the characteristic vertical zonation in these communities reflects the relative tolerance of the representative species. The detrimental effects of salinity are largely attributed to problems in nutrient uptake and wilting induced by low external water potentials (Munns *et al.*, 1983). In salt marsh species such as *Spartina alternifolia*, sodium is excluded at the roots thereby preventing the normal problem of potassium deficiency associated with saline soils (Smart and Barko, 1980). Salt-excreting glands are also present in the leaves of *S. alternifolia* and other salt marsh species, and the evaporated crystals of salt which form on their surfaces are subsequently washed off by the rain or tide (Anderson, 1974). Similarly, the nutrient balance in the growing tissue of seablite (*Suaeda maritima*) is maintained by the translocation of potassium out of older leaves before they are shed (Gorham and Wyn Jones, 1983). Succulence is another characteristic adaptation of salt marsh plants; it normally increases throughout the growing season so that tissue salt concentrations remain constant. Salt uptake may also be limited because of reduced transpiration as a result of this xeromorphic habit (Flowers, 1985). Halophytes must adapt to the high external water potential which develops in saline soils in order for water to flow from the roots to the shoots. This requires an increase in tissue solute concentrations so that the osmotic potential

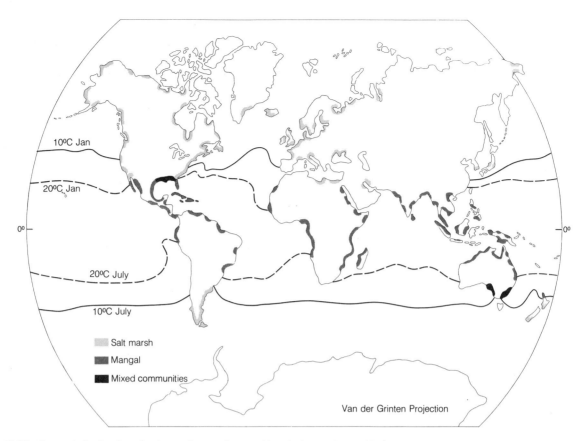

Figure 12.34 General distribution of salt marshes and mangal in relation to the 20 °C mean annual surface water temperature isotherm. (After Chapman, 1977.) (Redrawn from V. J. Chapman, ed. *Ecosystems of the World, Vol. 1. Wet Coastal Ecosystems*; published by Elsevier Science Publishers, 1977.)

of the plant exceeds that of the external solution. Osmoregulation is mostly achieved by the accumulation of inorganic salts. These are typically stored in the cell vacuoles to prevent problems of toxicity, with an equivalent reduction in water potential of the cytoplasm produced by the synthesis of organic solutes such as proline and betaine (Adam, 1990).

The frequency of flooding decreases as the level of the marsh is built up through accretion and the highest parts may only be covered by the occasional high spring tide. Although this reduces the input of salt, the salinity of the interstitial waters in the upper marsh may be highly variable because of the effects of evaporation and rainfall (Beeftink, 1977). High rates of evapotranspiration during drier periods can increase soil salinities in more elevated parts of the marsh and lead to the formation of salt crusts. This can result in an inversion of zonation in estuarine marshes which are flooded only by brackish water. Elsewhere flooding may be too infrequent to counteract leaching by rainwater. Soil salinities in the lower marshes are relatively constant and rarely exceed that of the tidal waters. The water drains slowly back to the sea during the ebb tide but the fine sediment usually remains waterlogged and deficient in oxygen. Anaerobic conditions and associated changes in soil chemistry vary with the microtopography of the salt marsh and this is reflected in the distribution of the plants, with some species being limited to the better drained creek banks and more elevated parts of the marsh (Armstrong *et al.*, 1985). Several species have adapted to the waterlogged conditions by developing aerenchymatous tissue to conduct oxygen to the roots (Anderson, 1974) but this does not always meet their respiration requirements. Growth in species such as *S. alternifolia* is still related to soil drainage (Howes *et al.*, 1981), and to survive under anaerobic conditions the metabolism of the plants is altered such that toxic products do not accumulate (Mendelssohn *et al.*, 1981). Oxidation of inorganic soil toxins also occurs in the rhizosphere of salt marsh plants (Mendelssohn and Postek, 1982). However, the distribution of these plants appears to be unrelated to variations in the levels of iron and manganese in the soil (Rozema *et al.*, 1985), although some species may be adversely affected by high sulphide concentrations (Havill *et al.*, 1985).

Comparatively few species of plant can tolerate the peculiar conditions of the salt marsh environment but there are still pronounced regional differences in floristic composition (Adam, 1990). The dominant species in arctic coastal marshes are grasses such as alkali grass

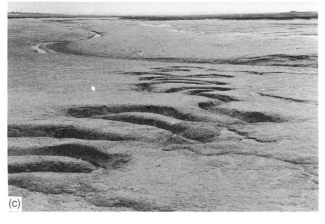

Figure 12.35 (a) A salt marsh dominated by *Salicornia* in the Bay of Fundy at high tide; (b) creek draining the salt marshes of the Dee estuary, Wales, at low tide; (c) gradual accumulation of sediment may eventually raise the level of the barren mudflats permitting the establishment of pioneer vascular plants. ((a) R. A. Wright; (b) and (c), O. W. Archibold.)

(*Puccinellia phryganodes*) and *P. maritima*. Sedges (*Carex* spp.) and rushes (*Juncus* spp.) are common in the upper marshes, and other species such as seaside arrow-grass *Triglochlin maritima*, common scurvy-grass (*Cochlearia officinialis*) and *Stellaria humifusa* may occur abundantly at some locations. However, the total number of species is very low and because of the short growing season all

are perennials. Reproduction is mainly by vegetative spread: seed production has never been observed in arctic populations of *P. phryganodes*, and the colonization of new sites is thought to depend on ice rafting of stolons (Sørensen, 1953). Salt marsh development is limited along open coasts by ice abrasion, and extensive coastal marshes occur only in sheltered inlets and bays (Jefferies, 1977). Melting of sea ice during the growing season reduces salinity in coastal waters below that of normal sea water, and zonation in the plant cover reflects the salinity of the floodwater rather than successional development in a saline environment.

The transition to temperate marshes is accompanied by a gradual increase in species diversity and a more pronounced vertical zonation. In Europe the characteristic species of the lower marshes are summer annuals such as the saltworts (*Salicornia* spp.) and *Suaeda maritima*. The seeds are distributed by the tides and most germinate in the spring (Beeftink, 1985). Normally there is no persistent seed bank, although seeds of other halophytes may accumulate in the soil (Bakker, 1985). Unlike many salt marsh species, germination in *Salicornia* is not prevented by high salinities and some even germinate while floating (Ungar, 1978). However, establishment on mudflats is often limited by poor root growth: many seedlings are uprooted by the tide and most perish. *Puccinella maritima*, sea lavender (*Limonium vulgare*) and sea aster (*Aster tripolium*) occur at higher levels on the marshes, and the dwarf shrub sea purslane (*Halimione portulacoides*) is conspicuous along well-drained creeks. Above the mean high tide-mark the vegetation cover is mainly of rushes and grasses. Mud rush (*Juncus gerardi*) is a common species, especially on wet sandy soils, but is replaced by *Festuca rubra* in drier sites and fiorin (*Agrostis stolonifera*) where soil salinity is lower. The presence of sea wormwood (*Artemisia maritima*), *Plantago maritima* and other halophytic species adds to the floristic diversity of the upper salt marsh communities (Chapman, 1976).

Spartina alternifolia was recorded in Britain in 1829 after being introduced from North America in ships' ballast. Prior to this the only species of cord grass native to Europe was *S. maritima* and this was comparatively rare (Marchant and Goodman, 1969a). *S. alternifolia* is now also virtually extinct in Britain (Marchant and Goodman, 1969b) but *S. anglica*, a recently evolved hybrid of *S. alternifolia* and *S. maritima*, has spread rapidly and become the dominant species in many salt marshes. It is a large, vigorous plant and has formed dense, monospecific stands, especially on the lower tidal mudflats that were previously unvegetated. *Spartina* is replaced by grasses such as *Puccinella maritima* where the stand is opened up by grazing and trampling or where growth is suppressed by accumulated litter. Extensive die-back of *Spartina* has occurred because of toxic conditions in the very fine-textured sediment that has been trapped by the plants (Adam, 1990).

The most extensive marshes in North America occur in the south-eastern United States where the gently sloping shorelines are protected by barrier islands. The dominant species in the intertidal zone throughout the eastern seaboard is cord grass (*Spartina alternifolia*): it occurs in two disinct forms. The tall form of *S. alternifolia* can reach 2–3 m in height and is found along the banks of tidal creeks: the dwarf form grows less than 50 cm high and occupies more elevated regions of the marshes (Valiela *et al.*, 1978). This distribution is attributed to differences in soil aeration and sulphide concentrations (DeLaune *et al.*, 1983). *S. alternifolia* is replaced by *Juncus gerardi* and other rushes near the high tide-mark. Seashore salt grass (*Distichlis spicata*) and cord grass (*S. patens*) appear on the highest parts of the marshes, with common reed (*Phragmites communis*) forming the landward margin in places where there is fresh water (Reimold, 1977). Above-ground production in stands of *S. alternifolia* can be high as 3250 g C m^{-2} yr^{-1} in some marshes in the south-eastern United States, although this is exceeded in stands of black rush (*Juncus roemerianus*) and *S. patens* (Table 12.3). Numerous salt marshes and tidal flats occur around the shores of the Gulf of Mexico. They are floristically similar to those on the Atlantic coast and are comprised mainly of *S. alternifolia*, *D. spicata* and *Juncus roemerianus*. However, once the sediment is stabilized the marsh plants are often replaced by mangroves (R. C. West, 1977).

Spartina foliosa is the dominant low-marsh species in California, but unlike the eastern seaboard, *Spartina* is absent elsewhere along the temperate Pacific coast of North America. Here the principal species in more saline areas include *Salicornia virginica*, *Triglochlin maritima* and lesser sea-spurry (*Spergularia marina*). At higher levels on the marshes the cover is mainly of *Distichlis spicata* and associated herbs such as sea milkwort (*Glaux maritima*) and *Plantago maritima*. Sedge marshes dominated by *Carex lyngbyei* occur in less frequently flooded areas and along tidal creeks: this becomes the principal vegetation type on more northerly tidal mudflats but is eventually replaced by *Puccinellia phryganodes* in Alaska (Macdonald, 1977).

Salt marsh communities similar to those of Europe and North America occur to a limited extent in Japan, South America and Australasia. Species such as *Suaeda maritima*, *Triglochlin maritima* and *Aster tripolium* are present in the lower salt marsh zones in Japan, but here the distinctive species is *Zoysia sinica* (Adam, 1990). In South America the most extensive salt marshes occur along the coasts of Uruguay and Argentina. They are dominated by South American species of *Spartina*, *Distichlis* and *Juncus* with shrubby species of *Salicornia*, *Suaeda* and *Atriplex* becoming more prominent further south in Patagonia (R. C. West, 1977). Similarly, many of the tidal flats in temperate parts of Australia support a cover of low shrubs interspersed with salt-tolerant grasses and herbs (Saenger *et al.*, 1977). This type of vegetation is characteristic of the salt marshes of the eastern Mediterranean and the Red Sea. Shrubby *Halocnemon strobilaceum* typically grows in areas which are recurrently washed by sea water, with species such as *Arthrocnemum glaucum*, *Halophyllum perfoliata* and *Zygophyllum album* extending landwards across the salt plains of the adjacent desert (Zahran, 1977).

Algae growing on the surface of the mud are also an important component of salt marsh communities. Green algae, such as *Enteromorpha* and *Rhizoclonium*, are common on sandy substrates with blue-green algae such as

Table 12.3 Primary production for selected salt marsh species in different coastal locations around North America (after Kennish, 1986)

Species	Location	Net above-ground production (g dry weight m^{-2} yr^{-1})	Net below-ground production (g dry weight m^{-2} yr^{-1})
Carex spp.	Arctic Coast	25–300	
Distichlis spicata	Gulf Coast	4000	
	Pacific Coast	750–1500	
	Atlantic Coast		1070–3400
Juncus gerardii	Maine	485	
	Atlantic Coast		1620–4290
Juncus roemerianus	North Carolina	754	
	Georgia	2200	
	Gulf Coast	4250	1360–7600
Puccinellia phryganodes	Arctic Coast	63–175	
Salicornia spp.	Pacific Coast	1113–2500	
	Atlantic Coast		430–1430
Spartina alternifolia	Atlantic Coast	500–2000	550–4200
	Gulf Coast	3250	279–6000
Spartina foliosa	Pacific Coast	1000–2125	
Spartina patens	Gulf Coast	7500	
	Atlantic Coast		310–3270

(Source: M. J. Kennish, adapted from Table 4, p. 163–5, *Ecology of Estuaries, Vol. 1. Physical and Chemical Aspects*; published by CRC Press, 1986.)

Oscillatoria and *Nostoc* present on muddy soils, and red algae, for example *Bostrychia*, often attached to the stems of higher plants (Chapman, 1977). The contribution of the algae to primary production varies with the composition of the marsh. A value of 105 g C m^{-2} yr^{-1} is reported for epibenthic algae in stands of *Spartina alternifolia* in the eastern United States: this is equivalent to 25% of vascular plant production (van Raalte *et al.*, 1976). In marshes dominated by *S. foliosa* net production by algae is 276 g C m^{-2} yr^{-1} compared with 340 g C m^{-2} yr^{-1} for vascular plants (Zedler, 1980). Snails and other macroconsumers may feed on the algae (Pace *et al.*, 1979) but macrophyte herbivory is typically very light in salt marshes and mostly results from insect activity. For example, leaf-chewing and sap-sucking insects consume about 5% of the plant material in stands of *S. alternifolia*, although annual production losses through leaf damage and early senescence can be much greater (Pfeiffer and Wiegert, 1981). Other species such as ducks and geese tend to feed on *S. alternifolia* because of its low content of soluble phenolics (Buchsbaum *et al.*, 1984) and this can have a considerable impact on the structure and composition of salt marsh communities (Smith, 1983).

As much as 80% of the above-ground primary production of *S. alternifolia* and practically all of the below-ground production is incorporated into the detritus food web (Valiela *et al.*, 1985). Soluble compounds are quickly leached from the litter with 5–40% weight loss occurring within a few weeks. Most of the material that remains is broken down more slowly by microbial activity and after 12 months only refractory substances such as lignins are left, which gradually accumulate as thin layers of peats. A small amount of the detritus is utilized by macroconsumers such as crabs and shrimps but they mostly feed on the colonies of decomposer organisms and egest the bulk of the dead plant material. Others feed on benthic diatoms or phytoplankton which are more readily assimilated than decomposer microbes (Montague *et al.*, 1981). Detritus feeders tend to have low assimilation efficiencies and continually rework the sediments in search of food. Their principal role in salt marsh energetics is to accelerate decomposition by breaking down detritus into smaller fragments and by concentrating organic matter in waste products. The chemical composition of the detritus affects the rate at which it is broken down. Thus soluble phenolic acids inhibit detritivore activity (Valiela *et al.*, 1979) and the high lignin content in cell walls increases resistance to microbial attack.

Aerobic micro-oganisms colonize suspended particulate matter and are also present in the thin, oxidized layer at the surface of the sediment, but most of the detritus is broken down within the sediment under anaerobic conditions (Wiebe *et al.*, 1981). The nitrogen content of the detritus tends to increase over time because of the higher protein content of bacterial biomass. However, much of the nitrogen is present in compounds that are not readily available for plant growth (Rice, 1982) and is ultimately lost through denitrification processes associated with detrital nitrate reduction (Whitney *et al.*, 1981). Some nitrogen is fixed by bacteria in the rhizosphere of *Spartina* and by algae in marsh pools and creeks (Teal *et al.*, 1979). However, this is not normally sufficient to compensate for losses by denitrification and tidal export, and additions of nitrogen through rain or groundwater may be necessary to sustain high productivity (Valiela *et al.*, 1978). Nutrient export through tidal flushing provides an important functional relationship between marshes and adjacent coastal waters. Similarly, estuarine marshes act as nursery grounds for many coastal organisms which in turn export nutrients and energy when they migrate to the sea.

12.10.4 Tropical mangrove formations

In the tropics and subtropics the stable intertidal mudflats usually support stands of mangroves. The best developed mangrove communities, or **mangal**, are associated with estuaries and deltas, although they do not grow extensively at the mouths of the Amazon and Congo (Tomlinson, 1986). The dominant plants are woody species of *Rhizophora* (red mangrove) and *Avicennia* (black mangrove), some of which grow to heights of 30 m. Most species of mangrove are widely distributed but none are pantropical and they are conveniently divided into two groups on the basis of longitude. Some 40 species are represented in the eastern group which covers the Indian and Pacific Oceans (30–180 °E) with regional diversities ranging from four species in Samoa to 32 species in South East Asia and the Western Pacific. The western group, comprised of only eight species, includes West Africa and the Americas and effectively extends from 15 °E to 120 °W: maximum diversity occurs in the western Caribbean and Central America, where seven species are reported. Mangroves grow where sea surface temperatures remain above 24–27 °C year round, but their latitudinal distribution is clearly influenced by the pattern of ocean currents. For example, the southern limit of mangal occurs at about 25 °S in the warm waters of Brazil compared with only 4 °S on the Pacific Coast, where temperatures are lowered by the Peruvian Current (R. C. West, 1977). In this way the range of mangal extends into temperate latitudes with the extreme limit for *Kandelia candel* occurring at about 31 °N in Japan, while *Avicennia marina* grows to 38 °S in New Zealand (Figure 12.34).

Mangal is best developed in coastal areas where naturally high rates of sedimentation are enhanced by the growth of the mangroves themselves. Characteristic vegetation zones normally develop as the land is built up and tidal flooding becomes less frequent. The width of each zone depends on the tidal range and slope of the shore (Macnae, 1968) but distinct zonal patterns are often

Figure 12.36 (a) Characteristic development of prop-roots on *Rhizophora* growing in Costa Rica; (b) looping form of roots in *Bruguiera* exposed at low tide. ((a) K. A. Hudson; (b) O. W. Archibold.)

Figure 12.37 (a) Pneumatophores of *Avicennia* emerging from mudflats in the Philippines; (b) a young grey mangrove (*Avicennia marina*) with pneumatophore roots colonizing tidal flats in Queensland. ((a) G. A. J. Scott, (b) G. W. and R. Wilson.)

disrupted by differences in local topography, run-off channels and sediment stability (Tomlinson, 1986) and by freshwater seepage (Semeniuk, 1983). The floristic composition of each zone is primarily determined by species tolerance to flooding and salinity. Mangroves grow on fine-grained soils that are to varying degrees low in oxygen and poorly consolidated. Most species of mangrove have extensive shallow root systems which are variously adapted to the conditions in the mud. Species of *Rhizophora* that colonize the lowest tidal flats are characteristically equipped with abundant prop-roots that arise from the trunk and lower branches (Figure 12.36). Even though additional roots develop as sedimentation continues, the trees cannot spread vegetatively in this way. In most species of *Avicennia* the lateral roots are equipped with pneumatophores which rise 30 cm or so above the surface of the soil (Figure 12.37); in *Bruguiera* 'knees' develop where the roots periodically loop above the substrate, and in *Heritiera* there are small plank buttresses (Tomlinson, 1986). All of these specialized structures are perforated with lenticels which connect to aerenchymatous tissue and so improve gas exchange in the root system. Respiratory consumption of air in the

aerenchyma creates a negative pressure in the roots when the lenticels are closed during tidal flooding: the lenticels open at low tide and air is sucked in by the pressure imbalance.

Tolerance of salinity in mangroves is related to the control of tissue water potentials through specialized leaves and stems. All genera of mangroves have rather thick, evergreen leaves and exhibit to varying degrees xeromorphic characteristics such as slightly sunken stomata, succulence and surface waxes (Stace, 1966). Similarly, the vessels in mangrove wood are very narrow and densely distributed. Such an arrangement is thought to reduce conduction losses through gas embolism, the frequency of which is theoretically increased by the high osmotic potential of the sea water. Salt is mostly excluded from mangroves by an ultrafiltration process in the roots (Scholander, 1968), although some is taken up and ultimately must be secreted or accumulated. Salt secretion in species such as *Avicennia marina* occurs by means of salt glands on the leaf surfaces (Fahn and Shimony, 1977). In salt-accumulating species, sodium and chloride are deposited in the older leaves and bark of the stems and pneumatophores. Leaf storage is generally accompanied by increased succulence. In some species salt is transported to the older senescing leaves and potassium and phosphorus is withdrawn prior to shedding; in deciduous species such as *Xylocarpus* annual leaf fall allows for excess salt removal before the start of each growing season (Saenger, 1982).

In addition to the controlling influence of salinity and tidal exposure, the zonal distribution of mangroves has often been associated with successional development in mangal. Throughout the Indo-Pacific region species of *Avicennia* form pure stands along the seaward fringe of the mangal (Macnae, 1968). *A. marina* is the pioneer species in areas where the substrate is relatively firm: *A. alba* grows on soft muds and *A. eucalyptifolia* is associated with coral islands and reefs. A similar role is reported for *Aegiceras corniculatum* in areas where salinity is lowered by freshwater seepage. Seedlings and saplings do not grow well in the shade and most die beneath the canopy of parent trees. Similarly, progression seaward is limited because the seedlings get smothered by silt, although under some conditions a narrow fringe of *Sonneratia alba* may be present beyond the *Avicennia* zone. Mature *Avicennia* trees are eventually overtopped and killed by *Rhizophora* which establishes beneath them. The *Rhizophora* zone occurs in areas which are less frequently flooded. The most widely distributed species is *R. mucronata*, with *R. apiculata* appearing further inland where salinities are lower. All of these species may develop into tall trees beneath which is an open understory of saplings, or less commonly, smaller mangroves such as *Ceriops tagal* and *Bruguiera parviflora*. *Bruguiera gymnorrhiza* is the characteristic species of the late stages in mangal succession where soil levels are raised above the influence of the tides, and the transition to terrestrial associations is marked by the increasing abundance of inland species.

Zonation is less pronounced in the Americas, perhaps because there are fewer species of mangroves here. *Rhizophora mangle* is generally regarded as the pioneer species in this region, and this seaward zone is fringed with seedlings and saplings growing in the mud below low-tide level (Chapman, 1976). Small turfy red seaweeds and blue-green algae cover the prop roots of the *Rhizophora* trees, and other algae grow in the mud. *Avicennia germinans* is the dominant species further inshore and ideally this is followed by successive zones of *Laguncularia racemosa* and *Conocarpus erecta*. However, there is increasing evidence that mangrove zonation reflects the influence of geomorphological activities on soil type and drainage rather than classical successional development. For example, pioneer stands of *Avicennia* usually develop on soils with a higher sand content, *Rhizophora* may colonize organic mucks and peats, and *Pelliciera rhizophorae* may become established on poorly oxidized clays (R. C. West, 1977). Similarly, in deltaic sites the distribution of mangroves is related to microtopography, with *A. germinans* on the levees, *R. mangle*

bordering the channels, and mixed stands of *R. mangle* and *L. racemosa* present in sites that are slowly subsiding (Thom, 1967). Interspecific competition does not occur in young mixed stands, and because the shade-intolerant *Laguncularia* has a faster growth rate than *Rhizophora* both species will continue to coexist for many years. However, seedlings of shade-tolerant *Rhizophora* become increasingly abundant beneath established canopies which in time favours the development of pure stands (Ball, 1980). Structural and floristic diversity in mangrove forests may therefore depend on frequent disturbance by hurricanes or other external forces (Craighead and Gilbert, 1962).

Vegetative reproduction is extremely rare in mangroves and propagation depends on seedling establishment. Most species are hermaphrodite, and although there is some evidence of self-compatability in *Rhizophora*, cross-pollination is the rule (Tomlinson, 1986). Pollination in *Rhizophora* is mainly by wind, but as with most species of mangroves the flowers are also frequently visited by bees. Birds are attracted to the large red flowers of *Bruguiera gymnorrhiza*, and in *Sonneratia* the flowers open in the evening and are visited by bats and moths. Mangroves are dispersed by water and all of their propagules will float for varying periods of time. Dispersal is principally by fruit or seeds but mangroves such as *Rhizophora* and *Bruguiera* are viviparous and germination occurs before the fruit is shed. Considerable mortality occurs while the seedlings are attached to the tree: in *R. mangle* less than 7% of the flower buds may produce mature seedlings (Gill and Tomlinson, 1971), most being lost though fungal and insect attack. Successful dispersal requires that the immersed propagules remain viable until they reach a suitable site but longevity varies between species. For example, the seedlings of *Rhizophora mangle* can remain viable for more than a year, compared with only 35 days for the one-seeded fruits of *Laguncularia racemosa* (Rabinowitz, 1978a). Similarly, seedlings of *Pelliciera rhizophorae* will normally sink after a few days whereas those of *Avicennia germinans* can remain afloat for many months until they rot. An obligate dispersal period is also required before the root systems develop: this can be as long as 40 days in *Rhizophora* and a further 15 days may then elapse before permanent rooting is achieved. Tidal sorting on the basis of interspecific propagule characteristics has been suggested as a mechanism for mangrove zonation (Rabinowitz, 1978b). However, the assumption that large propagules become stranded as water depth decreases and small propagules are restricted to shallower water because of limited root growth of the seedlings is not supported by the zonal patterns of the species.

The mosaic of vegetation types that occurs within mangrove communities results in variable productivity within a general range of 350 to 500 g C m^{-2} yr^{-1} (Mann, 1982). Primary production in species such as *R. mangle* is mainly influenced by the structure of the canopy. The

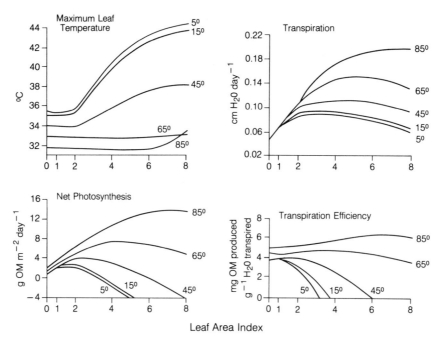

Figure 12.38 The influence of leaf-area index and leaf angle (° from horizontal) on daily maximum leaf temperatures, net photosynthesis, transpiration and transpiration efficiency in red mangrove (*Rhizophora mangle*) canopies in southern Florida. (After Miller, 1972.) (Reprinted with permission from P. C. Miller, Bioclimate, leaf temperature, and primary production in red mangrove canopies in south Florida, *Ecology*, 1972, **53**, 42.)

Table 12.4 Distribution of biomass in selected stands of mangroves

Species	Location	Leaves	Trunkwood and branches	Prop roots	Roots and pneumatophores	Source
			(t dry weight ha^{-1})			
Rhizophora brevistyla	Panama	3.5	159.3	116.4	189.8	Golley *et al.*, 1975
Rhizophora mangle	Puerto Rico	5.4	40.7	–	14.4	Golley *et al.*, 1962
	Florida	3.7	76.9	14.6	–	Snedaker and Pool, 1973
Avicennia nitida	Florida	0.1	3.5	–	–	
Avicennia marina	Australia	13.5	114.9	–	153.8	Briggs, 1977
Rhizophora apiculata	Thailand	7.4	90.2	61.2	–	Christensen, 1978

rate of net photosynthesis increases when the leaves are steeply inclined: this lowers the temperature of the leaves and allows more light to penetrate the canopy. Transpiration efficiency also increases with leaf inclination, even though transpiration rates are higher (Figure 12.38). The structure of the canopy is therefore adapted to maximize production under conditions of unlimited water supply (Miller, 1972). Differences in net photosynthesis in *Rhizophora*, *Avicennia* and *Laguncularia* have also been attributed to the salinity of the flood water (Carter *et al.*, 1973) and nutrient availability (Onuf *et al.*, 1977). Mangroves grow comparatively slowly with the larger species attaining girths of 60–70 cm after 50 years (Durant, 1941). Leaves typically account for less than 10% of the total above-ground biomass in mangal (Table 12.4). They have an average life span of about 11 months and leaf fall occurs throughout the year. The mean shed-

ding rate is equivalent to about 30% of the total leaf complement but generally increases during the summer months, when air temperatures and incident light levels are at a maximum. The leaves are replaced continuously, with maximum production often coinciding with periods of heavy rainfall or less frequent tidal flooding (Tomlinson, 1986).

Leaf material accounts for as much as 90% of the litterfall in some stands (Table 12.5). Decomposition is relatively rapid with daily mass losses of about 1% (Tam *et al.*, 1990). Nitrogen, phosphorus and potassium are quickly leached from the litter and this is mostly exported by tidal activity as dissolved organic matter (van der Valk and Attiwill, 1984). A considerable amount of dead leaf material can also be carried out with the tide (Boto and Bunt, 1981). Consequently, the amount of litter that accumulates in a mangrove community varies inversely

Table 12.5 Total annual dry weight yield of litter in various mangrove associations in Australia (after Bunt, 1982)

Dominant mangrove species	Leaves	Stipules	Twigs	Reproductive tissue	Other debris	Total
			(g dry weight m^{-2} yr^{-1})			
Avicennia marina	556.5	0.0	57.6	16.4	101.5	732.0
Bruguiera gymnorhiza	467.6	21.2	119.0	165.4	87.9	861.1
Ceriops tagal	419.1	1.8	55.6	31.2	244.8	752.5
Lumnitzera littorea	604.4	0.0	34.6	6.6	17.2	662.8
Rhizophora apiculata	600.4	98.5	88.8	250.9	76.2	1114.8
Rhizophora lamarkii	428.0	67.3	87.6	68.8	71.6	723.3
Rhizophora stylosa	558.9	82.8	87.1	131.5	74.5	934.8
Xylocarpus granatum	453.7	10.7	148.8	92.0	101.5	806.7

(Source: B. F. Clough, ed., *Mangrove Ecosystems in Australia: Structure Function and Management*; published by Australian Institute of Marine Science, 1982, reproduced with permission.)

with tidal flushing, and *in situ* decomposition becomes the dominant process only in the more protected inland areas (Lugo and Snedaker, 1974). Typically the amount of litter remains fairly constant and most is lost within a year. The leaves of *Avicennia marina* are mechanically weakened and begin to fragment within a month and ultimately all that remains is a small amount of refractory material (van der Valk and Attiwill, 1984). Some litter may be consumed by crabs and other invertebrates but colonization by bacteria, fungi and protozoans is eventually necessary for complete mineralization. The rate of microbial decomposition is related to temperature, moisture and oxygen levels within the substrate and is affected by the tidal regime (Steinke and Ward, 1987). An optimal moisture content of 50% is reported for microbial activity in mangrove sediments and this is characteristically reached at low tide in regularly flooded soils (Hesse, 1961). Microbial activity is also affected by litter quality. Thus the delay in decomposition of *Rhizophora mangle* leaf litter is attributed to its high tannin content (Cundell *et al.*, 1979) and subsequent decomposition is slowed by its comparatively high C:N ratio (Twilley *et al.*, 1986).

Very little of the leaf biomass is consumed by herbivores: depending on the tidal regime, some of the production enters the detritus food web within the mangrove community and varying amounts are exported to adjacent coastal systems (Odum and Heald, 1975). The detritivores are classed according to the size of material they consume. Shredders and grinders, such as amphipods and crabs, mostly chew large leaf fragments, and the finer leaf debris which settles on to the surface of the mud is consumed by fish and other deposit feeders. The finest particles which remain suspended in the water column are removed by filter feeders such as mussels. The nitrogen content of mangrove detritus increases with the growth of bacteria and fungi (Rice and Tenore, 1981) and in organically enriched sediment because of nitrogen fixation in the rhizosphere (Zuberer and Silver, 1978). In this way the protein content of freshly fallen leaves of *R. mangle* increases during decomposition, which consequently raises its value as a food for detritivores and sustains a diverse heterotrophic community (Heald, 1971).

Mangal provides a wide range of habitats for terrestrial and marine animals ranging from the tree canopy and rot holes in the trunks and branches to the tidal muds and water courses. Many species of bird are associated with mangal but few are restricted to this environment: they mostly feed in the creeks and pools and on the abundant insect fauna (Macnae, 1968). Flying foxes, monkeys and arboreal snakes are typically present in Asia, where the mangroves often merge with lowland rain forest. These animals are typically separated from marine organisms by their habitat requirements, although intermingling of the two populations is facilitated by tidal fluctuations. Thus, the crab-eating macaque (*Macaca uris*) mostly feeds on crabs and molluscs which it collects in the shallow tidal water. A pronounced vertical zonation similar to that on rocky shores is characteristic of some molluscs and barnacles that live on the prop roots and trunks of the trees. Crabs are mostly found at higher levels on the shore where aeration of the muddy sediment is often improved by their burrows. Some are carnivorous but most feed on detritus. The filter-feeding gastropods and bivalves are similarly most abundant in the seaward fringe of mangal, while species such as penaeid prawns migrate with the tides. Mudskippers are perhaps the most distinctive species associated with the mangrove forests. These small amphibious fish are found in the seaward fringe of mangal throughout South East Asia, Africa and northern Australia. At low tide they move across the mud flats by using their pelvic fins as crutches or by a jumping motion achieved by pushing their tails into the mud. All species of mudskippers swim and breathe like other fish when submerged. Oxygenated water held in the buccal cavity is similarly passed over the gills in species such as *Periophthalmus chrysospilos*, which climb into the trees just ahead of the rising tide (Macnae, 1968). As well as contributing to the marine food web through the production of detritus, mangroves are also used commercially for fuelwood, timber and other products. Thus mangroves, unlike most coastal ecosystems, have traditionally been viewed as an important economic resource and sylvicultural management practices are used throughout Asia to maximize wood production (Christensen, 1983).

12.11 HUMAN IMPACT ON COASTAL AND MARINE ECOSYSTEMS

Apart from the utilization of mangroves, commercial exploitation of coastal ecosystems was comparatively limited until their value for tourism and recreation was recognized. In some parts of the world seaweeds are collected for use in foods such as dulce and laver, and they are commercially important as a source of carrageenin, agar, algin and other products. Similarly, salt marshes and sand dunes are used to graze sheep and cattle, while shellfish are collected from estuarine mudflats throughout the world. The aesthetic quality of coastlines has resulted in a great increase in permanent and seasonal human residents and the effects are becoming increasingly evident, especially in fragile ecosystems such as sand dunes. At the same time commercial and industrial development has had a great effect on coastal regions, and tremendous quantities of waste materials have been dumped into the water without concern or knowledge of their environmental impact. Sewage and other organic wastes account for the greatest volume of discharge into coastal waters and estuaries. Oxygen levels are severely depleted where circulation is relatively poor, and further degradation depends on slower anaerobic processes. Many of the native plant and animal species are eliminated under these conditions. The effects of industrial discharges vary with the nature of the materials involved. The release of cooling water from power stations and factories usually has little impact on organisms beyond the immediate point of discharge, although in tropical areas heat dissipation may be slow and lethal water temperatures can persist over much greater areas. Inert particulate matter such as colliery waste may clog the feeding and respiratory structures of animals and plant photosynthesis may be reduced in the turbid water. Heavy metals such as mercury and lead and various organic chemicals such as PCBs (polychlorinated biphenyls) are not subject to bacterial attack and can accumulate in the tissues of plants and animals with harmful effects. Shipwrecks and marine accidents also contribute to the problem of pollution, with the most dramatic effects occurring as crude oil washes ashore. In addition, some coastal habitat has been modified through reclamation. The most extensive reclamation has occurred around the Ijsselmeer in The Netherlands, but numerous other projects in all parts of the world are progressively converting estuaries and bays to farmland, or to sites for residential or commercial development.

A variety of materials such as dredging spoil, fly ash and sewage sludge are dumped offshore in designated areas. Similarly, more dangerous materials such as radioactive wastes have, until recently, been dumped in deep water in the Atlantic and Pacific Oceans. The contaminated materials are packed in specially designed concrete-lined steel drums for disposal at depths of up to 4400 m. Ultimately the containers will rupture and slowly release their contents. The risk of contamination of the pelagic fishes is reduced because they do not feed on the sparse abyssal species, but it is possible that radioactive materials may eventually be brought to the surface by currents and incorporated into the food chain in this way. However, the level of contamination is considered to be minimal (Clark, 1989).

The ocean food webs terminate with the demersal and pelagic fishes and these have long been exploited as a source of food. About 95% of the world fishery catch comes from near-shore and reef areas, and these are also the nursery grounds for many species which spend their adult life in deeper water. The world catch of marine fish rose to 91 million tonnes in 1989 (Tolba, 1992) and what was originally thought to be an inexhaustible resource is now seriously threatened by overfishing. In 1992 the Government of Canada enacted a 2-year fishing ban on cod in order to reverse the depletion of stocks in the North Atlantic, and quotas have similarly been imposed for many other important species. Not only is there a decline in the total number of fish caught, but the size of the fish is also reduced by overfishing and the full potential of the resource is lost. Excessive exploitation can lead to the ultimate collapse of the breeding population and, as in the case of the whales, extraordinary efforts are needed in order to save the species from extinction. Environmental standards for the world's oceans were set out in the Law of the Sea in 1982, and beyond the 200-mile territorial limits this vast international resource is under the protection of the United Nations.

CHAPTER 13

The prospect of change

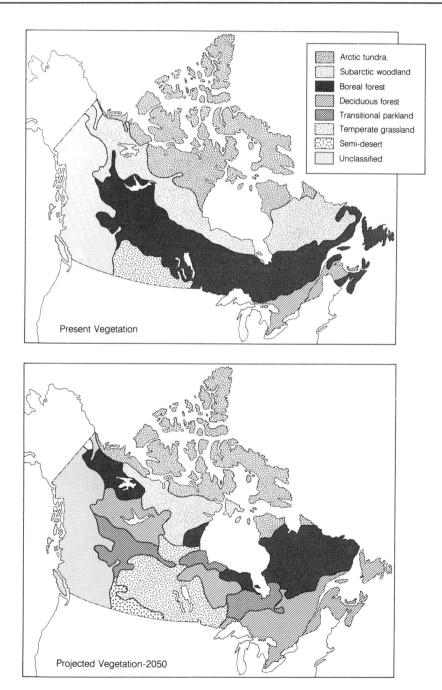

Figure 13.1 Present distribution of the major ecosystems in Canada and the potential response by the year 2050 to climate change as a result of a two-fold increase in atmospheric CO_2 levels. (After Rizzo and Wiken, 1992.) (Reproduced with permission from B. Rizzo and E. Wiken, Assessing the sensitivity of Canada's ecosystems to climate change, *Climatic Change*, 1992, **21**, 43, 50.)

13.1 THE DEMAND FOR NATURAL RESOURCES

The ability to adapt has allowed plants to survive in all but the most hostile of environments, and the broad vegetation patterns represented by the major biomes reflect the diversity of habitat conditions which have controlled this evolutionary process. Natural ecosystems are highly dynamic and evolve in response to environmental stimuli, but whereas in the past this process was conditioned only by natural causes, humans and their technology are now important agents of biogeographic change. The world's population increased from about 3.6 billion in 1970 to 5.5 billion in 1992 and is expected to reach 6 billion by 1998, perhaps rising to 11.6 billion by 2150 (Tolba and El-Kholy, 1992). Population growth has been accompanied by the development of mechanized agriculture and urban-industrial societies. This has greatly increased the demand for natural resources and created the problem of undesirable wastes. However, the relationship between population growth, resource development and the state of the environment varies regionally. Developed countries are mainly faced with problems of deterioration or depletion of global resources. Developing countries are more concerned with over-exploitation of their own ecological resources as their growing populations move into marginal lands: this has resulted primarily in a loss of forest cover, desertification and other forms of land degradation.

Approximately 17 million ha of forest was reportedly destroyed in Africa, Latin America and Asia in the period 1981–90: this is equivalent to a rate of 0.9% yr^{-1}. Some of the land is cleared for agriculture or simply to provide fuelwood. Where the forests are being exploited for timber the rate of extraction is normally much higher than the rate of natural regeneration, and regrowth is often prevented by the influx of slash-and-burn agriculturalists into logged-over areas. Consequently, less than 10% of the world's tropical forests is currently under sustainable management (Poore *et al.*, 1989). It is estimated that about 300 million ha of agricultural land will become unproductive during the period 1975–2000 and that this will be replaced by bringing new areas into cultivation (Buringh and Dudal, 1987). The problem is most acute in the developing countries of Africa and Asia but it is not restricted to these regions. Considerable losses have occurred in areas such as Australia and North America, mostly as a result of overgrazing of the rangelands. The growth of cities has similarly contributed to the loss of farmland: in North America, for example, the rate of conversion is about 2.5 million ha every decade. In addition to forests and rangeland, significant losses are also reported in wetland ecosystems, mainly because of drainage, landfill or pollution. General figures on land conversion can be misleading: in Europe, for example, forest area has increased in recent years. How-ever, this is mainly because of large-scale planting of species-poor coniferous monocultures: natural woodland continues to decline (Groombridge, 1992).

13.2 THE IMPACT OF CLIMATE CHANGE

Urbanization and the growing demand for goods and services has brought an enormous increase in industrial activity and consequently a growing demand for mineral and energy resources. The problems associated with industrial development vary with the type of operation and the environment in which it is carried out. The effects of mining and extraction are relatively localized but the waste products of processing and manufacturing have a much greater impact on the environment. Atmospheric pollution is the most widespread of these adverse effects. Past concerns about atmospheric quality mostly focused on the detrimental effects of sulphur dioxide and other gases on the vegetation in the immediate vicinity of industrial sites, but more recent studies emphasize the global effects of pollution such as acid rain, the depletion of the ozone layer and atmospheric warming.

Oxides of sulphur and nitrogen are the principal agents of acid deposition. Conifers and lichens are especially sensitive to these pollutants, which have also caused extensive damage to freshwater ecosystems. The problem of **acidification** is most pronounced in eastern North America and western Europe but it is also a threat in some tropical regions where the soils have a low buffering capacity (Rodhe and Herrera, 1988). Depletion of the **ozone layer** is largely attributed to the increased concentration of **chlorofluorocarbons** (**CFCs**) in the atmosphere. The primary impact of ozone depletion is to increase the amount of UV-B radiation that penetrates the Earth's atmosphere. Preliminary studies indicate that some plants can adapt to enhanced UV-B conditions by chemically altering the attenuation capacity of the epidermis (Robberecht and Caldwell, 1983) but often these conditions affect leaf development and growth is ultimately reduced (Teramura, 1983). Thus the composition of an ecosystem could eventually reflect the competitive ability of the species according to their sensitivity to UV-B.

Ozone depletion has also been linked to **global warming** because of the ability of CFCs to absorb IR radiation and enhance the greenhouse effect. The concentration of CFCs in the atmosphere is increasing at the rate of 4% yr^{-1} despite dramatic changes in their production and use. The other principal '**greenhouse gases**' that are affected by human activities are carbon dioxide, methane and nitrous oxide. Carbon dioxide is increasing at a rate of 0.5% yr^{-1}, mostly as a result of fossil fuel consumption and deforestation. Atmospheric concentrations of methane are increasing at the rate of 0.9% yr^{-1}. Most of the methane that enters the atmosphere is produced by anaerobic bacteria in natural wetlands: this is en-

Table 13.1 Climate change scenarios for the year 2050 predicted by various general circulation models (after Smith *et al.*, 1992)

| | Predicted change | |
General circulation model	Mean increase in global temperature (°C)	Mean increase in global precipitation (%)
Oregon State University (OSU)	2.8	7.8
Geophysical Fluid Dynamics Laboratory (GFDL)	4.0	8.7
Goddard Institute for Space Studies (GISS)	4.2	11.0
UK Meteorological Office (UKMO)	5.2	15.0

(Reprinted with permission from T. M. Smith, R. Leemans and H. H. Shugart, *Climatic Change*, **21**, Sensitivity of terrestrial carbon storage to CO_2-induced climate change: comparison of four scenarios based on general circulation models; published by Kluwer Academic Publishers, 1992.)

hanced by the creation of rice paddies. Conversely, emissions are reduced through drainage of wetlands. Cattle production is an additional source of methane. Nitrous oxides are produced by the consumption of fossil fuels and approximately 50% of the total human-made emissions are generated by the transport sector. The concentration of nitrous oxides in the atmosphere continues to increase at a rate of 0.25% yr^{-1}, mainly because of the greater number of vehicles in use (Tolba and El-Kholy, 1992).

Global temperatures increased by about 1 °C between 1861 and 1984 (Jones *et al.*, 1986). **General circulation models (GCMs)** used to predict climate change suggest that global warming will continue with temperatures rising by 2.8–5.2 °C by the year 2050 (Table 13.1). The greatest warming trend is expected in the polar regions where average winter temperatures could increase by 4–16 °C. It is anticipated that the effect will be less marked in oceanic regions because of the large heat capacity of the water but very different weather patterns and higher precipitation are predicted over land masses. Global precipitation is expected to increase by 8–15%, with the greatest change occurring in polar and tropical regions (Smith *et al.*, 1992). The increase in water vapour would itself contribute to the greenhouse effect but it is not known how this change in the amount and distribution of the cloud cover would affect the global radiation budget (Rind *et al.*, 1991).

The potential environmental impacts of climate change are many and varied. Sea level is currently rising by 2.4 mm yr^{-1} and is likely to increase with the accelerated melting of glaciers and ice caps (Peltier and Tushingham, 1989). Similarly, an increased rate of thawing is predicted for permafrost. On a global scale, the distribution of plant and animal species is largely controlled by climate and it is assumed that changes in temperature and precipitation will affect the distribution of the major ecosystems (Emanuel *et al.*, 1985). In Canada, for example, it is predicted that by the year 2050 the area of arctic and subarctic ecosystems could be reduced by 18% and the boreal forest reduced by 14%. Conversely, changing climatic conditions would favour a 19% increase in the area of grasslands, with temperate forests increasing by 11% and semi-desert regions increasing by 2% (Figure 13.1). These are potential changes that are predicted by cli-

matic conditions: the character of the vegetation would ultimately depend on species migration rates and subsequent changes in soil conditions and other environmental factors (Rizzo and Wiken, 1992). Based on current trends it is predicted that temperatures in central North America will rise by 2–4 °C in winter and 2–3 °C in summer. Precipitation could increase by 0–15% in winter but decrease by 5–10% in summer with a consequent reduction in soil moisture of 15–20%. This would have a dramatic effect on the agricultural potential of the region, with changes in land use subsequently transmitted to the natural ecosystems.

Similar predictions on a global scale suggest that the areal coverage of tundra and desert will decrease, while grasslands and forests will expand in response to CO_2-induced climate change (Smith *et al.*, 1992). The transition from tundra to boreal forest accounts for most of the predicted increase in the global extent of forest. Expansion of the grasslands would occur in the southern boreal forest, where tree cover is expected to decline because of drier conditions. However, higher precipitation is expected to cause a transition from desert to grassland or forest in temperate and tropical regions. The degree of change varies according to the different general circulation models that are used (Table 13.2). Predicted responses in forest cover are the most variable, with dry forest increasing by 0 to 71% of present area, and moist forest changing by − 10 to + 11%. This uncertainty is further illustrated by the composite changes which are predicted for the moist forests (Figure 13.2). The greatest decline in area is predicted for subtropical forests but this is mostly replaced by an increase in the area of tropical forest. This, together with the replacement of tundra and desert by more productive ecosystems, suggests that terrestrial carbon storage would increase by 0.4–9.5%. The potential reduction in atmospheric CO_2 of 4–85 ppm that results from this is incorporated into the GCM scenarios.

13.3 BIODIVERSITY AND SPECIES EXTINCTION

Natural ecosystems consist of integrated, self-regulating communities of organisms in equilibrium with their

Table 13.2 Changes in the areal extent of major biome-types under present climatic conditions and as predicted by various general circulation models of climate change (after Smith *et al.*, 1992)

Biome-type	Current area ($\times 1000 \, km^2$)	Change in area predicated by GSM ($\times 1000 \, km^2$)			
		OSU	GFDL	GISS	UKMO
Tundra	939	− 302	− 515	− 314	− 573
Desert	3699	− 619	− 630	− 962	− 980
Grassland	1923	380	969	694	810
Dry forest	1816	4	608	487	1296
Moist forest	5172	561	− 402	120	− 519

(Reprinted with permission from T. M. Smith, R. Leemans and H. H. Shugart, *Climatic Change*, **21**, Sensitivity of terrestrial carbon storage to CO_2-induced climate change: comparison of four scenarios based on general circulation models; published by Kluwer Academic Publishers, 1992.)

physical and chemical environment. Consequently, any changes in individual species populations affect the stability of the entire ecosystem. Through the process of natural selection, species are variously adapted to conditions existing in the ecosystem. Species regulate natural fluctuations in their environment through homeostatic mechanisms, so that productivity and performance are

thought to be enhanced by positive feedback under conditions of stability (Connell and Orias, 1964). In addition, the stability of ecosystems is presumed to increase with maturity and become more resilient to disturbance, primarily because of greater efficiency in resource use (DeAngelis, 1980) and better modulation of environmental extremes by dominant species with wide ecological

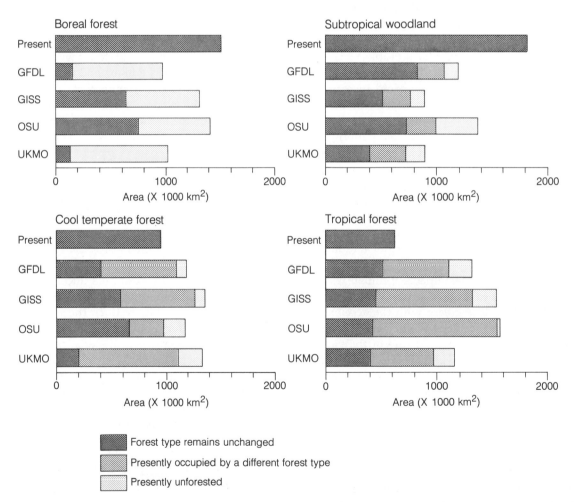

Figure 13.2 Areal coverage of mesic forest types under present climatic conditions and four climate change scenarios. (After Smith *et al.*, 1992.) (Reproduced with permission from T. M. Smith, R. Leemans and H. H. Shugart, Sensitivity of terrestrial carbon storage to CO_2-induced climate change: comparison of four scenarios based on general circulation models, *Climatic Change*, 1992, **21**, 379.)

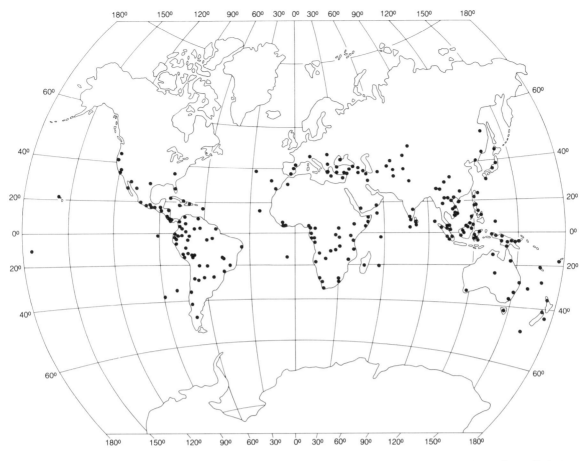

Figure 13.3 World distribution of centres of plant diversity. (After Groombridge, 1992.) (Reproduced with permission from B. Groombridge, ed., *Global Biodiversity: Status of the Earth's Living Resources*; published by Chapman & Hall, 1992.)

Van der Grinten Projection

amplitudes (Langford and Buell, 1969). Stability is determined by the ability of the component species to persist within the normal range of variation that occurs within the environment. High species diversity is characteristic of environments that provide favourable and constant conditions. The process is self-augmenting in that the addition of new species promotes the evolution of other life forms. However, overspecialization and small species populations increase their sensitivity to environmental stress, and the structure and function of the entire ecosystem must necessarily adjust to these changes. Community respiration tends to increase under conditions of stress and biomass accumulation is consequently reduced. Similarly, nutrient turnover increases but is less efficient, so that more nutrients are lost from an ecosystem under stress. Typically species diversity declines; there is a decrease in the size and life span of the residual organisms; weedy plants with readily dispersed propagules become more abundant; and the ecosystem is dominated by fewer, more tolerant species (Odum, 1985).

Species diversity is generally highest in warm, moist regions and decreases with latitude and aridity. Some species are widely distributed but many others are restricted to specific locations of unusually high diversity. Approximately 50 000 species, or 20% of all flowering plants, are endemic to only 18 locations representing 0.5% of the Earth's land surface. Fourteen of these '**hot spots**' are in tropical forests and the remainder are in mediterranean regions: they are considered areas of high conservation priority because of their unique flora (Myers, 1990). On a broader scale, the World Conservation Union (IUCN) plant conservation programme recognizes some 250 **Centres of Plant Diversity (CPD)**. These areas are particularly rich in plant life and would, if protected, safeguard the majority of wild plants in the world. An area is designated as a CPD on the basis of botanical importance rather than the degree of threat to the ecosystem, and normally contains a large number of endemic species amongst its comparatively rich flora (Groombridge, 1992). Most CPDs are in the plant-rich tropics where it is usually not possible to identify threatened plant species individually (Figure 13.3).

Habitat destruction and fragmentation are presently recognized as the most serious cause of species extinction.

Table 13.3 Number and status of endemic plant species on selected oceanic islands (after Groombridge, 1992)

	Endemic species				
	Extinct	Endangered	Vulnerable	Rare	Total
Cuba	25	322	294	142	3233
Jamaica	0	76	137	122	827
Hawaiian Is.	87	121	23	92	731
Canaries	1	127	120	129	593
Mauritius	21	75	45	55	236
Socotra	1	33	15	52	267
Ogasawara	4	22	21	21	152
Galapagos Is.	2	8	11	54	148
Juan Fernandez	1	54	38	0	123
Reunion	1	13	14	14	120
Madiera	0	17	29	34	118
Marquesa	1	18	13	7	105
Cyprus	0	7	9	28	90
Lord Howe Is.	1	3	1	72	84
Balearic Is.	1	8	11	23	70
Seychelles	0	17	26	7	63

(Source: B. Groombridge, ed., *Global Biodiversity: Status of the Earth's Living Resources*; published by Chapman & Hall, 1992.)

Table 13.4 Threatened animal species (after Groombridge, 1992)

	Number of species threatened	Approximate number of described species	Percentage of species threatened
Vertebrates			
Mammals	507	4 300	11.8
Birds	1029	9 700	10.6
Reptiles	169	4 800	3.5
Amphibians	57	4 000	1.4
Fishes	713	20 000	3.6
Invertebrates			
Molluscs	409	50 000	0.8
Corals and sponges	154	9 000	1.7
Insects	1083	750 000	0.1
Spiders	18	68 000	< 0.1
Crustaceans	126	42 000	0.3
Others	187	28 000	0.7

(Source: B. Groombridge, ed., *Global Biodiversity: Status of the Earth's Living Resources*; published by Chapman & Hall, 1992.)

Estimated rates of extinction based on floristic richness and projected deforestation rates indicate that 2–8% of tropical forest species will be lost between 1990 and 2015 (Reid, 1992). Island floras are equally sensitive, and 30% of plants threatened with extinction are island endemic (Groombridge, 1992). Damage to most island floras occurred during the period of European exploration and colonization, usually because of the introduction of goats and other livestock. Subsequent introductions of invasive, competitive plants also contributed to the loss of native species (Table 13.3). Approximately 23 000 plant species and subspecies of plants are threatened worldwide and over 800 species are known to have become extinct in historic times. However, such information is strongly biased geographically and many undocumented species could undoubtedly be added to these numbers.

A total of 4452 animal species were listed as threatened in the 1990 IUCN Red List (Table 13.4). The percentage of threatened species is highest for mammals (11.8%), birds (10.6%) and fish (3.5%). In comparison the list of threatened insects represents less than 0.1% of recorded species. Vertebrates may be more vulnerable to extinction because they are larger and require more resources. However, many invertebrates have very localized ranges, which makes them very susceptible to habitat loss. As with plants, the major threat to animal species is loss or fragmentation of habitat through cultivation, logging, pastoral development and settlement. Other causes include over-exploitation for commercial or subsistence reasons, accidental or deliberate introductions of competitive or predatory species, deliberate eradication of pest species, and disease. The majority of threatened

Table 13.5 Number of threatened vertebrate species (and percentage of locally known species) ranked for the leading eight countries in each class (after Groombridge, 1992)

Mammals		Birds		Reptiles		Amphibians		Fishes	
Madagascar	53 (50)	Indonesia	135 (9)	USA	25 (–)	USA	22 (–)	USA	164 (–)
Indonesia	49 (10)	Brazil	123 (10)	India	17 (4)	Italy	7 (21)	Mexico	98 (–)
Brazil	40 (10)	China	83 (8)	Mexico	16 (2)	Mexico	4 (1)	Indonesia	29 (–)
China	40 (10)	India	72 (7)	Bangladesh	14 (12)	Australia	3 (2)	South Africa	28 (–)
India	39 (12)	Colombia	69 (4)	Indonesia	13 (3)	India	3 (1)	Philippines	21 (–)
Australia	38 (13)	Peru	65 (4)	Malaysia	12 (4)	New Zealand	3 (100)	Australia	16 (–)
Zaire	31 (7)	Ecuador	64 (4)	Brazil	11 (2)	Seychelles	3 (25)	Canada	15 (–)
Tanzania	30 (10)	Argentina	53 (–)	Colombia	10 (3)	Spain	3 (12)	Thailand	13 (–)

(Reproduced with permission from B. Groombridge, ed., *Global Dioversity: Status of the Earth's Living Resources*; published by Chapman & Hall, 1992.)

mammalian species live in tropical countries such as Madagascar, Indonesia and Brazil (Table 13.5) and are primarily affected by destruction of tropical forest (Figure 13.4). Threatened bird species are similarly distributed, although flightless and ground-nesting species are especially vulnerable on oceanic islands due to introduced

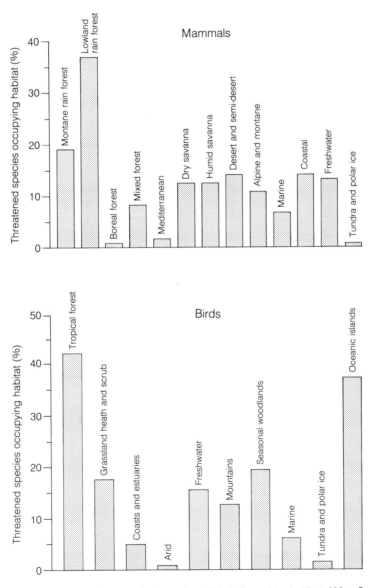

Figure 13.4 Distribution of threatened species of mammals (upper) and birds (lower) by habitat. (After Groombridge, 1992.) (Reproduced with permission from B. Groombridge, ed., *Global Biodiversity: Status of the Earth's Living Resources*; published by Chapman & Hall, 1992.)

Table 13.6 Land-use zone classification for Canadian National Parks (after Tolba and El-Kholy, 1992)

Zone I	Species preservation: these areas contain rare and endangered habitats and are strictly protected with access either prohibited or controlled.
Zone II	Wilderness: This is the best representation of the natural region of the park (60–90% of area). The objective is resource preservation, and use is dispersed with only limited facilities.
Zone III	Natural environment: access is primarily non-motorized, and the area acts typically as a buffer zone.
Zone IV	Recreation: major overnight facilities, e.g. camp sites, are concentrated in this area.
Zone V	Park services: this area is characterized by highly modified landscapes, but usually comprise less than 1% of park area.

(Source: M. K. Tolba and O. A. El-Kholy, eds, *The World Environment 1972–1992: Two Decades of Change*; published by United Nations Environment Programme, Information and Public Affairs Division, and Chapman & Hall, 1992.)

predators such as rats and mongooses. In contrast, many of the threatened species of reptiles, amphibians and fish are associated with temperate regions.

13.4 THE VALUE OF BIODIVERSITY

The economic value of plants for food, medicine and raw materials is immense but the importance of plant biodiversity is still poorly understood. Approximately 3000 plant species are regarded as sources of food: about 200 of these have been domesticated, but only 15–20 are crops of major economic importance (Groombridge, 1992). The development of high-yielding cultivars tends to produce genetically uniform crops and these are becoming more widely distributed at the expense of their wild relatives. The wild plants are an important genetic resource and are most commonly used in crop-breeding programmes to improve resistance to pests and diseases. Consequently, improvement in yield and increasing reliance on agrochemicals is contributing to the loss of genetic diversity of the world's food plants, which could affect future breeding programmes. Similarly, over 21 000 plants with medicinal uses are listed by the World Health Organization: of these only 5000 species have been thoroughly investigated as potential sources of new drugs. The most widely used medicinal plants are temperate species; little work has been done on the biochemical potential of tropical plants. Medicinal plants are still mainly harvested from the wild. Quinine (*Cinchona* spp.) was one example of a medicinal plant developed as a major crop species. The original supply was derived from wild *Cinchona* trees in the Andes but plantations were successfully established in Indonesia in the 1860s because of concerns, even then, that it might become extinct. About 100 species of medicinal plant are now cultivated in China, where over 700 000 tons are processed annually for traditional prescriptions (Xiao, 1991).

Many plant-derived pharmaceuticals are now synthetically manufactured, but the use of medicinal plants continues to grow and is expected to exceed $US500 billion by the year 2000 (Principe, 1991). About 120 clinically useful prescription drugs are derived from 95 species of higher plants, 39 of which originate from tropical forests. Only 23 tropical forest species are presently used in drugs marketed in the United States: their import value currently exceeds $US20 million, but this represents less than 20% of all plant-derived medicines in use in the United States (Farnsworth and Soejarto, 1991). However, 1300 forest species are used by indigenous peoples in the north-west Amazon for medicines, poisons and narcotics, and at least 6500 species are used in herbal remedies in East and South East Asia. Ethnobotanists estimate that 35 000–70 000 plant species are used medicinally in this way throughout the world, and most of these originate from the tropical forests. Few of these plants have been examined in detail, and although commercial use of tropical species is relatively small, there is an urgent need to preserve and document them before they are lost.

The economic value of agricultural, timber and medicinal products can be estimated fairly accurately but it is much more difficult to assess the full value of plant resources unless their contribution to society is fully appreciated. It may be possible to calculate the timber value of a tree species, for example, but its value in watershed protection, recycling of oxygen and climate control is largely ignored. However, the continual loss of natural environments has led to a growing awareness of the non-consumptive value of these shrinking resources. In economic terms, the process of conversion ceases once the perceived value of the wild resources cannot be increased. The growth of ecotourism is one example of the current trend away from traditional resource exploitation. The natural environment, often enhanced by a distinctive local culture, is the focus of ecotourism and so the unspoiled nature of exotic ecosystems is beginning to have a real economic value. For example, it is estimated that each lion in an African National Park has an annual visitor-attraction value of $US27 000 and each elephant herd is worth $US610 000 (Tolba and El-Kholy, 1992). Nature tourism is the leading foreign-exchange earner in countries such as Nepal, Kenya and Costa Rica, with total worldwide income in developing countries estimated at $US2–12 billion. However, careful management is needed to minimize environmental degradation and so maintain demand. Access to natural sites often leads to loss of plant and animal species. For this reason

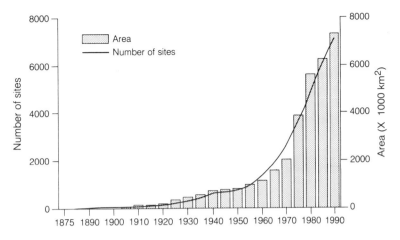

Figure 13.5 Global trends in the number and area of protected sites during the period 1875–1990. (After Groombridge, 1992.) (Reproduced with permission from B. Groombridge, ed., *Global Biodiversity: Status of the Earth's Living Resources*; published by Chapman & Hall, 1992.)

the Canadian Parks Service has established a 'Visitor Activity Management Process' which defines objectives, opportunities and constraints for visitor use. In this way the National Parks are classified according to five land-use zones to keep tourist impact at acceptable levels (Table 13.6).

13.5 CONSERVATION OF NATURAL RESOURCES

Conservation action is typically incorporated into the legal systems of established governments. Although policies vary from country to country, protection of plant species is normally legislated through restrictions on collection and trade, by prohibiting habitat destruction and by controlling the introduction of exotics. Similar legislation covers wild animal species, although international treaties are more common, especially for marine species. These policies have been complemented by the establishment of protected areas. The initial purpose of protected areas was generally to preserve spectacular scenery and to provide recreational areas, but in recent years the concept has grown to encompass the habitats of endangered species and ecosystems of great biodiversity (Groombridge, 1992). The IUCN recognizes 10 classes of protected areas, each of which have different management objectives. Included in this system are 96 **World Heritage Sites**, so designated because of their outstanding aesthetic or scientific qualities, and 300 **Biosphere Reserves** established during the UNESCO–MAB (Man and the Biosphere) Programme primarily for research and environmental monitoring. The total number of protected areas has increased considerably since 1972 (Figure 13.5). Approximately 8500 sites are currently listed, covering about 7.8 million km² or 5.2% of the Earth's land surface. The creation of Greenland National Park in

1974 (972 000 km²) and Great Barrier Reef Marine Park in the 1980s (340 000 km²) has had a considerable effect on the total protected area, so that more than 35% is contained within a few large reserves. More than 50% of the sites are smaller than 100 km², which suggests that fragmentation may ultimately limit conservation success. In addition, the major biomes are not uniformly protected, with lakes and temperate grasslands being especially under-represented (Table 13.7). Protected areas covering 6000 km² in Antarctica are excluded from these data.

Conservation of plant species is most readily achieved through habitat protection as part of the protected areas network. In Australia, for example, about 50% of nationally threatened species are in permanent protected areas: figures as high as 75% are reported for southern Africa and Britain, and 100% for Czechoslovakia (Groombridge, 1992). These countries have had long-standing programmes to identify and conserve threatened species but in many parts of the world threatened species are outside of protected areas. The habitat protection approach does not require detailed knowledge of the status of the species within protected areas and so is particularly useful in floristically diverse biomes such as tropical forests. The major limitation is that specific habitat conditions for many species will not be available unless extensive areas are set aside. Botanic gardens also play an important role in conservation. However, relatively few botanic gardens are located within the tropics (Table 13.8), although it is here that most new gardens have been created in recent years. Most botanic gardens now recognize that priority should be given to wild species, and collections of local native flora are gradually replacing exotic species. Such activities are co-ordinated at the international level by Botanic Gardens Conservation International (BGCI). Currently about 10 000 species of rare and threatened plants are now in cultivation, although the number of individuals is considered inadequate for

Table 13.7 Distribution and coverage of protected areas by biome type (after Groombridge, 1992)

	Protected areas		Biome area (km^2)	% of total area
	Number	Area (km^2)		
Tropical humid forests	501	522 000	10 513 000	4.96
Tropical dry forests	807	818 300	17 313 000	4.73
Subtropical and wet temperate forests	935	366 100	3 928 000	9.32
Tropical grasslands	56	198 200	4 265 000	4.65
Warm-winter deserts	296	957 700	24 280 000	3.94
Cold-winter deserts	139	364 700	9 250 000	3.94
Evergreen sclerophyllous woodland	786	177 400	3 757 000	4.72
Temperate grassland	196	70 000	8 977 000	0.78
Deciduous broad-leaf forests	1509	357 000	11 249 000	3.17
Boreal forest	440	487 000	17 026 000	2.86
Tundra	81	1 643 400	22 017 000	7.46
Mixed mountain systems	1265	819 600	10 633 000	7.71
Mixed island and marine systems	501	246 300	3 244 000	7.59
Lakes	18	6600	518 000	1.28

(Reproduced with permission from B. Groombridge, ed., *Global Biodiversity: Status of the Earth's Living Resources*; published by Chapman & Hall, 1992.)

Table 13.8 Botanic gardens and those with membership in Botanic Gardens Conservation International (BGCI) (after Groombridge, 1992)

	Number of botanic gardens	Number of BGCI members
Asia		
USSR (former)	160	1
India	68	7
China	66	4
Japan	59	0
Malaysia	9	5
Other countries	75	23
Europe		
Germany	73	12
France	66	18
United Kingdom	60	31
Italy	48	10
The Netherlands	39	5
Czechoslovakia	34	1
Other countries	213	41
North and Central America		
United States	266	53
Mexico	30	7
Canada	18	7
Other countries	42	13
South America		
Colombia	13	4
Brazil	11	4
Chile	9	1
Argentina	9	1
Other countries	23	6
Africa		
South Africa	17	11
Egypt	5	1
Nigeria	5	1
Kenya	5	1
Other countries	37	12
Oceania		
Australia	60	22
New Zealand	17	6
Papua New Guinea	4	2
Other countries	3	1

(Source: B. Groombridge, ed., *Global Biodiversity: Status of the Earth's Living Resources*; published by Chapman & Hall, 1992.)

conservation. In addition to their collections of live plants, many botanic gardens maintain extensive collections of viable seed. The advantage is that they represent a readily stored, genetically diverse source of plant material, although they must be carefully monitored and tested for viability. However, it is estimated that 50 000 plant species produce seeds that either have no natural dormancy or cannot survive conditions of storage. *In vitro* storage of germplasm and subsequent growth of plants from pieces of meristematic tissue under laboratory conditions can sometimes be used to maintain species with recalcitrant seeds. Reintroduction of extinct or endangered plants may then be possible. A total of 29 species have been reintroduced in this way, but long-term success is relatively low and this approach should not be regarded as a substitute for habitat conservation.

13.6 UNDERSTANDING THE ENVIRONMENT

The functional role of ecosystems in maintaining global metabolism is expressed in the **Gaia hypothesis** (Lovelock, 1988). This concept suggests that the biosphere modulates geochemical cycles and so maintains a homeostatic equilibrium which is now threatened by the vast scale of human activities. However, some species must inevitably be lost if environmental conditions are changing. The geologic record attests to species extinction as a result of changing physiography and climate. Consequently, attempts to restore presently disturbed ecosystems to their previous state could lead to further loss of resilience unless the factors causing stress are removed. Advances in research have led to a greater understanding of the holistic nature of the environment and have demonstrated the importance of global monitoring systems. For example, the World Meteorological Organization (WMO) has recently established the Global

Table 13.9 Major international environmental agreements since 1971 (after Tolba and El-Kholy, 1992)

Species and habitat:

1971	Convention on Wetlands of International Importance Especially as Waterfowl Habitat.
1972	Convention Concerning the Protection of the World Cultural and Natural Heritage.
1973	Convention on International Trade in Endangered Species.
1979	Convention on the Conservation of Migratory Species of Wild Animals.
1980	Convention on the Conservation of Antarctic Marine Living Resources.
1991	Protocol to the Antarctic Treaty on Conservation.

Marine Pollution:

1972	Convention on the Prevention of Marine Pollution by Dumping of Wastes and Other Matter.
1973	International Convention for the Prevention of Pollution from Ships and the Protocol of 1978 Relating Thereto.
1982	UN Convention on the Law of the Sea.
1990	International Convention on Oil Pollution Preparedness, Response and Co-operation.

Pollution of the Atmosphere:

1979	Convention on Long-Range Transboundary Air Pollution.
1985	Helsinki Protocol on the Reduction of Sulphur Emissions or Their Transboundary Fluxes by at Least 30 Per Cent.
1988	Sofia Protocol on the Reduction of Nitrogen Oxides or Their Transboundary Fluxes.
1985	Vienna Convention for the Protection of the Ozone Layer.
1987	Montreal Protocol on Substances that Deplete the Ozone Layer.

(Source: M. K. Tolba and O. A. El-Kholy, eds, *The World Environment 1972–1992: Two Decades of Change*; published by United Nations Environment Programme, Information and Public Affairs Division, and Chapman & Hall, 1992.)

Atmospheric Watch (GAW) which monitors the chemical composition of the atmosphere and precipitation. Similarly, the Global Environmental Monitoring System (GEMS) established as part of the United Nations Environment Programme (UNEP) now monitors atmosphere and climate, environmental pollutants and renewable resources in 142 countries. The environment changes continually and the fundamental alternatives that control the biosphere are evolution or extinction. However, with the vast scale of human intervention, the adverse changes that are predicted are causing major concerns.

Some hold the view of '**Nature forgiving**' – that conditions will revert to the original state once the stress is removed. Another view is that of '**Nature unforgiving**' – that environmental impacts are cumulative and difficult to reverse. A third view is that of '**Nature resilient**' – that ecological systems are strengthened by environmental stress, enabling them to survive more extreme conditions until a critical threshold is exceeded. Central to this third approach is the idea that ecosystems should be managed in ways that permit them to adapt to changing conditions and so promote sustainable development (Tolba and El-Kholy, 1992). The development of sound environmental policies requires extensive documentation of causes and effects of past and present activities. The increasing amount of environmental research provides a better understanding of human impact on ecosystems but also demonstrates the great uncertainty associated with predictive modelling that arises because of the complex nature of the biosphere.

In the past the environment received little attention from government and industry, but growing public pressure backed by scientific research has focused attention on these issues. The traditional approach to environmental degradation has been to try to correct problems after the fact. This reactive approach is being progressively superseded by an attitude of anticipation and prevention and is reflected, for example, in greater concerns for pollution control, the screening of potentially hazardous substances before they are marketed commercially and the more efficient use of energy. The submission of an **environmental impact assessment** (**EIA**) now customarily precedes any large-scale development and is designed to identify and predict the effect of the project on the biophysical environment. Public hearings enable interested parties to identify environmental concerns, as well as social issues that might arise. This type of consultative process has resulted in much greater public participation in the development and implementation of environmental policies. Membership in environmental organizations has increased dramatically in recent years: this has had a major influence on the attitude of business and industry as more companies begin to realize the economic benefits of developing acceptable environmental standards. In the industrialized countries this has resulted in the emergence of a new consumer society that is very much concerned with the environment, and consequently there is increasing pressure to develop an economic system that favours conservation and sustainable use of resources. The primary environmental concern in developing countries is management of land and fresh water in order to maintain food production and the export of resources which support their economies. A number of countries have developed national environmental plans which are designed to achieve sustainable development, but implementation has proved difficult (Tolba and El-Kholy, 1992).

The inability of national policies to protect the global environment adequately is being rectified through international agreements co-ordinated within the UN and other world organizations (Table 13.9). However, there still remains the need for comprehensive data on many environmental problems. Without this it is difficult to assess the degree of damage that has been done, or to determine the optimum way to halt and repair this degradation.

Today it is estimated that every second over 200 tonnes of carbon dioxide are released and 750 tonnes of topsoil are lost; while every day 47,000 hectares of forest are destroyed, over 16,000 hectares of land are turned to desert [and] between 100 and 300 species become extinct.

(Tolba and El-Kholy 1992 *The World Environment 1972–1992: two decades of challenge*, p. 811).

How resilient the natural ecosystems of the world are to such pressures is a matter of conjecture.

Glossary

aapa mire Minerotrophic wet mire in boreal regions consisting of circular or elongated mounds covered with dwarf shrubs and *Sphagnum* and mainly sedges in the intervening depressions.

abiotic Non-living.

acid rain Precipitation with abnormally low pH mainly due to sulpate emissions from industrial processes.

acclimation Reversible change in physiological tolerance through gradual adjustment to a change in environmental conditions.

acrotelm The upper, periodically aerated and partly living layer in a *Sphagnum* bog.

active layer Shallow surface layer of the ground in regions of permafrost that is subject to seasonal thawing.

acumen Tapering, point-like extension of the distal end of a broad-leaf blade.

adaptation A heritable change in an organism's structure, function or behaviour whereby it is better suited to its environment.

adaptive hyperthermia Capacity of some desert mammals to survive abnormally high body temperatures.

adiabatic process In meteorology a change in sensible temperature caused by expansion (cooling) or compression (warming) of air as it rises or descends in the atmosphere.

adventitious Applied to a structure not arising in its normal place, e.g. a root from a stem.

aerenchyma Spongy parenchymatous tissue with thin walls and large air spaces that permits air to diffuse through stems and roots (especially in aquatic plants).

aerial root A root that originates above ground level.

aerobic Having molecular oxygen present in the environment.

albedo Percentage of solar radiation that is reflected from a surface.

Alfisols One of 10 soil orders in the US Soil Taxonomy System, comprising relatively young acid soils with a clay-enriched B horizon usually formed under deciduous forest in humid, temperate and subtropical climates.

allelopathy Adverse influence of living plants (excluding micro-organisms) on each other through chemical exudates.

allochthonous Applied to material (plant debris, rocks, etc.) which originated elsewhere and has been transported to its present location (contrasted with autochthonous).

allophane An amorphous hydrated aluminosilicate of variable composition.

allowable cut The amount of wood that can be harvested from a forest to meet management objectives.

alluvial fan Cone-shaped deposit of alluvium (usually associated with desert landscapes) laid down by a stream where it emerges from an upland on to a plain.

alluvium Material, such as clay, sand and gravel, deposited by rivers and streams.

amino acids Organic acids containing both amino (NH_2) and carboxyl (COOH) groups which are the fundamental constituents of protein.

ammonification The biochemical process whereby NH_3-nitrogen is released from organic compounds through decomposition.

amorphous material A mineral that has no definite crystalline structure.

amphicarpy Production of two types of fruits by a single plant.

anadromous Living in salt water but breeding in fresh water.

anaerobic Having no molecular oxygen present in the environment

angiosperm Flowering plant in which the seeds produced from ovules are enclosed in fruits developed from matured ovaries.

Andosols Soils developed on volcanic ash.

annual A plant that completes its life cycle from germination to death in one year.

anthocyanin The pigment usually responsible for pink, red or purple colours in plant tissues.

anticyclone A region of high atmospheric sea-level pressure.

aphotic zone The deeper parts of seas and lakes where light intensity is insufficient for photosynthesis.

apomixis Reproduction by seeds formed without fertilization.

Arcto-Tertiary Flora Forests of conifers and deciduous trees that covered higher latitudes of the northern hemisphere in mid-Cenozoic times (15–30 million years BP).

aril Fleshy covering of seeds usually developed from the funicle or stalk of the ovule and generally associated with tropical plants.

argillic horizon Soil horizon rich in clay minerals.

Aridisols One of 10 soil orders in the US Soil Taxonomy System, associated with arid regions and characterized by low organic status and accumulation of calcium and sodium salts.

arroyo A steep-sided, flat-floored stream channel in arid regions.

association A plant community of definite floristic composition, dominated by particular species and growing under uniform habitat conditions.

autotroph Organism capable of producing organic compounds from inorganic materials by using energy from sunlight or oxidation processes, e.g. plants with chlorophyll and certain bacteria.

awn Stiff, bristle-like extension characteristic of the chaff-like lower bract enclosing the florets of some grasses.

bacterioplankton Planktonic bacteria suspended in the water column of lakes and oceans.

Batesian mimicry Resemblance of a harmless animal to a poisonous or distasteful one which is often conspicuously marked in order to gain protection from predators.

base cation Positively charged soil ions such as potassium and magnesium that function as plant nutrients.

bathypelagic zone Deep portion of the ocean between 1000 and 4000 m not including the bottom.

benthos Organisms which live on or in the bottom of the oceans, lakes or rivers.

biennial Plant which completes its life cycle in two years, growing vegetatively in the first year and flowering in the second.

biogenic decalcification Precipitation of $CaCO_3$ in lakes induced by rapid photosynthetic utilization of dissolved inorganic carbon.

biogenic sediments Sediments of biological origin, such as the calcareous and silicous oozes on the ocean floor.

biogeography The study of the geographic distribution of plants and animals.

biogeographic realm A major division of the Earth according to the evolutionary affinities of the constituent flora and fauna.

biological oxygen demand (BOD) Oxygen used up by the decay of organic materials in aquatic environments.

biological spectrum Percentage distribution of plants in a region according to the life-form classes of Raunkiaer.

bioluminescence The biochemical production of light by living organisms.

biomass Total mass of living organisms within a defined trophic level, population or community.

biome A major terrestrial climax community defined by the physiognomic similarity of the dominant plants.

biosphere The biologically inhabited portion of the Earth, its atmosphere and seas.

biota All of the species of plants and animals in a specified region.

bipolar distribution Discontinuous distribution of a taxon in the high latitudes of the northern and southern hemispheres.

blanket bog Extensive area of acid peatland characteristic of broad, poorly drained upland regions in wet climates.

bloodworms Oligochaete worms and the aquatic larvae of chironomid midges characteristic of organically polluted rivers.

bloom The sudden appearance, of brief duration, of dense growths of algae in waterbodies.

bog An area of wet acid peat which is unaffected by nutrient-rich groundwater.

bordered pit Pits with an overarched rim surrounding the membrane or thin part of the cell wall (characteristic of gymnosperm tracheids).

bottom-up processes Regulation of population structure in an (aquatic) ecosystem by the productivity of the autotrophic organisms.

bundle sheath The sheath of parenchymatous cells that surrounds the vascular bundles of leaves.

bulliform cells Large, thin-walled cells associated with leaf rolling in grasses.

bunch grass A grass with a dense, tufted growth habit produced by many stems arising from the root crown.

C₃ plants Plants in which the first product of photosynthesis after carbon dioxide fixation is phosphoglyceric acid (PGA) which contains three carbon atoms.

C₄ plants Plants in which the first product of photosynthesis after carbon dioxide fixation is oxaloacetic acid (OAA) which contains four carbon atoms.

caliche A soil layer that is more or less cemented by calcium or

magnesium salts precipitated from the soil solution.

CAM plants (Crassulacean acid metabolism) Plants in which carbon dioxide taken up during the night is stored as malic acid until fixed using the C_3 pathway the following day – mostly succulent xerophytes.

carbon:nitrogen ratio The ratio (as a percentage) of the weight of organic carbon to the weight of total nitrogen in soil or organic material.

cation A positively charged ion.

cation exchange capacity (CEC) The total amount of exchangeable cations that a soil can absorb.

cataphyll Small scale-like leaf often serving for protection.

catotelm Lower, anaerobic layers in a peat deposit which is dead except for the aerenchymatous roots of angiosperms.

cauliflory The production of flowers directly from the trunk or major branches of woody plants.

cavitation The rupture of the water column in xylem vessels and tracheids.

cellulose Complex carbohydrate that is the principal component of cell walls.

chamaephyte A woody plant in Raunkiaer's life-form classification with perennating buds located 0–25 cm above the soil surface.

Chernozem An order of soils with dark-coloured, organic-rich surface horizons and high base saturation that has developed principally under subhumid grasslands.

chemoautotrophic bacteria Bacteria which produce organic material by oxidizing reduced inorganic compounds to obtain energy and utilizing carbon dioxide as a source of carbon.

chemoheterotrophic bacteria Bacteria which produce organic material by oxidizing or reducing organic material as a source of energy and carbon.

Chenopodiaceae A family of herbaceous plants, often with reduced fleshy or hairy leaves, that are typically tolerant of saline soils.

chinook A warm, dry wind that descends the eastern slopes of the Rocky Mountains of North America.

climax community Relatively stable plant community that is perpetuated through adequate reproduction under prevailing environmental conditions.

clear-cut Logging practice in which all merchantable trees are removed from an area.

co-dominant One of two or more dominant species in a stand.

cold hardiness The capacity of an organism to tolerate low temperatures.

colluvium Accumulated deposits of rock and soil moved downslope by the force of gravity.

community An assemblage of organisms characterized by a distinctive combination of species and co-occuring in space and time.

competition Interaction between organisms with similar resource requirements.

compound leaves Leaf with the blade divided into leaflets.

conspecific Pertaining to individuals or populations belonging to the same species.

cool season grass A temperate grass which grows mainly during the cooler part of the year, usually in spring.

coppicing A common woodland management practice in which trees or shrubs are felled near the ground, promoting adventitious shoots to arise from the stump.

cosmopolitan Organisms with essentially world-wide distribution.

crepuscular Active at dawn or dusk.

Cryosols An order of soils in the Canadian taxonomic system with mineral or organic horizons that are perennially frozen material within 1 m of the surface.

cryptogam A lower plant such as a moss, fern or liverwort that does not produce seeds.

cryptophyte A plant in Raunkiaer's life-form classification with perennating buds covered with soil or water.

culm The stem of grass and bamboos which is usually hollow except at the nodes.

cultural eutrophication Detrimental enrichment of aquatic ecosystems through input of nitrates and phosphates by human activities.

current annual increment (CAI) Amount of growth of a tree or forest stand that occurs in a specific year or annually at a specific age.

decreaser A species that decreases in population density under continued grazing.

deforestation Removal of trees from a forested area without adequate replanting.

deep-supercooling The ability of water to remain in a liquid state at temperatures many degrees below freezing.

demersal Living on the seabed.

denitrification Bacterial conversion of organic nitrogenous compounds to gaseous nitrogen.

depression In meteorology a region of low atmospheric pressure.

desertification Degradation of rangeland and cropland in subhumid areas through inappropriate agricultural activities, especially during times of drought.

detritivore Organism that feeds on dead or decaying organic matter.

diagnostic horizon A distinctive and rigorously defined soil horizon that is used to classify soils into orders within the Soil Taxonomy.

diaheliotropism Movement of leaf blades in dicotyledonous plants in such a way that they remain perpendicular to direct solar radiation (solar tracking).

diapause A period of suspended growth and reduced metabolism in the life cycle of an insect which increases resistance to adverse environmental conditions.

diatoms Unicellular algae, some of which are colonial with siliceous and typically highly ornamented cell walls.

diel Referring to the 24-hr period of day and night.

diffuse porous Wood in which an equal and random distribution of large xylem vessel members is produced throughout the growing season.

dimictic In lake classification, referring to two seasonal periods of water circulation.

dimorphic Having two distinct forms.

dinoflagellates Motile biflagellate algae, often having cellulose cell walls comprised of discrete, sculptured plates.

dioecious Seed plants in which male and female flowers are borne on separate individuals.

dipterocarp A member of the Dipterocarpaceae, a family of tall tropical trees native to Malesia which produces two winged fruits.

disjunct A geographically isolated population or species beyond the characteristic range of the taxon.

dispersal The actual transfer or movement of organisms or their disseminules from one place to another.

diurnal Referring to daytime, in contrast to nocturnal.

dormancy A condition in organisms in which metabolic processes are relatively quiescent.

drip tip The long, attenuate tip characteristic of many leaves in tropical lowland forests.

ecological inversion surface Hypothetical plane within a canopy defined by the height where the actual microclimate crosses the average conditions within a community.

ecological tolerance The range of environmental conditions in which an organism can function.

ecotourism Visiting a location primarily to observe and enjoy the natural environment.

ecotype A population within a species that is genetically adapted to a particular habitat.

elfin forest Stunted forest at higher elevations in moist tropical regions.

endemic Confined to a specific region and with a comparatively restricted range.

Entisols One of 10 soil orders in the US Soil Taxonomy System, comprising mineral soils lacking well-developed horizons.

environmental impact assessment (EIA) A report based on detailed studies that discloses the environmental consequences and course of action associated with a development project.

ephemeral Short-lived.

epigeal Applied to germination in which the cotyledons grow above the soil surface.

epilimnion In a lake, the uppermost, warm-water layer that lies above the thermocline.

epipedon A diagnostic soil horizon that has formed at the surface.

epipelagic zone Open water zone normally to a depth of 200 m.

epiphyte A plant that grows attached to another plant but derives no sustenance from it.

epontic flora Algae growing within the basal layer of sea ice.

euphotic zone The zone in aquatic ecosystems having sufficient light for photosynthesis.

eutrophic Referring to a waterbody that is rich in plant nutrients.

evolutionary convergence The independent evolution of similar structures in organisms that are unrelated.

exothermic Applied to a chemical reaction in which energy is released as heat.

facilitation model A postulated process in vegetation succession whereby habitat modification by an established species favours its replacement by other species.

facultative Having the ability to live under various conditions.

fell-field A ground surface in tundra regions that is littered with rock fragments.

fen An area of waterlogged peat that is fed by nutrient-rich groundwater and supports mainly sedges, grasses and brown mosses.

fire cycle Average period of time between successive fires.

fire intensity The rate of heat release per unit time and unit length of fire front.

Fire Weather Index (FWI) A numerical rating, based on meteorological observations, of fire intensity in a standard fuel type.

flark A shallow seepage pond associated with minerotrophic mires.

floristic diversity The number of plant species that occur in a community or area.

floristic region A region of the Earth defined by its distinctive flora.

foliovore An animal that feeds on plant leaves.

forb A broad-leaf herbaceous plant.

forest-line The latitudinal or altitudinal limit of regular reproduction in trees.

formation A regionally extensive assemblage of natural vegetation with a recurring physiognomy determined by its dominant life forms.

formation-type A group of geographically widespread plant formations of similar physiognomy and occurring under similar environmental conditions.

frugivore An animal that feeds on fruits.

garua Coastal fog in South America associated with the cold Peruvian current.

Gelbstoff A general term for soluble organic materials that give marine waters a characteristic yellow hue.

General Circulation Model (GCM) Mathematical model used in the prediction of climate change.

geographic isolation The spatial segregation of two potentially interbreeding populations.

gibbsite A clay mineral that is one of the main constituents of laterite.

gilgai Microrelief in clayey soils produced by moisture-related expansion and contraction.

gley A waterlogged soil characterized by grey, blue or green coloration and mottles produced by the reduction of iron and other elements.

gregarious flowering Simultaneous flowering of members of a population.

halophyte A plant that grows naturally in saline environments.

haptophyte Plants such as seaweeds which are attached to a surface by a hapteron or root-like holdfast.

heliotropism An auxin-controlled growth response in plants to a directional light stimulus.

hemicryptophyte A herbaceous plant in Raunkiaer's life-form classification with perennating buds located at the soil surface, often protected by dead tissues.

hemiepiphyte An epiphytic plant that roots in the soil for part of its life

hermatypic Reef-building, referring to corals.

holoplankton An organism that is planktonic for all of its life cycle.

homeostasis The maintenance of constancy within a biological system because of the ability of organisms to make adjustments.

humivorous Feeding on humus.

hydrarch Referring to plant succession that begins in aquatic or wet habitats and progresses towards drier conditions.

hydroperiodism The flowering response of a plant to the periodic stimulus of water.

hypolimnion In lakes, the colder, non-circulating layer of water that lies below the thermocline.

Hypsithermal A period when temperatures were higher than at present, normally referring to the post-glacial Climatic Optimum 5000–8500 years BP.

igapó An open forest community that is partially submerged for 6–8 months of the year.

illuviation Accumulation of clays and elements such as iron and aluminium in the lower soil horizons following downward transport in suspension or solution.

Inceptisols One of 10 soil orders in the US Soil Taxonomy System, comprising young soils with weakly developed horizons.

increaser A plant species already present in an area that increases in abundance with overgrazing.

individualistic Referring to the concept that plant communities represent an assemblage of plant species with overlapping habitat requirements that has arisen because of fortuitous propagule availability.

inflorescence A cluster of flowers in any arrangement on a shoot.

inhibition model A postulated process in vegetation succession whereby a change in floristic composition is prevented until an established species dies out.

inselberg An isolated residual hill rising abruptly from an area of low relief typically in arid regions.

insolation The reception of solar radiation at a surface.

interception Precipitation held in a plant canopy and returned to the atmosphere by evaporation.

interspecific Between two distinct species.

intraspecific Between individuals of the same species.

invader A newly established species in an overgrazed area.

involucre A whorl of bracts or modified leaves beneath a flower or inflorescence.

kampfzone Belt of stunted trees near the upper limit of tree growth.

kaolinite A hydrous aluminosilicate clay mineral with no permanent ionic charge.

Krantz anatomy The distinctive arrangement of specialized tissue surrounding the leaf veins in C_4 plants.

Kreb's cycle (citric acid cycle) A series of enzyme-controlled reactions associated with aerobic respiration in which pyruvic acid is broken down to carbon dioxide and water, producing adenosine triphosphate (ATP) as an energy source for further biochemical synthesis.

krummholz Scrubby, stunted growth form of trees at the limit of growth in temperature mountains and arctic regions.

ladder fuel Combustible materials, such as lichen and dead branches, which facilitate the spread of fire into a tree canopy.

lag deposit A residual deposit of coarse rock fragments commonly found in arid regions.

lagg The depressed margin of a raised bog.

lammas shoot Elongation of a newly formed terminal bud in temperate trees that results in shoot growth abnormally late in the season.

laterite A residual soil material composed chiefly of iron and aluminium sesquioxides occurring as a granular or concretionary layer in tropical soils.

latosols Acidic tropical soils of low organic status characterized by the complete disintegration of parent material, leaching of silica, and reddish in colour due to the accumulation of iron and aluminium sesquioxides.

layering Propagation by a branch in contact with the ground developing roots and growing independently after detachment from the parent plant.

leaf conductance Flux of water vapour out of a leaf that is regulated by turgor-induced leaf movements and stomatal control.

life form The characteristic form or appearance of a species at maturity.

lignotuber Swollen root crown burl that serves as a regenerative organ in certain shrubs and trees in fire-prone environments

lithosols Poorly developed stony soils.

littoral zone In lakes, the shallow water zone occupied by rooted plants; in the sea, the intertidal zone.

macrophyte A large aquatic plant.

marsh Periodically inundated mineral ground characterized by a zonal or mosaic vegetation cover of submerged and floating aquatic plants, emergent sedges and grasses, and typically surrounded by shrubs and trees.

Massenerhebung effect Lowering of the tree limit on isolated mountain peaks and coastal mountains compared with more massive interior ranges.

mean annual increment (MAI) In forestry, the average annual increase in growth that has occurred up to a given age.

mean turnover time In plankton population studies, the average time needed to replace the total population.

meristem A region of active cell division in a plant.

meroplankton Plankton that is seasonal in appearance, or an organism that is planktonic for only part of its life cycle.

mesopelagic zone The dimly illuminated zone of the sea extending from about 200 to 1000 m.

mesophyte A plant having moderate moisture requirements.

Miocene A period of Earth history from 26 to 7 million years BP.

minerotrophic Applied to wet peatlands, such as fens, which are supplied with mineral-rich groundwater.

Mollisols One of 10 soil orders in the US Soil Taxonomy System, comprising grassland soils characterized by thick, organic-rich surface horizons and high base status.

monocarpic Flowering only once, then dying.

monoecious Seed plants bearing separate male and female flowers on the same individual.

monopodial Having one main axis of growth.

montmorillonite An aluminosilicate clay mineral having an expandable, water-permeable lattice structure which gives it very plastic properties.

mutualism Any kind of interspecies relationship that is beneficial to, but not essential for, the wellbeing of the participating organisms.

mycorrhiza The symbiotic association of the root of a higher plant with a fungus which assists in mineral recycling and growth.

myrmecophyte A plant that has structures adapted for the shelter or sustenance of ants or termites.

nanoplankton Very small phytoplankton 2–20 µm in size.

nekton Strong-swimming animals of the pelagic zone.

neritic That portion of the sea overlying the continental shelf.

nitrification The conversion of organic nitrogen (ammonium) compounds into nitrites and nitrates by aerobic soil bacteria.

nitrogen fixation The conversion of atmospheric nitrogen into organic nitrogen compounds, mainly by micro-organisms in the soil or in nodules of certain plants, especially legumes.

nurse plant A plant that affords protection to the developing seedlings of another species.

obligate Restricted to a particular way of life.

oceanic province That portion of the ocean beyond the continental shelf.

oligomictic Referring to thermally stable water which rarely circulates.

ombrophilous Referring to plants that endure much rain.

oligophotic Referring to low light intensity.

opportunistic Adapted to utilize a range of environmental resources depending on availability.

organic drift Small organisms and organic debris carried downstream in flowing water.

organismic Referring to the concept that a plant community has qualities reflecting a higher level of organization than is inherently present in the constituent organisms.

orographic In meteorology, referring to precipitation derived from air forced to rise over regions of highland.

Orthids A suborder of Aridisols lacking a clayey B horizon, but with an overall profile depth of at least 25 cm.

orthotropic Gravitational growth response which results in the longer axis of a shoot developing more or less vertically.

osmoregulation The homeostatic adjustment in osmotic concentration of internal fluids in living organisms.

osmotic potential The change in free energy or chemical potential of water produced by solutes.

overturn The wind-induced circulation of water and nutrients in lakes and oceans caused by the mixing of the thermal strata.

Oxisols One of 10 soil orders in the US Soil Taxonomy System, comprising intensely weathered tropical and subtropical soils with low cation exchange capacity.

palsa A mound of perennially frozen peat and mineral soil associated with bogs or fens.

paludification The process of bog expansion caused by a gradual rise in the water table through peat accumulation.

palynology The study of pollen and spores, especially in lake sediments, as an indicator of past environments and plant communities.

parthenogenesis The development of an embryo from an unfertilized egg.

patterned ground Various surface features, such as stone polygons, especially characteristic of arctic tundra environments.

perennial A plant the persists for more than two years and when mature will normally flower annually.

permafrost Permanently frozen ground.

phanerophyte A woody plant in Raunkiaer's life-form classification with perennating buds located more than 25 cm above ground, especially trees and shrubs.

phenology The study of the timing of periodic biological phenomena.

photoinhibition The debilitating effect of high light intensity on the photosynthetic capacity of green plants.

photoperiodism The response of plants (and animals) to the relative duration of light and darkness.

photorespiration Light-activated cellular respiration that occurs concurrently with active photosynthesis, especially in C_3 plants.

phototaxis The directional movement of an organism in response to a light stimulus.

phyllode A flattened leaf stalk which has assumed the form and function of a leaf.

physiognomy The general appearance or morphology of a community as determined by the life form of the dominant plants.

phytophagous Herbivorous.

phytotoxin A substance poisonous to green plants.

picoplankton Planktonic organisms $< 2\,\mu m$.

pioneer species An early colonizer of a newly available site.

plagiotropic Gravitational response which results in more or less horizontal shoot growth.

plankton Aquatic plants and animals, many of them microscopic, which are characteristically moved passively by winds, waves and currents.

Pleistocene A period of Earth history from 2 million to 10 000 years BP associated with the most recent glacial episodes.

plinthite Iron-rich material present in tropical and subtropical soils which hardens to rock-like consistency with repeated wetting and drying.

pneumatophore Erect, aerial outgrowth from the roots of some plants which grow in water or tidal areas, giving submerged tissues a direct connection to the atmosphere.

polymictic Referring to lakes in which water circulation occurs more or less continually with only short periods of stagnation.

polyploid Any organism or cell with more than two complete sets of chromosomes.

primary succession The natural development of a plant cover in an area previously devoid of vegetation.

primary production The assimilation of organic matter by a plant community.

primary productivity The amount of organic matter made in a given time by the green plants in an ecosystem.

prop root A root originating from the stem above ground level and entering the ground at a distance from the main stem.

pseudocarp False fruit in which the ripened ovary and its contents are combined with another structure such as the receptacle.

pyrophyte A plant characterized by unusual resistance to (or dependence on) fire.

Quaternary The most recent period of geologic time from 2 million years BP to the present.

radial oxygen loss Diffusion of oxygen from plant roots into anaerobic soils.

raised bog An acid peatland dominated by *Sphagnum* mosses that is typically convex and gently sloping from its centre to the ditch or waterway (lagg) that surrounds it.

raptor A predatory animal, normally referring to birds of prey.

recurrence surface A distinct stratum in a peat deposit resulting from a period of drying and oxidation of surface materials.

red tide The colouring of offshore marine water by dense phytoplankton populations.

red belt Altitudinal band of conifers in the Rocky Mountains with needles killed by desiccating chinook winds.

refugium An area unaffected by significant environmental changes experienced by a region as a whole, thus favouring the survival of relic species (normally associated with the events of the last glaciation).

regosols Weakly developed soils, particularly those on recently deposited or unstable parent materials.

rhizome A creeping underground perennial stem that seasonally produces roots and leafy shoots.

rhizophyte A rooted plant.

ring porous Wood with large xylem vessels mostly confined to early (spring) wood.

rotation age In forestry, the age at which optimum wood volume is attained for a defined combination of site conditions and utilization standards.

ruderal A plant characteristic of waste places.

ruminant A mammal which must chew the cud (regurgitated food) to assist passage of food through its complex, multi-chambered stomach.

saprolite A surface layer of deeply weathered and leached rock material in a clayey matrix overlying bedrock.

scansorial Referring to an organism that is adapted for climbing.

scarification Abrasion of a seed coat mechanically or with an inorganic acid rendering it permeable to water and gases, thus preparing it for germination.

sclerophyll A plant with tough, leathery evergreen leaves.

secondary metabolites Compounds produced from primary carbohydrates, amino acids and lipids, responsible for interspecific differences in plant colour, scent and resistance to pathogens and phytophagous organisms.

secondary species Subordinate species in a community that are typically representative of disturbed sites.

secondary succession The natural development of a plant cover following destruction of the established vegetation.

seed bank The population of dormant seeds in the soil and within the vegetation canopy.

sequential flowering Successive flowering in related plant species to minimize competition for pollinating or seed dispersal organisms.

serotinous Late developing, but normally referring to the sealed cones of gymnosperms which release seeds only after a fire.

sesquioxide An oxide containing oxygen and metallic elements in the ratio 3:2, in soils chiefly Fe_2O_3 and Al_2O_3.

site index A numerical evaluation of the productivity of a forest site based on the height of trees after a specified period (usually 100 years) of growth.

snowbank community Plant community associated with sites in which snow melts late in the year.

sod-forming grass A grass in which roots and rhizomes form a matted layer near the soil surface.

soil horizon An approximately horizontal layer of soil with distinctive properties reflecting the formation processes.

soil order The highest category in the Soil Taxonomy within the Comprehensive Soil Classification System of the US Department of Agriculture and distinguished on the basis of diagnostic horizons indicative of different soil-forming processes.

soil profile Vertical section of a soil through all of its horizons and extending into the parent material.

soil structure The combination or arrangement of primary soil particles into secondary units or peds of different sizes and shapes.

soil texture The relative proportions of sand (0.05–2.0 mm diam.), silt (0.05–0.002 mm diam.) and clay (< 0.002 mm diam.) sized mineral particles in a soil.

solifluction The gradual downslope movement of surface materials characteristic of, though not confined to, regions subject to alternate periods of freezing and thawing.

solonetz A type of soil in which the surface horizons of varying friability are underlain by compact, alkaline materials which usually exhibit a distinctive columnar structure.

Spodosol One of 10 soil orders in the US Soil Taxonomy System, comprising acid soils of low cation-exchange capacity with subsurface (B) horizons enriched with aluminium and iron sesquioxide and amorphous organic material (usually fulvic acid).

stem flow Precipitation that reaches the soil by way of plant stems and trunks.

stolon A horizontal stem growing along the surface of the ground from which adventitious roots and new plants may arise.

sublittoral zone That part of a lake which is too deep for rooted plants to grow, or the zone in the sea lying below the intertidal zone and extending to the edge of the continental shelf.

succession The process of vegetation change that culminates in a climax community.

succulent A plant, usually perennial, with fleshy, water-storing tissues typically associated with an arid or saline environments.

sukhovey Dust storm.

sustainable development Utilizing a renewable resource at a rate that is not detrimental to its productivity.

swamp A forested wetland.

sympodial Growth axis formed of successive dichotomous branches of unequal length.

synusia A group of plants all of the same general life form occurring together in the same habitat.

terra rossa Red soils developed on limestone due to the accumulation of insoluble iron oxides.

terrestrialization The development of mires by the overgrowth of lakes.

Tertiary Period of geological time covering 66 million to 2 million years BP.

thermal front Zone of abrupt temperature change between water-bodies of different mixing histories.

thermocline The layer in a thermally stratified body of water within which there is a rapid decrease in temperature with depth.

thermokarst Subsidence features associated with thawing of permafrost when the thermal regime has been altered through vegetation and ground disturbance.

thermotropism A growth response to the directional stimulus of heat.

therophyte An herbaceous annual plant in Raunkiaer's life-form classification which survives unfavourable periods as seeds.

throughflow Precipitation which drips to the soil from a vegetation canopy.

tiller A shoot arising from the base of a stem, as in many grasses.

timber-line The latitudinal or altitudinal limit of tall, erect trees.

tolerance model A postulated process in vegetation succession whereby habitat modification by an established species has little effect on other species, change in community composition being essentially controlled by the life span of the plants.

top-down processes Regulation of population structure in an (aquatic) ecosystem by the feeding intensity of the highest trophic level.

tracheid An elongated tapering xylem cell with abundant lignified pits but no perforations in the end walls, adapted for conduction and support.

translocation Transfer of materials in solution from one part of a plant to another.

trichome A hair-like outgrowth from an epidermal cell.

tunicate Enveloped with loose membranes.

turnover time In plankton studies, the time needed for the replacement of the total population.

Ultisols One of 10 soil orders in the US Soil Taxonomy System, comprising deeply weathered red and yellow clay-enriched soils of low base status in humid temperate and subtropical climates.

ungulate A hoofed mammal.

Vertisols One of 10 soil orders in the US Soil Taxonomy System, comprising clay-rich soils that swell and crack in seasonally wet and dry tropical and subtropical environments.

vivipary Germination of seeds while the fruit is still attached to the parent plant.

wadi A steep-sided, flat-floored stream channel in arid regions.

warm season grass A temperate grass which grows mainly during the warmer part of the year, usually late spring and summer.

water potential A thermodynamic parameter used to describe the energy status of water in plants, defined in terms of the chemical potential of the plant solute compared with the chemical potential of pure water; the unit of measurement is MPa.

whole tree harvesting Mechanized harvesting system in which trees are cut off at ground level and trunk and branches reduced to small wood chips.

Xeralf A suborder of Alfisols formed under xeric water regimes.

xerophyte A plant adapted to grow in arid places.

xylem pressure potential Negative hydrostatic pressure of water under tension in the non-living cells of the xylem

xylophagous Wood-eating.

xylopodium Swollen, woody subterranean parts characteristic of savanna shrubs and trees.

References

Abbot, I. and van Heurck, P. (1985) Comparison of insects and vertebrates as removers of seed and fruit in a Western Australian forest. *Australian Journal of Ecology*, **10**, 165–8.

Abdel Wahab, A. M. (1975) Nitrogen fixation by *Bacillus* strains isolated from the rhizosphere of *Ammophila arenaria*. *Plant and Soil*, **42**, 703–8.

Abee, A. and Lavender, D. (1972) Nutrient cycling in throughfall and litterfall in 450-year-old Douglas fir stands, in *Research on Coniferous Forest Ecosystems: First Year Progress in the Coniferous Forest Biome, US/IBP*, (eds J. F. Franklin, L. J. Dempster and R. H. Waring), USDA, Pacific Northwest Forest and Range Experiment Station, Portland, pp. 133–43.

Achituv, Y. and Dubinsky, Z. (1990) Evolution and zoogeography of coral reefs, in *Ecosystems of the World Vol 25. Coral Reefs*, (ed. Z. Dubinsky), Elsevier, Amsterdam, pp. 1–9.

Ackerman, T. L. and Bamberg, S. A. (1974) Phenological studies in the Mojave Desert at Rock Valley (Nevada test site), in *Phenology and Seasonality Modelling*, (ed. H. Lieth), New York, Springer, pp. 215–26.

Acocks, J. P. H. (1975) Veld types of South Africa. *Memoirs of the Botanical Survey of South Africa*, **40**, 1–128.

Adam, P. (1990) *Saltmarsh Ecology*, Cambridge University Press, Cambridge.

Adams, M. S. and Strain, B. R. (1968) Photosynthesis in stems and leaves of *Cercidium floridum*: spring and summer diurnal field response and relation to temperature. *Oecologia Plantarum*, **3**, 285–97.

Adams, M. S., Strain, B. R. and Ting, I. P. (1967) Photosynthesis in chlorophyllous stem tissue and leaves of *Cercidium floridum*: accumulation and distribution of ^{14}C from $^{14}CO_2$. *Plant Physiology*, **42**, 1797–9.

Adams, S., Strain, B. R. and Adams, M. S. (1970) Water-repellent soils, fire, and annual plant cover in a desert scrub community of southeastern California. *Ecology*, **51**, 696–700.

Adams, S. M. and DeAngelis, D. L. (1987) Indirect effects of early bass-shad interactions on predator population structure and food web dynamics, in *Predation: Direct and Indirect Impacts on Aquatic Communities*, (eds W. C. Kerfoot and A. Sih), University Press of New England, Hanover, pp. 103–17.

Addicott, F. T. (1982) *Abscission*, University of California Press, Berkeley.

Ahlgren, C. E. (1957) Phenological observations of nineteen native tree species in northeastern Minnesota. *Ecology*, **38**, 622–8.

Albertson, F. W. and Tomanek, G. W. (1965) Vegetation changes during a 30-year period in grassland communities near Hays, Kansas. *Ecology*, **46**, 714–20.

Albertson, F. W. and Weaver, J. E. (1944) Nature and degree of recovery of grassland from the great drought of 1933 to 1940. *Ecological Monographs*, **14**, 393–479.

Aleksandrova, V. D. (1988) *Vegetation of the Soviet Polar Deserts*, Cambridge University Press, Cambridge.

Alexander, V., Billington, M. and Schell, D. M. (1978) Nitrogen fixation in arctic and alpine tundra, in *Vegetation and Production Ecology of an Alaskan Arctic Tundra*, (ed. L. L. Tieszen), Springer, New York, pp. 539–58.

Allan, H. H. (1937) A consideration of the 'biological spectra' of New Zealand. *Journal of Ecology*, **25**, 116–52.

Allan, J. D. (1983) Predator–prey relationships in streams, in *Stream Ecology*, (eds J. R. Barnes and G. W. Minshall), Plenum Press, New York, pp. 191–229.

Allen, L. H., Lemon, E. and Müller, L. (1972) Environment of a Costa Rican forest. *Ecology*, **53**, 102–111.

Allen, W. A., Gausman, H. W. and Richardson, A. J. (1973) Willstätter-Stoll theory of leaf reflectance evaluated by ray tracing. *Applied Optics*, **12**, 2448–53.

Alvim, P. T. (1960) Moisture stress as a requirement for flowering of coffee. *Science*, **132**, 354.

Alvim, P. T. (1964) Tree growth periodicity in tropical climates, in *The Formation of Wood in Forest Trees*, (ed. M. H. Zimmermann), Academic Press, New York, pp. 479–95.

Alvim, P. T. and Alvim, R. (1978) Relation of climate to growth periodicity in tropical trees, in *Tropical Trees as Living Systems*, (eds P. B. Tomlinson and M. H. Zimmermann), Cambridge University Press, Cambridge, pp. 445–64.

Amen, R. D. (1966) The extent and role of seed dormancy in alpine plants. *Quarterly Review of Biology*, **41**, 271–81.

Amos, A. F., Langseth, M. G. and Markl, R. G. (1972) Visible oceanic saline fronts, in *Studies in Physical Oceanography*, Vol. 1 (ed. A. L.Gordon), Gordon and Breach, New York, pp. 49–62.

Anderson, A. B. (1981) White-sand vegetation of Brazilian Amazonia. *Biotropica*, **13**, 199–210.

Anderson, C. E. (1974) A review of structure in several North Carolina salt marsh plants, in *Ecology of Halophytes*, (eds R. J. Reimold and W. H. Queen), Academic Press, New York, pp. 307–44.

Anderson, H. E. (1970) Forest fuel ignitability. *Fire Technology*, **6**, 312–19, 322.

Anderson, J. A. R. (1983) The tropical peat swamps of western Malesia, in *Ecosystems of the World Vol. 4B. Mires: Swamp, Bog, Fen and Moor – Regional Studies*, (ed. A. J. P. Gore), Elsevier, Amsterdam, pp. 181–99.

Anderson, J. M. and Swift, M. J. (1983) Decomposition in tropical forests, in *Tropical Rain Forest: Ecology and Management*, (eds S. L. Sutton, T. C. Whitmore and A. C. Chadwick), Blackwell Scientific Publications, Oxford, pp. 287–309.

Anderson, K. L. (1965) Time of burning as it affects soil moisture in an ordinary upland bluestem prairie in the Flint Hills. *Journal of Range Management*, **18**, 311–16.

Anderson, K. L., Smith E. F. and Owensby, C. E. (1970) Burning bluestem prairie. *Journal of Range Management*, **23**, 81–92.

Andersen, F. O. (1976) Primary production in a shallow water lake with special reference to a reed swamp. *Oikos*, **27**, 243–50.

Andreev, V. N. and Aleksandrova, V. D. (1981) Geobotanical division of the Soviet Arctic, in *Tundra Ecosystems: A Comparative Analysis*, (eds L. C. Bliss, O. W. Heal and J. J. Moore), Cambridge University Press, Cambridge, pp. 25–34.

Andrews, J. C. and Gentien, P. (1982) Upwelling as a source of nutrients for the Great Barrier Reef ecosystem: a solution to Darwin's question? *Marine Ecology Progress Series*, **8**, 257–69.

Andrzejewska, L. and Gyllenberg, G. (1980) Small herbivore subsystem, in *Grasslands, Systems Analysis and Man*, (eds A. I. Breymeyer and G. M. van Dyne), Cambridge University Press, Cambridge, pp. 201–67.

Angel, M. V. (1984) Detrital organic fluxes through pelagic ecosystems, in *Flows of Energy and Materials in Marine Ecosystems*, (ed. M J. R. Fasham), Plenum Press, New York, pp. 475–516.

Aoki, M., Yabuki, K. and Koyama, H. (1975) Micrometeorology and assessment of primary production of a tropical rain forest in west Malaysia. *Journal of Agricultural Meterology*, **31**, 115–24.

ap Rees, T., Jenkin, L. E. T., Smith, A. M. and Wilson, P. M. (1987) The metabolism of flood-tolerant plants, in *Plant Life in Aquatic and Amphibious Habitats*, (ed. R. M. M. Crawford), Blackwell Scientific Publications, Oxford, pp. 227–38.

Archibold, O. W. (1989) Seed banks and vegetation processes in coniferous forests, in *Ecology of Soil Seed Banks*, (eds M. A. Leck, V. T. Parker and R. L. Simpson), Academic Press, San Diego, pp. 107–22.

Ares, J. (1976) Dynamics of the root system of blue grama. *Journal of Range Management*, **29**, 208–13.

Armond, P. A., Schreiber, U. and Björkman, O. (1978) Photosynthetic acclimation to temperature in the desert shrub, *Larrea divaricata*. *Plant Physiology*, **61**, 411–15.

Armstrong, W. (1967a) The use of polarography in the assay of oxygen diffusing from roots in anaerobic media. *Physiologia Plantarum*, **20**, 540–53.

Armstrong, W. (1967b) The oxidising activity of roots in waterlogged soils. *Physiologia Plantarum*, **20**, 920–26.

Armstrong, W. (1979) Aeration in higher plants. *Advances in Botanical Research*, **7**, 225–332.

Armstrong, W. and Boatman, D. J. (1967) Some field observations relating the growth of bog plants to conditions of soil aeration. *Journal of Ecology*, **55**, 101–10.

Armstrong, W., Wright, E. J., Lythe, S. and Gaynard, T. J. (1985) Plant zonation and the effects of the spring-neap tide cycle on soil aeration in a Humber salt marsh. *Journal of Ecology*, **73**, 323–39.

Arnott, J. T. (1979) *Germination and Seedling Establishment*, Canadian Forestry Service Publication 1339, 55–66.

Arrhenius, G. (1963) Pelagic sediments, in *The Sea: Ideas and Observations on Progress in the Study of the Seas*, Vol. 3 (ed. M. N. Hill), Wiley Interscience, New York, pp. 655–727.

Arriaga, L. (1988) Gap dynamics of a tropical cloud forest in northeastern Mexico. *Biotropica*, **20**, 178–84.

Aschmann, H. (1973a) Distribution and peculiarity of mediterranean ecosystems, in *Mediterranean Type Ecosystems: Origin and Structure*, (eds F. di Castri and H. A. Mooney), Springer, New York, pp. 11–19.

Aschmann, H. (1973b) Man's impact on the several regions with mediterranean climates, in *Mediterranean Type Ecosystems: Origin and Structure*, (eds F. di Castri and H. A. Mooney), Springer, New York, pp. 363–71.

Ashton, P. S. (1969) Speciation among tropical forest trees: some deductions in the light of recent evidence. *Biological Journal of the Linnean Society*, **1**, 155–96.

Ashton, P. S., Givnish, T. J. and Appanah, S. (1988) Staggered flowering in the Dipterocarpaceae: new insights into floral induction and the evolution of mast fruiting in the aseasonal tropics. *American Naturalist*, **132**, 44–66.

Attwell, R. I. G. (1966) Oxpeckers, and their associations with mammals in Zambia. *Puku*, **4**, 17–48.

Aubuchon, R. R., Thompson, D. R. and Hinckley, T. M. (1978) Environmental influences on photosynthesis within the crown of a white oak. *Oecologia*, **35**, 295–306.

Auclair, A. N. D., Bouchard, A. and Pajaczkowski, J. (1976) Productivity relations in a *Carex*-dominated ecosystem. *Oecologia*, **26**, 9–31.

Auclair, A. N. D. and Rencz, A. N. (1982) Concentration, mass, and distribution of nutrients in a subarctic *Picea mariana–Cladonia alpestris* ecosystem. *Canadian Journal of Forest Research*, **12**, 947–68.

Auer, V. (1960) The Quaternary history of Fuego-Patagonia. *Royal Society of London Proceedings*, Series B, **152**, 507–16.

Augspurger, C. K. (1983) Offspring recruitment around tropical trees: changes in cohort distance with time. *Oikos*, **40**, 189–96.

Auld, T. D. (1986a) Population dynamics of the shrub *Acacia suaveolens* (Sm.) Willd.: dispersal and the dynamics of the soil seed-bank. *Australian Journal of Ecology*, **11**, 235–54.

Auld, T. D. (1986b) Population dynamics of the shrub *Acacia suaveolens* (Sm.) Willd.: fire and the transition to seedlings. *Australian Journal of Ecology*, **11**, 373–85.

Axelrod, D. I. (1958) Evolution of the Madro-Tertiary flora. *Botanical Review*, **24**, 433–509.

Axelrod, D. I. (1966) Origin of deciduous and evergreen habits in temperate forests. *Evolution*, **20**, 1–15.

Axelrod, D. I. (1973) History of the mediterranean ecosystem in California, in *Mediterranean Type Ecosystems: Origin and Structure*, (eds F. di Castri and H. A. Mooney), Springer, New York, pp. 225–77.

Axelrod, D. I. (1975) Evolution and biogeography of Madrean-Tethyan sclerophyll vegetation. *Annals of the Missouri Botanical Garden*, **62**, 280–334.

Axelrod, D. I. (1985) Rise of the grassland biome, central North America. *Botanical Review*, **51**, 163–201.

Axelrod, D. I. and Raven, P. H. (1978) Late Cretaceous and Tertiary vegetation history of Africa, in *Biogeography and Ecology of Southern Africa*, (ed. M. J. A. Werger), Junk, The Hague, pp. 77–130.

Ayyad, M. A. (1981) Soil–vegetation–atmosphere interactions, in *Arid-land Ecosystems: Structure, Functioning and Management*, (eds D. W. Goodall, R. A. Perry and K. M. W. Howes), Cambridge, Cambridge University Press, **2**, pp. 9–31.

Ayyad, M. A. and Ghabbour, S. I. (1986) Hot desert of Egypt and the Sudan, in *Ecosystems of the World Vol. 12B. Hot Deserts and Arid Shrublands*, (eds M. Evanari, I. Noy-Meir and D. W. Goodall), Elsevier, Amsterdam, pp. 149–202.

Azcon, R., Barea, J. M. and Hayman, D. S. (1976) Utilization of rock phosphate in alkaline soils by plants inoculated with mycorrhizal fungi and phosphate-solubilizing bacteria. *Soil Biology and Biochemistry*, **8**, 135–8.

Azevedo, J. and Morgan, D. L. (1974) Fog precipitation in coastal California forests. *Ecology*, **55**, 1135–41.

Babb, T. A. and Bliss, L. C. (1974) Effects of physical disturbance on arctic vegetation in the Queen Elizabeth Islands. *Journal of Applied Ecology*, **11**, 549–62.

Babb, T. A. and Whitfield, D. W. A. (1977) Mineral nutrient cycling and limitation of plant growth in the Truelove Lowland ecosystem, in *Truelove Lowland, Devon Island, Canada: A High Arctic Ecosystem*, (ed. L. C. Bliss), The University of Alberta Press, Edmonton, pp. 589–606.

Backéus, I. (1972) Bog vegetation re-mapped after sixty years. *Oikos*, **23**, 384–93.

Bagyaraj, D. J. (1989) Mycorrhizas, in *Ecosystems of the World Vol. 14B. Tropical Rain Forest Ecosystems: Structure and Function*, (eds H. Lieth and M. J. A. Werger), Elsevier, Amsterdam, pp. 537–46.

Baig, M. N. and Tranquillini, W. (1976) Studies on upper timberline: morphology and anatomy of Norway spruce (*Picea abies*) and stone pine (*Pinus cembra*) needles from various habitat conditions. *Canadian Journal of Botany*, **54**, 1622–32.

Bailey, C. G. and Riegert, P. W. (1973) Energy dynamics of *Encoptolophus sordidus costalis* (Scudder) (Orthoptera: Acrididae) in a grassland ecosystem. *Canadian Journal of Zoology*, **51**, 91–100.

Baker, H. G. (1973a) Evolutionary relationships between flowering plants and animals in American and African tropical forests, in *Tropical Forest Ecosystems in Africa and South America: A Comparative Review*, (eds B. J. Meggers, E. S. Ayensu and W. D. Duckworth), Smithsonian Institution Press, Washington, pp. 145–59.

Baker, H. G. (1973b) A structural model of forces in buttressed tropical rain forest trees, Appendix, *Biotropica*, **5**, 93.

Baker, H. G. (1978) Chemical aspects of the pollination biology of woody plants in the tropics, in *Tropical Trees as Living Systems*, (eds P. B. Tomlinson and M. H. Zimmermann), Cambridge University Press, Cambridge, pp. 57–82.

Baker, H. G. (1989) Sources of the naturalized grasses and herbs in California grasslands, in *Grassland Structure and Function: California Annual Grassland*, (eds L. F. Huenneke and H. A. Mooney), Kluwer Academic Publishers, Dordrecht, pp. 29–38.

Baker, H. G., Bawa, K. S., Frankie G. W. and Opler P. A. (1983) Reproductive biology of plants in tropical forests, in *Ecosystems of the World Vol. 14A. Tropical Rain Forest Ecosystems*, (ed. F. B. Golley), Elsevier, Amsterdam, pp. 183–215.

Baker, H. S. (1950) *Principles of Silviculture*, McGraw-Hill, New York.

Bakker, J. P. (1985) The impact of grazing on plant communities, plant populations and soil conditions on salt marshes. *Vegetatio*, **62**, 391–98.

Baldwin, M., Kellogg, C. E. and Thorp, J. (1938) Soil classification, in *Soils and Men*, United States Department of Agriculture, Yearbook of Agriculture 1938, 979–1001.

Baldwin, I. T. and Schultz, J. C. (1983) Rapid changes in tree leaf chemistry induced by damage: evidence for communication between plants. *Science*, **221**, 277–8.

Ball, M. C. (1980) Patterns of secondary succession in a mangrove forest of southern Florida. *Oecologia*, **44**, 226–35.

Båmstedt, U. (1981) Seasonal energy requirements of macrozooplankton from Kosterfjorden, Western Sweden. *Kieler Meeresforschungen Sonderheft*, **5**, 140–52.

Barbault, R. (1983) Reptiles in savanna ecosystems, in *Ecosystems of the World Vol. 13. Tropical Savannas*, (ed. F. Bourlière), Elsevier, Amsterdam, pp. 325–36.

Barber, K. E. (1981) *Peat Stratigraphy and Climatic Change*, Balkema, Rotterdam.

Barber, R. T. and Smith, R. L. (1981) Coastal upwelling ecosystems, in *Analysis of Marine Ecosystems*, (ed. A. R. Longhurst), Academic Press, London, pp. 31–68.

Barbour, M. G. (1978) Salt spray as a microenvironmental factor in the distribution of beach plants at Point Reyes, California. *Oecologia*, **32**, 213–24.

Barbour, M. G. (1981) Plant–plant interactions, in *Arid-land Ecosystems: Structure, Functioning and Management*, Vol. 2 (eds D. W. Goodall, R. A. Perry and K. M. W. Howes), Cambridge University Press, Cambridge, pp. 33–49.

Barnes, R. S. K. and Hughes, R. N. (1988) *An Introduction to Marine Ecology*, 2nd edn, Blackwell Scientific Publications, Oxford.

Barr, B. M. and Braden, K. E. (1988) *The Disappearing Russian Forest: A Dilemma in Soviet Resource Management*, Rowman and Littlefield, London.

Barr, W. (1991) *Back From the Brink: The Road to Muskox Conservation in the Northwest Territories*. Komatik Series, Number 3, The Arctic Institute of North America, Calgary.

Bartell, S. M., Brenkert, A. L., O'Neill, R. V. and Gardner, R. H. (1988) Temporal variation in regulation of production in a pelagic food web model, in *Complex Interactions in Lake Communities*, (ed. S. R. Carpenter), Springer, New York, pp. 101–18.

Bartholomew, G. A. and Dawson, W. R. (1968) Temperature regulation in desert mammals, in *Desert Biology*, Vol. 1 (ed. G. W. Brown), Academic Press, New York, pp. 395–421.

Bartsch, A. F. (1948) Biological aspects of stream pollution. *Sewage Works Journal*, **20**, 292–302.

Basilier, K. (1980) Fixation and uptake of nitrogen in *Sphagnum* blue-green algal associations. *Oikos*, **34**, 239–42.

Basilier, K., Granhall, U. and Stenström, T.-A. (1978) Nitrogen fixation in wet minerotrophic moss communities of a subarctic mire. *Oikos*, **31**, 236–46.

Basinger, J. F. (1991) The fossil forests of the Buchanan Lake Formation (early Tertiary), Axel Heiberg Island, Canadian Arctic Archipelago: preliminary floristics and paleoclimate, in *Tertiary Fossil Forests of the Geodetic Hills, Axel Heiberg Island, Arctic Archipelago*, (eds R. L. Christie and N. J. McMillan), Geological Society of Canada, Ottawa, pp. 39–65.

Batanouny, K. H. (1979) Vegetation along the Jeddah–Mecca road: pattern and process as affected by human impact. *Journal of Arid Environments*, **2**, 21–30.

Batanouny, K. H. (1983) Human impact on desert vegetation, in *Man's Impact on Vegetation*, (eds W. Holzner, M. J. A. Werger and I. Ikusima), Junk, The Hague, pp. 139–49.

Batchelder, R. B. (1967) Spatial and temporal patterns of fire in the tropical world. *Proceedings of the Tall Timbers Fire Ecology Conference*, **6**, 171–90.

Battarbee, R. W. (1984) Diatom analysis and the acidification of lakes. *Philosophical Transactions of the Royal Society of London*, **305b**, 451–77.

Battarbee, R. W., Flower, R. J., Stevenson, A. C. and Rippey, B. (1985) Lake acidification in Galloway: a palaeoecological test of competing hypotheses. *Nature*, **314**, 350–52.

Battarbee, R. W., Flower, R. J., Stevenson, A. C., *et al.* (1988) Diatom and chemical evidence for reversibility of acidification of Scottish lochs. *Nature*, **332**, 530–32.

Batzli, G. O. (1981) Populations and energetics of small mammals in the tundra ecosystem, in *Tundra Ecosystems: A Comparative Analysis*, (eds L. C. Bliss, O. W. Heal and J. J. Moore), Cambridge University Press, Cambridge, pp. 377–96.

Bawa, K. S. (1977) The reproductive biology of *Cupania guatemalensis* Radlk. (Sapindaceae). *Evolution*, **31**, 52–63.

Bawa, K. S. (1980) Mimicry of male by female flowers and intrasexual competition for pollinators in *Jacaratia dolichaula* (D. Smith) Woodson (Caricaceae). *Evolution*, **34**, 467–74.

Bawa, K. S. (1983) Patterns of flowering in tropical plants, in *Handbook of Experimental Pollination Biology*, (eds C. E. Jones and R. J. Little), Scientific and Academic Editions, New York, pp. 394–410.

Bawa, K. S. and Crisp, J. E. (1980) Wind-pollination in the understory of a rain forest in Costa Rica. *Journal of Ecology*, **68**, 871–6.

Bawa, K. S. and Opler, P. A. (1975) Dioecism in tropical forest trees. *Evolution*, **29**, 167–79.

Baynton, H. W., Biggs, W. G., Hamilton, H. L., *et al.* (1965) Wind structure in and above a tropical forest. *Journal of Applied Meterology*, **4**, 670–75.

Bazilevich, N. I., Bykov, B. A. and Kurochkina, L. J. (1981) Cycles of mineral elements, in *Arid-land Ecosystems: Structure, Functioning and Management*, Vol. 2 (eds D. W. Goodall, R. A. Perry and K. M. W. Howes), Cambridge University Press, Cambridge, pp. 183–8.

Beadle, L. C. (1974) *The Inland Waters of Tropical Africa*, Longman, London.

Beaman, J. H. and Andresen, J. W. (1966) The vegetation, floristics and phytogeography of the summit of Cerro Potosi, Mexico. *The American Midland Naturalist*, **75**, 1–33.

Beard, J. B. (1973) *Turfgrass: Science and Culture*, Prentice-Hall, Englewood Cliffs.

Beard, J. S. (1953) The savanna vegetation of northern tropical America. *Ecological Monographs*, **23**, 149–215.

Beatley, J. C. (1967) Survival of winter annuals in the northern Mojave Desert. *Ecology*, **48**, 745–50.

Beatley, J. C. (1969) Dependence of desert rodents on winter annuals and precipitation. *Ecology*, **50**, 721–4.

Beatley, J. C. (1970) Perennation in *Astragalus lentiginosus* and *Tridens pulchellus* in relation to rainfall. *Madroño*, **20**, 326–32.

Beatley, J. C. (1974) Phenological events and their environmental triggers in Mojave Desert ecosystems. *Ecology*, **55**, 856–63.

Beatley, J. C. (1980) Fluctuations and stability in climax shrub and woodland vegetation of the Mojave, Great Basin and transition deserts of southern Nevada. *Israel Journal of Botany*, **28**, 149–68.

Becker, P. and Wong, M. (1985) Seed dispersal, seed predation, and juvenile mortality of *Aglaia* sp. (Meliaceae) in lowland dipterocarp rainforest. *Biotropica*, **17**, 230–37.

Bedoian, W. H. (1978) Human use of the pre-Saharan ecosystem and its impact on desertization, in *Social and Technological Management in Dry Lands: Past, Present, Indigenous and Imposed*, (ed. N. L. Gonzalez), Westview, Boulder, 61–109.

Beebe, J. D. and Hoffman, G. R. (1968) Effects of grazing on vegetation and soils in southeastern South Dakota. *American Midland Naturalist*, **80**, 96–110.

Beeftink, W. G. (1977) The coastal salt marshes of Western and Northern Europe: an ecological and phytosociological approach, in *Ecosystems of the World Vol. 1. Wet Coastal Ecosystems*, (ed. V. J. Chapman), Amsterdam, Elsevier, pp. 109–155.

Beeftink, W. G. (1985) Population dynamics of annual *Salicornia* species in the tidal salt marshes of the Oosterschelde, The Netherlands. *Vegetatio*, **61**, 127–36.

Beek, K. J. and Bramao, D. L. (1968) Nature and geography of South American soils, in *Biogeography and Ecology in South America*, Vol. 1 (eds E. J. Fittkau, J. Illies, H. Klinge, G. H. Schwabe and H. Sioli), Junk, The Hague, pp. 82–112.

Belay, A. (1981) An experimental investigation of inhibition of phytoplankton photosynthesis at lake surfaces. *New Phytologist*, **89**, 61–74.

Bell, F. C. (1979) Precipitation, in *Arid-land Ecosystems: Structure, Functioning and Management*, Vol. 1 (eds

D. W. Goodall, R. A. Perry and K. M. W. Howes), Cambridge University Press, Cambridge, pp. 373–92.

Bell, K. L. and Bliss, L. C. (1978) Root growth in a polar semidesert environment. *Canadian Journal of Botany*, **56**, 2470–90.

Bell, K. L. and Bliss, L. C. (1980) Plant reproduction in a high arctic environment. *Arctic and Alpine Research*, **12**, 1–10.

Bell, R. H. V. (1970) The use of the herb layer by grazing ungulates in the Serengeti, in *Animal Populations in Relation to Their Food Resources*, (ed. A. Watson), Blackwell, Oxford, pp. 111–24.

Bell, R. H. V. (1971) A grazing ecosystem in the Serengeti, *Scientific American*, **25**, 86–93.

Bennett, H. H. and Allison, R. V. (1928) *The Soils of Cuba*, Tropical Plant Research Foundation, Washington, DC.

Bentley, B. L. (1977) The protective function of ants visiting the extrafloral nectaries of *Bixa orellana* (Bixaceae). *Journal of Ecology*, **65**, 27–38.

Bentley, P. J. and Blumer, W. F. C. (1962) Uptake of water by the lizard, *Moloch horridus*. *Nature*, **194**, 699–700.

Berg, R. Y. (1975) Myrmecochorous plants in Australia and their dispersal by ants. *Australian Journal of Botany*, **23**, 475–508.

Bergquist, A. M., Carpenter, S. R. and Latino, J. C. (1985) Shifts in phytoplankton size structure and community composition during grazing by contrasting zooplankton assemblages. *Limnology and Oceanography*, **30**, 1037–45.

Berliner, R. and Orshan, G. (1971) Effects of photoperiod on seasonal dimorphism, flowering and fruiting of *Sarcopoterium spinosum* (L.) Sp. *Israel Journal of Botany*, **20**, 199–202.

Bernard, J. M. and Solsky, B. A. (1977) Nutrient cycling in a *Carex lacustris* wetland. *Canadian Journal of Botany*, **55**, 630–38.

Bernard, J. M. and Bernard, F. A. (1977) Winter standing crop and nutrient contents in five central New York wetlands. *Bulletin of the Torrey Botanical Club*, **104**, 57–9.

Bews, J. W. (1929) *The World's Grasses*, Russell and Russell, New York.

Bickerstaff, A., Wallace, W. L. and Evert, F. (1981) *Growth of Forests in Canada Part 2. A Quantitative Description of the Land Base and the Mean Annual Increment*, Information Report PI-X-1. Environment Canada: Canadian Forestry Service, Ottawa.

Biederbeck, V. O., Lowe, W. E., Paul, E. A. *et al.* (1974) *Soil Microorganisms: II. Decomposition of Cellulose and Plant Residues. Canadian Committee for IBP*, Matador Project Technical Report 39, University of Saskatchewan, Saskatoon.

Bienfang, P. K. and Takahashi, M. (1983) Ultraplankton growth rates in a subtropical ecosystem. *Marine Biology*, **76**, 213–18.

Billen, G. and Lancelot, C. (1988) Modelling benthic nitrogen cycling in temperate coastal ecosystems, in *Nitrogen Cycling in Coastal Marine Environments*, (eds T. H. Blackburn and J. Sørensen), Wiley, Chichester, pp. 341–78.

Billings, W. D. (1974a) Adaptations and origins of alpine plants. *Arctic and Alpine Research*, **6**, 129–42.

Billings, W. D. (1974b) Arctic and alpine vegetation: plant adaptations to cold summer climates, in *Arctic and Alpine*

Environments, (eds J. D. Ives and R. G. Barry), Metheun, London, pp. 403–43.

Billings, W. D. and Bliss, L. C. (1959) An alpine snowbank environment and its efects on vegetation, plant development, and productivity. *Ecology*, **40**, 388–97.

Billings, W. D. and Mooney, H. A. (1968) The ecology of arctic and alpine plants. *Biological Reviews*, **43**, 481–529.

Bird, D. F. and Kalff, J. (1986) Bacterial grazing by planktonic lake algae. *Science*, **231**, 493–5.

Bird, J. B. (1974) Geomorphic processes in the Arctic, in *Arctic and Alpine Environments*, (eds J. D. Ives and R. G. Barry), Metheun, London, pp. 703–20.

Birks, H. J. B. (1980) The present flora and vegetation of the moraines of the Klutlan Glacier, Yukon Territory, Canada: a study in plant succession. *Quaternary Research*, **14**, 60–86.

Birks, J. B. (1989) Holocene isochrone maps and patterns of tree-spreading in the British Isles. *Journal of Biogeography*, **16**, 503–40.

Biswell, H. H. (1974) Effects of fire on chaparral, in *Fire and Ecosystems*, (eds T. T. Kozlowski and C. E. Ahlgren), Academic Press, New York, pp. 321–64.

Björkman, O. (1968) Carboxydismutase activity in shade-adapted and sun-adapted species of higher plants. *Physiologia Plantarum*, **21**, 1–10.

Black, M. A., Maberly, S. C. and Spence, D. H. N. (1981) Resistance to carbon dioxide fixation in four submerged freshwater macrophytes. *New Phytologist*, **89**, 557–68.

Black, R. A. and Bliss, L. C. (1978) Recovery sequence of *Picea mariana–Vaccinium uliginosum* forests after burning near Inuvik, North-west Territories, Canada. *Canadian Journal of Botany*, **56**, 2020–30.

Black, R. A. and Bliss, L. C. (1980) Reproductive ecology of *Picea mariana* (Mill.) BSP., at tree line near Inuvik, North-west Territories, Canada. *Ecological Monographs*, **50**, 331–54.

Blackburn, M. (1981) Low latitude gyral regions, in *Analysis of Marine Ecosystems* (ed. A. R. Longhurst), Academic Press, London, pp. 3–29.

Blais, J. R. (1983) Trends in the frequency, extent, and severity of spruce budworm outbreaks in eastern Canada. *Canadian Journal of Forest Research*, **13**, 539–47.

Blasco, F. (1983) The transition from open forest to savanna in continental southeast Asia, in *Ecosystems of the World Vol. 13. Tropical Savannas*, (ed. F. Bourlière), Elsevier, Amsterdam, pp. 167–81.

Bliss, L. C. (1963) Alpine plant communities of the Presidential Range, New Hampshire. *Ecology*, **44**, 678–97.

Bliss, L. C. (1977) General summary, Truelove Lowland ecosystem, in *Truelove Lowland, Devon Island, Canada: A High Arctic Ecosystem*, (ed. L. C. Bliss), The University of Alberta Press, Edmonton, pp. 657–675.

Bliss, L. C. (1981) North American and Scandinavian tundras and polar desert, in *Tundra Ecosystems: A Comparative Analysis*, (eds L. C. Bliss, O. W. Heal and J. J. Moore), Cambridge University Press, Cambridge, pp. 8–24.

Bliss, L. C. (1983) Modern human impact in the Arctic, in *Man's Impact on Vegetation*, (eds W. Holzner, M. J. A. Werger and I. Ikusima), Junk, The Hague, pp. 213–25.

Blydenstein, J. (1968) Burning and tropical American savannas. *Proceedings Tall Timbers Fire Ecology Conference*, **8**, 1–14.

Boardman, N. K. (1977) Comparative photosynthesis of sun and shade plants. *Annual Review of Plant Physiology*, **28**, 355–77.

Böcher, T. W. (1963) Phytogeography of middle west Greenland. *Meddelelser om Grønland*, **148**, 1–289.

Bodkin, P. C., Spence, D. H. N. and Weeks, D. C. (1980) Photoreversible control of heterophylly in *Hippuris vulgaris* L. *New Phytologist*, **84**, 533–42.

Bodley, J. H. and Benson, F. C. (1980) Stilt-root walking by an Iriarteoid palm in the Peruvian Amazon. *Biotropica*, **12**, 67–71.

Boelter, D. H. (1964) Water storage characteristics of several peats *in situ*. *Soil Science Society of America Proceedings*, **28**, 433–5.

Boerboom. J. H. A. (1964) Microclimatological observations in the Wassenaar dunes. *Mededelingen van de Landbouwhogeschool te Wageningen*, **64**, 1–28.

Boerboom, J. H. A. and Wiersum, K. F. (1983) Human impact on tropical moist forest, in *Man's Impact on Vegetation*, (eds W. Holzner, M. J. A. Werger and I. Ikusima), Junk, The Hague, pp. 83–106.

Bogdan, A. V. (1977) *Tropical Pasture and Fodder Plants*, Longman, London.

Bokhari, U. G. (1977) Regrowth of western wheatgrass utilizing ^{14}C-labelled assimilates stored in belowground parts. *Plant and Soil*, **48**, 115–27.

Bokhari, U. G. and Singh, J. S. (1974) Effects of temperature and clipping on growth, carbohydrate reserves, and root exudation of western wheatgrass in hydroponic culture. *Crop Science*, **14**, 790–94.

Bokhari, U. G. and Singh, J. S. (1975) Standing state and cycling of nitrogen in soil–vegetation components of prairie ecosystems. *Annals of Botany*, **39**, 273–85.

Bonaccorso, F. J. and Humphrey, S. R. (1984) Fruit bat dynamics: their role in maintaining tropical forest diversity, in *Tropical Rain-forest*, (eds A. C. Chadwick and S. L. Sutton), Leeds Philosophical and Literary Society, Leeds, pp. 169–83.

Bond, W. (1980) Fire and senescent fynbos in the Swartberg, Southern Cape. *South African Forestry Journal*, **114**, 68–71.

Bond, W. J. (1985) Canopy-stored seed reserves (serotiny) in Cape Proteaceae. *South African Journal of Botany*, **51**, 181–6.

Boney, A. D. (1965) Aspects of the biology of the seaweeds of economic importance. *Advances in Marine Biology*, **3**, 105–253.

Boorman, L. A. (1977) Sand-dunes, in *The Coastline*, (ed. R. S. K. Barnes), Wiley, London, pp. 161–97.

Borchert, J. R. (1950) The climate of the central North American grassland. *Annals of the Association of American Geographers*, **40**, 1–39.

Borman, F. H. and Likens, G. E. (1979) *Pattern and Process in a Forested Ecosystem*, Springer, New York.

Borner, H. (1960) Liberation of organic substances from higher plants and their role in the soil sickness problem. *Botanical Review*, **26**, 393–424.

Botch, M. S. and Masing, V. V. (1983) Mire ecosystems in the U. S. S. R., in *Ecosystems of the World Vol. 4B. Mires: Swamp, Bog, Fen and Moor – Regional Studies*, (ed. A. J. P. Gore). Elsevier, Amsterdam, pp. 95–152.

Boto, K. G. and Bunt, J. S. (1981) Tidal export of particulate organic matter from a Northern Australian mangrove system. *Estuarine, Coastal and Shelf Science*, **13**, 247–55.

Boucher, C. and Moll, E. J. (1981) South African mediterranean shrublands, in *Ecosystems of the World Vol. 11. Mediterranean-type Shrublands*, (eds F. di Castri, D. W. Goodall and R. L. Specht), Elsevier, Amsterdam, pp. 233–48.

Bourgeron, P. S. (1983) Spatial aspects of vegetation structure, in *Ecosystems of the World Vol. 14A. Tropical Rain Forest Ecosystem: Structure and Function*, (ed. F. B. Golley), Elsevier, Amsterdam, pp. 29–47.

Bourlière, F. and Hadley, M. (1970) The ecology of tropical savannas. *Annual Review of Ecology and Systematics*, **1**, 125–52.

Bowes, G. (1987) Aquatic plant photosynthesis: strategies that enhance carbon gain, in *Plant Life in Aquatic and Amphibious Habitats* (ed. R. M. M. Crawford), Blackwell Scientific Publications, Oxford, pp. 79–98.

Boyce, S. G. (1954) The salt spray community. *Ecological Monographs*, **24**, 29–67.

Boyd, C. E. (1970) Production, mineral accumulation and pigment concentrations in *Typha latifolia* and *Scirpus americanus*. *Ecology*, **51**, 285–90.

Boyd, C. E. (1971) The dynamics of dry matter and chemical substances in a *Juncus effusus* population. *American Midland Naturalist*, **86**, 28–45.

Boyer, M. L. H. and Wheeler, B. D. (1989) Vegetation patterns in spring-fed calcareous fens: calcite precipitation and constraints on fertility. *Journal of Ecology*, **77**, 597–609.

Boylen, C. W. and Sheldon, R. B. (1976) Submergent macrophytes: growth under winter ice cover. *Science*, **194**, 841–2.

Bradley, W. G. and Mauer, R. A. (1971) Reproduction and food habits of Merriam's kangaroo rat, *Dipodomys merriami*. *Journal of Mammalogy*, **52**, 497–507.

Bräendle, R. and Crawford, R. M. M. (1987) Rhizome anoxia tolerance and habitat specialization in wetland plants, in *Plant Life in Aquatic and Amphibous Habitats*, (ed. R. M. M. Crawford), Blackwell Scientific Publications, Oxford, pp. 397–410.

Bragg, T. B. and Hulbert, L. C. (1976) Woody plant invasion of unburned Kansas bluestem prairie. *Journal of Range Management*, **29**, 19–24.

Braithwaite, R. W. and Gullen, P. K. (1978) Habitat selection by small mammals in a Victorian heathland. *Australian Journal of Ecology*, **3**, 109–27.

Brand, D. G. and Penner, M. E. (1991) Regeneration and growth of Canadian Forests, in *Canada's Timber Resources*, (ed. D. G. Brand), Information Report PI-X-101, Forestry Canada, Ottawa, pp. 51–9.

Brandon, P. C. (1967) Temperature features of enzymes affecting Crassulacean acid metabolism. *Plant Physiology*, **42**, 977–84.

Braun, E. L. (1964) *Deciduous Forests of Eastern North America*, Hafner, New York.

Bray, J. R. (1964) Primary consumption in three forest canopies. *Ecology*, **45**, 165–7.

Bray, J. R. and Gorham, E. (1964) Litter production in forests of the world. *Advances in Ecological Research*, **2**, 101–57.

Briggs, S. V. (1977) Estimates of biomass in a temperate mangrove community. *Australian Journal of Ecology*, **2**, 369–73.

Brinkmann, W. L. F. (1989) Hydrology, in *Ecosystems of the World Vol. 14B. Tropical Rain Forest Ecosystems: Biogeographical and Ecological Studies*, (eds H. Lieth and M. J. A. Werger), Elsevier, Amsterdam, pp. 89–98.

Bristow, J. M. (1969) The effects of carbon dioxide on the growth and development of amphibious plants. *Canadian Journal of Botany*, **47**, 1803–7.

Bristow, J. M. (1974) Nitrogen fixation in the rhizosphere of freshwater angiosperms. *Canadian Journal of Botany*, **52**, 217–21.

Brock, T. C. M. and Bregman, R. (1989) Periodicity in growth, productivity, nutrient content and decomposition of *Sphagnum recurvum* var. *mucronatum* in a fen woodland. *Oecologia*, **80**, 44–52.

Brokaw, N. (1982) Treefalls: frequency, timing, and consequences, in *The Ecology of a Tropical Forest*, (eds E. G. Leigh, A. S. Rand and D. M. Windsor), Smithsonian Institution Press, Washington, pp. 101–8.

Brongersma-Sanders, M. (1957) Mass mortality in the sea. *Geological Society of America Memoir 67*, vol. 1, 941–1010.

Brookfield, H. and Byron, Y. (1990) Deforestation and timber extraction in Borneo and the Malay Peninsula: the record since 1965. *Global Environmental Change*, **1**, 42–56.

Brookman-Amissah, J., Hall, J. B., Swaine, M. D. and Attakorah, J. Y. (1980) A re-assessment of a fire protection experiment in north-eastern Ghana savanna. *Journal of Applied Ecology*, **17**, 85–99.

Brooks, J. L. and Dodson, S. I. (1965) Predation, body size, and composition of plankton. *Science*, **150**, 28–35.

Brown, A. C. and McLachlan, A. (1990) *Ecology of Sandy Shores*, Elsevier, Amsterdam.

Brown, D. H. (1982) Mineral nutrition, in *Bryophyte Ecology*, (ed. A. J. E. Smith), Chapman & Hall, London, pp. 383–444.

Brown, J. H. and Davidson, D. W. (1977) Competition between seed-eating rodents and ants in desert ecosystems. *Science*, **196**, 880–82.

Brown J. H., Reichman, O. J. and Davidson, D. W. (1979) Granivory in desert ecosystems. *Annual Review of Ecology and Systematics*, **10**, 201–27.

Brown, J. S. and Venable, D. L. (1986) Evolutionary ecology of seed-bank annuals in temporally varying environments. *The American Naturalist*, **127**, 31–47.

Brown, K. S. and Brown, G. G. (1992) Habitat alteration and species loss in Brazilian forests, in *Tropical Deforestation and Species Extinction*, (eds T. C. Whitmore and J. A. Sayer), Chapman & Hall, London, pp. 119–42.

Brown, R. H. (1978) A difference in N use efficiency in C_3 and C_4 plants and its implications in adaptation and evolution. *Crop Science*, **18**, 93–8.

Brown, S. and Lugo, A. E. (1984) Biomass of tropical forests: a new estimate based on forest volumes. *Science*, **223**, 1290–93.

Brown, W. L. (1960) Ants, acacias and browsing mammals. *Ecology*, **41**, 587–92.

Brum, G. D. (1973) Ecology of the saguaro (*Carnegiea gigantea*): phenology and establishment in marginal populations. *Madroño*, **22**, 195–204.

Brunig, E. F. (1983) Vegetation structure and growth, in *Ecosystems of the World Vol. 14A. Tropical Rain Forest Ecosystems, Structure and Function*, (ed. F. B. Golley), Elsevier, Amsterdam, pp. 49–75.

Brynard, A. M. (1964) The influence of veld burning on the vegetation and game of the Kruger National Park, in *Ecological Studies in Southern Africa*, (ed. D. H. S. Davis), Junk, The Hague, pp. 371–93.

Bryson, R. A. (1966) Air masses, streamlines, and the boreal forest. *Geographical Bulletin*, **8**, 228–69.

Bryson, R. A. and Hare, F. K. (eds.) (1974) *World Survey of Climatology Vol. 11. Climates of North America*, Elsevier, Amsterdam,

Bucher, E. H. (1982) Chaco and caatinga-South American arid savannas, woodlands and thickets, in *Ecology of Tropical Savannas*, (eds P. J. Huntley and B. H. Walker), Springer, Berlin, pp. 48–79.

Buchsbaum, R., Valiela, I. and Swain, T. (1984) The role of phenolic compounds and other plant constituents in feeding by Canada geese in a coastal marsh. *Oecologia*, **63**, 343–9.

Buckley, R. C., Corlett, R. T. and Grubb, P. J. (1980) Are the xeromorphic trees of tropical upper montane rain forests drought-resistant. *Biotropica*, **12**, 124–36.

Buddemeier, R. W. and Kinzie, R. A. (1976) Coral growth. *Oceanography and Marine Biology Annual Review*, **14**, 183–225.

Bullock, S. H. and Bawa, K. S. (1981) Sexual dimorphism and the annual flowering pattern in *Jacaratia dolichaula* (D. Smith) Woodson (Caricaceae) in a Costa Rican rain forest. *Ecology*, **62**, 1494–504.

Bunnel, F. L., MacLean, S. F. and Brown, J. (1975) Barrow, Alaska, USA, in *Ecological Bulletins 20: Structure and Function of Tundra Ecosystems*, (eds T. Rosswall and O. W. Heal), Swedish Natural Science Research Council, Stockholm, pp. 73–124

Bunt, J. S. (1982) Studies of mangrove litter fall in tropical Australia, in *Mangrove Ecosystems in Australia: Structure, Function and Management*, (ed. B. F. Clough), Australian Institute of Marine Science, Canberra, pp. 223–37.

Burbridge, N. T. (1960) The phytogeography of the Australian region. *Australian Journal of Botany*, **8**, 75–209.

Burgess, P. F. (1971) The effect of logging on hill dipterocarp forests. *Malay Nature Journal*, **24**, 231–7.

Burgess, T. L. and Shmida, A. (1988) Succulent growth forms in arid environments, in *Arid Lands Today and Tomorrow*, (eds E. E. Whitehead, C. F. Hutchinson, B. N. Timmermann and R. G. Varady), Westview Press, Boulder, pp. 383–95.

Buringh, P. and Dudal, R. (1987) Agricultural land use in space and time, in *Land Transformation in Agriculture*, (eds M. G. Wolman and F. G. A. Fournier), Wiley, Chichester, pp. 9–43.

Burk, J. H. (1978) Seasonal and diurnal water potentials in selected chaparral shrubs. *American Midland Naturalist*, **99**, 244–6.

Burke, M. J., Gusta, L. V., Quamme, H. A., *et al.* (1976) Freezing and injury in plants. *Annual Review of Plant Physiology*, **27**, 507–28.

Burrows, C. J. (1977) Alpine grasslands and snow in the Arthur's Pass and Lewis Pass regions, South Island, New Zealand. *New Zealand Journal of Botany*, **15**, 665–86.

Butland, G. J. (1957) The human geography of southern Chile. *Institute of British Geographers Publication No. 24*, George Philip and Son, London.

Cade, T. J. and Maclean, G. L. (1967) Transport of water by adult sandgrouse to their young. *The Condor*, **69**, 323–43.

Caldwell, M. M. and Camp, L. B. (1974) Belowground productivity of two cool desert communities. *Oecologia*, **17**, 123–30.

Caldwell, M. M., White, R. S., Moore, R. T. and Camp, L. B. (1977) Carbon balance, productivity, and water use of cold-winter desert shrub communities dominated by C_3 and C_4 species. *Oecologia*, **29**, 275–300.

Callaghan, L. T. V. (1973) Studies on the factors affecting the primary production of bi-polar *Phleum alpinum*, in *Primary Production and Production Processes, Tundra Biome*, (eds L. C. Bliss and F. E. Wielgolaski), Tundra Biome Steering Committee, Edmonton, pp. 153–67.

Campbell, A. J. and Tanton, M. T. (1981) Effects of fire on the invertebrate fauna of soil and litter of a eucalypt forest, in *Fire and the Australian Biota*, (eds A. M. Gill, R. H. Groves and I. R. Noble), Australian Academy of Science, Canberra, pp. 215–41.

Campbell, E. O. (1983) Mires of Australasia, in *Ecosystems of the World Vol. 4B. Mires: Swamp, Bog, Fen and Moor – Regional Studies*, (ed. A. J. P. Gore), Elsevier, Amsterdam, pp. 153–80.

Cannell, M. G. R. (1982) *World Forest Biomass and Primary Production Data*, Academic Press, London.

Capone, D. G. and Carpenter, E. J. (1982) Nitrogen fixation in the marine environment. *Science*, **217**, 1140–42.

Carignan, R. and Kalff, J. (1980) Phosphorus sources for aquatic weeds: water or sediments? *Science*, **207**, 987–9.

Carleton, T. J. (1982) The pattern of invasion and establishment of *Picea mariana* (Mill.) BSP. into the subcanopy layers of *Pinus banksiana* Lamb. dominated stands. *Canadian Journal of Forest Research*, **12**, 973–84.

Carlisle, A., Brown, A. H. F. and White, E. J. (1966) The organic matter and nutrient elements in the precipitation beneath a sessile oak (*Quercus petraea*) canopy. *Journal of Ecology*, **54**, 87–98.

Carlucci, A. F. and Cuhel, R. L. (1977) Vitamins in the South Polar Seas: distribution and significance of dissolved and particulate vitamin B_{12}, thiamine, and biotin in the southern Indian Ocean, in *Adaptations Within Antarctic Ecosystems*, (ed. G. A. Llano), National Academy of Sciences, Washington, pp. 115–28.

Carney, H. J. and Elser, J. J. (1990) Strength of zooplankton–phytoplankton coupling in relation to lake trophic state, in *Large Lakes*, (eds M. M. Tilzer and C. Serruya), Springer, Berlin, pp. 615–31.

Caron, D. A., Pick, F. R. and Lean, D. R. S. (1985) Chroococcoid cyanobacteria in Lake Ontario: vertical and seasonal distributions during 1982. *Journal of Phycology*, **21**, 171–5.

Carpenter, S. R., Kitchell, J. F. and Hodgson, J. R. (1985) Cascading trophic interactions and lake productivity. *BioScience*, **35**, 634–9.

Carroll, E. J. and Ashton, D. H. (1965) Seed storage in soils of several Victorian plant communities. *The Victorian Naturalist*, **82**, 102–10.

Carter, J. V. and Brenner, M. L. (1985) Plant growth regulators and low temperature stress, in *Encyclopedia of Plant Physiology Vol II. Hormonal Regulation of Development III*, (eds R. P. Pharis and D. M. Reid), Springer, Berlin, pp. 418–43.

Carter, M. R., Burns, L. A., Cavinder, T. R. *et al.* (1973) *Ecosystems Analysis of the Big Cypress Swamp and Estuaries*, United States Environmental Protection Agency, Athens.

Cartwright, D. E. and Woods, A. J. (1963) Measurements of upper and lower tidal currents at Banc de la Chapelle. *Deutsche Hydrographische Zeitschrift*, **16**, 64–76.

Casparie, W. A. (1972) Bog development in south-eastern Drenthe (The Netherlands). *Vegetatio*, **25**, 1–271.

Caswell, H. and Reed, F. C. (1976) Plant–herbivore interactions: the indigestibility of C_4 bundle sheath cells by grasshoppers. *Oecologia*, **26**, 151–6.

Caswell, H., Reed, F., Stephenson, S. N. and Werner, P. A. (1973) Photosynthetic pathways and selective herbivory: a hypothesis. *American Naturalist*, **107**, 465–80.

Catling, P. C. and Newsome, A. E. (1981) Responses of the Australian vertebrate fauna to fire: an evolutionary approach, in *Fire and the Australian Biota*, (eds A. M. Gill, R. H. Groves and I. R. Noble), Australian Academy of Science, Canberra, pp. 367–93.

Cattelino, P. J., Noble, I. R., Slatyer, R. O. and Kessel, S. R. (1979) Predicting the multiple pathways of plant succession. *Environmental Management*, **3**, 41–50.

Chabot, B. F. (1979) Metabolic and enzymatic adaptations to low temperature, in *Comparative Mechanisms of Cold Adaptation*, (eds L. S. Underwood, L. L. Tieszen, A. B. Callahan and G. E. Folk), Academic Press, New York, pp. 283–301.

Chabot, B. F. and Hicks, D. J. (1982) The ecology of leaf life spans. *Annual Review of Ecology and Systematics*, **13**, 229–59.

Chalk, L. and Akpalu, J. D. (1963) Possible relation between the anatomy of the wood and buttressing. *Commonwealth Forestry Review*, **42**, 53–8.

Chambers, J. C., MacMahon, J. A. and Haefner, J. H. (1991) Seed entrapment in alpine ecosystems: effects of soil particle size and diaspore morphology. *Ecology*, **72**, 1668–77.

Chan, A. T. (1978) Comparative physiological study of marine diatoms and dinoflagellates in relation to irradiance and cell size. I. Growth under continuous light. *Journal of Phycology*, **14**, 396–402.

Chan, H. T. and Appanah, S. (1980) Reproductive biology of some Malaysian dipterocarps I: Flowering biology. *The Malaysian Forester*, **43**, 132–43.

Chapin, F. S. (1978) Phosphate uptake and nutrient utilization by Barrow tundra vegetation, in *Vegetation and Production Ecology of an Alaskan Arctic Tundra*, (ed. L. L. Tieszen), Springer, New York, pp. 483–507.

Chapin, F. S. and Shaver, G. R. (1981) Changes in soil properties and vegetation following disturbance of Alaskan arctic tundra. *Journal of Applied Ecology*, **18**, 605–17.

Chapin, F. S. and Shaver, G. R. (1985) Arctic, in *Physiological Ecology of North American Plant Communities*, (eds

B. F. Chabot and H. A. Mooney), Chapman & Hall, New York, pp. 16–40.

Chapin, F. S., Barsdate, R. J. and Barél, D. (1978) Phosphorus cycling in Alaskan coastal tundra: a hypothesis for the regulation of nutrient cycling. *Oikos*, **31**, 189–99.

Chapin, F. S., Johnson, D. A. and McKendrick, J. D. (1980) Seasonal movement of nutrients in plants of differing growth form in an Alaskan tundra ecosystem: implications for herbivory. *Journal of Ecology*, **68**, 189–209.

Chapman, V. J. (1976) *Coastal Vegetation*, 2nd edn, Pergamon Press, Oxford.

Chapman, V. J. (1977) Introduction, in *Ecosystems of the World Vol. 1. Wet Coastal Ecosystems*, (ed. V. J. Chapman), Amsterdam, Elsevier, pp. 1–29.

Charley, J. L. and West, N. E. (1975) Plant-induced soil chemical patterns in some shrub-dominated semi-desert ecosystems in Utah. *Journal of Ecology*, **63**, 945–63.

Charter, J. R. and Keay, R. W. J. (1960) Assessment of the Olokemeji fire-control experiment (investigation 254) 28 years after institution. *Nigerian Forestry Information Bulletin*, **3**, 1–32.

Chazdon, R. L. and Fetcher, N. (1984) Photosynthetic light environments in a lowland tropical rain forest in Costa Rica. *Journal of Ecology*, **72**, 553–64.

Chernov, Y. I. (1985) *The Living Tundra*, Cambridge University Press, Cambridge.

Chew, R. M. and Chew, A. E. (1970) Energy relationships of the mammals of a desert shrub (*Larrea tridentata*) community. *Ecological Monographs*, **40**, 1–21.

Christensen, B. (1978) Biomass and primary production of *Rhizophora apiculata* Bl. in a mangrove in southern Thailand. *Aquatic Botany*, **4**, 43–52.

Christensen, B. (1983) Mangroves – what are they worth? *Unasylva*, **35**, 2–15.

Christensen, N. L. (1988) Vegetation of the southeastern coastal plain, in *North American Terrestrial Vegetation*, (eds M. G. Barbour and W. D. Billings), Cambridge University Press, Cambridge, pp. 317–63.

Christensen, N. L. and Muller, C. H. (1975) Effects of fire on factors controlling plant growth in *Adenostoma* chaparral. *Ecological Monographs*, **45**, 29–55.

Christensen, P., Recher, H. and Hoare, J. (1981) Responses of open forests (dry sclerophyll forests) to fire regimes, in *Fire and the Australian Biota*, (eds A. M. Gill, R. H. Groves and I. R. Noble), Australian Academy of Science, Canberra, pp. 273–310.

Christensen, P. E. and Kimber, P. C. (1975) Effect of prescribed burning on the flora and fauna of the south-west Australian forests. *Proceedings of the Ecological Society of Australia*, **9**, 85–107.

Christie, E. K. and Detling, J. K. (1982) Analysis of interference between C_3 and C_4 grasses in relation to temperature and soil nitrogen supply. *Ecology*, **63**, 1277–84.

Church, N. S. (1960) Heat loss and the body temperatures of flying insects. I. Heat loss by evaporation of water from the body. *Journal of Experimental Biology*, **37**, 171–85.

Clark, D. A. and Clark, D. B. (1984) Spacing dynamics of a tropical rain forest tree: evaluation of the Janzen–Connell model. *The American Naturalist*, **124**, 769–88.

Clark, D. P. (1971) Flights after sunset by the Australian plague locust, *Chortoicetes terminifera* (Walk.), and their significance in dispersal and migration. *Australian Journal of Zoology*, **19**, 159–76.

Clark, F. E. and Paul, E. A. (1970) The microflora of grassland. *Advances in Agronomy*, **23**, 375–435.

Clark, F. E., Cole, C. V. and Bowman, R. A. (1980) Nutrient cycling, in *Grasslands, Systems Analysis and Man*, (eds A. I. Breymeyer and G. M. van Dyne), Cambridge University Press, Cambridge, pp. 659–712.

Clark, R. B. (1989) *Marine Pollution*, 2nd edn, Clarendon Press, Oxford.

Clarke, G. L. and Denton, E. J. (1962) Light and animal life, in *The Sea: Ideas and Observations on Progress in the Study of the Seas*, Vol. 1 (ed. M. N. Hill), Wiley Interscience, New York, pp. 456–68.

Clarke, G. L. and Hubbard, C. J. (1959) Quantitative records of the luminescent flashing of oceanic animals at great depths. *Limnology and Oceanography*, **4**, 163–80.

Clayton, W. D. (1958) Secondary vegetation and the transition to savanna near Ibadan, Nigeria. *Journal of Ecology*, **46**, 217–38.

Clements, F. E. (1916) *Plant Succession: An Analysis of the Development of Vegetation*, Carnegie Institution of Washington, Washington.

Cloudsley-Thompson, J. L. (1977) *Man and the Biology of Arid Zones*, Edward Arnold, London.

Clymo, R. S. (1967) Control of cation concentrations, and in particular of pH, in *Sphagnum* dominated communities, in *Chemical Environment in the Aquatic Habitat*, (eds H. L. Golterman and R. S. Clymo), Noord-Hollandische, Amsterdam, pp. 273–84.

Clymo, R. S. (1965) Experiments on breakdown of *Sphagnum* in two bogs. *Journal of Ecology*, **53**, 747–59.

Clymo, R. S. (1973) The growth of *Sphagnum*: some effects of environment. *Journal of Ecology*, **61**, 849–69.

Clymo, R. S. (1983) Peat, in *Ecosystems of the World Vol. 4A. Mires: Swamp, Bog, Fen and Moor – General Studies*, (ed. A. J. P. Gore), Elsevier, Amsterdam, pp. 159–224.

Clymo, R. S. and Hayward, P. M. (1982) The ecology of *Sphagnum*, in *Bryophyte Ecology*, (ed. A. J. E. Smith), Chapman & Hall, London, pp. 229–89.

Clymo, R. S. and Reddaway, E. J. F. (1971) Productivity of *Sphagnum* (bog-moss) and peat accumulation. *Hidrobiologia*, **12**, 181–92.

Coaldrake, J. E. (1970) The brigalow, in *Australian Grasses*, (ed. R. M. Moore), Australian National University Press, Canberra, pp. 123–40.

Coaldrake, J. E. (1979) The natural grasslands of Australasia, in *Ecology of Grasslands and Bamboolands in the World*, (ed. M. Numata), Junk, The Hague, pp. 133–40.

Cody, M. L. and Mooney, H. A. (1978) Convergence versus nonconvergence in Mediterranean-climate ecosystems. *Annual Review of Ecology and Systematics*, **9**, 265–321.

Cody, M. L., Fuentes, E. R., Glantz, W., Hunt, J. H. and Moldenke, A. R. (1977) Convergent evolution in the consumer organisms of mediterranean Chile and California, in *Convergent Evolution in Chile and California*, (ed. H. A. Mooney), Dowden, Hutchinson and Ross, Stroudsburg, pp. 144–92.

Coe, M. (1972) The South Turkana expedition. Scientific papers IX. Ecological studies of the small mammals of South Turkana. *Geographical Journal*, **138**, 316–38.

Coe, M. J. (1969) Microclimate and animal life in the Equatorial mountains. *Zoologica Africana*, **4**, 101–28.

Cole, D. W. and Rapp, M. (1981) Elemental cycling in forest ecosystems, in *Dynamic Properties of Forest Ecosystems*, (ed. D. E. Reichle), Cambridge University Press, Cambridge, pp. 341–409.

Cole, M. M. (1960) Cerrado, caatinga and pantanal: the distribution and origin of the savanna vegetation of Brazil. *Geographical Journal*, **126**, 168–79.

Cole, M. M. (1963) Vegetation and geomorphology in Northern Rhodesia: an aspect of the distribution of the savanna of central Africa. *Geographical Journal*, **129**, 290–310.

Cole, M. M. (1986) *The Savannas: Biogeography and Geobotany*, Academic Press, London.

Cole, C. V., Innis, G. S. and Stewart, J. W. B. (1977) Simulation of phosphorus cycling in semiarid grasslands. *Ecology*, **58**, 1–15.

Colebrook, J. M. (1982) Continuous plankton records: seasonal variations in the distribution and abundance of plankton in the North Atlantic Ocean and North Sea. *Journal of Plankton Research*, **4**, 435–62.

Coleman, D. C. and Sasson, A. (1980) Decomposer subsystem, in *Grasslands, Systems Analysis and Man*, (eds A. I. Breymeyer and G. M. van Dyne), Cambridge University Press, Cambridge, pp. 609–55.

Coles, S. M. (1979) Benthic microalgal populations on intertidal sediments and their role as precursors to salt marsh development, in *Ecological Processes in Coastal Environments*, (eds R. L. Jefferies and A. J. Davy), Blackwell Scientific Publications, Oxford, pp. 25–42.

Coley, P. D. (1983) Herbivory and defensive characteristics of tree species in a lowland tropical forest. *Ecological Monographs*, **53**, 209–33.

Coley, P. D., Bryant, J. P. and Chapin, F. S. (1985) Resource availability and plant antiherbivore defense. *Science*, **230**, 895–9.

Colinvaux, P. (1973) *Introduction to Ecology*, Wiley, New York.

Collins, B. S. and Pickett, S. T. A. (1987) Influence of canopy opening on the environment and herb layer in a northern hardwoods forest. *Vegetatio*, **70**, 3–10.

Collins, N. J., Baker, J. H. and Tilbrook, P. J. (1975) Signy Island, Maritime Antarctic, in *Structure and Function of Tundra Ecosystems*, (eds T. Rosswall and O. W. Heal), Swedish Natural Science Research Council, Stockholm, pp. 345–74.

Collins, N. M. (1989) Termites, in *Ecosystems of the World Vol. 14B. Tropical Rainforest Ecosystems: Biogeographical and Ecological Studies*, (eds H. Lieth and M. J. A. Werger), Elsevier, Amsterdam, pp. 455–71.

Collins, V. G., D'Sylva, B. T. and Latter, P. M. (1978) Microbial populations in peat, in *Production Ecology of British Moors and Montane Grasslands*, (eds O. W. Heal and D. F. Perkins), Springer, Berlin, pp. 94–112.

Connell, J. H. (1971) On the role of natural enemies in preventing competitive exclusion in some marine animals and in rain forest trees, in *Dynamics of Populations*, (eds P. J. den Boer and G. R. Gradwell),

Centre for Agricultural Publishing and Documentation, Wageningen, pp. 298–312.

Connell, J. H. (1978) Diversity in tropical rain forests and coral reefs. *Science*, **199**, 1302–10.

Connell, J. H. and Orias, E. (1964) The ecological regulation of species diversity. *American Naturalist*, **98**, 399–414.

Connell, J. H. and Slatyer, R. O. (1977) Mechanisms of succession in natural communities and their role in community stability and organisation. *American Naturalist*, **111**, 1119–44.

Connor, H. E. (1964) Tussock grassland communities in the Mackenzie Country, South Canterbury, New Zealand. *New Zealand Journal of Botany*, **2**, 325–51.

Cook, A. G., Critchley, B. R., Critchley, U., *et al.* (1980) Effects of cultivation and DDT on earthworm activity in a forest soil in the sub-humid tropics. *Journal of Applied Ecology*, **17**, 21–9.

Cooke, G. D., Welch, E. B., Peterson, S. A. and Newroth, P. R. (1986) *Lake and Reservoir Restoration*, Butterworths, Boston.

Coombs, J., Hall, D. O. and Chartier, P. (1983) *Plants as Solar Collectors*, Reidel, Dordrecht.

Cooper, J. P. and Tainton, N. M. (1968) Light and temperature requirements for the growth of tropical and temperate grasses. *Herbage Abstracts*, **38**, 167–76.

Coppock, D. L., Detling, J. K., Ellis, J. E. and Dyer, M. I. (1983) Plant–herbivore interactions in a North American mixed-grass prairie I. Effects of black-tailed prairie dogs on intraseasonal aboveground plant biomass and nutrient dynamics and plant species diversity. *Oecologia*, **56**, 1–9.

Corner, E. J. H. (1956) *Wayside Trees of Malaya*, 2nd edn, Government Printing Office, Singapore.

Cosgrove, D. J. (1967) Metabolism of organic phosphates in soil, in *Soil Biochemistry*, (eds A. D. McLaren and G. H. Petersen), Marcel Dekker, New York, pp. 216–28.

Costin, A. B. (1967) Alpine ecosystems of the Australasian region, in *Arctic and Alpine Environments*, (eds H. E. Wright and W. H. Osburn), University of Indiana Press, Bloomington, pp. 55–87.

Coulson, J. C. and Butterfield, J. (1978) An investigation of the biotic factors determining the rates of plant decomposition on blanket bog. *Journal of Ecology*, **66**, 631–50.

Coulson, J. C. and Whittaker, J. B. (1978) Ecology of moorland animals, in *Production Ecology of British Moors and Montane Grasslands*, (eds O. W. Heal and D. F. Perkins), Springer, Berlin, pp. 52–93.

Coulson, R. N. and Witter, J. A. (1984) *Forest Entomology*, Wiley, New York.

Coultas, C. L. and Duever, M. J. (1984) Soils of cypress swamps, in *Cypress Swamps*, (eds K. C. Ewel and H. T. Odum), University of Florida Press, Gainesville, pp. 51–9.

Coupland, R. T. (1950) Ecology of mixed prairie in Canada. *Ecological Monographs*, **20**, 271–315.

Coupland, R. T. (1958) The effects of fluctuations in weather upon the grasslands of the Great Plains. *Botanical Review*, **24**, 273–317.

Coupland, R. T. (1974a) Fluctuations in North American grassland vegetation, in *Handbook of Vegetation Science*

Part VIII: Vegetation Dynamics, (ed. R. Knapp), Junk, The Hague, pp. 235–41.

Coupland, R. T. (1974b) *Producers: IV. Underground Plant Parts. Canadian Committee for IBP, Matador Project Technical Report 41*, University of Saskatchewan, Saskatoon.

Coupland, R. T. (1979) The nature of grassland, in *Grassland Ecosystems of the World: Analysis of Grasslands and Their Uses*, (ed. R. T. Coupland), Cambridge University Press, Cambridge, pp. 23–9.

Coupland, R. T. and Abouguendia, Z. (1974) *Producers: V. Dynamics of Shoot Development in Grasses and Sedges. Canadian Committee for IBP, Matador Project Technical Report 51*, University of Saskatchewan, Saskatoon.

Coupland, R. T. and Johnson, R. E. (1965) Rooting characteristics of native grassland species in Saskatchewan. *Journal of Ecology*, **53**, 475–507.

Court, A. (1974) The climate of the conterminous United States, in *World Survey of Climatology Vol. 11: Climates of North America*, (eds R. A. Bryson and F. K. Hare), Elsevier, Amsterdam, pp. 193–343.

Coutinho, L. M. (1990) Fire in the ecology of the Brazilian cerrado, in *Fire in the Tropical Biota*, (ed. J. G. Goldammer), Springer, Berlin, pp. 82–105.

Cowles, S. (1982) Preliminary results investigating the effect of lichen ground cover on the growth of black spruce. *Le Naturaliste Canadien*, **109**, 573–81.

Cowling, R. M. and Lamont, B. B. (1985) Seed release in *Banksia*: the role of wet–dry cycles. *Australian Journal of Ecology*, **10**, 169–71.

Cowling, R. M., Lamont, B. B. and Enright, N. J. (1990) Fire and management of south-western Australian banksias. *Proceedings of the Ecological Society of Australia*, **16**, 177–83.

Cowling, R. M., Lamont, B. B. and Pierce, S. M. (1987) Seed bank dynamics of four co-occurring *Banksia* species. *Journal of Ecology*, **75**, 289–302.

Craig, P. C. (1984) Fish use of coastal waters of the Alaskan Beaufort Sea: a review. *Transactions of the American Fisheries Society*, **113**, 265–82.

Craighead, F. C. and Gilbert, V. C. (1962) The effects of Hurricane Donna on the vegetation of southern Florida. *The Quarterly Journal of the Florida Academy of Sciences*, **25**, 1–28.

Crawford, C. S. (1976) Feeding-season production in the desert millipede *Orthoporus ornatus* (Girard) (Diplododa). *Oecologia*, **24**, 265–76.

Crawford, C. S. (1979) Desert detritivores: a review of life history patterns and trophic roles. *Journal of Arid Environments*, **2**, 31–42.

Crawford, C. S. and Cloudsley-Thompson, J. L. (1971) Water relations and desiccation-avoiding behaviour in the vinegaroon *Mastigoproctus giganteus* (Arachnida: Uropygi). *Entomologica Experimentalis et Applicata*, **14**, 99–106.

Crawford, R. M. M. (1989) *Studies in Plant Survival*, Blackwell Scientific Publications, Oxford.

Crawford, R. M. M., Wishart, D. and Campbell, R. M. (1970) A numerical analysis of high altitude scrub vegetation in relation to soil erosion in the eastern cordillera of Peru. *Journal of Ecology*, **58**, 173–91.

Crawford, R. M. M. and Wishart, D. (1966) A multivariate analysis of the development of dune slack vegetation in relation to coastal accretion at Tentsmuir, Fife. *Journal of Ecology*, **54**, 729–43.

Cresswell, C. F., Ferrar, P., Grunow, J. O. *et al.* (1982) Phytomass, seasonal phenology and photosynthetic studies, in *Ecology of Tropical Savannas*, (eds B. J. Huntley and B. H. Walker), Springer, Berlin, pp. 476–97.

Crider, F. J. (1955) *Root-growth Stoppage Resulting from Defoliation of Grass*, US Department of Agriculture Technical Bulletin 1102.

Critchfield, H. J. (1966) *General Climatology*, Prentice-Hall, Englewood Cliffs.

Croat, T. B. (1979) The sexuality of the Barro Colorado Island flora. *Phytologia*, **42**, 319–48.

Crome, F. H. J. and Irvine, A. K. (1986) 'Two bob each way': the pollination and breeding system of the Australian rain forest tree *Syzygium cormiflorum* (Myrtaceae). *Biotropica*, **18**, 115–25.

Cryer, M., Peirson, G. and Townsend, C. R. (1986) Reciprocal interactions between roach, *Rutilus rutilus*, and zooplankton in a small lake: prey dynamics and fish growth and recruitment. *Limnology and Oceanography*, **31**, 1022–38.

Cuatrecasas, J. (1968) Paramo vegetation and its life forms, in *Geo-ecology of the Mountainous Regions of the Tropical Americas*, (ed. C. Troll), pp. 163–86.

Cuatrecasas, J. (1979) Growth forms of the Espeletiinae and their correlation to vegetation types of the high Tropical Andes, in *Tropical Botany*, (eds K. Larsen and L. B. Holm-Nielsen), Academic Press, London, pp. 397–410.

Cumberland, K. B. (1962) Moas and men: New Zealand about A.D. 1250. *Geographical Review*, **52**, 151–73.

Cummins, K. W. (1988) The study of stream ecosystems: a functional view, in *Concepts of Ecosystem Ecology*, (eds L. R. Pomeroy and J. J. Alberts), Springer, New York, pp. 247–62.

Cummins, K. W. and Klug, M. J. (1979) Feeding ecology of stream invertebrates. *Annual Review of Ecology and Systematics*, **10**, 147–72.

Cundell, A. M., Brown, M. S., Standford, R. and Mitchell, R. (1979) Microbial degradation of *Rhizophora mangle* leaves immersed in the sea. *Estuarine and Coastal Marine Science*, **9**, 281–6.

Currie, D. J. and Kalff, J. (1984) A comparison of the abilities of freshwater algae and bacteria to acquire and retain phosphorus. *Limnology and Oceanography*, **29**, 298–310.

Curtis, J. T. and Partch, M. L. (1948) Effect of fire on the competition between blue grass and certain prairie plants. *American Midland Naturalist*, **39**, 437–43.

Cutter, E. G. (1978) *Plant Anatomy. Part 1: Cells and Tissues*, 2nd edn, Arnold, London.

Dahl, B. E. and Hyder, D. N. (1977) Developmental morphology and management implications, in *Rangeland Plant Physiology*, (ed. R. E. Sosebee), Society for Range Management, Denver, pp. 257–90.

Dahl, E. (1975) Flora and plant sociology in Fennoscandian tundra areas, in *Fennoscandian Tundra Ecosystems: Part 1 Plants and Microorganisms*, (ed. F. E. Wielgolaski), Springer, New York, pp. 62–7.

Dahlman, R. C. and Kucera, C. L. (1965) Root productivity and turnover in native prairie. *Ecology*, **46**, 84–9.

Dalrymple, R. L. and Dwyer, D. D. (1967) Root and shoot growth of five range grasses. *Journal of Range Management*, **20**, 141–5.

Dalsted, K. J., Sather-Blair, S., Worcester, B. K. and Klukas, R. (1981) Application of remote sensing to prairie dog management. *Journal of Range Management*, **34**, 218–23.

Damman, A. W. H. (1978) Distribution and movement of elements in ombrotrophic peat bogs. *Oikos*, **30**, 486.

Damman, A. W. H. (1988) Regulation of nitrogen removal and retention in *Sphagnum* bogs and other peatlands. *Oikos*, **51**, 291–305.

Daniel, T. W. and Schmidt, J. (1972) Lethal and nonlethal effects of the organic horizons of forested soils on the germination of seeds from several associated conifer species of the Rocky Mountains. *Canadian Journal of Forest Research*, **2**, 179–84.

Danserau, P. and Segadas-Vianna, F. (1952) Ecological study of the peat bogs of eastern North America. *Canadian Journal of Botany*, **30**, 490–520.

Datta, S. C., Evanari, M. and Gutterman, Y. (1970) The heteroblasty of *Aegilops ovata* L., *Israel Journal of Botany*, **19**, 463–83.

Daubenmire, R. (1943) *Vegetational zonation in the Rocky Mountains*. The Botanical Review, **9**, 325–93.

Daubenmire, R. (1954) Alpine timberlines in the Americas and their interpretation. *Butler University Botanical Studies*, **11**, 119–36.

Daubenmire, R. (1966) Vegetation: identification of typal communities. *Science*, **151**, 291–8.

Daubenmire, R. (1968) Ecology of fire in grasslands. *Advances in Ecological Research*, **5**, 209–66.

Daubenmire, R. (1972a) Standing crops and primary production in savanna derived from semideciduous forest in Costa Rica. *Botanical Gazette*, **133**, 395–401.

Daubenmire, R. (1972b) Phenology and other characteristics of tropical semi-deciduous forest in north-western Costa Rica. *Journal of Ecology*, **60**, 147–70.

Daubenmire, R. F. (1943) Vegetational zonation in the Rocky Mountains. *The Botanical Review*, **9**, 325–93.

Daubenmire, R. F. (1967) *Plants and Environment*, 2nd edn, Wiley, New York.

Davidson, D. W. (1977) Foraging ecology and community organization in desert seed-eating ants. *Ecology*, **58**, 725–37.

Davies, D. D., Nascimento, K. H. and Patil, K. D. (1974) The distribution and properties of NADP malic enzyme in flowering plants. *Phytochemistry*, **13**, 2417–25.

Davis, C. B. (1973) 'Bark-striping' in *Arctostaphylos* (Ericaceae). *Madroño*, **22**, 145–9.

Davis, C. B. and van der Valk, A. G. (1983) Uptake and release of nutrients by living and decomposing *Typha glauca* Godr. tissues at Eagle Lake, Iowa. *Aquatic Botany*, **16**, 75–89.

Davis, M. B. (1981) Quaternary history and the stability of forest communities, in *Forest Succession*, (eds D. C. West, H. H. Shugart and D. B. Botkin), Springer, New York, pp. 132–53.

Davis, M. B. (1983) Holocene vegetational history of the eastern United States, in *Late-Quaternary Environments of the United States Vol. 2: The Holocene*, (ed. H. E. Wright), University of Minnesota Press, Minneapolis, pp. 166–81.

Dayton, P. K. (1973) Dispersion, dispersal, and persistence of the annual intertidal alga, *Postelsia palmaeformis* Ruprecht. *Ecology*, **54**, 433–8.

Dayton, P. K. (1975) Experimental evaluation of ecological dominance in a rocky intertidal algal community. *Ecological Monographs*, **45**, 137–59.

Deacon, H. J. (1983) The comparative evolution of mediterranean-type ecosystems: a southern perspective, in *Mediterranean-type Ecosystems: The Role of Nutrients*, (eds F. J. Kruger, D. T. Mitchell and J. U. M. Jarvis), Springer, Berlin, pp. 3–40.

DeAngelis, D. L. (1980) Energy flow, nutrient cycling, and ecosystem resilience. *Ecology*, **61**, 764–71.

DeAngelis, D. L., Gardner, R. H. and Shugart, H. H. (1981) Productivity of forest ecosystems studied during the IBP: the woodlands data set, in *Dynamic Properties of Forest Ecosystems*, (ed. D. E. Reichle), Cambridge University Press, Cambridge, pp. 567–672.

DeBano, L. F. and Conrad C. E. (1978) The effect of fire on nutrients in a chaparral ecosystem. *Ecology*, **59**, 489–97.

DeBano, L. F., Eberlein, G. E. and Dunn, P. H. (1979) Effects of burning on chaparral soils: I. Soil nitrogen. *Soil Society of America Journal*, **43**, 504–9.

DeBano, L. F., Savage, S. M. and Hamiliton, D. A. (1976) The transfer of heat and hydrophobic substances during burning. *Soil Society of America Journal*, **40**, 779–82.

Debazac, E. F. (1983) Temperate broad-leaved evergreen forests of the Mediterranean region and middle east, in *Ecosystems of the World Vol. 10. Temperate Broad-leaved Evergreen Forests*, (ed. J. D. Ovington), Elsevier, Amsterdam, pp. 107–23.

DeBernardi, R., Giussani, G. and Grimaldi, E. (1984) Lago Maggiore, in *Ecosystems of the World Vol. 23. Lakes and Reservoirs*, (ed. F. B. Taub), Elsevier, Amsterdam, pp. 247–66.

DeLaune, R. D., Smith, C. J. and Patrick, W. H. (1983) Relationship of marsh elevation, redox potential, and sulfide to *Spartina alternifolia* productivity. *Soil Science Society of America Journal*, **47**, 930–5.

Delcourt, P. A. and Delcourt, H. R. (1981) Vegetation maps for eastern North America: 40,000 yr B.P. to the present, in *Geobotany II*, (ed. R. C. Romans), Plenum, New York, pp. 123–65.

D'Elia, C. F. (1988) The cycling of essential elements in coral reefs, in *Concepts of Ecosystem Ecology*, (eds L. R. Pomeroy and J. J. Alberts), Springer, New York, pp. 195–230.

Delisle, G. P. and Dubé, D. E. (1983) One and one-half centuries of fire in Wood Buffalo National Park. Northern Forest Research Centre Forestry Report No. 28. Environment Canada: Canadian Forestry Service, Ottawa, p. 12.

del Moral, R., Willis, R. J. and Ashton, D. H. (1978) Suppression of coastal heath vegetation by *Eucalyptus baxteri*. *Australian Journal of Botany*, **26**, 203–19.

Dement, W. A. and Mooney, H. A. (1974) Seasonal variation in the production of tannins and cyanogenic glucosides in the chaparral shrub, *Heteromeles arbutifolia*. *Oecologia*, **15**, 65–76.

DeMott, W. R. (1983) Seasonal succession in a natural *Daphnia* assemblage. *Ecological Monograph*, **53**, 321–340.

DeMott, W. R. (1989) The role of competition in zooplankton succession, in *Plankton Ecology* (ed. U. Sommer), Springer, Berlin, pp. 195–252.

Denslow, J. S. (1980) Gap partitioning among tropical rainforest trees. *Biotropica*, **12** (supplement), 47–55.

Denslow, J. S. (1987) Tropical rainforest gaps and trees species diversity. *Annual Review of Ecology and Systematics*, **18**, 431–51.

Detling, J. K., Dyer, M. I., Procter-Gregg, C. and Winn, D. T. (1980) Plant–herbivore interactions: examination of potential effects of bison saliva on regrowth of *Bouteloua gracilis* (H. B. K.) Lag. *Oecologia*, **45**, 26–31.

Detling, J. K., Dyer, M. I. and Winn, D. T. (1979) Net photosynthesis, root respiration, and regrowth of *Bouteloua gracilis* following simulated grazing. *Oecologia*, **41**, 127–34.

Detling, J. K. and Painter, E. L. (1983). Defoliation responses of western wheatgrass populations with diverse histories of prairie dog grazing. *Oecologia*, **57**, 65–71.

di Castri, F. (1981) Mediterranean-type shrublands of the world, in *Ecosystems of the World Vol. 11. Mediterranean-type Shrublands*, (eds F. di Castri, D. W. Goodall and R. L. Specht), Elsevier, Amsterdam, pp. 1–52.

Dickinson, C. H. (1983) Micro-organisms in peatlands, in *Ecosystems of the World Vol. 4A. Mires: Swamp, Bog, Fen and Moor – General Studies*, (ed. A. J. P. Gore), Elsevier, Amsterdam, pp. 225–45.

Dickinson, C. H. and Dodd, J. L. (1976) Phenological pattern in the shortgrass prairie. *American Midland Naturalist*, **96**, 367–78.

Dickinson, C. H. and Maggs, G. H. (1974) Aspects of the decomposition of *Sphagnum* leaves in an ombrophilous mire. *New Phytologist*, **73**, 1249–57.

Dickmann, D. I. (1971) Photosynthesis and respiration by developing leaves of cottonwood (*Populus deltoides* Bartr.). Botanical Gazette, **132**, 253–9.

Dierberg, F. E. and Brezonik, P. L. (1981) Nitrogen fixation (acetylene reduction) associated with decaying leaves of pond cypress (*Taxodium distichum* var. *nutans*) in a natural and a sewage-enriched cypress dome. *Applied and Environmental Microbiology*, **41**, 1413–18.

Dierberg, F. E. and Brezonik, P. L. (1984) The effect of wastewater on the surface water and groundwater quality of cypress domes, in *Cypress swamps*, (eds K. C. Ewel and H. T. Odum), University of Florida Press, Gainesville, pp. 83–101.

Digby, P. S. B. (1953) Plankton production in Scoresby Sound, East Greenland. *Journal of Animal Ecology*, **22**, 289–322.

Dirzo, R. (1984) Insect–plant interactions: some ecophysiological consequences of herbivory, in *Physiological Ecology of Plants of the Wet Tropics*, (eds E. Medina, H. A. Mooney and C. Vasquez-Yanes), Junk, The Hague, pp. 209–24.

Disraeli, D. J. (1984) The effect of sand deposits on the growth and morphology of *Ammophila breviligulata*. *Journal of Ecology*, **72**, 145–54.

Dittus, W. P. J. (1985) The influence of cyclones on the dry evergreen forest of Sri Lanka. *Biotropica*, **17**, 1–14.

Dix, R. L. (1960) The effects of burning on the mulch structure and species composition of grasslands in western North Dakota. *Ecology*, **41**, 49–56.

Dixon, J. and Louw, G. (1978) Seasonal effects on nutrition, reproduction and aspects of thermoregulation in Namaqua sandgrouse (*Pterocles namaqua*). *Madoqua*, **11**, 19–29.

Dodd, J. L. and Lauenroth, W. K. (1979) Analysis of the response of a grassland ecosystem to stress, in *Perspectives in Grassland Ecology*, (ed. N. R. French), Springer, New York, pp. 43–8.

Doing, H. (1985) Coastal fore-dune zonation and succession in various parts of the world. *Vegetatio*, **61**, 65–75.

Dolph, G. E. and Dilcher, D. L. (1980) Variation in leaf size with respect to climate in Costa Rica. *Biotropica*, **12**, 91–9.

Doman, E. R. (1968) Prescribed burning and brush type conversion in California National Forests. *Tall Timbers Fire Ecology Conference, Proceedings*, **7**, 225–33.

Dorst, J. and Dandelot, P. (1970) *A Field Guide to the Larger Mammals of Africa*. Houghton Mifflin, Boston.

Douglas, G. W. and Bliss, L. C. (1977) Alpine and high subalpine plant communities of the North Cascades Range, Washington and British Columbia. *Ecological Monographs*, **47**, 113–50.

Dowding, P., Chapin, F. S., Wielgolaski, F. E. and Kilfeather P. (1981) Nutrients in tundra ecosystems, in *Tundra Ecosystems: A Comparative Analysis*, (eds L. C. Bliss, O. W. Heal and J. J. Moore), Cambridge University Press, Cambridge, pp. 647–83.

Doyle, G. J. (1973) Primary production estimates of native blanket bog and meadow vegetation growing on reclaimed peat at Glenamoy, Ireland, in *Primary Production and Production Processes, Tundra Biome*, (eds L. C. Bliss and F. E. Wielgolaski), University of Alberta, Edmonton, pp. 141–51.

Dragoun, F. J. and Kuhlman, A. R. (1968) Effect of pasture management practices on runoff. *Journal of Soil and Water Conservation*, **23**, 55–7.

Dregne, H. E. (1983) *Desertification of Arid Lands*. Harwood, Chur.

Drew, M. C. (1979) Root development and activities, in *Arid-land Ecosystems: Structure, Functioning and Management*, Vol. 1 (eds D. W. Goodall, R. A. Perry and K. M. W. Howes) Cambridge University Press, Cambridge, pp. 573–606.

Dring, M. J. (1982) *The Biology of Marine Plants*, Arnold, London.

Dromgoole, F. I. (1980) Desiccation resistance of intertidal and subtidal algae. *Botanica Marina*, **23**, 149–59.

Ducklow, H. W., Kirchman, D. L. and Rowe, G. T. (1982) Production and vertical flux of attached bacteria in the Hudson River plume of the New York Bight as studied with floating sediment traps. *Applied and Environmental Microbiology*, **43**, 769–76.

Duever, M. J. and Riopelle, L. A. (1983) Successional sequences and rates on tree islands in the Okefenokee Swamp. *American Midland Naturalist*, **110**, 186–93.

Dugdale, R. C. and Goering, J. J. (1967) Uptake of new and regenerated forms of nitrogen in primary productivity. *Limnology and Oceanography*, **12**, 196–206.

Dunbar, M. J. (1970) On the fishery potential of the sea waters of the Canadian North. *Arctic*, **23**, 150–74.

Dunbar, M. J. (1973) Stability and fragility in arctic ecosystems. *Arctic*, **26**, 179–85.

Durant, C. C. L. (1941) The growth of mangrove species in Malaya. *Malayan Forester*, **10**, 8–15.

Duvigneaud, P. and Denaeyer-De Smet, S. (1970) Biological cycling of minerals in temperate deciduous forests, in *Analysis of Temperate Forest Ecosystems*, (ed. D. E. Reichle), Springer, Berlin, pp. 199–225.

Dwyer, D. D. and Pieper, R. D. (1967) Fire effects on blue grama–pinyon–juniper rangeland in New Mexico. *Journal of Range Management*, **20**, 359–62.

Dyer, M. I. and Bokhari, U. G. (1976) Plant–animal interactions: studies of the effects of grasshopper grazing on blue grama grass. *Ecology*, **57**, 762–72.

Dyksterhuis, E. J. (1949) Condition and management of range land based on quantitative ecology. *Journal of Range Management*, **2**, 104–15.

Dykyjová, D. (1971) Productivity and solar energy conversion in reedswamp stands in comparison with outdoor mass cultures of algae in the temperate climate of Central Europe. *Photosynthetica*, **5**, 329–40.

Dykyjová, D. (1978) Nutrient uptake by littoral communities of helophytes, in *Pond Littoral Ecosystems*, Springer, Berlin, pp. 257–77.

Eden, M. J. (1974) Palaeoclimatic influences and the development of savanna in southern Venezuela. *Journal of Biogeography*, **1**, 95–109.

Edmisten, J. (1970) Preliminary studies of the nitrogen budget of a tropical rain forest, in *A Tropical Rain Forest: A Study of Irradiation and Ecology at ElVerde, Puerto Rico*, (eds H. T. Odum and R. F. Pigeon), US Atomic Energy Commission, Washington, H211–5.

Edmonson, W. T. and Lehman, J. T. (1981) The effect of changes in the nutrient income on the conditions of Lake Washington. *Limnology and Oceanography*, **26**, 1–29.

Edney, E. B. (1966) Animals of the desert, in *Arid Lands – A Geographical Appraisal*, (ed. E. S. Hills), Methuen, London, pp. 181–218.

Edney, E. B. (1974) Desert arthropods, in *Desert Biology*, Vol. 2 (ed. G. W. Brown), Academic Press, New York, pp. 311–84.

Edmunds, M. (1974) *Defence in Animals*, Longman, Harlow.

Edmunds, M. (1978) On the association between *Myrmarachne* spp. (Salticidae) and ants. *Bulletin of the British Arachnological Society*, **4**, 149–60.

Edwards, C. A., Reichle, D. E. and Crossley, D. A. (1970) The role of soil invertebrates in turnover of organic matter and nutrients, in *Analysis of Temperate Forest Ecosystems*, (ed. D. E. Reichle), Springer, Berlin, pp. 147–72.

Edwards, P. J. (1982) Studies of mineral cycling in a montane rain forest in New Guinea V. Rates of cycling in throughfall and litter fall. *Journal of Ecology*, **70**, 807–27.

Edwards, P. J. and Grubb, P. J. (1977) Studies of mineral cycling in a montane rain forest in New Guinea I. The distribution of organic matter in the vegetation and soil. *Journal of Ecology*, **65**, 943–69.

Edwards, P. J., Wratten, S. D. and Greenwood, S. (1986) Palatability of British trees to insects: constitutive and induced defences. *Oecologia*, **69**, 316–19.

Edwards, R. Y., Soos, J. and Ritcey, R. W. (1960) Quantitative observations on epidendric lichens used as food by caribou. *Ecology*, **41**, 425–31.

Egler, F. E. (1954) Vegetation science concepts I. Initial floristic composition, a factor in old-field vegetation development. *Vegetatio*, **4**, 412–17.

Ehleringer, J. (1981) Leaf absorptances of Mohave and Sonoran Desert plants. *Oecologia*, **49**, 366–70.

Ehleringer, J. and Björkman, O. (1977) Quantum yields for CO_2 uptake in C_3 and C_4 plants. *Plant Physiology*, **59**, 86–90.

Ehleringer, J. and Forseth I. (1980) Solar tracking by plants. *Science*, **210**, 1094–8.

Ehleringer, J. R. (1978) Implications of quantum yield differences on the distributions of C_3 and C_4 grasses. *Oecologia*, **31**, 255–67.

Ehleringer, J. R. and Björkman, O. (1978) Pubescence and leaf spectral characteristics in a desert shrub, *Encelia farinosa*. *Oecologia*, **36**, 151–62.

Ehleringer, J. R. and Mooney, H. A. (1978) Leaf hairs: effects on physiological activity and adaptive value to a desert shrub. *Oecologia*, **37**, 183–200.

Eis, S. (1973) Cone production of Douglas fir and grand fir and its climatic requirements. *Canadian Journal of Forest Research*, **3**, 61–70.

Eiten, G. (1972) The cerrado vegetation of Brazil. *The Botanical Review*, **38**, 201–341.

Eiten, G. (1978) Delimitation of the cerrado concept. *Vegetatio*, **36**, 169–78.

Elgabaly, M. M. (1977) Salinity and waterlogging in the Near-East region. *Ambio*, **6**, 36–9.

Ellenberg, H. (1988) *Vegetation Ecology of Central Europe*, Cambridge University Press, Cambridge.

Elliott-Fisk, D. L. (1983) The stability of the northern Canadian tree limit. *Annals of the Association of American Geographers*, **73**, 560–76.

Elmore, J. L. (1983) Factors influencing *Diaptomus* distributions: an experimental study in subtropical Florida. *Limnology and Oceanography*, **28**, 522–32.

El-Sayed, S. Z. (1978) Primary productivity and estimates of potential yields of the Southern Ocean, in *Polar Research: To the Present, and the Future* (ed. M. A. McWhinnie), Westview Press, Boulder, pp. 141–60.

Emanuel, W. R., Shugart, H. H. and Stevenson, M. P. (1985) Climatic change and the broad-scale distribution of terrestrial ecosystem complexes. *Climatic Change*, **7**, 29–43.

Emiliani, C. (1955) Pleistocene temperatures. *Journal of Geology*, **63**, 538–78.

Emiliani, C. and Flint, R. F. (1963) The Pleistocene record, in *The Sea: Ideas and Observations on Progress in the Study of the Seas*, Vol. 3 (ed. M. N. Hill), Wiley Interscience, New York, pp. 888–927.

Endean, R. (1973) Population explosions of *Acanthaster planci* and associated destruction of hermatypic corals in the Indo-West Pacific region, in *Biology and Geology of Coral Reefs Vol. 2*, (eds O. A. Jones and R. Endean), Academic Press, New York, pp. 389–438.

Eppley, R. W. and Peterson, B. J. (1979) Particulate organic matter flux and planktonic new production in the deep ocean. *Nature*, **282**, 677–80.

Eppley, R. W., Renger, E. H., Venrick, E. L. and Mullin, M. M. (1973) A study of plankton dynamics and nutrient cycling in the central gyre of the North Pacific Ocean. *Limnology and Oceanography*, **18**, 534–51.

Erickson, R. (1965) *Orchids of the West*, Paterson Brokensha, Perth.

Ericson, D. B., Ewing, M., Wollin, G. and Heezen, B. C. (1961) Atlantic deep-sea sediment cores. *Geological Society of America Bulletin*, **72**, 193–286.

Ernst, W. (1975) Variation in the mineral contents of leaves of trees in miombo woodland in south central Africa. *Journal of Ecology*, **63**, 801–7.

Essig, F. B. (1973) Pollination in some New Guinea palms. *Principes*, **17**, 75–83.

Etherington, J. R. (1967) Studies of nutrient cycling and productivity in oligotrophic ecosystems. I. Soil potassium and wind-blown sea-spray in a South Wales dune grassland. *Journal of Ecology*, **55**, 743–52.

Evanari, M., Shanan, L. and Tadmor, N. (1982) *The Negev – The Challenge of a Desert*, 2nd edn, Harvard University Press, Cambridge.

Evanari, M., Bamberg, S., Schulze, E. D. *et al.* (1975) The biomass production of some higher plants in Near-Eastern and American deserts, in *Photosynthesis and Productivity in Different Environments*. (ed. J. P. Cooper), Cambridge University Press, Cambridge, pp. 121–7.

Everett, K. R., Vassiljevskaya, V. D., Brown, J. and Walker, B. D. (1981) Tundra and analogous soils, in *Tundra Ecosystems: A Comparative Analysis*, (eds L. C. Bliss, O. W. Heal and J. J. Moore), Cambridge University Press, Cambridge, pp. 139–79.

Eyre, S. R. (1968) *Vegetation and Soils*, 2nd edn, Aldine Publishing Company, Chicago.

Faegri, K. and van der Pijl, L. (1979) *The Principles of Pollination Ecology*, 3rd edn, Pergamon, Oxford.

Fahn, A. and Shimony, C. (1977) Development of the glandular and non-glandular leaf hairs of *Avicennia marina* (Forsskål) Vierh. *Botanical Journal of the Linnean Society*, **74**, 37–46.

Fahnenstiel, G. L., Sicko-Goad, L., Scavia, D. and Stoermer, E. F. (1986) Importance of picoplankton in Lake Superior. *Canadian Journal of Fisheries and Aquatic Sciences*, **43**, 235–40.

Fareed, M. and Caldwell, M. M. (1975) Phenological patterns of two alpine tundra plant populations on Niwot Ridge, Colorado. *Northwest Science*, **49**, 17–23.

Farnsworth, N. R. and Soejarto, D. D. (1991) Global importance of medicinal plants, in *The Conservation of Medicinal Plants*, (eds O. Akerele, V. Heywood and H. Synge), Cambridge University Press, Cambridge, pp. 25–51.

Farrish, K. W. and Grigal, D. F. (1988) Decomposition in an ombrotrophic bog and a minerotrophic fen in Minnesota. *Soil Science*, **145**, 353–8.

Federer, C. A. (1976) Differing diffusive resistance and leaf development may cause differing transpiration among hardwoods in spring. *Forest Science*, **22**, 359–64.

Feeny, P. (1976) Plant apparency and chemical defense, in *Recent Advances in Phytochemistry. Vol. 10. Biochemical Interaction Between Plants and Insects*, (eds J. W. Wallace and R. L. Mansell), Plenum Press, New York, pp. 1–40.

Feinsinger, P. (1978) Ecological interactions between plants and hummingbirds in a successional tropical community. *Ecological Monographs*, **48**, 269–87.

Fenchel, T. (1984) Suspended marine bacteria as a food source, in *Flows of Energy and Materials in Marine Ecosystems*, (ed. M. J. R. Fasham), Plenum Press, New York, pp. 301–15.

Fernandez, O. A. and Caldwell, M. M. (1975) Phenology and dynamics of root growth of three cool semi-desert shrubs under field conditions. *Journal of Ecology*, **63**, 703–14.

Fetcher, N., Oberbauer, S. F. and Strain, B. R. (1985) Vegetation effects on microclimate in lowland tropical forest in Costa Rica. *International Journal of Biometeorology*, **29**, 145–55.

Fiala, K. and Kvĕt, J. (1971) Dynamic balance between plant species in South Moravian reedswamps, in *The Scientific Management of Animal and Plant Communities for Conservation*, (eds E. Duffey and A. S. Watt), Blackwell Scientific Publications, Oxford, pp. 241–69.

Fireman, M. and Hayward, H. E. (1952) Indicator significance of some shrubs in the Escalante Desert, Utah. *The Botanical Gazette*, **114**, 143–55.

Fisher, R. F. (1979) Possible allelopathic effects of reindeer-moss (*Cladonia*) on jack pine and white spruce. *Forest Science*, **25**, 256–60.

Fisher, T. R. (1979) Plankton and primary production in aquatic systems of the Central Amazon Basin. *Comparative Biochemistry and Physiology*, **62A**, 31–8.

Fittkau, E. J. and Klinge, H. (1973) On biomass and trophic structure of the central Amazonian rain forest ecosystem. *Biotropica*, **5**, 2–14.

Fitzgerald, B. M. (1981) Predatory birds and mammals, in *Tundra Ecosystems: A Comparative Analysis*, (eds L. C. Bliss, O. W. Heal and J. J. Moore), Cambridge University Press, Cambridge pp. 485–508.

Fitzgerald, R. D. and Bailey, A. W. (1984) Control of aspen regrowth by grazing with cattle. *Journal of Range Management*, **37**, 156–8.

Fitzpatrick, E. A. (1979) Radiation, in *Arid-land Ecosystems: Structure, Functioning and Management Vol. 1*, (eds D. W. Goodall, R. A. Perry and K. M. W. Howes), Cambridge University Press, Cambridge, pp. 347–71.

Fitzpatrick, E. A. and Nix, H. A. (1970) The climatic factor in Australian grassland ecology, in *Australian Grasslands*, (ed. R. M. Moore), Australian National University Press, Canberra, pp. 3–26.

Flanagan, P. W. and Scarborough, A. M. (1974) Physiological groups of decomposer fungi on tundra plant remains, in *Soil Organisms and Decomposition in Tundra*, (eds A. J. Holding, O. W. Heal, S. F. MacLean and P. W. Flanagan), Tundra Biome Steering Committee, Stockholm, pp. 159–81.

Flanagan, P. W. and Veum, A. K. (1974) Relationships between respiration, weight loss, temperature and moisture in organic residues on tundra, in *Soil Organisms and Decomposition in Tundra*, (eds A. J. Holding, O. W. Heal, S. F. MacLean and P. W. Flanagan), Tundra Biome Steering Committee, Stockholm, pp. 249–77.

Fleming, T. H. and Heithaus, E. R. (1981) Frugivorous bats, seed shadows, and the structure of tropical forests. *Biotropica*, **13** (Supplement) 45–53.

Flenley, J. (1979) *The Equatorial Rain Forest: A Geological History*, Butterworths, London.

Fleming, C. A. (1962) New Zealand biogeography: a paleontologist's approach. *Tuatara*, **10**, 53–108.

Fletcher, J. E., Sorensen, D. L. and Porcella, D. B. (1978) Erosional transfer of nitrogen in desert ecosystems, in *Nitrogen in Desert Ecosystems Vol. 9*, (eds N. E. West and J. Skujiņš), Dowden, Hutchinson and Ross, Stroudsburg, pp. 171–81.

Fletcher, J. M. (1974) Annual rings in modern and medieval times, in *The British Oak*, (eds M. G. Morris and F. H. Perring), The Botanical Society of the British Isles, Faringdon, pp. 80–97.

Flint, R. F. (1971) *Glacial and Quaternary Geology*, Wiley, New York.

Flowers, T. J. (1985) Physiology of halophytes. *Plant and Soil*, **89**, 41–56.

Fodor, S. S. (1965) Phytogeographical zonation of the alpine vegetation of Transcarpathia, in *Studies on the Flora and Vegetation of High-mountain Areas*, (ed. V. N. Sukachev), Israel Program for Scientific Translations, Jerusalem, pp. 83–97.

Fogden, M. P. L. (1972) The seasonality and population dynamics of equatorial forest birds in Sarawak. *The Ibis*, **114**, 307–41.

Fogg, G. E. (1975) Biochemical pathways in unicellular plants, in *Photosynthesis and Productivity in Different Environments*, (ed. J. P. Cooper), Cambridge University Press, Cambridge, pp. 437–57.

Fogg, G. E. (1980) Phytoplanktonic primary production, in *Fundamentals of Aquatic Ecosystems*, (eds R. K. Barnes and K. H. Mann), Blackwell Scientific Publications, Oxford, pp. 24–45.

Fogg, G. E. (1982) Nitrogen cycling in sea waters. *Philosophical Transactions of the Royal Society of London*, Series B, **296**, 511–20.

Foldats, E. and Rutkis, E. (1975) Ecological studies of chapparo (*Curatella americana* L.) and manteco (*Byrsonima crassifolia* H. B. K.) in Venezuela. *Journal of Biogeography*, **2**, 159–78.

Fonteyn, P. J. and Mahall, B. E. (1978) Competition among desert perennials. *Nature*, **275**, 544–5.

Forestry Canada (1992) *Selected Forestry Statistics Canada 1991, Information Report E-X-46*. Policy and Economics Directorate, Ottawa.

Forman, R. T. T. (1975) Canopy lichens with blue-green algae: a nitrogen source in a Colombian rain forest. *Ecology*, **56**, 1176–84.

Forrest, G. I. and Smith, R. A. H. (1975) The productivity of a range of blanket bog vegetation types in the northern Pennines. *Journal of Ecology*, **63**, 173–202.

Forseth, I. N., Ehleringer, J. R., Werk, K. S. and Cook, C. S. (1984) Field water relations of Sonoran Desert annuals. *Ecology*, **65**, 1436–44.

Foster, D. R. and Glaser, P. H. (1986) The raised bogs of south-eastern Labrador, Canada: classification, distribution, vegetation and recent dynamics. *Journal of Ecology*, **74**, 47–71.

Foster, N. W. (1974) Annual macroelement transfer from *Pinus banksiana* Lamb. forest to soil. *Canadian Journal of Forest Research*, **4**, 470–76.

Foster, R. B. (1977) *Tachigalia versicolor* is a suicidal neotropical tree. *Nature*, **268**, 624–6.

Foster, R. B. (1982) The seasonal rhythm of fruitfall on Barro Colorado Island, in *The Ecology of a Tropical Forest*, (eds E. G. Leigh, A. S. Rand and D. M. Windsor), Smithsonian Institution Press, Washington, pp. 151–72.

Foth, H. D. and Schafer, J. W. (1980) *Soil Geography and Land Use*, Wiley, New York.

Fowells, H. A. (1965) *Sylvics of Forest Trees of the United States*, Agriculture Handbook No. 271. United States Department of Agriculture, Forest Service, Washington.

Fowler, S. V. and Lawton, J. H. (1985) Rapidly induced defenses and talking trees: the devil's advocate position. *The American Naturalist*, **126**, 181–95.

Fox, J. F. (1983) Germinable seed banks of interior Alaskan tundra. *Arctic and Alpine Research*, **15**, 405–11.

Frank, R. M. and Safford, L. O. (1970) Lack of viable seeds in the forest floor after clearcutting. *Journal of Forestry*, **68**, 776–8.

Frankie, G. W. and Baker, H. G. (1974) The importance of pollinator behaviour in the reproductive biology of tropical trees. *Annales del Instituto de Biologia, Serie Botanica*, **45**, 1–10.

Frankie, G. W., Baker, H. G. and Opler, P. A. (1974) Comparative phenological studies of trees in tropical wet and dry forests in the lowlands of Costa Rica. *Journal of Ecology*, **62**, 881–913.

Frankie, G. W., Opler, P. A. and Bawa, K. S. (1976) Foraging behaviour of solitary bees: implications for outcrossing of a neotropical forest tree species. *Journal of Ecology*, **64**, 1049–57.

Frankie, G. W., Haber, W. A., Opler, P. A. and Bawa, K. S. (1983) Characteristics and organization of the large bee pollination system in the Costa Rican dry forest, in *Handbook of Experimental Pollination Biology*, (eds C. E. Jones and R. J. Little), Scientific and Academic Editions, New York, pp. 411–47.

Franklin, J. F. (1988) Pacific northwest forests, in *North American Terrestrial Vegetation*, (eds M. G. Barbour and W. D. Billings), Cambridge University Press, Cambridge, pp. 103–30.

Franklin, J. F. and Dyrness, C. T. (1969) *Vegetation of Oregon and Washington*. USDA Forest Service Research Paper PNW-80. Portland, Pacific Northwest Forest and Range Experiment Station.

Franklin, J. F. and Hemstrom, M. A. (1981) Aspects of succession in the coniferous forests of the Pacific Northwest, in *Forest Succession: Concepts and Application*, (eds D. C. West, H. H. Shugart and D. B. Botkin), Springer, New York, pp. 213–29.

Franklin, J. F., Moir, W. H., Douglas, G. W. and Wiberg, C. (1971) Invasion of subalpine meadows by trees in the Cascade Range, Washington and Oregon. *Arctic and Alpine Research*, **3**, 215–24.

Freckman, D. W. and Mankau, R. (1977) Distribution and trophic structure of nematodes in desert soils. *Ecological Bulletins (Stockholm)*, **25**, 511–14.

Freeland, R. O. (1952) Effect of age of leaves upon the rate of photosynthesis in some conifers. *Plant Physiology*, **27**, 685–90.

Frei, J. K. and Dodson, C. H. (1972) The chemical effect of certain bark substrates on the germination and early growth of epiphytic orchids. *Bulletin of the Torrey Botanical Club*, **99**, 301–7.

French, N. R. (1979) Principal subsystem interactions in grasslands, in *Perspectives in Grassland Ecology*, (ed. N. R. French), Springer, New York, pp. 173–90.

French, N. R., Grant, W. E., Grodzinski, W. and Swift, D. M. (1976) Small mammal energetics in grassland ecosystems. *Ecological Monographs*, **46**, 201–20.

Frenzel, B. (1968) The Pleistocene vegetation of northern Eurasia. *Science*, **161**, 637–49.

Friedman, J., Orshan, G. and Ziger-Cfir, Y. (1977) Suppression of annuals by *Artemisia herba-alba* in the Negev Desert of Israel. *Journal of Ecology*, **65**, 413–26.

Fritz, E. (1931) The role of fire in the redwood region. *Journal of Forestry*, **29**, 939–50.

Fritzell, E. K. (1989) Mammals in prairie wetlands, in *Northern Prairie Wetlands*, (ed. A. van der Valk), Iowa State University Press, Ames, pp. 268–301.

Fry, C. H. (1983) Birds in savanna ecosystems, in *Ecosystems of the World Vol. 13. Tropical Savannas*, (ed. F. Bourlière), Elsevier, Amsterdam, pp. 337–57.

Fuentes, E. R. and Etchegaray, J. (1983) Defoliation patterns in matorral ecosystems, in *Mediterranean-type Ecosystems: The Role of Nutrients*, (eds F. J. Kruger, D. T. Mitchell and J. U. M. Jarvis), Springer, Berlin, pp. 525–42.

Furley, P. A. and Ratter, J. A. (1988) Soil resources and the plant communities of the central Brazilian cerrado and their development. *Journal of Biogeography*, **15**, 97–108.

Galinato, M. I. and van der Valk, A. G. (1986) Seed germination traits of annuals and emergents recruited during drawdowns in the Delta Marsh, Manitoba, Canada. *Aquatic Botany*, **26**, 89–102.

Galoux, A., Benecke, P., Gietl, G. *et al.* (1981) Radiation, heat, water and carbon dioxide balances, in *Dynamic Properties of Forest Ecosystems*, (ed. D. E. Reichle), Cambridge University Press, Cambridge, pp. 87–204.

Gameson, A. L. H. and Wheeler, A. (1977) Restoration and recovery of the Thames estuary, in *Recovery and Restoration of Damaged Ecosystems*, (eds J. Cairns, K. L. Dickson and E. E. Herricks), University Press of Virginia, Charlottesville, pp. 72–101.

Ganf, G. G. (1980) Phytoplankton, in *The Functioning of Freshwater Ecosystems*, (eds E. D. Le Cren and R. H. Lowe-McConnell), Cambridge University Press, Cambridge, pp. 188–94.

Ganf, G. G. and Horne, A. J. (1975) Diurnal stratification, photosynthesis and nitrogen-fixation in a shallow, equatorial Lake (Lake George, Uganda). *Freshwater Biology*, **5**, 13–39.

Garcia-Moya, E. and McKell, C. M. (1970) Contribution of shrubs to the nitrogen economy of a desert-wash plant community. *Ecology*, **51**, 81–8.

Gardner, C. A. (1957) The fire factor in relation to the vegetation of Western Australia. *The Western Australian Naturalist*, **5**, 166–73.

Gartland, J. S., McKey, D. B., Waterman, P. G., Mbi, C. N. and Struhsaker, T. T. (1980) A comparative study of the phytochemistry of two African rain forests. *Biochemical Systematics and Ecology*, **8**, 401–22.

Garwood, N. C. (1989) Tropical soil seed banks: a review, in *Ecology of Soil Seed Banks*, (eds M. A. Leck, V. T. Parker and R. L. Simpson), Academic Press, San Diego, pp. 149–209.

Gates, D. M. (1980) *Biophysical Ecology*, Springer, New York.

Gates, D. M., Alderfer, R. and Taylor, E. (1968) Leaf temperatures of desert plants. *Science*, **159**, 954–5.

Gaudet, J. J. (1976) Nutrient relationships in the detritus of a tropical swamp. *Archiv fur Hydrobiologia*, **78**, 213–39.

Gaudet, J. J. (1977) Uptake, accumulation, and loss of nutrients by papyrus in tropical swamps. *Ecology*, **58**, 415–22.

Gautier-Hion, J.-M. Duplantier, R., Quris, R. *et al.* (1985) Fruit characters as a basis of fruit choice and seed dispersal in a tropical forest vertebrate community. *Oecologia*, **65**, 324–37.

Geis, J. W., Tortorelli, R. L. and Boggess, W. R. (1971) Carbon dioxide assimilation of hardwood seedlings in relation to community dynamics in central Illinois. *Oecologia*, **7**, 276–89.

Gentilli, J. (ed.) (1971) *World Survey of Climatology Vol. 13. Climates of Australia and New Zealand*, Elsevier, Amsterdam.

Gentry, A. H. (1974) Flowering phenology and diversity in tropical Bignoniaceae. *Biotropica*, **6**, 64–8.

George, D. G. and Harris, G. P. (1985) The effect of climate on long-term change in the crustacean zooplankton biomass of Lake Windermere, U. K. *Nature*, **316**, 536–9.

George, M. F., Becwar, M. R. and Burke, M. J. (1982) Freezing avoidance by deep undercooling of tissue water in winter-hardy plants. *Cryobiology*, **19**, 628–39.

George, M. F., Burke, M. J., Pellett, H. M. and Johnson, A. G. (1974) Low temperature exotherms and woody plant distribution. *HortScience*, **9**, 519–22.

Gibbs, R. J. (1970) Circulation in the Amazon River estuary and adjacent Atlantic Ocean. *Journal of Marine Research*, **28**, 113–23.

Gill, A. M. (1981) Adaptive responses of Australian vascular plant species to fires, in *Fire and the Australian Biota*, (eds A. M. Gill, R. H. Groves and I. R. Noble), Australian Academy of Science, Canberra, pp. 243–72.

Gill, A. M. and Tomlinson, P. B. (1971) Studies on the growth of red mangrove (*Rhizophora mangle* L.). 2. Growth and differentiation of aerial roots. *Biotropica*, **3**, 63–77.

Gillon, Y. (1983) The invertebrates of the grass layer, in *Ecosystems of the World Vol. 13. Tropical Savannas*, (ed. F. Bourlière), Elsevier, Amsterdam, pp. 289–311.

Gillon, Y. and Gillon, D. (1967) Recherches ecologiques dans la Savane de Lamto (Cote D'Ivoire): cycle annuel des effectifs et des biomasses d'arthropodes de la strate herbacee. *La Terre et la Vie*, **21**, 262–77.

Givnish, T. J. (1978) On the adaptive significance of compound leaves, with particular reference to tropical trees, in *Tropical Trees as Living Systems*, (eds P. B. Tomlinson and M. H. Zimmermann), Cambridge University Press, Cambridge, pp. 351–80.

Gjærevoll, O. and Bringer, K.-G. (1965) Plant cover of the alpine regions. *Acta Phytogeographica Suecica*, **50**, 257–68.

Glantz, W. E., Thorington, R. W., Giacalone-Madden, J. and Heaney, L. R. (1982) Seasonal food use and demographic trends in *Sciurus granatensis*, in *The Ecology of a Tropical Forest*, (eds E. G. Leigh, A. S. Rand and D. M. Windsor), Smithsonian Institution Press, Washington, pp. 239–52.

Glazovskaya, M. A. (1984) *Soils of the World Vol. II. Soil Geography*, A. A. Balkema, Rotterdam,

Gleason, H. A. (1926) The individualistic concept of the plant association. *Bulletin of the Torrey Botanical Club*, **53**, 7–26.

Glennie, K. W. (1987) Desert sedimentary environments, present and past – a summary. *Sedimentary Geology*, **50**, 135–65.

Gliwicz, Z. M. and Pijanowska, J. (1989) The role of predation in zooplankton succession, in *Plankton Ecology*, (ed. U. Sommer), Springer, Berlin, pp. 253–96.

Gloser, J. (1977) Characteristics of CO_2 exchange in *Phragmites communis* Trin. derived from measurements *in situ*. *Photosynthetica*, **11**, 139–47.

Goering, J. J. and Parker, P. L. (1972) Nitrogen fixation by epiphytes on sea grasses. *Limnology and Oceanography*, **17**, 320–23.

Goering, J. J., Dugdale, R. C. and Menzel, D. W. (1966) Estimates of *in situ* rates of nitrogen uptake by *Trichodesmium* sp. in the tropical Atlantic Ocean. *Limnology and Oceanography*, **11**, 614–20.

Goldman, B. and Talbot, F. H. (1976) Aspects of the ecology of coral reef fishes, in *Biology and Geology of Coral Reefs*, Vol. 3 (eds O. A. Jones and R. Endean), Academic Press, New York, pp. 125–54.

Goldman, C. R. and Horne, A. J. (1983) *Limnology*, McGraw-Hill, New York.

Goldman, C. R., Morgan, M. D., Threlkeld, S. T. and Angeli, N. (1979) A population dynamics analysis of the cladoceran disappearance from Lake Tahoe, California–Nevada. *Limnology and Oceanography*, **24**, 289–97.

Golley, F., Odum, H. T. and Wilson, R. F. (1962) The structure and metabolism of a Puerto Rican red mangrove forest in May. *Ecology*, **43**, 9–19.

Golley, F. B. (1983) Nutrient cycling and nutrient conservation, in *Ecosystems of the World Vol. 14A. Tropical Rain Forest Ecosystems, Structure and Function*, (ed. F. B. Golley), Elsevier, Amsterdam, pp. 137–56.

Golley, F. B., McGinnis, J. T., Clements, R. G., Child, G. I. and Duever, M. J. (1975) *Mineral Cycling in a Tropical Moist Forest Ecosystem*, University of Georgia Press, Athens.

Good, R. (1964) *The Geography of Flowering Plants*, 3rd edn, William Clowes, London.

Goodfellow, S. and Barkham, J. P. (1974) Spectral transmission curves for a beech (*Fagus sylvatica* L.) canopy. *Acta Botanica Neerlandica*, **23**, 225–30.

Goodwin, T. W. and Mercer, E. I. (1983) *Introduction to Plant Biochemistry*, 2nd edn, Pergamon Press, Oxford.

Gopal, B. (1987) *Water Hyacinth*, Elsevier, Amsterdam.

Gordon, A. G. (1983) Nutrient cycling dynamics in differing spruce and mixedwood ecosystems in Ontario and the effects of nutrient removals through harvesting, in *Resources and Dynamics of the Boreal Zone*, (eds R. W. Wein, R. R. Riewe and I. R. Methven), Association of Canadian Universities for Northern Studies, Ottawa, pp. 97–118.

Gore, A. J. P. and Allen, S. E. (1956) Measurement of exchangeable and total cation content for H^+, Na^+, K^+, Mg^{++}, Ca^{++} and iron, in high level blanket peat. *Oikos*, **7**, 48–55.

Goreau, T. F. (1961) On the relation of calcification to primary productivity in reef building organisms, in *The Biology of Hydra and of Some Other Coelenterates*, (eds H. M. Lenhoof and W. F. Loomis), University of Miami Press, Coral Gables, pp. 269–85.

Gorham, E. (1956) The ionic composition of some bog and fen waters in the English Lake District. *Journal of Ecology*, **44**, 142–52.

Gorham, J. and Wyn Jones, R. G. (1983) Solute distribution in *Suaeda maritima*. *Planta*, **157**, 344–9.

Gosz, J. R., Likens, G. E. and Bormann, F. H. (1972) Nutrient content of litter fall on the Hubbard Brook experimental watershed, New Hampshire. *Ecology*, **53**, 769–84.

Gould, E. (1978) Foraging behaviour of Malaysian nectar-feeding bats. *Biotropica*, **10**, 184–93.

Gower, S. T. (1987) Relations between mineral nutrient availability and fine root biomass in two Costa Rican tropical wet forests: a hypothesis. *Biotropica*, **19**, 171–5.

Grace, J. and Marks, T. C. (1978) Physiological aspects of bog production at Moor House, in *Production Ecology of British Moors and Montane Grasslands*, (eds O. W. Heal and D. F. Perkins), Springer, Berlin, pp. 38–51.

Graetz. R. D. (1981) Plant–animal interactions, in *Arid-land Ecosystems: Structure, Functioning and Management*, Vol. 2 (eds D. W. Goodall, R. A. Perry and K. M. W. Howes), Cambridge University Press, Cambridge, pp. 85–103.

Grainger, E. H. (1979) Primary production in Frobisher Bay, Arctic Canada, in *Marine Production Mechanisms*, (ed. M. J. Dunbar), Cambridge University Press, Cambridge, pp. 9–30.

Granhall, U. and Lid-Torsvik, V. (1975) Nitrogen fixation by bacteria and free-living blue-green algae in tundra areas, in *Fennoscandian Tundra Ecosystems: Part 1. Plants and Microorganisms*, (ed. F. E. Wielgolaski), Springer, New York, pp. 305–15.

Grant, W. E. and Birney, E. C. (1979) Small mammal community structure in North American grasslands. *Journal of Mammalogy*, **60**, 23–36.

Grassle, J. F. (1986) The ecology of deep-sea hydrothermal vent communities. *Advances in Marine Biology*, **23**, 301–62.

Gray, J. T. (1982) Community structure and productivity in *Ceanothus* chaparral and coastal sage scrub of southern California. *Ecological Monographs*, **52**, 415–35.

Gray, R. and Bonner, J. (1948) An inhibitor of plant growth from the leaves of *Encelia farinosa*. *American Journal of Botany*, **35**, 52–7.

Gray, J. T. and Schlesinger, W. H. (1981) Nutrient cycling in mediterranean type ecosystems, in *Resource Use by Chaparral and Matorral*, (ed. P. C. Miller), Springer, New York, pp. 259–85.

Green, B. H. (1968) Factors influencing the spatial and temporal distribution of *Sphagnum immbricatum* Hornsch. ex Russ. in the British Isles. *Journal of Ecology*, **56**, 47–58.

Green, M. S. and Etherington, J. R. (1977) Oxidation of ferrous iron by rice (*Oryza sativa* L.) roots: a mechanism for waterlogging tolerance. *Journal of Experimental Botany*, **28**, 678–90.

Greenberg, A. E. (1964) Plankton of the Sacramento River. *Ecology*, **45**, 40–49.

Greenberg, R. and Gradwohl, J. (1980) Leaf surface specializations of birds and *Arthropods* in a Panamanian forest. *Oecologia*, **46**, 115–24.

Greene, C. H. (1985) Planktivore functional groups and patterns of prey selection in pelagic communities. *Journal of Plankton Research*, **7**, 35–40.

Greenland, D. J. and Kowal, J. M. L. (1960) Nutrient content of the moist tropical forest of Ghana. *Plant and Soil*, **12**, 154–74.

Gregory, S. V. (1983) Plant–herbivore interactions in stream systems, in *Stream Ecology*, (eds J. R. Barnes and G. W. Minshall), Plenum Press, New York, pp. 157–89.

Grier, C. C. (1978) A *Tsuga heterophylla–Picea sitchensis* ecosystem of coastal Oregon: decomposition and nutrient balances of fallen logs. *Canadian Journal of Forest Research*, **8**, 198–206.

Grier, C. C. and Logan, R. S. (1977) Old-growth *Pseudotsuga menziesii* communities of a western Oregon watershed: biomass distribution and production budgets. *Ecological Monographs*, **47**, 373–400.

Grier, C. C., Vogt, K. A., Keyes, M. R. and Edmonds, R. L. (1981) Biomass distribution and above- and below-ground production in young and mature *Abies amabalis* zone ecosystems of the Washington Cascades. *Canadian Journal of Forest Research*, **11**, 155–67.

Grieve, B. J. and Hellmuth, E. O. (1970) Eco-physiology of Western Australian plants. *Oecologia Plantarum*, **5**, 33–67.

Griffiths, J. F. (1972) *World Survey of Climatology Vol. 10. Climates of Africa*, (ed. H. E. Landsberg), Elsevier, Amsterdam.

Grime, J. P. (1979) *Plant Strategies and Vegetation Processes*, Wiley, Chichester.

Groombridge, B. (ed.) (1992) *Global Biodiversity: Status of the Earth's Living Resources*, Chapman & Hall, London.

Grubb, P. J. (1977) Control of forest growth and distribution on wet tropical mountains: with special reference to mineral nutrition. *Annual Review of Ecology and Systematics*, **8**, 83–107.

Grubb, P. J. and Whitmore, T. C. (1966) A comparison of montane and lowland rain forest in Ecuador II. The climate and its effects on the distribution and physiognomy of the forests. *Journal of Ecology*, **54**, 303–33.

Grubb, P. J., Lloyd, J. R., Pennington, T. D. and Whitmore, T. C. (1963) A comparison of montane and lowland rain forest in Ecuador. *Journal of Ecology*, **51**, 567–601.

Guerrant, E. O. and Fiedler, P. L. (1981) Flower defenses against nectar-pilferage by ants. *Biotropica*, **13** (suppl), 25–33.

Guilcher, A. (1963) Estuaries, deltas, shelf, slope, in *The Sea: Ideas and Observations on Progress in the Study of the Seas*. Vol. 3 (ed. M. N. Hill), Wiley Interscience, New York, pp. 620–54.

Gusta, L. V., Tyler, N. J. and Hwei-Hwang Chen, T. (1983) Deep undercooling in woody taxa growing north of the −40 °C isotherm. *Plant Physiology*, **72**, 122–8.

Guthrie, R. D. (1982) Mammals of the mammoth steppe as paleoenvironmental indicators, in *Paleoecology of Beringia*, (eds D. M. Hopkins, J. V. Matthews, C. E. Schweger and S. B. Young), Academic Press, New York, pp. 307–26.

Gutterman, Y. (1972) Delayed seed dispersal and rapid germination as survival mechanisms of the desert plant *Blepharis persica* (Burm.) Kuntze. *Oecologia*, **10**, 145–9.

Hadley, E. B. and Kieckhefer, B. J. (1963) Productivity of two prairie grasses in relation to fire frequency. *Ecology*, **44**, 389–95.

Hadley, N. F. (1972) Desert species and adaptation. *American Scientist*, **60**, 338–47.

Haffer, J. (1969) Speciation in Amazonian forest birds. *Science*, **165**, 131–37.

Hall, D. J., Threlkeld, S. T., Burns, C. W. and Crowley, P. H. (1976) The size–efficiency hypothesis and the size structure of zooplankton communities. *Annual Review of Ecology and Systematics*, **7**, 177–208.

Hall, E. A. A., Specht, R. L. and Eardley, C. M. (1964) Regeneration of the vegetation on Koonamore Vegetation Reserve, 1926–1962. *Australian Journal of Botany*, **12**, 205–64.

Halldal, P. (1974) Light and photosynthesis of different marine algal groups, in *Optical Aspects of Oceanography*, (eds N. G. Jerlov and E. Steemann Nielsen), Academic Press, London, pp. 345–60.

Hallé, F., Oldeman, R. A. A., and Tomlinson, P. B. (1978) *Tropical Trees and Forests*, Springer, Berlin.

Hämet-Ahti, L. (1983) Human impact on closed boreal forest (taiga), in *Man's Impact on Vegetation*, (eds W. Holzner, M. J. A. Werger and I. Ikusima), Junk, The Hague, pp. 201–11.

Hamilton, A. (1989) African forests, in *Ecosystems of the World Vol. 14B. Tropical Rain Forest Ecosystems*, (eds H. Lieth and M. J. A. Werger), Elsevier, Amsterdam, pp. 155–182.

Hamilton, W. J. and Seely, M. K. (1976) Fog basking by the Namib Desert beetle, *Onymacris unguicularis*. *Nature*, **262**, 284–5.

Hamilton, W. J., Buskirk, R. and Buskirk, W. H. (1977) Intersexual dominance and differential mortality of Gemsbok *Oryx gazella* at Namib Desert waterholes. *Madoqua*, **10**, 5–19.

Hammer, U. T. (1980) Geographical variations, in *The Functioning of Freshwater Ecosystems* (eds E. D. Le Cren and R. H. Lowe-McConnell), Cambridge University Press, Cambridge, pp. 235–46.

Hanes, T. L. (1965) Ecological studies on two closely related chaparral shrubs in southern California. *Ecological Monographs*, **35**, 213–35.

Hanes, T. L. (1971) Succession after fire in the chaparral of southern California. *Ecological Monographs*, **41**, 27–52.

Hanes, T. L. (1977) California chaparral, in *Terrestrial Vegetation of California*, (eds M. G. Barbour and J. Major), Wiley, New York, pp. 417–69.

Hanes, T. L. (1981) California chaparral, in *Ecosystems of the World Vol. 11. Mediterranean-type Shrublands*, (eds F. di Castri, D. W. Goodall and R. L. Specht), Elsevier, Amsterdam, pp. 139–74.

Hardy, F. (1945) The soils of South America. *Plants and Plant Science in Latin America*, **2**, 322–6.

Hardy, Y., Mainville, M. and Schmitt, D. M. (1986) *An Atlas of Spruce Budworm Defoliation in Eastern North America, 1938–80*. USDA Forest Service, Cooperative State Research Service, Miscellaneous Publication No. 1449.

Hare, F. K. (1950) Climate and zonal divisions of the boreal forest formation in eastern Canada. *The Geographical Review*, **40**, 615–35.

Hare, F. K. and Hay, J. E. (1974) The climate of Canada and Alaska, in *World Survey of Climatology Vol. 11. Climates of North America*, (eds R. A. Bryson and F. K. Hare), Elsevier, Amsterdam, pp. 49–192.

Hare, F. K. and Ritchie, J. C. (1972) The boreal bioclimates. *The Geographical Review*, **62**, 333–65.

Harling, G. (1979) The vegetation types of Ecuador – a brief survey, in *Tropical Botany*, (eds K. Larsen and L. B. Holm-Nielsen), Academic Press, London, pp. 165–74.

Harmon, M. E., Franklin, J. F., Swanson, F. J. *et al.* (1986) Ecology of coarse woody debris in temperate ecosystems. *Advances in Ecological Research*, **15**, 133–302.

Harris, G. P. (1986) *Phytoplankton Ecology*, Chapman & Hall, London.

Harris, S. A. (1986) *The Permafrost Environment*, Croom Helm, London.

Harrison, A. T., Small, E. and Mooney, H. A. (1971) Drought relationships and distribution of two mediterranean climate California plant communities. *Ecology*, **52**, 869–75.

Harrison, J. L. (1962) The distribution of feeding habits among animals in a tropical rain forest. *Journal of Animal Ecology*, **31**, 53–63.

Harrison, W. G. (1978) Experimental measurements of nitrogen remineralization in coastal waters. *Limnology and Oceanography*, **23**, 684–94.

Hartman, W. L. (1988) Historical changes in the major fish resources of the Great Lakes, in *Toxic Contaminants and Ecosystem Health: A Great Lakes Focus* (ed. M. S. Evans), Wiley, New York, pp. 103–31.

Hartshorn, G. S. (1978) Tree falls and tropical forest dynamics, in *Tropical Trees as Living Systems*, (eds P. B. Tomlinson and M. H. Zimmermann), Cambridge University Press, Cambridge, pp. 617–38.

Harvey, R. A. and Mooney, H. A. (1964) Extended dormancy of chaparral shrubs during severe drought. *Madroño*, **17**, 161–3.

Haslam, S. M. (1978) *River Plants*, Cambridge University Press, Cambridge.

Hasle, G. R. (1956) Phytoplankton and hydrography of the Pacific part of the Antarctic Ocean. *Nature*, **177**, 616–17.

Haukioja, E. (1981) Invertebrate herbivory at tundra sites, in *Tundra Ecosystems: A Comparative Analysis*, (eds L. C. Bliss, O. W. Heal and J. J. Moore), Cambridge University Press, Cambridge, pp. 547–5.

Haukioja, E., Suomela, J. and Neuvonen, S. (1985) Long-term inducible resistance in birch foliage: triggering cues and efficacy on a defoliator. *Oecologia*, **65**, 363–9.

Haverty, M. I. and Nutting, W. L. (1975) Natural wood preferences of desert termites. *Annals of the Entomological Society of America*, **68**, 533–6.

Havill, D. C., Ingold, A. and Pearson, J. (1985) Sulphide tolerance in coastal halophytes. *Vegetatio*, **62**, 279–85.

Hayes, D. C. and Seastedt, T. R. (1987) Root dynamics of tallgrass prairie in wet and dry years. *Canadian Journal of Botany*, **65**, 787–91.

Heal, O. W. and French, D. D. (1974) Decomposition of organic matter in tundra, in *Soil Organisms and Decomposition in Tundra*, (eds A. J. Holding, O. W. Heal, S. F. MacLean and P. W. Flanagan), Tundra Biome Steering Committee, Stockholm, pp. 279–309.

Heal, O. W., Latter, P. M. and Howson, G. (1978) A study of the rates of decomposition of organic matter, in *Production Ecology of British Moors and Montane Grasslands*, (eds O. W. Heal and D. F. Perkins), Springer, Berlin, pp. 137–59.

Heal, O. W., Flanagan P. W., French, D. D. and MacLean, S. F. (1981) Decomposition and accumulation of organic matter, in *Tundra Ecosystems: A Comparative Analysis*, (eds L. C. Bliss, O. W. Heal and J. J. Moore), Cambridge University Press, Cambridge, pp. 587–633.

Heald, E. J. (1971) The production of organic detritus in a south Florida estuary. *Sea Grant Technical Bulletin No. 6*, University of Miami Sea Grant Program, Miami.

Hecky, R. E. and Kilham, P. (1988) Nutrient limitation of phytoplankton in freshwater and marine environments: a review of recent evidence on the effects of enrichment. *Limnology and Oceanography*, **33**, 796–822.

Hedberg, O. (1964) Features of Afro-alpine plant ecology. *Acta Phytogeographica Suecica*, **49**, 1–144.

Hedberg, O. (1969) Evolution and speciation in a tropical high mountain flora. *Biological Journal of the Linnean Society*, **1**, 135–48.

Hedegart, T. (1976) Breeding systems, variation and genetic improvement of teak (*Tectona grandis* L. f.), in *Tropical Trees. Variation, Breeding and Conservation*, (eds J. Burley and B. T. Styles), Academic Press, London, pp. 109–21.

Heinselman, M. L. (1975) Boreal peatlands in relation to environment, in *Coupling of Land and Water Systems*, (ed. A. D. Hasler), Springer, New York, pp. 93–103.

Heinrich, A. K. (1962) The life histories of plankton animals and seasonal cycles of plankton communities in the oceans. *Journal du Conseil. Conseil International pour l'Exploration de la Mer*, **27**, 15–24.

Heithaus, E. R., Fleming, T. H. and Opler, P. A. (1975) Foraging patterns and resource utilization in seven species of bats in a seasonal tropical forest. *Ecology*, **56**, 841–54.

Hejný, S., Kvet, J. and Dykyjová, D. (1981) Survey of biomass and net production of higher plant communities in fishponds. *Folia Geobotanica et Phytotaxonomica*, **16**, 73–84.

Hellawell, J. M. (1986) *Biological Indicators of Freshwater Pollution and Environmental Management*, Elsevier, London.

Hellmers, H. and Ashby, W. C. (1958) Growth of native and exotic plants under controlled temperatures and in the San Gabriel Mountains California. *Ecology*, **39**, 416–28.

Hellmers, H., Horton, J. S., Juhren, G. and O'Keefe, J. (1955) Root systems of some chaparral plants in southern California. *Ecology*, **36**, 667–78.

Hemond, H. F. (1983) The nitrogen budget of Thoreau's Bog. *Ecology*, **64**, 99–109.

Hendrickson, O. Q., Chatarpaul, L. and Burgess, D. (1989) Nutrient cycling following whole-tree and conventional harvest in northern mixed forest. *Canadian Journal of Forest Research*, **19**, 725–35.

Henwood, K. (1973) A structural model of forces in buttressed tropical rain forest trees. *Biotropica*, **5**, 83–93.

Herdendorf, C. E. (1990) Distribution of the world's large lakes, in *Large Lakes* (eds M. M. Tilzer and C. Serruya), Springer, Berlin, pp. 3–38.

Heslop-Harrison, Y. (1978) Carnivorous plants. *Scientific American*, **238**, 104–15.

Hesse, P. R. (1961) The decomposition of organic matter in a mangrove swamp soil. *Plant and Soil*, **14**, 249–63.

Hesthagen, T. (1986) Fish kills of Atlantic salmon (*Salmo salar*) and brown trout (*Salmo trutta*) in an acidified river of SW Norway. *Water, Air, and Soil Pollution*, **30**, 619–28.

Heusser, C. J. (1972) Palynolgy and phytogeographical significance of a late-Pleistocene refugium near Kalaloch, Washington. *Quaternary Research*, **2**, 189–201.

Heusser, C. J. (1974) Vegetation and climate of the southern Chilean Lake District during and since the last interglaciation. *Quaternary Research*, **4**, 290–315.

Higgins, D. G. and Ramsey, G. S. (1992) *Canadian Forest Fire Statistics: 1988–1990*. Forestry Canada, Petawawa National Forestry Institute Information Report PI-X-107E/F.

Highsmith, R. C., Riggs, A. C. and D'Antonio, C. M. (1980) Survival of hurricane-generated coral fragments and a disturbance model of reef calcification/growth rates. *Oecologia*, **46**, 322–9.

Hills, L. V., Klovan, J. E. and Sweet, A. R. (1974) *Juglans eocinerea* n. sp., Beaufort Formation (Tertiary), southwestern Banks Island, Arctic Canada. *Canadian Journal of Botany*, **52**, 65–90.

Hinneri, S., Sonesson, M. and Veum, A. K. (1975) Soils of Fennoscandian IBP tundra ecosystems, in *Fennoscandian Tundra Ecosystems – Part 1. Plants and Microorganisms*, (ed. F. E. Wielgolaski), Springer, New York, pp. 31–40.

Hiratsuka, Y. (1987) *Forest Tree Diseases of the Prairie Provinces*. Canadian Forestry Service Information Report NOR–X–286.

Hobbie, J. E. (1984) Polar limnology, in *Ecosystems of the World Vol. 23. Lakes and Reservoires* (ed. F. B. Taub), Elsevier, Amsterdam, pp. 63–105.

Hockey, P. A. R., Siegfried, W. R., Crowe, A. A. and Cooper, J. (1983) Ecological structure and energy requirements of the sandy beach avifauna of southern Africa, in *Sandy Beaches as Ecosystems*, (eds A. McLachlan and T. Erasmus), Junk, The Hague, pp. 507–21.

Hockings, B. (1970) Insect associations with the swollen thorn acacias. *Transactions of the Royal Entomological Society of London*, **122**, 211–55.

Hoffmann, A. (1972) Morphology and histology of *Trevoa trinervis* (Rhamnaceae), a drought deciduous shrub from the Chilean matorral. *Flora*, **161**, 527–38.

Hoffmann, A. and Kummerow, J. (1978) Root studies in the Chilean matorral. *Oecologia*, **32**, 57–69.

Hoffmann, A. J. and Hoffmann, A. E. (1976) Growth pattern and seasonal behaviour of buds of *Colliguaya odorifera*, a shrub from the Chilean mediterranean vegetation. *Canadian Journal of Botany*, **54**, 1767–74.

Hoffmann, R. S. and Taber, R. D. (1967) Origin and history of Holarctic tundra ecosystems, with special reference to their vertebrate faunas, in *Arctic and Alpine Environments*, (eds H. E. Wright and W. H. Osburn), Indiana University Press, Bloomington, pp. 143–70.

Hofstetter, R. H. (1983) Wetlands in the United States, in *Ecosystems of the World Vol. 4B. Mires: Swamp, Bog, Fen and Moor – Regional Studies*, (ed. A. J. P. Gore), Elsevier, Amsterdam, pp. 201–44.

Hogg, B. M. and Hudson, H. J. (1966) Micro-fungi on leaves of *Fagus sylvatica* I. The micro-fungal succession. *Transactions of the British Mycological Society*, **49**, 185–92.

Holaday, A. S., Salvucci, M. E. and Bowes, G. (1983) Variable photosynthesis/photorespiration ratios in *Hydrilla* and other submersed aquatic macrophyte species. *Canadian Journal of Botany*, **61**, 229–36.

Holdgate, M. W. (1961) Vegetation and soils in the south Chilean Islands. *Journal of Ecology*, **49**, 559–80.

Holding, A. J. (1981) The microflora of tundra, in *Tundra Ecosystems: A Comparative Analysis*, (eds L. C. Bliss, O. W. Heal and J. J. Moore), Cambridge University Press, Cambridge, pp. 561–85.

Holdridge, L. R., Grenke, W. C., Hatheway, W. H., Liang, T. and Tosi, J. A. (1971) *Forest Environments in Tropical Life Zones*, Pergamon Press, Oxford.

Holm-Hansen, O., El-Sayed, S. Z., Franceschini, G. A. and Cuhel, R. L. (1977) Primary production and the factors controlling phytoplankton growth in the Southern Ocean, in *Adaptations Within Antarctic Ecosystems*, (ed. G. A. Llano), National Academy of Sciences, Washington, pp. 11–50.

Holway, J. G. and Ward, R. T. (1965) Phenology of alpine plants in northern Colorado. *Ecology*, **46**, 73–83.

Honer, T. G. and Bickerstaff, A. (1985) *Canada's Forest Area and Wood Volume Balance 1977–81: An Appraisal of Change Under Present Levels of Management*, Canadian Forestry Service, Pacific Forestry Centre BC-X-272.

Honjo, S. (1978) Sedimentation of materials in the Sargasso Sea at a 5,367 m deep station. *Journal of Marine Research*, **36**, 469–92.

Hope, G. S. (1976) The vegetational history of Mt Wilhelm, Papua New Guinea. *Journal of Ecology*, **64**, 627–63.

Hopkins, B. (1966) Vegetation of the Olokemeji Forest Reserve, Nigeria. IV. The litter and soil with special reference to their seasonal changes. *Journal of Ecology*, **54**, 687–703.

Hopkins, B. (1970a) Vegetation of the Olokemeji Forest Reserve, Nigeria. VI. The plants on the forest site with special reference to their seasonal growth. *Journal of Ecology*, **58**, 765–93.

Hopkins, B. (1970b) Vegetation of the Olokemeji Forest Reserve, Nigeria. VII. The plants on the savanna site with special reference to their seasonal growth. *Journal of Ecology*, **58**, 795–825.

Hopkins, H. C. (1984) Floral biology and pollination ecology of the Neotropical species of *Parkia*. *Journal of Ecology*, **72**, 1–23.

Hopkins, H.C and Hopkins, M. J. G. (1982) Predation by a snake of a flower-visiting bat at *Parkia nitida* (Leguminosae: Mimosoideae). *Brittonia*, **34**, 225–7.

Horn, H. S. (1971) *The Adaptive Geometry of Trees*, Princeton University Press, Princeton.

Horn, H. S. (1975) Forest succession. *Scientific American*, **232**, 90–98.

Horne, A. J. and Goldman, C. R. (1972) Nitrogen fixation in Clear Lake, California. I. Seasonal variation and the role of heterocysts. *Limnology and Oceanography*, **17**, 678–92.

Horton, J. S. and Kraebel, C. J. (1955) Development of vegetation after fire in the chamise chaparral of southern California. *Ecology*, **36**, 244–62.

Hough, R. A. (1974) Photorespiration and productivity in submersed aquatic vascular plants. *Limnology and Oceanography*, **19**, 912–27.

Howe, H. F. (1979) Fear and frugivory. *The American Naturalist*, **114**, 925–31.

Howe, H. F. (1981) Dispersal of a Neotropical nutmeg (*Virola sebifera*) by birds. *The Auk*, **98**, 88–98.

Howe, H. F. (1982) Fruit production and animal activity in two tropical trees, in *The Ecology of a Tropical Forest*, (eds E. G. Leigh, A. S. Rand and D. M. Windsor) Smithsonian Institution Press, Washington, pp. 189–99.

Howe, H. F. and Estabrook, G. F. (1977) On intraspecific competition for avian dispersers in tropical trees. *The American Naturalist*, **111**, 817–32.

Howe, H. F. and Smallwood, J. (1982) Ecology of seed dispersal. *Annual Review of Ecology and Systematics*, **13**, 201–28.

Howe, H. F., Schupp, E. W. and Westley, L. C. (1985) Early consequences of seed dispersal for a Neotropical tree (*Virola surinamensis*). *Ecology*, **66**, 781–91.

Howe, H. F. and Vande Kerckhove, G. A. (1981) Removal of wild nutmeg (*Virola surinamensis*) crops by birds. *Ecology*, **62**, 1093–106.

Howe, H. S. (1977) Bird activity and seed dispersal of a tropical wet forest tree. *Ecology*, **58**, 539–50.

Howell, D. J. (1977) Time sharing and body partitioning in bat–plant pollination systems. *Nature*, **270**, 509–10.

Howes, B. L., Howarth, R. W., Teal, J. M. and Valiela, I. (1981) Oxidation–reduction potentials in a salt marsh: spatial patterns and interactions with primary production. *Limnology and Oceanography*, **26**, 350–60.

Hubbell, S. P. (1979) Tree dispersion, abundance, and diversity in a tropical dry forest. *Science*, **203**, 1299–309.

Hubbell, S. P. (1980) Seed predation and the coexistence of tree species in tropical forests. *Oikos*, **35**, 214–29.

Hubert, B. A. (1977) Estimated productivity of muskox on Truelove Lowland, in *Truelove Lowland, Devon Island, Canada: A High Arctic Ecosystem*, (ed. L. C. Bliss), The University of Alberta Press, Edmonton, pp. 467–91.

Huiskes, A. H. L. (1979) Biological flora of the British Isles: *Ammophila arenaria* (L.) Link (*Psamma arenaria* (L.) Roem. et Schult.; *Calamagrostis arenaria* (L.) Roth). *Journal of Ecology*, **67**, 363–82.

Hulbert, L. C. (1988) Causes of fire effects in tallgrass prairie. *Ecology*, **69**, 46–58.

Hulme, M. A., Ennis, T. J. and Lavallée, A. (1983) Current status of *Bacillus thuringiensis* for spruce budworm control. *The Forestry Chronicle*, **59**, 58–61.

Huntley, B. J. and Morris, J. W. (1982) Structure of the Nylsvley savanna, in *Ecology of Tropical Savannas*, (eds B. J. Huntley and B. H. Walker), Springer, Berlin, pp. 433–55.

Huntley, B. J. and Walker, B. H. (1982) Introduction, in *Ecology of Tropical Savannas*, (eds B. J. Huntley and B. H. Walker), Springer, Berlin, pp. 1–2.

Hustich, I. (1979) Ecological concepts and biogeographical zonation in the north: the need for a generally accepted terminology. *Holarctic Ecology*, **2**, 208–17.

Hutchinson, G. E. (1967) *A Treatise on Limnology Vol. 2. Introduction to Lake Biology and the Limnoplankton*, Wiley, New York.

Hutchinson, G. E. (1973) Eutrophication. *American Scientist*, **61**, 269–79.

Hutchinson, J. N. (1980) The record of peat wastage in the East Anglian Fenlands at Holme Post, 1848–1978 A.D. *Journal of Ecology*, **68**, 229–49.

Hutchison, V. H. and Larimer, J. L. (1960) Reflectivity of the integuments of some lizards from different habitats. *Ecology*, **41**, 199–209.

Huttel, C. (1975) Root distribution and biomass in three Ivory Coast rain forest plots, in *Tropical Ecological Systems*, (eds F. B. Golley and E. Medina), Springer, New York, pp. 123–30.

Huxley, C. (1980) Symbiosis between ants and epiphytes. *Biological Reviews*, **55**, 321–40.

Hynes, H. B. N. (1970) *The Ecology of Running Waters*, University of Toronto Press, Toronto.

Ignaciuk, R. and Lee, J. A. (1980) The germination of four annual strand-line species. *New Phytologist*, **84**, 581–91.

Ijzerman, R. (1931) *Outline of the Geology and Petrology of Surinam*, Martinus Nijhoff, The Hague.

Ikonnikov, S. S. (1965) Recent data on the flora of Pamir, in *Studies on the Flora and Vegetation of High Mountain Areas*, (ed. V. N. Sukachev), Israel Program for Scientific Translations, Jerusalem, pp. 250–56.

Ingram, H. A. P. (1978) Soil layers in mires: function and terminology. *Journal of Soil Science*, **29**, 224–7.

Ingram, H. A. P. (1983) Hydrology, in *Ecosystems of the World Vol. 4A. Mires: Swamp, Bog, Fen and Moor – General Studies*, (ed. A. J. P. Gore), Elsevier, Amsterdam, pp. 67–158.

Ives, J. D. (1974a) Permafrost, in *Arctic and Alpine Environments*, (eds J. D. Ives and R. G. Barry), Methuen, London, pp. 159–94.

Ives, J. D. (1974b) Biological refugia and the nunatak hypothesis, in *Arctic and Alpine Environments*, (eds J. D. Ives and R. G. Barry), Methuen, London pp. 605–36.

Ives, W. G. H. and Wong, H. R. (1988) *Tree and Shrub Insects of the Prairie Provinces*, Information Report NOR-X-292, Ministry of Supply and Services, Canada.

Izhaki, I. and Safriel, U. N. (1990) The effect of some mediterranean scrubland frugivores upon germination patterns. *Journal of Ecology*, **78**, 56–65.

Jackson, J. F. (1981) Seed size as a correlate of temporal and spatial patterns of seed fall in a Neotropical forest. *Biotropica*, **13**, 121–30.

Jacobs, M. (1974) Botanical panorama of the Malesian archipelago (vascular plants), in *Natural Resources of Humid Tropical Asia*, Unesco, Paris, pp. 263–94.

Janos, D. P. (1983) Tropical mycorrhizas, nutrient cycles and plant growth, in *Tropical Rain Forest: Ecology and Management*, (eds S. L. Sutton, T. C. Whitmore and A. C. Chadwick), Blackwell Scientific Publications, Oxford, pp. 327–45.

Janson, C. H. (1983) Adaptation of fruit morphology to dispersal agents in a Neotropical forest. *Science*, **219**, 187–8.

Janson, C. H., Terborgh, J. and Emmons, L. H. (1981) Non-flying mammals as pollinating agents in the Amazonian forest. *Biotropica*, **13**, (suppl.), 1–6.

Janzen, D. H. (1968) Reproductive behaviour in the Passifloraceae and some of its pollinators in Central America. *Behaviour*, **32**, 33–48.

Janzen, D. H. (1970) Herbivores and the number of tree species in tropical forests. *The American Naturalist*, **104**, 501–28.

Janzen, D. H. (1971a) Seed predation by animals. *Annual Review of Ecology and Systematics*, **2**, 465–92.

Janzen, D. H. (1971b) Euglossine bees as long-distance pollinators of tropical plants. *Science*, **171**, 203–5.

Janzen, D. H. (1975a) *Ecology of Plants in the Tropics*, Edward Arnold, London.

Janzen, D. H. (1975b) Behaviour of *Hymenaea courbaril* when its predispersal seed predator is absent. *Science*, **189**, 145–6.

Janzen, D. H. (1976) Why bamboos wait so long to flower. *Annual Review of Ecology and Systematics*, **7**, 347–91.

Janzen, D. H. (1978) Seeding patterns of tropical trees, in *Tropical Trees as Living Systems*, (eds P. B. Tomlinson and M. H. Zimmermann), Cambridge University Press, Cambridge, pp. 83–128.

Janzen, D. H. (1980) Specificity of seed-attacking beetles in a Costa Rican deciduous forest. *Journal of Ecology*, **68**, 929–52.

Jarman, P. J. and Sinclair, A. R. E. (1979) Feeding strategy and the pattern of resource-partitioning in ungulates, in *Serengeti, Dynamics of an Ecosystem*, (eds A. R. E. Sinclair and M. Norton-Griffiths), University of Chicago Press, Chicago, pp. 130–63.

Jarvis, P. G. and Leverenz, J. W. (1983) Productivity of temperate deciduous and evergreen trees, in *Physiological Plant Ecology IV. Vol. 12D. Ecosystem Processes: Mineral Cycling, Productivity and Man's Influence*, (eds O. L. Lange, P. S. Nobel, C. B. Osmond and H. Ziegler), Springer, Berlin, pp. 232–80.

Jarvis, P. G. and Sandford, A. P. (1986) Temperate forests, in *Photosynthesis in Contrasting Environments*, (eds N. R. Baker and S. P. Long), Elsevier, Amsterdam, pp. 199–236.

Jassby, A. D. (1975) The ecological significance of sinking to planktonic bacteria. *Canadian Journal of Microbiology*, **21**, 270–74.

Jassby, A. D. and Goldman, C. R. (1974) Loss rate from a lake phytoplankton community. *Limnology and Oceanography*, **19**, 618–27.

Jefferies, R. L. (1977) The vegetation of salt marshes at some coastal sites in Arctic North America. *Journal of Ecology*, **65**, 661–72.

Jefferies, R. L. (1981) Osmotic adjustment and the response of halophytic plants to salinity. *BioScience*, **31**, 42–6.

Jeffers, J. N. R. and Boaler, S. B. (1966) Ecology of a miombo site, Lupa North Forest Reserve, Tanzania. I. Weather and plant growth, 1962–64. *Journal of Ecology*, **54**, 447–63.

Jeník, J. (1978) Roots and root systems in tropical trees: morphologic and ecologic aspects, in *Tropical Trees as Living Systems*, (eds P. B. Tomlinson and M. H. Zimmermann), Cambridge University Press, Cambridge, pp. 323–49.

Jennings, D. H. (1968) Halophytes, succulence and sodium in plants – a unified theory. *New Phytologist*, **67**, 899–911.

Jennings, D. H. (1976) The effects of sodium chloride on higher plants. *Biological Reviews*, **51**, 453–86.

Jensen, V. (1974) Decomposition of angiosperm tree leaf litter, in *Biology of Plant Litter Decomposition*, Vol. 1 (eds C. H. Dickinson and G. J. F. Pugh), Academic Press, London, pp. 69–104.

Jerlov, N. G. (1976) *Marine Optics*, Elsevier, Amsterdam.

Johansson, D. R. (1989) Vascular epiphytism in Africa, in *Ecosystems of the World Vol. 14B. Tropical Rain Forest Ecosystems: Biogeographical and Ecological Studies*, (eds H. Lieth and M. J. A. Werger), Elsevier, Amsterdam, pp. 183–94.

Johansson, L.-G. and Linder, S. (1980) Photosynthesis of *Sphagnum* in different microhabitats on a subarctic mire, in *Ecology of a Subarctic Mire*, (ed. M. Sonesson), Stockholm, Swedish Natural Science Research Council, pp. 181–90.

Johns, A. D. (1988) Effects of 'selective' timber extraction on rain forest structure and composition and some consequences for frugivores and folivores. *Biotropica*, **20**, 31–7.

Johnson, D. A. and Caldwell, M. M. (1975) Gas exchange of four arctic and alpine tundra plant species in relation to atmospheric and soil moisture stress. *Oecologia*, **21**, 93–108.

Johnson, D. A. and Tieszen, L. L. (1976) Aboveground biomass allocation, leaf growth, and photosynthesis patterns in tundra plant forms in arctic Alaska. *Oecologia*, **24**, 159–73.

Johnson, D. W., Cole, D. W., Bledsoe, C. S. *et al.* (1982) Nutrient cycling in forests of the Pacific Northwest, in *Analysis of Coniferous Forest Ecosystems in the Western United States*, (ed. R. L. Edmonds), Hutchinson Ross Publishing Company, Stroudsburg, pp. 186–232.

Johnson, E. A. (1981) Vegetation organization and dynamics of lichen woodland communities in the Northwest Territories, Canada. *Ecology*, **62**, 200–215.

Johnson, K. A. and Whitford, W. G. (1975) Foraging ecology and relative importance of subterranean termites in Chihuahuan Desert ecosystems. *Environmental Entomology*, **4**, 66–70.

Johnson, L. C. and Damman, A. W. H. (1991) Species-controlled *Sphagnum* decay on a South Swedish raised bog. *Oikos*, **61**, 234–42.

Johnson, L. C., Damman, A. W. H. and Malmer, N. (1990) *Sphagnum* macrostructure as an indicator of decay and compaction in peat cores from an ombrotrophic South Swedish peat-bog. *Journal of Ecology*, **78**, 633–47.

Johnson, M. P. (1967) Temperature dependent leaf morphogenesis in *Ranunculus flabellaris*. *Nature*, **214**, 1354–5.

Johnson, P. L. and Billings, W. D. (1962) The alpine vegetation of the Beartooth Plateau in relation to cryopedogenic processes and patterns. *Ecological Monographs*, **32**, 105–35.

Johnson, W. S., Gigon, A., Gulmon, S. L. and Mooney, H. A. (1974) Comparative photosynthetic capacities of intertidal algae under exposed and submerged conditions. *Ecology*, **55**, 450–53.

Johnston, C. S., Jones, R. G. and Hunt, R. D. (1977) A seasonal carbon budget for a laminarian population in a Scottish sea-loch. *Helgoländer Wissenschaftliche Meeresuntersuchungen*, **30**, 527–45.

Johnston, J. P. (1941) Height–growth periods of oak and pine reproduction in the Missouri Ozarks. *Journal of Forestry*, **39**, 67–8.

Jones, C. F. (1930) Agricultural regions of South America, instalment VII. *Economic Geography*, **6**, 1–36.

Jones, E. W. (1955) Ecological studies on the rain forest of southern Nigeria IV. The plateau forest of the Okomu Forest Reserve. *Journal of Ecology*, **43**, 564–94.

Jones, E. W. (1956) Ecological studies on the rain forest of southern Nigeria IV (continued). The plateau forest of the Okomu Forest Reserve. *Journal of Ecology*, **44**, 83–117.

Jones, H. E. (1971) Comparative studies of plant growth and distribution in relation to waterlogging. II. An experimental study of the relationship between transpiration and the uptake of iron in *Erica cinerea* L. and *E. tetralix* L. *Journal of Ecology*, **59**, 167–78.

Jones, M. B. (1986) Wetlands, in *Photosynthesis in Contrasting Environments*, (eds N. R. Baker and S. P. Long), Elsevier, Amsterdam, pp. 103–38.

Jones, M. B. and Muthuri, F. M. (1985) The canopy structure and microclimate of papyrus (*Cyperus papyrus*) swamps. *Journal of Ecology*, **73**, 481–91.

Jones, P. D., Wigley, T. M. L. and Wright, P. B. (1986) Global temperature variations between 1861 and 1984. *Nature*, **322**, 430–4.

Jordan, C., Golley, F., Hall, J. and Hall, J. (1980) Nutrient scavenging of rainfall by the canopy of an Amazonian rain forest. *Biotropica*, **12**, 61–6.

Jordan, C. F. (1985) *Nutrient Cycling in Tropical Forest Ecosystems*, Wiley, Chichester.

Jordan, C. F. and Escalante, G. (1980) Root biomass in an Amazonian rain forest. *Ecology*, **61**, 14–18.

Jordan, C. F. and Uhl, C. (1978) Biomass of a 'tierra firme' forest of the Amazon Basin. *Oecologia Plantarum*, **13**, 387–400.

Josens G. (1983) The soil fauna of tropical savannas. III. The termites, in *Ecosystems of the World Vol. 13. Tropical Savannas*, (ed. F. Bourliere), Elsevier, Amsterdam, pp. 505–24.

Jow, W. M., Bullock, S. H. and Kummerow, J. (1980) Leaf turnover rates of *Adenostoma fasciculatum* (Rosaeae). *American Journal of Botany*, **67**, 256–61.

Joyce, C. (1993) Taxol: search for a cancer drug. *BioScience*, **43**, 133–6.

Juhren, M., Went, F. W. and Phillips, E. (1956) Ecology of desert plants. IV. Combined field and laboratory work on germination of annuals in the Joshua Tree National Monument, California. *Ecology*, **37**, 318–30.

Julander, O. (1945) Drought resistance in range and pasture grasses. *Plant Physiology*, **20**, 573–99.

Junk, W. J. (1983) Ecology of swamps on the middle Amazon, in *Ecosystems of the World Vol. 4B. Mires: Swamp, Bog, Fen and Moor – Regional Studies*, (ed. A. J. P. Gore), Elsevier, Amsterdam, pp. 269–94.

Kabzems, A., Kosowan, A. L. and Harris, W. C. (1986) *Mixedwood Section in an Ecological Perspective*. Technical Bulletin No. 8, Canada–Saskatchewan Forest Resource Development Agreement.

Kadlec, R. H. (1987) Northern natural wetland water treatment systems, in *Aquatic Plants for Water Treatment and Resource Recovery*, (eds K. R. Reddy and W. H. Smith), Magnolia Publishing, Orlando, pp. 83–98.

Kain, J. M. (1979) A view of the genus *Laminaria*. *Oceanography and Marine Biology An Annual Review*, **17**, 101–61.

Kalff, J. and Welch, H. E. (1974) Phytoplankton production in Char Lake, a natural polar lake, and in Meretta Lake, a polluted polar lake, Cornwallis Island, Northwest Territories. *Journal of the Fisheries Research Board of Canada*, **31**, 621–36.

Kaminsky, R. (1981) The microbial origin of the allelopathic potential of *Adenostoma fasciculatum* H & A. *Ecological Monographs*, **51**, 365–82.

Kangas, P. C. and Hannan, G. L. (1985) Vegetation on muskrat mounds. *American Midland Naturalist*, **113**, 392–6.

Karban, R. (1983) Induced responses of cherry trees to periodical cicada oviposition. *Oecologia*, **59**, 226–31.

Karr, J. R. (1989) Birds, in *Ecosystems of the World Vol 14B. Tropical Rainforest Ecosystems*, (eds H. Lieth and M. J. A. Werger), Elsevier, Amsterdam, pp. 401–16.

Kassas, M. (1966) Plant life in deserts, in *Arid lands – A Geographical Appraisal*, (ed. E. S. Hills), Methuen, London, pp. 145–80.

Kassas, M. and Batanouny, K. H. (1984) Plant ecology, in *Sahara Desert*, (ed. J. L. Cloudsley-Thompson), Pergamon, Oxford, pp. 77–90.

Kato, R., Tadaki, Y. and Ogawa, H. (1978) Plant biomass and growth increment studies in Pasoh Forest. *Malayan Nature Journal*, **30**, 211–24.

Kayll, A. J. (1963) *A Technique for Studying the Fire Tolerance of Living Tree Trunks*. Canadian Department of Forestry Publication No. 1012, Ottawa.

Kearney, M. S. (1982) Recent seedling establishment at timberline in Jasper National Park, Alta. *Canadian Journal of Botany*, **60**, 2283–7.

Kearney, T. H. and Shantz, H. L. (1912) The water economy of dry-land crops, in *Yearbook of the United States Department of Agriculture – 1911*, Washington, pp. 351–62.

Keast, A. (1980) Synthesis: ecological basis and evolution of the Neoarctic–Neotropical bird migration system, in *Migrant Birds of the Neotropics*, (eds A. Keast and E. S. Morton), Smithsonian Institution Press, Washington, pp. 559–76.

Keay, R. W. J. (1949) An example of Sudan zone vegetation in Nigeria. *Journal of Ecology*, **37**, 335–64.

Keel, S. H. K. and Prance, G. T. (1979) Studies of the vegetation of a white-sand black-water igapó (Rio Negro, Brazil). *Acta Amazonica*, **9**, 645–55.

Keeley, J. E. (1977) Seed production, seed populations in soil, and seedling production after fire for two congeneric pairs of sprouting and nonsprouting chaparral shrubs. *Ecology*, **58**, 820–9.

Keeley, J. E. (1979) Population differentiation along a flood frequency gradient: physiological adaptations to flooding in *Nyssa sylvatica*. *Ecological Monographs*, **49**, 89–108.

Keeley, J. E. (1987a) Role of fire in seed germination of woody taxa in California chaparral. *Ecology*, **68**, 434–43.

Keeley, J. E. (1987b) Ten years of change in seed banks of the chaparral shrubs, *Arctostaphylos glauca* and *A. glandulosa*. *The American Midland Naturalist*, **117**, 446–8.

Keeley, J. E. (1987c) The adaptive radiation of photosynthetic modes in the genus *Isoetes* (Isoetaceae), in *Plant Life in Aquatic and Amphibious Habitats*, (ed. R. M. M. Crawford), Blackwell Scientific Publications, Oxford, pp. 113–28.

Keeley, J. E. and Hays, R. L. (1976) Differential seed predation on two species of *Arctostaphylos* (Ericaceae). *Oecologia*, **24**, 71–81.

Keeley, J. E. and Keeley, S. C. (1984) Postfire recovery of California coastal sage scrub. *The American Midland Naturalist*, **111**, 105–17.

Keeley, J. E. and Keeley, S. C. (1987) Role of fire in the germination of chaparral herbs and suffrutescents. *Madroño*, **34**, 240–49.

Keeley, J. E. and Zedler, P. H. (1978) Reproduction of chaparral shrubs after fire: a comparison of sprouting and seeding strategies. *The American Midland Naturalist*, **99**, 142–61.

Keeley, J. E., Morton, B. A., Pedrosa, A. and Trotter, P. (1985) Role of allelopathy, heat and charred wood in the germination of chaparral herbs and suffrutescents. *Journal of Ecology*, **73**, 445–58.

Keeley, S. C. and Johnson, A. W. (1977) A comparison of the pattern of herb and shrub growth in comparable sites in Chile and California. *The American Midland Naturalist*, **97**, 120–32.

Keeley, S. C. and Pizzorno, M. (1986) Charred wood stimulated germination of two fire-following herbs of the California chaparral and the role of hemicellulose. *American Journal of Botany*, **73**, 1289–97.

Keeley, S. C., Keeley, J. E., Hutchinson, S. M. and Johnson, A. W. (1981) Postfire succession of the herbaceous flora in the southern California chaparral. *Ecology*, **62**, 1608–21.

Kellman, M. (1979) Soil enrichment by neotropical savanna trees. *Journal of Ecology*, **67**, 565–77.

Kellman, M., Miyanishi, K. and Hiebert, P. (1985) Nutrient retention by savanna ecosystems. II. Retention after fire. *Journal of Ecology*, **73**, 953–62.

Kellman, M. C. (1970) On the nature of secondary plant succession, *Proceedings of the Canadian Association of Geographers*, University of Manitoba, pp. 193–8.

Kellman, M. C. (1988) Disturbance x habitat interactions: fire regimes and the savanna/forest balance in the Neotropics. *Canadian Geographer*, **32**, 80–82.

Kelly, P. M., Jones, P. D., Sear, C. B., Cherry, B. S. G. and Tavakol, R. K. (1982) Variations in surface air temperatures: Part 2. Arctic regions, 1881–1980. *Monthly Weather Review*, **110**, 71–83.

Kelly, V. R. and Parker, V. T. (1990) Seed bank survival and dynamics in sprouting and nonsprouting *Arctostaphylos* species. *The American Midland Naturalist*, **124**, 114–23.

Kemp, P. R. (1989) Seed banks and vegetation processes in deserts, in *Ecology of Soil Seed Banks*, (eds M. A. Leck, V. T. Parker and R. L. Simpson), Academic, San Diego, pp. 257–281.

Kemp, P. R. and Gardetto, P. E. (1982) Photosynthetic pathway types of evergreen rosette plants (Liliaceae) of the Chihuahuan Desert. *Oecologia*, **55**, 149–56.

Kemp, P. R. and Williams, G. J. (1980) A physiological basis for niche separation between *Agropyron smithii* (C_3) and *Bouteloua gracilis* (C_4). *Ecology*, **61**, 846–58.

Kendall, R. L. (1969) An ecological history of the Lake Victoria basin. *Ecological Monographs*, **36**, 121–76.

Kennan, T. C. D. (1972) The effects of fire on two vegetation types at Matapos, Rhodesia. *Proceedings Tall Timbers Fire Ecology Conference*, **11**, 53–98.

Kennish, M. J. (1986) *Ecology of Estuaries Vol. 1. Physical and Chemical Aspects*, CRC Press, Baton Rouge.

Kerfoot, O. (1963) The root systems of tropical forest trees. *Commonwealth Forestry Review*, **42**, 19–26.

Kerfoot, W. C., Levitan, C. and DeMott, W. R. (1988) *Daphnia*–phytoplankton interactions: density-dependent shifts in resource quality. *Ecology*, **69**, 1806–25.

Kerlinger, P., Lein, M. R. and Sevick, B. J. (1985) Distribution and population fluctuations of wintering Snowy Owls (*Nyctea scandiaca*) in North America. *Canadian Journal of Zoology*, **63**, 1829–34.

Kershaw, K. A. (1977) Studies on lichen-dominated systems. XX. An examination of some aspects of the northern boreal lichen woodlands in Canada. *Canadian Journal of Botany*, **55**, 393–410.

Kershaw, K. A. (1985) *Physiological Ecology of Lichens*, Cambridge University Press, Cambridge.

Kevan, P. G. (1975) Sun-tracking solar furnaces in high arctic flowers: significance for pollination and insects. *Science*, **189**, 723–6.

Kienholz, R. (1941) Seasonal course of height growth in some hardwoods in Connecticut. *Ecology*, **22**, 249–58.

Kilander, S. (1965) Alpine zonation in the southern part of the Swedish Scandes. *Acta Phytogeographica Suecica*, **50**, 78–84.

King, H. G. C. and Heath, G. W. (1967) The chemical analysis of small samples of leaf material and the relationship between the disappearance and composition of leaves. *Pedobiologia*, **7**, 192–7.

Kingdon, J. (1990) *Island Africa. The evolution of Africa's rare animals and plants*, Collins, London.

Kira, T. (1975) Primary production of forests, in *Photosynthesis and Productivity in Different Environments*, (ed. J. P. Cooper), Cambridge University Press, Cambridge, pp. 5–40.

Kira, T. (1977) Introduction, Forest vegetation of Japan, in *JIBP Synthesis Vol. 16. Primary Productivity of Japanese Forests*, (eds T. Shidei and T. Kira), University of Tokyo Press, Tokyo, pp. 1–9.

Kira, T. and Yabuki, K. (1978) Primary production rates in the Minamata forest, in *JIBP Synthesis Vol. 18. Biological Production in a Warm-temperate Evergreen Oak Forest of Japan*, (eds T. Kira, Y. Ono and T. Hosokawa), University of Tokyo Press, Tokyo, pp. 131–8.

Kirk, J. T. O. (1983) *Light and Photosynthesis in Aquatic Ecosystems*, Cambridge University Press, Cambridge.

Kirk, J. T. O. (1986) Freshwater environments, in *Photosynthesis in Contrasting Environments*, (eds N. R. Baker and S. P. Long), Elsevier, Amsterdam, pp. 295–336.

Kittredge, J. (1948) *Forest Influences*, McGraw-Hill, New York.

Kittredge, J. (1955) Litter and forest floor of the chaparral in parts of the San Dimas Experimental Forest, California. *Hilgardia*, **23**, 563–96.

Kjellberg, B., Karlsson, S. and Kerstensson, I. (1982) Effects of heliotropic movements of flowers of *Dryas octopetala* L. on gynoecium temperature and seed development. *Oecologia*, **54**, 10–13.

Kleinfeldt, S. E. (1978) Ant-gardens: the interaction of *Codonanthe crassifolia* (Gesneriaceae) and *Crematogaster longispina* (Formicidae). *Ecology*, **59**, 449–56.

Klinge, H. (1965) Podzol soils in the Amazon basin. *Journal of Soil Science*, **16**, 95–103.

Klinge, H. (1973a) Root mass estimation in lowland tropical rain forests of central Amazonia, Brazil. I. Fine root masses of a pale yellow latosol and a giant humus podzol. *Tropical Ecology*, **14**, 29–38.

Klinge, H. (1973b) Root mass estimation in lowland tropical rain forests of central Amazonia, Brazil. II. 'Coarse root mass' of trees and palms in different height classes. *Anais du Academia Brasileira de Ciencas*, **45**, 595–609.

Klinge, H., Rodrígues, W. A., Brunig, E. and Fittkau, E. J. (1975) Biomass and structure in a central Amazonian rain forest, in *Tropical Ecological Systems*, (eds F. B. Golley and E. Medina), Springer, New York, pp. 115–22.

Klopatek, J. M. (1978) Nutrient dynamics of freshwater riverine marshes and the role of emergent macrophytes, in *Freshwater Wetlands*, (eds R. E. Good, D. F. Whigham and R. L. Simpson), Academic, New York, pp. 195–215.

Klopatek, J. M. and Stearns, F. W. (1978) Primary productivity of emergent macrophytes in a Wisconsin freshwater marsh ecosystem. *Am. Mid. Nat.*, **100**, 320–32.

Koblentz-Mishke, O. J., Volkovinsky, V. V. and Kabanova, J. G. (1970) Plankton primary production of the world ocean, in *Scientific Exploration of the South Pacific*, (ed. W. S. Wooster), National Academy of Sciences, Washington, pp. 183–93.

Koelmeyer, K. O. (1959) The periodicity of leaf change and flowering in the principal forest communities of Ceylon. *Ceylon Forester*, **4**, 157–89, 308–64.

Koerselman, W., de Caluwe, H. and Kieskamp, W. M. (1989) Denitrification and dinitrogen fixation in two quaking fens in the Vechtplassen area, The Netherlands. *Biogeochemistry*, **8**, 153–65.

Koford, C. B. (1957) The vicuña and the puna. *Ecological Monographs*, **27**, 153–219.

Koller, D. and Roth, N. (1964) Studies on the ecological and physiological significance of amphicarpy in *Gymnarrhena micrantha* (Compositae). *American Journal of Botany*, **51**, 26–35.

Korte, P. A. and Fredrickson, L. H. (1977) Loss of Missouri's lowland hardwood ecosystem. *Transactions of the North American Wildlife and Natural Resource Conference*, **42**, 31–41.

Kotov, M. I. (1965) High-mountain vegetation of the southeastern part of Central Tien Shan, in *Studies on the Flora and Vegetation of High Mountain Areas*, (ed. V. N. Sukachev), Israel Program for Scientific Translations, Jerusalem, pp. 229–36.

Kovda, V. A., Samoilova, E. M., Charley, J. L. and Skujiņš, J. J. (1979) Soil processes in arid lands, in *Arid-land Ecosystems: Structure, Functioning and Management*, Vol. 1 (eds D. W. Goodall, R. A. Perry and K. M. W. Howes), Cambridge University Press, Cambridge, pp. 439–70.

Kozlowski, T. T. (1982) Water supply and tree growth. Part I. Water deficits. *Forestry Abstracts*, **43**, 57–95.

Krajina, V. J. (1969) Ecology of forest trees in British Columbia, in *Ecology of Western North America*, Vol. 2 (eds V. J. Krajina and R. C. Brooke), University of British Columbia, Vancouver, pp. 1–146.

Kramer, P. J. and Kozlowski, T. T. (1960) *Physiology of Trees*, McGraw-Hill, New York.

Kratz, T. K. and deWitt, C. B. (1986) Internal factors controlling peatland–lake ecosystem development. *Ecology*, **67**, 100–7.

Kratz, T. K., Frost, T. M. and Magnuson, J. J. (1987) Inferences from spatial and temporal variability in ecosystems: long-term zooplankton data from lakes. *American Naturalist*, **129**, 830–46.

Krause, D. and Kummerow, J. (1977) Xeromorphic structure and soil moisture in the chaparral. *Oecologia Plantarum*, **12**, 133–48.

Krebs, J.S and Barry, R. G. (1970) The arctic front and the tundra–taiga boundary in Eurasia. *The Geographical Review*, **60**, 548–54.

Kriebel, H. B. and Wang, C.-W. (1962) The interaction between provenance and degree of chilling in bud-break of sugar maple. *Silvae Genetica*, **11**, 125–30.

Krog, J. (1955) Notes on temperature measurements indicative of special organization in arctic and subarctic plants for utilization of radiated heat fom the sun. *Physiologia Plantarum*, **8**, 836–9.

Krueger, K. W. and Ruth, R. H. (1969) Comparative photosynthesis of red alder, Douglas fir, Sitka spruce, and western hemlock seedlings. *Canadian Journal of Botany*, **47**, 519–27.

Krueger, K. W. and Trappe, J. M. (1967) Food reserves and seasonal growth of Douglas fir seedlings. *Forest Science*, **13**, 192–202.

Kruger, F. J. (1979) South African heathlands, in *Ecosystems of the World Vol. 9A. Heathlands and Related Shrublands*, (ed. R. L. Specht), Elsevier, Amsterdam, pp. 19–80.

Kruger, F. J. and Bigalke, R. C. (1984) Fire in fynbos, in *Ecological Effects of Fire in South African Ecosystems*, (eds P. de V. Booysen and N. M. Tainton), Springer, Berlin, pp. 67–114.

Kucera, C. L. and Ehrenreich, J. H. (1962) Some effects of annual burning on central Missouri prairie. *Ecology*, **43**, 334–6.

Kulman, H. M. (1971) Effects of insect defoliation on growth and mortality of trees. *Annual Review of Entomology*, **16**, 289–324.

Kummerow, J. (1973) Comparative anatomy of sclerophylls of mediterranean climatic areas, in *Mediterranean-type Ecosystems: Origin and Structure*, (eds F. di Castri and H. A. Mooney), Springer, New York, pp. 157–67.

Kummerow, J. (1981) Structure of roots and root systems, in *Ecosystems of the World Vol. II. Mediterranean-type Shrublands*, (eds F. di Castri, D. W. Goodall and R. L. Specht), Elsevier, Amsterdam, pp. 269–88.

Kummerow, J. (1983) Comparative phenology of mediterranean-type plant communities, in *Mediterranean-type Ecosystems: the Role of Nutrients*, (eds F. J. Kruger, D. T. Mitchell and J. U. M. Jarvis), Springer, Berlin, pp. 300–17.

Kummerow, J. and Mangan, R. (1981) Root systems in *Quercus dumosa* Nutt. dominated chaparral in southern California. *Oecologia Plantarum*, **2**, 177–88.

Kummerow, J., Krause, D. and Jow, W. (1977) Root systems of chaparral shrubs. *Oecologia*, **29**, 163–77.

Kummerow, J., Montenegro, G. and Krause, D. (1981) Biomass, phenology, and growth, in *Resource Use by Chaparral and Matorral*, (ed. P. C. Miller), Springer, New York, pp. 69–96.

Kuvaev, V. B. (1965) Distribution pattern of the vegetation in western Verkhoyansk mountain area, in *Studies on the Flora and Vegetation of High Mountain Areas*, (ed. V. N. Sukachev), Israel Program for Scientific Translations, Jerusalem, pp. 67–82.

Lacey, J. R. and van Poolen, H. W. (1981) Comparison of herbage production on moderately grazed and ungrazed western ranges. *Journal of Range Management*, **34**, 210–12.

Ladd, H. S. (1977) Types of coral reefs and their distribution, in *Biology and Geology of Coral Reefs Vol. IV. Geology 2* (eds O. A. Jones and R. Endean), Academic Press, New York, pp. 1–19.

Ladefoged, K. (1952) The periodicity of wood formation. *Danske Biologiske Skrifter*, **7**, 1–98.

Laing, C. C. (1958) Studies in the ecology of *Ammophila breviligulata* I. Seedling survival and its relation to population increase and dispersal. *Botanical Gazette*, **119**, 208–16.

Lamb, I. M. (1970) Antarctic terrestrial plants and their ecology, in *Antarctic Ecology*, Vol. 2 (ed. M. W. Holdgate), Academic Press, London, pp. 733–51.

Lamont, B. (1980) Blue-green algae in nectar of *Banksia* aff. *Sphaerocarpa. The Western Australian Naturalist*, **14**, 193–4.

Lamont, B. B. (1983) Strategies for maximizing nutrient uptake in two mediterranean ecosystems of low nutrient status, in *Mediterranean Type Ecosystems: The Role of Nutrients*, (eds F. J. Kruger, D. T. Mitchell and J. U. M. Jarvis), Springer, Berlin, pp. 246–73.

Lamotte, M. (1975) The trophic structure and function of a tropical savannah ecosystem, in *Tropical Ecological Systems*, (eds F. B. Golley and E. Medina), Springer, New York, pp. 179–222.

Lamotte, M. (1983) Amphibians in savanna ecosystems, in *Ecosystems of the World Vol. 13. Tropical Savannas*, (ed. F. Bourlière), Elsevier, Amsterdam, pp. 313–23.

Lampert, W., Fleckner, W., Rai, H. and Taylor, B. E. (1986) Phytoplankton control by grazing zooplankton: a study on the spring clear-water phase. *Limnology and Oceanography*, **31**, 478–90.

Lamprey, H. F. (1963) Ecological separation of the large mammal species in the Tarangire Game Reserve, Tanganyika. *East African Wildlife Journal*, **1**, 63–92.

Lang, G. E. and Knight, D. H. (1979) Decay rates for boles of tropical trees in Panama. *Biotropica*, **11**, 316–7.

Langenheim, J. H. (1962) Vegetation and environmental patterns in the Crested Butte area, Gunnison County, Colorado. *Ecological Monographs*, **32**, 265–85.

Langenheim, J. H. (1984) The roles of plant secondary chemicals in wet tropical ecosystems, in *Physiological Ecology of Plants of the Wet Tropics*, (eds E. Medina, H. A. Mooney and C. Vazquez-Yanes), Junk, The Hague, pp. 189–208.

Langford, A. N. and Buell, M. F. (1969) Integration, identity and stability in the plant association. *Advances in Ecological Research*, **6**, 83–135.

Lanly, J. P. (1982) *Tropical Forest Resources*, Food and Agriculture Organization of the United Nations, Rome.

Larcher, W. (1969) The effect of environmental and physiological variables on the carbon dioxide gas exchange of trees. *Photosynthetica*, **3**, 167–98.

Larcher, W. (1980) *Physiological Plant Ecology*, 2nd edn, Springer, Berlin.

Larcher, W., Cernusca, A., Schmidt, L., *et al.* (1975) Mt. Patscherkofel, Austria, in *Ecological Bulletins 20: Structure and Function of Tundra Ecosystems*, (eds T. Rosswall and O. W. Heal), Swedish Natural Science Research Council, Stockholm, pp. 125–39.

Larsen, J. A. (1971) Vegetational relationships with air mass frequencies: boreal forest and tundra. *Arctic*, **24**, 177–94.

Larsen, J. A. (1980) *The Boreal Ecosystem*, Academic Press, New York.

Larsen, J. A. (1982) *Ecology of the Northern Lowland Bogs and Conifer Forests*, Academic Press, New York.

Lassoie, J. P., Hinckley, T. M. and Grier, C. C. (1985) Coniferous forests of the Pacific Northwest, in *Physiological Ecology of North American Plant Communities*, (eds B. F. Chabot and H. A. Mooney), Chapman & Hall, New York, pp. 127–61.

Lathrop, E. W. and Archbold, E. F. (1980) Plant response to Los Angeles aqueduct construction in the Mojave Desert. *Environmental Management*, **4**, 137–48.

Latif, A. F. A. (1984) Lake Nasser – the new man-made lake in Egypt (with reference to Lake Nubia), in *Ecosystems of the World Vol. 23. Lakes and Reservoirs*, (ed. F. B. Taub), Elsevier, Amsterdam, pp. 385–410.

Launchbaugh, J. L. (1964) Effects of early spring burning on yields of native vegetation, *Journal of Range Management*, **17**, 5–6.

Lavelle, P. (1983) The soil fauna of tropical savannas. II. The earthworms, in *Ecosystems of the World Vol. 13. Tropical Savannas*, (ed. F. Bourlière), Elsevier, Amsterdam, pp. 485–504.

Lawlor, D. W. (1987) *Photosynthesis: Metabolism, Control, and Physiology*, Longman, Harlow.

Lawson, G. W., Armstrong-Mensah, K. O. and Hall, J. B. (1970) A catena in tropical moist semi-deciduous forest near Kade, Ghana. *Journal of Ecology*, **58**, 371–98.

Lawson, G. W., Jeník, J. and Armstrong-Mensah, K. O. (1968) A study of a vegetation catena in Guinea savanna at Mole Game Reserve (Ghana). *Journal of Ecology*, **56**, 505–22.

Lee, D. W. (1987) The spectral distribution of radiation in two Neotropical rainforests. *Biotropica*, **19**, 161–6.

Lee, D. W. and Lowry, J. B. (1980) Young-leaf anthocyanin and solar ultraviolet. *Biotropica*, **12**, 75–6.

Lee, J. A. and Ignaciuk, R. (1985) The physiological ecology of strandline plants. *Vegetatio*, **62**, 319–26.

Lee, K. E. and Wood, T. G. (1971) Physical and chemical effects on soils of some Australian termites, and their pedological significance. *Pedobiologia*, **11**, 376–409.

Lee, R. (1978) *Forest Microclimatology*, Columbia University Press, New York.

Leetham, J. W. and Milchunas, D. G. (1985) The composition and distribution of soil microarthropods in the shortgrass steppe in relation to soil water, root biomass, and grazing by cattle. *Pedobiologia*, **28**, 311–25.

Le Fèvre, J. (1986) Aspects of the biology of frontal systems. *Advances in Marine Biology*, **23**, 163–299.

Lehman, J. T. (1984) Grazing, nutrient release, and their impacts on the structure of phytoplankton communities, in *Trophic Interactions Within Aquatic Ecosystems*, (eds D. G. Meyers and J. R. Strickler), American Association for the Advancement of Science, Washington, pp. 49–72.

Lehman, J. T. and Scavia, D. (1982) Microscale patchiness of nutrients in plankton communities. *Science*, **216**, 729–30.

Leitch, J. A. (1989) Politicoeconomic overview of prairie potholes, in *Northern Prairie Wetlands* (ed. A. van der Volk), Iowa State University Press, Ames, pp. 2–14.

Le Houérou, H. N. (1974) Fire and vegetation in the Mediterranean Basin. *Tall Timbers Fire Ecology Conference, Proceedings*, **13**, 237–77.

Le Houérou, H. N. (1977) The nature and causes of desertification, in *Desertification*, (ed. M. H. Glantz), Westview Press, Boulder, pp. 17–38.

Le Houérou, H. N. (1979) North Africa, in *Arid-land Ecosystems: Structure, Functioning and Management*, Vol. I (eds D. W. Goodall, R. A. Perry and K. M. W. Howes), Cambridge University Press, Cambridge, pp. 83–107.

Le Houérou, H. N. (1986) The desert and arid zones of northern Africa, in *Ecosystems of the World Vol. 12B. Hot Deserts and Arid Shrublands, B*, (eds M. Evanari, I. Noy-Meir and D. W. Goodall), Elsevier, Amsterdam, pp. 101–47.

Le Houérou, H. N. (1981) Impact of man and his animals on mediterranean vegetation, in *Ecosystems of the World Vol. 11. Mediterranean-type Shrublands*, (eds F. di Castri, D. W. Goodall and R. L. Specht), Elsevier, Amsterdam, pp. 479–521.

Leigh, E. G. and Windsor, D. M. (1982) Forest production and regulation of primary consumers on Barro Colorado Island, in *The Ecology of a Tropical Forest*, (eds E. G. Leigh, A. S. Rand and D. M. Windsor), Smithsonian Institution Press, Washington, pp. 111–22.

Leighton, M. and Leighton, D. R. (1983) Vertebrate responses to fruiting seasonality within a Bornean rain forest, in *Tropical Rain Forest: Ecology and Management*, (eds S. L. Sutton, T. C. Whitmore and A. C. Chadwick), Blackwell Scientific Publications, Oxford, pp. 181–96.

Leistner, O. A. (1979) Southern Africa, in *Arid-land Ecosystems: Structure, Functioning and Management*, Vol. 1 (eds D. W. Goodall, R. A. Perry and K. M. W. Howes), Cambridge University Press, Cambridge, pp. 109–43.

Lemon, R. C. (1968) Effects of fire on an African plateau grassland. *Ecology*, **49**, 316–22.

Levenson, J. B. (1981) Woodlots as biogeographic islands in southeastern Wisconsin, in *Forest Island Dynamics in Man-dominated Landscapes*, (eds R. L. Burgess and D. M. Sharpe), Springer, New York, pp. 13–39.

Leverenz, J. W. and Jarvis, P. G. (1979) Photosynthesis in Sitka spruce. VIII. The effects of light flux density and direction on the rate of net photosynthesis and the stomatal conductance of needles. *Journal of Applied Ecology*, **16**, 919–32.

Leverenz, J. W. and Jarvis, P. G. (1980) Photosynthesis in Sitka spruce (*Picea sitchensis* (Bong.) Carr.). X. Acclimation to quantum flux density within and between trees. *Journal of Applied Ecology*, **17**, 697–708.

Levieux, J. (1983) The soil fauna of tropical savannas. IV. The ants, in *Ecosystems of the World Vol. 13. Tropical Savannas*, (ed. F. Bourlière) Elsevier, Amsterdam, pp. 525–40.

Lewis, A. R. (1988) Buttress arrangement in *Pterocarpus officinalis* (Fabaceae): effects of crown asymmetry and wind. *Biotropica*, **20**, 280–85.

Lewis, J. B. (1981) Coral reef ecosystems, in *Analysis of Marine Ecosystems* (ed. A. R. Longhurst), Academic Press, London, pp. 127–58.

Lewis, J. B. (1987) Measurements of groundwater seepage flux onto a coral reef: spatial and temporal variations. *Limnology and Oceanography*, **32**, 1165–9.

Lewis Smith, R. I. (1988) Destruction of antarctic terrestrial ecosystems by a rapidly increasing fur seal population. *Biological Conservation*, **45**, 55–72.

Lieberman, D. and Lieberman, M. (1987) Forest tree growth and dynamics at La Selva, Costa Rica (1969–82). *Journal of Tropical Ecology*, **3**, 347–58.

Lieberman, D., Lieberman, M., Hartshorn, G. and Peralta, R. (1985a) Growth rates and age–size relationships of tropical wet forest trees in Costa Rica. *Journal of Tropical Ecology*, **1**, 97–109.

Lieberman, D., Lieberman, M., Peralta, R. and Hartshorn, G. S. (1985b) Mortality patterns and stand turnover rates in a wet tropical forest in Costa Rica. *Journal of Ecology*, **73**, 915–24.

Lieffers, V. J. and Rothwell, R. L. (1987a) Effects of drainage on substrate temperature and phenology of some trees and shrubs in an Alberta peatland. *Canadian Journal of Forest Research*, **17**, 97–104.

Lieffers, V. J. and Rothwell, R. L. (1987b) Rooting depth of peatland black spruce and tamarack in relation to depth of water table. *Canadian Journal of Botany*, **65**, 817–21.

Lieth, H. (1975) Historical survey of primary productivity research, in *Primary Productivity of the Biosphere*, (ed. H. Leith and R. H. Whittaker), Springer, New York, pp. 7–16.

Likens, G. E., Bormann, F. H., Johnson N. M., Fisher, D. W. and Pierce, R. S. (1970) Effects of forest cutting and herbicide treatment on nutrient budgets in the Hubbard Brook watershed-ecosystem. *Ecological Monographs*, **40**, 31–47.

Limbach, W. E., Oechel, W. C. and Lowell, W. (1982) Photosynthetic and respiratory responses to temperature and light of three Alaskan tundra growth forms. *Holoarctic Ecology*, **5**, 150–7.

Linder, S. and Rook, D. A. (1984) Effects of mineral nutrition on carbon dioxide exchange and partitioning of carbon in trees, in *Nutrition of Plantation Forests*, (eds G. D. Bowen and E. K. S. Nambiar), Academic Press, London, pp. 211–36.

Lindsay, D. C. (1978) The role of lichens in antarctic ecosystems. *The Bryologist*, **81**, 268–76.

Livingstone, D. A. (1975) Late Quaternary climatic change in Africa. *Annual Review of Ecology and Systematics*, **6**, 249–80.

Llano, G. A. (1970) Habitats and vegetation, in *Antarctic Ecology*, Vol. 2 (ed. M. W. Holdgate), Academic Press, London, p. 864.

Loach, K. (1970) Shade tolerance in tree seedlings II. Growth analysis of plants raised under artifical shade. *New Phytologist*, **69**, 273–86.

Lockwood, J. G. (1986) The causes of drought with particular reference to the Sahel. *Progress in Physical Geography*, **10**, 111–19.

Lodge, D. M., Barko, J. W., Strayer, D. *et al.* (1988) Spatial heterogeneity and habitat interactions in lake communities, in *Complex Interactions in Lake Communities*, (ed. S. R. Carpenter), Springer, New York, pp. 181–208.

Loehr, R. C., Ryding, S.-O. and Sonzogni, W. C. (1989) Estimating the nutrient load to a waterbody, in *The Control of Eutrophication of Lakes and Reservoirs*, (eds S.-O. Ryding and W. Rast) Unesco, Paris, pp. 115–46.

Longman, K. A. (1985) Tropical forest trees, in *Handbook of Flowering*, Vol. 1 (ed. A. H. Halevy), CRC Press, Boca Raton, pp. 23–39.

Longman, K. A. and Jeník, J. (1987) *Tropical Forest and Its Environment*, 2nd edn, Longman Scientific and Technical, Harlow.

Looman, J. (1976) Productivity of permanent bromegrass pastures in the parklands of the prairie provinces. *Canadian Journal of Plant Science*, **56**, 829–35.

Loomis, R. S., Williams, W. A. and Duncan, W. G. (1967) Community architecture and the productivity of terrestrial plant communities, in *Harvesting the Sun*, (eds A. San Pietro, F. J. Greer and T. J. Army), Academic Press, New York, pp. 291–308.

Lopes, A. S. and Cox, F. R. (1977) A survey of the fertility status of surface soils under 'cerrado' vegetation in Brazil. *Soil Science Society of America Journal*, **41**, 742–7.

Lossaint, P. (1973) Soil–vegetation relationships in mediterranean ecosystems of southern France, in *Mediterranean Type Ecosystems: Origin and Structure*, (eds F. di Castri and H. A. Mooney), Springer, New York, pp. 199–210.

Louw, G. N. and Seely, M. K. (1982) *Ecology of Desert Organisms*, Longman, London.

Löve, A. and Löve, D. (1974) Origin and evolution of the arctic and alpine floras, in *Arctic and Alpine Environments*, (eds J. D. Ives and R. G. Barry), Methuen, London, pp. 571–603.

Loveless, A. R. (1961) A nutritional interpretation of sclerophylly based on differences in the chemical composition of sclerophyllous and mesophytic leaves. *Annals of Botany*, **25**, 168–84.

Lovelock, J. (1988) *The Ages of Gaia*, Norton, New York.

Loveridge, J. P. (1968a) The control of water loss in *Locusta migratoria migratorioides* R. and F. I. Cuticular water loss. *Journal of Experimental Biology*, **49**, 1–13.

Loveridge, J. P. (1968b) The control of water loss in *Locusta migratoria migratorioides* R. and F. II. Water loss through the spiracles. *Journal of Experimental Biology*, **49**, 15–29.

Lowe, R. G. (1963) The height of buttresses in relation to size of stem and crown. *Journal of the West African Science Association*, **8**, 6–17.

Ludlow, M. M. and Jarvis, P. G. (1971) Photosynthesis in Sitka spruce (*Picea sitchensis* (Bong.) Carr.). I. General characteristics. *Journal of Applied Ecology*, **8**, 925–53.

Ludwig, J. A. (1987) Primary productivity in arid lands: myths and realities. *Journal of Arid Environments*, **13**, 1–7.

Lugo, A. E. and Snedaker, S. C. (1974) The ecology of mangroves. *Annual Review of Ecology and Systematics*, **5**, 39–64.

Lund, J. W. G. (1965) The ecology of the freshwater phytoplankton. *Biological Reviews*, **40**, 231–93.

Lund, J. W. G., Mackereth, F. J. H. and Mortimer, C. H. (1963) Changes in depth and time of certain chemical and physical conditions and of the standing crop of *Asterionella formosa* Hass. in the north basin of Windermere in 1947. *Philosophical Transactions of the Royal Society*, **246**, 255–89.

Lüning, K. (1979) Growth strategies of three *Laminaria* species (Phaeophyceae) inhabiting different depth zones in the sublittoral region of Helgoland (North Sea). *Marine Ecology Progress Series*, **1**, 195–207.

Lüning, K. and Dring, M. J. (1979) Continuous underwater light measurement near Helgoland (North Sea) and its significance for characteristic light limits in the sublittoral region. *Helgoländer Wissenschaftliche Meeresuntersuchungen*, **32**, 403–24.

Lydolph, P. E. (1977) *World Survey of Climatology Vol. 7. Climates of the Soviet Union*, Elsevier, Amsterdam.

Lynch, M. (1980) The evolution of cladoceran life histories. *Quarterly Review of Biology*, **55**, 23–42.

Macdonald, K. B. (1977) Plant and animal communities of Pacific North American salt marshes, in *Ecosystems of the World Vol. I. Wet Coastal Ecosystems*, (ed. V. J. Chapman), Amsterdam, Elsevier, pp. 166–91.

MacGeachy, J. K. and Stearn, C. W. (1976) Boring by macro-organisms in the coral *Montastrea annularis* on Barbados reefs. *Internationale Revue der Gesamten Hydrobiologie*, **61**, 715–45.

MacLean, S. F. (1974) Primary production, decomposition, and the activity of soil invertebrates in tundra ecosystems; a hypothesis, in *Soil Organisms and Decomposition in Tundra*, (eds A. J. Holding, O. W. Heal, S. F. MacLean and P. W. Flanagan), Tundra Biome Steering Committee, Stockholm, pp. 197–206.

MacLean, S. F. (1981) Introduction: invertebrates, in *Tundra Ecosystems: A Comparative Analysis*, (eds L. C. Bliss, O. W. Heal and J. J. Moore), Cambridge University Press, Cambridge, pp. 509–16.

MacLean, S. F., Fitzgerald, B. M. and Pitelka, F. A. (1974) Population cycles in arctic lemmings: winter reproduction and predation by weasels. *Arctic and Alpine Research*, **6**, 1–12.

MacLulich, D. A. (1937) Fluctuations in the Numbers of the Varying Hare (Lepus americanus). *University of Toronto Studies, Biological Series No. 43*. University of Toronto Press, Toronto.

MacMahon, J. A. and Wagner, F. H. (1985) The Sonoran and Chihuahuan Deserts of North America, in *Ecosystems of the World Vol. 12A. Hot Desert and Arid Shrublands, A*, (eds M. Evanari, I. Noy-Meir and D. W. Goodall), Elsevier, Amsterdam, pp. 105–202.

Macnae, W. (1968) A general account of the fauna and flora of mangrove swamps and forests in the Indo-West-Pacific region. *Advances in Marine Biology*, **6**, 73–270.

McAuliffe, J. R. (1984) Sahuaro-nurse tree associations in the Sonoran Desert: competitive effects of sahuaros. *Oecologia*, **64**, 319–21.

McCarthy, J. J. (1980). Nitrogen, in *The Physiological Ecology of Phytoplankton*, (ed. I. Morris), University of California Press, Berkeley, pp. 191–233.

McCormick, J. F. and Platt, R. B. (1980) Recovery of an Appalachian forest following the chestnut blight. *The American Midland Naturalist*, **104**, 264–73.

McCown, B. H. (1978) The interactions of organic nutrients, soil nitrogen, and soil temperature and plant growth and survival in the arctic environment, in *Vegetation and Production Ecology of an Alaskan Arctic Tundra*, (ed. L. L. Tieszen), Springer, New York, pp. 435–56.

McDonald, P. M. and Littrell, E. E. (1976) The bigcone Douglas fir–canyon live oak community in southern California. *Madroño*, 23, 310–20.

McGinnies, W. (1984) Chemically thinning blue grama range for increased forage and seed production. *Journal of Range Management*, **37**, 412–414.

McGinnies, W. G. (1968) Vegetation of desert environments, in *Deserts of the World*, (eds W. G. McGinnies, B. J. Goldman and P. Paylore), University of Arizona Press, Tucson, pp. 381–566.

McGraw, J. B. and Vavrek, M. C. (1989) The role of buried viable seeds in arctic and alpine plant communities, in *Ecology of Soil Seed Banks*, (eds M. A. Leck, V. T. Parker and R. L. Simpson), Academic Press, San Diego, pp. 91–105.

McIntyre, A. D. (1977) Sandy foreshores, in *The Coastline*, (ed. R. S. K. Barnes), Wiley, London, pp. 31–47.

McKey, D. (1975) The ecology of coevolved seed dispersal systems, in *Coevolution of Animals and Plants*, (eds L. E. Gilbert and P. H. Raven), University of Texas Press, Austin, pp. 159–91.

McLaughlin, S. B. and McConathy, R. K. (1979) Temporal and spatial patterns of carbon allocation in the canopy of white oak. *Canadian Journal of Botany*, **57**, 1407–13.

McLean, A. (1969) Fire resistance of forest species as influenced by root systems. *Journal of Range Management*, **22**, 120–2.

McManmon, M. and Crawford, R. M. M. (1971) A metabolic theory of flooding tolerance: the significance of enzyme distribution and behaviour. *New Phytologist*, **70**, 299–306.

McMillan, C. (1959) The role of ecotypic variation in the distribution of the central grassland of North America. *Ecological Monographs*, **29**, 285–308.

McMillen, G. G. and McClendon, J. H. (1979) Leaf angle: an adaptive feature of sun and shade leaves. *Botanical Gazette*, **140**, 437–42.

McNaughton, S. J. (1974) Development control of net productivity in *Typha latifolia* ecotypes. *Ecology*, **55**, 864–69.

McNaughton, S. J. (1985) Ecology of a grazing ecosystem: the Serengeti. *Ecological Monographs*, **55**, 259–94.

McPherson, J. K. and Muller, C. H. (1969) Allelopathic effects of *Adenostoma fasciculatum*, 'chamise', in the California chaparral. *Ecological Monographs*, **39**, 177–98.

McPherson, J. K., Chou, C. H. and Muller, C. H. (1971) Allelopathic constituents of the chaparral shrub *Adenostoma fasciculatum*. *Phytochemistry*, **10**, 2925–33.

McQueen, D. J., Post, J. R. and Mills, E. L. (1986) Trophic relationships in freshwater pelagic ecosystems. *Canadian Journal of Fisheries and Aquatic Sciences*, **43**, 1571–81.

McRill, M. and Sagar, G. R. (1973) Earthworms and seeds. *Nature*, **243**, 482.

Mabberley, D. J. (1992) *Tropical Rain Forest Ecology*, 2nd edn, Blackie, Glasgow.

Maberly, S. C. and Spence, D. H. N. (1983) Photosynthetic inorganic carbon use by freshwater plants. *Journal of Ecology*, **71**, 705–24.

Mackney, D. (1961) A podzol development sequence in oakwoods and heath in central England. *Journal of Soil Science*, **12**, 23–40.

Macan, T. T. (1961) Factors that limit the range of freshwater animals. *Biological Reviews*, **36**, 151–98.

Maddock, L. (1979) The 'migration' and grazing succession, in *Serengeti: Dynamics of an Ecosystem*, (eds A. R. E. Sinclair and M. Norton-Griffiths), University of Chicago Press, Chicago, pp. 104–29.

Magalhaes, A. C. and Angelocci, L. R. (1976) Sudden alterations in water balance associated with flower bud opening in coffee plants. *Journal of Horticultural Science*, **51**, 419–23.

Mahall, B. E. and Bormann, F. H. (1978) A quantitative description of the vegetative phenology of herbs in a northern hardwood forest. *Botanical Gazette*, **139**, 467–81.

Maher, W. J. (1973) *Birds: I. Population Dynamics. Canadian Committee for IBP, Matador Project Technical Report 34*, University of Saskatchewan, Saskatoon.

Maikawa, E. and Kershaw, K. A. (1976) Studies on lichen-dominated systems. XIX. The postfire recovery sequence of black spruce–lichen woodland in the Abitau Lake Region, N. W. T. *Canadian Journal of Botany*, **54**, 2679–87.

Main, A. R. (1981) Fire tolerance of heathland animals, in *Ecosystems of the World Vol. 9B. Heathlands and Related Shrublands*, (ed. R. L. Specht), Elsevier, Amsterdam, pp. 85–90.

Malaisse, F., Freson, R., Goffinet, G. and Malaisse-Mousset, M. (1975) Litter fall and litter breakdown in miombo, in *Tropical Ecological Systems*, (eds F. B. Golley and E. Medina), Springer, New York, pp. 137–52.

Malle, K.-G. (1985) Metallgehalt und schwebstoffgehalt im Rhein II. *Zeitschrift fur Wasser und Abwasser Forschung*, **18**, 207–9.

Mallik, A. U. and Newton, P. F. (1988) Inhibition of black spruce seedling growth by forest-floor substrates of central Newfoundland. *Forest Ecology and Management*, **23**, 273–83.

Malmer, N. (1986) Vegetational gradients in relation to environmental conditions in northwestern European mires. *Canadian Journal of Botany*, **64**, 375–83.

Malmer, N. (1988) Patterns in the growth and the accumulation of inorganic constituents in the *Sphagnum* cover on ombrotrophic bogs in Scandinavia. *Oikos*, **53**, 105–20.

Malmer, N. and Holm, E. (1984) Variations in the C/N-quotient of peat in relation to decomposition rate and age determination with ^{120}Pb. *Oikos*, **43**, 171–82.

Malmer, N. and Sjörs, H. (1955) Some determinations of elementary constituents in mire plants and peat. *Botaniska Notiser*, **108**, 46–80.

Mandelli, E. F. and Burkholder, P. R. (1966) Primary productivity in the Gerlache and Bransfield Straits of Antarctica. *Journal of Marine Research*, **24**, 15–27.

Mann, K. H. (1973) Seaweeds: their productivity and strategy for growth. *Science*, **182**, 975–81.

Mann, K. H. (1982) *Ecology of Coastal Waters*, University of California Press, Berkeley.

Mann, K. H. and Chapman, A. R. O. (1975) Primary production of marine macrophytes, in *Photosynthesis and Productivity in Different Environments*, (ed. J. P. Cooper), Cambridge University Press, Cambridge, pp. 207–23.

Mann, L. K., Johnson, D. W., West, D. C. *et al.* (1988) Effects of whole-tree and stem-only clearcutting on post harvest hydrologic losses, nutrient capital, and regrowth. *Forest Science*, **34**, 412–28.

Manokaran, N. and Kochummen, K. M. (1987) Recruitment, growth and mortality of tree species in a lowland dipterocarp forest in Peninsular Malaysia. *Journal of Tropical Ecology*, **3**, 315–30.

Marchand, P. J. and Roach, D. A. (1980) Reproductive strategies of pioneering alpine species: seed production, dispersal, and germination. *Arctic and Alpine Research*, **12**, 137–46.

Marchant, C. J. and Goodman, P. J. (1969a) Biological flora of the British Isles. *Spartina maritima* (Curtis) Fernald. *Journal of Ecology*, **57**, 287–91.

Marchant, C. J. and Goodman, P. J. (1969b) Biological flora of the British Isles. *Spartina alternifolia* Loisel. *Journal of Ecology*, **57**, 291–95.

Marcuzzi, G. (1979) *European Ecosystems*, Junk, The Hague.

Mares, M. A., Morello, J. and Goldstein, G. (1985) The Monte Desert and other subtropical semi-arid biomes of Argentina, with comments on their relation to North American arid areas, in *Ecosystems of the World Vol. 12A. Hot Desert and Arid Shrublands, A*, (eds M. Evanari, I. Noy-Meir and D. W. Goodall), Elsevier, Amsterdam, pp. 203–37.

Margaris, N. S. (1975) Effect of photoperiod on seasonal dimorphism of some mediterranean plants. *Berichte der Schweizerischen Botanischen Gesellschaft*, **85**, 96–102.

Margaris, N. S. (1976) Structure and dynamics in a phryganic (east Mediterranean) ecosystem. *Journal of Biogeography*, **3**, 249–59.

Margaris, N. S. (1977) Physiological and biochemical observations in seasonal dimorphic leaves of *Sarcopoterium spinosum* and *Phlomis fruticosa*. *Oecologia Plantarum*, **12**, 343–50.

Mark, A. F. (1965) Flowering, seeding, and seedling establishment of narrow-leaved snow tussock, *Chionochloa rigida*. *New Zealand Journal of Botany*, **3**, 180–93.

Mark, A. F. and Bliss, L. C. (1970) The high-alpine vegetation of Central Otago, New Zealand. *New Zealand Journal of Botany*, **8**, 381–451.

Marks, P. L. (1974) The role of pin cherry (*Prunus pensylvanica* L.) in the maintenance of stability in northern hardwood ecosystems. *Ecological Monographs*, **44**, 73–88.

Marks, P. L. (1975) On the relation between extension growth and sucessional status of deciduous trees of the northeastern United States. *Bulletin of the Torrey Botanical Club*, **102**, 172–7.

Marquis, D. A. (1975) Seed storage and germination under northern hardwood forests. *Canadian Journal of Forest Research*, **5**, 478–84.

Marsh, J. A. (1970) Primary production of reef-building calcareous red algae. *Ecology*, **51**, 255–63.

Marshall, J. K. (1965) *Corynephorus canescens* (L.) P. Beauv. as a model for the *Ammophila* problem. *Journal of Ecology*, **53**, 447–63.

Martin, A. R. H. (1966) The plant ecology of the Grahamstown Nature Reserve: II. Some effects of burning. *Journal of South African Botany*, **32**, 1–39.

Martin, N. J. and Holding, A. J. (1978) Nutrient availability and other factors limiting microbial activity in the blanket peat, in *Production Ecology of British Moors and Montane Grasslands*, (eds O. W. Heal and D. F. Perkins), Springer, Berlin, pp. 113–35.

Martineau, R. (1984) *Insects Harmful to Forest Trees*. Environment Canada: Multiscience Publications in cooperation with the Canadian Forestry Service, Ottawa.

Mason, B. J. (1976) Towards the understanding and prediction of climatic variations. *Quarterly Journal of the Royal Meteorological Society*, **102**, 473–98.

Mason, C. F. and Bryant, R. J. (1975) Production, nutrient content and decomposition of *Phragmites communis* Trin. and *Typha angustifolia* L. *Journal of Ecology*, **63**, 71–95.

Matthews, E. G. (1976) *Insect Ecology*, University of Queensland Press, St Lucia.

Matthews, J. D. (1963) Factors affecting the production of seed by forest trees. *Forestry Abstracts*, **24**, i–xiii.

Mattson, W. J., Lawrence, R. K., Haack, R. A., Herms, D. A. and Charles P.-J. (1988) Defensive strategies of woody plants against different insect-feeding guilds in relation to plant ecological strategies and intimacy of association with insects, in *Mechanisms of Woody Plant Defenses Against Insects*, (eds W. J. Mattson, J. Levieux, C. Bernard-Dagan), Springer, New York, pp. 3–38.

Maunder, W. J. (1971) Elements of New Zealand's climate, in *World Survey of Climatology Vol. 13. Climates of Australia and New Zealand*, (ed. J. Gentilli), Elsevier, Amsterdam, pp. 229–384.

May, L. H. (1960) The utilization of carbohydrate reserves in pasture plants after defoliation. *Herbage Abstracts*, **30**, 239–45.

May, R. M. (1976) Patterns in multi-species communities, in *Theoretical Ecology*, (ed. R. M. May), Blackwell Scientific Publications, Oxford, pp. 142–62.

Mayer, A. M. and Poljakoff-Mayber, A. (1982) *The Germination of Seeds*, 3rd edn, Pergamon, Oxford.

Mayhew, W. W. (1968) Biology of desert amphibians and reptiles, in *Desert Biology*, Vol. 1 (ed. G. W. Brown), Academic Press, New York, pp. 195–356.

Medina, E. (1971) Effect of nitrogen supply and light intensity during growth on the photosynthetic capacity and carboxydismutase activity of leaves of *Atriplex patula* ssp. *hastata. Carnegie Institution Yearbook 1970*, pp. 551–9.

Medina, E. (1982) Physiological ecology of neotropical savanna plants, in *Ecology of Tropical Savannas*, (eds B. J. Huntley and B. H. Walker), Springer, Berlin, pp. 308–35.

Medina, E. (1983) Adaptations of tropical trees to moisture stress, in *Ecosystems of the World Vol. 14A. Tropical Rain Forest Ecosystems Structure and Function*, (ed. F. B. Golley), Elsevier, Amsterdam, pp. 225–37.

Medina, E. Garcia, V. and Cuevas, E. (1990) Sclerophylly and oligotrophic environments, relationships between leaf structure, mineral nutrient content, and drought resistance in tropical rain forests of the upper Rio Negro region. *Biotropica*, **22**, 51–64.

Medway, Lord (1972) Phenology of a tropical rain forest in Malaya. *Biological Journal of the Linnean Society*, **4**, 117–46.

Meguro, H., Ito, K. and Fukushima, H. (1967) Ice flora (bottom type): a mechanism of primary production in Polar seas and the growth of diatoms in sea ice. *Arctic*, **20**, 114–33.

Meidner, H and Sheriff, D. W. (1976) *Water and Plants*, Wiley, New York.

Meigs, P. (1953) World distribution of arid and semi-arid homoclimates, in *Reviews of Research on Arid Zone Hydrology*, Arid Zone Programme, 1, UNESCO, Paris, pp. 203–10.

Meijer, W. (1959) Plantsociological analysis of montane rainforest near Tjibodas, West Java. *Acta Botanica Neerlandica*, **8**, 277–91.

Meinzer, F., Seymour, V. and Goldstein, G. (1983) Water balance in developing leaves of four tropical savanna woody species. *Oecologia*, **60**, 237–43.

Meinzer, F. C., Goldstein, G. H. and Rundel, P. W. (1985) Morphological changes along an altitude gradient and their consequences for an Andean giant rosette plant. *Oecologia*, **65**, 278–83.

Menaut, J. C. (1983) The vegetation of African savannas, in *Ecosystems of the World Vol. 13. Tropical Savannas*, (ed. F. Bourlière), Elsevier, Amsterdam, pp. 109–49.

Menaut, J. C. and César, J. (1979) Structure and primary productivity of Lamto savannas, Ivory Coast. *Ecology*, **60**, 1197–210.

Mendelssohn, I. A., McKee, K. L. and Patrick, W. H. (1981) Oxygen deficiency in *Spartina alternifolia* roots: metabolic adaptation to anoxia. *Science*, **214**, 439–41.

Mendelssohn, I. A. and Postek, M. T. (1982) Elemental analysis of deposits on the roots of *Spartina alternifolia* Loisel. *American Journal of Botany*, **69**, 904–12.

Menzel, D. W. and Spaeth, J. P. (1962) Occurrence of vitamin B$_{12}$ in the Sargasso Sea. *Limnology and Oceanography*, **7**, 151–62.

Meybeck, M. (1979) Concentrations des eaux fluviales en éléments majeurs et apports en solution aux océans. *Revue de Géologie Dynamique et de Géographie Physique*, **21**, 215–46.

Meybeck, M., Chapman, D. V. and Helmer, R. (eds) (1989) *Global Freshwater Quality*, Blackwell, Oxford.

Meyer, J. L. and Schultz, E. T. (1985) Tissue condition and growth rate of corals associated with schooling fish. *Limnology and Oceanography*, **30**, 157–66.

Mielke, H. W. (1989) *Patterns of Life*. Unwin Hyman, London.

Mierle, G., Clark, K. and France, R. (1986) The impact of acidification on aquatic biota in North America: a comparison of field and laboratory results. *Water, Air, and Soil Pollution*, **31**, 593–604.

Millar, C. S. (1974) Decomposition of coniferous leaf litter, in *Biology of Plant Litter Decomposition*, (eds C. H. Dickinson and G. J. F. Pugh), Academic Press, London, pp. 105–28.

Millar, J. B. (1976) Wetland Classification in Western Canada: a Guide to Marshes and Shallow Open Water Wetlands in the Grasslands and Parklands of the Prairie Provinces. *Canadian Wildlife Service Report No. 37*, Environment Canada, Ottawa.

Miller, A. (1976) The climate of Chile, in *World Survey of Climatology Vol. 12. Climates of Central and South America*. (ed. W. Schwerdtfeger), Elsevier, Amsterdam, pp. 113–45.

Miller, G. A. (1986) Pubescence, floral temperature and fecundity in species of *Puya* (Bromeliaceae) in the Ecuadorian Andes. *Oecologia*, **70**, 155–60.

Miller, P. C. (1972) Bioclimate, leaf temperature, and primary production in red mangrove canopies in South Florida. *Ecology*, **53**, 22–45.

Miller, P. C. and Ng, E. (1977) Root:shoot biomass ratios in shrubs in southern California and central Chile. *Madroño*, **24**, 215–23.

Miller, P. C. and Poole, D. K. (1979) Patterns of water use by shrubs in southern California. *Forest Science*, **25**, 84–98.

Miller, P. C., Bradbury, D. E., Hajek *et al.* (1977) Past and present environment, in *Convergent Evolution in Chile and California*, (ed. H. A. Mooney), Dowden, Hutchinson and Ross, Stroudsburg, pp. 27–72.

Mills, E. L. and Forney, J. L. (1988) Trophic dynamics and development of freshwater pelagic food webs, in *Complex Interactions in Lake Communities*, (ed. S. R. Carpenter), Springer, New York, pp. 11–30.

Mills, J. N. (1986) Herbivores and early postfire succession in southern California chaparral. *Ecology*, **67**, 1637–49.

Minnich, R. A. (1983) Fire mosaics in southern California and northern Baja California. *Science*, **219**, 1287–94.

Mispagel, M. E. (1978) The ecology and bioenergetics of the acridid grasshopper, *Bootettix punctatus* on creosotebush, *Larrea tridentata*, in the northern Mojave Desert. *Ecology*, **59**, 779–88.

Misra, R. (1983) Indian savannas, in *Ecosystems of the World Vol. 13. Tropical Savannas*, (ed. F. Bourlière), Elsevier, Amsterdam, pp. 151–66.

Mitchell, J. E. and Pfadt, R. E. (1974) A role of grasshoppers in a shortgrass prairie ecosystem. *Environmental Entomology*, **3**, 358–60.

Mitsch, W. J. and Gosselink, J. G. (1986) *Wetlands*, Van Nostrand Reinhold, New York.

Miyawaki, A. and Sasaki, Y. (1985) Floristic changes in the *Castanopsis cuspidata* var. *sieboldii*-forest communities along the Pacific Ocean coast of the Japanese islands. *Vegetatio*, **59**, 225–34.

Moebus, K. and Johnson, K. M. (1974) Exudation of dissolved organic carbon by brown algae. *Marine Biology*, **26**, 117–25.

Monk, C. D. (1966) An ecological significance of evergreenness. *Ecology*, **47**, 504–5.

Monson, R. K. and Williams, G. J. (1982) A correlation between photosynthetic temperature adaptation and seasonal phenology patterns in the shortgrass prairie. *Oecologia*, **54**, 58–62.

Monson, R. K., Littlejohn, R. O. and Williams, G. J. (1983) Photosynthetic adaptation to temperature in four species from the Colorada shortgrass steppe: a physiological model for coexistence. *Oecologia*, **58**, 43–51.

Monson, R. K., Sackschewsky, M. R. and Williams, G. J. (1986) Field measurements of photosynthesis, water-use efficiency, and growth in *Agropyron smithii* (C_3) and *Bouteloua gracilis* (C_4) in the Colorada shortgrass steppe. *Oecologia*, **68**, 400–9.

Montague, C. L., Bunker. S. M., Haines, E. B., Pace, M. L. and Wetzel, R. L. (1981) Aquatic macroconsumers, in *The Ecology of a Salt Marsh*, (eds L. R. Pomeroy and R. G. Wiegert), Springer, New York, pp. 69–85.

Montenegro, G., Rivera, O. and Bas, F. (1978) Herbaceous vegetation in the Chilean matorral. *Oecologia*, **36**, 237–44.

Montenegro, G., Hoffmann, A. J., Aljaro, M. E. and Hoffmann, A. E. (1979) *Satureja gilliesii*, a poikilohydric shrub from the Chilean mediterranean vegetation. *Canadian Journal of Botany*, **57**, 1206–13.

Montes, R. and Medina, E. (1977) Seasonal changes in nutrient content of leaves of savanna trees with different ecological behaviour. *Geo-Eco-Trop*, **4**, 295–307.

Montgomery, R. F. and Askew, G. P. (1983) Soils of tropical savannas, in *Ecosystems of the World Vol. 13. Tropical Savannas*, (ed. F. Bourlière), Elsevier, Amsterdam, pp. 63–78.

Mooney, H. A. (1981) Primary production in mediterranean-climate regions, in *Ecosystems of the World Vol 11. Mediterranean-type Shrublands*, (eds F. di Castri, D. W. Goodall and R. L. Specht), Elsevier, Amsterdam, pp. 249–55.

Mooney, H. A. (1983) Carbon-gaining capacity and allocation patterns of mediterranean-climate plants, in *Mediterranean-type Ecosystems: The Role of Nutrients*, (eds F. J. Kruger, D. T. Mitchell and J. U. M. Jarvis), Springer, Berlin, pp. 103–19.

Mooney, H. A. and Chu, C. (1974) Seasonal carbon allocation in *Heteromeles arbutifolia*, a California evergreen shrub. *Oecologia*, **14**, 295–306.

Mooney, H. A. and Dunn, E. L. (1970) Convergent evolution of mediterranean-climate evergreen sclerophyll shrubs. *Evolution*, **24**, 292–303.

Mooney, H. A. and Gulmon, S. L. (1979) Environmental and evolutionary constraints on the photosynthetic characteristics of higher plants, in *Topics in Plant Population Biology*, (eds O. T. Solbrig, S. Jain, G. B. Johnson and P. H. Raven), Columbia University Press, New York, pp. 316–37.

Mooney, H. A. and Miller, P. C. (1985) Chaparral, in *Physiological Ecology of North American Plant Communities*, (eds B. F. Chabot and H. A. Mooney), Chapman & Hall, New York, pp. 213–31.

Mooney, H. A. and Rundel, P. W. (1979) Nutrient relations of the evergreen shrub, *Adenostoma fasciculatum*, in the California chaparral. *Botanical Gazette*, **140**, 109–13.

Mooney, H. A. and Strain, B. R. (1964) Bark photosynthesis in ocotillo. *Madroño*, **17**, 230–3.

Mooney, H. A., Björkman, O. and Collatz, G. J. (1978) Photosynthetic acclimation to temperature in the desert shrub *Larrea divaricata*. *Plant Physiology*, **61**, 406–10.

Mooney, H. A., Ehleringer, J. and Björkman, O. (1977a) The energy balance of leaves of the evergreen shrub *Atriplex hymenelytra*. *Oecologia*, **29**, 301–10.

Mooney, H. A., Harrison, A. T. and Morrow, P. A. (1975) Environmental limitations of photosynthesis on a California evergreen shrub. *Oecologia*, **19**, 293–301.

Mooney, H. A., Parsons, D. J. and Kummerow, J. (1974) Plant development in mediterranean climates, in *Phenology and Seasonality Modelling*, (ed. H. Lieth), Springer, New York, pp. 255–67.

Mooney, H. A., Troughton, J. H. and Berry, J. A. (1974) Arid climates and photosynthetic systems. *Carnegie Institution Year Book 73*, 793–805.

Mooney, H. A., Dunn, E. L., Shropshire, F. and Song, L. (1970) Vegetation comparisons between the mediterranean climatic areas of California and Chile. *Flora*, **159**, 480–96.

Mooney, H. A., Kummerow, J., Johnson, A. W. *et al.* (1977b) The producers – their resources and adaptive responses, in *Convergent Evolution in Chile and California*, (ed. H. A. Mooney), Dowden, Hutchinson and Ross, Stroudsburg, pp. 85–143.

Moore, H. E. and Uhl, N. W. (1973) The monocotyledons: their evolution and comparative biology: IV. Palms and the origin and evolution of monocotyledons. *Quarterly Review of Biology*, **48**, 414–36.

Moore, J. J., Dowding, P. and Healy, B. (1975) Glenamoy, Ireland, in *Structure and Function of Tundra Ecosystems*, (eds T. Rosswall and O. W. Heal), Swedish Natural Science Research Council, Stockholm, pp. 321–43.

Moore, L. A. and Willson, M. F. (1982) The effect of microhabitat, spatial distribution, and display size on dispersal of *Lindera benzoin* by avian frugivores. *Canadian Journal of Botany*, **60**, 557–60.

Moore, M. R. and Vankat, J. L. (1986) Responses of the herb layer to the gap dynamics of a mature beech–maple forest. *American Midland Naturalist*, **115**, 336–47.

Moore, R. M. (1970) South-eastern temperate woodlands and grasslands, in *Australian Grasslands*, (ed. R. M. Moore), Australian National University Press, Canberra, pp. 169–90.

Moore, T. R. (1984) Litter decomposition in a subarctic spruce–lichen woodland, eastern Canada. *Ecology*, **65**, 299–308.

Moreau, R. E. (1972) *The Palaearctic–African Bird Migration Systems*, Academic Press, London.

Morgan, D. C. and Smith, H. (1981) Non-photosynthetic responses to light quality, in *Encyclopedia of Plant Physiology Vol. 12A. Physiological Plant Ecology I. Responses to the Physical Environment*, Springer, Berlin, pp. 109–34.

Mori, S. A., Boom, B. M., de Carvalho, A. M. and dos Santos, T. S. (1983) Southern Bahian moist forest. *The Botanical Review*, **49**, 155–232.

Morison, C. G. T., Hoyle, A. C. and Hope-Simpson, J. F. (1948) Tropical soil–vegetation catenas and mosaics. A case study in the south-western part of the Anglo-Egyptian Sudan. *Journal of Ecology*, **36**, 1–84.

Morris, J. W., Bezuidenhout, J. J. and Furniss, P. R. (1982) Litter decomposition, in *Ecology of Tropical Savannas*, (eds B. J. Huntley and B. H. Walker), Springer, Berlin, pp. 535–53.

Morris, J. T. (1991) Effects of nitrogen loading on wetland ecosystems with particular reference to atmospheric deposition. *Annual Review of Ecology and Systematics*, **22**, 257–79.

Morrow, P. A. (1983) The role of sclerophyllous leaves in determining insect grazing damage, in *Mediterranean-type Ecosystems: The Role of Nutrients*, (eds F. J. Kruger, D. T. Mitchell and J. U. M. Jarvis), Springer, Berlin, pp. 509–24.

Moss, B. (1980) *Ecology of Fresh Waters*, Blackwell Scientific Publications, Oxford.

Mott, J. J. (1972) Germination studies on some annual species from an arid region of Western Australia. *Journal of Ecology*, **60**, 293–304.

Mott, J. J. and Andrew, M. H. (1985) The effect of fire on the population dynamics of native grasses in tropical savannas of north-west Australia. *Proceedings of the Ecological Society of Australia*, **13**, 231–9.

Mueller, I. M. (1941) An experimental study of rhizomes of certain prairie plants. *Ecological Monographs*, **11**, 165–88.

Mugasha, A. G., Pluth, D. J., Higginbotham, K. O. and Takyi, S. K. (1991) Foliar responses of black spruce to thinning and fertilization on a drained shallow peat. *Canadian Journal of Forest Research*, **21**, 152–63.

Muller, C. H. (1940) Plant succession in the *Larrea–Flourensia* climax. *Ecology*, **21**, 206–12.

Muller, C. H. (1953) The association of desert annuals with shrubs. *American Journal of Botany*, **40**, 53–60.

Muller, W. H. and Muller, C. H. (1956) Association patterns involving desert plants that contain toxic products. *American Journal of Botany*, **43**, 354–61.

Mullette, K. J. (1978) Studies of the lignotubers of *Eucalyptus gummifera* (Gaertn. and Hochr.). I. The nature of the lignotuber. *Australian Journal of Botany*, **26**, 9–13.

Mulroy, T. W. and Rundel, P. W. (1977) Annual plants: adaptations to desert environments. *BioScience*, **27**, 109–14.

Mundie, J. H. (1974) Optimization of the salmonid nursery stream. *Journal of the Fisheries Research Board of Canada*, **31**, 1827–37.

Munns, R., Greenway, H. and Kirst, G. O. (1983) Halotolerant eukaryotes, in *Physiological Plant Ecology III Vol. 12C, Responses to the Chemical and Biological Environment*, (eds O. L. Lange, P. S. Nobel, C. B. Osmond and H. Ziegler), Springer, Berlin, pp. 59–135.

Murphy, P. G. (1975) Net primary productivity in tropical terrestrial ecosystems, in *Primary Productivity of the Biosphere*, (eds H. Lieth and R. H. Whittaker), Springer, New York, pp. 217–31.

Murphy, P. G. and Lugo, A. E. (1986) Structure and biomass of a subtropical dry forest in Puerto Rico. *Biotropica*, **18**, 89–96.

Muscatine, L. and D'Elia, C. F. (1978) The uptake, retention, and release of ammonium by reef corals. *Limnology and Oceanography*, **23**, 725–34.

Mutch, R. W. (1970) Wildland fires and ecosystems – a hypothesis. *Ecology*, **51**, 1046–51.

Myers, J. H. and Williams, K. S. (1984) Does tent caterpillar attack reduce the food quality of red alder foliage. *Oecologia*, **62**, 74–9.

Myers, N. (1980) The present status and future prospects of tropical moist forests. *Environmental Conservation*, **7**, 101–14.

Myers, N. (1990) The biodiversity challenge: Expanded hot-spots analysis. *The Environmentalist*, **10**, 243–55.

Nahal, I. (1981) The mediterranean climate from a biological viewpoint, in *Ecosystems of the World Vol. 11. Mediterranean-type Shrublands*, (eds F. di Castri, D. W. Goodall and R. L. Specht), Elsevier, Amsterdam, pp. 63–86.

Nauwerck, A., Duncan, A., Hillbricht-Ilkowska, A. and Larsson, P. (1980) Zooplankton, in *The Functioning of Freshwater Ecosystems*, (eds E. D. Le Cren and R. H. Lowe-McConnell), Cambridge University Press, Cambridge, pp. 251–85.

Naveh, Z. (1974) The ecology of fire in Israel. *Tall Timbers Fire Ecology Conference, Proceedings*, **13**, 131–70.

Naveh, Z. (1975) The evolutionary significance of fire in the Mediterranean region. *Vegetatio*, **29**, 199–208.

Naveh, Z. and Whittaker, R. H. (1979) Structural and floristic diversity of shrublands and woodlands in northern Israel and other Mediterranean areas. *Vegetatio*, **41**, 171–90.

Navez, A. E. (1930) On the distribution of tabular roots in *Ceiba* (Bombacaceae). *Proceedings of the National Academy of Sciences*, **16**, 339–44.

Neely, R. K. and Baker, J. L. (1989) Nitrogen and phosphorus dynamics and the fate of agricultural runoff, in *Northern Prairie Wetlands*, (ed. A. van der Valk), Iowa State University Press, Ames, pp. 93–131.

Neely, R. K. and Davis, C. B. (1985) Nitrogen and phosphorus fertilization of *Sparganium eurycarpum* Engelm. and *Typha glauca* Godr. stands. II. Emergent plant decomposition. *Aquatic Botany*, **22**, 363–75.

Neill, W. E. (1975) Experimental studies of microcrustacean competition, community composition and efficiency of resource utilization. *Ecology*, **56**, 809–26.

Nelson, J. F. and Chew, R. M. (1977) Factors affecting seed reserves in the soil of a Mojave Desert ecosystem, Rock Valley, Nye County, Nevada. *American Midland Naturalist*, **97**, 300–320.

Nemoto, T. and Harrison, G. (1981) High latitude ecosystems, in *Analysis of Marine Ecosystems*, (ed. A. R. Longhurst), Academic Press, London, pp. 95–126.

Ng, F. S. P. (1966) Age at first flowering in dipterocarps. *The Malayan Forester*, **29**, 290–5.

Ng, F. S. P. (1978) Strategies of establishment in Malayan forest trees, in *Tropical Trees as Living Systems*, (eds P. B. Tomlinson and M. H. Zimmermann), Cambridge University Press, Cambridge, pp. 129–62.

Ng, F. S. P. (1980) Germination ecology of Malaysian woody plants. *The Malaysian Forester*, **43**, 406–37.

Ng, F. S. P. and Loh, H. S. (1974) Flowering-to-fruiting periods of Malaysian trees. *The Malaysian Forester*, **37**, 127–32.

Nichols, H. (1975) *Palynological and Paleoclimatic Study of the Late Quaternary Displacements of the Boreal Forest–Tundra*

Ecotone in Keewatin and Mackenzie, N. W. T., Canada. Occasional Paper 15, Institute of Arctic and Alpine Research, Boulder.

Nichols, H. (1976) Historical aspects of the northern Canadian treeline. *Arctic*, **29**, 38–47.

Nicholson, D. I. (1965) A study of virgin forest near Sandakan North Borneo. *Proceedings of the Symposium on Humid Tropics Vegetation*, Unesco, Kuching, pp. 67–87.

Nicholson, P. H. (1981) Fire and the Australian aborigine – an enigma, in *Fire and the Australian Biota*, (eds A. M. Gill, R. H. Groves and I. R. Noble), Australian Academy of Science, Canberra, pp. 55–76.

Nishioka, M. and Kirita, H. (1978) Decomposition cycles, in *JIBP Synthesis Vol. 18. Biological Production in a Warm-temperate Evergreen Oak Forest of Japan*, (eds T. Kira, Y. Ono and T. Hosokawa), University of Tokyo Press, Tokyo, pp. 231–88.

Nix, H. A. (1983) Climate of tropical savannas, in *Ecosystems of the World Vol. 13. Tropical Savannas*, (ed. F. Bourlière), Elsevier, Amsterdam, pp. 37–62.

Njoku, E. (1963) Seasonal periodicity in the growth and development of some forest trees in Nigeria I. Observations on mature trees. *Journal of Ecology*, **51**, 617–24.

Nobel, P. S. (1976) Water relations and photosynthesis of a desert CAM plant, *Agave deserti*. *Plant Physiology*, **58**, 576–82.

Nobel, P. S. (1977) Water relations and photosynthesis of a barrel cactus, *Ferocactus acanthodes*, in the Colorado desert. *Oecologia*, **27**, 117–33.

Nobel, P. S. (1980a) Interception of photosynthetically active radiation by cacti of different morphology. *Oecologia*, **45**, 160–66.

Nobel, P. S. (1980b) Morphology, surface temperatures, and northern limits of columnar cacti in the Sonoran Desert. *Ecology*, **61**, 1–7.

Nobel, P. S. (1980c) influences of minimum stem temperatures on ranges of cacti in southwestern United States and central Chile. *Oecologia*, **47**, 10–5

Nobel, P. S. (1981) Spacing and transpiration of various sized clumps of a desert grass, *Hilaria rigida*. *Journal of Ecology*, **69**, 735–42.

Nobel, P. S. (1982a) Orientations of terminal cladodes of platyopuntias. *Botanical Gazette*, **143**, 219–24.

Nobel, P. S. (1982b) Low-temperature tolerance and cold hardening of cacti. *Ecology*, **63**, 1650–6.

Nobel, P. S. (1985) Desert succulents, in *Physiological Ecology of North American Plant Communities*, (eds B. F. Chabot and H. A. Mooney), Chapman & Hall, New York, pp. 181–97.

Nobel, P. S. (1987) Photosynthesis and productivity of desert plants, in *Progress in Desert Research*, (eds L. Berkofsky and M. G. Wurtele), Rowman and Littlefield, New Jersey, pp. 41–66.

Nobel, P. S. and Smith, S. D. (1983) High and low temperature tolerances and their relationships to distribution of agaves. *Plant, Cell and Environment*, **6**, 711–19.

Noirot, C. (1970) The nests of termites, in *Biology of Termites*, (eds K. Krishna and F. M. Weesner), Academic Press, New York, pp. 73–125.

Norton, B. E. and McGarity, J. W. (1965) The effect of burning of native pasture on soil temperature in northern New South Wales. *Journal of the British Grassland Society*, **20**, 101–5.

Noy-Meir, I. (1973) Desert ecosystems: environment and producers. *Annual Review of Ecology and Systematics*, **4**, 25–51.

Noy-Meir, I. (1974) Desert ecosystems: higher trophic levels. *Annual Review of Ecology and Systematics*, **5**, 195–214.

Numata, M., Miyawaki, A. and Itow, D. (1972) Natural and semi- natural vegetation in Japan. *Blumea*, **20**, 435–81.

Nye, P. H. (1955) Some soil-forming processes in the humid tropics. IV. The action of the soil fauna. *Journal of Soil Science*, **6**, 73–83.

Nye, P. H. and Tinker, P. B. (1977) *Solute Movement in the Soil–Root System*, Blackwell, Oxford.

Nye, P. H. and Greenland, D. J. (1960) *The Soil Under Shifting Cultivation*. Technical Communication No. 51, Commonwealth Agricultural Bureaux, Farnham Royal.

O'Brien, R. T. (1978) Proteolysis and ammonification in desert soils, in *Nitrogen in Desert Ecosystems*, (eds N.E West and J. J. Skujiņš), Dowden, Hutchinson and Ross, Stroudsburg, pp. 50–59.

Ode, D. J., Tieszen, L. L. and Lerman, J. C. (1980) The seasonal contribution of C_3 and C_4 plant species to primary production in a mixed prairie. *Ecology*, **61**, 1304–11.

Odum, E. P. (1969) The strategy of ecosystem development. *Science*, **164**, 262–70.

Odum, E. P. (1971) *Fundamentals of Ecology*, 3rd edn, Saunders, Philadelphia.

Odum, E. P. (1985) Trends expected in stressed ecosystems. *BioScience*, **35**, 419–22.

Odum, H. T. and Odum, E. P. (1955) Trophic structure and productivity of a windward coral reef community on Eniwetok Atoll. *Ecological Monographs*, **25**, 291–320.

Odum, W. E. and Heald, E. J. (1975) The detritus-based food web of an estuarine mangrove community, in *Estuarine Research, Vol. 1. Chemistry, Biology and the Estuarine System*, (ed. L. E. Cronin), Academic Press, New York, pp. 265–86.

Oechel, W. C. and Lawrence, W. (1981) Carbon allocation and utilization, in *Resource Use by Chaparral and Matorral*, (ed. P. C. Miller), Springer, New York, pp. 185–235.

Oechel, W. C. and Sveinbjörnsson, B. (1978) Primary production processes in arctic bryophytes at Barrow, Alaska, in *Vegetation and Production Ecology of an Alaskan Arctic Tundra*, (ed. L. L. Tieszen), Springer, New York, pp. 269–98.

Oechel, W. C. and van Cleve, K. (1986) The role of bryophytes in nutrient cycling in the taiga, in *Forest Ecosystems in the Alaskan Taiga*, (eds K. van Cleve, F. S. Chapin, P. W. Flanagan, L. A. Viereck and C. T. Dyrness), Springer, New York, pp. 121–37.

Oechel, W. C., Lawrence, W., Mustafa, J. and Martinez, J. (1981) Energy and carbon acquisition, in *Resource Use by Chaparral and Matorral*, (ed. P. C. Miller), Springer, New York, pp. 151–83.

Ogden, J. C. and Lobel, P. S. (1978) The role of herbivorous fishes and urchins in coral reef communities. *Environmental Biology of Fishes*, **3**, 49–63.

Ohiagu, C. E. (1979) Nest and soil populations of *Trinervitermes* spp. with particular reference to *T. geminatus* (Wasmann), (Isoptera), in southern Guinea savanna near Mokwa, Nigeria. *Oecologia*, **40**, 167–78.

Ohiagu, C. E. and Wood, T. G. (1979) Grass production and decomposition in southern Guinea savanna, Nigeria. *Oecologia*, **40**, 155–65.

Ohle, W. (1956) Bioactivity, production, and energy utilization of lakes. *Limnology and Oceanography*, **1**, 139–49.

Ohlson, M. and Dahlberg, B. (1991) Rate of peat increment in hummock and lawn communities on Swedish mires during the last 150 years. *Oikos*, **61**, 369–78.

Okusanya, O. T. (1979a) An experimental investigation into the ecology of some maritime cliff species. II. Germination studies. *Journal of Ecology*, **67**, 293–304.

Okusanya, O. T. (1979b) An experimental investigation into the ecology of some maritime cliff species. III. Effect of sea water on growth. *Journal of Ecology*, **67**, 579–90.

Olson, J. S. (1975) Productivity of forest ecosystems, in *Productivity of World Ecosystems*, National Academy of Sciences, Washington, pp. 33–43.

Olson, R. J. (1980) Nitrate and ammonium uptake in Antarctic waters. *Limnology and Oceanography*, **25**, 1064–74.

Ondok, J. P. and Gloser, J. (1978) Net photosynthesis and dark respiration in a stand of *Phragmites communis* Trin. calculated by means of a model II. Results. *Photosynthetica*, **12**, 337–43.

Onuf, C. P., Teal, J. M. and Valiela, I. (1977) Interactions of nutrients, plant growth and herbivory in a mangrove ecosystem. *Ecology*, **58**, 514–26.

Onyeanusi, A. E. (1989) Large herbivore grass offtake in Masai Mara National Reserve: implications for the Serengeti–Mara migrants. *Journal of Arid Environments*, **16**, 203–9.

Open University Course Team (1989a) *Ocean Chemistry and Deep-sea Sediments*, Pergamon Press, Oxford.

Open University Course Team (1989b) *Seawater: Its Composition, Properties and Behaviour*, Pergamon Press, Oxford.

Opler, P. A. and Bawa, K. S. (1978) Sex ratios in tropical forest trees. *Evolution*, **32**, 812–21.

Opler, P. A., Baker, H. G. and Frankie, G. W. (1980a) Plant reproductive characteristics during secondary succession in Neotropical lowland forest ecosystems. *Biotropica*, **12** (supplement), 40–6.

Opler, P. A., Frankie, G. W. and Baker, H. G. (1976) Rainfall as a factor in the release, timing, and synchronization of anthesis by tropical trees and shrubs. *Journal of Biogeography*, **3**, 231–6.

Opler, P. A., Frankie, G. W. and Baker, H. G. (1980b) Comparative phenological studies of treelet and shrub species in tropical wet and dry forests in the lowlands of Costa Rica. *Journal of Ecology*, **68**, 167–88.

Oppenheimer, J. R. (1982) *Cebus capucinus*: home range, population dynamics and interspecific relationships, in *The Ecology of a Tropical Forest*, (eds E. G. Leigh, A. S. Rand and D. M. Windsor), Smithsonian Institution Press, Washington, pp. 253–72.

Orians, G. H., Gates, R. G., Mares, M. A. *et al.* (1977) Resource utilization systems, in *Convergent Evolution in Warm Deserts*, (eds G. H. Orians and O. T. Solbrig), Dowden, Hutchinson and Ross, Stroudsburg, pp. 164–224.

Orshan, G. (1963) Seasonal dimorphism of desert and mediterranean chamaephytes and its significance as a factor in their water economy, in *The Water Relations of Plants*, (eds A. J. Rutter and F. H. Whitehead), Wiley, New York, pp. 206–22.

Orshan, G. (1972) Morphological and physiological plasticity in relation to drought. *Wildland Shrubs – Their Biology and Utilization, Proceedings International Symposium, Utah State University*, 245–54.

Orshan, G. (1983) Approaches to the definition of mediterranean growth forms, in *Mediterranean-type Ecosystems: The Role of Nutrients*, (eds F. J. Kruger, D. T. Mitchell and J. U. M. Jarvis), Springer, Berlin, pp. 86–100.

Orshan, G. (1986) The deserts of the Middle East, in *Ecosystems of the World Vol. 12B. Hot Deserts and Arid Shrublands, B*, (eds M. Evanari, I. Noy-Meir and D. W. Goodall), Elsevier, Amsterdam, pp. 1–28.

Osburn, W. S. (1974) Radioecology, in *Arctic and Alpine Environments*, (eds J. D. Ives and R. G. Barry), Methuen, London, pp. 875–903.

Osmond, C. B. (1974) Leaf anatomy of Australian saltbushes in relation to photosynthetic pathways. *Australian Journal of Botany*, **22**, 39–44.

Osmond, C. B. (1979) Ion uptake, transport and excretion, in *Arid-land Ecosystems: Structure, Functioning and Management*, Vol. 1 (eds D. W. Goodall, R. A. Perry and K. M. W. Howes), Cambridge University Press, Cambridge, pp. 607–25.

Østbye, E., Berg, A., Blehr, O. *et al.* (1975) Hardangervidda, Norway, in *Ecological Bulletins 20, Structure and Function of Tundra Ecosystems*, (eds T. Rosswall and O. W. Heal), Swedish Natural Science Research Council, Stockholm, pp. 225–64.

Ovington, J. D. and Olson, J. S. (1970) Biomass and chemical content of El Verde lower montane rain forest plants, in *A Tropical Rain Forest: A Study of Irradiation and Ecology at El Verde, Puerto Rico*, (eds H. T. Odum and R. F. Pigeon), United States Atomic Energy Commision, Washington, pp. H53–75.

Owadally, A. W. (1979) The dodo and the tambalacoque tree. *Science*, **203**, 1363–4.

Owen, J. (1974) A contribution to the ecology of the African baobab (*Adansonia digitata* L.). *Savanna*, **3**, 1–12.

Owensby, C. E., Hyde, R. M. and Anderson, K. L. (1970) Effects of clipping and supplemental nitrogen and water on loamy upland bluestem range. *Journal of Range Management*, **23**, 341–6.

Owen-Smith, N. (1982) Factors influencing the consumption of plant products by large herbivores, in *Ecology of Tropical Savannas*, (eds B. J. Huntley and B. H. Walker), Springer, Berlin, pp. 359–404.

Paasche, E. (1980) Silicon, in *The Physiological Ecology of Phytoplankton*, (ed. I. Morris), University of California Press, Berkeley, pp. 259–84.

Paasche, E. (1988) Pelagic primary production in nearshore waters, in *Nitrogen Cycling in Coastal Marine Environments*, (eds T. H. Blackburn and J. Sørensen), Wiley, Chichester, pp. 33–57.

Pace, M. L., Shimmel, S. and Darley, W. M. (1979) The effect of grazing by a gastropod, *Nassarius obsoletus*, on the benthic microbial community of a salt marsh mudflat. *Estuarine and Coastal Marine Science*, **9**, 121–34.

Packer, J. G. (1974) Differentiation and dispersal in alpine floras. *Arctic and Alpine Research*, **6**, 117–28.

Paerl, H. W. and Kellar, P. E. (1978) Significance of bacterial–*Anabaena* (Cyanophyceae) associations with respect to N_2 fixation in freshwater. *Journal of Phycology*, **14**, 254–60.

Paillet, F. L. (1984) Growth-form and ecology of American chestnut sprout clones in northeastern Massachusetts. *Bulletin of the Torrey Botanical Club*, **111**, 316–28.

Pakarinen, P. (1978) Production and nutrient ecology of three *Sphagnum* species in southern Finnish raised bogs. *Annales Botanici Fennici*, **15**, 15–26.

Parker, V. T. and Kelly, V. R. (1989) Seed banks in California chaparral and other mediterranean climate shrublands, in *Ecology of Soil Seed Banks*, (eds M. A. Leck, V. T. Parker and R. L. Simpson), Academic Press, San Diego, pp. 231–55.

Paton, D. F. and Hosking, W. J. (1970) Wet temperate forests and heaths, in *Australian Grasslands*, (ed. R. M. Moore), Australian National University Press, Canberra, pp. 141–58.

Patric, J. H. and Hanes, T. L. (1964) Chaparral succession in a San Gabriel Mountain area of California. *Ecology*, **45**, 353–60.

Patrick, W. H. and Mahapatra, I. C. (1968) Transformation and availability to rice of nitrogen and phosphorus in waterlogged soils. *Advances in Agronomy*, **20**, 323–359.

Patriquin, D. and Knowles, R. (1972) Nitrogen fixation in the rhizosphere of marine angiosperms. *Marine Biology*, **16**, 49–58.

Patten, D. T. and Dinger, B. E. (1969) Carbon dioxide exchange patterns of cacti from different environments. *Ecology*, **50**, 686–8.

Patton, W. K. (1976) Animal associates of living reef corals, in *Biology and Geology of Coral Reefs*, Vol. 3 (eds O. A. Jones and R. Endean), Academic Press, New York, pp. 1–36.

Paul, E. A., Clark, F. E. and Biederbeck, V. O. (1979) Micro-organisms, in *Grassland Ecosystems of the World: Analysis of Grasslands and their Uses*, (ed. R. T. Coupland), Cambridge University Press, Cambridge, pp. 87–96.

Payette, S. and Filion, L. (1985) White spruce expansion at the tree line and recent climatic change. *Canadian Journal of Forest Research*, **15**, 241–51.

Payette, S., Deshaye, J. and Gilbert, H. (1982) Tree seed populations at the treeline in Rivière aux Feuilles area, northern Quebec, Canada. *Arctic and Alpine Research*, **14**, 215–21.

Pearsall, W. H. (1956) Two blanket-bogs in Sutherland. *Journal of Ecology*, **44**, 493–516.

Pearsall, W. H. (1950) *Mountains and Moorlands*, Collins, London.

Pearse, P. H., Lang, A. J. and Todd, K. L. (1986) The backlog of unstocked forest land in British Columbia and the impact of reforestation programs. *The Forestry Chronicle*, **62**, 514–21.

Peckarsky, B. L. (1980) Predator–prey interactions between stoneflies and mayflies: behavioural observations. *Ecology*, **61**, 932–43.

Peden, D. G., van Dyne, G. M., Rice, R. W. and Hansen, R. M. (1974) The trophic ecology of *Bison bison* L. on shortgrass plains. *Journal of Applied Ecology*, **11**, 489–97.

Pederson, R. L. and van der Valk, A. G. (1984) Vegetation change and seed banks in marshes: ecological and management implications. *Transactions of the North American Wildlife and Natural Resources Conference*, **49**, 271–80.

Pedrós-Alió, C. (1989) Toward an autecology of bacterioplankton, in *Plankton Ecology*, (ed. U. Sommer), Springer, Berlin, pp. 297–336.

Pedrós-Alió, C. and Brock, T. D. (1983) The impact of zooplankton feeding on the epilimnetic bacteria of a eutrophic lake. *Freshwater Biology*, **13**, 227–39.

Peet, R. K. (1981) Changes in biomass and production during secondary forest succession, in *Forest Succession*, (eds D. C. West, H. H. Shugart and D. B. Botkin), Springer, New York, pp. 324–38.

Peet, R. K. (1988) Forests of the Rocky Mountains, in *North American Terrestrial Vegetation*, (eds M. G. Barbour and W. D. Billings), Cambridge University Press, Cambridge, pp. 63–101.

Peltier, W. R. and Tushingham, A. M. (1989) Global sea level rise and the greenhouse effect: might they be connected? *Nature*, **244**, 806–10.

Pemadasa, M. A. and Lovell, P. H. (1974) Factors affecting the distribution of some annuals in the dune system at Aberffraw, Anglesey. *Journal of Ecology*, **62**, 403–16.

Pennington, W. (1969) *The History of British Vegetation*, The English Universities Press, London.

Perry, R. A. (1970) Arid shrublands and grasslands, in *Australian Grasses*, (ed. R. M. Moore), Australian National University Press, Canberra, pp. 246–59.

Perry, R. A. and Lazarides, M. (1962) Vegetation of the Alice Springs Area, in *General Report on the Lands of the Alice Springs Area, Northern Territory, 1956–57, Land Research Series No. 6*, CSIRO, Melbourne, pp. 208–36.

Perry, T. O. and Wang, C. W. (1960) Genetic variation in the winter chilling requirement for date of dormancy break for *Acer rubrum*. *Ecology*, **41**, 790–94.

Petch, T. (1930) Buttress roots. *Annals of the Royal Botanic Gardens, Peradeniya*, **11**, 277–85.

Petersen, H. (1982) Structure and size of soil animal populations. *Oikos*, **39**, 306–29.

Petersen, R. C. and Cummins, K. W. (1974) Leaf processing in a woodland stream. *Freshwater Biology*, **4**, 343–68.

Pfeiffer, W. J. and Wiegert, R. G. (1981) Grazers on *Spartina* and their predators, in *The Ecology of a Salt Marsh*, (eds L. R. Pomeroy and R. G. Wiegert), Springer, New York, pp. 87–112.

Phillips, D. H. and Burdekin, D. A. (1982) *Diseases of Forest and Ornamental Trees*, Macmillan, London.

Phillips, D. L. and MacMahon, J. A. (1981) Competition and spacing patterns in desert shrubs. *Journal of Ecology*, **69**, 97–115.

Phillips, W. S. (1963) Depths of roots in soil, *Ecology*, **44**, 424.

Phleger, F. B. (1977) Soils of marine marshes, in *Ecosystems of the World Vol. 1. Wet Coastal Ecosystems*, (ed. V. J. Chapman), Elsevier, Amsterdam, pp. 69–77.

Pielou, E. C. (1979) *Biogeography*, Wiley, New York.

Pieterse, P. J. and Cairns, A. L. P. (1986) The effect of fire on an *Acacia longifolia* seed bank in the south- western Cape. *South African Journal of Botany*, **52**, 233–6.

Pignatti, S. (1979) Plant geographical and morphological evidences in the evolution of the Mediterranean flora (with particular reference to the Italian representatives). *Webbia*, **34**, 243–55.

Pignatti, S. (1983) Human impact in the vegetation of the Mediterranean basin, in *Man's Impact on Vegetation*, (eds W. Holzner, M. J. A. Werger and I. Ikusima), Junk, The Hague, pp. 151–61.

Pike, L. H. (1978) The importance of epiphytic lichens in mineral cycling. *The Bryologist*, **81**, 247–57.

Pisano, E. (1983) The Magellanic tundra complex, in *Ecosystems of the World Vol. 4B. Mires: Swamp, Bog, Fen and Moor – Regional Studies*, (ed. A. J. P. Gore), Elsevier, Amsterdam, pp. 295–329.

Pollet, F. C. and Bridgewater, P. B. (1973) Phytosociology of peatlands in central Newfoundland. *Canadian Journal of Forest Research*, **3**, 433–42.

Polunin, N. (1960) *Introduction to Plant Geography*, Longman, London.

Polunin, N. C. V. (1984) The decomposition of emergent macrophytes in fresh water. *Advances in Ecological Research*, **14**, 115–66.

Polunin, O. and Walters, M. (1985) *A Guide to the Vegetation of Britain and Europe*, Oxford University Press, Oxford.

Ponnamperuma, F. N. (1972) The chemistry of submerged soils. *Advances in Agronomy*, **24**, 29–96.

Poole, D. K. and Miller, P. C. (1975) Water relations of selected species of chaparral and coastal sage communities. *Ecology*, **56**, 1118–28.

Poole, D. K., Roberts, S. W. and Miller, P. C. (1981) Water utilization, in *Resource Use by Chaparral and Matorral*, (ed. P. C. Miller), Springer, New York, pp. 123–49.

Poore, M. E. D. (1968) Studies in Malaysian rain forest I. The forest on Triassic sediments in Jengka Forest Reserve. *Journal of Ecology*, **56**, 143–96.

Poore, D., Burgess, P., Palmer, J., *et al.* (1989) *No Timber Without Trees*, Earthscan Publications, London.

Popma, J., Bongers, F., Martínez-Ramos, M. and Veneklaas, E. (1988) Pioneer species distribution in treefall gaps in Neotropical rain forest; a gap definition and its consequences. *Journal of Tropical Ecology*, **4**, 77–88.

Porsild, A. E. (1958) Geographical distribution of some elements in the flora of Canada. *Geographic Bulletin*, **11**, 57–77.

Porsild, A. E., Harington, C. R. and Mulligan, G. A. (1967) *Lupinus arcticus* Wats. Grown from seeds of Pleistocene age. *Science*, **158**, 113–14.

Porter, J. W., Battey, J. F. and Smith, G. J. (1982) Perturbation and change in coral reef communities. *Proceedings of the National Academy of Science*, **79**, 1678–81.

Prance, G. T. (1982) Forest refuges, evidence from woody angiosperms, in *Biological Diversification in the Tropics*, (ed. G. T. Prance), Columbia University Press, New York, pp. 137–58.

Prance, G. T. (1989) American Tropical forests, in *Ecosystems of the World Vol. 14B. Tropical Rain Forest Ecosystems*, (eds H. Lieth and M. J. A. Werger), Elsevier, Amsterdam, pp. 99–132.

Pratt, T. K. and Stiles, E. W. (1983) How long fruit-eating birds stay in the plants where they feed: implications for seed dispersal. *The American Naturalist*, **122**, 797–805.

Prézelin, B. B. (1976) The role of peridinin–chlorophyll a-proteins in the photosynthetic light adaption of the marine dinoflagellate, *Glenodinium* sp. *Planta*, **130**, 225–33.

Principe, P. P. (1991) Valuing the biodiversity of medicinal plants, in *The Conservation of Medicinal Plants*, (eds O. Akerele, V. Heywood and H. Synge), Cambridge University Press, Cambridge, pp. 79–124.

Pritchard, D. W. (1955) Estuarine circulation patterns. *Proceedings of the American Society of Civil Engineers*, **81**, Separate Paper 717.

Prohaska, F. (1976) The climate of Argentina, Paraguay and Uruguay, in *World Survey of Climatology Vol. 12. Climates of Central and South America*, (ed. W. Schwerdtfeger), Elsevier, Amsterdam, pp. 13–112.

Provasoli, L. (1963) Organic regulation of phytoplankton fertility, in *The Sea: Ideas and Observations on Progress in the Study of the Seas*, Vol. 2 (ed. M. N. Hill), Wiley Interscience, New York, pp. 165–219.

Pruitt, W. O. (1959) Snow as a factor in the winter ecology of the Barren Ground caribou (*Rangifer arcticus*). *Arctic*, **12**, 159–79.

Pugh, K. B., Andrews, A. R., Gibbs, C. F., *et al.* (1974) Some physical, chemical, and microbiological characteristics of two beaches of Anglesey. *Journal of Experimental Marine Biology and Ecology*, **15**, 305–33.

Pulle, A. (1906) *An Enumeration of the Vascular Plants Known from Surinam, Together With Their Distribution and Synonymy*. E. J. Brill, Leiden.

Putz, F. E. and Milton, K. (1982) Tree mortality rates on Barro Colorado Island, in *The Ecology of a Tropical Forest*, (eds E. G. Leigh, A. S. Rand and D. M. Windsor), Smithsonian Institution Press, Washington, pp. 95–100.

Pyke, G. H. and Waser, N. M. (1981) The production of dilute nectars by hummingbird and honeyeater flowers. *Biotropica*, **13**, 260–70.

Pyne, S. J. (1986) 'These conflagrated prairies': a cultural fire history of the grasslands, in *The Prairie: Past, Present and Future. Proceedings of the Ninth North American Prairie Conference*, (eds G. K. Clambey and R. H. Pemble), Tri-College University Centre for Environmental Studies, Fargo, pp. 131–7.

Quick, C. R. and Quick, A. S. (1961) Germination of ceanothus seeds. *Madroño*, **16**, 23–30.

Quinn, J. A. and Hervey, D. F. (1970) Trampling losses and travel by cattle on sandhills range. *Journal of Range Management*, **23**, 50–55.

Rabinowitz, D. (1978a) Dispersal properties of mangrove propagules. *Biotropica*, **10**, 47–57.

Rabinowitz, D. (1978b) Early growth of mangrove seedlings in Panama, and an hypothesis concerning the relationship of dispersal and zonation. *Journal of Biogeography*, **5**, 113–33.

Ramsay, J. M. and Rose Innes, R. (1963) Some quantitative observations on the effects of fire on the Guinea savanna vegetation of northern Ghana over a period of eleven years. *African Soils*, **8**, 41–86.

Ranney, J. W., Bruner, M. C. and Levenson, J. B. (1981) The importance of edge in the structure and dynamics of forest islands, in *Forest Island Dynamics in Man-dominated Landscapes*, (eds R. L. Burgess and D. M. Sharpe), Springer, New York, pp. 67–95.

Rapp, M. and Lossaint, P. (1981) Some aspects of mineral cycling in the garrigue of southern France, in *Ecosystems of the World Vol. 11. Mediterranean-type Shrublands*, (eds F. di Castri, D. W. Goodall and R. L. Specht), Elsevier, Amsterdam, pp. 289–301.

Rauh, W. (1985) The Peruvian–Chilean deserts, in *Ecosystems of the World Vol. 12A. Hot Deserts and Arid Shrublands, A*, (eds M. Evanari, I. Noy-Meir and D. W. Goodall), Elsevier, Amsterdam, pp. 239–67.

Raunkiaer, C. (1934) *The Life Forms of Plants and Statistical Plant Geography*, Clarendon Press, Oxford.

Rauzi, F. and Hanson, C. L. (1966) Water intake and runoff as affected by intensity of grazing. *Journal of Range Management*, **19**, 351–6.

Rauzi, F. and Smith, F. M. (1973) Infiltration rates: three soils and three grazing levels in north-eastern Colorado. *Journal of Range Management*, **26**, 126–9.

Raven, J. A. and Richardson, K. (1986) Marine environments, in *Photosynthesis in Contrasting Environments*, (eds N. R. Baker and S. P. Long), Elsevier, Amsterdam, pp. 337–98.

Raven, P. H. (1963) Amphitropical relationships in the floras of North and South America. *The Quarterly Review of Biology*, **38**, 151–77.

Raven, P. H. (1973) The evolution of mediterranean floras, in *Mediterranean Type Ecosystems: Origin and Structure*, (eds F. di Castri and H. A. Mooney), Springer, New York, pp. 213–24.

Rawitscher, F. (1948) The water economy of the vegetation of the 'Campos Cerrados' in southern Brazil. *Journal of Ecology*, **36**, 238–68.

Rawson, D. S. (1939) Some physical and chemical factors in the metabolism of lakes. *American Association for the Advancement of Science, Publications*, **10**, 9–26.

Reardon, P. O. and Huss, D. L. (1965) Effects of fertilization on a little bluestem community. *Journal of Range Management*, **18**, 238–41.

Reardon, P. O., Leinweber, C. L. and Merril, L. B. (1974) Response of sideoats grama to animal saliva and thiamine. *Journal of Range Management*, **27**, 400–1.

Reddy, K. R. and DeBusk, W. F. (1987) Nutrient storage capabilities of aquatic and wetland plants, in *Aquatic Plants for Water Treatment and Resource Recovery*, Magnolia, Orlando, pp. 337–57.

Redford, K. H. and da Fonseca, G. A. B. (1986) The role of gallery forests in the zoogeography of the cerrado's non-volant mammalian fauna. *Biotropica*, **18**, 126–35.

Redmann, R. E. (1978) Plant and soil water potentials following fire in a northern mixed grassland. *Journal of Range Management*, **31**, 443–5.

Redmann, R. E. (1985) Adaptation of grasses to water stress – leaf rolling and stomate distribution. *Annals of the Missouri Botanical Garden*, **72**, 833–42.

Rees, A. R. (1964) Some observations on the flowering behaviour of *Coffea rupestris* in southern Nigeria. *Journal of Ecology*, **52**, 1–7.

Reich, P. B. and Borchert, R. (1988) Changes with leaf age in stomatal function and water status of several tropical tree species. *Biotropica*, **20**, 60–69.

Reichman, O. J. (1979) Desert granivore foraging and its impact on seed densities and distributions. *Ecology*, **60**, 1085–92.

Reichman, O. J. (1984) Spatial and temporal variation of seed distributions in Sonoran Desert soils. *Journal of Biogeography*, **11**, 1–11.

Reichman, O. J. and Oberstein, D. (1977) Selection of seed distribution types by *Dipodomys merriami* and *Perognathus amplus*. *Ecology*, **58**, 636–43.

Reichman, O. J., Prakash, I. and Roig, V. (1979) Food selection and consumption, in *Arid-land Ecosystems: Structure, Functioning and Management*, Vol. 1 (eds D. W. Goodall, R. A. Perry and K. M. W. Howes), Cambridge University Press, Cambridge, pp. 681–716.

Reid, W. V. (1992) How many species will there be?, in *Tropical Deforestation and Species Extinction*, (eds T. C. Whitmore and J. A. Sayer), Chapman & Hall, London, pp. 55–73.

Reimold, R. J. (1977) Mangals and salt marshes of eastern United States, in *Ecosystems of the World Vol. 1. Wet Coastal Ecosystems*, (ed. V. J. Chapman), Elsevier, Amsterdam, pp. 157–66.

Remmert, H. (1980) *Arctic Animal Ecology*, Springer, Berlin.

Repetto, R. (1990) Deforestation in the tropics. *Scientific American*, **262**, 36–42.

Retzer, J. L. (1974) Alpine soils, in *Arctic and Alpine Environments*, (eds J. D. Ives and R. G. Barry), Methuen, London, pp. 771–802.

Reukema, D. L. (1982) Seedfall in a young-growth Douglas fir stand: 1950–1978. *Canadian Journal of Forest Research*, **12**, 249–54.

Reynolds, C. S. (1984) *The Ecology of Freshwater Phytoplankton*, Cambridge University Press, Cambridge.

Reynolds, D. N. (1984) Populational dynamics of three annual species of alpine plants in the Rocky Mountains. *Oecologia*, **62**, 250–5.

Rhoades, D. F. (1979) Evolution of plant chemical defense against herbivores, in *Herbivores, Their Interaction With Secondary Plant Metabolites*, (eds G. A. Rosenthal and D. H. Janzen), Academic Press, New York, pp. 3–54.

Rhoades, F. M. (1981) Biomass of epiphytic lichens and bryophytes on *Abies lasiocarpa* on a Mt Baker lava flow, Washington. *The Bryologist*, **84**, 39–47.

Rhoades, D. F. and Cates, R. G. (1975) Toward a general theory of plant antiherbivore chemistry, in *Recent Advances in Phytochemistry Vol. 10. Biochemical Interaction Between Plants and Insects*, (eds J. W. Wallance and R. L. Mansell), Plenum Press, New York, pp. 168–213.

Rice, D. L. (1982) The detritus nitrogen problem: new observations and perspectives from organic geochemistry. *Marine Ecology Progress Series*, **9**, 153–62.

Rice, D. L. and Tenore, K. R. (1981) Dynamics of carbon and nitrogen during the decomposition of detritus derived from estuarine macrophytes. *Estuarine, Coastal and Shelf Science*, **13**, 681–90.

Rich, P. H., Wetzel, R. G. and Thuy, N. V. (1971) Distribution, production and role of aquatic macrophytes in a southern Michigan marl lake. *Freshwater Biology*, **1**, 3–21.

Richards, P. W. (1936) Ecological observations on the rain forest of Mount Dulit, Sarawak. Part I. *Journal of Ecology*, **24**, 1–37.

Richards, P. W. (1952) *The Tropical Rain Forest*, Cambridge University Press, Cambridge.

Richards, P. W. (1973) Africa, the 'odd man out', in *Tropical Forest Ecosystems in Africa and South America: A Comparative Review*, (eds B. J. Meggers, E. S. Ayensu and W. D. Duckworth), Smithsonian Institution Press, Washington, pp. 21–6.

Richards, P. W. (1983) The three-dimensional structure of tropical rain forest, in *Tropical Rain Forest: Ecology and Management*, (eds S. L. Sutton, T. C. Whitmore and A. C. Chadwick), Blackwell Scientific Publications, Oxford, pp. 3–10.

Richardson, C. J. and Davis, J. A. (1987) Natural and artificial wetland ecosystems: ecological opportunities and limitations, in *Aquatic Plants for Water Treatment and Resource Recovery*, (eds K. R. Reddy and W. H. Smith), Magnolia Publishing, Orlando, pp. 819–54.

Richardson, C. J. and Marshall, P. E. (1986) Processes controlling movement, storage, and export of phosphorus in a fen peatland. *Ecological Monographs*, **56**, 279–302.

Richardson, K., Griffiths, H., Reed, M. L., *et al.* (1984) Inorganic carbon assimilation in the Isoetids, *Isoetes lacustris* L. and *Lobelia dortmanna* L. *Oecologia*, **61**, 115–21.

Richardson, S. D. (1957) The effect of leaf age on the rate of photosynthesis in detached leaves of tree seedlings. *Acta Botanica Neerlandica*, **6**, 445–57.

Richardson, S. D. (1966) *Forestry in Communist China*, The Johns Hopkins Press, Baltimore,

Richter, W. (1984) A structural approach to the function of buttresses of *Quararibea asterolepis*. *Ecology*, **65**, 1429–35.

Ricklefs, R. E. (1990) *Ecology*, 3rd edn, Freeman, New York.

Riewe, R. R. (1981) Changes in Eskimo utilisation of arctic wildlife, in *Tundra Ecosystems: A Comparative Analysis*, (eds L. C. Bliss, O. W. Heal and J. J. Moore), Cambridge University Press, Cambridge, pp. 721–30.

Riley, J. P. and Chester, R. (1971) *Introduction to Marine Chemistry*, Academic Press, London.

Rind, D., Chiou, E.-W., Chu, W. *et al.* (1991) Positive water vapour feedback in climate models confirmed by satellite data. *Nature*, **349**, 500–503.

Risser, P. G. (1988) Abiotic controls on primary productivity and nutrient cycles in North American grasslands, in *Concepts of Ecosystem Ecology*, (eds L. R. Pomeroy and J. J. Alberts), Springer, New York, pp. 115–29.

Risser, P. G., Birney, E. C., Blocker, H. D. *et al.* (1981) *The True Prairie Ecosystem*, Hutchinson Ross Publishing Company, Stroudsburg.

Ritchie, J. C. (1987) *Postglacial Vegetation of Canada*, Cambridge University Press, Cambridge.

Ritchie, J. C. (1982) The modern and late-Quaternary vegetation of the Doll Creek area, north Yukon, Canada. *New Phytologist*, **90**, 563–603.

Rizzo, B. and Wiken, E. (1992) Assessing the sensitivity of Canada's ecosystems to climatic change. *Climatic Change*, **21**, 37–55.

Robarts, R. D. (1984) Factors controlling primary production in a hypertrophic lake (Hartbeespoort Dam, South Africa). *Journal of Plankton Research*, **6**, 91–105.

Robberecht, R. and Caldwell, M. M. (1983) Protective mechanisms and acclimation to solar ultraviolet-B radiation in *Oenothera stricta*. *Plant Cell and Environment*, **6**, 477–85.

Robertson, J. H. (1939) A quantitative study of true-prairie vegetation after three years of extreme drought. *Ecological Monographs*, **9**, 431–91.

Robertson, A. and Scavia, D. (1984) North American Great Lakes, in *Ecosystems of the World Vol. 11. Lakes and Reservoirs*, (ed. F. B. Taub), Elsevier, Amsterdam, pp. 135–76.

Robins, J. K. and Susut, J. P. (1974) *Red Belt in Alberta*, Northern Forest Research Centre, Information Report NOR-X-99, Edmonton.

Robinson, G. A. (1970) Continuous plankton records: variation in the seasonal cycle of phytoplankton in the North Atlantic. *Bulletin of Marine Ecology*, **6**, 333–45.

Rochefort, L., Vitt, D. H. and Bayley, S. E. (1990) Growth, production, and decomposition dynamics of *Sphagnum* under natural and experimentally acidified conditions. *Ecology*, **71**, 1986–2000.

Rodhe, W. (1969) Crystallization of eutrophication concepts in northern Europe, in *Eutrophication, Causes, Consequences, Correctives*, National Academy of Sciences, Washington, pp. 50–64.

Rodhe, H. and Herrera, R. (1988) (eds.) *Acidification in Tropical Countries*, Wiley, Chichester.

Rodin, L. E. and Bazilevich, N. I. (1967) *Production and Mineral Cycling in Terrestrial Vegetation*, Oliver and Boyd, Edinburgh.

Roe, F. G. (1970) *The North American Buffalo: A Critical Study of the Species in its Wild State*, 2nd edn, University of Toronto Press, Toronto.

Rose, A. B. and Platt, K. (1987) Recovery of northern fiordland alpine grasslands after reduction in the deer population. *New Zealand Journal of Ecology*, **10**, 23–33.

Rose, A. B. and Platt, K. H. (1990) Age-states, population structure, and seedling regeneration of *Chionochloa pallens* in Canterbury alpine grasslands, New Zealand. *Journal of Vegetation Science*, **1**, 89–96.

Rose, A. B., Harrison, J. B. J. and Platt, K. H. (1988) Alpine tussockland communities and vegetation–landform–soil relationships, Wapiti Lake, Fiordland, New Zealand. *New Zealand Journal of Botany*, **26**, 525–40.

Rose Innes, R. (1972) Fire in West African vegetation. *Proceedings of the Tall Timbers Fire Ecology Conference*, **11**, 147–73.

Rosseland, B. O. (1986) Ecological effects of acidification on tertiary consumers. Fish population responses. *Water, Air, and Soil Pollution*, **30**, 451–60.

Rossiter, R. C. and Ozanne, P. G. (1970) South-western temperate forests, woodlands, and heaths, in *Australian Grasslands*, (ed. R. M. Moore), Australian National University Press, Canberra, pp. 199–218.

Rosswall, T. and Granhall, U. (1980) Nitrogen cycling in a subarctic ombrotrophic mire, in *Ecology of a Subarctic Mire*, (ed. M. Sonesson), Swedish Natural Science Research Council, Stockholm, pp. 209–34.

Rosswall, T., Flower-Ellis, J. G. K., Johansson, L. G., Jonsson, S., Rydén, B. E. and Sonesson, M. (1975) Stordalen (Abisko), Sweden, in *Ecological Bulletins 20: Structure and Function of Tundra Ecosystems*, (eds T. Rosswall and O. W. Heal), Swedish Natural Science Research Council, Stockholm, pp. 265–94.

Roth, I. (1984) *Stratification of Tropical Forests as Seen in Leaf Structure*, Junk, The Hague.

Rowe, F. W. E. and Vail, L. (1984) Crown-of-Thorns: GBR *Not Under Threat*. *Search*, **11**, 211–13.

Rowe, J. S. (1956) Uses of undergrowth plant species in forestry. *Ecology*, **37**, 461–73.

Rowe, J. S. (1977) Forest Regions of Canada, *Canadian Forestry Service Publication No. 1300*, Ottawa.

Rowe, J. S. (1983) Concepts of fire effects on plant individuals and species, in *The Role of Fire in Northern Circumpolar Ecosystems*, (eds R. W. Wein and D. A. MacLean), Wiley, Chichester, pp. 135–54.

Rowe, J. S. and Coupland, R. T. (1984) Vegatation of the Canadian plains. *Prairie Forum*, **9**, 231–48.

Rozema, J., Luppes, E. and Broekman, R. (1985) Differential response of salt-marsh species to variation of iron and manganese. *Vegetatio*, **62**, 293–301.

Rundel, P. W. (1981a) The matorral zone of central Chile, in *Ecosystems of the World Vol. 11. Mediterranean-type Shrublands*, (eds F. di Castri, D. W. Goodall and R. L. Specht), Elsevier, Amsterdam, pp. 175–201.

Rundel, P. W. (1981b) Fire as an ecological factor, in *Physiological Plant Ecology I: Responses to the Physical Environment*, (eds O. L. Lange, P. S. Nobel, C. B. Osmond and H. Ziegler), Springer, Berlin, pp. 501–38.

Rundel, P. W. (1983) Impact of fire on nutrient cycles in mediterranean-type ecosystems with reference to chaparral, in *Mediterranean Type Ecosystems: The Role of Nutrients*, (eds F. J. Kruger, D. T. Mitchell and J. U. M. Jarvis), Springer, Berlin, pp. 192–207.

Rundel, P. W. and Parsons, D. J. (1979) Structural changes in chamise (*Adenostoma fasciculatum*) along a fire-induced age gradient. *Journal of Range Management*, **32**, 462–6.

Rundel, P. W. and Parsons, D. J. (1980) Nutrient changes in two chaparral shrubs along a fire-induced age gradient. *American Journal of Botany*, **67**, 51–8.

Runkle, J. R. (1982) Patterns of disturbance in some old-growth mesic forests of eastern North America. *Ecology*, **63**, 1533–46.

Rutherford, M. C. (1980) Annual plant production-precipitation relations in arid and semi-arid regions. *South African Journal of Science*, **76**, 53–6.

Rutherford, M. C. (1982) Woody plant biomass distribution in *Burkea africana* savannas, in *Ecology of Tropical Savannas*, (eds B. J. Huntley and B. H. Walker), Springer, Berlin, pp. 120–41.

Ruuhijärvi, R. (1983) The Finnish mire types and their regional distribution, in *Ecosystems of the World Vol. 4B. Mires: Swamp, Bog, Fen and Moor – Regional Studies*, (ed. A. J. P. Gore), Elsevier, Amsterdam, pp. 47–67.

Ryan, J. K. and Hergert, C. R. (1977) Energy budget for *Gynaephora groenlandica* (Homeyer) and *G. rossii* (Curtis) (Lepidoptera: Lymantriidae) on Truelove Lowland, in *Truelove Lowland, Devon Island, Canada: A High Arctic Ecosystem*, (ed. L. C. Bliss), The University of Alberta Press, Edmonton, pp. 395–409.

Rycroft, D. W., Williams, D. A. and Ingram, H. A. P. (1975) The transmission of water through peat. I. Review. *Journal of Ecology*, **63**, 535–56.

Ryding, S.-O. and Rast, W. (eds.) (1989) *The Control of Eutrophication of Lakes and Reservoirs*, Unesco, Paris.

Ryther, J. H. (1963) Geographic variations in productivity, in *The Sea: Ideas and Observations on Progress in the Study of the Seas*, Vol. 2 (ed. M. N. Hill), Wiley Interscience, New York, pp. 347–80.

Ryther, J. H. and Menzel, D. W. (1959) Light adaptation by marine phytoplankton. *Limnology and Oceanography*, **4**, 492–7.

Ryvarden, L. (1971) Studies in seed dispersal I. Trapping of diaspores in the alpine zone at Finse, Norway. *Norwegian Journal of Botany*, **18**, 215–26.

Ryvarden, L. (1975) Studies in seed dispersal II. Winter-dispersed species at Finse, Norway. *Norwegian Journal of Botany*, **22**, 21–4.

Saenger, P. (1982) Morphological, anatomical and reproductive adaptations of Australian mangroves, in *Mangrove Ecosystems in Australia*, (ed. B. F. Clough), Australian Institute of Marine Science, Canberra, pp. 153–91.

Saenger, P., Specht, M. M., Specht, R. L. and Chapman, V. J. (1977) Mangal and coastal salt-marsh communities in Australasia, in *Ecosystems of the World Vol. 1. Wet Coastal Ecosystems*, (ed. V. J. Chapman), Elsevier, Amsterdam, pp. 293–345.

Sakai, A. and Weiser, C. J. (1973) Freezing resistance of trees in North America with reference to tree regions. *Ecology*, **54**, 118–26.

Salati, E. and Vose, P. B. (1984) Amazon Basin: a system in equilibrium. *Science*, **225**, 129–38.

Saldarriaga, J. G. (1987) Recovery following shifting cultivation, in *Amazonian Rain Forests*, (ed. C. F. Jordan), Springer, New York, pp. 24–33.

Salonen, K., Jones, R. I. and Arvola, L. (1984) Hypolimnetic phosphorus retrieval by vertical migrations of lake phytoplankton. *Freshwater Biology*, **14**, 431–8.

Sanchez, P. A. and Boul, S. W. (1975) Soils of the tropics and the world food crisis. *Science*, **188**, 598–603.

Sanchez, P. A., Villachica, J. H. and Bandy, D. E. (1983) Soil fertility dynamics after clearing a tropical rainforest in Peru. *Soil Science Society of America Journal*, **47**, 1171–8.

Sand-Jensen, K. (1987) Environmental control of bicarbonate use among freshwater and marine macrophytes, in *Plant Life in Aquatic and Amphibious Habitats*, (ed. R. M. M. Crawford), Blackwell Scientific Publications, Oxford, pp. 99–112.

San José, J. J. and Medina, E. (1975) Effect of fire on organic matter production and water balance in a tropical savanna, in *Tropical Ecological Systems*, (eds F. B. Golley and E. Medina), Springer, New York, pp. 251–64.

San José, J. J., Berrade, F. and Ramirez, J. (1982) Seasonal changes of growth, mortality and disappearance of belowground root biomass in the *Trachypogon* savanna grass. *Acta Oecologica/Oecologia Plantarum*, **3**, 347–58.

Sarmiento, G. (1983) The savannas of tropical America, in *Ecosystems of the World Vol. 13. Tropical Savannas*, (ed. F. Bourlière), Elsevier, Amsterdam, pp. 245–88.

Sarmiento, G. (1984) *The Ecology of Neotropical Savannas*. Harvard University Press, Cambridge.

Sarmiento, G. and Monasterio, M. (1969) Studies on the savanna vegetation of the Venezuelan llanos I. The use of association-analysis. *Journal of Ecology*, **57**, 579–98.

Sarmiento, G. and Monasterio, M. (1975) A critical consideration of the environmental conditions associated with the occurrence of savanna ecosystems in tropical America, in *Tropical Ecological Systems*, (eds F. B. Golley and E. Medina), Springer, New York, pp. 223–50.

Sarmiento, G. and Monasterio, M. (1983) Life forms and phenology, in *Ecosystems of the World Vol. 13. Tropical Savannas*, (ed. F. Bourlière), Elsevier, Amsterdam, pp. 79–108.

Sarmiento, G., Goldstein, G. and Meinzer, F. (1985) Adaptive strategies of woody species in neotropical savannas. *Biological Reviews*, **60**, 315–55.

Sarukhan, J. (1978) Studies on the demography of tropical trees, in *Tropical Trees as Living Systems*, (eds P. B. Tomlinson and M. H. Zimmermann), Cambridge University Press, Cambridge, pp. 163–84.

Saunders, G. W. (1969) Some aspects of feeding in zooplankton, in *Eutrophication: Causes, Consequences, Correctives*, National Academy of Sciences, Washington, pp. 556–73.

Saunders, G. W. (1980) Organic matter and decomposers, in *The Functioning of Freshwater Ecosystems*, (eds E. D. Le Cren and R. H. Lowe-McConnell), Cambridge University Press, Cambridge, pp. 341–92.

Savidge, G. and Foster, P. (1978) Phytoplankton biology of a thermal front in the Celtic Sea. *Nature*, **271**, 155–6.

Savile, D. B. O. (1972) *Arctic Adaptations in Plants*, Canada Department of Agriculture, Monograph No. 6, Ottawa.

Sazima, M. and Sazima, I. (1978) Bat pollination of the passion flower, *Passiflora mucronata* in southeastern Brazil. *Biotropica*, **10**, 100–9.

Scagel, R. F., Bandoni, R. J., Rouse, G. E. *et al.* (1965) *An Evolutionary Survey of the Plant Kingdom*, Wadsworth, Belmont.

Scavia, D. and Fahnenstiel, G. L. (1988) From picoplankton to fish: complex interactions in the Great Lakes, *Complex Interactions in Lake Communities*, (ed. S. R. Carpenter), Springer, New York, pp. 85–97.

Schaefer, R. (1973) Microbial activity under seasonal conditions of drought in mediterranean climates, in *Mediterranean Type Ecosystems: Origin and Structure*, (eds F. di Castri and H. A. Mooney), Springer, New York, pp. 191–8.

Schaffer, W. M. and Gadgil, M. D. (1975) Selection for optimal life histories in plants, in *Ecology and Evolution of Communities*, (ed. M. L. Cody and J. M. Diamond), Harvard, Cambridge, pp. 142–57.

Schantz, E. J. (1960) Biochemical studies on paralytic shellfish poisons. *Annals New York Academy of Sciences*, **90**, 843–55.

Schechter, Y. and Galai, C. (1980) The Negev – a desert reclaimed, in *Desertification*, (eds M. R. Biswas and A. K. Biswas), Pergamon Press, Oxford, pp. 255–308.

Scheibling, R. E. and Stephenson, R. L. (1984) Mass mortality of *Strongylocentrotus droebachiensis* (Echinodermata: Echinoidea) off Nova Scotia, Canada. *Marine Biology*, **78**, 153–64.

Schell, D. M. and Alexander, V. (1973) Nitrogen fixation in arctic coastal tundra in relation to vegetation and micro-relief. *Arctic*, **26**, 130–7.

Schimel, D., Stillwell, M. A. and Woodmansee, R. G. (1985) Biogeochemistry of C, N, and P in a soil catena of the shortgrass steppe. *Ecology*, **66**, 276–82.

Schimper, A. F. W. (1903) *Plant-geography Upon a Physiological Basis*, Clarendon, Oxford.

Schindler, J. E. (1971) Food quality and zooplankton nutrition. *Journal of Animal Ecology*, **40**, 589–95.

Schlesinger, W. H. (1978) Community structure, dynamics and nutrient cycling in the Okefenokee cypress swamp–forest. *Ecological Monographs*, **48**, 43–65.

Schlesinger, W. H. and Gill, D. S. (1980) Biomass, production, and changes in the availability of light, water, and nutrients during the development of pure stands of the chaparral shrub, *Ceanothus megacarpus*, after fire. *Ecology*, **61**, 781–9.

Schlesinger, W. H. and Hasey, M. H. (1981) Decomposition of chaparral shrub foliage: losses of organic and inorganic constituents from deciduous and evergreen leaves. *Ecology*, **62**, 762–74.

Schmidt-Nielsen, B. (1954) Water conservation in small desert rodents, in *Biology of Deserts*, (ed. J. L. Cloudsley-Thompson), The Institute of Biology, London, pp. 173–81.

Schmidt-Nielsen, K. (1964) *Desert Animals. Physiological Problems of Heat and Water*, Clarendon, Oxford.

Schmidt-Nielsen, K. (1972) *How Animals Work*, Cambridge University Press, Cambridge.

Schmidt-Nielsen, K., Schmidt-Nielsen, B., Jarnum, S. A. and Houpt, T. R. (1957) Body temperature of the camel and its relation to water economy. *American Journal of Physiology*, **188**, 103–12.

Schneider, R. L. and Scharitz, R. R. (1986) Seed bank dynamics in a southeastern riverine swamp. *American Journal of Botany*, **73**, 1022–30.

Schodde, R. (1981) Bird communities of the Australian mallee: composition, derivation, distribution, structure and seasonal cycles, in *Ecosystems of the World Vol. 11. Mediterranean-type Shrublands*, (eds F. di Castri, D. W. Goodall and R. L. Specht), Elsevier, Amsterdam, pp. 387–415.

Scholander, P. F. (1968) How mangroves desalinate seawater. *Physiologia Plantarum*, **21**, 251–61.

Schonbeck, M. W. and Norton, T. A. (1979a) An investigation of drought avoidance in intertidal fucoid algae. *Botanica Marina*, **22**, 133–44.

Schonbeck, M. W. and Norton, T. A. (1979b) The effects of brief periodic submergence on intertidal fucoid algae. *Estuarine and Coastal Marine Science*, **8**, 205–11.

Schonbeck, M. W. and Norton, T. A. (1980) Factors controlling the lower limits of fucoid algae on the shore. *Journal of Experimental Biology and Ecology*, **43**, 131–50.

Schopf, J. M. (1970) Relation of floras of the southern hemisphere to continental drift. *Taxon*, **19**, 657–74.

Schultz, A. M. (1969) A study of an ecosystem: the arctic tundra, in *The Ecosystem Concept in Natural Resource Management*, (ed. G. M. Van Dyne), Academic Press, New York.

Schultz, A. M., Launchbaugh, J. L. and Biswell, H. H. (1955) Relationship between grass density and brush seedling survival. *Ecology*, **36**, 226–38.

Schulze, B. R. (1972) South Africa, in *World Survey of Climatology Vol 10. Climates of Africa*, (ed. J. F. Griffiths), Elsevier, Amsterdam, pp. 501–86.

Schwarz, A. G. and Redmann, R. E. (1988) C_4 grasses from the boreal forest region of northwestern Canada. *Canadian Journal of Botany*, **66**, 2424–30.

Schwegman, J. E. and Anderson, R. C. (1986) Effect of eleven years of fire exclusion on the vegetation of a southern Illinois barren remnant, in *The Prairie: Past, Present and Future. Proceedings of the Ninth North American Prairie Conference*, (eds G. K. Clambey and R. H. Pemble), Tri-College University Centre for Environmental Studies, Fargo, pp. 146–8.

Schwerdtfeger, W. (ed.) (1976) *World Survey of Climatology Vol 12. Climates of Central and South America*. Elsevier, Amsterdam.

Schwintzer, C. R. (1978) Nutrient and water levels in a small Michigan bog with high tree mortality. *Am. Mid. Nat.*, **100**, 441–51.

Schwintzer, C. R. and Williams, G. (1974) Vegetation changes in a small Michigan bog. *Am. Mid. Nat.*, **92**, 447–59.

Scott, B. D. and Jitts, H. R. (1977) Photosynthesis of phytoplankton and zooxanthellae on a coral reef. *Marine Biology*, **41**, 307–15.

Scott, J. A., French, N. R. and Leetham, J. W. (1979) Patterns of consumption in grasslands, in *Perspectives in Grassland Ecology*, (ed N. R. French), Springer, New York, pp. 89–105.

Sculthorpe, C. D. (1967) *The Biology of Aquatic Vascular Plants*, Arnold, London.

Seely, M. K. and Hamilton, W. J. (1976) Fog catchment sand trenches constructed by tenebrionid beetles, *Lepidochora*, from the Namib Desert. *Science*, **193**, 484–6.

Semeniuk, V. (1983) Mangrove distribution in northwestern Australia in relationship to regional and local freshwater seepage. *Vegetation*, **53**, 11–31.

Sephton, D. H. and Harris, G. P. (1984) Physical variability and phytoplankton communities: VI. Day to day changes in primary productivity and species abundance. *Archiv für Hydrobiologie*, **102**, 155–75.

Serruya, C. and Pollinger, U. (1983) *Lakes of the Warm Belt*, Cambridge University Press, Cambridge.

Shanks, A. L. and Trent, J. D. (1980) Marine snow: sinking rates and potential role in vertical flux. *Deep-Sea Research*, **27A**, 137–43.

Shantz, H. L. and Marbut C. F. (1923) *Vegetation and Soils of Africa*, AMS Press, New York.

Sharifi, M. R., Nilsen, E. T. and Rundel, P. W. (1982) Biomass and net primary productivity of *Prosopis glandulosa* (Fabaceae) in the Sonoran Desert of California. *American Journal of Botany*, **69**, 760–7.

Shaver, G. R. (1981) Mineral nutrient and nonstructural carbon utilisation, in *Resource Use by Chaparral and Matorral*, (ed. P. C. Miller), Springer, New York, pp. 237–57.

Shaver, G. R. (1983) Mineral nutrient and nonstructural carbon pools in shrubs from mediterranean-type ecosystems of California and Chile, in *Mediterranean Type Ecosystems: The Role of Nutrients*, (eds F. J. Kruger, D. T. Mitchell and J. U. M. Jarvis), Springer, Berlin, pp. 286–99.

Shaver, G. R. and Billings, W. D. (1975) Root production and root turnover in a wet tundra ecosystem, Barrow, Alaska. *Ecology*, **56**, 401–9.

Shaver, G. R. and Billings, W. D. (1977) Effects of daylength and temperature on root elongation in tundra graminoids. *Oecologia*, **28**, 57–65.

Shaw, N. H. and Norman, J. T. (1970) Tropical and sub-tropical woodlands and grasslands, in *Australian Grasslands*, (ed. R. M. Moore), Canberra, Australian National University Press, pp. 112–22.

Sheppe, W. (1972) The annual cycle of small mammal populations on a Zambian floodplain. *Journal of Mammalogy*, **53**, 445–60.

Sheppe, W. and Osborne, T. (1971) Patterns of use of a flood plain by Zambian mammals. *Ecological Monographs*, **41**, 179–205.

Shmida, A. (1985) Biogeography of the desert flora, in *Ecosystems of the World. Vol 12A: Hot Deserts and Arid Shrublands*, (eds M. Evanari, I. Noy-Meir and D. W. Goodall), Elsevier, Amsterdam, pp. 23–77.

Shmida, A. and Whittaker, R. H. (1981) Pattern and biological microsite effects in two shrub communities, southern California. *Ecology*, **62**, 234–51.

Shreve, F. (1942) The desert vegetation of North America. *The Botanical Review*, **8**, 195–246.

Shreve, F. and Hinkley, A. L. (1937) Thirty years of change in desert vegetation. *Ecology*, **18**, 463–78.

Sieburth, J. McN. (1969) Studies on algal substances in the sea. III. The production of extracellular organic matter by littoral marine algae. *Journal of Experimental Marine Biology and Ecology*, **3**, 290–309.

Signor, P. W. (1990) The geologic history of diversity. *Annual Review of Ecology and Systematics*, **21**, 509–39.

Silvester, W. B. (1977) Dinitrogen fixation by plant associations excluding legumes, in *A Treatise on Dinitrogen Fixation, Section IV: Agronomy and Ecology*, (eds R. W. F. Hardy and A. H. Gibson), Wiley, New York, pp. 141–90.

Silvertown, J. W. (1982) *Introduction to Plant Population Ecology*, Longman, London.

Simard, A. J. (1973) *Forest Fire Weather Zones of Canada*, Environment Canada, Forestry Service, Ottawa.

Simonson, R. W. (1959) Outline of a generalised theory of soil genesis. *Soil Science of Society of America Proceedings*, **23**, 152–6.

Simpson, G. G. (1947) Holarctic mammalian faunas and continental relationships during the Cenozoic. *Bulletin of the Geological Society of America*, **58**, 613–88.

Sims, P. L. and Coupland, R. T. (1979) Producers, in *Grassland Ecosystems of the World: Analysis of Grasslands and Their Uses*, (ed. R. T. Coupland), Cambridge University Press, Cambridge, pp. 49–72.

Sims, P. L. and Singh, J. S. (1978) The structure and function of ten western North American grasslands II, intra-seasonal dynamics in primary producer compartments. *Journal of Ecology*, **66**, 547–72.

Sims, P. L. and Singh, J. S. (1978) The structure and function of ten western North American grasslands III. Net primary production, turnover and efficiencies of energy capture and water use. *Journal of Ecology*, **66**, 573–97.

Sims, P. L. and Singh, J. S. (1978) The structure and function of ten western North American grasslands IV. Compartmental transfers and energy flow within the ecosystem. *Journal of Ecology*, **66**, 983–1009.

Sims, P. L., Singh, J. S. and Lauenroth, W. K. (1978) The structure and function of ten western North American grasslands I. Abiotic and vegetational characteristics. *Journal of Ecology*, **66**, 251–85.

Sinclair, A. R. E. (1975) The resource limitation of trophic levels in tropical grassland ecosystems. *Journal of Animal Ecology*, **44**, 497–520.

Sinclair, W. A., Lyon, H. H. and Johnson, W. T. (1987) *Diseases of Trees and Shrubs*, Comstock Publishing Associates, Ithaca.

Sjörs, H. (1983) Mires of Sweden, in *Ecosystems of the World Vol 4B. Mires: Swamp, Bog, Fen and Moor – Regional Studies*, (ed. A. J. P. Gore), Elsevier, Amsterdam, pp. 69–94.

Sladen, W. J. L., Menzie, C. M. and Reichel, W. L. (1966) DDT residues in Adelie penguins and a crabeater seal from Antarctica. *Nature*, **210**, 670–73.

Slingsby, P. and Bond, W. J. (1985) The influence of ants on the dispersal distance and seedling recruitment of *Leucospermum conocarpodendron* (L.) Buek (Protaceae). *South African Journal of Botany*, **51**, 30–4.

Small, E. (1972a) Ecological significance of four critical elements in plants of raised *Sphagnum* peat bogs. *Ecology*, **53**, 498–503.

Small, E. (1972b) Photosynthetic rates in relation to nitrogen recycling as an adaption to nutrient deficiency in peat bog plants. *Canadian Journal of Botany*, **50**, 2227–33.

Smart, R. M. and Barko, J. W. (1980) Nitrogen nutrition and salinity tolerance of *Distichlis spicata* and *Spartina alternifolia*. *Ecology*, **61**, 630–38.

Smayda, T. J. (1980) Phytoplankton species succession, in *The Physiological Ecology of Phytoplankton*, (ed. I. Morris), University of California Press, Berkeley, pp. 493–570.

Smeins, F. E. and Olsen, D. E. (1970) Species composition and production of a native northwestern Minnesota tall grass prairie. *Am. Mid. Nat.*, **84**, 398–410.

Smirnoff, N. and Crawford, R. M. M. (1983) Variation in the structure and response to flooding of root aerenchyma in some wetland plants. *Annals of Botany*, **51**, 237–49.

Smith, A. P. (1972) Buttressing of tropical trees: a descriptive model and new hypotheses. *The American Naturalist*, **106**, 32–46.

Smith, A. P. (1979) Function of dead leaves in *Espeletia schultzii* (Compositae) an Andean caulescent rosette species. *Biotropica*, **11**, 43–7.

Smith, A. P. and Young, T. P. (1987) Tropical alpine plant ecology. *Annual Review of Ecology and Systematics*, **18**, 137–58.

Smith, K. C. (1989) *The Science of Photobiology*, 2nd edn, Plenum Press, New York.

Smith, R. A. H. and Forrest, G. I. (1978) Field estimates of primary production, in *Production Ecology of British Moors and Montane Grasslands*, (eds O. W. Heal and D. F. Perkins), Springer, Berlin, pp. 17–37.

Smith, S. D. and Nobel, P. S. (1986) Deserts, in *Photosynthesis in Contrasting Environments*, (eds N. R. Baker and S. P. Long), Elsevier, Amsterdam, pp. 13–62.

Smith, S. D., Didden-Zopfy, B. and Nobel, P. S. (1984) High-temperature responses of North American cacti. *Ecology*, **65**, 643–51.

Smith, S. V. (1984) Phosphorus versus nitrogen limitation in the marine environment. *Limnology and Oceanography*, **29**, 1149–60.

Smith, S. V. (1988) Mass balance in coral reef-dominated areas, in *Coastal-Offshore Ecosystem Interactions*, (ed. B.-O. Jansson), Springer, Berlin, pp. 209–26.

Smith, S. V., Kimmerer, W. J. and Walsh, T. W. (1986) Vertical flux and biogeochemical turnover regulate nutrient limitation of net organic production in the North Pacific Gyre. *Limnology and Oceanography*, **31**, 161–7.

Smith, S. V., Kimmerer, W. J., Laws, E. A., Brock, R. E. and Walsh, T. W. (1981) Kaneohe Bay sewage diversion experiment: perspectives on ecosystem responses to nutritional perturbation. *Pacific Science*, **35**, 279–395.

Smith, T. J. (1983) Alteration of salt marsh plant community composition by grazing snow geese. *Holarctic Ecology*, **6**, 204–10.

Smith, T. M., Leemans, R. and Shugart, H. H. (1992) Sensitivity of terrestrial carbon storage to CO_2-induced climate change: comparison of four scenarios based on general circulation models. *Climatic Change*, **21**, 367–84.

Smith, W. K. and Geller, G. N. (1980) Leaf and environmental parameters influencing transpiration: theory and field measurements. *Oecologia*, **46**, 308–13.

Smith, W. K., Young, D. R., Carter, G. A., Hadley, J. L. and McNaughton, G. M. (1984) Autumn stomatal closure in six conifer species of the central Rocky Mountains. *Oecologia*, **63**, 237–42.

Smythe, N. (1970) Relationships between fruiting seasons and seed dispersal methods in a Neotropical forest. *The American Naturalist*, **104**, 25–35.

Smythe, N., Glantz, W. E. and Leigh, E. G. (1982) Population regulation in some terrestrial frugivores, in *The Ecology of a Tropical Forest*, (eds E. G. Leigh, A. S. Rand and D. M. Windsor), Smithsonian Institution Press, Washington, pp. 227–38.

Snedaker, S. C. and Pool, D. J. (1973) Mangrove forest types and biomass, in *The Role of Mangrove Ecosystems in the Maintenance of Environmental Quality and a High Productivity of Desirable Fisheries*, Center for Aquatic Sciences, Gainsville, pp. 1–13.

Snow, D. W. (1965) A possible selective factor in the evolution of fruiting seasons in tropical forest. *Oikos*, **15**, 274–81.

Snow, D. W. (1981) Tropical frugivorous birds and their food plants: a world survey. *Biotropica*, **13**, 1–14.

Soholt, L. F. (1973) Consumption of primary production by a population of kangaroo rats (*Dipodomys merriami*) in the Mojave Desert. *Ecological Monographs*, **43**, 357–76.

Soil Survey Staff (1975) *Soil Taxonomy: A basic system of soil classification for making and interpreting soil surveys*. United States Department of Agriculture Handbook 436.

Solbrig, O. T., Barbour, M. A., Cross, J. *et al.* (1977) The strategies and community patterns of desert plants, in *Convergent Evolution in Warm Deserts*, (eds G. H. Orians and O. T. Solbrig), Dowden, Hutchinson and Ross, Stroudsburg, pp. 67–106.

Solbrig, O. T., Cody, M. L., Fuentes, E. R. *et al.* (1977) The origin of the biota, in *Convergent Evolution in Chile and California*, (ed. H. A. Mooney), Dowden, Hutchinson and Ross, Stroudsburg, pp. 13–26.

Sollins, P., Grier, C. C., McCorison, F. M. *et al.* (1980) The internal element cycles of an old-growth Douglas fir ecosystem in western Oregon. *Ecological Monographs*, **50**, 261–85.

Sommer, U. (1981) The role of r- and K-selection in the succession of phytoplankton in Lake Constance. *Acta Oecologica/Oecologia Generalis*, **2**, 327–42.

Sommer, U. (1989) The role of competition for resources in phytoplankton succession, in *Plankton Ecology*, (ed. U. Sommer), Springer, Berlin, pp. 57–106.

Sorensen, F. C. and Ferrell, W. K. (1973) Photosynthesis and growth of Douglas fir seedlings when grown in different environments. *Canadian Journal of Botany*, **51**, 1689–98.

Sørensen, T. (1941) Temperature relations and phenology of the northeast Greenland flowering plant. *Meddelelser om Grønland*, **125**, 1–305.

Sørensen, T. (1953) A revision of the Greenland species of *Puccinellia* Parl. with contributions to our knowledge of the Arctic *Puccinellia* flora in general. *Meddelelser om Grønland*, **136**, 1–179.

Soriano, A. (1979) Distribution of grasses and grasslands of South America, in *Ecology of Grasslands and Bamboolands in the World*, (ed. M. Numata), Junk, The Hague, pp. 84–91.

Sorokin, Y. I. (1978) Microbial production in the coral-reef community. *Archiv für Hydrobiologie*, **83**, 281–323.

Sorokin, Y. I. (1981) Microheterotrophic organisms in marine ecosystems, in *Analysis of Marine Ecosystems*, (ed. A. R. Longhurst), Academic Press, London, pp. 293–42.

Sournia, A. (1976) Oxygen metabolism of a fringing reef in French Polynesia. *Helgoländer Wissenschaftliche Meeresuntersuchungen*, **28**, 401–10.

Southward, A. J. and Southward, E. C. (1978) Recolonization of rocky shores in Cornwall after use of toxic dispersants to clean up the *Torrey Canyon* spill. *Journal of the Fisheries Research Board of Canada*, **35**, 682–706.

Spain, A. V. (1984) Litterfall and the standing crop of litter in three tropical Australian rainforests. *Journal of Ecology*, **72**, 947–61.

Sparling, J. H. (1967) Assimilation rates of some woodland herbs in Ontario. *Botanical Gazette*, **128**, 160–8.

Specht, R. L. (1957) Dark Island heath (Ninety-mile Plain, South Australia). IV. Soil moisture patterns produced by rainfall interception and stem-flow. *Australian Journal of Botany*, **5**, 137–50.

Specht, R. L. (1969) A comparison of the sclerophyllous vegetation characteristic of mediterranean type climates in France, California, and southern Australia. *Australian Journal of Botany*, **17**, 277–92.

Specht, R. L. (1973) Structure and functional response of ecosystems in the mediterranean climate of Australia, in *Mediterranean Type Ecosystems: Origin and Structure*, (eds F. di Castri and H. A. Mooney), Springer, New York, pp. 113–20.

Specht, R. L. (1981) Mallee ecosystems in southern Australia, in *Ecosystems of the World Vol. 11. Mediterranean-type Shrublands*, (eds F. di Castri, D. W. Goodall and R. L. Specht), Elsevier, Amsterdam, pp. 203–31.

Specht, R. L. and Moll, E. J. (1983) Mediterranean-type heathlands and sclerophyllous shrublands of the world: an overview, in *Mediterranean Type Ecosystems: The Role of Nutrients*, (eds F. J. Kruger, D. T. Mitchell and J. U. M. Jarvis), Springer, Berlin, pp. 41–65.

Specht, R. L. and Rayson, P. (1957a) Dark Island heath (Ninety-mile Plain, South Australia). I. Definition of the ecosystem. *Australian Journal of Botany*, **5**, 52–85.

Specht, R. L. and Rayson, P. (1957b) Dark Island heath (Ninety-mile Plain, South Australia). III. The root systems. *Australian Journal of Botany*, **5**, 103–14.

Specht, R. L., Rayson, P. and Jackman, M. E. (1958) Dark Island heath (Ninety-mile Plain, South Australia). VI. Pyric succession: changes in composition, coverage, dry weight, and mineral nutrient status. *Australian Journal of Botany*, **6**, 59–88.

Speight, M. C. D. and Blackith, R. E. (1983) The animals, in *Ecosystems of the World Vol. 4A. Mires: Swamp, Bog, Fen and Moor – General Studies*, (ed. A. J. P. Gore), Elsevier, Amsterdam, pp. 349–65.

Spence, J. R. (1990) Seed rain in grassland, herbfield, snowbank, and fellfield in the alpine zone, Craigieburn Range, South Island, New Zealand. *New Zealand Journal of Botany*, **28**, 439–50.

Spetzman, L. A. (1959) Vegetation of the Arctic Slope of Alaska. *United States Geological Survey Professional Paper 302–B*, 19–58.

Stace, C. A. (1966) The use of peridermal characters in phylogenetic considerations. *New Phytologist*, **65**, 304–18.

Stace, H. C. T., Hubble, G. D., Brewer, R., *et al.* (1968) *A Handbook of Australian Soils*, Rellim Technical Publications, Glenside.

Stanek, W. (1977) Ontario clay belt peatlands – are they suitable for forest drainage. *Canadian Journal of Forest Research*, **7**, 656–65.

Stanton, N. L. (1988) The underground in grasslands. *Annual Review of Ecology and Systematics*, **19**, 573–89.

Stark, N. (1970) The nutrient content of plants and soils from Brazil and Surinam. *Biotropica*, **2**, 51–60.

Stark, N. M. and Jordan, C. F. (1978) Nutrient retention by the root mat of an Amazonian rain forest. *Ecology*, **59**, 434–7.

Starker, T. J. (1934) Fire resistance in the forest. *Journal of Forestry*, **32**, 462–7.

Stearns, F. (1978) Management potential: summary and recommendations, in *Freshwater Wetlands: Ecological Processes and Management Potential*, (eds R. E. Good,

D. F. Whigham and R. L. Simpson), New York, Academic Press, pp. 357–63.

Stebbins, G. L. (1971) *Chromosomal Evolution in Higher Plants*, Addison-Wesley, London.

Steers, J. A. (1977) Physiography, in *Ecosystems of the World Vol. 1. Wet Coastal Ecosystems*, (ed. V. J. Chapman), Elsevier, Amsterdam, pp. 31–60.

Stein, R. A. and Magnuson, J. J. (1976) Behavioural response of crayfish to a fish predator. *Ecology*, **57**, 751–61.

Stein, R. A., Threlkeld, S. T., Sandgren, C. D. *et al.* (1988) Size structured interactions in lake communities, in *Complex Interactions in Lake Communities*, (ed. S. R. Carpenter), Springer, New York, pp. 161–79.

Steinke, T. D. and Ward, C. J. (1987) Degradation of mangrove leaf litter in the St Lucia Estuary as influenced by season and exposure. *South African Journal of Botany*, **53**, 323–8.

Steinmann, F. and Bräendle, R. (1984) Carbohydrate and protein metabolism in the rhizomes of the bullrush (*Schoenplectus lacustris* (L.) Palla in relation to natural development of the whole plant. *Aquatic Botany*, **19**, 53–63.

Stephenson, T. A. and Stephenson, A. (1954) Life between tide-marks in North America IIIB. Nova Scotia and Prince Edward Island: the geographical features of the region. *Journal of Ecology*, **42**, 46–70.

Sterner, R. W. (1986) Herbivores' direct and indirect effects on algal populations. *Science*, **231**, 605–7.

Steward, K. K. and Ornes, W. H. (1975) The autecology of sawgrass in the Florida Everglades. *Ecology*, **56**, 162–71.

Stiles, F. G. (1975) Ecology, flowering phenology, and hummingbird pollination of some Costa Rican *Heliconia* species. *Ecology*, **56**, 285–301.

Stocker, O. and Holdheide, W. (1938) Die assimilation Helgoländer Gezeitenalgen während der Ebbezeit. *Zeitschrift für Botanik*, **32**, 1–59.

Stockner, J. G. and Antia, N. J. (1986) Algal picoplankton from marine and freshwater ecosystems: a multidisciplinary perspective. *Canadian Journal of Fisheries and Aquatic Sciences*, **43**, 2472–503.

Stockner, J. G. and Shortreed, K. S. (1989) Algal picoplankton production and contribution to food-webs in oligotrophic British Columbia lakes. *Hydrobiologia*, **173**, 151–66.

Stoddart, D. R. (1969) Ecology and morphology of recent coral reefs. *Biological Reviews*, **44**, 433–98.

Stokes, P. M. (1986) Ecological effects of acidification on primary producers in aquatic systems. *Water, Air, and Soil Pollution*, **30**, 421–38.

Stone, E. C. (1951) The stimulative effect of fire on the flowering of the golden brodiaea (*Brodiaea ixioides* Wats. var. *lugens* Jeps.). *Ecology*, **32**, 534–7.

Stone, E. C. and Juhren, G. (1953) Fire stimulated germination. *California Agriculture*, **7**, 13–14.

Stoner, W. A. and Miller, P. C. (1975) Water relations of plant species in the wet coastal tundra at Barrow, Alaska. *Arctic and Alpine Research*, **7**, 109–24.

Strahler, A. H. and Strahler, A. N. (1992) *Modern Physical Geography*, 4th edn, Wiley, New York.

Stratton, D. A. (1989) Longevity of individual flowers in a Costa Rican cloud forest: ecological correlates and phylogenetic constraints. *Biotropica*, **21**, 308–15.

Strong, A. E. and Eadie, B. J. (1978) Satellite observations of calcium carbonate precipitations in the Great Lakes. *Limnology and Oceanography*, **23**, 877–87.

Succow, M. and Lange, E. (1984) The mire types of the German Democratic Republic, in *European Mires*, (ed. P. D. Moore), Academic Press, London, pp. 149–75.

Svedarsky, W. D., Buckley, P. E. and Feiro, T. A. (1986) The effect of 13 years of annual burning on an aspen-prairie ecotone in northwestern Minnesota, in *The Prairie: Past, Present and Future. Proceedings of the Ninth North American Prairie Conference*, (eds G. K. Clambey and R. H. Pemble), Tri-College University Centre for Environmental Studies, Fargo, pp. 118–22.

Sverdrup, H. U., Johnson, M. W. and Fleming, R. H. (1942) *The Oceans: Their Physics, Chemistry, and General Biology*, Prentice-Hall, New York.

Svoboda, J. (1977) Ecology and primary production of raised beach communities, Truelove Lowland, in *Truelove Lowland, Devon Island, Canada: A High Arctic Ecosystem*, (ed. L. C. Bliss), The University of Alberta Press, Edmonton, pp. 185–216.

Swaine, M. D. and Hall, J. B. (1983) Early succession on cleared forest land in Ghana. *Journal of Ecology*, **71**, 601–27.

Swaine, M. D. and Whitmore, T. C. (1988) On the definition of ecological species groups in tropical rain forests. *Vegetatio*, **75**, 81–6.

Swaine, M. D., Lieberman, D. and Putz, F. E. (1987) The dynamics of tree populations in tropical forest: a review. *Journal of Tropical Ecology*, **3**, 359–66.

Swank, S. E. and Oechel, W. C. (1991) Interactions among the effects of herbivory, competition, and resource limitation on chaparral herbs. *Ecology*, **72**, 104–15.

Swanson, G. A. and Duebbert, H. F. (1989) Wetland habitats of waterfowl in the Prairie pothole region, in *Northern Prairie Wetlands*, (ed. A. van der Valk), Iowa State University Press, Ames, pp. 228–67.

Swift, M. J. and Anderson, J. M. (1989) Decomposition, in Ecosystems of the World Vol. 14B. Tropical Rain Forest Ecosystems: Biogeographical and Ecological Studies, (eds H. Leith and M. J. A. Werger), Elsevier, Amsterdam, pp. 547–69.

Syers, J. K. and Walker, T. W. (1969a) Phosphorus transformations in a chronosequence of soils developed on wind-blown sand in New Zealand. I. Total and organic phosphorus. *Journal of Soil Science*, **20**, 57–64.

Syers, J. K. and Walker, T. W. (1969b) Phosphorus transformations in a chronosequence of soils developed on wind-blown sand in New Zealand. II. Inorganic phosphorus. *Journal of Soil Science*, **20**, 318–24.

Sykes, J. M. and Bunce, R. G. H. (1970) Fluctuations in litter-fall in a mixed deciduous woodland over a three-year period 1966–68. *Oikos*, **21**, 326–9.

Szarek, S. R. (1979) Primary production in four North American deserts: indices of efficiency. *Journal of Arid Environments*, **2**, 187–209.

Szarek, S. R. and Ting, I. P. (1975) Physiological responses to rainfall in *Opuntia basilaris* (Cactaceae). *American Journal of Botany*, **62**, 602–9.

Szczepanski, A. (1973) Chlorophyll in the assimilation parts of helophytes. *Polskie Archiwum Hydrobiologii*, **20**, 67–71.

Tait, R. V. (1972) *Elements of Marine Biology: An Introductory Course*, 2nd edn, Butterworths, London.

Tallis, J. H. (1983) Changes in wetland communities, in *Ecosystems of the World Vol. 4A. Mires: Swamp, Bog, Fen and Moor – General Studies*, (ed. A. J. P. Gore), Elsevier, Amsterdam, pp. 311–47.

Tam, N. F. Y., Vrijmoed, L. L. P. and Wong, Y. S. (1990) Nutrient dynamics associated with leaf decomposition in a small subtropical mangrove community in Hong Kong. *Bulletin of Marine Science*, **47**, 68–78.

Tamm, C. O. (1954) Some observations on the nutrient turn-over in a bog community dominated by *Eriophorum vaginatum* L. *Oikos*, **5**, 186–94.

Tansley, A. G. (1953) *The British Islands and Their Vegetation*, Cambridge, Cambridge University Press.

Tasker, R. and Smith, H. (1977) The function of phytochrome in the natural environment – V. Seasonal changes in radiant energy quality in woodlands. *Photochemistry and Photobiology*, **26**, 487–91.

Taylor, C. J. (1960) *Synecology and Silviculture in Ghana*, Nelson, Edinburgh.

Taylor, C. R. (1969) The eland and the oryx. *Scientific American*, **220**, 88–95.

Taylor, H. C. (1978) Capensis, in *Biogeography and Ecology of Southern Africa*, (ed. M. J. A. Werger), Junk, The Hague, pp. 171–229.

Teal, J. M., Valiela, I. and Berlo, D. (1979) Nitrogen fixation by rhizosphere and free-living bacteria in salt marsh sediments. *Limnology and Oceanography*, **24**, 126–32.

Tedrow, J. C. F. (1977) *Soils of the Polar Landscapes*, Rutgers University Press, New Brunswick.

Teeri, J. A. and Stowe, L. G. (1976) Climatic patterns and the distribution of C_4 grasses in North America. *Oecologia*, **23**, 1–12.

Temple, S. A. (1977) Plant–animal mutualism: coevolution with dodo leads to near extinction of plant. *Science*, **197**, 885–6.

Teramura, A. H. (1983) Effects of ultraviolet-B radiation on the growth and yield of crop plants. *Physiologia Plantarum*, **58**, 415–27.

Tevis, L. (1958) Interrelations between the harvester ant *Veromessor pergandei* (Mayr) and some desert ephemerals. *Ecology*, **39**, 695–704.

Thirgood, J. V. (1981) *Man and the Mediterranean Forest*, Academic Press, London.

Thom, B. G. (1967) Mangrove ecology and deltaic geomorphology: Tabasco, Mexico. *Journal of Ecology*, **55**, 301–43.

Thomas, D. B. (1979) Patterns of abundance of some tenebrionid beetles in the Mojave Desert. *Environmental Entomology*, **8**, 568–74.

Thomas, P. A. and Wein, R. W. (1985) The influence of shelter and the hypothetical effect of fire severity on the postfire establishment of confiers from seed. *Canadian Journal of Forest Research*, **15**, 148–55.

Thompson, C. H. (1981) Podzol chronosequences on coastal dunes of eastern Australia. *Nature*, **291**, 59–61.

Thompson, K. and Hamilton, A. C. (1983) Peatlands and swamps of the African continent, in *Ecosystems of the World Vol 4B. Mires: Swamp, Bog, Fen and Moor* (ed. A. J. P. Gore), Elsevier, Amsterdam, pp. 331–73.

Thompson, J. N. (1980) Treefalls and colonization patterns of temperate forest herbs. *The American Midland Naturalist*, **104**, 176–84.

Thompson, J. T. (1977) Ecological deterioration: local-level rule-making and enforcement problems in Niger, in *Desertification*, (ed. M. H. Glantz), Westview, Boulder, pp. 57–79.

Thornthwaite, C. W. (1948) An approach toward a rational classification of climate. *Geographical Review*, **38**, 55–94.

Thornton, J. A. (1987) Aspects of eutrophication management in tropical/sub-tropical regions, *Journal of the Limnological Society of Southern Africa*, **13**, 25–43.

Threlkeld, S. T. (1986) Resource-mediated demographic variation during the midsummer succession of a cladoceran community. *Freshwater Biology*, **16**, 673–83.

Thrower, N. J. W. and Bradbury, D. E. (1973) The physiography of the mediterranean lands with special emphasis on California and Chile, in *Mediterranean Type Ecosystems: Origin and Structure*, (eds F. di Castri and H. A. Mooney), Springer, New York, pp. 37–52.

Tieszen, L. L. (1978) Photosynthesis in the principal Barrow, Alaska species: a summary of field and laboratory responses, in *Vegetation and Production Ecology of an Alaskan Arctic Tundra*, (ed. L. L. Tieszen), Springer, New York, pp. 241–68.

Tieszen, L. L., Senyimba, M. M., Imbamba, S. K. and Troughton, J. H. (1979) The distribution of C_3 and C_4 grasses and carbon isotope discrimination along an altitudinal and moisture gradient in Kenya. *Oecologia*, **37**, 337–50.

Tieszen, L. L., Lewis, M. C., Miller, P. C., *et al.* (1981) An analysis of processes of primary production in tundra growth forms, in *Tundra Ecosystems: A Comparative Analysis*, (eds L. C. Bliss, O. W. Heal and J. J. Moore), Cambridge University Press, Cambridge, pp. 285–356.

Tillman, U. and Lampert, W. (1984) Competitive ability of differently sized *Daphnia* species: an experimental test. *Freshwater Ecology*, **2**, 311–23.

Tilman, D., Kilham, S. S. and Kilham, P. (1982) Phytoplankton community ecology, the role of limiting nutrients. *Annual Review of Ecology and Systematics*, **13**, 349–72.

Tilzer, M. M. (1973) Diurnal periodicity in the phytoplankton assemblage of a high mountain lake. *Limnology and Oceanography*, **18**, 15–30.

Tilzer, M. M. (1984) Estimation of phytoplankton loss rates from daily photosynthetic rates and observed biomass changes in Lake Constance. *Journal of Plankton Research*, **6**, 309–24.

Tilzer, M. M. and Goldman, C. R. (1978) Importance of mixing, thermal stratification and light adaptation for phytoplankton productivity in Lake Tahoe (California–Nevada). *Ecology*, **59**, 810–21.

Tilzer, M., Pyrina, I. L. and Westlake, D. F. (1980) Phytoplankton, in *The Functioning of Freshwater Ecosystems*, (eds E. D. Le Cren and R. H. Lowe-McConnell), Cambridge University Press, Cambridge, pp. 163–70.

Timoney, K. P., La Roi, G. H., Zoltai, S. C. and Robinson, A. L. (1992) The high subarctic forest-tundra of northwestern Canada: position, width, and vegetation gradients in relation to climate. *Arctic*, **45**, 1–9.

Tinley, K. L. (1982) The influence of soil moisture balance on ecosystem patterns in southern Africa, in *Ecology of Tropical Savannas*, (eds B. J. Huntley and B. H. Walker), Springer, Berlin, pp. 175–92.

Tolba, M. K. (1992) *Saving Our Planet: Challenges and Hopes*, Chapman & Hall, London.

Tolba, M. K. and El-Kholy, O. A. (eds.) (1992) *The World Environment 1972–1992: Two Decades of Challenge*, Chapman & Hall, London.

Toledo, V. M. (1977) Pollination of some rain forest plants by non-hovering birds in Veracruz, Mexico. *Biotropica*, **9**, 262–7.

Tomanek, G. W. and Albertson, F. W. (1957) Variations in cover, composition, production, and roots of vegetation on two prairies in western Kansas. *Ecological Monographs*, **27**, 267–81.

Tomaselli, R. (1981) Main physiognomic types and geographic distribution of shrub systems related to mediterranean climates, in *Ecosystems of the World Vol. 11. Mediterranean-type Shrublands*, (eds F. di Castri, D. W. Goodall and R. L. Specht), Elsevier, Amsterdam, pp. 95–106.

Tomlinson, G. H. (1990) *Effects of Acid Deposition on the Forests of Europe and North America*, CRC Press, Boca Raton.

Tomlinson, P. B. (1983) Structural elements of the rain forest, in *Ecosystems of the World Vol. 14A. Tropical Rain Forest Ecosystems*, (ed. F. B. Golley), Elsevier, Amsterdam, pp. 9–28.

Tomlinson, P. B. (1986) *The Botany of Mangroves*, Cambridge University Press, Cambridge.

Toms, R. G. (1985) River pollution control since 1974. *Water Pollution Control*, **84**, 178–86.

Torres, J. C., Gutiérrez, J. R. and Fuentes, E. R. (1980) Vegetative responses to defoliation of two Chilean matorral shrubs. *Oecologia*, **46**, 161–3.

Tothill, J. C. (1969) Soil temperatures and seed burial in relation to the performance of *Heteropogon contortus* and *Themeda australis* in burnt native woodland pastures in eastern Queensland. *Australian Journal of Botany*, **17**, 269–75.

Trabaud, L. (1981) Man and fire: impacts on mediterranean vegetation, in *Ecosystems of the World Vol. 11. Mediterranean-type Shrublands*, (eds F.di Castri, D. W. Goodall and R. L. Specht). Elsevier, Amsterdam, pp. 523–37.

Tranquillini, W. (1979) *Physiological Ecology of the Alpine Timberline*, Springer, Berlin.

Trapnell, C. G. (1959) Ecological results of woodland burning experiments in Northern Rhodesia. *Journal of Ecology*, **49**, 129–68.

Trapnell, C. G., Friend, M. T., Chamberlain, G. T. and Birch, H. F. (1976) The effects of fire and termites on a Zambian woodland soil. *Journal of Ecology*, **64**, 577–88.

Trlica, M. J. and Schuster, J. L. (1969) Effects of fire on grasses of the Texas High Plains. *Journal of Range Management*, **22**, 329–33.

Trollope, W. S. W. (1982) Ecological effects of fire in South African savannas, in *Ecology of Tropical Savannas*, (eds B. J. Huntley and B. H. Walker), Springer, Berlin, pp. 292–306.

Trollope, W. S. W. (1984) Fire in savanna, in *Ecological Effects of Fire in South African Ecosystems*, (eds P. de V. Booysen and N. M. Tainton), Springer, Berlin, pp. 149–75.

Trumble, H. C. and Woodroffe K. (1954) The influence of climatic factors on the reaction of desert shrubs to grazing by sheep, in *Biology of Deserts*, (ed. J. L. Cloudsley-Thompson), Institute of Biology, London, pp. 129–47.

Tseplyaev, V. P. (1965) *The Forests of the U. S. S.R.*, Israel Program for Scientific Translations, Jerusalem.

Tucker, C. J., Dregne, H. E. and Newcomb, W. W. (1991) Expansion and contraction of the Sahara Desert from 1980 to 1990. *Science*, **253**, 299–301.

Tunstall, B. R., Walker, J. and Gill, A. M. (1976) Temperature distribution around synthetic trees during grass fires. *Forest Science*, **22**, 269–76.

Turner F. B. and Randall, D. C. (1987) The phenology of desert shrubs in southern Nevada. *Journal of Arid Environments*, **13**, 119–28.

Turner, F. B. and Randall, D. C. (1989) Net production by shrubs and winter annuals in southern Nevada. *Journal of Arid Environments*, **17**, 23–36.

Turner, I. M. (1990) Tree seedling growth and survival in a Malaysian rain forest. *Biotropica*, **22**, 146–54.

Turner, J. and Singer, M. J. (1976) Nutrient distribution and cycling in a sub-alpine coniferous forest ecosystem. *Journal of Applied Ecology*, **13**, 295–301.

Turner, J. R. G. (1982) How do refuges produce biological diversity? Allopatry and parapatry, extinction and gene flow in mimetic butterflies, in *Biological Diversification in the Tropics*, (ed. G. T. Prance), Columbia University Press, New York, pp. 309–35.

Twilley, R. R., Lugo, A. E. and Patterson-Zucca, C. (1986) Litter production and turnover in basin mangrove forests in southwest Florida. *Ecology*, **67**, 670–83.

Uhl, C. and Jordan, C. F. (1984) Succession and nutrient dynamics following forest cutting and burning in Amazonia. *Ecology*, **65**, 1476–90.

Uhl, C. and Vieira, I. C. G. (1989) Ecological impacts of selective logging in the Brazilian Amazon: a case study from the Paragominas Region of the State of Para. *Biotropica*, **21**, 98–106.

Uhl, N. W. and Moore, H. E. (1977) Correlations of inflorescence, flower structure, and floral anatomy with pollination in some palms. *Biotropica*, **9**, 170–90.

Umamaheswararo, M. and Sreeramulu, T. (1964) An ecological study of some intertidal algae of the Visakhapatnam coast. *Journal of Ecology*, **52**, 595–616.

Ungar, I. A. (1978) Halophyte seed germination. *Botanical Review*, **44**, 233–64.

Urban, N. R. and Eisenreich, S. J. (1988) Nitrogen cycling in a forested Minnesota bog. *Canadian Journal of Botany*, **66**, 435–49.

Urquhart, C. and Gore, A. J. P. (1973) The redox characteristics of four peat profiles. *Soil Biology and Biochemistry*, **5**, 659–72.

US Department of Agriculture (1948) *Woody-plant Seed Manual*. United States Department of Agriculture, Forest Service, Miscellaneous Publication 654, Washington, D. C.

Valentine, H. T., Wallner, W. E. and Wargo, P. M. (1983) Nutritional changes in host foliage during and after defoliation and their relation to the weight of gypsy moth pupae. *Oecologia*, **57**, 298–302.

Valiela, I., Teal, J. M. and Deuser, W. G. (1978) The nature of growth forms in the salt marsh grass *Spartina alternifolia*. *American Naturalist*, **112**, 461–70.

Valiela, I., Koumjian, L., Swain, T. *et al.* (1979) Cinnamic acid inhibition of detritus feeding. *Nature*, **280**, 55–7.

Valiela, I., Teal, J. M., Volkmann, S. *et al.* (1978) Nutrient and particulate fluxes in a salt marsh ecosystem: tidal exchanges and inputs by precipitation and groundwater. *Limnology and Oceanography*, **23**, 798–812.

Valiela, I., Teal, J. M., Allen, S. D. *et al* (1985) Decomposition in salt marsh ecosystems: the phases and major factors affecting disappearance of above-ground organic matter. *Journal of Experimental Marine Biology and Ecology*, **89**, 29–54.

Van, T. K., Haller, W. T. and Bowes, G. (1976) Comparison of the photosynthetic characteristics of three submersed aquatic plants. *Plant Physiology*, **58**, 761–8.

van Cleve, K., Oliver, L., Schlentner, R., Viereck, L. A. and Dyrness, C. T. (1983a) Productivity and nutrient cycling in taiga forest ecosystems. *Canadian Journal of Forest Research*, **13**, 747–66.

van Cleve, K., Dyrness, C. T., Viereck, L. A., *et al.* (1983b) Taiga ecosystems in interior Alaska. *BioScience*, **33**, 39–44.

van der Hammen, T. (1974) The Pleistocene changes of vegetation and climate in tropical South America. *Journal of Biogeography*, **1**, 3–26.

van der Pijl, L. (1955) Some remarks on myrmecophytes. *Phytomorphology*, **5**, 190–200.

van der Pijl, L. and Dodson, C. H. (1966) *Orchid Flowers*, University of Miami Press, Coral Gables.

van der Valk, A. G. (1974) Mineral cycling in coastal foredune plant communities in Cape Hatteras National Seashore. *Ecology*, **55**, 1349–58.

van der Valk, A. G. (1977) The role of leaves in the uptake of nutrients by *Uniola paniculata* and *Ammophila breviligulata*. *Chesapeake Science*, **18**, 77–9.

van der Valk, A. G. (1986) The impact of litter and annual plants on recruitment from the seed bank of a lacustrine wetland. *Aquatic Botany*, **24**, 13–26.

van der Valk, A. G. and Attiwill, P. M. (1984) Decomposition of leaf and root litter of *Avicennia marina* at Westernport Bay, Victoria, Australia. *Aquatic Botany*, **18**, 205–21.

van der Valk, A. G. and Davis, C. B. (1978) The role of seed banks in the vegetation dynamics of prairie glacial marshes. *Ecology*, **59**, 322–35.

van der Valk, A. G. and Davis, C. B. (1979) A reconstruction of the recent vegetational history of a prairie marsh, Eagle Lake, Iowa, from its seed bank. *Aquatic Botany*, **6**, 29–51.

van Donselaar-Ten Bokkel Huinink, W. A. E. (1966) Structure, root systems and periodicity of savanna plants and vegetations in northern Surinam. *Wentia*, **17**, 1–162.

van Eerdt, M. M. (1985) The influence of vegetation on erosion and accretion in salt marshes of the Oosterschelde, The Netherlands. *Vegetatio*, **62**, 367–73.

Vanni, M. J. (1986) Competition in zooplankton communities: suppression of small species by *Daphnia pulex.*, *Limnology and Oceanography*, **31**, 1039–56.

van Raalte, C. D., Valiela, I. and Teal, J. M. (1976) Production of epibenthic salt marsh algae: light and nutrient limitation. *Limnology and Oceanography*, **21**, 862–72.

van Steenis, C. G. G. J. (1968) Frost in the tropics, in *Proceedings of the Symposium on Recent Advances in Tropical Ecology*, (eds R. Misra and B. Gopal), Part 1, The International Society for Tropical Ecology, Faridabad, pp. 154–67.

van Valen, L. (1975) Life, death, and energy of a tree. *Biotropica*, **7**, 260–9.

van Wagner, C. E. (1974) *Structure of the Canadian Forest Fire Weather Index*. Canadian Forestry Service Publication No. 1333. Department of the Environment, Ottawa.

van Wagner, C. E. (1978) Age–class distribution and the forest fire cycle. *Canadian Journal of Forest Research*, **8**, 220–27.

van Wagner, C. E. (1983) Fire behaviour in northern conifer forests and shrublands, in *The Role of Fire in Northern Circumpolar Ecosystems*, (eds R. W. Wein and D. A. MacLean), Wiley, Chichester, pp. 65–95.

van Wilgen, B. W. (1981) Some effects of fire on fynbos plant community composition and structure at Jonkershoek, Stellenbosch. *South African Forestry Journal*, **118**, 42–55.

van Zinderen Bakker, E. M. (1978) Quaternary vegetation changes in southern Africa, in *Biogeography and Ecology of Southern Africa*, (ed. M. J. A. Werger), Junk, The Hague, pp. 131–43.

Vareschi, V. (1962) La Quema como factor ecologico en los llanos. *Sociedad Venezolana de Ciencias Naturales*, **23**, 9–26.

Vasek, F. C. (1980a) Creosote bush: long-lived clones in the Mojave Desert. *American Journal of Botany*, **67**, 246–55.

Vasek, F. C. (1980b) Early successional stages in Mojave Desert scrub vegetation. *Israel Journal of Botany*, **28**, 133–48.

Vaughan, T. A. (1986) *Mammalogy*, 3rd edn, Saunders, Philadelphia.

Vázquez-Yanes, C. (1974) Studies on the germination of seeds of *Ochroma lagopus* Swartz. *Turrialba*, **24**, 176–9.

Vázquez-Yanes, C. and Smith, H. (1982) Phytochrome control of seed germination in the tropical rain forest pioneer trees *Cecropia obtusifolia* and *Piper auritum* and its ecological significance. *New Phytologist*, **92**, 477–85.

Veblen, T. T., Schlegel, F. M. and Oltremari, J. V. (1983) Temperate broad-leaved evergreen forests of South America, in *Ecosystems of the World Vol. 10. Temperate Broad-leaved Evergreen Forests*, (ed. J. D. Ovington), Elsevier, Amsterdam, pp. 5–31.

Verhoeven, J. T. A. (1986) Nutrient dynamics in minerotrophic peat mires. *Aquatic Botany*, **25**, 117–37.

Verhoeven, J. T. A., Maltby, E. and Schmitz, M. B. (1990) Nitrogen and phosphorus mineralization in fens and bogs. *Journal of Ecology*, **78**, 713–26.

Verry, E. S. (1975) Streamflow chemistry and nutrient yields from upland-peatland watersheds in Minnesota. *Ecology*, **56**, 1149–57.

Vesey-FitzGerald, D. F. (1960) Grazing succession among East African game animals. *Journal of Mammalogy*, **41**, 161–72.

Vesk, M. and Jeffrey, S. W. (1977) Effect of blue-green light on photosynthetic pigments and chloroplast structure in unicellular marine algae from six classes. *Journal of Phycology*, **13**, 280–8.

Viereck. L. A. (1983) The effects of fire in black spruce ecosystems of Alaska and northern Canada, in *The Role of Fire in Northern Circumpolar Ecosystems*, (eds R. W. Wein and D. A. MacLean), Wiley, Chichester, pp. 201–20.

Vincent, W. F., Wurtsbaugh, W., Neale, P. J. and Richerson, P. J. (1986) Polymixis and algal production in a tropical lake: latitudinal effects on the seasonality of photosynthesis. *Freshwater Biology*, **16**, 781–803.

Vine, P. J. (1974) Effects of algal grazing and aggressive behaviour of the fishes *Pomacentrus lividus* and *Acanthurus sohal* on coral-reef ecology. *Marine Biology*, **24**, 131–6.

Vitt, D. H., Achuff, P. and Andrus, R. E. (1975) The vegetation and chemical properties of patterned fens in the Swan Hills, north central Alberta. *Canadian Journal of Botany*, **53**, 2776–95.

Vogl, R. J. (1974) Effects of fire on grasslands, in *Fire and Ecosystems*, (eds T. T. Kozlowski and C. E. Ahlgren), Academic Press, New York, pp. 139–94.

Vogt, K. A., Grier, C. C. and Vogt, D. J. (1986) Production, turnover, and nutrient dynamics of above- and below-ground detritus of world forests. *Advances in Ecological Research*, **15**, 303–77.

Vossbrinck, C. R., Coleman, D. C. and Woolley, T. A. (1979) Abiotic and biotic factors in litter decomposition in a semiarid grassland. *Ecology*, **60**, 265–71.

Vowinckel, T., Oechel, W. C. and Boll, W. G. (1975) The effect of climate on the photosynthesis of *Picea mariana* at the subarctic tree line. 1. Field measurements. *Canadian Journal of Botany*, **53**, 604–20.

Wager, H. G. (1938) Growth and survival of plants in the arctic. *Journal of Ecology*, **26**, 390–410.

Waibel, L. (1948) Vegetation and land use in the Planalto Central of Brazil. *Geographical Review*, **38**, 529–54.

Waisel, Y., Liphschitz, N. and Kuller, Z. (1972) Patterns of water movement in trees and shrubs. *Ecology*, **53**, 520–3.

Waksman, S. A. and Purvis, E. R. (1932) The microbiological population of peat. *Soil Science*, **34**, 95–109.

Walker, B. D. and Peters, T. W. (1977) Soils of Truelove Lowland and Plateau, in *Truelove Lowland, Devon Island, Canada: A High Arctic Ecosystem*, (ed. L. C. Bliss), University of Alberta Press, Edmonton, pp. 31–62.

Walker, D. (1970) Direction and rate in some British post-glacial hydroseres, in *Studies in the Vegetational History of the British Isles*, (eds D. Walker and R. G. West), Cambridge University Press, Cambridge, pp. 117–39.

Walker, D. (1982) Speculations on the origin and evolution of Sunda–Sahul rain forests, in *Biological Diversification in the Tropics*, (ed. G. T. Prance), Columbia University Press, New York, pp. 554–75.

Walker, D. and Walker, P. M. (1961) Stratigraphic evidence of regeneration in some Irish bogs. *Journal of Ecology*, **49**, 169–85.

Walker, J. and Gillison, A. N. (1982) Australian savannas, in *Ecology of Tropical Savannas*, (eds B. J. Huntley and B. H. Walker), Springer, Berlin, pp. 5–24.

Wallace, A., Romney, E. M. and Hunter, R. B. (1978a) Nitrogen cycle in the northern Mojave Desert: implications and predictions, in *Nitrogen in Desert Ecosystems*, (eds N. E. West and J. Skujiņš), Dowden, Hutchinson and Ross, Stroudsburg, pp. 207–18.

Wallace, A., Romney, E. M., Kleinkopf, G. E. and Soufi, S. M. (1978b) Uptake of mineral forms of nitrogen by desert plants, in *Nitrogen in Desert Ecosystems*, (eds N. E. West and J. Skujiņš), Dowden, Hutchinson and Ross, Stroudsburg, pp. 130–51.

Wallace, C. C., Watt, A. and Bull, G. D. (1986) Recruitment of juvenile corals onto coral tables preyed upon by *Acanthaster planci*. *Marine Ecology Progress Series*, **32**, 299–306.

Wallace, L. L. and Dunn, E. L. (1980) Comparative photosynthesis of three gap phase successional tree species. *Oecologia*, **45**, 331–40.

Wallen, B. (1980) Changes in structure and function of *Ammophila* during primary succession. *Oikos*, **34**, 227–38.

Wallen, B., Falkengren-Grerup, U. and Malmer, N. (1988) Biomass, productivity and relative rate of photosynthesis of *Sphagnum* at different water levels on a South Swedish peat bog. *Holarctic Ecology*, **11**, 70–6.

Wallen, C. C. (ed.) (1970) *World Survey of Climatology Vol. 5. Climates of Northern and Western Europe.* Elsevier, Amsterdam.

Walsh, J. J. (1971) Relative importance of habitat variables in predicting the distribution of phytoplankton at the ecotone of the Antarctic upwelling ecosystem. *Ecological Monographs*, **41**, 291–309.

Walsh, J. J. (1981) Shelf-sea ecosystems, in *Analysis of Marine Ecosystems*, (ed. A. R. Longhurst), Academic Press, London, pp. 159–96.

Walsh, J. J. (1983) Death in the sea: enigmatic phytoplankton losses. *Progress in Oceanography*, **12**, 1–86.

Walsh, T. and Barry, T. A. (1958) The chemical composition of some Irish peats, *Proceedings of the Royal Irish Academy*, **59**, 305–28.

Walter, H. (1985) *Vegetation of the Earth*, 3rd edn, Springer, Berlin.

Walter, H. (1986) The Namib Desert, in *Ecosystems of the World Vol. 12B. Hot Deserts and Arid Shrublands*, (eds M. Evanari, I. Noy-Meir and D. W. Goodall), Elsevier, Amsterdam, pp. 245–82.

Walter, H. and Box, E. O. (1983a) Caspian Lowland biome, in *Ecosystems of the World Vol. 5. Temperate Deserts and Semi-deserts*, (ed. N. E. West), Elsevier, Amsterdam, pp. 9–41.

Walter, H. and Box, E. O. (1983b) Middle Asian deserts, in *Ecosystems of the World Vol. 5. Temperate Deserts and Semi-deserts*, (ed. N. E. West), Elsevier, Amsterdam, pp. 79–104.

Walter, H. and Box, E. O. (1983c) The Karakum Desert, an example of a well-studied eu-biome, in *Ecosystems of the World Vol. 5. Temperate Deserts and Semi-deserts*, (ed. N. E. West), Elsevier, Amsterdam, pp. 105–59.

Walter, H. and Box, E. O. (1983d) The deserts of central Asia, in *Ecosystems of the World Vol. 5. Temperate Deserts and*

Semi-deserts, (ed. N. E. West), Elsevier, Amsterdam, pp. 193–236.

Wanders, J. B. W. (1976a) The role of benthic algae in the shallow reef of Curaçao (Netherlands Antilles). I. Primary productivity in the coral reef. *Aquatic Botany*, **2**, 235–70.

Wanders, J. B. W. (1976b) The role of benthic algae in the shallow reef of Curaçao (Netherlands Antilles). II. Primary productivity of the *Sargassum* beds on the north-east coast submarine plateau. *Aquatic Botany*, **2**, 327–35.

Wang, C.-W. (1961) *The Forests of China*, Maria Moors Cabot Foundation Publ. 5, Harvard University, Cambridge.

Warburg, M. R. (1966) On the water economy of several Australian geckos, agamids, and skinks. *Copeia*, **2**, 230–5.

Ward, P. (1965) Feeding ecology of the Black-faced Dioch *Quelea quelea* in Nigeria. *Ibis*, **107**, 173–214.

Ward, P. (1971) The migration patterns of *Quelea quelea* in Africa. *Ibis*, **113**, 275–97.

Wardle, P. (1963) Evolution and distribution of the New Zealand flora, as affected by Quaternary climates. *New Zealand Journal of Botany*, **1**, 3–17.

Wardle, P. (1968) Engelmann spruce (*Picea engelmannii* Engel.) at its upper limits on the Front Range, Colorado. *Ecology*, **49**, 483–95.

Wardle, P. (1974) Alpine timberlines, in *Arctic and Alpine Environments*, (eds J. D. Ives and R. G. Barry), Methuen, London, pp. 371–402.

Wardle, P., Bulfin, M. J. A. and Dugdale, J. (1983) Temperate broad-leaved evergreen forests of New Zealand, in *Ecosystems of the World Vol. 10. Temperate Broad-leaved Evergreen Forests*, (ed. J. D. Ovington), Elsevier, Amsterdam, pp. 33–71.

Waring, R. H. and Running, S. W. (1978) Sapwood water storage: its contribution to transpiration and effect upon water conductance through the stems of old-growth Douglas fir. *Plant, Cell and Environment*, **1**, 131–40.

Wassen, M. J., Barendregt, A., Palczynski, A., *et al.* (1990) The relationship between fen vegetation gradients, groundwater flow and flooding in an undrained valley mire at Biebrza, Poland. *Journal of Ecology*, **78**, 1106–22.

Watts, I. E. M. (1969) Climates of China and Korea, in *Climates of Northern and Eastern Asia*, Elsevier, Amsterdam, pp. 1–117.

Watts, W. A. and Stuiver, M. (1980) Late Wisconsin climate of northern Florida and the origin of species-rich deciduous forest. *Science*, **210**, 325–7.

Waughman, G. J. (1980) Chemical aspects of the ecology of some south German peatlands. *Journal of Ecology*, **68**, 1025–46.

Weakly, H. E. (1943) A tree-ring record of precipitation in western Nebraska. *Journal of Forestry*, **41**, 816–19.

Weaver, H. (1974) Effects of fire on temperate forests: western United States, in *Fire and Ecosystems*, (eds T. T. Kozlowski and C. E. Ahlgren), Academic Press, New York, pp. 279–319.

Weaver, J. E. (1950) Effects of different intensities of grazing on depth and quantity of roots of grasses. *Journal of Range Management*, **3**, 100–113.

Weaver, J. E. (1954) *North American Prairie*, Johnsen Publishing Company, Lincoln.

Weaver, J. E. (1958) Classification of root systems of forbs of grassland and a consideration of their significance. *Ecology*, **39**, 391–401.

Weaver, J. E. (1968) *Prairie Plants and their Environment: A Fifty-Year Study in the Midwest*, University of Nebraska Press, Lincoln.

Weaver, J. E. and Albertson, F. W. (1943) Resurvey of grasses, forbs, and underground plant parts at the end of the Great Drought. *Ecological Monographs*, **13**, 63–117.

Weaver, J. E. and Albertson, F. W. (1956) *Grasslands of the Great Plains*, Johnsen Publishing Company, Lincoln.

Weaver, J. E. and Darland, R. W. (1949) Soil–root relationships of certain native grasses in various soil types. *Ecological Monographs*, **19**, 301–38.

Webb, K. L., DuPaul, W. D., Wiebe, W. *et al.* (1975) Enewetak (Eniwetok) Atoll: aspects of the nitrogen cycle on a coral reef. *Limnology and Oceanography*, **20**, 198–210.

Webb, W., Szarek, S., Lauenroth, W. and Smith, M. (1978) Primary productivity and water use in native forest, grassland, and desert ecosystems. *Ecology*, **59**, 1239–47.

Webber, P. J. (1974) Tundra primary productivity, in *Arctic and Alpine Environments*, (eds J. D. Ives and R. G. Barry), Methuen, London, pp. 445–73.

Webber, P. J. (1978) Spatial and temporal variation of the vegetation and its production, Barrow, Alaska, in *Vegetation and Production Ecology of an Alaskan Arctic Tundra*, (ed. L. L. Tieszen), Springer, New York, pp. 37–112.

Webber, P. J., Miller, P. C., Chapin, F. S. and McCown, B. H. (1980) The vegetation: pattern and succession, in *An Arctic Ecosystem*, (eds J. Brown, P. C. Miller, L. L. Tieszen and F. L. Bunnell), Dowden, Hutchinson and Ross, Stroudsburg, pp. 186–218.

Wee, Y. C. (1978) Vascular epiphytes of Singapore's wayside trees. *The Gardens' Bulletin, Singapore*, **31**, 114–26.

Wein, R. W. and Moore, J. M. (1977) Fire history and rotations in the New Brunswick Acadian Forest. *Canadian Journal of Forest Research*, **7**, 285–94.

Weiser, C. J. (1970) Cold resistance and injury in woody plants. *Science*, **169**, 1269–78.

Weisser, P., Weisser, J., Schrier, K. and Robres, L. (1975) Discovery of a subterranean species of *Neochilenia* (= *Chileorebutia, Thelocephalia*) in the Atacama Desert, Chile and notes about its habitat. *Excelsa*, **5**, 97–99, 104.

Weller, G. and Holmgren, B. (1974) The microclimates of the arctic tundra. *Journal of Applied Meterology*, **11**, 854–62.

Wellington, A. B. and Noble, I. R. (1985) Seed dynamics and factors limiting recruitment of the mallee *Eucalyptus incrassata* in semi-arid, south-eastern Australia. *Journal of Ecology*, **73**, 675–66.

Wells, J. W. (1957) Coral reefs. *Geological Society of America, Memoir 67*, vol. 1, 609–31.

Wells, P. V. (1969) The relation between mode of reproduction and extent of speciation in woody genera of the California chaparral. *Evolution*, **23**, 264–7.

Wells, P. V. (1970) Postglacial vegetational history of the Great Plains. *Science*, **167**, 1574–82.

Went, F. W. (1942) The dependence of certain annual plants on shrubs in southern California deserts. *Bulletin of the Torrey Botanical Club*, **69**, 100–14.

Went, F. W. (1949) Ecology of desert plants. II. The effect of rain and temperature on germination and growth. *Ecology*, **30**, 1–13.

Went, F. W. (1953) The effects of rain and temperature on plant distribution in the desert, in *Desert Research, International Symposium, Jerusalem*, 230–40.

Went, F. W. (1955) The ecology of desert plants. *Scientific American*, **192**, 68–75.

Went, F. W. (1969) A long term test of seed longevity. II. *Aliso*, **7**, 1–12.

Went, F. W. (1979) Germination and seedling behaviour of desert plants, in *Arid-land Ecosystems: Structure, Functioning and Management*, Vol. 1 (eds D. W. Goodall, R. A. Perry and K. M. W. Howes), Cambridge University, Cambridge, pp. 477–89.

Went, F. W. and Westergaard, M. (1949) Ecology of desert plants. III. Development of plants in the Death Valley National Monument, California. *Ecology*, **30**, 26–38.

Werger, M. J. A. (1978) The Karoo-Namib region, in *Biogeography and Ecology of Southern Africa*, (ed. M. J. A. Werger), Junk, The Hague, pp. 231–302.

Werger, M. J. A. (1983) Tropical grasslands, savannas, woodlands: natural and manmade, in *Man's Impact on Vegetation*. (eds W. Holzner, M. J. A. Werger and I. Ikusima), Junk, The Hague, pp. 107–37.

Werger, M. J. A. (1986) The Karoo and southern Kalahari, in *Ecosystems of the World Vol. 12B. Hot Deserts and Arid Shrublands*, (eds M. Evanari, I. Noy-Meir and D. W. Goodall), Elsevier, Amsterdam, pp. 283–359.

Werger, M. J. A. and Coetzee, B. J. (1978) The Sudano-Zambezian region, in *Biogeography and Ecology of Southern Africa*, (ed. M. J. A. Werger), Junk, The Hague, pp. 301–462.

West, C. (1985) Factors underlying the late seasonal appearance of the lepidopterous leaf-mining guild on oak. *Ecological Entomology*, **10**, 111–20.

West, N. E. (1978) Physical inputs of nitrogen to desert ecosystems, in *Nitrogen in Desert Ecosystems*, (eds N. E. West and J. Skujiņš), Dowden, Hutchinson and Ross, Stroudsburg, pp. 165–70.

West, N. E. (1981) Nutrient cycling in desert ecosystems, in *Arid-land Ecosystems: Structure, Functioning and Management*, Vol. 2 (eds D. W. Goodall, R. A. Perry and K. M. W. Howes), Cambridge University, Cambridge, pp. 301–24.

West, N. E. (1988) Intermontane deserts, shrub steppes, and woodlands, in *North American Terrestrial Vegetation*, (eds M. G. Barbour and W. D. Billings), Cambridge University Press, Cambridge, pp. 209–30.

West, N. E. and Klemmedson, J. O. (1978) Structural distribution of nitrogen in desert ecosystems, in *Nitrogen in Desert Ecosystems*, (eds N. E. West and J. J. Skujiņš), Dowden, Hutchinson and Ross, Stroudsburg, pp. 1–19.

West, R. C. (1977a) Tidal salt-marsh and mangal formations of Middle and South America, in *Ecosystems of the World Vol. 1. Wet Coastal Ecosystems*, (ed. V. J. Chapman), Elsevier, Amsterdam, pp. 193–213.

West, R. G. (1977b) *Pleistocene Geology and Biology: With Especial Reference to the British Isles* (2nd edn), Longman, London.

Westhoff, V. (1971) The dynamic structure of plant communities in relation to the objectives of conservation, in *The Scientific Management of Animal and Plant Communities for Conservation*, (eds E. Duffey and A. S. Watt), Blackwell Scientific Publications, Oxford, pp. 3–14.

Westlake, D. F. (1967) Some effects of low-velocity currents on the metabolism of aquatic macrophytes. *Journal of Experimental Botany*, **18**, 187–205.

Westlake, D. F. (1980) Macrophytes, in *The Functioning of Freshwater Ecosystems*, (eds E. D. Le Cren and R. H. Lowe-McConnell), Cambridge University Press, Cambridge, pp. 177–82.

Westman, W. E. (1981) Seasonal dimorphism of foliage in Californian coastal sage scrub. *Oecologia*, **51**, 385–88.

Westman, W. E. (1983) Plant community structure – spatial partitioning of resources, in *Mediterranean-type Ecosystems: The Role of Nutrients*, (eds F. J. Kruger, D. T. Mitchell and J. U. M. Jarvis), Springer, Berlin, pp. 417–45.

Wetzel, R. G. (1975) *Limnology*, Saunders, Philadelphia.

Whitcomb, R. F., Robbins, C. S., Lynch, J. F. *et al.* (1981) Effects of forest fragmentation on avifauna of the eastern deciduous forest, in *Forest Island Dynamics in Man-dominated Landscapes*, (eds R. L. Burgess and D. M. Sharpe), Springer, New York, pp. 125–205.

White, F. (1977) The underground forests of Africa: a preliminary review. *The Gardens' Bulletin, Singapore*, **29**, 57–71.

White, J. (1980) Demographic factors in populations of plants, in *Demography and Evolution in Plant Populations*, (ed. O. T. Solbrig), University of California Press, Berkeley, pp. 21–48.

White, R. G., Bunnell, F. L., Gaare, E., Skogland, T. and Hubert B. (1981) Ungulates on arctic ranges, in *Tundra Ecosystems: A Comparative Analysis*, (eds L. C. Bliss, O. W. Heal and J. J. Moore), Cambridge University Press, Cambridge, pp. 397–483.

Whitford, H. N. (1906) The vegetation of the Lamao Forest Reserve. *The Philippine Journal of Science*, **1**, 373–431.

Whitford, W. G. (1978) Foraging in seed-harvester ants *Pogonomyrmex* spp. *Ecology*, **59**, 185–9.

Whitford, W. G., Johnson, P. and Ramirez, J. (1976) Comparative ecology of the harvester ants *Pogonomyrmex barbatus* (F. Smith) and *Pogonomyrmex rugosus* (Emery). *Insectes Sociaux*, **23**, 117–32.

Whitmore, T. C. (1962) Bark morphology as an aid to forest recognition and taxonomy in Dipterocarpaceae. *Flora Malesiana Bulletin*, **18**, 1017–9.

Whitmore, T. C. (1974) Change with time and the role of cyclones in tropical rain forest on Kolombangara, Solomon Islands. *Commonwealth Forestry Institute, Paper No. 46*.

Whitmore, T. C. (1984) *Tropical Rain Forests of the Far East*, 2nd edn, Clarendon, Oxford.

Whitmore, T. C. (1989a) Southeast Asian tropical forests, in *Ecosystems of the World Vol. 14B. Tropical Rain Forest Ecosystems: Biogeographical and Ecological Studies*, (eds H. Lieth and M. J. A. Werger), Elsevier, Amsterdam, pp. 195–218.

Whitmore, T. C. (1989b) Canopy gaps and the two major groups of forest trees. *Ecology*, **70**, 536–8.

Whitmore, T. C. and Sayer, J. A. (eds) (1992) *Tropical Deforestation and Species Extinction*, Chapman & Hall, London.

Whitney, D. M., Chalmers, A. G., Haines, E. B. *et al.* (1981) The cycles of nitrogen and phosphorus, in *The Ecology of a Salt Marsh*, (eds L. R. Pomeroy and R. G. Weigert), Springer, New York, pp. 163–81.

Whittaker, R. H. (1965) Dominance and diversity in land plant communities. *Science*, **147**, 250–60.

Whittaker, R. H. (1970) *Communities and Ecosystems*, Collier-Macmillan, London.

Whittaker, R. H. (1977) Evolution of species diversity in land communities. *Evolutionary Biology*, **10**, 1–67.

Whittaker, R. H. and Likens, G. E. (1975) The biosphere and Man, in *Primary Productivity of the Biosphere*, (ed. H. Lieth and R. H. Whittaker), Springer, New York, pp. 305–28.

Whittaker, R. H. and Niering, W. A. (1975) Vegetation of the Santa Catalina Mountains, Arizona. V. Biomass, production, and diversity along the elevation gradient. *Ecology*, **56**, 771–90.

Whittaker, R. H. and Woodwell, G. M. (1969) Structure, production and diversity of the oak–pine forest at Brookhaven, New York. *Journal of Ecology*, **57**, 155–74.

Wickman, F. E. (1951) The maximum height of raised bogs and a note on the motion of water in soligenous mires. *Geologiska Föreningens i Stockholm Förhandlingar*, **73**, 413–22.

Widden, P. (1977) Microbiology and decomposition on Truelove Lowland, in *Truelove Lowland, Devon Island, Canada: A High Arctic Ecosystem*, (ed. L. C. Bliss), The University of Alberta Press, Edmonton, pp. 505–30.

Wiebe, W. J., Christian, R. R., Hansen, J. A. *et al.* (1981) Anaerobic respiration and fermentation, in *The Ecology of a Salt Marsh*, (eds L. R. Pomeroy and R. G. Wiegert), Springer, New York, pp. 137–59.

Wiebes, J. T. (1979) Co-evolution of figs and their insect pollinators. *Annual Review of Ecology and Systematics*, **10**, 1–12.

Wielgolaski, F. E., Bliss, L. C., Svoboda, J. and Doyle, G. (1981) Primary production of tundra, in *Tundra Ecosystems: A Comparative Analysis*, (eds L. C. Bliss, O. W. Heal and J. J. Moore), Cambridge University Press, Cambridge, pp. 187–225.

Wielgolaski, F. E. and Kärenlampi, L. (1975) Plant phenology of Fennoscandian tundra areas, in *Fennoscandian Tundra Ecosystems*, (ed. F. E. Wielgolaski), Springer, New York, pp. 94–102.

Wielgolaski, F. E., Kjelvik, S. and Kallio, P. (1975) Mineral content of tundra and forest tundra plants in Fennoscandia, in *Fennoscandian Tundra Ecosystems*, (ed. F. E. Wielgolaski), Springer, New York, pp. 316–32.

Wiens, J. A. (1974) Climatic instability and the 'ecological saturation' of bird communities in North American grasslands. *Condor*, **76**, 385–400.

Wilkinson, C. R. and Sammarco, P. W. (1983) Effects of fish grazing and damselfish territoriality on coral reef algae. II. Nitrogen fixation. *Marine Ecology Progress Series*, **13**, 15–9.

Willard, B. E. and Marr, J. W. (1970) Effects of human activities on alpine tundra ecosystems in Rocky Mountain National Park, Colorado. *Biological Conservation*, **2**, 257–65.

Willard, J. R. (1974) *Soil Invertebrates: VIII. A Summary of Populations and Biomass. Canadian Committee for IBP, Technical Report 56*, University of Saskatchewan, Saskatoon.

Williams, I. J. M. (1972) A revision of the genus *Leucadendron* (Protaceae). *Contributions from the Bolus Herbarium 3*. The Bolus Herbarium, Cape Town.

Williams, O. B. (1979) Ecosystems of Australia, in *Arid-land Ecosystems: Structure, Functioning and Management*, Vol. 1 (eds D. W. Goodall, R. A. Perry and K. M. W. Howes), Cambridge University, Cambridge, pp. 145–212.

Williams, O. B. and Calaby, J. H. (1985) The hot deserts of Australia, in *Ecosystems of the World Vol. 12A. Hot Deserts and Arid Shrublands, A*, (eds M. Evanari, I. Noy-Meir and D. W. Goodall), Elsevier, Amsterdam, pp. 269–312.

Williams, W. D. (1984) Australian lakes, in *Ecosystems of the World Vol. 23. Lakes and Reservoirs*, (ed. F. B. Taub), Elsevier, Amsterdam, pp. 499–519.

Willis, A. J. (1965) The influence of mineral nutrients on the growth of *Ammophila arenaria*. *Journal of Ecology*, **53**, 735–45.

Willis, A. J. and Jefferies, R. L. (1963) Investigations on the water relations of sand-dune plants under natural conditions, in *The Water Relations of Plants*, (eds A. J. Rutter and F. H. Whitehead), Wiley, New York, pp. 168–89.

Wilton, W. C. (1963) Black spruce seedfall immediately following fire. *Forestry Chronicle*, **39**, 477–9.

Winberg, G. G. (1980) General characteristics of freshwater ecosystems based on Soviet IBP studies, in *The Functioning of Freshwater Ecosystems*, (eds E. D. Le Cren and R. H. Lowe-McConnell), Cambridge University Press, Cambridge, pp. 481–91.

Winkler, E. M. (1977) Insolation of rock and stone, a hot item. *Geology*, **5**, 188–9.

Winkler, E. M. and Singer, P. C. (1972) Crystallisation pressure of salts in stone and concrete. *Geological Society of America Bulletin*, **83**, 3509–13.

Winkworth, R. E. (1971) Longevity of buffel grass seed sown in an arid Australian range. *Journal of Range Management*, **24**, 141–5.

Winter, D. F., Banse, K. and Anderson, G. C. (1975) The dynamics of phytoplankton blooms in Puget Sound, a fjord in the northwestern United States. *Marine Biology*, **29**, 139–76.

Winter, W. H. (1987) Using fire and supplements to improve cattle production from monsoon tallgrass pastures. *Tropical Grasslands*, **21**, 71–80.

Witkamp, M. and van der Drift, J. (1961) Breakdown of forest litter in relation to environmental factors. *Plant and Soil*, **15**, 295–311.

Wolfe, J. A. (1972) An interpretation of Alaskan Tertiary floras, in *Floristics and Paleofloristics of Asia and Eastern North America*, (ed. A. Graham), Elsevier, Amsterdam, pp. 201–33.

Wood, T. W. W. (1970) Wind damage in the forest of western Samoa. *The Malayan Forester*, **33**, 92–9.

Wood, W. B. (1990) Tropical deforestation: balancing regional development demands and global environmental concerns. *Global Environmental Change*, **1**, 23–41.

Woodhouse, R. M., Williams, J. G. and Nobel, P. S. (1980) Leaf orientation, radiation interception, and nocturnal acidity increases by the CAM plant *Agave deserti* (Agavaceae). *American Journal of Botany*, **67**, 1179–85.

Woodmansee, R. G. (1978) Additions and losses of nitrogen in grassland ecosystems. *BioScience*, **28**, 448–53.

Woodmansee, R. G. (1979) Factors influencing input and output of nitrogen in grasslands, in *Perspectives in Grassland Ecology*, (ed. N. R. French), Springer, New York, pp. 117–34.

Woods, F. W. and Shanks, R. E. (1959) Natural replacement of Chestnut by other species in the Great Smoky Mountains National Park. *Ecology*, **40**, 349–61.

Wratten, S. D., Edwards, P. J. and Dunn, I. (1984) Wound-induced changes in the palatability of *Betula pubescens* and *B. pendula*. *Oecologia*, **61**, 372–5.

Wright, D. I. and O'Brien, W. J. (1984) Model analysis of the feeding ecology of a freshwater planktivorous fish, in *Trophic Interactions Within Aquatic Ecosystems*, (eds D. G. Meyers and J. R. Strickler), American Association for the Advancement of Science, Washington, pp. 243–67.

Wright, H. A. (1974) Effect of fire on southern mixed prairie grasses. *Journal of Range Management*, **27**, 417–19.

Wright, H. A. and Bailey, A. W. (1982) *Fire Ecology*, Wiley, New York.

Wright, H. A., Bunting, S. C. and Neuenschwander, L. F. (1976) Effect of fire on honey mesquite. *Journal of Range Management*, **29**, 467–71.

Wright, J. B. (ed.) (1977) *Oceanography*, Open University Press, Milton Keynes.

Wright, S. J. (1983) The dispersion of eggs by a bruchid beetle among *Scheelea* palm seeds and the effect of distance to the parent palm. *Ecology*, **64**, 1016–21.

Wyatt-Smith, J. (1966) Ecological studies on Malayan forests. *Malayan Forestry Department Research Pamphlet 52*.

Wycherley, P. R. (1973) The phenology of plants in the humid tropics. *Micronesica*, **9**, 75–96.

Xiao, P. (1991) The Chinese approach to medicinal plants – their utilization and conservation, in *The Conservation of Medicinal Plants*, (eds O. Akerele, V. Heywood and H. Synge), Cambridge University Press, Cambridge, pp. 305–13.

Yamasaki, S. (1984) Role of plant aeration in zonation of *Zizania latifolia* and *Phragmites australis*. *Aquatic Botany*, **18**, 287–97.

Yarie, J. (1981) Forest fire cycles and life tables: a case study from interior Alaska. *Canadian Journal of Forest Research*, **11**, 554–62.

Yeaton, R. I. (1978) A cyclical relationship between *Larrea tridentata* and *Opuntia leptocaulis* in the northern Chihuahuan Desert. *Journal of Ecology*, **66**, 651–6.

Yeaton, R. I., Travis, J. and Gilinsky, E. (1977) Competition and spacing in plant communities: the Arizona upland association. *Journal of Ecology*, **65**, 587–95.

Yentsch, C. S. (1980) Light attenuation and phytoplankton photosynthesis, in *The Physiological Ecology of Phytoplankton*, (ed. I. Morris), University of California Press, Berkeley, pp. 95–127.

Young, D. R. and Smith, W. K. (1979) Influence of sunflecks on the temperature and water relations of two subalpine understory congeners. *Oecologia*, **43**, 195–205.

Young, S. B. (1982) The vegetation of land-bridge Beringia, in *Paleoecology of Beringia*, (eds D. M. Hopkins, J. V. Matthews, C. E. Schweger, S. B. Young), Academic Press, New York, pp. 179–91.

Young, T. C., DePinto, J. V., Flint, S. E., *et al.* (1982) Algal availability of phosphorus in municipal wastewater. *Journal of the Water Pollution Control Federation*, **54**, 1505–16.

Yurtsev, B. A. (1972) Phytogeography of northeastern Asia and the problem of Transberingian floristic interrelations, in *Floristics and Paleofloristics of Asia and Eastern North America*, (ed. A. Graham), Elsevier, Amsterdam, pp. 19–54.

Zahran, M. A. (1977) Africa A. Wet formations of the African Red Sea coast, in *Ecosystems of the World Vol. 1. Wet Coastal Ecosystems*, (ed. V. J. Chapman), Elsevier, Amsterdam, pp. 215–31.

Zahran, M. A., Abdel Wahid, A. A. and El-Demerdash, M. A. (1979) Economic potential of *Juncus* plants, in *Arid Land Plant Resources*, (eds J. R. Goodin and D. K. Northington), Texas Tech University, Lubbock, pp. 244–60.

Zaret, T. M. (1980) *Predation and Freshwater Communities*, Yale University Press, New Haven.

Zavitkovski, J. and Newton, M. (1968) Ecological importance of snowbrush *Ceanothus velutinus* in the Oregon Cascades. *Ecology*, **49**, 1134–45.

Zedler, J. B. (1980) Algal mat productivity: comparisons in a salt marsh. *Estuaries*, **3**, 122–31.

Zedler, P. H. (1981) Vegetation change in chaparral and desert communities in San Diego County, California, in *Forest Succession, Concepts and Application*, (eds D. C. West, H. H. Shugart and D. B. Botkin), Springer, New York, pp. 406–30.

Zinke, P. J. (1973) Analogies between the soil and vegetation types of Italy, Greece, and California, in *Mediterranean Type Ecosystems: Origin and Structure*, (eds F. di Castri and H. A. Mooney), Springer, New York, pp. 61–80.

Zlotin, R. I. and Khodashova, K. S. (1980) *The Role of Animals in Biological Cycling of Forest–Steppe Ecosystems*, Dowden, Hutchinson and Ross, Stroudsburg.

Zoltai, S. C. (1975) Structure of subarctic forests on hummocky permafrost terrain in northwestern Canada. *Canadian Journal of Forest Research*, **5**, 1–9.

Zoltai, S. C. and Pollett, F. C. (1983) Wetlands in Canada: their classification, distribution, and use, in *Ecosystems of the World Vol. 4B. Mires: Swamp, Bog, Fen and Moor – Regional Studies*, (ed. A. J. P. Gore), Elsevier, Amsterdam, pp. 245–68.

Zoltai, S. C. and Vitt, D. H. (1990) Holocene climatic change and the distribution of peatlands in western interior Canada. *Quaternary Research*, **33**, 231–40.

Zuberer, D. A. and Silver, W. S. (1978) Biological dinitrogen fixation (acetylene reduction) associated with Florida mangroves. *Applied and Environmental Microbiology*, **35**, 567–75.

Index

Page numbers appearing in **bold** refer to figures and page numbers appearing in *italic* refer to tables.